Theoretical Astrophysics
Volume III: Galaxies and Cosmology

This timely volume provides comprehensive coverage of all aspects of cosmology and extragalactic astronomy at an advanced level. Beginning with an overview of the key observational results and necessary terminology, it goes on to cover important topics including the theory of galactic structure and galactic dynamics, structure formation, cosmic microwave background radiation, formation of luminous galaxies in the universe, intergalactic medium, and active galactic nuclei. Topics are developed in a contemporary fashion, with emphasis on currently active research areas.

This self-contained text has a modular structure, and contains over one hundred worked exercises. It can be used alone or in conjunction with the previous two accompanying volumes (Volume I: *Astrophysical Processes*, and Volume II: *Stars and Stellar Systems*). The textbook develops all aspects of extragalactic astronomy and cosmology in a detailed and pedagogical way and will be invaluable to researchers and graduate students of extragalactic astronomy, astrophysics, and theoretical physics.

Thanu Padmanabhan is a Senior Professor and Dean of Core Academic Programmes at the Inter-University Centre for Astronomy and Astrophysics in Pune, India. His research interests are quantum theory, gravitation, cosmology, and structure formation in the universe. He has published more than one hundred technical papers in these areas, and this is his sixth book to be published by Cambridge University Press. He also has written more than one hundred popular science articles and several regular columns on astronomy, recreational mathematics, and history of science for international journals and newspapers.

Professor Padmanabhan is an elected fellow of the Indian Academy of Sciences, the National Academy of Sciences, the Indian National Science Academy, and a member of the International Astronomical Union. He has received numerous awards including the Shanti Swarup Bhatnagar Prize in Physics (1996) and the Millenium Medal (2000) awarded by the Council for Scientific and Industrial Research, India.

THEORETICAL ASTROPHYSICS

Volume III: Galaxies and Cosmology

T. PADMANABHAN

Inter-University Centre for Astronomy and Astrophysics
Pune, India

CAMBRIDGE
UNIVERSITY PRESS

PUBLISHED BY THE PRESS SYNDICATE OF THE UNIVERSITY OF CAMBRIDGE
The Pitt Building, Trumpington Street, Cambridge, United Kingdom

CAMBRIDGE UNIVERSITY PRESS
The Edinburgh Building, Cambridge CB2 2RU, UK
40 West 20th Street, New York, NY 10011-4211, USA
477 Williamstown Road, Port Melbourne, VIC 3207, Australia
Ruiz de Alarcón 13, 28014 Madrid, Spain
Dock House, The Waterfront, Cape Town 8001, South Africa

http://www.cambridge.org

First published 2002

Printed in the United States of America

Typeface Times Roman 11/13 pt.   *System* LATEX 2_ε [TB]

A catalog record for this book is available from the British Library.

Library of Congress Cataloging in Publication Data

ISBN 0 521 56242 2 hardback
ISBN 0 521 56630 4 paperback

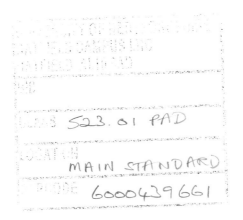

To all my friends

THEORETICAL ASTROPHYSICS

– in three volumes –

VOLUME I: PHYSICAL PROCESSES

Chapters 1: Order of Magnitude Astrophysics; 2: Dynamics; 3: Special Relativity, Electrodynamics and Optics; 4: Basics of Electromagnetic Radiation; 5: Statistical Mechanics; 6: Radiative Processes; 7: Spectra; 8: Neutral Fluids; 9: Plasma Physics; 10: Gravitational Dynamics; 11: General Theory of Relativity; 12: Basics of Nuclear Physics.

VOLUME II: STARS AND STELLAR SYSTEMS

Chapters 1: Overview: Stars and Stellar Systems; 2: Stellar Structure; 3: Stellar Evolution; 4: Supernova; 5: White Dwarfs, Neutron Stars and Black Holes; 6: Pulsars; 7: Binary Stars and Accretion; 8: Sun and Solar System; 9: Interstellar Medium; 10: Globular Clusters.

VOLUME III: GALAXIES AND COSMOLOGY

Chapters 1: Overview: Galaxies and Cosmology; 2: Galactic Structure and Dynamics; 3: Friedmann Model of the Universe; 4: Thermal History of the Universe; 5: Structure Formation; 6: Cosmic Microwave Background Radiation; 7: Formation of Baryonic Structures; 8: Active Galactic Nuclei; 9: Intergalactic Medium and Absorption Systems; 10: Cosmological Observations.

Contents

Preface

From where has this Creation sprung ? Who holds or does not hold ?
He who is its Surveyer in the highest heaven, He alone knows
And yet maybe He doth not know ?

— Rig Veda, Verse 10.129.7.

During the past decade or so, theoretical astrophysics has emerged as one of the most active research areas in physics. This advance has also been reflected in the greater interdisciplinary nature of research that has been carried out in this area in recent years. As a result, those who are learning theoretical astrophysics with the aim of making a research career in this subject need to assimilate a considerable amount of concepts and techniques, in different areas of astrophysics, in a short period of time. Every area of theoretical astrophysics, of course, has excellent textbooks that allow the reader to master that *particular* area in a well-defined way. Most of these textbooks, however, are written in a traditional style that focusses on one area of astrophysics (say stellar evolution, galactic dynamics, radiative processes, cosmology, etc.). Because different authors have different perspectives regarding their subject matter, it is not very easy for a student to understand the key unifying principles behind several different astrophysical phenomena by studying a plethora of separate textbooks, as they do not link up together as a series of core books in theoretical astrophysics covering everything that a student would need. A few books, which *do* cover the whole of astrophysics, deal with the subject at a rather elementary (first-course) level.

What we require is clearly something analogous to the famous Landau–Lifshitz course in theoretical physics, but focussed to the subject of theoretical astrophysics at a fairly advanced level. In such a course, one could present all the key physical concepts (eg. radiative processes, fluid mechanics, plasma physics, etc.) from a unified perspective and then apply them to different astrophysical situations.

This book is the third in a set of three volumes that are intended to do exactly that. The three volumes form one single coherent unit of study, using which a student can acquire mastery over all the traditional astrophysical topics. What is more, these volumes will emphasise the unity of concepts and techniques in different branches of astrophysics. The interrelationship among different areas and common features in the analysis of different theoretical problems will be stressed throughout. Because many of the basic techniques need to be developed only once, it is possible to achieve significant economy of presentation and crispness of style in these volumes.

Needless to say, there are some basic "boundary conditions" one has to respect in such an attempt to cover the whole of theoretical astrophysics in approximately 3×600 pages. Not much space is available to describe the nuances in greater length or to fill in details of algebra. For example, I have made conscious choices as to which parts of the algebra can be left to the reader and which parts need to be worked out explicitly in the text, and I have omitted detailed discussions of elementary concepts and derivations. However, I do *not* expect the reader to know anything about astrophysics. All astrophysical concepts are developed *ab initio* in these volumes. The approach used in these three volumes is similar to that used by Genghis Khan, namely (*i*) cover as much area as possible, (*ii*) capture the important points, and (*iii*) be utterly ruthless!

To cut out as much repetition as possible, the bulk of the physical principles are presented at one go in Vol. I and are applied in the other two volumes to different situations. These three volumes also concentrate on *theoretical* aspects. Observation and phenomenology are, of course, discussed in Vols. II and III to the extent necessary to make the motivation clear. However, I do not have the space to discuss how these observations are made, the errors, reliability, etc., of the observations or the astronomical techniques. (Maybe there should be a fourth volume describing observational astrophysics!)

The target audience for this three-volume work will be fairly large and comprises: (1) students in the first year of their Ph.D. Program in theoretical physics, astronomy, astrophysics, and cosmology; (2) research workers in various fields of theoretical astrophysics, cosmology etc.; and (3) teachers of graduate courses in theoretical astrophysics, cosmology, and related subjects. In fact, anyone working in or interested in some area of astronomy or astrophysics will find something useful in these volumes. They are also designed in such a way that parts of the material can be used in modular form to suit the requirements of different people and different courses.

Let me briefly highlight the features that are specific to Vol. III. The reader is assumed to be familiar with the material covered in Vol. I, having either studied that volume (which is the recommended procedure!) or through independent courses in basic physics. This volume also uses several topics developed in

Vol. II; for example, the discussion of accretion disks (in Chap. 7 of Vol. II) and the interstellar medium (Chap. 9 of Vol. II) finds application in the study of active galactic nuclei and the intergalactic medium. The spirit of the three coherent volumes is to avoid repetition as much as possible, and hence I have merely referred to the relevant parts of Vol. I or Vol. II whenever some input is required. Given this background, it was fairly easy to order the topics of Vol. III in a logical sequence, and I have broadly followed the same format as that of Vol. II. Chapter 1 provides a broad overview of galaxies and cosmology and rapidly introduces several observational and theoretical concepts that are developed in detail in the later chapters. This is done to tackle the interdependency of concepts (and jargon) that prevents the development of topics in a fully streamlined fashion. I expect the reader to rapidly go through Chap. 1 in the first reading and come back to it as and when required. Chapter 2 introduces several aspects of galactic structure and dynamics, building on the observational inputs already covered in Chap. 1. Chapters 3–5 form the core of cosmology and structure formation. The logical structure is somewhat similar to the one I adopted in my 1993 book *Structure Formation in the Universe*, but I have updated the contents quite significantly, keeping in mind the nature of these volumes as well as recent developments. For example, I have included a discussion of models with a nonzero cosmological constant that are enjoying considerable popularity at present. I have also added a somewhat simple discussion of different gauges that are used in the study of structure formation. Chapter 6 deals with cosmic microwave background radiation, which will continue to attract attention in the coming years. Even as this chapter was being written, data from CMBR experiments were revised, necessitating my rewriting of the chapter. Quite obviously, the details and the comparison with observation will not survive for long, but I have tried to present the fundamental issues in such a way that they will have a much longer shelf life. Chapter 7 deals with the actual formation of luminous galaxies in the universe, which is probably one of the most complex topics under active investigation. I have refrained from quoting results of numerical simulations mainly because I do not believe the hydrosimulations have yet reached a stage to warrant inclusion in a basic textbook like this. Instead, I have tried to give semianalytic arguments and qualitative descriptions to give the reader a flavour of our current understanding of this difficult subject. Chapter 8 takes a detour to describe the important features of active galactic nuclei. I have drawn heavily on the material developed in earlier volumes (in particular, as regards radiative processes and physics of accretion disks) and have tried to give a coherent picture of active galactic nuclei in different wave bands. Chapter 9 describes the physics of intergalactic medium and absorption systems; once again, the emphasis is on relatively modern topics like the role of star formation and the production of extragalactic background light. Finally, Chap. 10 provides a rapid overview of observations relevant to cosmology and the techniques used to measure different cosmological parameters. This chapter also includes a discussion of the evidence for dark matter in different scales.

This volume also provided a tough challenge as regards the discussion of phenomenological input and a few words regarding my policy are in order. I have followed essentially the same philosophy as I used in Vol. II and have tried to avoid the two pitfalls: (1) Drowning the reader in accurate but unclassified sea of astronomical data just because accurate data are available or (2) ignoring the phenomenological input and treating the subject as a branch of applied mathematics – a criticism I have heard as regards my book *Structure Formation in the Universe*. As in Vol. II, I have tried to describe the necessary observational issues (but not observational techniques) and provide a minimum of observational data whenever it is relevant. I have also tried to motivate theoretical developments based on specific observational inputs, especially when a more fundamental approach would be unwarranted or facetious. At the same time I have tried to bring some amount of method and order to the presentation of the topics so that the reader will be able to grasp how a theoretical astrophysicist goes about the task of developing the models. (Once again, this makes Chap. 1 somewhat different in form and content compared with other chapters. I hope this is not too much of a distraction.)

All this required the exercise of my judgement in deciding the choice of topics, their emphasis, and the proper blend of phenomenology, observations, and theoretical rigour. It is impossible to satisfy everyone as regards the "correctness" of such decisions, and I have tried to do some optimisation so as to provide maximum benefit to the reader. The two topics that I probably would have liked to discuss at length but had to forego are the physics of cosmic rays and the stucture and dynamics of galaxy clusters. Except for these, I believe I have done justice to most of the topics at a level appropriate for these volumes.

Any one of these topics is fairly vast and often requires a full textbook to do justice to it, whereas I have devoted approximately 60 pages to discuss each of them! I would like to emphasise that such a crisp, condensed discussion is not only possible but also constitutes a basic matter of policy in these volumes. After all, the idea *is* to provide the student with the essence of several textbooks in one place. It should be clear to lecturers that these materials can be easily regrouped to serve different graduate courses at different levels, especially when complemented by other textbooks.

Because of the highly pedagogical nature of the material covered in this volume, I have not given detailed references to original literature except on rare occasions when a particular derivation is not available in standard textbooks. The annotated list of references given at the end of the book cites several other textbooks that I found very useful. Some of these books, of course, contain extensive bibliographies and references to original literature. The selection of core books cited here clearly reflects the personal bias of the author, and I apologise to anyone who feels their work or contribution has been overlooked.

Several people have contributed to the making of these volumes. The idea for these volumes originated over a dinner with J. P. Ostriker in late 1994, while I was visiting Princeton. I was lamenting to Jerry about the lack of a comprehensive set of books covering all of theoretical astrophysics and Jerry said, "Why don't *you* write them?" He was very enthusiastic and supportive of the idea and gave extensive comments and suggestions on the original outline I produced the next week. I am grateful to him for the comments and for the moral support that I needed to launch into such a project. I sincerely hope the volumes do not disappoint him.

Adam Black (who was with Cambridge University Press at that time) took up the proposal with his characteristic enthusiasm and initiative. I should also thank him for choosing six excellent (anonymous) referees for this proposal whose support and comments helped to mould it into the proper framework. The processing of this volume was handled by Simon Mitton of Cambridge University Press, and I thank him for the effort he has put in.

Many of my friends and colleagues carried out the job of reading the earlier drafts and providing comments. Many colleagues, especially Jaichan Hwang, Chanda Jog, Ofer Lahav, Niranjan Sambhus, Shiv Sethi, T. Souradeep, R. Srianand, K. Subramanian, and Yogesh Wadedekar, have made comments on selected chapters. Some of the figures and data were provided by C.W. Churchill, Tapas Kumar Das, George Djorgovski, Bill Keel, John Peacock, Niranjan Sambhus, Shiv Sethi, and C.C. Steidel, and I thank them for their help.

I have been a regular visitor to the Astronomy department of Caltech during the past several years, and the work on the volumes has benefitted tremendously through my discussions and interactions with the students and staff of the Caltech Astronomy department. I would like to specially thank Roger Blandford, Peter Goldreich, Shri Kulkarni, Sterl Phinney, and Tony Readhead for several useful discussions and for sharing with me their insights and experience in the teaching of astrophysics. Part of the work on this volume was done while I was visiting the Institute of Astronomy, Cambridge, England, in May 2001.

This project would not have been possible but for the dedicated support from Vasanthi Padmanabhan, who not only did the entire TEXing and formatting but also produced most of the figures – often writing the necessary programs for the same. I thank her for the help in all three volumes. It is a pleasure to acknowledge the library and other research facilities available at IUCAA, which were useful in this task.

T. Padmanabhan

1

Overview: Galaxies and Cosmology

1.1 Introduction

Attempts to understand extragalactic objects and the universe by using the laws of physics lead to difficulties that have no parallel in the application of the laws of physics to systems of a more moderate scale. The key difficulty arises from the fact that our universe exhibits temporal evolution and is not in steady state. Thus different epochs in the past evolutionary history of the universe are unique (and have occurred only once), and the current state of the universe is a direct consequence of the conditions that were prevalent in the past. For example, most of the galaxies in the universe have formed sometime in the past during a particular phase in the evolution of the universe. This is in contrast to star formation within a galaxy that we can observe directly and study by using standard statistical methods.

In principle, we should be able to see the events that took place in the universe in the past because of the finite light travel time. By observing sufficiently far-away regions of the universe, we will be able to observe the universe as it was in the past. Although technological innovation will eventually allow us to directly observe and understand all the past events in the history of the universe (especially when neutrino astronomy and gravitational wave astronomy start complementing photon-based observations), we are far from such a satisfactory state of affairs at present. Direct observational evidence today spans only a tiny fraction in the past history of the universe and is not available for sufficiently early epochs. Hence the straightforward approach of starting with known initial conditions for the laws of physics (expressed as a differential equation, say) and integrating them forward in time cannot be adopted to the study of the universe.

An alternative procedure is to start with the current state of the universe and integrate the same equations backwards in time in order to understand its past history. Even in this attempt, progress is not easy because data available at the present epoch are insufficient. The primary problem is what was stressed in Vol. II, Chap. 1: Observational data of adequate quality and quantity become scarce as we probe larger and larger scales. Further, we have no direct laboratory evidence regarding nearly 90% of the matter that is present in the universe; there is also some indirect evidence to suggest that nearly 60% of the matter present in the universe today obeys a fairly exotic equation of state.

These difficulties – which are unique when we attempt to apply the laws of physics to an evolving universe – require us to proceed in a multifaceted manner. Our approach will be to develop a broad paradigm describing the evolution of the universe and the formation of structures in it and iterate the details by constantly comparing the theoretical predictions with observational data. This paradigm is based on the idea that the universe was reasonably homogeneous, isotropic, and fairly featureless – except for small fluctuations in the energy density – at sufficiently early times. It is then possible to set up the equations that describe a model for the universe and integrate them forward in time. The results will depend on the composition of the universe, its current expansion rate, and the initial spectrum of density perturbations. Varying these parameters allows us to construct a library of evolutionary models for the universe that can then be compared with observations in order to restrict the parameter space.

Our approach in many of the chapters in this volume are based on the preceding paradigm of *parameterised cosmology*. The aim will be to deduce as many features of the observed universe as possible from a small set of parameters. Such an approach has proved to be extremely successful in the past two decades, mainly because of the advances in technology that allow good-quality observations. Some of the observations planned during the next two decades hold the hope of determining fairly accurately the parameters that characterize the universe, thereby reducing the problem to one of integration of the relevant equations.

It is possible to consider the study of extragalactic astronomy and cosmology from a broader perspective and ask why the parameters describing the universe have the values that are attributed to them. In other words, why does the observed universe follow one template out of a class of models that can be constructed based on the known laws of physics? Such a question – although intuitively appealing – has no mathematically rigorous and unique formulation and hence will be ignored in our discussion.

A completely different issue will be whether the laws of physics can be used to reduce the number of independent parameters and assumptions in any cosmological model. This is certainly possible once our knowledge of high-energy interactions of particles gets better. At present direct laboratory evidence exists for particle interactions only at energies less than about 100 GeV, and particle-physics models describing higher energies do not have the level of certainty required for making definite *predictions* about the evolution of the universe. Eventually, when our understanding of high-energy particle physics improves to an adequate level, it can be applied to the early phases of the universe. We stress the fact that the procedure of applying laws of physics to understand the behaviour of the universe is hindered *only* because we are ignorant about the relevant laws of physics at sufficiently high energies.[1] (The superscripted numbers throughout the book refer to items in the Notes and References chapter at the end of the book.)

1.2 Evolution of the Universe

Observations suggest that the universe at large scales is homogeneous and isotropic. The fractional fluctuations $(\delta\rho/\rho)_R$ in the mass (and energy) density ρ (which is due to the existence of structures like galaxies, clusters etc.), within a randomly placed sphere of radius R, decrease with R as a power law. This suggests that we can model the universe as being made up of a smooth background with an average density, superposed with fluctuations in the density that are large at small scales but decrease with scale. At sufficiently large scales, the universe may be treated as being homogeneous and isotropic with a uniform density.

It was shown in Vol. 1, Chap. 1, that the only large-scale motion compatible with homogeneity and isotropy is the one with the velocity field of the form $\dot{\mathbf{r}}(t) = \mathbf{v}(t) = f(t)\mathbf{r}$. This allows us to describe the position $\mathbf{r}$ of any material body in the universe in the form $\mathbf{r} = a(t)\mathbf{x}$, where $a(t)$ is another arbitrary function related to $f(t)$ by $f(t) = (\dot{a}/a)$ and $\mathbf{x}$ is a constant for any given material body in the universe. It is conventional to call $\mathbf{x}$ and $\mathbf{r}$ the *comoving* and the *proper* coordinates of the body and $a(t)$ the *expansion factor*. (Even though, if $\dot{a} < 0$, it acts as a contraction factor.)

The dynamics of the universe is entirely determined by the function $a(t)$. The simplest choice will be $a(t) = $ constant, in which case there will be no motion in the universe and all matter will be distributed uniformly in a static configuration. It is, however, clear that such a configuration will be violently unstable when the mutual gravitational forces of the bodies are taken into account. Any such instability will eventually lead to the random motion of particles in localized regions, thereby destroying the initial homogeneity. Observations, however, indicate that this is not true and that the relation $\mathbf{v} = (\dot{a}/a)\mathbf{r}$ does hold in the observed universe with $\dot{a} > 0$. In that case, the dynamics of $a(t)$ can be qualitatively understood along the following lines. Consider a particle of *unit* mass at the location r with respect to some coordinate system. Equating the sum of its kinetic energy $v^2/2$ and gravitational potential energy $[-GM(r)/r]$ that is due to the attraction of matter inside a sphere of radius r, to a constant, we find that $a(t)$ should satisfy the condition

$$\frac{1}{2}\dot{a}^2 - \frac{4\pi G\rho(t)}{3}a^2 = \text{constant}, \tag{1.1}$$

where ρ is the mean density of the universe; that is,

$$\frac{\dot{a}^2}{a^2} + \frac{k}{a^2} = \frac{8\pi G}{3}\rho(t), \tag{1.2}$$

where k is a constant. Although the preceding argument to determine this equation is fallacious, Eq. (1.2) happens to be exact and arises from the proper application of Einstein's theory of relativity to a homogeneous and isotropic distribution of matter with ρ interpreted as the energy density. We shall now describe some simple aspects of such an evolution that will be taken up for detailed study in the later chapters.

Observations suggest that our universe today (at $t = t_0$) is governed by Eq. (1.2) with $(\dot{a}/a)_0 \equiv H_0 = 0.3 \times 10^{-17} h \text{ s}^{-1}$, where $h \approx (0.5\text{–}1)$. This is equivalent to $H_0 = 100h \text{ km s}^{-1} \text{ Mpc}^{-1}$, where 1 Mpc $\approx 3 \times 10^{24}$ cm is a convenient unit for cosmological distances. (We will also use the units 1 kpc $= 10^{-3}$ Mpc and 1 pc $= 10^{-6}$ Mpc in our discussion.) From H_0 we can form the time scale $t_{\text{univ}} \equiv H_0^{-1} \approx 10^{10} h^{-1}$ yr and the length scale $c H_0^{-1} \approx 3000 h^{-1}$ Mpc; t_{univ} characterizes the evolutionary time scale of the universe and $c H_0^{-1}$ is of the order of the largest length scales currently accessible in cosmological observations. The relation $\mathbf{v} = f(t)\mathbf{r} = (\dot{a}/a)\mathbf{r} = H_0 \mathbf{r}$ is called *Hubble's law*, and H_0 is called *Hubble's constant*. From H_0 we can also construct a quantity with the dimensions of density, called the *critical density*:

$$\rho_c = \frac{3 H_0^2}{8 \pi G} = 1.88 h^2 \times 10^{-29} \text{ gm cm}^{-3} = 2.8 \times 10^{11} h^2 \, M_\odot \text{ Mpc}^{-3}$$
$$= 1.1 \times 10^4 h^2 \text{ eV cm}^{-3} = 1.1 \times 10^{-5} h^2 \text{ protons cm}^{-3}. \tag{1.3}$$

(The last two "equalities" should be interpreted in terms of conversion of mass into energy by a factor c^2 and the conversion of mass into number of baryons by a factor m_p^{-1}, where m_p is the proton mass.) It is conventional to measure all other mass and energy densities in the universe in terms of the critical density. If ρ_i is the mass or the energy density associated with a particular species, then we define a density parameter Ω_i through the ratio $\Omega_i \equiv (\rho_i/\rho_c)$. In general, both ρ_i and ρ_c can be defined at any given epoch in the universe and not necessarily at the present moment $t = t_0$; by convention, ρ_c is always defined in terms of the present value of the Hubble constant, although ρ_i could, in general, be a function of time: $\rho_i = \rho_i(t)$. In this case, Ω_i will also depend on time and we define $\Omega_i(t) \equiv \rho_i(t)/\rho_c$.

It is obvious from Eq. (1.2) that the numerical value of k can be absorbed into the definition of $a(t)$ by rescaling it so that we can treat k as having one of the three values $(0, -1, +1)$. The choice among these three values for k is decided by Eq. (1.2) depending on the value of Ω; we see that $k = 1, 0$ or -1, depending on whether Ω is greater than, equal to, or less than unity. The fact that k is proportional to the total energy of the dynamical system described by Eq. (1.2) shows that $a(t)$ will have a maximum value followed by a contracting phase to the universe if $k = 1$ and $\Omega > 1$.

To determine the nature of the cosmological model we need to determine the value of Ω for the universe, taking into account all forms of energy densities that exist at present. Further, to determine the form of $a(t)$ we need to determine how the energy density of any given species varies with time. We now briefly describe the issues involved in this task.

If a particular kind of energy density is described by an equation of state of the form $p = w\rho$, where p is the pressure and w is a constant, then the equation for energy conservation in an expanding background, $d(\rho a^3) = -pd(a^3)$, can be integrated to give $\rho \propto a^{-3(1+w)}$. Equation (1.2) can be now written in the form

$$\frac{\dot{a}^2}{a^2} = H_0^2 \sum_i \Omega_i \left(\frac{a_0}{a}\right)^{3(1+w_i)} - \frac{k}{a^2}, \tag{1.4}$$

where each of these species is identified by density parameter Ω_i and the equation of state is characterized by w_i. The most familiar forms of energy densities are those due to pressureless matter with $w_i = 0$ (that is, nonrelativistic matter with rest-mass-energy density ρc^2 dominating over the kinetic-energy density, $\rho v^2/2$) and radiation with $w_i = (1/3)$. The density parameter contributed today by visible, nonrelativistic, baryonic matter in the universe is $\Omega_B \approx (0.01-0.2)$ and the density parameter that is due to radiation is $\Omega_R \approx 2 \times 10^{-5}$. Unfortunately, models for the universe with just these two constituents for the energy density are in violent disagreement with observations. As we shall see in later chapters, it appears to be necessary to postulate (1) the existence of pressureless ($w = 0$) *non*baryonic dark matter that does not couple with radiation and has a density of at least $\Omega_{DM} \approx 0.3$; because it does not emit light, it is called *dark matter*; (2) an exotic form of matter (called either *cosmological constant* or *vacuum-energy density*) with an equation of state $p = -\rho$ (that is, $w = -1$) that has a density parameter of $\Omega_V \approx 0.7$. The evidence for the existence of nonbaryonic dark matter seems to be fairly definitive whereas the evidence for the existence of cosmological constant is somewhat less definitive. Keeping this in mind, we will concentrate on two typical cosmological models throughout this volume. The first one will have $\Omega_V = 0$ and $0 \leq \Omega_{DM} \leq 1$; the second one will have $\Omega_V + \Omega_{DM} = 1$.

Figure 1.1 provides an inventory of the density contributed by different forms of matter in the universe, and these entries will be discussed in different sections of this chapter. The x axis is actually a combination of Ω and the Hubble parameter h because different components are measured by different techniques. (Usually $n = 1$ or 2; numerical values are for $h = 0.7$.) The top two positions in the contribution to Ω are from a cosmological constant and nonbaryonic dark matter. It is unfortunate that we do not have laboratory evidence for the existence of the first two dominant contributions to the energy density in the universe. This feature alone could make most of the cosmological paradigm described in this book irrelevant at a future date. Alternatively, laboratory detection of a nonbaryonic dark-matter candidate will be an important discovery in establishing the standard paradigm of structure formation.

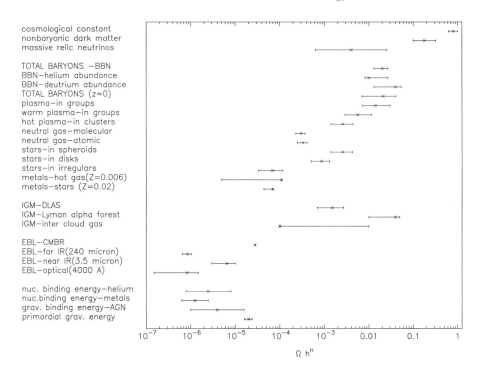

Fig. 1.1.   Cosmic inventory of energy densities. See text for description.

Exercise 1.1

Determining the matter content: Let us assume that the universe contains material with several different equations of state, each characterized by a constant value $w = p/\rho$. Introduce the parameter $\alpha \equiv 3(1 + w)$ and the function $\Omega(\alpha)$ that describes the amount of energy density contributed by a species with a given value of α. Explain how the knowledge of the function $a(t)$ can be used to determine $\Omega(\alpha)$. [Answer: We first note that the term (k/a^2) can be thought of as contributed by a hypothetical species of matter with $w = -(1/3)$. Hence Eq. (1.4) can be written in the form

$$\frac{\dot{a}^2}{a^2} = H_0^2 \sum_i \Omega_i \left(\frac{a_0}{a}\right)^{3(1+w_i)}, \tag{1.5}$$

with a term having $w_i = -(1/3)$ added to the sum. In the continuum limit, this equation can be rewritten as

$$\left(\frac{dq}{d\tau}\right)^2 = \int_{-\infty}^{\infty} d\alpha \, \Omega(\alpha) e^{-\alpha q}, \tag{1.6}$$

where $(a/a_0) = \exp(q)$ and $\tau = H_0 t$. The function $\Omega(\alpha)$ is assumed to have finite support or to decrease fast enough for the expression on the right-hand side to converge. If the observations determine the function $a(t)$, then the left-hand side can be expressed as a function of q. An inverse Laplace transform of this equation will then determine the form of $\Omega(\alpha)$, thereby determining the composition of the universe, as long as all matter can be described by an equation of state of the form $p = w\rho$.]

The evolution of the universe, with the energy content as described above, is straightforward to determine and we shall illustrate it for a simple model with $\Omega_{DM} + \Omega_B + \Omega_R \approx 1; \Omega_V = 0$. If neither particles nor photons are created or destroyed during the expansion, then the number density of particles or photons will decrease as a^{-3} as a increases. In the case of photons, the wavelength will also get stretched during expansion with $\lambda \propto a$; because the energy density of material particles is nmc^2 whereas that of photons of frequency ν is $nh\nu = (nhc/\lambda)$, it follows that the energy densities of radiation and matter vary as $\rho_{rad} \propto a^{-4}$ and $\rho_{matter} \propto a^{-3}$. Combining with the result $\rho_{rad} \propto T^4$ for thermal radiation, it follows that any thermal spectrum of photons in the universe will have its temperature varying as $T \propto a^{-1}$. In the past, when the universe was smaller, it would also have been (1) denser, (2) hotter, and – at sufficiently early epochs – (3) dominated by radiation-energy density.

The light emitted at an earlier epoch by an object will reach us today with the wavelengths stretched because of the expansion. If the light was emitted at $a = a_e$ and received today (when $a = a_0$), the wavelength will change by the factor $(1 + z_e) = (a_0/a_e)$, where z_e is called the *redshift*, which corresponds to the epoch of emission a_e. Because the observed luminosity L of a source is proportional to $(p_\gamma c) d^3 p_\gamma \propto \nu^3 d\nu \propto (1 + z)^{-4}$, where $p_\gamma = (\epsilon/c) = (h\nu/c)$ is the photon momentum, it will decrease as $(1 + z)^{-4}$.

When the temperature of the universe is higher than the temperatures corresponding to the atomic ionisation energy, the matter content in the universe will be a high-temperature plasma. Further, when the temperature of the universe is higher than the binding energy of the nuclei ($\sim$MeV), none of the heavy elements (helium and the metals) could have existed in the universe. Starting from such a hot initial plasma stage, the universe cools as it expands and nucleosynthesis of some amount of deuterium, helium and lithium takes place when $k_B T \lesssim$ MeV. This process does not proceed to form any other heavier elements in significant quantities. This is because – for the observed range of matter and radiation-energy densities – the universe expands too fast to allow the synthesis of heavier metals. The primordial abundance of helium and deuterium is therefore a sensitive test of the different parameters of the universe and will be explored in detail in Chap. 4. The three terms in Fig. 1.1 marked BBN give the constraints arising from *big bang nucleosynthesis*.

In the early hot phase, the radiation will be in thermal equilibrium with matter; as the universe cools below $k_B T \simeq (\epsilon_a/10)$, where ϵ_a is the binding energy of atoms, the electrons and ions will combine to form neutral atoms and radiation will decouple from matter. This occurs at $T_{dec} \simeq 3 \times 10^3$ K. As the universe expands further, these photons will exist in the form of thermal background radiation with a temperature that scales as $T \propto (1/a)$. It turns out that the major component of the extragalactic background light (EBL) that exists today is in the microwave band and can be fitted very accurately by a thermal spectrum at a temperature of ~ 2.7 K. It seems reasonable to interpret this radiation as a relic arising from the early hot phase of the evolving universe. The intensity per logarithmic band of frequency, νB_ν, for this radiation peaks at a wavelength of 1 mm and the maximum intensity is 5.3×10^{-7} W m^{-2} rad^{-2} over the entire sky. The intensity per square arcsecond of the sky is approximately 1.33×10^{-17} W m^{-2} arcsec^{-2}. The energy density that is due to this radiation today will be $\rho_\gamma \simeq (k_B T)^4/(\hbar c)^3 \simeq 5.7 \times 10^{-13}$ ergs cm^{-3}, which corresponds to a *mass* density of $(\rho_\gamma/c^2) = 5.7 \times 10^{-34}$ gm cm^{-3} (this is marked as the entry EBL-CMBR in Fig. 1.1; CMBR stands for *cosmic microwave background radiation*). Taking the matter density today as $\rho_0 = 10^{-30}$ gm cm^{-3}, we find that $\rho_\gamma \simeq 5.7 \times 10^{-4} \rho_0$; radiation (with $\rho_\gamma \propto a^{-4}$) would have dominated over matter (with $\rho \propto a^{-3}$) when the redshift was larger than $z_{\rm eq} \equiv (\rho/\rho_\gamma) \approx 1.7 \times 10^3$.

1.3 Formation of Dark-Matter Halos

The considerations of the last section were independent of the explicit form of $a(t)$. We now turn to the solutions of Eq. (1.2) that determine $a(t)$ and the issue of the formation of structures. The simplest solution to Eq. (1.2) will occur for $k = 0$ if we take the matter density in the universe to decrease as a^{-3} with expansion. Then we get $a(t) = (t/t_0)^{2/3}$ with $t_0^{-2} = (6\pi G\rho_0)$, and $a(t)$ is normalised to $a = 1$ at the present epoch $t = t_0$.

Such a totally uniform universe, of course, will never lead to any of the inhomogeneous structures seen today. However, if the universe has even the slightest inhomogeneity in the past, then gravitational instability can amplify the density perturbations. To see how this comes about in the simplest context, consider Eq. (1.2) written in the equivalent form as

$$\ddot{a} = -\frac{4\pi G\rho_0}{3a^2} = -\left(\frac{2}{9t_0^2}\right)\frac{1}{a^2}, \tag{1.7}$$

where we have put $\rho = (\rho_0 a_0^3/a^3)$ and differentiated Eq. (1.2) once with respect to t. If we perturb $a(t)$ slightly to $a(t) + \delta a(t)$ such that the corresponding fractional density perturbation is $\delta \equiv (\delta\rho/\rho) = -3(\delta a/a)$, we find that δa satisfies the equation

$$\frac{d^2}{dt^2}\delta a = \left(\frac{4}{9t_0^2}\right)\frac{\delta a}{a^3} = \frac{4}{9}\frac{\delta a}{t^2}. \tag{1.8}$$

This equation has the growing solution $\delta a \propto t^{4/3} \propto a^2$. Hence the density perturbation $\delta = -3(\delta a/a)$ grows as $\delta \propto a$. When the perturbations have grown sufficiently, their self-gravity will start dominating and the matter can collapse to form a gravitationally bound system. The dark matter will form virialised, gravitationally bound structures with different masses and radii. The baryonic matter will cool by radiating energy, sink to the centres of the dark-matter halos, and form galaxies. We now discuss some of the features of such a case for structure formation, starting with the formation of dark-matter haloes. The formation of galaxies will be discussed in the next section.

To describe the growth of structures in the universe, it is convenient to use the spatial Fourier transform $\delta_\mathbf{k}(t)$ of the density contrast $\delta(t, \mathbf{x}) \equiv [\rho(t, \mathbf{x}) - \rho_{\mathrm{bg}}]/\rho_{\mathrm{bg}}$, where $\rho_{\mathrm{bg}}(t)$ is the smooth background density. We treat the density fluctuation $\delta_\mathbf{k}(t)$ as a realisation of a random processes. Then we can define the power spectrum of fluctuations at a given wave number k by $P(k, t) \equiv \langle |\delta_\mathbf{k}(t)|^2 \rangle$, where the averaging symbol denotes that we are treating $P(k, t)$ as a statistical quantity averaged over an ensemble of possibilities; statistical isotropy of the universe implies that the power spectrum can depend on only the magnitude $|\mathbf{k}|$ of the wave number. The power per logarithmic band in k is given by

$$\Delta_k^2(t) = \frac{k^3 |\delta_k(t)|^2}{2\pi^2} = \frac{k^3 P(k, t)}{2\pi^2}. \tag{1.9}$$

For a smoothly varying power spectrum, this quantity is related to the mean-square fluctuation in density (or mass) at the scale $R \approx k^{-1}$ in the universe by

$$\Delta_k^2 = \left(\frac{\delta\rho}{\rho} \right)^2_{R \simeq k^{-1}} = \left(\frac{\delta M}{M} \right)^2_{R \simeq k^{-1}} \cong \sigma^2(R, t). \tag{1.10}$$

Since we can associate a mass scale $M = (4\pi/3)\rho_{\mathrm{bg}}(t_0)R^3$ with a length scale R, one can also treat σ^2 as a function of mass scale: $\sigma^2 = \sigma^2(M, t)$. We shall see in Chap. 5 that the power spectrum of fluctuations in the universe is fairly smooth and hence can be approximated by a power law in k locally at any given time so that $P(k) \propto k^n$. From the result derived above, $\delta \propto a$, it follows that

$$\Delta_k^2(t) \propto a^2 k^{n+3}, \qquad \sigma^2(R, t) \propto a^2 R^{-(n+3)} \tag{1.11}$$

as long as $\sigma \ll 1$, with n being a slowly varying function of scale k or R.

The pattern of density fluctuations is thus characterised by the power spectrum $P(k, t)$ at any given time. The gravitational potential that is due to a density perturbation $\delta\rho = \bar{\rho}\delta$ in a region of size R will be $\phi \propto (\delta M/R) \propto \bar{\rho}\delta R^2$. In an expanding universe $\bar{\rho} \propto a^{-3}$ and $R \propto a$, and the perturbation δ grows as $\delta \propto a$ [see the discussion following Eq. (1.8)], making ϕ constant in time. In particular, the fluctuations that existed in the universe at the time when radiation decoupled from matter would have left their imprint on the radiation field. Because photons climbing out of a potential well of size ϕ will lose energy and undergo a redshift $(\Delta\nu/\nu) \approx (\phi/c^2)$, we would expect to see a temperature anisotropy in the microwave radiation of the order of $(\Delta T/T) \approx (\Delta\nu/\nu) \approx (\phi/c^2)$. The largest potential wells would have left their imprint on the cosmic background radiation at the time of decoupling of radiation and matter. We shall see later that the galaxy clusters constitute the deepest gravitational potential wells in the universe from which the escape velocities are $v_{\text{clus}} \approx (GM/R)^{1/2} \approx 10^3$ km s^{-1}. This will lead to a temperature anisotropy of $\Delta T/T \approx (v_{\text{clus}}/c)^2 = 10^{-5}$. Such a temperature perturbation has indeed been observed in the microwave background radiation, vindicating the case for structure formation.

The entry marked gravitational binding energy in Fig. 1.1 is essentially a measure of $(v/c)^2$ for the largest scales that are gravitationally bound. Equivalently, it can be thought of as the amount of power in the gravitational potential per logarithmic band in Fourier space. Its value can be determined from the temperature anisotropies in CMBR and will be discussed in Chap. 6.

When $\sigma(R, t) \to 1$, that particular scale characterized by R will go nonlinear and matter at that scale will collapse and form a bound structure. Because this occurs when the density contrast σ reaches some critical value $\sigma_c \approx 1$, it follows from relations (1.11) that the scale that goes nonlinear at any given time t in the past (corresponding to a redshift z) obeys the relation

$$R_{\text{NL}}(t) \propto a(t)^{2/(n+3)} = R_{\text{NL}}(t_0)(1+z)^{-2/(n+3)}. \tag{1.12}$$

Equivalently, structures with mass $M \propto R_{\text{NL}}^3$ will form at a redshift z where

$$M_{\text{NL}}(z) = M_{\text{NL}}(t_0)(1+z)^{-6/(n+3)}. \tag{1.13}$$

Such virialised, gravitationally bound structures – once formed – will remain frozen at a mean density $\bar{\rho}$, which is approximately $f_c \simeq 200$ times the background density of the universe at the redshift of formation, z (see Chap. 5). Taking the background density of the universe at redshift z to be $\rho_{\text{bg}}(z) = \rho_c \Omega(1+z)^3$, we find that the mean density $\bar{\rho}$ of an object that would have collapsed at redshift z is given by $\bar{\rho} \simeq \Omega \rho_c f_c (1+z)^3$. We define the *circular velocity* v_c for such a collapsed body as

$$v_c^2 \equiv \frac{GM}{r} \equiv \frac{4\pi G}{3}\bar{\rho}r^2. \tag{1.14}$$

If $\bar{\rho}$ is eliminated in terms of v_c, the redshift of formation of an object can be expressed in the form

$$(1+z) \simeq 5.8 \left(\frac{200}{\Omega f_c}\right)^{1/3} \frac{(v_c/200 \text{ km s}^{-1})^{2/3}}{(r/h^{-1} \text{ Mpc})^{2/3}}. \tag{1.15}$$

It is interesting that such a fairly elementary calculation leads to an acceptable result regarding the redshift for the formation of first structures. If we consider small-scale halos (approximately a few kiloparsecs), the formation redshift can go up to, say, 20. This calculation also introduces the notion of *hierarchical clustering* in which smaller scales go nonlinear and virialise earlier on and the merging of these smaller structures leads to hierarchically bigger and bigger structures. Of course, the process is supplemented by the larger scales going nonlinear by themselves but the importance of mergers cannot be ignored in galaxy-formation scenarios.

It is in fact possible to work out all the properties of collapsed structures from the preceding formalism. To do this, we begin by noting that $\sigma(M, t)$ denotes the *typical* value of density fluctuations that exist in the universe at a given time t. In a statistical description, we could think of regions having a density contrast of, say, ν times larger than the typical value $\sigma(M, t)$ occurring with a probability $\mathcal{P}(\nu)$ that could be approximated as a Gaussian of unit variance in most models. Such a $\nu\sigma$ fluctuation will collapse at a redshift z, which is determined by the condition $\nu\sigma(M, t) = \sigma_c = \mathcal{O}(1)$. Because $\sigma(M, t) \propto a(t)$, it follows that $\sigma(M, t) = \sigma_0(M)(1 + z)^{-1}$, where $\sigma_0(M)$ is the fiducial value of the density fluctuation today. This leads to the first result, namely the redshift $z_{\text{coll}}(M)$ at which any given mass scale collapses in terms of $\sigma_0(M)$:

$$(1 + z_{\text{coll}}) = \frac{\nu\sigma_0(M)}{\sigma_c}. \tag{1.16}$$

Next we use the fact that the density of structures collapsing at redshift z is f_c times larger than the background density at z; hence $\rho = \rho_c f_c(1 + z)^3$ in an $\Omega = 1$ universe. Given M and ρ, we can compute the radius by $R^3 = (3M/4\pi\rho)$, the circular velocity by $v_c = (GM/R)^{1/2}$, and the gravitational potential energy by $U = -(3/5)(GM^2/R)$. The only input needed to compute these quantities is $\sigma_0(M)$, and we obtain by straightforward algebra the following results:

$$\rho = 5.6 \times 10^{13} h^2 \left(\frac{f_c}{200}\right) \left(\frac{\nu}{\sigma_c}\right)^3 \sigma_0^3(M) \, M_\odot \, \text{Mpc}^{-3}, \tag{1.17}$$

$$R = 2.57 h^{-2/3} \left(\frac{M}{10^{11} M_\odot}\right)^{1/3} \left(\frac{f_c}{200}\right)^{-1/3} \left(\frac{\sigma_c}{\nu}\right) \sigma_0^{-1}(M) \, \text{Mpc}, \tag{1.18}$$

$$v_c = 75.5 h^{1/3} \left(\frac{M}{10^{11} M_\odot}\right)^{1/3} \left(\frac{f_c}{200}\right)^{1/6} \left(\frac{\sigma_c}{\nu}\right)^{-1/2} \sigma_0^{1/2}(M) \, \text{km s}^{-1}, \tag{1.19}$$

$$U = 6.72 \times 10^{57} h^{2/3} \left(\frac{M}{10^{11} M_\odot}\right)^{5/3} \left(\frac{f_c}{200}\right)^{1/3} \left(\frac{\sigma_c}{\nu}\right)^{-1} \sigma_0(M) \, \text{ergs}. \tag{1.20}$$

The free-fall time scale for a constant density sphere of radius R and mass M is $t_{\text{ff}} = (\pi/2\sqrt{2})(R^3/GM)^{1/2}$ (see Vol. II, Chap. 3, Eq. 3.13) and is approximately

$$t_{\text{ff}} = 0.77 h^{-1} \left(\frac{f_c}{200}\right)^{-1/2} \left(\frac{\sigma_c}{\nu}\right)^{3/2} \sigma_0^{-3/2}(M) \, \text{Gyr}. \tag{1.21}$$

From virial theorem, we know that the collapse of a protogalaxy from a marginally bound initial state ($E \simeq 0$; $T_i \simeq |U|$) to a virialised final state ($T_f \simeq |U|/2$) will lead to the release of a comparable amount of energy. If this is released over a free-fall time scale of $t_{\rm ff} \simeq 10^8$ yr, then the resulting luminosity is approximately $L = U/t_{\rm ff}$, which scales as

$$L = 7 \times 10^7 h^{5/3} \left(\frac{M}{10^{11} \, M_\odot} \right)^{5/3} \left(\frac{f_c}{200} \right)^{5/6} \left(\frac{\sigma_c}{\nu} \right)^{-5/2} \sigma_0^{5/2}(M) \, L_\odot.$$

(1.22)

Figure 1.2 shows M, ν_c, and L in terms of the collapsed redshift for a cosmological model in which $\sigma_0(M)$ is adequately approximated by the fitting function

$$\sigma_0(M) = \frac{7.2 m^{-0.035}}{1 + 0.82 m^{0.23}}, \qquad m = \frac{M}{10^{11} \, M_\odot}, \tag{1.23}$$

in the range $10^7 \lesssim (M/M_\odot) \lesssim 10^{14}$. We have also taken $\sigma_c = 1.68$ and $f_c = 170$; these numerical values as well as the nature of $\sigma_0(M)$ will be justified in Chap. 5, but their orders of magnitude should be obvious from physical considerations. The three curves (from bottom to top) in each of the frames are for $\nu = 1, 2, 3$, which roughly correspond to collapsed objects with fractional abundances of $(2\pi)^{-1/2} \exp(-\nu^2/2) = (0.24, 0.054, 4.4 \times 10^{-3})$. Figure 1.2 shows that the mass scales from 10^7 to $10^{12} \, M_\odot$ collapse and virialise, forming nonlinear self-gravitating structures in the redshift range $z = (10{-}1)$. Further, it is clear that, although a 1σ fluctuation at the mass scale of, say, $10^{11} \, M_\odot$ will collapse at $z \approx 1$, a 3σ fluctuation at the same scale can collapse as early as $z \approx 6$. Because the probability for a 3σ fluctuation relative to a 1σ fluctuation is down by a factor of $(0.0044/0.24) = 0.018$, we would expect $\sim 2\%$ of the structures having a mass of $10^{11} \, M_\odot$ to form as early as $z = 6$, although copious production of such structures will occur only at $z \approx (1{-}2)$. This analysis shows that galaxy formation is an extended process rather than a single event in the models we will consider.

Exercise 1.2
Putting asunder what gravity has put together: We saw in Vol. II, Chap. 4, that a supernova explosion releases $\sim 10^{51}$ ergs of energy. Show that this hydrodynamic energy can, in principle, unbind the baryonic gas in the dark-matter halos at $z > 5$. Take the binding energy of baryons to be a factor $\Omega_B/\Omega_{\rm DM} \approx 0.15$ less than the binding energy U computed above.

1.4 Galaxy Formation

The description in the previous section applies only to dark matter that is not directly coupled to radiation. The baryons inside the dark-matter halos can radiate energy, cool, and sink to the centres of the dark-matter halos. As the halos merge, the baryonic structures can survive provided they can cool and condense sufficiently within these halos. We now estimate the conditions for this.

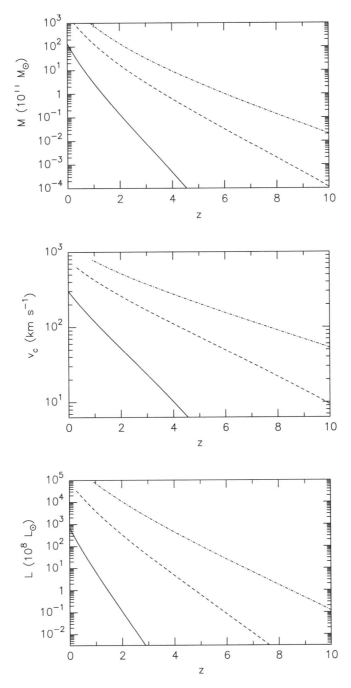

Fig. 1.2.   Properties of self-gravitating halos formed by gravitational collapse of over-dense regions at different redshifts. See text for details.

The cooling of a plasma occurs mainly through two processes. The first is the radiation emitted during recombination of electrons and ions, which, if it escapes the plasma, can be a source of *recombination cooling*. From the discussion in Vol. I, Chap. 6, Section 6.12, we know that the recombination rate varies as $n^2 T^{-1/2}$. The second is the bremsstrahlung cooling with an energy-loss rate proportional to $n^2 T^{1/2}$ (see Vol. I, Chap. 6, Section 6.12). For systems with temperature $k_B T \simeq (GMm_p/R)$, which is much higher than the ionisation potential $\alpha^2 m_e c^2$, the dominant cooling mechanism is thermal bremsstrahlung. The cooling time for this process is

$$t_{\text{cool}} \simeq \frac{nk_B T}{(d\mathcal{E}/dt\,dV)} = \left(\frac{\hbar}{m_e c^2}\right)\left(\frac{1}{n\lambdabar_e^3}\right)\left(\frac{k_B T}{m_e c^2}\right)^{1/2}\frac{1}{\alpha^3}, \qquad (1.24)$$

and the time scale for gravitational collapse is

$$t_{\text{grav}} \simeq \left(\frac{GM}{R^3}\right)^{-1/2}. \qquad (1.25)$$

The condition for efficient cooling, $t_{\text{cool}} < t_{\text{grav}}$, coupled to $k_B T \simeq GMm_p/R$, leads to the constraint $R < R_g$, with

$$R_g \simeq \alpha^3 \alpha_G^{-1} \lambdabar_e \left(\frac{m_p}{m_e}\right)^{1/2} \simeq 74 \text{ kpc}, \qquad (1.26)$$

where $\alpha_G \equiv (Gm_p^2/\hbar c) \approx 6 \times 10^{-39}$ is the gravitational (equivalent) of the fine-structure constant. The analysis assumed that $k_B T > \alpha^2 m_e c^2$; for $R \simeq R_g$ this constraint is equivalent to the condition $M > M_g$, with

$$M_g \simeq \alpha_G^{-2}\alpha^5 \left(\frac{m_p}{m_e}\right)^{1/2} m_p \simeq 3 \times 10^{44} \text{ gm}. \qquad (1.27)$$

This result suggests that systems that have a mass of approximately 3×10^{44} gm and a radius of $\sim$70 kpc could rapidly cool, fragment, and form gravitationally bound structures. Most galaxies have masses around this region, and this is one possible scenario for forming galaxies. Note that the mass and length scales in Eq. (1.27) and (1.26) arise entirely from the fundamental physical constants.

The original (maximum) radius of the cooling plasma estimated above is $\sim$70 kpc. After the matter has cooled and contracted, the final radius is more like (10–20) kpc, which is the typical radii of large galaxies. For $M_g \simeq 3 \times 10^{44}$ gm and $R_g \simeq 20$ kpc, the density is $\rho_{\text{gal}} \simeq 10^{-25}$ gm cm^{-3}, which is $\sim$10^5 times larger than the current mean density, $\rho_0 \simeq 10^{-30}$ gm cm^{-3}, of the universe. If we assume that high-density regions with $\bar{\rho} \gtrsim 100\bar{\rho}_{\text{univ}}$ collapsed to form the galaxies, then the galaxy formation must have taken place when the density of the universe was $\sim$1000 times larger; the value of $a(t)$ would have been 10 times smaller, and the redshift of the galaxy formation should have been $z_{\text{gal}} \lesssim 9$. Further, if the protogalactic plasma condensations were almost touching each other at the time of formation, these centres (which would have been at a separation of ~150 kpc) would have now moved apart to a distance of $150(1 + z_{gal})$ kpc ≈ 1500 kpc $= 1.5$ Mpc. This is indeed the mean separation between the large galaxies today. The nearest galaxy with a radius of ~10 kpc, at a distance of 1 Mpc, would subtend an angle of $\theta_{gal} \approx 10^{-2}$ rad $\approx 30'$. A galaxy at a distance of ~4000 Mpc will subtend $0.5''$.

Still larger structures than galaxies, called *galaxy clusters*, with masses of $\sim10^{47}$ gm, a radius of ~3 Mpc, and a mean density of 10^{-27} gm cm^{-3}, exist in the universe as gravitationally bound systems. Our preceding argument shows that the gas in these structures could not yet have cooled and will have a virial temperature of $T \approx (GMm_p/Rk_B) \approx 4 \times 10^7$ K. In fact, Fig. 1.2 shows that clusters are just about collapsing under self-gravity at the present epoch. We also see this more directly by comparing the age of the universe with the free-fall collapse time. The cosmic time t and the redshift z can be related to each other once the form of $a(t)$ is known. In the simple case of $\Omega = 1$, with $a(t) = (t/t_0)^{2/3} = (1 + z)^{-1}$, we have

$$t(z) = t_0(1 + z)^{-3/2} = (2/3)H_0^{-1}(1 + z)^{-3/2} \approx 6.5h^{-1}(1 + z)^{-3/2} \text{ Gyr}, \tag{1.28}$$

whereas other models can change this value by a numerical factor of order unity. On the other hand, the free-fall time scale under gravity for a system of mass M and initial radius R_{init} is given by

$$t_{ff} = \left(\frac{\pi^2}{8G}\right)^{1/2} R_{init}^{3/2} M^{-1/2} \approx 0.5 \text{ Gyr} \left(\frac{R_{init}}{100 \text{ kpc}}\right)^{3/2} \left(\frac{M}{10^{12} M_\odot}\right)^{-1/2}$$
$$\approx 1.6 \text{ Gyr} \left(\frac{R_{init}}{10 \text{ Mpc}}\right)^{3/2} \left(\frac{M}{10^{15} M_\odot}\right)^{-1/2}. \tag{1.29}$$

Comparing Eqs. (1.28) and (1.29), we see that the peak epoch of galaxy formation should be in the range $z \approx (2\text{–}10)$, depending on the cosmological models, whereas cluster formation is probably still an ongoing process.

Let us next consider in more detail the processes that generate the luminosity of the galaxies. We saw earlier that if the typical binding energy of an average galaxy is radiated over the free-fall time then the luminosity is approximately $L \simeq 5 \times 10^9 L_\odot$. Much higher luminosities, however, can be achieved by nucleosynthesis. We know that ~30 MeV of energy is released for every baryon, which is converted to metals (see Vol. I, Chap. 12). If the total mass burnt in stars during protogalactic star bursts is M_* and the mass of metals is $M_Z \equiv ZM_*$ (which defines Z), the energy released is

$$E_{nucl} = 30 \text{ MeV} \left(\frac{M_Z}{m_p}\right) = 3 \times 10^{-2} Z M_* c^2. \tag{1.30}$$

Taking $Z \simeq Z_\odot \simeq 10^{-2}$, we find that $(E_{nucl}/M_*c^2) \equiv \Delta X \approx 3 \times 10^{-4}$. The energy released by this process will be

$$E_{\text{nucl}} \simeq M_* c^2 \Delta X \simeq 7.2 \times 10^{61} \text{ ergs} \left(\frac{M_*}{10^{11} \, M_\odot} \right) \left(\frac{\Delta X}{4 \times 10^{-4}} \right). \quad (1.31)$$

If *this* energy is released in a time scale $t_{\text{ff}} \simeq 10^8$ yr, the resulting luminosity is approximately $5 \times 10^{12} \, L_\odot$ and the mean star-formation rate (SFR) would be several hundred solar masses per year. Even if the process proceeded on a more extended, hierarchical time scale of 10^9 yr, the resulting luminosity would be comparable with that of a bright galaxy. This suggests that a generic protogalaxy will have a luminosity of $L_{\text{gal}} \simeq 10^{10}$–$10^{12} \, L_\odot$. Taking a luminosity distance of approximately $d \simeq H_0^{-1} c \simeq 2 \times 10^{28}$ cm, we find that the expected bolometric flux will be approximately $(10^{-14}$–$10^{-16})$ ergs cm^{-2} s^{-1}. Equating this bolometric luminosity to νF_ν, we find that the R-band magnitude will be ~25, or $F_\nu \approx 0.1$ mJy at $\lambda \approx 0.3$ mm. (For the definition of magnitudes in terms of physical units, see Vol. II, Chap. 1.)

The preceding discussion illustrates an important link among star formation, the production of metals, and the production of starlight. It is clear that

$$\frac{dM_Z}{dV \, dt} \approx Z \frac{dM_*}{dV \, dt}, \qquad \frac{dE}{dV \, dt} \approx 0.03 Z \frac{dM_*}{dV \, dt}, \qquad \frac{dE}{dV \, dt} \approx 0.03 \frac{dM_Z}{dV \, dt}. \quad (1.32)$$

Thus the density of starlight produced per second and the density of metals produced per second are proportional to the SFR. We shall have occasion to use this relation several times in different chapters.

The actual morphology of the galaxy that is formed will depend on the time scale for star formation in comparison with the free-fall time scale. If most of the early, Population II stars form over a time scale that is somewhat shorter than the free-fall time, then the protogalactic gas will be converted into a collisionless system of stars before the collapse is well underway. (For the definition of Pop I and Pop II stars, see Vol. II, Chap. 1.) In this picture, each galaxy forms in its own environment through an isolated collapse. Another variant of this case will be through mergers of a large number of protogalactic fragments – but in a time scale comparable with the free-fall time of the system – accompanied by rapid star-formation activity. Because the time scales are comparable, these two scenarios are hardly distinguishable from each other observationally. This will lead to spheroidal morphologies with stars moving in orbits with large eccentricity.

It is also possible to imagine an alternative extreme scenario in which the merger of large number of fragments occur over a time scale *comparable* with the Hubble time $t(z)$ [see Eq. (1.28)], with most of the stars having *already* formed within the merging subunits. This is quite different from the first example and will, in general, lead to a different final state. The gas, settling down within a dark-matter potential well, will gradually acquire the lowest energy configuration for a given amount of angular momentum – which will be a centrifugally supported rotating thin disk. This occurs before most of the gas is converted into stars, and hence this case – with a lower SFR – will lead to disklike systems. At high redshifts, when merging was more frequent, the formation of massive galactic disks would have been unlikely as they are almost invariably destroyed during encounters. As the merging rate decreases, the gradual assembly of disk-shaped galaxies increases.

This discussion suggests that, from the SFR and the collapse rate, we could get two broad morphologies, viz., spheroids or disks. Observations show that the morphology of most galaxies can be understood in terms of the relative proportion of spheroidal and disk components (see Section 1.5). Galaxies that have very little mass in the form of a disk component are called *ellipticals* whereas those with very little mass in the form of a spheroidal component are called *disks*. In reality, most of the galaxies fall in between containing a disk component, a spheroidal bulge near the centre, and a spherical halo of stars around the disk. From the preceding consideration, we expect the formation of halo and bulge components to occur at an earlier phase followed by an epoch of disk formation. In what follows we will use the term ellipticals to denote galaxies that are spheroidal in shape.

With the Hubble Space Telescope (HST) it is possible to image the distant galaxies with sufficient resolution so as to classify their morphology and thus compare it with the population of galaxies at $z = 0$. Such surveys show that elliptical galaxies have undergone only passive evolution (that is, the evolution of the galaxy luminosity that is due to the natural evolution of stars in the galaxy) in the redshift range ($0 \lesssim z \lesssim 1$). Over the same period of time, the disks of spiral galaxies show moderate evolution and an overall increase in the activity of star formation. The luminosity functions of spirals and ellipticals are similar in the redshift range ($0 < z < 1$), suggesting that galaxies like the Milky Way have already formed and were in place by $z = 1$. In sharp contrast, the irregular-shaped galaxies have undergone a spectacular amount of evolution in both luminosity and abundance.

The diversity of morphological structures seen in Hubble deep fields (HDFs) possibly suggests a similar diversity in their formation process. The compact spheroidal galaxies probably have formed as a result of gravitational collapse of a protogalaxy or by the merging of several subunits. Estimating the observed SFR from the UV radiation emitted by the massive stars, we may conclude that, if left undisturbed, they would have evolved to elliptical and spiral galaxies of medium mass today. The irregular and fragmented morphology, on the other hand, suggests intense star formation that occurs through interactions and merging events. We shall now discuss the morphology of galaxies in greater detail.

1.5 Morphological Classification of Galaxies

The classification of stars, for example, has a sound basis in terms of Hertzsprung–Russel (H-R) diagrams and evolutionary tracks. We do not have any similar, theoretically well-understood classification schemes for galaxies. Most of the schemes that exist in literature are subjective, physically ill-justified, and – at times – even misleading. We shall, nevertheless, give a brief description of the properties and classification of galaxies in a manner as logical as the current limited understanding will permit.

Because no fundamental description from first principles is possible, it is necessary to provide a classification of galaxies based on observed morphological features. This approach is complicated by at least two factors. First, the visual appearance of a galaxy strongly depends on the wave band in which it is observed. In the optical band, a spiral galaxy may extend to only, say, 15 kpc, but the 21-cm observations of neutral hydrogen might reveal a structure that is larger and extends to, say, 50 kpc. Second, one must remember that sky brightness puts a practical limitation on distinguishing the diffuse structure of the galaxy beyond a particular limit. The typical values of night-sky brightness in U, B, V, R, and I bands are 22.0, 22.7, 21.8, 20.9, and 19.9 magnitude arcsecond^{-2} (see Vol. II, Chap. 1). If the sky brightness is subtracted inaccurately, the luminosity profile of the galaxy will be incorrectly estimated in the outer regions – either with a sharper cutoff than warranted (because of overcorrecting for the sky brightness) or with a flattening profile (if the sky brightness has been underestimated). This is a serious problem, and special techniques are required for analysing the observations correctly when the galaxy brightness is close to the background-sky brightness. Hence even defining a radius or size of a galaxy is a nontrivial and often ill-defined endeavour.

However, if we want to proceed with this approach, it is best to select a particular wave band in which a wide variety of galactic observations is available. The visible band is a natural choice for this task. Galaxies, when observed in the visible band, seem to be made of two components, loosely called *bulge* and *disk*, with the understanding that there could exist galaxies in which one of the components may be completely absent. This classification also has *some* theoretical basis, as outlined in last section. The bulge dominates the central portions of the galaxies, is reasonably spherical in shape, and is supported against self-gravity by the orbital motion of the stars in the mean gravitational field. Most of the bulge stars are Population II and have a large radial component of orbital velocity (as to be expected in the case described in the last section). The disk, in contrast, is a flattened structure in which stars move in almost circular orbits and it contains young massive Population I stars, interstellar dust, and molecular clouds.

Different morphological types of galaxies correspond to combining the bulge and the disk components in different proportions. If the galaxy has virtually no disk, it is usually called an *elliptical* and is classified as E0, ..., Sa, ..., where

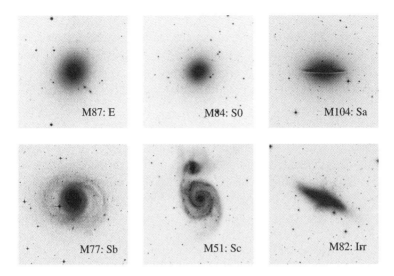

Fig. 1.3. Examples of galaxies having different morphologies. (Figure courtesy of John Peacock.)

the notation En denotes a shape in which the ratio (b/a) of a major to a minor axis is given by $(b/a) = (1 - n/10)$. If the bulge-to-disk contribution is of the order of unity and the disk is reasonably smooth (with no spiral structure, etc.), the galaxy is called *lenticular* and is denoted by S0. In the same configuration, if the disk shows an additional structure in the form of spiral arms, it is called a *spiral* galaxy and is denoted by Sa. If, further, the central region of spiral galaxy contains a prominent barlike structure, such galaxies are denoted by Sb. Galaxies that do not fall in any of these categories are called *irregulars*. This particular classification scheme is called the *Hubble classification*, and E0, E1, etc., are said to represent different *Hubble types* (see Fig. 1.3).

Historically, there was an idea that the sequence E0, E1, ..., S0, Sa/Sb could represent an evolutionary sequence. One often finds in literature references to early- and late-type galaxies, with the convention that E0 is the earliest and Sb is the latest. We will occasionally use this terminology with an understanding that it has *no* direct physical basis.

Because defining an outer radius for a galaxy is a nontrivial task, the next best thing we could do is to characterize the surface brightness of the galaxy as seen in the plane of the sky. If a galaxy is described in terms of a bulge and a disk component, we need to specify the surface-brightness profiles for each of them. We begin with the disk component.

The disk brightness profile usually has an exponential falloff, given by,

$$\Sigma(R) = \Sigma_0 \exp(-R/R_d), \tag{1.33}$$

where R is the two-dimensional radial coordinate in the plane of the disk and

Σ_0 and R_d are constants. Integrating over all R gives the total luminosity as $L_{\text{tot}} = 2\pi \Sigma_0 R_d^2$. The corresponding surface brightness projected on the sky, μ, is usually measured in V-band magnitude per square arcsecond. If the physical luminosity density is Σ (measured in $L_{V\odot}$ per square parsec, where $L_{V\odot}$ is the V-band luminosity of the Sun), then these two quantities are related by

$$\Sigma = 10^{(2/5)(26.4-\mu)} L_{V\odot}\,\text{pc}^{-2}. \tag{1.34}$$

For example, $\mu = 26.4$ V-mag arcsec^{-2} will correspond to $1\,L_{V\odot}$ pc^{-2}. The easiest way to derive such relations is to note that surface brightness is independent of the distance (see Vol. II, Chap. 1) when cosmological expansion is ignored and to calculate both μ and Σ at a convenient distance. We choose a distance

$$D = \frac{180}{\pi}(60)^2\,\text{pc} = 206,265\,\text{pc}, \tag{1.35}$$

at which a length of 1 pc will subtend an angle of 1 arcsec. At this distance, the Sun will have the apparent magnitude

$$m_V = M_{V\odot} + 5\log\frac{D}{10\,\text{pc}} = 4.83 + 5\log(20,626.5) = 26.4. \tag{1.36}$$

At this convenient distance 1 arcsec^2 will contain a luminosity $L = \Sigma\,L_{V\odot}(1\,\text{pc}^2)$ and will have a magnitude $m_V = 26.4 - (5/2)\log\Sigma$. Equating this to μ and solving for Σ, we get the relation given in Eq. (1.34). In terms of μ, Eq. (1.33) becomes

$$\mu(R) = \mu_0 + 1.086\left(\frac{R}{R_d}\right). \tag{1.37}$$

A wide class of spirals has an exponential form for the surface brightness, with the central surface brightness ranging from 21.5 to 23 B-mag arcsec^{-2}. Taking a value of $\mu_0 \simeq 21$ V-mag arcsec^{-2} and using Eq. (1.34), we get $\Sigma_0 = 170\,L_{V\odot}$ pc^{-2}. Thus the surface-density profile of these spirals can be expressed in the form

$$\Sigma(R) \approx 170 e^{-R/R_d} L_{V\odot}\,\text{pc}^{-2}. \tag{1.38}$$

In our galaxy, the brightness towards the galactic poles above the Sun (which is located at a radial distance of $R = R_0 \simeq 8$ kpc) is $\sim$24 V-mag arcsec^{-2}. This implies [through Eq. (1.34)] that $\Sigma(R_0)/2 = 9\,L_{V\odot}$ pc^{-2} with the factor half arising from the fact that there are two sides to the disk of the Milky Way. If the exponential law is applicable to the Milky Way, then we can immediately determine the scale length by using the value of $\Sigma(R_0)$; we get $R_d \approx 3.6$ kpc, which – as we shall see – is consistent with other observations as well.

In determining the properties of disk galaxies, it is important to keep in mind some observational selection effects in order to avoid spurious correlations. As an example of observational limitations, let us consider the distribution of central brightness of a sample of disk galaxies with an exponential profile and total luminosity $L_{tot} = 2\pi \Sigma_0 R_d^2$. Let R_c be the radius at which the surface brightness Σ falls to a prespecified value Σ_c. The radius R_c satisfies the equation $R_c/R_d = \ln(\Sigma_0/\Sigma_c)$. Writing $R_d = (L_{tot}/2\pi \Sigma_0)^{1/2}$, we get

$$R_c = \left(\frac{L_{tot}}{2\pi \Sigma_0}\right)^{1/2} \ln\left(\frac{\Sigma_0}{\Sigma_c}\right). \tag{1.39}$$

This equation shows that R_c has a distinct maximum as a function of the central surface brightness Σ_0. In a sample of galaxies we average over a luminosity function, thereby replacing L_{tot} with some average value without affecting the fact that R_c has a maximum as a function of Σ_0. The maximum value of R_c occurs at $\Sigma_0 = \Sigma_c e^2$ or – in terms of magnitudes – at $\mu_0 = \mu_c - 2.17$. It could happen that, because of observational limitations, a galaxy is included in the sample only if R_c (at which the surface brightness Σ falls to a prespecified value Σ_c) is greater than some minimum value R_{min}. Then the sample will preferentially contain galaxies with central surface brightness that cluster around the maximum value ($\mu_c - 2.17$). If $\mu_c \approx (24$–$25)$ B-mag arcsec^{-2}, then we may end up concluding that the central surface brightness of the galaxies is peaked around $\mu_0 \approx (21.83$–$22.83)$ B-mag arcsec^{-2}. Obviously, this result arises because of an observational selection effect.

Exercise 1.3

Practice with surface brightness: Consider a typical Sbc galaxy, NGC 5055, located at a distance of $D = 8$ Mpc and seen at an inclination $i = 60°$. (This is the angle between the normal to the plane of the galaxy and the line of sight.) R-band photometry along the major axis of this galaxy shows that the surface magnitude can be fitted reasonably well by the relation

$$\mu_R = 19.75 + 0.72 \left(\frac{\theta}{1 \text{ arcmin}}\right). \tag{1.40}$$

Determine the surface-luminosity profile in physical units. [Answer: Taking the apparent surface brightness to be $\Sigma \propto e^{-R/R_d}$ and using the relation between μ and Σ, we know that $\log \Sigma = -(2/5)\mu + \text{constant} = -(R/R_d)\log e + \text{constant}$ so that

$$\mu = \text{constant} + 1.086 \frac{R}{R_d}. \tag{1.41}$$

Comparing with the fit to the observation, we find that R_d subtends an angle of 1.51 arcmin. At a distance of 8 Mpc, 1 arcmin corresponds to $(60/206,265) \times 8$ Mpc $= 2.33$ kpc so that the physical scale length for our system is 3.5 kpc. We next need the relationship between Σ and μ evaluated in the R band. We can obtain this exactly as in the text by considering an object at a distance 206,265 pc (where 1 arcsec corresponds to 1 pc) and noting that the R-band magnitude of the Sun is $m_R = 5\log(206,265/10) + 4.74 = 26.31$. Hence the relation we now need is

$$\Sigma_R = 10^{(2/5)(26.31 - \mu_R)} L_{R\odot} \, \text{pc}^{-2}. \tag{1.42}$$

Using this relation and noting that the apparent projected density and the actual surface density are related by $\Sigma_{\text{app}} = (\Sigma_{\text{true}}/\cos i)$ we find the luminosity profile of the galaxy to be

$$\Sigma_R(R) = 190e^{-(R/3.5\,\text{kpc})} L_{R\odot} \, \text{pc}^{-2} \tag{1.43}$$

The total luminosity obtained by integration over all R is $L_{\text{tot}} = 2\pi \Sigma_0 R_d^2 = 1.5 \times 10^{10} L_{\odot}$.]

Let us next consider the bulge component that has a more extended distribution and that can be modelled by taking the falloff as $\exp[-(R/R_0)^\alpha]$ with $\alpha \simeq 0.25$. Because elliptical galaxies are dominated by a bulge component, this distribution also describes the surface-density profiles of ellipticals. The scale length R_0 is related to the size of the elliptical galaxy although, as stated earlier, defining the size of a galaxy itself is a nontrivial task. It is conventional in the description of elliptical galaxies to use three different definitions for the radius, all of which are motivated by practical considerations: (1) The Holmberg radius R_H is taken to be the length of a semimajor axis of an ellipsoid whose isophotal surface brightness is $\mu = 26.5$ B-mag arcsec^{-2}; (2) the core radius R_c is defined to be the one at which the surface brightness drops to half its central value, i.e., $\Sigma(R_c) = (1/2)\Sigma(0)$; and (3) the effective radius R_e is defined such that half the light is enclosed by this radius.

In terms of R_e, the surface brightness of the bulge (or for that matter of an elliptical galaxy that is made of essentially a bulge component) is well described by

$$I(R) = I_e \exp\left\{-7.67\left[\left(\frac{R}{R_e}\right)^{1/4} - 1\right]\right\}, \tag{1.44}$$

which is usually called the *de Vaucouleurs law*. The corresponding result, in units of magnitude per square arcsecond, is given by

$$\mu(R) = \mu_e + 8.3268\left[\left(\frac{R}{R_e}\right)^{1/4} - 1\right]. \tag{1.45}$$

The total luminosity obtained by integration over all R is $L_{\text{tot}} = 22.7 I_e R_e^2$.

In general, a galaxy will have both bulge and disk components. Figure 1.4 shows a typical B-band surface-brightness profile of an Sa0 galaxy. It is clear that there are two contributions with different radial dependences adding up to give the observed surface brightness. At large radius the brightness profile is

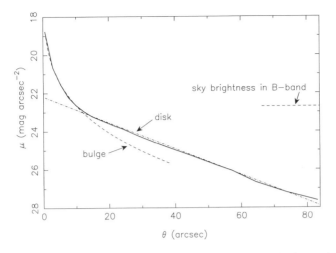

Fig. 1.4. Surface brightness of a typical Sa0 galaxy showing the bulge and disk contributions to the total light.

well fitted by

$$\mu_{\mathrm{disk}}(\theta) = 22.185 + 0.068 \left(\frac{\theta}{1 \text{ arcsec}} \right), \tag{1.46}$$

which is characteristic of an exponential disk. Near the origin, the intensity profile is well fitted by

$$\mu_{\mathrm{bulge}}(\theta) = 15.028 + 4.277 \left(\frac{\theta}{1 \text{ arcsec}} \right)^{1/4}, \tag{1.47}$$

which is characteristic of a bulge near the centre. If we assume this fit to extend all the way to origin, the central brightness that is due to the bulge is 15.028 magnitudes, which is significantly higher than the extrapolated central brightness that is due to the disk of 22.185 magnitudes.

The ratio of the bulge luminosity to the disk luminosity varies smoothly with the Hubble type steadily decreasing from S0 to Sm galaxies. Although the classification of galaxies based on the bulge and the disk components is purely phenomenological, it turns out that several other properties of the galaxies correlate reasonably with the Hubble type assigned based on the morphology. In fact, it is possible to use the bulge-to-disk ratio for a more quantitative comparison of different physical properties with the morphology.

Exercise 1.4

Practice with surface brightness: (a) Convert the fits given for the Sa0 galaxy in Fig. 1.4 into physical units, given that this particular galaxy is at a distance of ~60 Mpc. (b) Estimate the disk and the bulge scale lengths in physical units. (c) What is the total amount of luminosity in the bulge compared with that in the disk?

Let us next consider the spectral-energy distribution (SED) of a galaxy. Because the galaxy is made up of stars, gas and some amount of dust, it is – in principle – possible to express the total luminosity $L(\lambda)$ of the galaxy at wavelength λ at any given time in the form

$$L(\lambda) = \int dV \sum_i \rho_i L_i(\lambda) \exp(-\tau) + L_{\text{others}}(\lambda), \qquad (1.48)$$

where ρ_i is the density of stars of type i having a luminosity $L_i(\lambda)$ and τ is the correction for absorption of light at this wavelength within the galaxy; $L_{\text{others}}(\lambda)$ is due to nonstellar sources. (Because stars evolve over a period of time, this luminosity, in general, will vary with time.) It should be stressed that, although stars contribute the dominant amount of energy in a galaxy, there are also other forms of radiation that dominate different wave bands. For example, the radiation in the x-ray band is possibly due to individual accretion disks in binary stars whereas the radio luminosity arises from synchrotron radiation from ambient electrons in the galaxy.

Figure 1.5 shows the integrated SED of the galaxy NGC 7714. The UV light and the optical light come essentially from stars with different ranges of masses and corresponding temperatures. If the dominant contribution is from stars in the mass range $M_1 < M < M_2$ and if the temperature–mass relationship of stars is given by the function $T(M)$, then the energy radiated per logarithmic band of

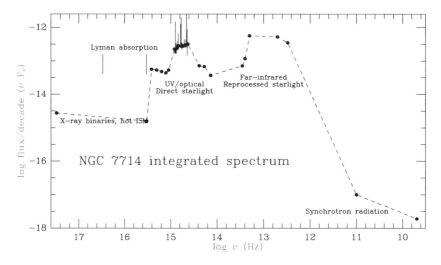

Fig. 1.5. SED of NGC 7714. Note that most of the light is contributed by the stars in the UV and the optical bands, whereas some amount of far-IR radiation arises because of reprocessing of starlight by dust. The high-energy radiation in the x-ray band and radio emission are subdominant. (Figure courtesy of Bill Keel.)

frequency will be

$$\nu I_\nu \propto \int_{M_1}^{M_2} dM \, \frac{\nu^4 n(M)}{\exp[h\nu/k_B T(M)] - 1}, \tag{1.49}$$

where $n(M)$ is the number density of stars in a mass range $(M, M + dM)$. As the stars evolve, the variables in this equation will change with time, making νI_ν change. However, two simple limits of this integral can be easily worked out. If the dominant contribution comes from a narrow range of masses, then the integral is dominated by a Planckian at a given temperature and the spectrum will be approximately thermal. On the other hand, if a broad range of masses contributes so that the limits M_1 and M_2 are irrelevant, then the resulting intensity will be an approximate power law if $T \propto M^\alpha$ and $n(M) \propto M^{-\beta}$. Evaluating the integral, we easily see that

$$\nu I_\nu \propto \nu^{4-(\beta-1)/\alpha} \tag{1.50}$$

in the $M_1 \to 0$, $M_2 \to \infty$ limit. These forms, however, are applicable over only a limited range. At high frequencies ($h\nu > 13.6$ eV, $\lambda < 912$ Å) absorption by hydrogen atoms will strongly suppress the intensity (which can be seen clearly in Fig. 1.5). At long wavelengths, it is necessary to take into account reradiation of energy by dust. A simple model for this reprocessing can be based on the assumption that a fraction f of the optical-UV luminosity is absorbed and reemitted in the far infrared (FIR) so that $L_{\mathrm{FIR}} = fL$. This FIR radiation will be in the form of a modified Planck spectrum with a SED $B_\nu \epsilon_\nu$, where B_ν is the Planckian intensity for a temperature in the range $T \approx (45\text{–}80)$ K and ϵ_ν is the dust emissivity that could be taken as a power law: $\epsilon_\nu \propto \nu^x$, where $x = 1$ for $\nu > 3 \times 10^{10}$ Hz (100 μm) and $x = 2$ for $\nu < 2 \times 10^{11}$ Hz (1 mm) with a linear extrapolation between the two limits. This will lead to the form shown in Fig. 1.5 at the FIR band.

The total integrated light from a galaxy, in a given band, can be used to assign the magnitudes (and colours) in that band. Such observations are complicated by the following factors: (1) the in situ absorption of light in the galaxy, (2) the differences in the absorption at different wavelengths (for example, the absorption in the K band at 2.2 μm, is only one tenth of that in the visual), and (3) the fact that the observed and the emitted wavelengths are different because of redshift arising at cosmological scales. When corrections are made for these effects to the extent possible, we find that the colours of the galaxy are broadly correlated with the galaxy type, as shown in Fig. 1.6. By and large, elliptical, lenticular, and early spiral galaxies have absolute blue magnitudes, $-22 < M_B < -16$, whereas later-type spirals and irregulars have $-18 < M_B < -12$.

This trend suggests, for example, that Sc s have greater fraction of massive main-sequence stars relative to earlier spirals, making Sc galaxies bluer than Sa and Sb galaxies. For successively later types of galaxies, a larger part of overall light is emitted in bluer wavelengths, implying an increasingly greater fraction

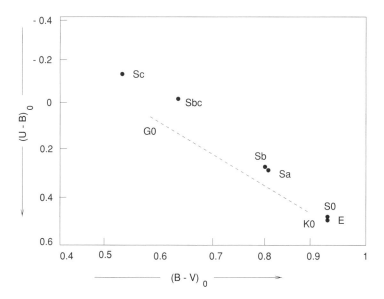

Fig. 1.6.   Colour–colour diagram for galaxy types. The blueness of the galaxy increases upwards on the *y* axis and to the left on the *x* axis. The dashed line gives the corresponding colour–colour values for main-sequence stars for comparison.

of younger, more massive, main-sequence stars. (Note that – in the astronomical terminology, which respects history more than it respects physics – galaxies of the *later* Hubble type are actually dominated by stars on the *earlier* part of the upper main sequence.) Irregulars are bluest of the Hubble sequence, with $(B - V) \simeq 0.4$. Many irregulars also get bluer towards their centres (unlike early-type spirals, which become redder), suggesting that irregulars have active star-forming regions in the centre. These ideas about the composition are confirmed by the (M/L) ratio for spirals – which decreases progressively as one moves to the later Hubble type – since upper main-sequence stars have lower mass-to-light ratios. The dashed line in Fig. 1.6 is the corresponding curve for the main-sequence stars.

At low redshifts ($z \approx 0$) the most dominant sink of baryons in galaxies is the spheroidal component. The ellipticals and the bulges of disk systems currently contain, in their stellar population, $\sim$63% of the baryonic mass that is in the galaxies in the form of stars, stellar remnants, or gas. The disks contain a total of only $\sim$21% of the baryonic mass, and the gas in the galaxies comprises $\sim$15% of the total baryonic mass. These and related components are given in Fig. 1.1 against appropriate entries.

1.6 The Evolution of Galaxies

Estimates of the age of the stars suggest that much of the star formation took place when the universe was $\sim$20% of its current age. If $t_0 \simeq 14$ Gyr, then most ellipticals and bulges of disk galaxies should have formed when the universe was ~ 3 Gyr old. To form a total stellar mass of $\sim 10^{11} \, M_\odot$ in a few gigayears requires star-formation rates of $\sim 10^2 \, M_\odot \, \text{yr}^{-1}$. For any reasonable initial mass function (IMF), we would expect larger number of low-mass stars to be formed compared with high-mass stars. However, because high-mass stars are significantly more luminous than the low-mass stars, they dominate the luminosity in the initial stages of star formation. After the first few million years of continuous star formation at a rate of, say, $10 \, M_\odot \, \text{yr}^{-1}$, a galaxy would have acquired a stellar mass of $10^7 \, M_\odot$. Of these, nearly 2×10^5 are massive stars that contribute a total luminosity of approximately $2 \times 10^{10} \, L_\odot$. This activity increases the luminosity of the galaxy by $\sim 20\%$, and its colour becomes bluer because of the radiation by massive stars. Eventually (after $\sim 10^6$ yr), these high-mass stars die out and the luminosity of the galaxy is dominated by low-mass stars.

As the stars evolve over different time scales and move in the H-R diagram, the physical properties of the galaxy will change with time. To study this process in detail, we begin with an IMF that characterizes the number of stars in a given mass range (which is produced at any given time) and some assumptions regarding the variation of star-formation rate in time. The IMF is usually taken to be a power law, $(dN/dM) \propto M^{-\alpha}$ with an index $\alpha \approx 2.35$ (see Vol. II, Chap. 3, Section 3.3). As regards the SFR, the two extreme limits that will bracket the possibilities are that (1) all the stars have formed in a single burst at some time in the past or (2) the SFR is constant over the relevant period of time. Standard stellar-evolution theory can then be used to evolve these stars forwards in time and to determine the light emitted by the collection of stars at any given instant. Figure 1.7 shows the results of one such theoretical study that uses a Salpeter IMF with index $\alpha = 2.35$ and metallicity $Z = 0.008$. The star-formation rate was assumed to be continuous at $1 \, M_\odot \, \text{yr}^{-1}$. The continuous curves are for $t = 2, 10, 20, 50, 100, 200, 500,$ and 900 Myr from bottom to top, and it is clear that overall luminosity increases because of the star formation. The two dashed curves at the extreme values of t are fits of the form

$$\lambda L_\lambda \propto \left(\frac{\lambda_0}{\lambda}\right)^n \frac{1}{\exp(\lambda_0/\lambda) - 1}. \tag{1.51}$$

At $t = 2$ Myr, the fitting parameters are given by $\lambda_0 = 3000$ Å and $n = 4$, which correspond to a Planckian distribution with temperature $T = 4.73 \times 10^4$ K. At $t = 900$ Myr, the fit is with $\lambda_0 = 1000$ Å and $n = 2.2$, which are fairly distorted from a pure Planckian. We saw in Section 1.5 [see relation (1.50)] that when a broad set of stars contributes to the luminosity, $\nu I_\nu \propto \lambda L_\lambda \propto \nu^{4-(\beta-1)/\alpha}$. Taking $T \propto M^{3/4}$ (corresponding to $L_* \propto M_*^5$ for the upper MS stars) and $n(M) \propto M^{-2.35}$, we have $\alpha = 3/4$, $\beta = 2.35$, giving $\lambda L_\lambda \propto \nu^{2.2} \propto \lambda^{-2.2}$. Also note that, for most of the range of $\lambda \approx (10^3 - 10^4)$ Å, the function $[\exp(\lambda_0/\lambda) - 1]^{-1}$ can be approximated as proportional to λ, making $L_\lambda \propto \lambda^{-2.2}$; in terms of frequencies,

$$L_\nu \propto \lambda^2 L_\lambda \propto \nu^{0.2} \approx \text{constant}. \tag{1.52}$$

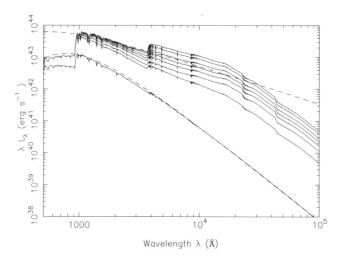

Fig. 1.7.   Results of population synthesis that uses a Salpeter IMF with index $\alpha = 2.35$ and metallicity $Z = 0.008$. The SFR was assumed to be continuous at $1\,M_\odot\,\text{yr}^{-1}$. The continuous curves are for $t = 2, 10, 20, 50, 100, 200, 500,$ and 900 Myr from bottom to top. The dashed curves are fits discussed in text. (Figure based on the data made available on the website http://stsci.edu/science/starburst99.)

These galaxies emit almost constant power per unit frequency in the range $10^3\,\text{Å} < \lambda < 10^4\,\text{Å}$; putting in all the numbers, we see that the flux is $\sim 10^{20.6}$ W Hz^{-1} for a SFR of $1\,M_\odot\,\text{yr}^{-1}\,\text{Mpc}^{-3}$.

Figure 1.8 shows the corresponding curves when the star formation was assumed to be instantaneous and the parameters adjusted so that the SED at the earliest moment is similar to the case discussed above. The curves from top to bottom are for $t = 2, 10, 20, 50, 100, 200, 500,$ and 900 Myr, and it is clear that the overall luminosity is decreasing with time. The aging away of blue O and B stars around one gigayear leads to the characteristic spectral shape shortwards of 4000 Å. The divergence of the predictions for the two cases in Figs. 1.7 and 1.8 show that very different SEDs for the galaxy can be expected, depending on the star-formation history.

Figure 1.9 gives the evolution of bolometric magnitude M_{bol} as a function of time for these models, with the left panel giving the results for continuous star formation and the right panel giving the results for instantaneous star formation. The solid curves and the dotted–dashed curves have $\alpha = 2.35$, and the upper cutoff for the initial mass function is 100 and $30\,M_\odot$, respectively. The long dashed curve has $\alpha = 3.30$ and a cutoff mass of $100\,M_\odot$.

This procedure for determining the SED of galaxies (called *population synthesis*), although simple in principle, introduces several uncertainties in practice. To begin with, the IMF is very poorly known and we need to take into account the fact that it could vary both in space and in time. There is no guarantee that all the parts in the galaxy have formed at the same time and – in fact, as discussed

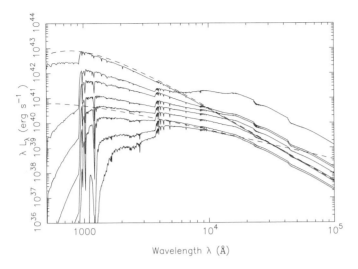

Fig. 1.8. Curves corresponding to those in Fig. 1.7 when the star formation was assumed to be instantaneous. The curves from top to bottom are for $t = 2, 10, 20, 50,$ $100, 200, 500,$ and 900 Myr. The dashed curves are fits discussed in text. (Figure based on the data made available on the website http://stsci.edu/science/starburst99.)

in Section 1.4 – there is a strong indication that different parts of the galaxy did form at different epochs. The uncertainties in stellar-evolution theory add further sources of error. It is also necessary to take into account the fact that, when a galaxy is observed in different wavelengths, we will be sampling different types of stellar populations. (In a galaxy like the Milky Way, K and M giants contribute

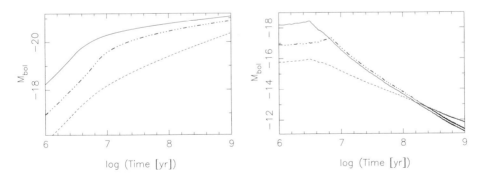

Fig. 1.9. The evolution of bolometric magnitude M_{bol} as a function of time for the two models described in Figs. 1.7 and 1.8. The left panel gives the results for continuous star formation, and the right panel gives the results for instantaneous star formation. The solid curves and the dotted–dashed curves have $\alpha = 2.35$ and the upper cutoff for the initial mass function is 100 and 30 $M_\odot$, respectively. The dashed curves have $\alpha = 3.30$ and a cutoff mass of 100 $M_\odot$. (Figure based on the data made available on the website http://stsci.edu/science/starburst99.)

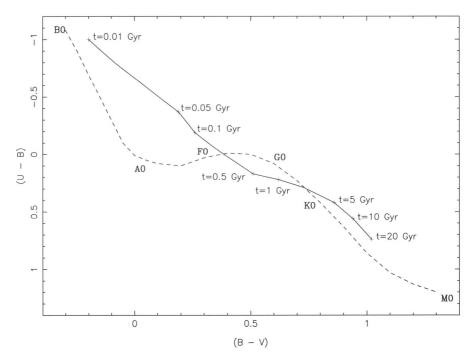

Fig. 1.10.   Population synthesis result for a Salpeter IMF and a single burst of star formation; the age is indicated near the solid curve. The dashed curve is the corresponding colour–colour plot for the main-sequence stars.

75% of the integrated I-band light but only 1% of the V-band light. In contrast, B stars contribute only 3% in the I band but 41% in the V band.)

To the lowest order of approximation, we will expect the galaxy to redden as its stars age if the star formation does not continuously replenish the evolution and disappearance of old stars. Simple models for stellar-evolution do lead to a conclusion in conformity with this. Figure 1.10 shows a population synthesis result for colour evolution for a Salpeter IMF and a burst of star formation. The colour–colour diagram for main-sequence stars is superposed on the figure for comparison.

Some of the preceding results regarding the luminosity evolution of galaxies can be qualitatively understood from stellar-evolution theory. We saw in Vol. II, Chap. 3, Section 3.6, that stars with $M \gtrsim 1.25\,M_\odot$ evolve in $t < 5$ Gyr. Thus in any stellar population that originated from a burst of star formation, there will be only stars with initial masses $M \lesssim 1.25\,M_\odot$ for $t > 5$ Gyr. Further, these low-mass stars emit most of their energy during the giant branch (GB) phase. Because the time these stars spend in the GB is fairly short, we can estimate the luminosity of this population by

$$L \approx \left(E_{GB} \frac{dN}{dM} \right)_{M_{GB}} \left| \frac{dM_{GB}}{dt} \right| \tag{1.53}$$

where $E_{GB}(M)$ is the total energy emitted in the GB, (dN/dM) is the number of stars in the population with mass in the range $(M, M + dM)$, and $M_{GB}(t)$ is the age at which a star of mass M turns off the main sequence. From our understanding of stellar evolution, we can take

$$\frac{M_{GB}(t)}{M_\odot} \simeq \left(\frac{t}{10\,\text{Gyr}}\right)^{-0.4}, \quad \frac{dN}{dM} \simeq K\left(\frac{M}{M_\odot}\right)^{-\alpha}, \tag{1.54}$$

with $\alpha \simeq 2.35$. Combining these, we get

$$L \simeq \frac{K M_\odot\, E_{GB}(M_{GB})}{(25\,\text{Gyr})} \left(\frac{M_{GB}}{M_\odot}\right)^{3.5-\alpha} \tag{1.55}$$

or, equivalently,

$$\frac{d\ln L}{d\ln t} = \left[\frac{d\ln E_{GB}}{dM_{GB}} + (3.5 - \alpha)\right]\frac{d\ln M_{GB}}{d\ln t} = 0.4\alpha - \left(1.4 + 0.4\frac{d\ln E_{GB}}{dM_{GB}}\right). \tag{1.56}$$

Because $(d\ln E_{GB}/d\ln M_{GB})$ is in the range $(0, 1)$ and $\alpha \approx 2.35$, we expect L to be a decreasing function of t; more detailed models suggest that $(d\ln L/d\ln t) \simeq 0.3\alpha - 1.6$, which is in reasonable agreement with the preceding analysis.

The preceding discussion highlights the importance of understanding the star-formation history of different types of galaxies in order to make useful predictions about them. Observationally the star formation in galaxies is studied by use of different diagnostics, all of which have different levels of reliability. We now briefly discuss some of them.

Recombination line emission (especially $H\alpha$) is characteristic of HII regions (see Vol. II, Chap. 9, Section 9.3), which are made of ionised gas around young star clusters. The amount of Lyman continuum photons emitted by any single massive young star can be predicted from the theory of stellar atmosphere for different spectral types and evolutionary stages. Roughly, the number of O7 stars needed to produce a given $H\beta$ luminosity is proportional to $H\beta$ luminosity, $L(H\beta)$. Given an IMF, we can relate the total mass of stars formed to the number of stars of spectral type O7. This allows us to relate the total SFR to the $H\beta$ luminosity (or to the $H\alpha$ luminosity, which is ~2.9 times stronger under low-density conditions). For a Salpeter IMF, we find that the SFR scales as

$$\text{SFR} \approx \left[\frac{L(H\alpha)}{1.12 \times 10^{41}\,\text{ergs s}^{-1}}\right] M_\odot\,\text{yr}^{-1}. \tag{1.57}$$

Different assumptions about the IMF, especially a change in the lower mass cutoff or slope of the lower main sequence, will lead to a factor of few changes in this result. Such studies show that the SFR is broadly correlated with the Hubble type and increases from E/S0 galaxies to Sc galaxies. In general, there exists a fairly tight correlation between the the average SFR in a gaseous disk Σ_{sfr} and the average surface density of gas Σ_{gas}, which can be stated in the form[2]

$$\Sigma_{sfr} = (2.5 \pm 0.7) \times 10^{-4} \left[\frac{\Sigma_{gas}}{1 \, M_\odot \, pc^{-2}} \right]^{1.4 \pm 0.15} M_\odot \, yr^{-1} \, kpc^{-2}. \quad (1.58)$$

This relation holds observationally over almost 5 orders of magnitude in Σ_{gas} and is called the *Schmidt law*.

Another way of measuring the SFR is to relate it to the UV intensity produced by the star. From stellar-evolution theory, we know that massive and young stars are very bright in the UV, whether or not they produce a discernable HII region. In principle, this measurement allows us to see star formation directly; however, dust extinction is a major source of uncertainty in using this method. Stars that produce UV radiation are also the same ones that produce most of the metals and disperse them in supernova bursts. Hence we expect a strong correlation between the UV background light from galaxies and the metallicities. (We shall say more about this in Chaps. 7 and 8.)

The studies of UV radiation background in the universe have allowed a very tentative inference of the SFR, which can be parameterised in the form

$$\frac{dM_*}{dt \, dV} = \dot{\rho}_* = A \frac{e^{az}}{e^{bz} + c}. \quad (1.59)$$

A naive conversion of the observed UV radiation into the SFR leads to the parameter values $A = 0.11 \, M_\odot \, yr^{-1} \, Mpc^{-3}$, $a = 3.4, b = 3.8$, and $c = 44.7$. This curve is shown by the thick solid curve in the top left frame of Fig. 1.11. Correcting the UV luminosity for dust enhances the estimate of SFR, especially at high redshifts but in a model-dependent way.[3] Such a study leads to the form shown in top right frame of Fig. 1.11 by the thick dashed curve, which has the parameters $A = 0.13 \, M_\odot \, yr^{-1} \, Mpc^{-3}$, $a = 2.2, b = 2.2$, and $c = 6.0$. At high redshifts, this fit gives essentially a constant SFR of $\sim 0.1 \, M_\odot \, yr^{-1} \, Mpc^{-3}$. Given the SFR as a function of z, we can obtain the total amount of gas that has been processed into stars by some epoch t from the integral

$$\rho_*(t) = \int_0^t dz \left(\frac{dt}{dz} \right) \dot{\rho}_*(z). \quad (1.60)$$

These curves are shown in the bottom left frame of Fig. 1.11 by the thick curves for the two cases in terms of $\Omega_*(t) = \rho_*(t)/\rho_c$. A somewhat more relevant quantity is the amount of gas that has been converted into stars per Hubble time, which is given by $(a/\dot{a})\dot{\rho}_*$. This quantity – which is not constrained to be monotonic, unlike $\rho_*(t)$ – is shown in the bottom right frame. The two bottom frames show that nearly half the stars are produced at $z \lesssim 1$, which is not apparent in the top frames; this result arises from the fact the the universe spends less time at higher redshifts.

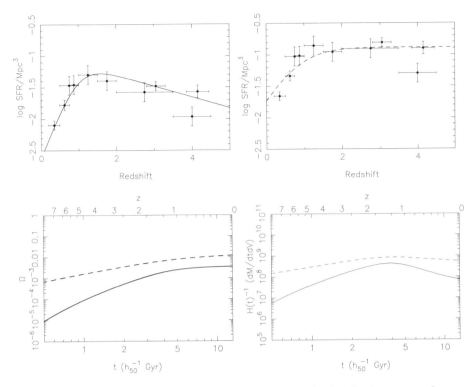

Fig. 1.11. Rate of production of stars in the universe. The data in the top two frames are based on the UV radiation background in the universe, and the top right frame has the data corrected for dust obscuration. The thick solid and dashed curves in the top two frames are parameterised fits to the data. The actual SFR is probably bracketed by these two curves. The thick dashed and solid curves in the bottom left frame give the integrated total star formation (obtained by integration of the curves in the top frame over time). The result is presented in terms of the density parameter Ω that is due to the stars. The two curves in the bottom right frame give the amount of star formation within a Hubble time in units of $M_\odot$ Mpc^{-3}. (Data courtesy of C.C. Steidel.)

We can also try to infer the SFR from the reprocessed FIR emission arising from the dust. A significant fraction of FIR emission arises from dust grains that are heated by the absorption of light in the UV/optical band from young stars and reradiate the energy as approximately blackbody emission in the FIR. The exact interpretation of energy balance among UV, optical, and IR bands in a given galaxy is quite involved (and uncertain) but we could form some broad conclusions regarding the global energy balance, as the total amount of energy removed by dust absorption must emerge in the FIR. In such a case, the luminosity in the $(10–300)\mu$ band must be equal to that in the $(912–3000)$Å band. There is some empirical evidence for a tight correlation between total FIR emission and Hα luminosity (with $L_{\text{FIR}} \approx 10^{2.2} L_{\text{H}\alpha}$), suggesting that FIR emission is correlated to the current SFR.

1.7 Properties of Disk Galaxies

Outside the central region, disk galaxies show the exponential luminosity profile [see Eq. (1.33)], although near the centre some minor deviations are often observed. When seen edge on, the luminosity of the disk arises from a very thin sheet well-fitted by the function $\mathrm{sech}^2(z/z_0)$, where z is the vertical coordinate normal to the plane of the disk (which is assumed to be located at $z = 0$; see Exercise 2.4, p. 93, for a theoretical basis). Combining these results, we may take the three-dimensional luminosity distribution of a disk as

$$L(R, z) = L_0 \exp(-R/R_d)\,\mathrm{sech}^2(z/z_0). \tag{1.61}$$

The centre brightnesses of many galaxies are typically in the range of approximately (22 ± 1) B-mag arcsec^{-2}. The disks also show a fairly sharp truncation in the radial direction beyond $R \approx (4.2 \pm 0.6)(R/R_d)$.

Late-type disk galaxies contain good deal of dust, in which the B-band optical depth in the central regions is approximately $\tau = (1 - 2)$ whereas the outer regions (near the 25 B-mag arcsec^{-2} isophote) are fairly transparent with $\tau = (0.1–0.5)$. Most of the interstellar gas and dust are found in a disk with a scale height significantly smaller than that of the stellar disk. The radial distribution of gas does not follow the exponential distribution of starlight; in particular, neutral-hydrogen profiles show a wide range of distribution.

When a particular disk galaxy shows a dominant bulge, the intensity distribution of the bulge is quite close to de Vaucouleur's law [Eq. (1.44)]. The bulge colours are similar to ellipticals of the same luminosity and – to the extent that colour gradients can be measured – they are also similar to those of ellipticals. The colours of galactic disks themselves do not show the systematic trends with radius and total luminosity (unlike, as we shall see later in Section 1.8, the ellipticals).

The central regions of disk galaxies have been investigated extensively because many of the galactic bulges are suspected to harbor a massive black hole in the centre. We obtain the radius of influence r_* of a black hole of mass M_{BH} on a stellar system with velocity dispersion σ_*^2 by equating the gravitational potential energy per unit mass (GM_{BH}/r) to the kinetic energy per unit mass $(\sigma_*^2/2)$:

$$r_* \simeq \mathcal{O}(1)\frac{GM_{\mathrm{BH}}}{\sigma_*^2} \simeq 4\left(\frac{M_{\mathrm{BH}}}{10^7\,M_\odot}\right)\left(\frac{\sigma_*}{100\,\mathrm{km\,s}^{-1}}\right)^{-2}\,\mathrm{pc}. \tag{1.62}$$

The angle subtended by this radius in a galaxy located at a distance D from us is

$$\theta \simeq \frac{r_*}{D} \approx 0.1''\left(\frac{M_{\mathrm{BH}}}{10^7\,M_\odot}\right)\left(\frac{\sigma_*}{100\,\mathrm{km\,s}^{-1}}\right)^{-2}\left(\frac{D}{10\,\mathrm{Mpc}}\right)^{-1}. \tag{1.63}$$

Figure 1.12 shows lines of constant M_{BH} on a $\sigma^2 - D$ plane for $\theta = 1''$, which is the limit of resolution from ground-based telescopes. A vast majority of galaxies

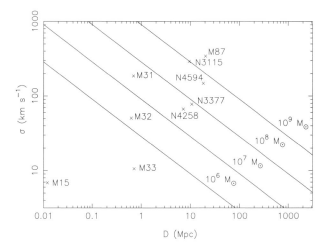

Fig. 1.12. Each of the lines indicates the detection limit for the effects of a black hole of mass M_{BH} in the center of a galaxy at distance D and having a stellar velocity dispersion σ. The region above the lines is not accessible from ground-based optical observations, which have an angular resolution of 1 arcsec.

are located at the right top corner of this figure (corresponding to $D > 20$ Mpc and $\sigma > 200$ km s^{-1}), and in such galaxies we cannot detect the effect of a black hole unless it is very massive. Some cases in which the existence of the central black hole is reasonably well established are indicated in the figure.

Although confirmed detection of the black hole is difficult, we can check for the consistency of the existence of a black hole of certain mass M_{BH} with the density profile near the origin. In this region, we can distinguish between two possible limiting behaviours. The first is a core with constant density and the luminosity profile of a form like, say, $I(r) = I_0[1 + (r/r_c)^2]^{-1}$ (where I_0 is the central brightness and r_c is the core radius). The second is a central region with a cusp in the density so that the intensity goes as $I \propto r^{-\alpha}$, which is indicative of central black hole. (We will discuss these in detail in Chap. 2, Section 2.5.)

It is important to realize that limitations of atmospheric seeing can artificially mimic the effect of a constant-density core region, so that high-resolution photometry is required for distinguishing between the two profiles mentioned above. The atmospheric seeing can be characterized by a point-spread function (PSF), $P(\mathbf{x})$, which gives the probability that a given photon will hit the imaging device at a point that is offset by a vector $\mathbf{x}$ compared with where it would have hit in the absence of atmospheric fluctuations. Hence the apparent surface brightness will be related to the true surface brightness by

$$I_{app}(\mathbf{R}) = \int d^2\mathbf{R}' P(\mathbf{R} - \mathbf{R}') I_{true}(\mathbf{R}'). \tag{1.64}$$

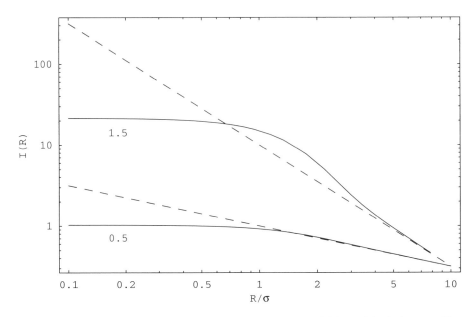

Fig. 1.13.   Illustration of the smoothing that is due to PSF, which could mimic a corelike region even though the true distribution of the intensity has a spike. The quantity $I_{app}(R)$ is plotted against R/σ in suitably scaled units for $I_{true}(R') \propto R'^{-\gamma}$ with $\gamma = 0.5$ and 1.5. The dashed curve is $I_{true}(R')$. The relative normalisation of the two sets of curves is arbitrary.

In the case of circularly symmetric distribution and a Gaussian PSF,

$$P(d) = \frac{1}{2\pi\sigma^2} \exp\left(-\frac{d^2}{2\sigma^2}\right), \tag{1.65}$$

the integral gives

$$I_{app}(R) = \int_0^\infty \frac{dR'\,R'}{\sigma^2} I_{true}(R') I_0\left(\frac{RR'}{\sigma^2}\right) \exp\left(-\frac{R^2 + R'^2}{2\sigma^2}\right), \tag{1.66}$$

where I_0 is the modified Bessel function. The asymptotic properties of I_0 lead to the result that if $I_{true} \propto R^{-\gamma}$ with $0 < \gamma < 1$, then I_{app} tends to a constant for $R \lesssim \sigma$. Further, the light removed from the central region is redistributed to outer radii, thereby increasing the surface brightness in the outer regions slightly (see Fig. 1.13). It is therefore necessary to have high-resolution images to decide observationally whether a cusp exists in the centre.

In many spiral galaxies, the inferred mass of the black hole (assuming it exists) is found to be well correlated with the absolute blue magnitude of the bulge. Figure 1.14 shows the correlation between the inferred mass of the black hole with the absolute magnitude of the spheroidal component of the galaxy – which is the bulge mass in the case of disks and the total mass in the case of ellipticals.

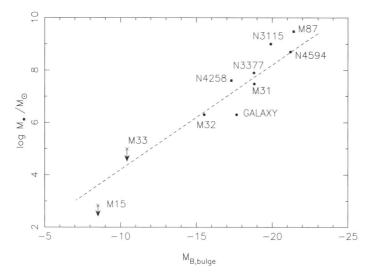

Fig. 1.14.   Correlation between the mass of the central black hole and the light emitted by the spheroidal bulge of the galaxy measured in terms of the blue magnitude.

Although the correlation is apparent, we must remember that brighter galaxies could be seen at larger distances for which it would not be possible to resolve the effect of a small-mass black hole on the stellar dynamics (see Fig. 1.12); hence we would preferentially see larger-mass black holes at brighter galaxies that could be seen from farther distances. The observed correlation, however, is stronger than could be obtained purely from this selection effect and is likely to be intrinsic. A linear fit in Fig. 1.14 between $\ln M_{BH}$ and the blue magnitude $M_{B,bulge}$ is equivalent to the relation $M_{BH} \simeq 0.006 m_{bulge}$, where m_{bulge} is the mass of the bulge.[4] (The entry corresponding to gravitational binding energy in relativistic form in Fig. 1.1 is estimated from the density of compact nuclei of galaxies, assuming they contain relativistic black holes.)

One of the most striking visual features of disk galaxies is the wide variety of spiral structures exhibited by many of them. The spiral arms are often sites of star formation and in such disks azimuthal colour variations are correlated to the spiral pattern with hot young stars and emission nebulae distributed along the ridge line of spiral arms. The overall structure is well defined, with a clear spiral pattern in bright-disk galaxies whereas the spirals are less well defined in low-luminosity disks. In addition to spirals, a large fraction of disk galaxies have *bars*, which are thin linear structures crossing the face of the galaxy. The bars have an axial ratio of 2:1 or greater with fairly constant surface brightness. The outer regions of disk galaxies are also often warped, with the edge-on galaxies showing a warped structure like an integral sign. The warps are not only quite common, but also exhibit fairly simple properties. They usually develop between R_{25} and $R_{26.5}$, with the line of nodes being approximately straight within $R < R_{26.5}$. (Here R_n

denotes the length of the semimajor axis of an ellipsoid whose isophotal surface brightness is $nB-$ mag arcsec^{-2}.) In many cases, the line of nodes takes the form of a loose leading spiral beyond $R > R_{26.5}$.

The disk galaxies show a fair amount of rotation, and the study of rotation velocities as a function of radial distance has been a valuable tool in mapping the density distribution in and around the disk. In all the cases, the rotation curves are either flat or gently rising at large radii, as far as the observations go. It has also been seen that the shape of the rotation curve exhibits some interesting correlations with other properties of the galaxies. For example, (1) the rotation curves tend to rise more rapidly near the centre and peak at higher maximum velocity (v_{max}) with increasing B-band luminosity; (2) among galaxies with equal B-band luminosities, spirals of the earlier type have larger values of v_{max}; and (3) within a given Hubble type, more luminous galaxies have larger values of v_{max}. The fact that galaxies of different Hubble types (with different bulge-to-disk luminosity ratios) show a similar pattern in rotation curves is taken to imply that rotational velocity at the outer reaches of the galaxy is possibly determined by the distribution of a dark-matter halo in these galaxies.

It is possible to obtain a simple scaling relation between the luminosity of the galaxy and v_{max} under certain conditions. Starting from the result $M \propto R v_{max}^2 / G$ and assuming that the (M/L) ratio for spirals is approximately constant, we get $L \propto v_{max}^2 R$. Further, if all spirals have approximately the same central surface brightness, we would expect (L/R^2) to be approximately constant. Eliminating R from these two relations, we find that the total luminosity of the disk scales with the rotation velocity as $L \propto v_{max}^4 = C v_{max}^4$, where C is a constant. This is called the *Tully–Fisher relation*. [The same result is obtained even if the central brightness $I(0)$ is not a constant, provided that $(M/L) \propto I(0)^{-1/2}$.] At a more fundamental level, we can attempt to relate v and M by using relation (1.19) for a given fluctuation spectrum $\sigma_0(M)$. For $\sigma_0(M) \propto R^{-(n+3)/2} \propto M^{-(n+3)/6}$, Eq. (1.19) gives $v \propto M^{(1-n)/12}$. At galactic scales the power spectrum has the approximate slope $n \approx -2$ [see Eq. (1.23); we will derive this in Chap. 5), giving $M \propto v^4$. Taking the logarithms to obtain the absolute magnitude, we get

$$M = M_\odot - 2.5 \log \left(\frac{L}{L_\odot} \right) = -10 \log v_{max} + \text{constant.} \qquad (1.67)$$

Observations show that the best-fit relation between absolute blue magnitude M_B and v_{max} indeed is of the form $M_B = -\alpha \log v_{max} + \beta$, where the pair (α, β) has the values (9.95, 3.15), (10.2, 2.71), (11.0, 3.31) for Sa, Sb, and Sc spirals, respectively. This is quite close to the relation $L \propto v_{max}^4$ obtained above.

Several other properties of the disk galaxies are strongly correlated with the Hubble type. We had already seen in Section 1.5 how the (B-V) colour of the galaxy decreases as we move to a later Hubble type. Because main-sequence stars B are short lived, they must have formed comparatively recently, requiring the bluer Sc galaxies to have fair amount of gas and dust. This is confirmed by observations with the ratio $\langle M_{gas}/M_{total} \rangle$ increasing steadily as we move from Sa to Scd. The relative amount of atomic and molecular hydrogen also change with a Hubble type, with the molecular hydrogen fraction decreasing from Sab to Scd and with the overall amount of molecular hydrogen varying from $10^6 \, M_\odot$ (for dwarf spirals) to $5 \times 10^{10} \, M_\odot$ (for massive spirals). This is possibly because Sa spirals are more centrally condensed, providing deeper gravitational wells in which the gas can collect and form molecules.

Because the amount of dust is approximately $(150-600)$ times lower than the mass of gas in the interstellar medium (ISM), the dust is also strongly correlated with Hubble type. The FIR luminosity of galaxies is primarily due to dust, and Infrared Astronomical Satellite (IRAS) observations show that (L_{FIR}/L_B) is larger in Sc compared with Sa or Sb, which is consistent with other observations.

Individual spiral galaxies also exhibit colour gradients with the bulges being redder than the disks. It was seen in Vol. II, Chap. 2, Section 2.4, that a star with higher metallicity (on the average) will be redder than a star with lower metallicity when all else remains equal. (This arises because greater metallicity implies more numbers of electrons per atom, on the average, and the transitions of these electrons contribute to opacity in stellar photospheres. Hence the opacity of a metal-rich star will be greater in trapping light more efficiently and causing the star to puff up. This causes T_{eff} to decrease, making the star redder.) Hence redness of the bulges suggests that they are more metal rich than the disks. In the Milky Way, for example,

$$\frac{d[\text{He/H}]}{dr} = -0.01 \pm 0.008 \, \text{dex kpc}^{-1};$$

$$\frac{d[\text{O/H}]}{dr} = -0.07 \pm 0.015 \, \text{dex kpc}^{-1}; \qquad (1.68)$$

$$\frac{d[\text{Fe/H}]}{dr} = -0.01 \text{ to } 0.05 \, \text{dex kpc}^{-1}$$

indicating the correctness of this idea. The square brackets used in the preceding equations are defined by relations of the kind

$$\left[\frac{\text{Fe}}{\text{H}} \right] \equiv \log_{10} \left(\frac{N_{Fe}}{N_H} \right) - \log_{10} \left(\frac{N_{Fe}}{N_H} \right)_\odot. \qquad (1.69)$$

This is the reason for the units dex kpc^{-1} that appear in Eqs. (1.68). Star formation, which is more in the gas-rich disk than in the bulge, also contributes to establishing a colour gradient because the disk produces young, hot, blue stars at a greater rate. The overall metallicity based on both [Fe/H] and [O/H] increases, with M_B showing that chemical enrichment was more efficient in luminous massive galaxies.

Another statistic that seems to be correlated with the Hubble type is the abundance of the globular clusters. The galaxies of the earlier type seem to have

Table 1.1. *Galaxy parameters – I*

Parameter	Sa	Sb	Sc	Sd/Sm	Im/Ir
M_B	-17 to -23	-17 to -23	-16 to -22	-15 to -20	-13 to -18
$(M/M_\odot)$	10^9–10^{12}	10^9–10^{12}	10^9–10^{12}	10^8–10^{10}	10^8–10^{10}
$\langle L_{bulge}/L_{total}\rangle_B$	0.3	0.13	0.05	—	—
Diameter D_{25}(kpc)	5–100	5–100	5–100	0.5–50	0.5–50
$\langle M/L_B\rangle$ $(M_\odot/L_\odot)$	6.2 ± 0.6	4.5 ± 0.4	2.6 ± 0.2	$\simeq 1$	$\simeq 1$
μ_{0B} (mag arcsec^{-2})	21.52 ± 0.39	21.52 ± 0.39	21.52 ± 0.39	22.61 ± 0.47	22.61 ± 0.47
$\langle B-V\rangle$	0.75	0.64	0.52	0.47	0.37
$\langle M_{gas}/M_{total}\rangle$	0.04	0.08	0.16	0.25 (Scd)	—
$\langle M_{H_2}/M_{HI}\rangle$	2.2 ± 0.6(Sab)	1.8 ± 0.3	0.73 ± 0.13	0.19 ± 0.10	—
$\langle S_N\rangle$	1.2 ± 0.2	1.2 ± 0.2	0.5 ± 0.2	0.5 ± 0.2	—

produced more globular clusters during their formative stages. To compare the abundance of globular clusters, it is usual to define a specific frequency of globular clusters by

$$S_N = N_t \frac{L_{15}}{L_V} = N_t 10^{0.4(M_V + 15)}, \tag{1.70}$$

where N_t is the total number of globular clusters in a galaxy, L_V is the luminosity of the galaxy, and L_{15} is the absolute magnitude of a galaxy with $M_V = -15$. The latter factor normalises the globular-cluster abundance to standard absolute magnitude. Table 1.1 shows that the specific frequency S_N increases as one moves from Sc to Sa. The value is even larger for ellipticals, implying that ellipticals have more globular clusters per unit luminosity.

It is also noted that the radius R_{25} of the disk (corresponding to a surface-brightness level of 25 B-mag arcsec^{-2}) correlates well with the absolute B magnitude:

$$\log\left(\frac{R_{25}}{1\,\text{kpc}}\right) = -0.249\,M_B - 4.0. \tag{1.71}$$

Combining with the relation between M_B and v_{max}, we can relate R_{25} to v_{max} and thus estimate the total mass $M \propto v_{max}^2 R$ once v_{max} is known. The observed mass is in the range $(10^9$–$10^{12})M_\odot$ and is only weakly correlated with the Hubble type of the galaxy.

The stellar distribution rotates slightly slower than the gas, but some of the galaxies contain a significant percentage of counterrotating disk stars. As regards the bulges, their rotation is similar to elliptical galaxies of the same luminosity and the rotation is fast enough to explain their observed flattening.

1.8 Properties of Elliptical Galaxies

We next consider the bulge component of galaxies and – in particular – ellipticals. In the simplest approximation, an elliptical galaxy is made of all bulge and no disk. When examined in detail, however, they show a wide variety of behaviour.[5]

To begin with, although the luminosity profile of the ellipticals are reasonably well fitted by de Vaucouleur's profile [Eq. (1.44)], the central region shows a more interesting structure. Current HST observations show that several elliptical galaxies do show a cuspy region with $I \propto r^{-\alpha}$, $\alpha \approx (0.5\text{–}0.3)$ up to the limit of HST resolution. We will see later in Chap. 2, Section 2.5, that the existence of a central black hole could lead to such a cusp in the density profile. There is also a weak correlation between the index α and the overall magnitude of the galaxy.

The standard procedure for the analysis of light distribution of an elliptical is to determine curves of constant intensity (called *isophotes*) and attempt to fit ellipses to these isophotes. If the intrinsic three-dimensional distribution of light in the galaxy is constant on elliptical surfaces with three different scale lengths (such a galaxy is usually called *triaxial*), then it is easy to show that the ellipticity and the orientation of the isophotes (which are obtained by two-dimensional projection onto the sky) will vary. (See Fig. 1.15 for a simple illustration of this effect.) The variation in the inclination of the major axis of the ellipses (with some reference direction) as a function of the isophote intensity (or radius) is usually called the *isophotal twist*. Isophotal twist provides a handle on analysing the triaxiality of the underlying distribution.

Another useful way of analysing the light distribution in an elliptical is to study the variation of brightness along the angular direction of an ellipse that fits the isophotes on the average. This is done by fitting an intensity profile of the

Fig. 1.15.   The left part of the figure gives three elliptical surfaces on which the intensity is constant. These elliptical surfaces are triaxial. The "shadow" on the right-hand side gives the projection of the intensity on a plane perpendicular to an axis that does *not* coincide with the principal axes of the ellipsoids. The figure shows that the major axes of the ellipses in the projected plane are not parallel to each other. (Figure courtesy of N. Sambus.)

Fig. 1.16.   Polar plot of the function $I(\theta)$ with $I_0 = 1$, $a_2 = 0.4$, and $a_4 = \pm 0.03$, and all other coefficients in Eq. (1.72) set to zero. The solid curve corresponds to $a_4 < 0$, and the dashed curve corresponds to $a_4 > 0$. It is clear that the sign of a_4 decides the shape of the boundary.

form

$$I(\theta) = I_0 + \sum_{n=1} (a_n \cos n\theta + b_n \sin n\theta) \tag{1.72}$$

and determining the coefficients a_n and b_n. If only the first term (I_0) on the right-hand side is retained, the intensity is constant on circles; higher-order terms denote deviation from circularity. By choosing the isophotes to be ellipses, we can make a_1, a_2, and a_3 and all of b_n small. Then the first nontrivial term in the summation – which indicates deviation from ellipticity – will be for $n = 4$, which usually varies in the range $-0.02 < a_4/I_0 < 0.04$. Galaxies with $a_4 < 0$ are called *boxy* because the isophotes are somewhat rectangular, whereas those with $a_4 > 0$ are called *disky* because the isophotes appear somewhat pointed (see Fig. 1.16). The importance of this classification arises from the fact that they seem to be well correlated with several other properties of the galaxies.

Exercise 1.5
Dimensional reduction: (a) Express the surface-brightness distribution $I(R)$ projected onto the sky in terms of the three-dimensional luminosity density $j(r)$ of a galaxy that is supposed to be spherically symmetric. Invert this relation to express j in terms of I. (b) Consider a class of $j(r)$ given by

$$j(r) = \frac{3 - \gamma}{4\pi} \frac{La}{r^\gamma (r + a)^{4-\gamma}}, \tag{1.73}$$

where L is the total luminosity, a is a characteristic scale length, and γ is an integer or half-integer. Determine the surface-brightness profile in terms of $s \equiv R/a$ for $\gamma = 0, 1, 2$. (c) Show that, for a general γ, $I(R)$ can be expressed in the form

$$I(R) \propto R^{1-\gamma} \int_0^\infty \frac{\cosh^{1-\gamma} \psi \, d\psi}{(a + R \cosh \psi)^{4-\gamma}}. \tag{1.74}$$

How does $I(R)$ behave near $R = 0$? {Hint: (a) Expressing $I(R)$ in terms of $j(r)$ is easy as it involves only integration along the line of sight z. We get

$$I(R) = \int_{-\infty}^{\infty} dz\, j(r) = 2 \int_{R}^{\infty} \frac{j(r)r\, dr}{\sqrt{r^2 - R^2}}. \tag{1.75}$$

Because this is an Abel integral equation, we can invert it to get

$$j(r) = -\frac{1}{\pi} \int_{r}^{\infty} \frac{dI}{dR} \frac{dR}{\sqrt{R^2 - r^2}}. \tag{1.76}$$

(b) For $\gamma = 0, 1, 2$ the surface-brightness profile in terms of $s = (R/a)$ varies as

$$I(s) \propto \begin{cases} \dfrac{-2 - 13s^2 + 3s^2(4 + s^2)|s^2 - 1|^{-1/2}c(s)}{4(s^2 - 1)^3} & \text{(for } \gamma = 0) \\[4mm] \dfrac{(2 + s^2)|s^2 - 1|^{-1/2}c(s) - 3}{2(s^2 - 1)^2} & \text{(for } \gamma = 1), \\[4mm] \dfrac{\pi}{4s} - \dfrac{1}{s^2 - 1}\left[\dfrac{1}{2} - \dfrac{(2 - s^2)c(s)}{|s^2 - 1|^{1/2}}\right] & \text{(for } \gamma = 2) \end{cases} \tag{1.77}$$

where $c(s) = \cosh^{-1}(1/s)$ for $s \leq 1$ and $\cos^{-1}(1/s)$ for $s \geq 1$. The model for $\gamma = 2$ is called a *Jaffe model*. (c) Relation (1.74) shows that the behaviour of $I(R)$ near $R = 0$ differs drastically depending on whether $\gamma > 1$ or $\gamma < 1$. When $\gamma > 1$, the integral is approximately independent of R for $R \ll 1$, making $I(R) \propto R^{-(\gamma-1)}$, which is cuspy near the origin. For $\gamma < 1$, $I(R)$ is approximately constant near the origin.}

Exercise 1.6

In the eyes of the beholder: Consider an oblate spheroidal galaxy symmetrical about the shortest axis z in which the density of stars $\rho(x, y, z)$ depends on only the combination $l^2 = A^{-2}(x^2 + y^2) + B^{-2}z^2$, with $A \geq B \geq 0$. This galaxy is viewed at an angle i with respect to the z axis. Show that the projection of the density on the sky will be elliptical with an axis ratio $q = [(B/A)^2 \sin^2 i + \cos^2 i]^{1/2}$. What is the corresponding result if the original galaxy is a prolate spheroid? What will be the distribution of the apparent axis ratio q for a fixed B/A when the viewing angle is assumed to vary randomly?

Morphologically, we can further subdivide ellipticals into five different groups, as indicated in Table 1.2. Of these, cDs and normal ellipticals have surface-brightness profiles that follow the $r^{1/4}$ law, and – as the mass of the galaxy decreases – there is a smooth transition to the exponential profile, which is apparent in the case of dwarf ellipticals (dEs) and dwarf spheroidals (dSphs). The latter two (dEs and dSphs) represent a class of objects that are fundamentally different from normal ellipticals in several ways. Their surface brightness, absolute B magnitudes, and metallicities are all much lower than those of normal ellipticals of the same mass. These differences are also reflected in the correlations exhibited by these objects. For example, the dependence of the effective radius r_e on the blue magnitude M_B differ quite significantly between normal ellipticals and dEs. The same is true for the dependence of the surface brightness on the absolute blue magnitude. Some of these differences could be related to the

Table 1.2. Galaxy parameters – II

Parameter	S0	cD	E	dE	dSph	BCD
M_B	-17 to -22	-22 to -25	-15 to -23	-13 to -19	-8 to -15	-14 to -17
$M(M_\odot)$	10^{10}–10^{12}	10^{13}–10^{14}	10^8–10^{13}	10^7–10^9	10^7–10^8	$\simeq 10^9$
Diameter D_{25}(kpc)	10–100	300–1000	1–200	1–10	0.1–0.5	<3
$\langle M/L_B \rangle$ $(M_\odot/L_\odot)$	$\simeq 10$	>100	10–100	$\simeq 10$	5–100	0.1–10
$\langle S_N \rangle$	$\simeq 5$	$\simeq 15$	$\simeq 5$	4.8 ± 1.0	—	—

fact that the dEs have low gravitational potential-well depths that prevent them from retaining a significant amount of gas; in fact, dSphs are virtually devoid of gas. In general, however, elliptical galaxies – especially E/S0 – do have a hot gas of mass 10^8–10^{10} $M_\odot$ at temperatures of $T \simeq 10^7$ K, which can be detected through x-ray emission. In addition, there is a warm (10^4 K) gas component of mass 10^4–10^5 $M_\odot$ observable in H_α and a cold (10^2 K) component of HI of mass 10^7–10^9 $M_\odot$ detectable at 21 cm. The CO emission indicates that there could also be another 10^7–10^9 $M_\odot$ of molecular hydrogen in these systems. Many elliptical galaxies also show dust lanes produced by cold interstellar material that seems to be kinematically distinct from the stellar component; for example, in many cases, the cold gas exhibits counterrotation.

The elliptical galaxies also become redder towards their centre; a factor of 10 decrease in radius changes (B–V) by 0.03 and (U–V) by 0.10. The metal lines become stronger with decreasing radius whereas H_β lines do not show a definite correlation. The logarithmic slope of the metal-abundance gradient is approximately -0.2 for a typical elliptical.

One might have thought, from the elliptical shape of a system that the oblateness (or prolateness) could possibly be related to the rotation. Such a simple picture, however, falls significantly short of explaining the observed features of many ellipticals. As we have seen, these galaxies are neither oblate or prolate but are triaxial and there is no single preferred axis of rotation. Further, a fraction of ellipticals show counterrotation of the stellar population in the centre as well as faint evidence for embedded disks. Actual measurement of rotational velocities of elliptical galaxies show that more luminous ones have mean rotational velocities $V_{\rm rot}$ that are much less than their velocity dispersion σ. If a galaxy of ellipticity ϵ is modelled as an oblate rotator with isotropic stellar velocity dispersion, then it can be shown that $(V_{\rm rot}/\sigma)^2 \simeq \epsilon(1-\epsilon)^{-1}$ (see Chap. 2). For example, a galaxy with $\epsilon = 0.4$ will require that $(V_{\rm rot}/\sigma) \simeq 0.8$ for rotational support. The actual ratio for most luminous ellipticals is in the range of 0.01. If we define a rotation parameter $\mathcal{R}$ as the ratio between the observed value of $(V_{\rm rot}/\sigma)$ and the theoretical value needed for rotational support, then we find that $\langle \mathcal{R} \rangle \approx 0.4$ for bright ellipticals and giant ellipticals. On the other hand, less luminous galaxies, with $-18 > M_B > -20.5$, have $\langle \mathcal{R} \rangle \simeq 0.9$, implying that they could be rotationally supported. The same is true of bulges of spiral galaxies. These observations suggest that more massive systems are supported by the velocity dispersion of stars whereas the less massive ones possess a fair amount of rotational support.

One significant systematic trend exhibited by the elliptical galaxies is the existence of a *fundamental plane*. Observations show that the effective radius R_e, the central velocity dispersion σ_0, and the mean surface brightness I_e (within the effective radius R_e) are connected by a relation of the form

$$R_e \propto \sigma_0^{1.49} I_e^{-0.86}. \tag{1.78}$$

From virial theorem, we would expect

$$R \propto \frac{M}{\sigma^2} \propto \frac{L}{\sigma^2}\left(\frac{M}{L}\right) \propto \frac{IR^2}{\sigma^2}\left(\frac{M}{L}\right), \tag{1.79}$$

so that $R \propto \sigma^2 I_e^{-1}(M/L)^{-1}$. If the mass-to-light ratio varies as $M/L \propto M^{0.2} \propto L^{0.25}$, we get $R \propto \sigma^{4/3} I^{-5/6}$, which could approximately account for the fundamental plane of ellipticals. The best-fit equation to the fundamental plane in terms of the magnitude μ_e is

$$\log(R_e) = 0.34\mu_e + 1.49 \log \sigma_0 - 6.60, \tag{1.80}$$

where R_e is measured in parsecs, μ_e in B magnitude per square arcsecond, and σ_0 in kilometers per second. Because σ_0 and μ_e can be measured directly from spectroscopy and photometry, we can use this relation to obtain the physical radius R_e. Comparison with the apparent angular radius in the sky will allow us to determine the distance to the galaxy, thereby providing a valuable observational tool. Figure 1.17 illustrates the fundamental plane. The same data set characterized by (R_e, μ_e, σ_0) is shown in projection onto three axes planes in the top two and bottom left figures. In the bottom right frame, the same data are projected onto the $(\log R_e, 1.49 \log \sigma_0 + 0.3427\mu_e)$ plane in which the scatter is minimum, demonstrating Eq. (1.80).

The fundamental plane also allows us to derive a simple relationship between diameters and velocity dispersions of these galaxies. Let D_n be the diameter within which the mean surface brightness is $I_n = 20.75\ B-\text{mag arcsec}^{-2}$ and assume that (1) the intensity profile is given by a function of the form $I(R) = I_e f(R/R_e)$ and that (2) most of the light interior to D_n comes from radii at which $f(x) \simeq x^{-\alpha}$ with $\alpha \approx 1.2$. It then follows that

$$I_n = 8I_e \left(\frac{R_e}{D_n}\right)^2 \int_0^{D_n/2r_e} dx\, x f(x), \tag{1.81}$$

which, for $f(x) = x^{-\alpha}$, gives $D_n \propto r_e I_e^{1/\alpha} \propto r_e I_e^{0.8}$. Using Eq. (1.78) for the

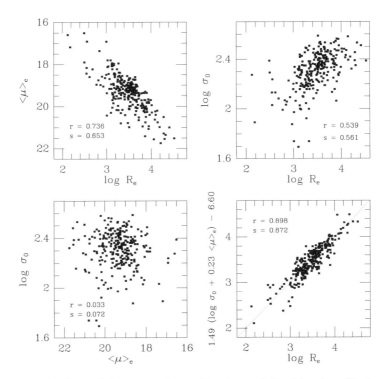

Fig. 1.17. Projections of data onto observable planes defined by the effective radius (actually semimajor axis) R_e in parsecs, the enclosed average surface brightness within the R_e isophote $\langle\mu\rangle_e$, and the central velocity dispersion σ_0, for a sample of 259 galaxies for which we have reliable data in hand. The lower right panel shows the fundamental plane fit for this sample. Pearson (r) and Spearman (s) correlation coefficients are given in each panel. (Figure courtesy of George Djorgovski.)

fundamental plane, we get

$$D_n \propto \sigma^{1.4} I_e^{0.07}, \tag{1.82}$$

which is called the *Faber–Jackson relation*. The weak dependence on I_e shows that there will be a tight correlation between D_n and σ, allowing these relations to be used for an estimate of distances.

Some of the properties of the spiral and the elliptical galaxies are summarised in Tables 1.1 and 1.2.

1.9 Milky Way Galaxy

Our own galaxy, the Milky Way, is obviously one of the best-studied examples of a spiral galaxy. In this section, several features of this galaxy are outlined, starting with a description of the components.

1.9.1 Components of the Milky Way

Broadly speaking, the Milky Way is a disk galaxy that has a distinct spiral structure, with the Sun being located at a distance of approximately $R_0 = (8 \pm 0.5)$ kpc from the centre of the galaxy. Viewed from the Earth, the centre of the Galaxy is in the direction of Sagittarius with the coordinates $\alpha = 17^h 42^m 29.3^s \pm 0.15^s$, $\delta = -28°59'18'' \pm 3''$ (see Vol. II, Chap. 1, Section 1.2, for a description of the coordinate system). The disk is actually made of different components that can be broadly divided as *young thin disk, old thin disk*, and *thick disk*. The thin disk, having a total mass of approximately $6 \times 10^{10}\, M_\odot$, has had an ongoing star formation for nearly 10^{10} yr and contains stars with a wide range of ages. From the ages, we can further divide the disk population of stars into (1) stars in the spiral arms, (2) young stars, (3) intermediate-age stars, and (4) old stars. The spiral arm has the youngest of stars and also includes HI and molecular clouds, HII regions, protostars, OB stars, and type I Cepheids, all of which trace the spiral pattern of Milky Way. The scale height of this material is roughly 100 pc, and they move in nearly circular orbits with a net velocity of $\sim$220 km s^{-1}. The other three populations are compared with the spiral arm population in Table 1.3. Note that the old thin disk has a scale height of nearly $z_{\text{thin}} \simeq 325$ pc whereas the thick disk has a scale height of $z_{\text{thick}} \simeq 1.4$ kpc. An empirical fit to the number of stars per unit volume based on the star-count data is

$$n(z, R) = n_0 \left(e^{-z/z_{\text{thin}}} + 0.02 e^{-z/z_{\text{thick}}} \right) e^{-R/h_R}, \qquad (1.83)$$

with $h_R = 3.5$ kpc and $n_0 \simeq 0.02$ pc^{-3}. (Such a fit combines stars of different types and is somewhat an oversimplification. In general, multiple components for each population of stars are required for describing the distribution accurately.)

Table 1.3. Components of the Milky Way

Parameter	Disks			Spheroids		
	Neutral Gas	Thin Disk	Thick Disk	Central Bulge	Stellar Halo	DM Halo
$(M/10^{10} M_\odot)$	0.5	6	0.2–0.4	1	0.1	55
$(L_B/10^{10} L_\odot)$	—	1.8	0.02	0.3	0.1	0
$M/L_B (M_\odot/L_\odot)$	—	3	—	3	$\sim$1	∞ (?)
Diameter (kpc)	50	50	50	2	100	>200
Form	$e^{-z/h}$	$e^{-z/h}$	$e^{-z/h}$	—	$r^{-3.5}$	$(a^2 + r^2)^{-1}$
Scale height (kpc)	0.16	0.325	1.4	0.4	3	$a = 2.8$
σ_w (km s^{-1})	5	20	60	120	90	220
$[Fe/H]$	>0.1	−0.5 to 0.3	−1.6 to −0.4	−1 to 1	−4.5 to −0.5	0 (?)
Age (Gyr)	0–17	<12	14–17	10–17	14–17	Pregalactic ?

The luminosity per unit volume of the old thin disk alone can be fitted to the form

$$L(R, z) = L_0 e^{-R/h_R} \operatorname{sech}^2(z/z_0),   \tag{1.84}$$

with $z_0 \simeq 2z_{\text{thin}}$ and $L_0 \simeq 0.05\, L_\odot\, \text{pc}^{-3}$.

In a disk galaxy such as ours, the thickness of the disk, $d \approx 200\, \text{pc}$, plays a vital role in determining the geometrical features relevant for the analysis of several observations, for example, the star count. To illustrate this feature, consider a model for the Milky Way in which stars are formed at a constant rate with an IMF

$$\frac{dN_0}{dM} \propto M^{-(1+x)}.   \tag{1.85}$$

We further assume that the luminosity of a star is related to its mass by $L \propto M^\alpha$, with $\alpha \simeq 4$, which is valid for stars more massive than $1\, M_\odot$; the lifetime of a star then varies as $t \propto (M/L) \propto M^{-(\alpha-1)}$. The number of stars actually shining in the main sequence at any given instant is the product of the SFR and the lifetime of the star. This gives

$$\frac{dN_*}{dM} \propto t\frac{dN_0}{dM} \propto M^{-(x+\alpha)}.   \tag{1.86}$$

Hence the number of stars per logarithmic interval in luminosity, which is shining in the main sequence, is given by

$$L\frac{dN_*}{dL} = M\frac{d\ln M}{d\ln L}\frac{dN_*}{dM} \propto M^{-(x+\alpha-1)} \propto L^{-\left(\frac{x+\alpha-1}{\alpha}\right)}.   \tag{1.87}$$

This result also gives the correct scaling for the number of stars with a mass greater than M or – equivalently – luminosity greater than L. For definiteness, we now take $\alpha \approx 4$, so that we get $N_*(> M) \propto M^{-(3+x)}$ and $N_*(> L) \propto L^{-(3+x)/4}$.

If the stars are distributed in a three-dimensional region around us, then the flux from stars at distance r will vary as $F \propto (L/r^2)$ so that the volume surveyed up to some limiting flux F will scale as $r^3 \propto (L/F)^{3/2}$. Hence the number of stars brighter than a flux F per octave in L detected in a three-dimensional survey will vary as $L^{3/2}L^{-(3+x)/4}$. This distribution is flat when $x = 3$ (corresponding to $dN/dL \propto L^{-5/2}$); if $x < 3$, then intrinsically luminous, distant stars will dominate the star count. On the other hand, if the region sampled is large compared with the scale height of the disk so that the distribution is two-dimensional, then the flux still varies as $F \propto (L/r^2)$ whereas the volume covered by the survey goes as $r^2 \propto (L/F)$. The number of stars brighter than F per octave in L now varies as $LL^{-(3+x)/4}$. This distribution is flat when $x = 1$ [corresponding to $(dN/dL) \propto L^{-2}$]; when $x > 1$, intrinsically faint, nearby stars dominate the star count.

There is some indication that the actual value of x is $x \simeq 1.35$ (see Vol. II, Chap. 3, Section 3.6). Because $x < 3$, the counts within a 200-pc local region are dominated by luminous stars. On the other hand, because $x > 1$, the intrinsically faint stars dominate from the region beyond 200 pc. A natural cutoff for faint stars occurs when the lifetime is comparable with $t_{\text{univ}} \simeq 10^{10}$ yr. For fainter stars, we cannot multiply by the lifetime t in relation (1.86) and we must keep

$$\frac{dN_*}{dM} \propto \frac{dN_0}{dM} \propto M^{-(1+x)}, \qquad L\frac{dN}{dL} \propto M^{-x} \propto L^{-x/4}. \qquad (1.88)$$

In this case, the quantity $LL^{-x/4}$ increases with L for $x = 1.35$. The distance of $r = 200$ pc corresponds to a magnitude difference of $m - M_{\text{bol}} = 5 \log \times (r/10\,\text{pc}) \simeq 6.5$ mag. So stars with $M_{\text{bol}} = 0$ will drop from the counts at $m_{\text{bol}} = 6.5$ whereas the G stars with $M_{\text{bol}} = 5$ will drop from the counts at $m_{\text{bol}} = 11.5$. At the end brighter than $m_{\text{bol}} = 6.5$, the OB and the A stars will dominate the count whereas at the end fainter than $m_{\text{bol}} = 11.5$ the main-sequence G stars will dominate the count. This analysis illustrates the nontrivial effects of disk geometry on observations.

Our galaxy also contains, in addition to the disk, distinct spheroidal populations in the forms of a central bulge, a stellar halo, and a dark-matter halo. The galactic bulge has a radius of ~1 kpc and contains a peanut-shaped bar in the centre that is substantially inclined to our line of sight. The ratio of minor to major axis of the bulge is ~0.6, and its surface brightness follows de Vaucouleur's law with $r_e \simeq 0.7$ kpc. Except along some specified directions, the density of material towards the centre of the galaxy is so high that there is an extinction of more than 28 mag. This makes direct observations, especially in the optical band, extremely difficult.

Our galaxy also contains another extended spheroidal halo essentially made of globular clusters and field stars that have large velocity components perpendicular to the galactic plane. Globular clusters are mostly found within a distance of ~33 kpc, although there are members who exist as far away as 10^2 kpc. The number-density profile of the globular clusters and field stars in the halo has the form $n_{\text{halo}}(r) = n_0 r^{-3.5}$.

Observations of the rotation curve of the galaxy based on 21-cm radiation suggest that there must exist another spherically distributed halo of matter extending out to at least 100 kpc and possibly farther. As we will see in Chap. 2, Section 2.3, it is difficult to uniquely specify the density profile of this dark-matter halo, although a distribution of the form $\rho(r) \propto (a^2 + r^2)^{-1}$ with $a = 2.8$ kpc seems to be consistent with observation. If this profile extends up to 100 kpc, then the dark matter will constitute ~90% of the mass within 100 kpc.

The Milky Way also contains, in addition to the stars, a significant amount of gas and dust clouds in the form of ISMs, which have been described in detail in Vol. II, Chap. 9. Molecular hydrogen and cool dust are predominantly found in the regions (3–8) kpc from galactic centre, whereas atomic hydrogen is found in the region (3–25) kpc away. The HII and the dust stick tightly to the plane of the galaxy with a low vertical scale height of ~90 pc whereas the scale height for atomic hydrogen is about 160 pc. The total mass of HI is approximately $4 \times 10^9 \, M_\odot$, and the mass of atomic hydrogen is $\sim 10^9 \, M_\odot$. In the solar neighbourhood, gas contributes a mass density of $\sim 0.04 \, M_\odot \, pc^{-3}$, of which atomic hydrogen contributes $\sim 77\%$. At $R \gtrsim 12$ kpc, the scale height of HI increases dramatically to a value of more than 800 pc and the gas develops a well-defined warp with a maximum angle of deviation of $\sim 15°$. There is also evidence for a very hot tenuous gas stretching up to distances of 50 kpc.

1.9.2 Metallicity

Each of the components of the Milky Way, described in the previous subsection, seem to have an identity other than just morphological. This evidence comes from studying the iron-to-hydrogen ratio in the atmospheres of stars that reside in these components. Because iron is synthesised in stars that have reached the end of their evolution and because new stars forming out of an enriched ISM will have greater iron abundance, we would expect the iron content to correlate with the age of the stellar population. We take the metallicity as being defined by the relation

$$\left[\frac{Fe}{H} \right] \equiv \log_{10} \left(\frac{N_{Fe}}{N_H} \right) - \log_{10} \left(\frac{N_{Fe}}{N_H} \right)_\odot \qquad (1.89)$$

(The square brackets on the left-hand side are an unnecessarily complicated – but conventional – notation, defined by the expression on the right-hand side.) We expect the value to vary from approximately -4.5 for old, extremely metal-poor stars to approximately $+1$ for young metal-rich stars. In the thin disk, the metallicity ranges between $-0.5 < [Fe/H] < 0.3$ whereas in the thick disk we have $-0.6 < [Fe/H] < -0.4$. The bulge has a wide range of stars, with $-1 < [Fe/H] < +1$, with a mean value near $+0.3$. This suggests that the stellar members of the thin disk are probably younger than those in the thick disk. It is likely that star formation in the thin disk began $(10–12) \times 10^9$ yr ago and is still going on. The star formation in the thick disk appears to have occurred several billion years before the onset of star formation in the thin disk. On the average, the metallicity of stars in the bulge is twice the solar value, suggesting that at least a portion of the bulge is young, perhaps only 10×10^9 yr old. As regards the globular clusters, the older metal-poor clusters have $[Fe/H] < -0.8$ and belong to an extended spherical halo of stars; the younger clusters with $[Fe/H] > -0.8$ form a much flatter distribution and could even be associated with the thick disk.

Although such a relationship between age and metallicity sounds reasonable, the following caveats must be kept in mind. To begin with, note that supernovas do not appear until $\sim 10^9$ yr after the beginning of star formation. Further, mixing of material in the ISM has its own time scale. It is therefore possible for a local region of the ISM to be enriched in iron after 10^9 yr whereas another similar region may not experience the same level of enrichment. Evidence based on the second generation of stars would make the iron-rich region appear younger, even though both regions have intrinsically the same age. To avoid this difficulty, we could attempt to measure the metallicity based on some other element abundance like [O/H] defined in the same way as in Eq. (1.89). The enrichment through oxygen can occur by means of a type II supernova, which appears after only 10^7 years. These observations reconfirm the idea that the stars of the thin disk are probably significantly younger than their thick-disk counterparts, and it is likely that the star formation in the thin disk started 10–12 Gyr ago and is still ongoing.

1.9.3 Kinematics

To characterise the motion of a star in the Milky Way, we need to specify its three components of velocity, conventionally denoted as

$$\Pi \equiv \frac{dR}{dt}, \qquad \Theta \equiv R\frac{d\theta}{dt}, \qquad Z \equiv \frac{dz}{dt} \qquad (1.90)$$

in a cylindrical coordinate system (R, θ, z). (Note that this coordinate system has θ measured from the line joining the Sun and the galactic centre and increases in the direction of galactic rotation. When viewed from the north galactic pole, the Milky Way rotates clockwise rather than counterclockwise, making this coordinate system left handed.) We next define a *local standard of rest* (LSR) as centred at a point that is instantaneously located on the Sun and moving in a circular orbit along the solar circle. Hence, by definition, the velocity components of the origin of the LSR are given by $\Pi_{LSR} \equiv 0$, $\Theta_{LSR} \equiv \Theta_0 = \Theta(R_0)$, and $Z_{LSR} \equiv 0$. The velocity of any star relative to the LSR is called the *peculiar velocity* and has the components

$$u = \Pi - \Pi_{LSR} = \Pi, \qquad v = \Theta - \Theta_{LSR} = \Theta - \Theta_0, \qquad w = Z - Z_{LSR} = Z. \qquad (1.91)$$

In a galaxy with perfect axisymmetry and reflection symmetry about $z = 0$, the average values of u and w for a collection of stars should vanish as we expect as many stars to be moving inwards as are moving outwards or towards the north galactic pole as towards the south galactic pole. (The deviation of the galaxy from perfect axisymmetry and reflection symmetry about $z = 0$ will, however, lead to some nonzero values for $\langle u \rangle$ and $\langle w \rangle$.) The mean value of v, however, will not be zero in general. If we consider a bunch of stars with $u = w = 0$, then such stars must be at either apogee or perigee when they move with the LSR. It follows that $\Theta < \Theta_0$ (making $v < 0$) if the star is at apogee and $\Theta > \Theta_0$ (making $v > 0$) if the star is at perigee. Because on the average there are more stars inside the solar radius than outside, this will make $\langle v \rangle < 0$. For the Sun, the components of the peculiar velocity are

$$u_\odot = -9\,\mathrm{km\,s^{-1}}, \qquad v_\odot = 12\,\mathrm{km\,s^{-1}}, \qquad w_\odot = 7\,\mathrm{km\,s^{-1}}. \qquad (1.92)$$

This motion is approximately with a speed 16.5 km s^{-1} towards $l = 53°$, $b = 25°$, which is a point in the constellation of Hercules. Once the solar motion is known, the velocities of stars relative to Sun can be transformed to peculiar motions relative to the LSR.

The velocity distribution of stars in the Milky Way plays a crucial role in several studies and also serves as a benchmark for comparison with other galaxies. By and large, the velocity distribution of stars (around a mean) can be taken to be a Gaussian in the three components v_1, v_2, v_z, where the first two are in the plane of the galaxy and the third component is perpendicular to the plane, with the corresponding three dispersions being σ_1, σ_2, and σ_z. Contour plots of stellar populations in the u–v plane will indicate the velocity dispersions of different classes of stars (see Fig. 1.18). This figure brings out two important trends. First, there is a distinct correlation between velocity dispersion and metallicity; when combined with the age–metallicity correlation, this suggests that the oldest stars in the galaxy have the widest range of peculiar velocities. This feature is consistent with our discussion in Section 1.6 about the formation of different components. A second feature is the clear asymmetry in the velocity ellipsoid along the v axis, known as *asymmetric drift*. Few stars are observed with $v > 65$ km s^{-1}, whereas there are several metal-poor stars with $v < -250$ km s^{-1}.

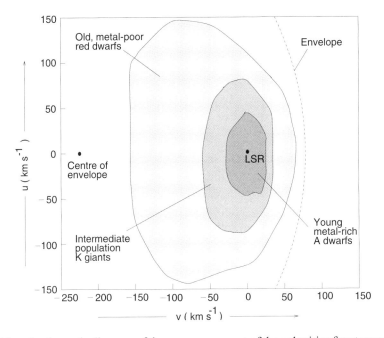

Fig. 1.18.   A schematic diagram of the u, v component of the velocities for stars near the Sun. The LSR is located at $(u, v) = (0, 0)$. The enveloping circle with its centre shown reveals the orbital velocity of the LSR.

The fact that the centre of the envelope with a radius of roughly 300 km s^{-1} is located near $v = -220$ km s^{-1} suggests that the orbital speed of the LSR is $\Theta(R_0) = 220$ km s^{-1}.

This result is consistent with the observations of the rotation curve of our galaxy. As shown in Vol. II, Chap. 9, Section 9.9, the 21-cm emission from neutral-hydrogen gas in our ISM can be used to map the velocity field of our galaxy. The discussion there showed that the radial velocity of gas can be written in the form

$$v_R = R_0 [\Omega(R) - \Omega_0] \sin l, \qquad R_0 = 8 \text{ kpc}, \qquad R_0 \Omega_0 = \Theta_0 = 200 \text{ km s}^{-1}, \tag{1.93}$$

where $\Omega(R)$ is the angular velocity of material at a distance R from the centre. As we move to larger distance, the maximum value of v_R occurs when R is a minimum, that is, at the tangent point (see Fig. 1.19). Hence the maximum value occurs at $R = R_0 |\sin l|$. Taking the rotational velocity $\Theta(R) = \Omega(R)R$, we get the relation

$$\Theta [R = R_0 |\sin l|] = v_{max}(l) + \Theta_0 |\sin l|. \tag{1.94}$$

Radio observations provide the maximum value of velocity $v_{max}(l)$ along different directions l. For example, the typical maximum values at $l = 60°, 45°, 30°$, and $15°$ are $v_{max} = 40, 80, 110,$ and 150 km s^{-1}, respectively. At these four points, $\Theta_0 |\sin l|$ has the values 190, 156, 110, and 57 km s^{-1} so that rotational velocity becomes $\Theta = v_{max} + \Theta_0 |\sin l| = (230, 236, 220,$ and 207 km s^{-1}). The radii for these four values of l are given by $R = R_0 |\sin l| = (6.9, 5.7, 4,$ and

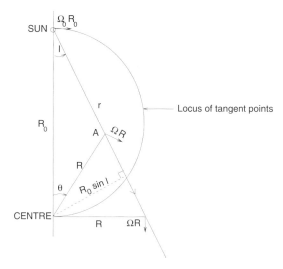

Fig. 1.19. Geometry illustrating the measurement of galactic rotation from the Doppler shift.

2.1 kpc). This shows that in a range from ~2 to 7 kpc, the rotation curve rises a little initially and remains reasonably flat. (A more precise analysis, of course, will have to correct for the peculiar motion of the Sun with respect to the local rotational velocity; this will lead to a correction of the order of 15 km s^{-1}.)

Exercise 1.7
Distance to a pulsar: A pulsar at $l = 33°$, $b = 0°$, shows a 21-cm absorption line that has a radial velocity of $v = 45$ km s^{-1}. Obtain a lower bound for the distance to the pulsar, assuming a flat rotation curve for our galaxy. [Answer. Assuming a flat rotation curve, the observed radial velocity can be written in the form $v_R = \Theta_0[(R_0/R) - 1]\sin l$. Using $l = 33°$, $v_R = 45$ km s^{-1}, $\Theta_0 = 220$ km s^{-1}, and $R_0 = 8$ kpc, we get $R = 0.7$, $R_0 = 5.8$ kpc. If the distance to the pulsar is d, then the two possible solutions for d are given by the roots of the quadratic equation $R^2 = R_0^2 + d^2 - 2R_0 d \cos l$. Knowing all other parameters, we can solve this to give $d = 2.9$ kpc and 10.6 kpc. The minimum distance to the pulsar is therefore 2.9 kpc.]

1.9.4 The Galactic Centre

The line of sight from the Sun to the galactic centre traverses a significant amount of material, making observations of the central region difficult in the optical wavelength. Hence the galactic centre is usually studied at wavelengths longer than 1 μm or at x-ray and γ-ray bands. The IR K band centred at 2.2 μm is also used because a large number of old Pop I stars of the K and the M types ($T_{eff} \approx 4000$ K) are readily visible in this band. These studies suggest that the mass density of stars rises as $r^{-1.8}$ (which is close to the r^{-2} behaviour of an isothermal sphere), up to ~2 pc. There is also an indication that, closer to the centre (at $r < 2$ pc), the stellar density rises substantially faster – at least as steep as $r^{-2.7}$ – suggesting that large amount of mass exists near the centre. It is estimated that nearly $(3-4) \times 10^6 \, M_\odot$ material exists within a sphere of a radius less than 0.5 pc.

Because the distribution of dust in the central region is nonuniform, it is possible to see farther along some directions towards the centre than along others. One such direction – called *Baade's window* – along $(l, b) = (1°, -3.9°)$ has been extensively used to probe the central regions of our galaxy. Using Baade's window, it is possible to monitor the variation in the brightnesses of several million stars in order to detect gravitational lensing – and consequent amplification in the intensity – of one star by another. (This process, called *microlensing*, will be discussed in detail in Chap. 3.) The optical depth τ for a lensing event will be proportional to the product of the surface density Σ_b of stars in the bulge and the effective depth L_b of the bulge. Let us consider how τ varies if we deform a prolate spheroidal bulge to a barlike structure while looking down along the long axis. If $q(< 1)$ is the axis ratio of the bulge, conservation of mass requires

$\Sigma_b q^2 L_b^2$ to remain constant, while conservation of volume requires $L_b^3 q^2$ to be a constant. Combining these, we find that

$$\tau \propto \Sigma_b L_b \propto L_b^2 \propto q^{-4/3}. \tag{1.95}$$

Comparing the observed value of τ with theoretically predicted values for models of the galaxy allows us to decide the nature of mass distribution near the centre and especially whether there is a bar in the centre of our galaxy. Most of the microlensing studies give an optical depth in the range of $\tau = (3-4) \times 10^{-6}$, which is roughly a factor of 3 lower than what is predicted for axisymmetric models of the Milky Way, thereby indicating a bar.[6] These are consistent with photometric observations in Baade's window if we take the long axis of the bar to be at $\sim 15°$ to the line joining the Sun to the centre of the galaxy.

1.10 Features of Active Galactic Nuclei

We have been discussing so far galaxies that are considered "normal" in the sense that their energy emission is mostly due to starlight. Another class of extragalactic sources that has attracted considerable amount of attention and is probably *not* powered by starlight is called *active galactic nuclei* (AGN). These objects display a wide variety of features, and their basic nature is still under considerable debate. We shall now provide a phenomenological introduction to some of the key properties of these objects and comment on different observational issues related to them.

The reason for the significant interest in AGN arises from the fact that many of them display spectacular behaviour. They produce very high luminosities (even 10^4 times the luminosity of a typical galaxy) in tiny volume (less than ~ 1 pc^3). Their spectra extend over a broad range of frequencies and, at least in one case, the luminosity per decade in frequency is roughly constant across 13 orders of magnitude in frequencies. Their line spectrum (in optical and UV) is quite strong, and the total flux in lines could be few to tens of percent of the continuum flux. Some of the lines have a width that indicates bulk velocities ranging up to 10^4 km s^{-1}. AGNs also show remarkable variety in their radio morphology and strong variability in different wave bands.

Such a wide variety of physical phenomena indicates fairly complex and diverse mechanisms of energy generation, and hence any simple attempt to classify AGNs is doomed to failure. What can be achieved is the limited goal of identifying a broad subclass of objects, each of which can be characterized by certain features. It must be remembered that there are exceptions to the classification schemes within each of the categories that can be defined. We begin by discussing some of the observable features of the AGN and the observational caveats related to these features. Later on, in Section 1.11, we shall describe different subclasses of objects, based on these observable features.[7]

1.10.1 Compact Sizes, Variability, and Continuum Emission

One of the simplest characterization of AGN is that they are very compact objects, emitting anywhere between 10^{42} and 10^{48} ergs s^{-1} of power. Because a typical field galaxy has a luminosity of 10^{44} ergs s^{-1}, the AGN have power outputs that range from 1% of that of a typical galaxy to 10^4 times as large. A power of 10^{47} ergs s^{-1} is equivalent to the conversion of the rest-mass energy at the rate of $\sim$2 solar mass per year. Alternatively, this is equivalent to the energy output of $\sim 10^7$ O-type stars confined to a volume of $\sim 10^{-6}$ pc^3.

For nearby AGNs, which can be seen embedded in a diffuse host galaxy, the optical image of the nuclear region appears as a bright point source whose flux can exceed the flux from the rest of the host galaxy. This suggests that, in most of the AGNs, which are too far away for the host galaxy to be seen, we can attribute the emission as arising from a compact nuclear region of the host galaxy, thereby justifying the name AGN.

There are, however, several observational caveats that we should keep in mind while considering this property. To begin with, our ability to detect the AGN and its host depends on the luminosity contrast between the two, which varies from case to case and also depends on the wave band of observation. For example, some AGN have very high x-ray luminosity compared with that of any normal galaxy; hence in the x-ray band, the AGN will appear as point sources. In contrast, the radio emission extends over a much larger region and often stretches beyond the host galaxy. Second, we must note that cosmological redshift effects have a bearing on this criterion. At $z \simeq 1$, a typical galaxy subtends $\sim$1 arcsec, which is at the resolution limit of most ground-based optical telescopes. In addition, the surface brightness of the host falls as $(1 + z)^{-4}$ (see Section 1.3) while the nucleus remains effectively as an unresolved point, thereby enhancing the contrast. Finally, for AGN with $z \gtrsim 1$, the light in the visible band originates in the UV band of the rest frame; because most galaxies are comparatively less luminous in the UV, we again achieve an artificial enhancement in the contrast. All these go to show that observing the host galaxy for the AGN is a fairly difficult task.

Exercise 1.8

Question of contrast: The AGN can sometimes even outshine the integrated light from the host by a large factor. To estimate this, assume that a diffraction-limited image is produced by a circular aperture of radius a in a telescope operating at wavelength λ. What is the angular radius inside which 99% of the light is contained? Because quasars can easily produce 100 times more luminosity than their host, the light in the far wings of the PSF can be very significant. How do you think this problem can be tackled? [Hint: The intensity profile in diffraction can be expressed in the form $I(\theta) = [2J_1(x)/x]^2$, where J_1 is the Bessel function and $x = 2\pi\theta a/\lambda$. From the asymptotic form $I \approx (4/\pi x^3)$, show that 99% of the light is within $\theta \approx 10(\lambda/a)$.]

The optical luminosities of most AGN vary at a level of $\sim 10\%$ in a time scale of few years, which is quite unlike those of normal galaxies. The relative change in the amplitude seems to increase at higher frequencies, with nearly a factor of 2 variation in x rays. Although this effect is most easily seen when the luminosity is plotted as a function of time, quantitative considerations of variability require us to obtain the power spectrum of the varying flux which, in turn, depends on both the total stretch of time t_{obs} for which the source is monitored as well as on the sampling interval Δt. To obtain the true power spectrum, we need the limits $t_{obs} \rightarrow \infty$ and $\Delta t \rightarrow 0$, which, of course, are impossible to obtain. Hence the measured rms fluctuations in the flux depend on the sampling details except when we are interested in time scales such that $\Delta t \ll t \ll t_{obs}$. Unlike variable stars, AGN do not show any periodicities or special time scales, and, in fact, the power spectra in the frequency space is fairly broadband.

The continuum radiation from the AGN is quite different from the spectra of normal galaxies. Roughly speaking, the luminosity of a typical galaxy arises from the luminosities of the individual stars that radiate as blackbodies. The typical stellar surface temperature varies by only a factor of ~ 10 among the stars that dominate the galactic power output. Because the bulk of the radiation in a blackbody is confined to a range that varies by only a factor of ~ 3 in frequency, a typical galaxy emits all its power within no more than one decade in frequency; in fact, it is usually less than a decade. (The only exception to this general rule arises because of the presence of interstellar dust that could absorb a significant fraction of original radiation in optical and UV and reradiate it in the FIR. This could lead to a secondary peak in the IR in dusty galaxies.) The continuum spectra of AGN, in contrast, are relatively flat in terms of νF_ν (see Fig. 1.20), all the way from mid–IR to hard x rays that span a range of 10^5 in frequencies. Very crudely, the broadband spectrum of an AGN continuum can be

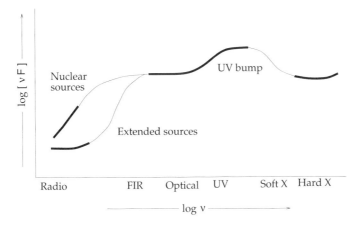

Fig. 1.20. Schematic spectrum of AGN.

characterized by a power law: $F_\nu = C\nu^{-\alpha}$, where $\alpha \lesssim 1$. (A flat distribution of energy in every decade of frequency corresponds to $\alpha \approx 1$.) Such a power-law spectrum has relatively far more energy at high frequencies compared with a blackbody spectrum. Consequently, the colour indices U–B and B–V for AGN will be quite different from those of stars; in a colour–colour plot they will lie well above the main-sequence stars. Compared with that of normal galaxies, the fraction of energy radiated in radio is approximately an order of magnitude larger and the fraction that is emitted in x rays could be even 3 to 4 orders of magnitude larger. As Fig. 1.20 shows, there are minor features and local maxima in the spectra, but no one frequency band dominates the energy output of AGN. The features in the spectrum could arise because of the superposition of different emission mechanisms operating at different energy scales. In particular, we would like to distinguish between thermal and nonthermal emission in these sources. Nonthermal emission is characterised by high luminosities in the radio band, high linear polarization, strong x-ray emission, rapid variability, and a power-law spectral signature. These features exist at different levels in different wave bands and sources.

The key ingredient of the theoretical model for AGN is the assumption that they are powered by a compact central engine made of a super massive black hole fed by an accretion disk. This is a natural mechanism for producing high luminosity from a compact region. If the radiation, emitted from a region of size R that is moving with speed v, varies at a time scale Δt, then we must have

$$R \leq \frac{c\Delta t}{\gamma} = c\Delta t\sqrt{1 - \frac{v^2}{c^2}}. \tag{1.96}$$

This constraint arises from the fact that, in the rest frame of the emitting region, causal processes can act coherently only over a size $R \leq c\Delta\tau$ because the information about the variation in light intensity can travel a distance $c\Delta\tau$ only in a proper time interval $\Delta\tau$. Using $\Delta\tau = \Delta t/\gamma$ leads to the inequality given in relation (1.96). For $\Delta t \approx 1$ h and $\gamma \approx 1$, we get $R \lesssim 10^{14}$ cm, which is smaller than the size of the solar system. Because the total luminosity of AGN is larger than that of an entire galaxy, we need a very efficient and compact mechanism for power generation. Of all the known radiation mechanisms, accretion onto a massive object is the most efficient one as it can easily convert rest-mass energy into radiation with an efficiency of a few percent. When a body of mass m falls from $r = \infty$ to $r \approx 5\,R_s$ where $R_s \simeq (2\,GM_{\mathrm{BH}}/c^2)$ is the Schwarzchild radius of a compact object of mass M_{BH}, it radiates the potential energy

$$U = \frac{GM_{\mathrm{BH}}m}{5\,R_s} = \frac{GM_{\mathrm{BH}}m}{10\,GM_{\mathrm{BH}}/c^2} = 0.1\,mc^2. \tag{1.97}$$

This oversimplified Newtonian calculation gives an efficiency η of 10% which is an order of magnitude more than the efficiency achieved in hydrogen fusion.

(Only direct particle–antiparticle annihilation can give rise to higher efficiencies.) With this efficiency, we can achieve a luminosity of $L = \eta \dot{M} c^2$ if the accretion rate is $\dot{M}$. In other words,

$$\dot{M} = \frac{L}{\eta c^2} = 2 \left(\frac{L_{46}}{\eta_{0.1}} \right) M_\odot \, \text{yr}^{-1}, \tag{1.98}$$

which is not difficult to achieve in the dense galactic centres. However, we know that accretion luminosity is bounded by the Eddington luminosity L_E (see Vol. II, Chap. 1, Section 1.6), which is related to the mass M_{BH} of the compact object by

$$L_E \simeq 1.5 \times 10^{38} \frac{M_{BH}}{M_\odot} \, \text{ergs s}^{-1}. \tag{1.99}$$

To achieve a luminosity of $L \simeq 5 \times 10^{46}$ ergs s^{-1}, we need $M_{BH} > 3.3 \times 10^8 \, M_\odot$. Note that the mass corresponding to a black hole with the Schwarzchild radius $R \simeq 10^{14}$ cm (which is the size of the central engine determined from variability) is indeed $3.3 \times 10^8 \, M_\odot$. This suggests that we might be able to get a consistent picture involving a massive black hole fuelled by accretion.

The real difficulty in accretion models is not with energetics but (1) with time scales and (2) in providing a mechanism that will make matter shed its angular momentum. To see the issue of time scales, note that the age of the universe at $z \approx 5$ is only approximately $(5\text{–}10) \times 10^8 h_{0.75}^{-1}$ yr. This is not very long compared with the galactic free-fall time [approximately $10^8 (r/10 \, \text{kpc})$ yr] or the black hole growth time defined by $\tau_{BH} = (\dot{M}/M_{BH})^{-1}$. Using $\dot{M} \propto L/\eta$ and $M_{BH} \propto L_E$, we find that $\tau_{BH} \approx 4 \times 10^7 \, \eta_{0.1}(L_E/L)$ yr. If a black hole has to grow from a original mass of $\sim 10 \, M_\odot$ to a final mass of $10^8 \, M_\odot$, it would take ~ 16 e-folding times; that is, the growth time is more like $16 \, \tau_{BH} \approx 6 \times 10^8 \, \eta_{0.1}(L_E/L)$ yr. The universe just may not have sufficient time at high redshifts to allow this. Low-efficiency accretion permits more rapid growth of a black hole (probably in its early stages), but this will require far more total mass to be accreted in order to provide the same amount of energy.

The situation is made worse by the fact that the angular momentum of the matter needs to be removed efficiently if accretion has to work. For matter in Keplerian orbits in a galaxy, the angular momentum for unit mass is $J/m = (GMr)^{1/2}$, where M is the mass interior to radius r. If we take $M \approx 10^{11} M_\odot$ and $r \approx 10$ kpc for estimating the initial angular momentum J_i and take $M = M_{BH} = 10^7 \, M_\odot$ and $r = 0.01$ pc for estimating the final angular momentum J_f, then $J_f \approx 10^{-5} J_i$, showing that almost all the angular momentum should be removed by some viscous drag if accretion has to be feasible.

Assuming this can (somehow) be achieved, an accretion disk will provide a nearly thermal radiation in the optical and the UV. At first sight this might sound surprising as stellar-mass compact objects in high-mass x-ray binary/low-mass x-ray binary (HMXB/LMXB) systems emit in x rays (see Vol. II, Chap. 1,

Section 1.6), and we might have expected super massive black holes to lead to higher-energy radiation. In reality, the increase in the Schwarzchild radius of the black hole works in the opposite direction, as can be seen by the following argument: If we take the luminosity emitted from regions near the black hole to be proportional to $L \propto (GM_{BH}\dot{M})/R_s$, where $R_s = (2\,GM_{BH}/c^2)$, then $L \propto \dot{M}$ and is independent of M_{BH}. If this energy is emitted from an area $A \propto R_s^2$, the flux will be $F \propto (L/A) \propto (\dot{M}/M_{BH}^2)$. Assuming the radiation is thermal with temperature T, we can write $F \propto T^4$, thereby leading to the estimate of the disk temperature:

$$T_{disk} = \left(\frac{3c^6 \dot{M}}{8\pi\sigma G^2 M_{BH}^2} \right)^{1/4}, \tag{1.100}$$

where the proportionality constants have been reintroduced. If the total emission is a fraction f of the Eddington luminosity and the efficiency of the process is η, we also have the relation

$$\eta\dot{M}c^2 = f\frac{4\pi Gc}{\kappa}M_{BH}, \tag{1.101}$$

[where $\kappa = 0.2(1 + X)$ cm^2 gm^{-1} is the electron-scattering opacity with $X \approx 0.7$], leading to

$$\dot{M} = \frac{f}{\eta}\frac{4\pi G}{\kappa c}M_{BH}. \tag{1.102}$$

Substituting into Eq. (1.100), we get

$$T_{disk} = \left(\frac{3c^5 f}{2\kappa\sigma GM_{BH}\eta} \right)^{1/4} \propto M_{BH}^{-1/4}. \tag{1.103}$$

This shows that the temperature of the disk actually decreases with the increasing mass of the black hole. As an example, consider a situation with $M_{BH} \simeq 10^8\, M_\odot$, $F \simeq 1, \eta \simeq 0.1$, and $L \simeq 1.5 \times 10^{46}$ ergs s^{-1}. This requires an accretion rate given by Eq. (1.102), which translates to $\dot{M} \simeq 2.6\, M_\odot$ yr^{-1}. The corresponding temperature obtained from Eq. (1.103) is $T_{disk} = 7.3 \times 10^5$ K. Such a spectrum will peak around 400 Å in the extreme-UV band. It is generally believed that such a component of thermal radiation could be responsible for the UV bump seen in the spectra of AGN.

Compared with that of optical, the origin of high-energy spectra is more complex and ill understood. Pair-production processes as well as inverse Compton scattering could contribute to high-energy radiation, although the details differ from source to source. It is easy to see that pair production should be important from the following arguments. The threshold energy for pair production is $h\nu \approx (m_e c^2)$; we can estimate the number n_γ of photons near the threshold by dividing the energy density $U \approx L_\gamma/4\pi R^2 c$ of gamma-ray photons within a region of radius R by the threshold energy $m_e c^2$. The optical depth for pair production is $\tau_{\gamma-\gamma} \simeq n_\gamma \sigma_T R$. The pair production will be important if $\tau_{\gamma-\gamma} \gtrsim 1$. This condition can be expressed in terms of the *compactness parameter* $\ell \equiv (L_\gamma/R)(\sigma_T/m_e c^3)$ as

$$\ell \equiv \frac{L_\gamma}{R} \sigma_T m_e c^3 \gtrsim 4\pi. \tag{1.104}$$

Reexpressing the compact parameter in terms of Eddington luminosity L_E and Schwarzchild radius R_s as $\ell = 2\pi(L_\gamma/L_E)(R_s/R)(m_p/m_e)$, we find that the condition for pair production becomes

$$\frac{L_\gamma}{L_E} \gtrsim \frac{2m_e}{m_p} \left(\frac{R}{R_s}\right) \approx 1.1 \times 10^{-3} \left(\frac{R}{R_s}\right). \tag{1.105}$$

Because $R \approx 10\,R_s$, it is possible to meet this condition in many AGN, suggesting that pair production can be an important effect.

In addition to the primary x-ray emission, we also expect reradiation of x rays from the K and L ionisation edges of iron, magnesium, silicon, and sulphur. In particular, the iron K-α line observed in many AGN is interpreted as arising from such a reflected radiation.

Finally we note that the energy radiated by AGN can be computed if the luminosity function of these sources is known. If $n(S, z)$ is the number of sources with bolometric flux S at a redshift z per steradian in the sky, then the integral of $Sn(S, z)$ over all S will give the intensity that is due to all sources per steradian. In a *static* universe we can obtain the corresponding energy density by multiplying by $(4\pi/c)$. In an expanding universe, we have to take into account the fact that energy emitted at an earlier epoch is redshifted away because of cosmic expansion. Hence the actual energy emitted has to be higher than a factor of $(1 + z)$ in an expanding universe in order to produce the same observed flux today. Therefore the energy density generated by the sources is given by

$$U = \frac{4\pi}{c} \int (1 + z)\, n(S, z)\, S\, dS\, dz$$

$$= \langle (1 + z) \rangle \frac{4\pi}{c} \int n(S, z)\, S\, dS = \langle (1 + z) \rangle U_{\text{source}}, \tag{1.106}$$

where $\langle (1 + z) \rangle$ is the mean redshift of the sources. The density parameter corresponding to this energy density is $\Omega = (U/\rho_c c^2)$.

The quasars in the optical B band (which dominates the output) give $\Omega \approx 1.1 \times 10^{-8} h^{-2} \langle (1 + z) \rangle$. (For comparison, the total amount of starlight in the B band from galaxies, released over a time scale H_0^{-1}, is $\Omega \approx 0.5 \times 10^{-6} h^{-2}$.) If all bright galaxies ($M < -21$) have harboured quasars, then by dividing the energy produced per unit volume U by the number density of bright galaxies, we can estimate the amount of energy that has to be emitted per host galaxy and hence the the remnant mass. This turns out to be $M_{\text{rem}} \approx 2 \times 10^6 h^{-3} M_\odot$. Alternatively, if only $\sim 1\%$ of the bright galaxies host the quasars, then the remnant mass has to be one hundred times larger: $M_{\mathrm{rem}} \approx 2 \times 10^8 h^{-3} M_\odot$. Because the total energy U radiated by AGN can be expressed in terms of $\dot{M}$, M_{BH}, and an efficiency factor, we can also estimate the remnant mass as $M_{\mathrm{rem}} \approx 8 \times 10^7 h^{-3} \eta^{-1} (t_Q/t_{\mathrm{dyn}}) M_\odot$, where t_Q is the time the quasar is "on" and t_{dyn} is the evolution time scale that will be of the order of the age of the universe. If $t_Q \approx t_{\mathrm{dyn}}$ and $\eta \approx 0.1$, then the remnant masses are large ($M_{\mathrm{rem}} \approx 10^9 h^{-3} \eta_{0.1} M_\odot$). In this case, only a small fraction of the galaxies harbour quasars and $(L/L_E) \lesssim 0.05 h \eta_{0.1}$. On the other hand, if $t_Q \ll t_{\mathrm{dyn}}$, then every bright galaxy could have a quasar with, say, $M_{\mathrm{rem}} \approx 10^7 h^{-3} \eta_{0.1} M_\odot$ if $t_q \approx 0.1 t_{\mathrm{dyn}}$. In this case, $(L/L_E) \lesssim 3 h \eta_{0.1}$.

1.10.2 Radio Emission and Jets

A class of AGN shows bright radio lobes and jets connecting the central source to the lobes. Because of the high resolution and sensitivity that can be achieved in the radio band, it is often possible to do milliarcsecond imaging of the AGN in this band and most of the *known* AGN are strong radio emitters. Even then, the radio-band luminosity never accounts for more than $\sim 1\%$ of bolometric luminosity and less biased surveys show that the predominance of radio emission could also be partly due to the selection effects.

In contrast to optical/UV emission, the radio spectrum is nonthermal and is most likely to originate from synchrotron radiation of relativistic electrons. The rate of synchrotron energy loss $\dot{E}_{\mathrm{synch}}$ by a single electron moving in a magnetic field with energy density U_{mag} is given by (see Vol. I, Chap. 6, Section 6.10)

$$-\dot{E}_{\mathrm{synch}} = 2\sigma_T \gamma^2 \beta^2 \sin^2 \theta U_{\mathrm{mag}} c, \qquad (1.107)$$

where $\cos\theta = \hat{\mathbf{v}} \cdot \hat{\mathbf{B}}$. The characteristic frequency of gyration of an electron in a magnetic field B is given by $\omega_g = (eB/\gamma m)$, where m is the mass of the electron and $\gamma = (1 - v^2/c^2)^{1/2}$. The characteristic time scale will then be ω_g^{-1}. However, because of relativistic beaming of the radiation, the characteristic time scale in which the radiation will sweep past a given observer is $(\gamma^3 \omega_g)^{-1/2}$ (see Vol. I, Chap. 6, Section 6.10). This result can be easily obtained as follows: Let us assume that the radiation is relativistically beamed into a cone of angle $\Delta\phi$. If the radius of the orbit is R, the electron takes a time $R\Delta\phi/v$ to cover the arc of length $R\Delta\phi$ but the light travel time across the segment is $R\Delta\phi/c$. Hence the observed time interval over which the radiation sweeps past the observer is

$$(\Delta t)_{\mathrm{obs}} = (R\Delta\phi)\left(\frac{1}{v} - \frac{1}{c}\right) = \left(\frac{R\Delta\phi}{c}\right)(\beta^{-1} - 1). \qquad (1.108)$$

Using $\beta = (R\omega_g/c)$ and assuming that $\gamma \gg 1$, we get $(\Delta t)_{\mathrm{obs}} = (\omega_g \gamma)^{-1}(1 - \beta^{-1}) \simeq \omega_g^{-1} \gamma^{-3}/2$. Hence the characteristic frequency in which the synchrotron radiation will peak is given by $\nu = \gamma^3 \nu_g$. The synchrotron emissivity from a bunch of electrons will be

$$\epsilon_\nu = 4\pi j_\nu = -\dot{E} N(E)\frac{dE}{d\nu}, \tag{1.109}$$

where $N(E)$ is the number of electrons in the energy range $(E, E + dE)$. For a power-law spectrum of electrons with $N(>\gamma) = N_0 \gamma^{1-x}$, where N_0 is the total electron number density, we have

$$N(E) = N_0(x-1)(m_e c^2)^{x-1} E^{-x}. \tag{1.110}$$

This leads to the synchrotron emissivity

$$\epsilon_\nu \cong N_0(x-1)\frac{ce^3}{4\pi^2 m}\left(\frac{e}{m}\right)^{(x-1)/2} B^{(x+1)/2}\nu^{(1-x)/2} \propto \nu^{-\alpha}, \tag{1.111}$$

if we take $\langle \sin^2\theta \rangle = 2/3$. Hence a power-law spectrum of electrons will lead to power-law synchrotron emission with $\alpha = (1/2)(x-1)$. This is considered to be the basic origin of the nonthermal power-law spectrum in the radio band of AGN.

1.10.3 Emission Lines

The first feature of AGN emission lines that distinguishes it from the spectra of most stars and galaxies is the equivalent widths of the lines, which are often as large as 100 Å. These lines are quite prominent compared with any other generic source known in the sky. Second, the emission lines show much less variation from one AGN to another, in contrast with many other physical properties. We almost always observe Lyman-α, Balmer, C IV 1549 doublet, [OIII] 5007, and iron K-α [6.4-keV] lines. The linewidths of the emission lines show a bimodal distribution with broad lines having widths corresponding to several thousand kilometres per second, and narrow lines having a width of only few hundred kilometres per second.

It is generally believed that the line radiation in AGN originates from clouds of different density and temperature that are in bulk orbital motion around the central engine (see Fig. 1.23) with the broad lines originating from high-density high-orbital-velocity clouds closer to the centre [called the *broad-line region* (BLR)] and the narrow-lines emerging from low-density lower-velocity clouds located at a larger distance from the centre [called the *narrow-line region* (NLR)]. The linewidths of, say, 5000 km s^{-1}would imply a temperature of $T \approx 10^9$ K if they are produced by random thermal motion; but the temperature inferred from various line ratios in the BLR is much smaller and hence the bulk motion is essential to interpret the linewidth. We shall discuss several aspects of the NLR and the BLR in Chap. 8.

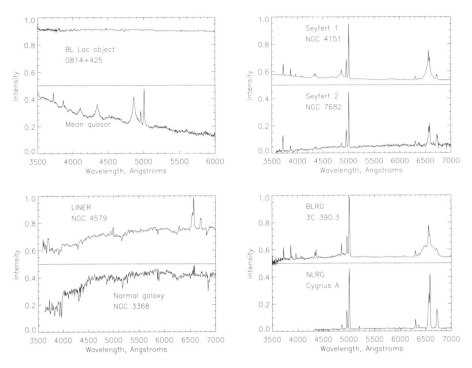

Fig. 1.21. The typical spectra of a wide class of AGN. (Figure courtesy of Bill Keel.)

1.10.4 Absorption Systems

Because the process of gravitational instability that leads to the condensation of galaxylike objects cannot be 100% efficient, it would leave some amount of matter uniformly distributed in between the galaxies. The light from distant quasars will have to pass through this matter and will contain signature of the state of such an *intergalactic medium* (IGM). The photons (with $v > v_I$) produced in the first-generation objects could cause a significant amount of ionisation of the IGM, especially the low-density regions. When a flux of photons (with $v > v_I$) impinges on a gas of neutral hydrogen with number density n_H, it will have an ionisation optical depth of $\tau = n_H \sigma_{bf} R$. Setting $\tau = 1$ gives a critical *column* density for ionisation to be $N_c \equiv n_H R = \sigma_{bf}^{-1} \simeq 10^{17}$ cm^{-2}. Regions with a hydrogen column density of $N_c \simeq nR$ greater than 10^{17} cm^{-2} will appear as patches of neutral regions in the ionised plasma of the IGM. Such regions can be studied by absorption of light from more distant sources, (especially through Lyman-α absorption corresponding to the transition between $n = 1$ and $n = 2$ levels) and are called *Lyman-α clouds*.These clouds probably contain a fair fraction of baryons in the redshift range $z \approx (2-4)$. (The density parameter contributed by these clouds are shown in Fig. 1.1 under the IGM.) We shall discuss them in detail in Chap. 9.

Table 1.4. *Properties of AGN*

Object	Emission Lines		Associated Galaxy		
	Broad	Narrow	Type	Luminosity	Evolution
PRG	SW	SW	E	S	S
WRG	W	W	E	<S	W
QSO (radio loud)	S	SW	E	S	S
QSO (radio quiet)	S	SW		<S	S
BL Lac	0	0W	E	<S	W
Sy 1	SW	SW	Sa-Sbc	$M_V < -20$	W
Sy 2	0	SW	Sa-Sbc	$M_V < -20$	?
LINERs	0	SW	E-Sbc	$M_V < -20$	?

Notes: Observational classification of AGN and related objects.[8] Notation in table. Spectral lines: S, strong; W, weak; 0, absent. QSO, quasi-stellar object; Sy 1, Seyfert type I galaxy; Sy 2, Seyfert type II galaxy; LINERs, low-ionisation nuclear-emission regions.

1.11 Taxonomy of Active Galactic Nuclei

Given the preceding properties, we can attempt to classify the different kinds of AGN into subsets that have certain definite features. This classification is, of course, not watertight, and we often come across sources that could be classified into different categories based on which property is given priority. Nevertheless, the taxonomy given below certainly helps in the study of AGN and serves as a standard backdrop for comparing models. The properties are summarised in Table 1.4 and the typical spectra are shown in Fig. 1.2.1.

1.11.1 Radio Galaxies

The radio galaxies (RGs) are usually identified with luminous ellipticals and are often associated with jets and radio-emitting lobes that could extend up to distances of from 10 kpc to 1 Mpc from the core region. Because of the extensive studies that have taken place in the radio band, these galaxies are further subdivided based on several features. One simple division depends on the luminosity at $\nu = 1.4$ GHz. If $L_{1.4} \lesssim 10^{25}$ W Hz^{-1}, they are called *weak-radio galaxies* (WRGs), and if $L_{1.4} \gtrsim 10^{25}$ W Hz^{-1}, they are called *powerful-radio galaxies* (PRGs). The spectrum of WRGs in radio bands is a power law with a rather steep index ($\alpha \gtrsim 0.4$) whereas PRGs have a relatively flat spectrum (with $\alpha \lesssim 0.4$ around 1 GHz).

Because many RGs have an extended jetlike structure ending in radio lobes, it is possible to classify RGs based on the ratio q of the separation between the two lobes to the overall size of the source. If $q \lesssim 0.5$, the RG is called *edge-darkened sources* or *Fanaroff-Riley type I* (FRI), whereas if $q \gtrsim 0.5$, then it is called *edge-brightened sources* or *Fanaroff-Riley type II* (FRII). The FRII-type RGs tend to have high radio luminosities with L (178 MHz) $> 10^{25}$ W Hz^{-1} sr^{-1} whereas FRI RGs have lower values. The FRII also more often exhibits the symmetrical morphology of a double source with hot spots. Among the FRIs, low-luminosity RGs with L (178 MHz) $\lesssim 3 \times 10^{23}$ W Hz^{-1} sr^{-1} are more symmetrical with an edge-darkened jet. The FRIs with L (178 MHz) $\gtrsim 3 \times 10^{23}$ W Hz^{-1} sr^{-1} have asymmetrical bright one-sided jets. The difference between the two FR types could be due to a subsonic motion of the jet head in FRI compared with supersonic motion in FRII, which leads to shock waves and strong radiation.

1.11.2 Quasars

Quasars are AGN with an angular size of $\lesssim 1''$ with broad emission lines and no strong radio morphology (so as to distinguish them from the RGs). Approximately 10% of the quasars are radio loud with $[\nu F_\nu \, (5 \text{ GHz})/\nu F_\nu \, (250 \text{ nm})] \gtrsim 10^{-3}$, whereas the remaining 90% are radio quiet. Table 1.5 compares the density of objects per square degree of sky at different bolometric magnitudes and shows how the quasi-stellar object (QSO) density increases at fainter magnitudes.

Table 1.5. *Space density of extragalactic objects*

Object	Luminosity	Density (Gpc^{-3})
Spiral Galaxies	$M_V < -20$	5×10^6
	$M_V < -22$	3×10^5
	$M_V < -23$	3×10^3
Elliptical galaxies	$M_V < -20$	1×10^6
(including S0)	$M_V < -22$	1×10^5
	$M_V < -23$	1×10^4
Clusters of galaxies		3×10^3
RGs	$L_{1.4 \text{ GHz}} > 10^{23.5}$ W Hz^{-1}	3×10^3
	$L_{1.4 \text{ GHz}} > 10^{25}$ W Hz^{-1}	10
Radio quasars	$L_{1.4 \text{ GHz}} > 10^{25}$ W Hz^{-1}	3
Radio-quiet quasars	$M_V < -23$	100
	$M_V < -25$	1
Sy 1	$M_V < -20$	4×10^4
Sy 2	$M_V < -20$	1×10^5
BL Lac	$L_{1.4 \text{ GHz}} > 10^{23.5}$ W Hz^{-1}	80

Note: Local space density of some types of objects.[8]

1.11.3 BL Lac Objects

These AGN show fairly extreme properties: They are highly variable in radio, optical, and x-ray bands (at time scales of $\lesssim 1$ day) with no broad optical emission lines; the optical radiation can be strongly polarized, with the degree of polarization being 5% to 40%. Their continuum luminosity has a maximum in the FIR region ($\sim 2\,\mu$m), where the luminosity could be $\nu F_\nu \gtrsim 10^{47}$ ergs s^{-1}.

1.11.4 Seyfert Type I Galaxies

These are mostly spiral galaxies which resemble low-luminosity radio-quiet quasars. Their spectra, however, contain strong permitted emission lines with broad wings (corresponding to velocities of ~ 5000 km s^{-1}) and narrower forbidden lines. The narrow lines are characteristic of emission from low-density $[n_e \simeq (10^3 - 10^6)$ cm$^{-3}]$ ionized gas. The absence of broad forbidden line emission suggests that they originate from a high-density ($n_e \approx 10^9$ cm^{-3}) region. In the radio band, they appear as flat-spectrum radio sources.

1.11.5 Seyfert Type II Galaxies

These are analogous to Seyfert type I but with equally wide permitted and forbidden lines and no broad wings. Typically the bolometric luminosity of SyII is lower than that of SyI by a factor of ~ 100, although they are more frequent than SyI with a ratio of ~ 2:1.

Both Seyfert types have a quasarlike nucleus, but their host galaxy is clearly detectable. Weak absorption lines that are due to late-type giant stars in the host galaxy are also observed in both the Seyferts.

1.11.6 Low-Ionisation Nuclear-Emission Regions

In contrast to Seyferts, low-ionisation nuclear-emission regions (LINERs) have strong lines of low ionisation of elements such as oxygen, sulphur, etc. They also have a strong OIII λ(4363) line, indicating a kinetic temperature of $T = \sim 40,000$ K. The width of the emission lines corresponds to approximately 200–400 km s^{-1}. In their broad emission lines, LINERs resemble Seyferts, but the narrow lines have a much lower degree of ionisation. These are the least luminous and at the same time most common of the AGN. One quantitative method for distinguishing LINERs from other AGN like Seyferts (which also have strong emission lines) is to compare the intensity ratios of two suitably chosen lines belonging to different elements but having fairly close frequencies. For example, consider the two ratios

$$R_1 = \frac{[\text{O III}]\lambda 5007}{\text{H}\beta\lambda 4861}, \qquad R_2 = \frac{[\text{N II}]\lambda 6583}{\text{H}\alpha\lambda 6563}. \qquad (1.112)$$

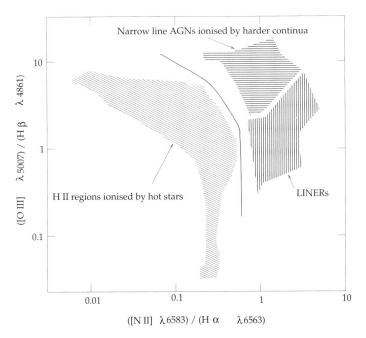

Fig. 1.22.   A schematic diagram showing how the ratio of line intensities can be used to distinguish between different sources of ionisation. The region on left and bottom of the solid curve represents the line ratios expected when the ionisation is due to starlike spectra, as in HII regions. The region to the right and top of the solid curve has line ratios characteristic of regions ionised by harder spectra, like those of AGN.

When different objects are plotted in the R_1-R_2 plane, they fall into different regions (see Fig. 1.22). Because the ratios are taken between lines which are close in wavelengths, we can minimise effects that are due to reddening. The diagonally shaded region shows HII regions that are ionised by hot stars. In contrast, the horizontally shaded region corresponds to narrow-line AGN that are ionised by continuum radiation that has a greater fraction of high-energy photons than a blackbody spectrum (for example, a power-law spectrum). The region shaded vertically can be distinguished from HII regions by higher values of R_2 and from Seyferts by lower values of R_1. Tables 1.4 and 1.5 summarise these properties.

Given the variety of these sources, it is not surprising that a considerable amount of effort was spent in obtaining correlations between properties of different classes of objects. For example, the following properties have been found: (1) The variability time scale is broadly correlated with the total luminosity of the source. (2) The radio-loud AGN have a larger Hα/Hβ ratio than the radio-quiet AGN. The host galaxies of radio-loud AGN are preferentially E or SO galaxies whereas the host galaxies of radio-quiet AGN are preferentially spirals. (3) The radio luminosity functions of the sources show a continuous variation along spirals $\rightarrow$ Seyferts $\rightarrow$ optical QSOs; ellipticals $\rightarrow$ radio quasars. (4) Nearly 15% of Seyferts have close companions (indicating strong interaction) in contrast to the 3% in a controlled sample of non-Seyferts; similarly, in a sample of interacting galaxies, nearly 13% show Seyfert nuclei in contrast to 5% among the sample of isolated galaxies. (5) Among radio-loud AGN, those with physical companions have nearly (5–10) times strong-radio emission than do isolated galaxies of the same type and absolute magnitude.

Exercise 1.9
Taxonomy practice: The best way to understand the classification of AGN is to try one's hand at some real-life examples. Looking up the data from the Web, classify the following extragalactic objects: (a) NGC 1265, (b) NGC 1275, (c) 3C 48, (d) 3C 111, and (e) 3C 279. [Answer: (a) This extended object at $z \simeq 0.021$ with small radio flux is considered to be a FRI RG. (b) This object has a compact core and diffuse emission. It is very radio bright [$F(178 \text{ MHz}) \simeq 70$ Jy] but shows [OIII] emission. This has been classified in the literature as both a Seyfert and a BL Lac object. (c) Forbidden and permitted lines allow this to be classified as a radio-loud quasar. (d) This broad-line RG with bright-radio luminosity is classified as a FRII object. (e) This shows a strong optical variation with very few lines and is often classified as a BL Lac object.]

The discussion of variety of observational features related to AGN shows that there are some common features present in any class of AGN; it is also clear that there are significant differences between the subclasses. Given such a diversity, it is not possible to state categorically whether all AGN can be thought of as being manifestations of the same basic phenomena or whether they represent entirely different classes of sources. Even the choice of radiation mechanism for different wave bands and different classes of objects is not uniquely determined. Such an unsatisfactory state of affairs forces us to start with a theoretical scenario that has at least some of the key ingredients of the AGN built in and to ask how best the scenario can explain the different phenomena. It should be stressed that the theoretical scenario would require conjectures that are neither derivable directly from first principles nor obtainable in a unique manner from the observations. In fact, some of the observed features require adding components to the theory that are not natural. In spite of such limitations, the theoretical approach outlined has the advantage of providing a backdrop against which observations and models can be compared and contrasted.

The key ingredient of the theoretical model for AGN is the assumption that they are powered by a compact central engine made of a super massive black hole fed by an accretion disk. This is a natural mechanism for producing high luminosity from a compact region. The accretion disk is relatively thin near the black hole and thickens as we proceed outwards. A schematic picture of such a model is shown in Fig. 1.23, in which the thicker part of the accretion disk is marked *torus*; the figure is a cross-section of an axially symmetric geometrical structure. There

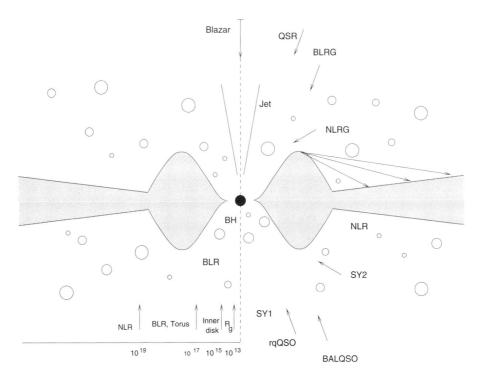

Fig. 1.23.   A schematic diagram describing the unified model for AGN. The figure is not to scale. The distances are given in units of $(M_{BH}/10^8 \, M_\odot)$cm.

are gas clouds with different physical conditions strewn all around this region. Those that are closer to the black hole are comparatively denser and move with larger bulk speeds and could lead to broad-line emission; those that are farther away are less dense and move with comparatively smaller speeds, leading to narrow-line emission. Under certain conditions, matter could be ejected in the form of powerful jets in a plane perpendicular to the accretion disk (marked *jet* in the figure). Given this geometrical model, we attempt to explain the wide variety of AGN phenomena as arising because of a combination of radiation processes and our viewing angle. Some tentative examples are also indicated in the figure.

This description (usually called a *unified model for AGN*) is probably esthetically quite pleasing because it attempts to relate a wide variety of phenomena with a small number of parameters. To avoid misunderstanding, we stress the fact that this is only a tentative model to describe the observed phenomena and its validity is not conclusively demonstrated.

1.12 Luminosity Function of Galaxies and Quasars

The abundance of objects like galaxies, quasars, etc., can be characterised in two different ways. The first one provides a count of objects with an apparent flux or magnitude per solid angle in the sky. Because the information is two-dimensional and we are not measuring the actual distance to the object, this observation is comparatively easy and is essentially limited by only the limiting flux or magnitude that can be obtained in a given wave band and by the angular resolution of the instrument. If $n(F_\nu) \equiv (dN/d\Omega\, dF_\nu)$ denotes the number of sources with the observed flux densities between F_ν and $F_\nu + dF_\nu$ per solid angle in the sky, we can immediately obtain the total intensity of radiation that is due to all the sources as

$$I_\nu = \int_0^\infty F_\nu\, n(F_\nu)\, dF_\nu. \tag{1.113}$$

If apparent magnitudes are used instead of fluxes, they can be related by $F_\nu \propto 10^{-0.4m_\nu}$, with the constant of proportionality depending on the passband in which observations are carried out. In this case, the integrand in Eq. (1.113) is proportional to $10^{-0.4m}(dN/d\Omega\, dm)$. Multiplying the intensity by $(4\pi/c)$ will give the energy density at the local neighbourhood that is due to these sources.

A more detailed description of the abundance of the sources is provided by the luminosity function $\Phi(L, z)\, dL$, which gives the number of sources per unit comoving volume in the luminosity band $L, L + dL$. Depending on the nature of the source, L could be either the luminosity in a given wave band or bolometric luminosity. This quantity can be obtained only from a survey that also has redshift information for the sources. Further, in order to be useful, there must exist sufficient number of objects at any given redshift bin. We shall now consider the source counts and the luminosity function for galaxies and quasars.

1.12.1 Galaxy Counts and Luminosity Function

We begin with a brief description of the source count of galaxies available from HDFs. These observations from the HST go for the faintest magnitudes currently available for galaxies.[9] The HDF survey has detected $\sim$3000 galaxies at U_{300}, B_{450}, V_{606}, and I_{814} and 1700 galaxies at J_{110} and H_{160}. At still longer wavelengths, there are $\sim$300 galaxies in the K band, 50 at (6.7–15) μm, and 9 at 3.2 μm. Nearly 200 of them now have redshifts, with the median redshift being approximately $z = 1$ at $R = 24$. The area of the sky covered by these observations is $\sim$5.3 arcmin^2 for HDF-N, which corresponds to (4.6 Mpc)^2 at $z = 3$ for a universe with $\Omega_{NR} = 0.3$, $\Omega_V = 0.7$, and 0.7 arcmin^2 in the case HDF-S. The top panel of Fig. 1.24 shows the counts of galaxies in different passbands from these surveys as well as from some ground-based observations. The bottom panel of Fig. 1.24 plots these counts in a more physically relevant manner by multiplying by the factor $10^{-0.4m}$. The area under the curve in the bottom panel indicates the total amount of intensity contributed by these galaxies. The figure suggests that when galaxies up to the limiting magnitudes are taken into consideration, the total integrated contribution of light has effectively converged. Actual

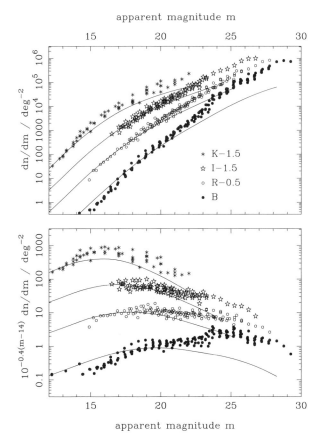

Fig. 1.24.   Top panel: Differential K, I, R, and B band galaxy counts as functions of magnitude from the HDF and some ground-based data. Bottom panel: The same data are replotted by including a factor of $10^{-0.4m}$ so as to directly provide the extragalactic background light per magnitude bin in different bands. The lines are based on no-evolution models in which the high redshift galaxy counts are predicted by transporting the current galaxy population to higher redshifts in a $\Omega_{\rm DM} = 1$ universe. (Figure courtesy of John Peacock.)

integration shows that νI_ν varies in the range (2.9–9.0) nW m^{-2} sr^{-1} in the wavelength band (3600–16,000)Å.[10] We shall discuss this in more detail in Chap. 9.

Let us next consider the luminosity function for galaxies, which has been investigated extensively for nearby galaxies ($z \approx 0$). We shall concentrate on these results and briefly mention the status of investigations at higher redshifts. While considering galaxies at $z = 0$, we treat the luminosity function as a function of a single variable, $\Phi(L)\,dL$, which gives the number of galaxies per unit volume in the luminosity band $L, L + dL$.

The determination of the luminosity function of galaxies could be attempted along the same lines as the determination of the stellar luminosity function.

We begin by measuring the apparent magnitudes of all the galaxies in some representative sample. Next we use some distance indicator, (e.g., Hubble law), to convert apparent magnitudes into absolute magnitudes or luminosities. Dividing the number of galaxies in a given luminosity bin centred around L by the volume of space V surveyed, we can obtain the function $\Phi(L)$.

In practice, we again encounter several serious difficulties: (1) To begin with, it is known that a fraction of the galaxies occur in clusters and superclusters. There is no assurance that the luminosity functions of field galaxies will be the same as those in clusters. It is necessary to make sure that we use fairly unexceptional regions of the sky while choosing the sample. (2) Because higher luminosity galaxies can be detected within larger volumes, we encounter the standard Malmquist bias (see Vol. II, Chap. 1, Section 1.8) in any flux-limited sample. We need to correct for this in obtaining the luminosity function. (3) Estimating the distances to galaxies is a difficult process, and Hubble law $\mathbf{v} = H_0\mathbf{r}$ is valid only if $\mathbf{v}$ is due to cosmological expansion. Local-density fluctuations will also contribute to $\mathbf{v}$. Hubble's law ignores such velocities of galaxies (called *peculiar velocities*) that are due to local concentrations of mass. This problem is quite acute for nearby galaxies for which the peculiar motions can dominate the Hubble expansion. Because low-luminosity galaxies can be observed in only the nearby region, this leads to severe uncertainties in the faint end of the galaxy luminosity function. (4) When the redshifts of the galaxies are significant, we need to apply further corrections because the wavelength band in which the apparent magnitude is measured will be different from the wavelength band in which the radiation is originally emitted. The observed wavelength band also selects the dominant class of stars that contribute to the light. For example, the B-band light is due to the small minority of a stellar population consisting of young stars, so there is no assurance that the B-band luminosity function and, say, the K-band luminosity function of the galaxies will be the same. (5) When the redshifts of the galaxies are significant, we also need to worry about the fact that high-redshift galaxies are seen during an earlier phase of the universe and might have evolved over the course of time. Separating out evolutionary effects in any luminosity function is a nontrivial and – in general – an ill-posed problem.

Given the uncertainties, it is clear that the observed galaxy luminosity function cannot be trusted at the same level as, say, the stellar luminosity function. Nevertheless, it provides an important quantitative tool in the study of galaxies. The luminosity function is reasonably well described by

$$\Phi(L) = \left(\frac{\Phi^*}{L^*}\right)\left(\frac{L}{L^*}\right)^{\alpha}\exp\left(-\frac{L}{L^*}\right), \tag{1.114}$$

which translates to

$$\Phi(M) = (0.4\ln 10)\,\Phi^*\,10^{0.4(\alpha+1)(M^*-M)}\exp\left[-10^{0.4(M^*-M)}\right] \tag{1.115}$$

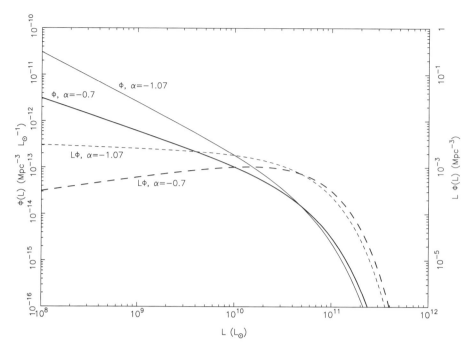

Fig. 1.25.   The luminosity function of galaxies. The numbers on the vertical axis on the left-hand side correspond to $\Phi(L)$ given by the Schecter luminosity function, indicated by the solid curve for two different slopes. The numbers on the vertical axis on the right-hand side correspond to $L\Phi(L)$, given by the dashed curves for two different slopes.

in terms of absolute luminosities. The parameter $\Phi^* = (1.6 \pm 0.3) \times 10^{-2}h^3$ Mpc^{-3} sets the overall normalisation; $\alpha = -(1.07 \pm 0.07)$ determines the slope of the luminosity function at the faint end; $L_B^* = (1.2 \pm 0.1) \times 10^{10}h^{-2}\,L_\odot$ [which corresponds to $M_B^* = (-19.7 \pm 0.1) + 5\log h$] is the characteristic luminosity above which the number of galaxies drops exponentially in the blue band. This function, called the *Schecter luminosity function*, is plotted in Fig. 1.25. (The K-band luminosity is similar except for the difference between M_B^* and M_K^*, which arises because of the B$-$K colours of the typical galaxies.) Because $\alpha \simeq -1$, the integral of $\Phi(L)$ over L can diverge (or be very large) at low L, suggesting a copious number of low-luminosity galaxies. The total luminosity density contributed by all galaxies, however, is finite:

$$l_{\text{tot}} = \int_0^\infty L\,\Phi(L)\,dL = \Phi_0 L^*\Gamma\,(2+\alpha). \qquad (1.116)$$

For $\alpha = -1$, $l_{\text{tot}} = \Phi_0 L^*$. There is a fair amount of scatter in the observed values of Φ^*, M^*, and α in the published literature, and the values quoted above should be treated with caution. Determination of these parameters have given Φ^* as varying in the range $(1.4-2.6) \times 10^{-2}h^3$ Mpc^{-3}, $M^*(B)$ varying in the range from -18.8 to -19.68, and α varying in the range $0.97-1.22$. The l_{tot}, however, does not vary all that much, and most observations are in the range $l_{\text{tot}} = (1.8-2.2) \times 10^8 h \, L_\odot \, \text{Mpc}^{-3}$.

There have been some attempts to study the luminosity function of galaxies at high redshifts in order to ascertain whether there is a systematic evolution. At $z \approx 3$, the luminosity function of galaxies can be fitted with $\alpha = -1.60 \pm 0.13$, $m_* = 24.48 \pm 0.15$, and $\Phi_* = 1.6 \times 10^{-2}$; at $z \approx 4$, the corresponding parameters are $\alpha \approx -1.60$, $m_* \approx 24.97$, and $\Phi_* = 1.3 \times 10^{-2}$ in the I band. It is difficult to make useful comments about galaxy evolution based on these results at present.[11]

Given the luminosity function for galaxies, we can estimate several useful quantities in a relatively straightforward manner. We shall now discuss some of them.

(1) Ignoring the effects of space–time curvature, we may relate the redshift z and the observed luminosity f of a galaxy to its distance r and intrinsic luminosity L by $z \simeq v/c \simeq (H_0 r/c)$ and $f = (L/4\pi r^2)$. Then the joint distribution function in z and f for galaxies is

$$\frac{dN}{d\Omega \, dz \, df} = \frac{1}{4\pi} \int_0^\infty 4\pi r^2 \, dr \, \Phi\left(\frac{L}{L_*}\right) \delta_D\left(z - \frac{H_0 r}{c}\right) \delta_D\left(f - \frac{L}{4\pi r^2}\right). \quad (1.117)$$

The factor $(1/4\pi)$ in front of the integral converts the number density to density per solid angle and the Dirac delta functions select the required redshift and flux. Using the known form of Φ and evaluating the integral, we get

$$\frac{dN}{d\Omega \, dz \, df} = \frac{4\pi}{L_*}\left(\frac{c}{H_0}\right)^5 z^4 \Phi(z^2/z_c^2), \qquad z_c^2 \equiv \frac{H_0^2 L_*}{4\pi f c^2}. \quad (1.118)$$

The redshift distribution at a fixed apparent luminosity f varies as $z^4 \Phi(z^2/z_c^2)$, which shows the effect of diminishing volume at low z (the z^4 factor).

(2) We can obtain the mean redshift of galaxies with a flux f by multiplying the Eq. (1.118) by z, integrating over all z, and normalising the expression properly. This gives

$$\langle z \rangle = z_c \frac{\int dy \, y^2 \Phi(y)}{\int dy \, y^{3/2} \Phi(y)} = z_c \frac{\Gamma(3 + \alpha)}{\Gamma(5/2 + \alpha)}. \quad (1.119)$$

(3) Integrating Eq. (1.118) over all z gives the number density of galaxies per unit flux interval. We get

$$\frac{dN}{d \ln f} = \frac{\Phi_*}{2}\left(\frac{L_*}{4\pi f}\right)^{3/2} \Gamma\left(\frac{5}{2} + \alpha\right). \quad (1.120)$$

(4) The mean luminosity per unit volume is given by

$$j = \int_0^\infty L \, \Phi\left(\frac{L}{L_*}\right) \frac{dL}{L_*} = \Gamma(2 + \alpha)\Phi_* L_* = (0.77 - 1.3) \times 10^8 h \, L_\odot \, \text{Mpc}^{-3}. \quad (1.121)$$

From this, we can estimate a typical number density of galaxies,

$$n_* \equiv \frac{j}{L_*} = \Gamma(2 + \alpha)\Phi_* = (6.7 - 14.9) \times 10^{-3} h^3 \text{ Mpc}^{-3}, \qquad (1.122)$$

and a mean separation between the galaxies,

$$d_* = n_*^{-1/3} = (4.1 - 5.3)h^{-1} \text{ Mpc}. \qquad (1.123)$$

(5) Multiplying the mass-to-light ratio $\mathcal{R} = (10 - 15)h(M_\odot/L_\odot)$ by the luminosity density j in Eq. (1.121) we get

$$\rho = j\mathcal{R} = (0.9 - 1.6) \times 10^9 h^2 \, M_\odot \text{ Mpc}^{-3} = (5.9 - 10.8) \times 10^{-32} h^2 \text{ g cm}^{-3}. \qquad (1.124)$$

(6) If the galaxies are assumed to be isothermal spheres of mass M and line-of-sight velocity dispersion σ, then their binding energies will be $U \approx 2M\sigma^2$. We can convert the velocity dispersion into luminosity by using the Faber–Jackson or the Tully–Fisher relation: $L/L^* = (\sigma/\sigma^*)^4$. Assuming a constant (M/L) ratio allows us to convert the mass to the luminosity and express U in terms of L. Integrating over the Schecter function then gives the binding energy per unit volume in galaxies as

$$U = 2\Phi^* L^* \sigma^{*2} \langle M/L \rangle \Gamma(5/2 + \alpha). \qquad (1.125)$$

σ^* is $\sim$240 km s^{-1} for ellipticals and 170 km s^{-1} for spirals. Taking an average value, we find that the binding energy of the visible parts of the galaxy is given by $1300h^2(\langle M/L \rangle/10)M_\odot c^2$ Mpc^{-3}. (Note that $M_\odot c^2$ Mpc$^{-3} = 6.1 \times 10^{20}$ ergs cm^{-3}.)

The luminosity function of galaxies in the cluster is similar and can be fitted to the Schecter form but with different values for the parameters. The Φ^* for cluster galaxies will certainly be higher because of the larger density of the galaxies in a cluster; studies show that even the other parameters are somewhat different: $M_B^* = (-19.5 \pm 0.1) + 5 \log h$ and $\alpha = (-1.27 \pm 0.04)$. The cutoff magnitude is not very different, but the slope at the faint end is considerably steeper. There has also been fair amount of investigation of the luminosity function of galaxies belonging to different morphological types like spirals, ellipticals, irregulars, etc. Studies show that there is a preponderance of ellipticals and S0s in the clusters compared with the field: as many as 40% of the galaxies in some clusters are ellipticals, whereas in others the proportion is nearly 15%. The elliptical fraction correlates with the morphology of the cluster, with those clusters that have a regular symmetric appearance having higher elliptical fractions.

1.12.2 Quasar Counts and Luminosity Function

Similar investigations have also resulted in the determination of the surface density of quasars in the sky as well as the quasar luminosity function.[12] Unlike galaxies, which radiate with a characteristic SED, quasars have a spectrum that extends over a wide range of frequencies. This has led to quasars being investigated in different bands from radio to x rays, and the counts as well as

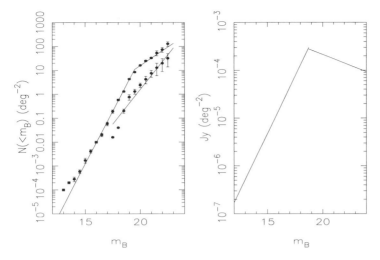

Fig. 1.26.   Left: The integral surface densities of quasars as functions of apparent B magnitude. Right: The same data are replotted, multiplying by a factor proportional to $10^{-0.4B}$ in order to obtain the intensity contributed by the quasars.

luminosity functions can, in principle, be defined in any of these bands. The most well-determined set of quasar counts, however, is in the B band in which the number density $N(<m_B)$ of quasars per square degree in the sky with magnitude less than m_B can be fitted by two broken power laws of the form

$$\log N(<m_B) = \begin{cases} 0.88m_B - 16.11 & \text{(for } m_B < 18.75) \\ 0.31m_B - 5 & \text{(for } m_B > 19.5) \end{cases}. \qquad (1.126)$$

This fit, as well as the data, is shown in the left panel of Fig. 1.26 along with another best-fit line,

$$\log N(<m_B) = 0.58m_B - 11.4, \qquad (1.127)$$

for quasars at $z > 2.2$. The right panel of Fig. 1.26 is a replot obtained by multiplying the count by $4.36 \times 10^{23} \times 10^{-0.4m_B - 20}$ in order to obtain the intensity in jansky per square degree. It is clear that in the B band the contribution of quasars is somewhat subdominant to those of galaxies by a factor of ~ 0.1. The actual amount contributed by quasars is ~ 4 nW m^{-2} sr^{-1}. However, as previously mentioned, quasars are broadband radiators and contribute a fair amount of background in high energies, for example in x-rays.

Let us next consider the luminosity function of quasars that is known at higher redshifts as well. The blue-band luminosity function can be expressed in the form

$$\phi(L, z)\, dL = \phi^* \left\{ \left[\frac{L}{L^*(z)} \right]^{-\alpha} + \left[\frac{L}{L^*(z)} \right]^{-\beta} \right\}^{-1} \frac{dL}{L^*(z)}. \qquad (1.128)$$

Using the relationships

$$M_B = M_B^*(z) - 2.5 \log\left(\frac{L}{L^*(z)}\right), \qquad \phi(M_B, z)\, dM_B = \phi(L, z)\left|\frac{dL}{dM_B}\right| dM_B,$$
(1.129)

we can express this in a more convenient form in terms of the blue magnitudes as follows:

$$\phi(M_B, z)\, dM_B = \frac{\phi^* \, dM_B}{10^{0.4[M_B - M_B^*(z)](\alpha+1)} + 10^{0.4[M_B - M_B^*(z)](\beta+1)}}, \quad (1.130)$$

where

$$M_B^*(z) = \begin{cases} A - 2.5k \, \log(1+z) & (\text{for } z < z_{\max}) \\ M_B^*(z_{\max}) & (\text{for } z > z_{\max}) \end{cases}, \quad (1.131)$$

with

$$\alpha = -3.9, \quad \beta = -1.5, \quad A = -20.9 + 5\log h, \quad k = 3.45,$$
$$z_{\max} = 1.9, \quad \phi^* = 5.2 \times 10^3 h^3 \, \text{Gpc}^{-3}\text{mag}^{-1}.$$
(1.132)

This function gives the number density of quasars with magnitudes between M_B and $M_B + dM_B$ at a redshift z [the inverse relation is based on $L = L_\odot 10^{0.4(M_\odot - M)}$ with $M_\odot = 5.48$ and $L_\odot = 2.36 \times 10^{33}$ ergs s^{-1}] and is plotted in Fig. 1.27 for different redshifts. There has been significant amount of effort to understand the luminosity function of quasars at $z > 3$; however, the situation is still very inconclusive.

The space density of quasars today is approximately $1.1 \times 10^{-3} h^3$ Mpc^{-3}, which is approximately a factor of $(0.02-0.1)$ lower than the space density of galaxies today. It is also obvious from the fitting function that quasars show strong

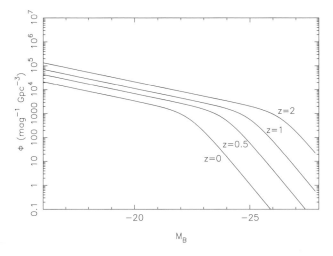

Fig. 1.27.   Quasar luminosity function. See text for discussion.

evolution over a redshift. Assuming that 1% of the bright galaxies ($M < -19$) contain quasars today and scaling the result to $z = 2.2$ by pure density evolution, we would estimate that there must be more than approximately 5 galaxies per Mpc^3 in order to accommodate all the quasars. This is much more than the number of bright galaxies per unit volume at that redshift, showing that pure density evolution will not work. We shall discuss these features in detail in Chap. 8.

1.13 Distribution of Matter

The nearest large galaxy to the Milky Way is the spiral M31, usually called the *Andromeda galaxy*, which is ~710 kpc away. Its mass is approximately $3 \times 10^{11} M_\odot$, and it has a size of ~50 kpc. Studies show that it is a spiral galaxy that we see from the Earth almost edge on. (It may be noted that galaxies are packed in the universe in a manner very different from the way the stars are distributed inside a galaxy: The distance from the Milky Way to the nearest large galaxy is only 20 galactic diameters, whereas the distance from the Sun to the nearest star is 30×10^6 times the diameter of these individual stars.)

There is some evidence to suggest that the Andromeda and the Milky Way galaxies are gravitationally bound to each other. They have a relative velocity – towards each other – of ~300 km s^{-1}. In fact, these are only two large members of a bunch of about 30 galaxies, all of which together constitute what is known as the *Local Group*. The entire Local Group can be contained within a spherical volume of approximately few megaparsecs in radius.

This kind of clustering of galaxies into groups is typical in the distribution of the galaxies in the universe. A careful study within a size of ~20 Mpc from our galaxy shows that only 10%–20% of the galaxies do *not* belong to any group; they are called *field* galaxies.[13]

Groups may typically consist of approximately 5–100 galaxies; a system with more than 100 galaxies is conventionally called a *cluster*. The sizes of groups range from a few hundred kiloparsecs to one or two megaparsecs. Clusters have a size of typically a few megaparsecs. We may approximate clusters and groups, just like galaxies, as gravitationally bound systems of effectively point particles. The large gravitational potential energy is counterbalanced by the large kinetic energy of random motion in the system. The line-of-sight velocity dispersion in groups is typically 200 km s^{-1} whereas that in clusters can be nearly 1000 km s^{-1}. There are several similarities between clusters of galaxies and stars in an elliptical galaxy. For example, the radial distribution of galaxies in a cluster can be adequately fitted by the $R^{1/4}$ law with an effective radius $R_e \simeq (1-2)h^{-1}$ Mpc. Approximately 10% of all galaxies are members of large clusters. In addition to galaxies, clusters also contain very hot intracluster gas at temperatures of 10^7–10^8 K.

The two large clusters nearest to the Milky Way are the Virgo cluster and the Coma cluster. The Virgo cluster, located ~15 Mpc away, has a diameter of

~3 Mpc and contains several thousand galaxies. This is a typical irregular cluster and does not exhibit a central condensation or a discernible shape. The Coma cluster, on the other hand, is almost spherically symmetric with a marked central condensation. It has an overall size of ~3 Mpc, and its central core is ~600 kpc in size. The core is populated with elliptical and spheroidal galaxies with a density nearly 30 times larger than that of our Local Group. These values are typical for large clusters. Coma is located at a distance of ~80 Mpc.

The distribution of galaxies around our Local Group has been studied extensively. It turns out that most of the galaxies nearby lie predominantly in a plane–called the *supergalactic plane* – that is approximately perpendicular to the plane of our own galaxy. The dense set of galaxies in this plane is called the Local Supercluster, and the Virgo cluster is nearly at the centre of this highly flattened disklike system. (The term *supercluster* is used to denote structures bigger than clusters.) Broadly speaking, the Local Supercluster consists of three components: ~20 percent of the brightest galaxies that form the core is the Virgo cluster itself; another 40% of galaxies lie in a flat disk with two extended, disjoint groups of galaxies; the remaining 40% are confined to a small number of groups scattered around. Nearly 80% of all matter in the Local Supercluster lies in a plane.[14] Studies of distant galaxies show that there are many superclusters in our universe separated by large voids.

To study such clustering of galaxies quantitatively, we need a good survey of the universe giving the coordinates of galaxies in the sky. Of the three coordinates needed to specify the position of the galaxy, the two angular coordinates are easy to obtain. There exist today several galaxy catalogues containing the angular positions of galaxies, in particular regions of the sky, complete up to a chosen depth.[15] (The Automated Plate Measuring (APM) galaxy survey has approximately 5×10^6 galaxies out to a depth of $600h^{-1}$ Mpc; the Lick catalogue has approximately 1.6×10^6 galaxies and depth of $200h^{-1}$ Mpc; the IRAS catalogue has more than 14,000 galaxies that are prominent in the IR band; these are a few of the major catalogues available today; see Vol. II, Chap. 1, Section 1.8, for a description of surveys in different wave bands.) If we know the redshift z of these galaxies as well, then we can attribute to it a line-of-sight velocity $v \cong zc$. If we further assume that this velocity is due to cosmic expansion, then we can assign to the galaxy a radial distance of $r \cong H_0^{-1}v$. This will provide us with the galaxy position (r, θ, ϕ) in the sky. The main difficulty in completing the survey lies in obtaining telescope time to make a systematic measurement of redshifts for the galaxies that are members of a catalogue. The Las Campanas Redshift Survey (LCRS), for example, has 26,418 galaxies in the R band reaching a depth of $\sim 30h^{-1}$ Mpc. Another major project, currently underway, is the Sloan Digital Sky Survey (SDSS), which will cover $\sim 10^{4\circ}$ of sky, $30°$ away from the galactic plane in five bands with limiting magnitudes of $U' = 22.3$, $G' = 23.3$, $R' = 23.1$, $I' = 22.3$, and $Z' = 20.8$. This survey is expected to detect approximately 5×10^7 galaxies and 10^8 starlike sources among which there will be $\sim 10^6$ AGN candidates that can be selected by colour techniques.

The amount of data we have today points to an interesting pattern in galaxy distribution.[16] The single most useful function characterizing the galaxy distribution is what is called the *two-point correlation function*: $\xi_{GG}(r)$. This function is defined by the relation

$$dP = \bar{n}^2[1 + \xi_{GG}(\mathbf{r}_1 - \mathbf{r}_2)]\, d^3\mathbf{r}_1\, d^3\mathbf{r}_2, \qquad (1.133)$$

where dP is the probability of finding two galaxies simultaneously in the regions $(\mathbf{r}_1, \mathbf{r}_1 + d^3\mathbf{r}_1)$ and $(\mathbf{r}_2, \mathbf{r}_2 + d^3\mathbf{r}_2)$ and $\bar{n}$ is the mean number density of galaxies in space. The homogeneity of the background universe guarantees that $\xi_{GG}(\mathbf{r}_1, \mathbf{r}_2) = \xi_{GG}(\mathbf{r}_1 - \mathbf{r}_2)$, and isotropy will further make $\xi_{GG}(\mathbf{r}) = \xi_{GG}(|\mathbf{r}|)$. From Eq. (1.133) it follows that $\xi_{GG}(r)$ measures the excess probability (over random) of finding a pair of galaxies separated by a distance $\mathbf{r}$; so if $\xi_{GG}(r) > 0$, we may interpret it as a clustering of galaxies over and above the random Poisson distribution. A considerable amount of effort was spent in the past decades in determining $\xi_{GG}(r)$ from observations. These studies show that

$$\xi_{GG}(r) \simeq \left(\frac{r}{r_0}\right)^{-\gamma}, \quad r_0 = (5 \pm 0.5)h^{-1}\,\mathrm{Mpc}, \quad \gamma = (1.77 \pm 0.04) \quad (1.134)$$

in the range $10h^{-1}\,\mathrm{kpc} \lesssim r \lesssim 10h^{-1}\,\mathrm{Mpc}$.

It is also possible, by use of the catalogues of rich clusters (like the Abell catalog, which contains 4,076 clusters), to compute the correlation function between galaxy clusters. The result turns out to be

$$\xi_{CC}(r) \simeq \left(\frac{r}{25h^{-1}\,\mathrm{Mpc}}\right)^{-1.8}. \qquad (1.135)$$

Comparing this result with the galaxy–galaxy correlation function ξ_{GG} given by Eq. (1.134) we see that clusters are more strongly correlated than the individual galaxies.[17] If firmly established, this result has important implications for the theories of structure formation. Unfortunately, $\xi_{CC}(r)$ is not as well established as ξ_{GG} and hence we must be cautious in interpreting results that depend on ξ_{CC}.

In recent years researchers have also resorted to less quantitative – but more appealing – diagnostics to demonstrate the clustering of galaxies. Many galaxy surveys – like the slices of the CfA survey – present striking *visual* patterns of galaxy distribution. The patterns are consistent with the interpretation that the universe contains several voids of a size of approximately $20h^{-1}$–$50h^{-1}$ Mpc. The CfA slices, in fact, suggest that galaxies are concentrated on sheetlike structures surrounding nearly empty voids.[18]

The distribution of matter seems to be reasonably uniform when observed at scales bigger than $\sim 100h^{-1}$ Mpc or so. For comparison, note that the size of the observed universe is $\sim 3000h^{-1}$ Mpc. Thus we may treat matter distribution in the universe as homogeneous while dealing with phenomena at scales larger than 100 Mpc or so.

1.14 Extragalactic Background Radiation

Many physical processes in the universe lead to the net emission of photons that in turn progressively degrade through their interaction with matter. As a result, we would expect certain amount of background radiation to exist in the universe at any given epoch. Such a radiation background can be separated into two well-defined categories. The first one is essentially a relic from the hotter, earlier phase of the universe, which is identified as CMBR. This will be discussed in detail in Chap. 6. The second category of extragalactic background arises from processes related to structure formation that eventually produces photons that survive as backgrounds in different wave bands. They contain valuable information about the high-redshift universe and structure formation and have been studied fairly extensively. We shall now briefly describe the nature of this (secondary) background radiation in different bands.

Figure 1.28 summarizes the current knowledge of EBL at different bands. The unit for νI_ν can be conveniently taken to be $\mathrm{nW\,m^{-2}\,sr^{-1}} = 10^{-6}\ \mathrm{ergs\ s^{-1}}$ $\mathrm{cm^{-2}\,sr^{-1}}$; multiplying this quantity by $4\pi/c$ will give the SED $U_\nu = (4\pi/c)(\nu I_\nu)$ with the dimensions of ergs per cubic centimetre. Dividing the energy density by $\rho_c c^2$ will give the density parameter contributed by the radiation background; roughly $1\ \mathrm{nW\,m^{-2}\,sr^{-1}}$ corresponds to $\Omega h_{50}^2 = 10^{-7}$. This quantity

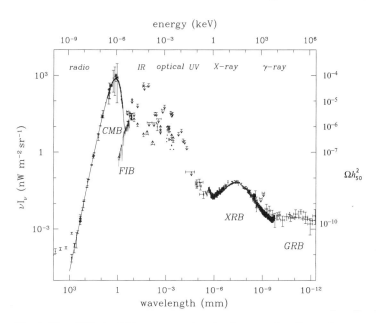

Fig. 1.28.   EBL in different wave bands. See text for discussion.

is plotted on the right-hand vertical axis and is also summarised in Fig. 1.28. It is clear from the figure that – if we ignore the CMBR that is of primordial origin – the maximum energy density is in the UV to IR band, which is what we will concentrate on.[19]

The majority of the EBL at UV to IR wavelengths originate from the starlight in the universe. In particular, the far-UV [(1000–1600)Å] light from the galaxies is produced by the same hot massive stars that also generate most of the metals. Thus the EBL in this band is a good tracer of the SFR and also traces the metallicity enhancement of the universe. Observations have converged at EBL fluxes in the range of 2–5 nW m^{-2} sr^{-1} in this band. (Of this, quasars and AGN probably contribute less than 0.2 nW m^{-2} sr^{-1}, with the remaining amount arising from the galaxies; we shall discuss the AGN in Chap. 8.) Comparing the cumulative flux of detected sources in the optical with the EBL flux suggests that nearly 50% of the flux arises from unresolved sources. These observations also suggest that the energy density does not vary too much over the spectral range $(0.1–10^3)\mu$.

The sources of EBL at optical (3000 Å –1 μ) and IR $(1–1000)\mu$ are more complex. The SED of a conventional stellar population peaks around $(1–1.5)\mu$ during most of the stellar ages; hence the optical EBL will include older populations at low redshifts and younger population at high redshifts. More importantly, even the normal stellar emission at the UV–optical will indirectly contribute to the IR along the following lines described in Section 1.5: During the period of active star formation, most of the light in a galaxy will be emitted in the optical and the UV bands and the spectrum (if not obscured by dust) will be almost flat ($I_\nu \approx$ constant) up to the Lyman continuum cutoff at 912 Å. This light is associated with the most luminous stars with the shortest lifetime. The SFR of 1 $M_\odot$ yr^{-1} leads to a luminosity of approximately 2.2 × 10^9 $L_\odot$. The presence of dust in galaxies leads to reprocessing of the optical–UV light into the FIR. This will lead to a correspondingly large amount of FIR radiation from the primeval galaxies.

The EBL is also related to the amount of metals produced in this phase. If the total bolometric intensity of starlight is taken to be ~50 nW m^{-2} sr^{-1}, then the corresponding energy density is $U = 2 \times 10^{-14}(1 + z)$ ergs cm^{-3} = 1 × 10$^{-8}(1 + z)$ MeV cm^{-3}, where the $(1 + z)$ factor corrects for the cosmological redshift, with z taken to be the mean epoch of emission. Each baryon that is converted to helium releases 25 MeV of energy, rising to a total of ~30 MeV per baryon for conversion from hydrogen to heavy elements. Dividing U by 30 MeV gives the mean baryon number density in metals involved in the production of background light as $n_Z \approx 4 \times 10^{-10}(1 + z)$ cm^{-3}. If we take the baryon number density to be $n_B = 1.1 \times 10^{-7}$ cm^{-3}, then the ratio of mass fraction in heavy elements that can produce the observed EBL is $Z \approx 4 \times 10^{-3}(1 + z)$. If the bulk of radiation is produced at $z = 2$ then $Z \approx 0.01$. (The entry corresponding to nuclear binding energy in Fig. 1.1 is computed in a similar manner, assuming $h_0 = 0.7$ and that intracluster plasma contains a metallicity of ~0.2 $Z_\odot$.)

Finally we note that the AGN also contributes a certain amount of background radiation. The amount of background light produced by any population of sources can be computed if the luminosity function of these sources is known. The steep spectrum radio sources with $S_{408} > 10$ mJy gives $\Omega \approx 7 \times 10^{-11} h^{-2} \langle (1+z) \rangle$; the quasars in the optical B band give $\Omega \approx 1.1 \times 10^{-8} h^{-2} \langle (1+z) \rangle$; the quasars in the x-ray band at 2 keV contribute $\Omega \approx 9 \times 10^{-11} h^{-2} \langle (1+z) \rangle$. For comparison, the total amount of starlight in the B band released over a time scale of H_0^{-1} is $\Omega \approx 0.5 \times 10^{-6} h^{-2}$. (All the preceding numbers are based on observations of νI_ν in certain bands and might require large bolometric corrections.) Also note that the total amount of energy released in supernovas over a time scale H_0^{-1} is $\sim 4000 h^2 (E/10^{51}$ erg) and the binding energy of visible part of the galaxies is $\sim 1300 h^2 (\langle M/L \rangle /10)$.

2

Galactic Structure and Dynamics

2.1 Introduction

In this chapter several aspects of galactic structure and evolution are discussed. Some of the ideas introduced in Chap. 1 are used and the contents of this chapter will be needed in Chaps. 7–10.

In the study of stellar structure and evolution in Vol. II, we could begin with a series of physically justifiable assumptions, derive the relevant equations describing the stars, and integrate these equations to understand the structure and evolution of stars. Such an approach is impossible in the case of galaxies for several reasons. To begin with, we do not understand how galaxies have formed. (It is true that there are several uncertainties in the case of star formation as well but they refer to details rather than to the fundamental process itself.) Second, observational data related to the galaxies are by no means statistically as well determined and abundant as data related to the stars. The reason essentially has to do with the fact that galaxies are located farther away from us and thus are more difficult to observe with the same level of accuracy. Third, galaxies, being collisionless systems of stars, are intrinsically more complicated compared with stars – which are made of collisional gas – from the point of view of mathematical description.[1]

Given these difficulties, it is better to divide the study of galactic structure and dynamics into several separate aspects and investigate each of them as though they are disconnected from each other. The latter assumption, of course, is only approximately true and we eventually will need to worry about how the individual aspects fit in the overall scheme of things. We shall address some of these broader questions in Chap. 7 and concentrate on more well-defined and conceptually simpler issues in this chapter.

A natural subdivision of the topics in the study of galaxies is based on different time scales and length scales that are involved in their description. The longest time scale that is operational with regard to galaxies is the time scale of formation, $t_{for} \approx \rho_*/\dot{\rho}_*$, where ρ_* is the density of stars in a galaxy and $\dot{\rho}_*$ is the rate of star

formation. Both these quantities could vary in space and time; but for typical
values, this time scale is 1–10 Gyr and is comparable with the cosmological time
scales in the redshift range $z \simeq (0\text{–}5)$, depending on the model chosen. At time
scales shorter than this, we may think of the galaxy as a well-defined physical
system that has already formed, and we can ignore the effects related to the
actual scenario for the formation of galaxy. The second time scale is the dynami-
cal one, $t_{\mathrm{dyn}} \approx (R/v)$, where R is the typical size of the galaxy and v is the typical
velocity of stars; this is $t_{\mathrm{dyn}} \simeq 0.1$ Gyr. This is, for example, the time scale that
would govern the process of violent relaxation (Vol. I, Chap. 10, Section 10.6)
if it was operational in the early stages of galaxy formation. Finally, the stars in
the galaxy evolve at a time scale that is typically 0.01–1 Gyr, depending on their
masses. Over this time scale, the gaseous and chemical contents of the galaxy
will change significantly. (The time scale for gravitational two-body relaxation,
$t_{\mathrm{relax}} \approx (N/\ln N)t_{\mathrm{dyn}}$, is considerably larger than the age of the universe and can
be ignored; see Vol. I, Chap. 10, Section 10.7.)

The preceding time scales are relevant for an isolated galaxy. If a galaxy is
in close proximity with one or more galaxies, then it is necessary to take into
account the effects of galaxy–galaxy encounters. As we shall see in Chap. 7, this
effect is significant because virtually every galaxy would have undergone a close
encounter with another galaxy some time during its past history.

The simplest (and mathematically most well-defined) question related to galaxies is the modelling of the steady-state distribution of stars in it. In addressing this
problem, we treat stars as structureless point particles interacting only through
gravity. Our description of galaxies begins with such a crude first approximation
in Sections 2.2–2.7. At the next level, we could incorporate some aspects of
stellar evolution and its effects on the chemical evolution of galaxies. This is
attempted in Section 2.8. Finally, in Section 2.9 we describe some simple effects
that could arise because of galaxy–galaxy interaction. Issues related to the formation and early evolution of galaxies as well as the effect of galaxies on the
IGM will be discussed in Chaps. 7 and 9.

2.2 Models for Galaxies in Steady State

In mathematical terms, the modelling of the steady-state distribution of stars in a
galaxy requires finding self-consistent solutions to the collisionless Boltzmann
equation (CBE) in which the smooth distribution of stars generates a gravitational
potential and the orbits admissible in this potential are populated by stars so as
to ensure self-consistency (see Vol. I, Chap. 10, Section 10.4).

The distribution of stars in a galaxy can be described by a function $f(\mathbf{x}, \mathbf{v}, t)$
that may be interpreted as the (relative) probability of finding a star in the phase
space in the interval $(\mathbf{x}, \mathbf{x} + d^3\mathbf{x}; \mathbf{v}, \mathbf{v} + d^3\mathbf{v})$ at time t. The smoothed-out mass

density of the stars at any point $\mathbf{x}$ will be

$$\rho(\mathbf{x}, t) = m \int f(\mathbf{x}, \mathbf{v}, t) \, d^3\mathbf{v}, \tag{2.1}$$

where m is the average mass of the stars. Such a smoothed-out density will produce a smooth gravitational potential $\phi(\mathbf{x}, t)$, where

$$\nabla^2 \phi = 4\pi G\rho. \tag{2.2}$$

We may assume that each star moves in this smooth gravitational field along some specific orbit. The conservation of the total number of stars, expressed by $(df/dt) = 0$, will reduce to

$$
\begin{aligned}
\frac{df}{dt} &= \frac{\partial f}{\partial \mathbf{x}} + \dot{\mathbf{v}} \cdot \frac{\partial f}{\partial \mathbf{v}} + \dot{\mathbf{x}} \cdot \frac{\partial f}{\partial \mathbf{x}} \\
&= \frac{\partial f}{\partial t} + \mathbf{v} \cdot \frac{\partial f}{\partial \mathbf{x}} - \nabla\phi \cdot \frac{\partial f}{\partial \mathbf{v}} = 0.
\end{aligned}
\tag{2.3}
$$

Models for galaxies are based on the solution to coupled equations (2.2) and (2.3).

It should be noted that the *actual* gravitational potential at any point $\mathbf{x}$, because of the stars, is

$$\phi_{\mathrm{act}}(\mathbf{x}, t) = -\sum_i \frac{Gm}{|\mathbf{x} - \mathbf{x}_i(t)|}, \tag{2.4}$$

where $\mathbf{x}_i(t)$ is the position of the ith star at time t. This will be different from the ϕ in Eq. (2.2), which was produced by the smooth density. Because of this difference, the actual trajectories of the stars will differ appreciably from the orbits of the smooth potential after sufficiently long time intervals. However, it can be easily shown (see Vol. I, Chap. 10, Section 10.7) that this time scale is very large compared with the orbital time scales in galaxies. Hence Eqs. (2.2) and (2.3) provide an excellent description of galaxies over reasonable time scales.

Because these equations allow a wide variety of solutions, different classes of models are possible for the galaxies. Unfortunately, there is no way of choosing any subset of them as appropriate to a galaxy without further input. Theoretically, we would have expected the initial conditions (at the time of formation of galaxies) and the process of violent relaxation (see Vol. I, Chap. 10, Section 10.6) to have led to a subset of all possible solutions to the CBE. However, it has not been possible to provide a general analysis of the time-dependent CBE, thereby determining the nature of final configurations to which the system evolves at late times. The best we can do at present is to provide a series of simplified time-independent models that serve as a backdrop for comparing theory and observations. We shall discuss a few such models in this section.

When the galaxy is in steady state, $f(t, \mathbf{x}, \mathbf{v}) = f(\mathbf{x}, \mathbf{v})$. To produce such a steady-state solution we can proceed as follows: Let $C_i = C_i(\mathbf{x}, \mathbf{v})$, $i = 1$,

2, ..., be a set of isolating integrals of motion for the stars moving in the potential ϕ (which, right now, is not known). It is obvious that any function $f(C_i)$ of the C_i will satisfy the steady-state Boltzmann equation: $(df/dt) = (\partial f/\partial C_i)$ $\dot{C}_i = 0$ because $\dot{C}_i$ is identically zero. If we can now determine ϕ from f self-consistently and populate the orbits of ϕ with stars, we have solved the problem. Let us consider some specific examples to see how this idea works.

In these calculations, it is convenient to shift the origin of $|\phi|$ by defining a new potential $\psi \equiv -\phi + \phi_0$, where ϕ_0 is a constant. (We will choose the value of ϕ_0 such that ψ vanishes at the "boundary" of the galaxy.) The new potential satisfies

$$\nabla^2 \psi = -4\pi G\rho, \qquad (2.5)$$

and the boundary condition $\psi \to \phi_0$ as $|\mathbf{x}| \to \infty$. We also define a "shifted" energy for the stars, $\epsilon = -E + \phi_0$; because $\phi_0 = \psi + \phi$, $\epsilon = -E + \psi + \phi = -(1/2)v^2 + \psi$.

The simplest galactic models are the ones in which $f(\mathbf{x}, \mathbf{v})$ depends on $\mathbf{x}$ and $\mathbf{v}$ only through the quantity ϵ so that $f = f(\epsilon) = f(\psi - \frac{1}{2}v^2)$. The density $\rho(\mathbf{x})$ corresponding to this distribution is

$$\rho(\mathbf{x}) = \int_0^{\sqrt{2\psi}} 4\pi v^2 \, dv f \left(\psi - \frac{1}{2}v^2 \right) = \int_0^{\psi} 4\pi \, d\epsilon f(\epsilon) \sqrt{2(\psi - \epsilon)}. \quad (2.6)$$

The limits of integration are chosen in such a way to pick only the stars bound in the galaxy's potential. The right-hand side is a known function of ψ, once $f(\epsilon)$ is specified. The Poisson equation,

$$\frac{1}{r^2}\frac{d}{dr}\left(r^2 \frac{d\psi}{dr} \right) = -4\pi G\rho = -16\pi^2 G \int_0^{\psi} d\epsilon f(\epsilon)\sqrt{2(\psi - \epsilon)}, \quad (2.7)$$

can now be solved – with some central value $\psi(0)$ and the boundary condition $\psi'(0) = 0$ – determining $\psi(r)$. Once $\psi(r)$ is known, all other variables can be computed.

Three different choices for $f(\epsilon)$ have been extensively used in the literature to describe spherically symmetric systems. We will summarise these models briefly.

2.2.1 Polytropes

The simplest form of $f(\epsilon)$ is a power law with $f(\epsilon) = A\epsilon^{n-3/2}$ for $\epsilon > 0$ and zero otherwise. Using Eq. (2.6), we see that this corresponds to density distributions of the form $\rho = B\psi^n$ (for $\psi > 0$) with $B = (2\pi)^{3/2} A\Gamma(n - 1/2)[\Gamma(n + 1)]^{-1}$. (Clearly, we need $n > 1/2$ to obtain finite density.) The Poisson equation now becomes

$$\frac{1}{l^2}\frac{d}{dl}\left(l^2 \frac{d\xi}{dl} \right) = \begin{cases} -\xi^n & \xi \geq 0 \\ 0 & \xi \leq 0 \end{cases}, \qquad (2.8)$$

where we have introduced the variables

$$L = [4\pi G \psi(0)^{n-1} B]^{-\frac{1}{2}}, \qquad l = (r/L), \qquad \xi = [\psi/\psi(0)]. \qquad (2.9)$$

This equation is the Lane–Emden equation that was discussed extensively in Vol. I, Chap. 10, Section 10.3. The case that is of interest in galactic modelling arises for $n = 5$. In this case, Eq. (2.8) has the simple solution

$$\xi = \left(1 + \frac{1}{3}l^2\right)^{-\frac{1}{2}}, \qquad (2.10)$$

which corresponds to a density profile of $\rho \propto [1 + (1/3)l^2]^{-5/2}$ and a total mass of $M = [\sqrt{3}L\psi(0)/G]$. This model (called the *Plummer model*) provides a reasonable description of some elliptical galaxies.

2.2.2 Isothermal Spheres

This model corresponds to the distribution function

$$f(\epsilon) = \frac{\rho_0}{(2\pi\sigma^2)^{3/2}} \exp\left(\frac{\epsilon}{\sigma^2}\right), \qquad (2.11)$$

parameterised by two constants, ρ_0 and σ. We can easily verify that the mean-square velocity $\langle v^2 \rangle$ is $3\sigma^2$ and that the density distribution is $\rho(r) = \rho_0 \exp(\psi/\sigma^2)$. The central density is $\rho_c = \rho_0 \exp[\psi(0)/\sigma^2]$. It is conventional to define a core radius and a set of dimensionless variables by

$$r_0 = \left(\frac{9\sigma^2}{4\pi G\rho_c}\right)^{1/2}, \qquad l = \frac{r}{r_0}, \qquad \xi = \frac{\rho}{\rho_c}. \qquad (2.12)$$

Then the Poisson equation can be rewritten in the form

$$\frac{1}{l^2}\frac{d}{dl}\left(\frac{l^2}{\xi}\frac{d\xi}{dl}\right) = -9\xi. \qquad (2.13)$$

This equation has to be integrated numerically [with the boundary condition $\xi(0) = 1$ and $\xi'(0) = 0$] to determine the density profile. It was shown (in Vol. I, Chap. 10, Section 10.3) that the solution has the asymptotic limit of $\xi \simeq (2/9l^2)$. Hence the exact solution has a mass profile with $M(r) \propto r$ at larger r; the model has to be cut off at some radius to provide a finite mass.

For $l \lesssim 2$, the numerical solution is well approximated by the function $\xi(l) = (1 + l^2)^{-\frac{3}{2}}$ (see Fig. 2.1). The projected, two-dimensional, surface density corresponding to this $\xi(l)$ is $S(R) = 2(1 + R^2)^{-1}$, where R is the projected radial distance. In other words, the (true) central surface density $\Sigma(0) \simeq 2\rho_c r_0$, where ρ_c is the central volume density and r_0 is the core radius. Because the mass density of the isothermal sphere varies as r^{-2} at large distances, the potential ϕ

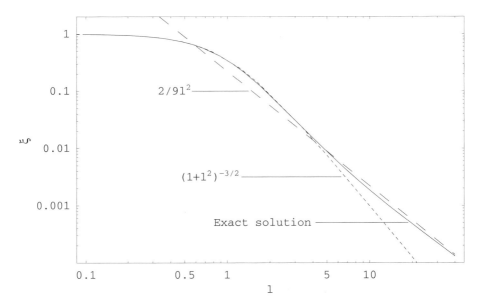

Fig. 2.1. The exact isothermal density profile $\xi(l)$ (solid curve) as well as the two asymptotic forms valid at small and large radii. For large radii, $\xi(l) \approx (2/9)l^{-2}$, which is a long-dashed straight line in the logarithmic plot, and for small radii $\xi(l) \approx (1+l^2)^{-3/2}$, which is shown by a short-dashed curve.

varies as $\ln r$. A disk of test stars embedded in such an isothermal sphere will have constant rotational velocities.

2.2.3 King Model

These are based on the distribution function with

$$f(\epsilon) = \frac{\rho_c}{(2\pi\sigma^2)^{3/2}}\left(e^{\epsilon/\sigma^2} - 1\right), \quad \epsilon \geq 0, \tag{2.14}$$

and $f(\epsilon) = 0$ for $\epsilon < 0$. Because $f(\epsilon)$ vanishes for $\epsilon < 0$, this model may be thought of as a truncated isothermal sphere. The density, when integrated from the origin numerically, will vanish at some radius r_t (called the tidal radius). The quantity $c = \log(r_t/r_0)$ is a measure of how concentrated the system is. Bright elliptical galaxies are well described by such a model with $[\psi(0)/\sigma^2] \simeq 10$ and $c \simeq 2.4$.

The preceding examples show how galactic models of reasonable description can be constructed as solutions to the CBE. Unfortunately, this particular approach has no dynamical content and is merely a tool to fit observations. In fact, the situation is worse than that; for a wide class of acceptable $\rho(r)$, it is always possible to obtain an $f(\epsilon)$ that will self-consistently reproduce the given density profile. To see this, we begin by noting that a given $\rho(r)$ uniquely determines a

$\psi(r)$ and hence the function $\rho(\psi)$. Writing Eq. (2.6) as

$$\frac{1}{\sqrt{8}\pi}\rho(\psi) = 2\int_0^\psi f(\epsilon)\sqrt{\psi - \epsilon}\, d\epsilon \qquad (2.15)$$

and differentiating both sides with respect to ψ, we get

$$\frac{1}{\sqrt{8}\pi}\frac{d\rho}{d\psi} = \int_0^\psi \frac{f(\epsilon)\, d\epsilon}{\sqrt{\psi - \epsilon}}. \qquad (2.16)$$

This equation (called Abel's integral equation) has the solution

$$f(\epsilon) = \frac{1}{\sqrt{8}\pi^2}\frac{d}{d\epsilon}\int_0^\epsilon \left(\frac{d\rho}{d\psi}\right)\frac{d\psi}{\sqrt{\epsilon - \psi}}, \qquad (2.17)$$

which determines $f(\epsilon)$. [Although this procedure gives an $f(\epsilon)$, there is no assurance that it will be positive definite; obviously, the method works only for those $\rho(r)$ for which $f(\epsilon) > 0$ for all ϵ.]

The above procedure leads to a galaxy model that will have an isotropic velocity dispersion. This follows from the fact that, in the above approach, $\rho(r)$ was used to determine an $f(\epsilon)$ that depends on only v^2. Because observations suggest that galaxies have an anisotropic velocity dispersion, it would be interesting to ask whether our procedure can be modified to cover them. If we allow the distribution function to depend on the angular momentum J as well as on ϵ, then a wider class of models can be constructed to fit the same $\rho(r)$. To see this, assume that f depends on ϵ and J^2 only through the combination

$$Q = \epsilon - \frac{J^2}{2R_0^2}, \quad R_0 = \text{constant}. \qquad (2.18)$$

We can easily integrate out the angular variables in the velocity space and obtain the density distribution

$$\rho(r) = \frac{2\pi\sqrt{8}}{(1 + r^2/R_0^2)}\int_0^\psi f(Q)\sqrt{\psi - Q}\, dQ \equiv \frac{2\pi\sqrt{8}}{(1 + r^2/R_0^2)}\mu(\psi). \qquad (2.19)$$

Comparing Eqs. (2.19) and (2.6), we see that

$$f(Q) = \frac{1}{\pi^2\sqrt{8}}\frac{d}{dQ}\int_0^Q \left(\frac{d\mu}{d\psi}\right)\frac{d\psi}{\sqrt{Q - \psi}}. \qquad (2.20)$$

Thus, given a $\rho(r)$, we determine $\psi(r)$ and another function $\mu(r) \equiv \rho(r)\,(1 + r^2/R_0^2)\,(2\pi\sqrt{8})^{-1}$. Eliminating r between $\mu(r)$ and $\psi(r)$, we obtain $\mu(\psi)$ and consequently $f(Q)$. This distribution function will lead to the density distribution $\rho(r)$ we started with. When $f = f(\epsilon)$, the velocity dispersion is isotropic with $\langle v_r^2 \rangle = \langle v_\theta^2 \rangle = \langle v_\phi^2 \rangle$. For a distribution function of the form $f(\epsilon, J^2)$, we have $v_\theta^2 = v_\phi^2 \neq v_r^2$, thereby providing some amount of anisotropy in the velocity dispersion.

2.2.4 Axisymmetric Systems

Similar modelling also works for a disklike system if the distribution function is taken to be of the form $f = f(\epsilon, J_z)$. We will mention a few examples: If the distribution function is taken to be

$$
f(\epsilon, J_z) = \begin{cases} A J_z^n \exp\left(\epsilon/\sigma^2\right), & (\text{for } J_z \geq 0) \\ 0, & (\text{for } J_z \leq 0) \end{cases}, \tag{2.21}
$$

then we obtain a surface density for the disk, which is

$$
\Sigma(R) = \frac{(n+1)\sigma^2}{2\pi G R} \equiv \frac{\Sigma_0 R_0}{R}. \tag{2.22}
$$

The circular velocity of stars, $v_c^2 = -R(\partial\psi/\partial R) = 2\pi G \Sigma_0 R_0$, is a constant for this model, called *Mestel's disk*. These parameters are related to the parameters in the distribution function by

$$
n = \left(v_c^2/\sigma^2\right) - 1, \qquad A = \Sigma_0 R_0 \left[2^{n/2}\sqrt{\pi}\,\Gamma(n+1/2)\sigma^{n+2}\right]^{-1}. \tag{2.23}
$$

The velocity dispersion in the radial direction σ^2 is a free parameter characterising the disk. The parameter n is a measure of the "coldness" of the disk.

A more complicated set of disk models (called the *Kalnajs disk*) can be obtained from the distribution function, which has the form

$$
f(\epsilon, J_z) = A \left[\left(\Omega_0^2 - \Omega^2\right)a^2 + 2(\epsilon + \Omega J_z)\right]^{-1/2} \tag{2.24}
$$

when the term in the square brackets is positive and zero otherwise. This function leads to the following surface density and potential:

$$
\Sigma(R) = \Sigma_0 \left(1 - \frac{R^2}{a^2}\right)^{1/2}, \qquad \phi(R) = \frac{\pi^2 G \Sigma_0}{4a} R^2 \equiv \frac{1}{2}\Omega_0^2 R^2, \tag{2.25}
$$

with Σ_0 and A related by $\Sigma_0 = 2\pi A a \sqrt{\Omega_0^2 - \Omega^2}$. The parameter Ω is free and describes the mean (systematic) rotational velocity of the system: $\langle v_\phi \rangle = \Omega R$ (see Exercise 2.3).

Exercise 2.1
Logarithmic potential: Another galactic model, which we shall have occasion to discuss extensively, has the gravitational potential of the form

$$
\phi = \Phi_L(R, z) \equiv \frac{1}{2}v_0^2 \ln\left(R_c^2 + R^2 + \frac{z^2}{q_\Phi^2}\right), \tag{2.26}
$$

where R_c, v_0, and q_Φ are constants. Show that the mass density $(4\pi G)^{-1}\nabla^2\phi$ corresponding to the potential in Eq. (2.26) is given by

$$
\rho_L(R, z) = \left(\frac{v_0^2}{4\pi G q_\Phi^2}\right)\frac{\left(2q_\Phi^2 + 1\right)R_c^2 + R^2 + 2\left(1 - \frac{1}{2}q_\Phi^{-2}\right)z^2}{\left(R_c^2 + R^2 + z^2 q_\Phi^{-2}\right)^2}. \tag{2.27}
$$

As we shall see in Section 2.4, this logarithmic potential helps us to understand several features of orbital structure in real galaxies. Find a distribution function that will lead to this potential self-consistently.

Exercise 2.2

Disks with negligible angular momentum: We intuitively associate highly flattened systems with systems that have a large angular momentum. Consider a disk galaxy in which all stars rotate in the same direction and are supported against gravity by centrifugal force. Is it possible to construct now another *disk* galaxy with the same spatial distribution of stars but with negligible angular momentum?

Exercise 2.3

Properties of disks: Verify the various results quoted in the text as regards the polytropes, isothermal spheres, King's model, Mestel's disk, and the Kalnajs disk. (a) Show that galaxy models with distribution function of the form $f(E)$ have isotropic velocity dispersion, that is, $\langle v_r^2 \rangle = \langle v_\theta^2 \rangle = \langle v_\phi^2 \rangle$. Also show that, if $f = f(E, J^2)$, then $\langle v_\theta^2 \rangle = \langle v_\phi^2 \rangle \neq \langle v_r^2 \rangle$. (b) Show that, for the isothermal sphere, $\langle v^2 \rangle = 3\sigma^2$. (c) Show that, for the Mestel's disk, $\langle v_r^2 \rangle = \sigma^2$ and $\langle v_\phi \rangle = \sqrt{2}\Gamma(n/2 + 1)\,[\Gamma(n/2 + 1/2)]^{-1}\sigma$. Hence show that, as n increases to large values, $[\langle v_\phi \rangle/\sigma]$ increases as n. (d) For the Kalnajs disk, show that $\langle v_\phi \rangle/r = \Omega$ is independent of r; thus the disk rotates like a rigid body. Also show that, for $\Omega \ll \Omega_0$, the system is "hot," with the support against gravitational force arising from random velocities, whereas, for $\Omega \approx \Omega_0$, the system is "cold" and centrifugally supported. (e) Consider a distribution function $f = f(E, J_z)$ that leads to a density $\rho(r, z)$. Let $f_\pm(E, J_z) = \frac{1}{2}[f(E, J_z) \pm f(E, -J_z)]$. Show that ρ is determined by f_+ and $\langle v_\phi \rangle$ is determined by f_-.

Exercise 2.4

Vertical density profile of a disk: Find the density profile of a thin self-gravitating disk of stars whose velocity distribution is everywhere Maxwellian and the distribution varies along only the z direction. [Answer. We take the one-dimensional phase-space density to be

$$f = f(E_z) = \rho_0 \left(2\pi\sigma_z^2\right)^{-1/2} \exp\left(-E_z/\sigma_z^2\right), \qquad E_z = \frac{v_z^2}{2} + \Phi(z). \quad (2.28)$$

In terms of the dimensionless variables,

$$\phi \equiv \frac{\Phi}{\sigma_z^2}, \qquad \zeta \equiv \frac{z}{z_0}, \qquad z_0 \equiv \frac{\sigma_z}{\sqrt{8\pi G\rho_0}}, \quad (2.29)$$

the Poisson equation can be reduced to the form

$$2\frac{d^2\phi}{d^2\zeta^2} = e^{-\phi}. \quad (2.30)$$

Solving this with the boundary condition $\phi(0) = 0 = (d\phi/d\zeta)_0$, we get the distribution

$$\rho(z) = \rho_0 \, \text{sech}^2\left(\frac{1}{2}\frac{z}{z_0}\right), \quad (2.31)$$

which describes the vertical density profile.]

2.3 Aspects of Stellar Orbits

It was pointed out in the beginning of Section 2.2 that it is not possible to derive the structure and gravitational potential of galaxies by starting from first principles and studying their formation. To choose among various solutions of the CBE, like the ones described in the last section, it is necessary to compare their properties with observations. These observations are invariably related to the motion of the stars in the galactic potential, and hence it is worthwhile to discuss the orbits of stars in some general class of potentials that will be appropriate for the description of the galaxies.

When the stars are treated as point particles and the motion (as well as the gravitational field) is treated as nonrelativistic, the equation of motion for any given star is simply $\ddot{\mathbf{x}} = -\nabla\Phi(\mathbf{x})$, which can be integrated once the form of Φ is known. The analytical solutions for the orbits, of course, are available for only a small subset of all potentials; but because they offer valuable insight into the nature of motion, we begin with these cases.

2.3.1 Spherically Symmetric Potentials

In the case of spherically symmetric potentials, there are two constants of motion; namely the total angular momentum per unit mass $\mathbf{L} = \mathbf{r} \times \mathbf{v}$ and the energy per unit mass E. The orbit is confined to a plane perpendicular to the direction of $\mathbf{L}$ and the trajectory, in terms of the polar coordinates in the plane, $[r(t), \psi(t)]$, is given by

$$r^2\dot{\psi} = L, \qquad \dot{r} = \frac{dr}{dt} = \pm\sqrt{2[E - \Phi(r)] - \frac{L^2}{r^2}} \qquad (2.32)$$

(see Vol. I, Chap. 2, Section 2.3). These equations reduce the problem to quadrature. In particular, the radial period is given by

$$T_r = 2\int_{r_1}^{r_2} \frac{dr}{\sqrt{2[E - \Phi(r)] - (L^2/r^2)}}, \qquad (2.33)$$

where r_1 and r_2 are the turning points at which $\dot{r} = 0$. During this period, the azimuthal angle changes by the amount

$$\Delta\psi = 2\int_{r_1}^{r_2} \frac{d\psi}{dr}\,dr = 2\int_{r_1}^{r_2} \frac{L}{r^2}\frac{dt}{dr}\,dr = 2L\int_{r_1}^{r_2} \frac{dr}{r^2\sqrt{2[E - \Phi(r)] - (L^2/r^2)}}. \qquad (2.34)$$

The corresponding azimuthal period is

$$T_\psi = \frac{2\pi}{\Delta\psi}\,T_r. \qquad (2.35)$$

Because $\Delta\psi/2\pi$ will not – in general – be a rational number, the orbit will not be closed and will fill the annular region between circles of radii r_1 and r_2.

Among all spherically symmetric potentials, there are two that form natural boundaries for more realistic potentials. The first is the Kepler potential $\Phi \propto r^{-1}$ that arises if we assume that the entire mass of the galaxy is concentrated at the centre and the star is moving under its influence. The second is the harmonic potential with $\Phi \propto r^2$ that arises if all the mass in the galaxy is distributed uniformly within a sphere of radius r. Because any realistic mass distribution will be centrally condensed, it is likely to be between these two limiting forms. To study the two limiting cases, it is convenient to have a form of $\Phi(r)$ that interpolates between the two and for which analytic solutions are available. One such special potential is

$$\Phi = -\frac{GM}{b + \sqrt{(r^2 + b^2)}}, \tag{2.36}$$

called the *isochrone potential*. When $b \to 0$, this leads to the Kepler potential whereas for $b \to \infty$ the leading order behavior is ($\Phi +$ constant) $\propto r^2$. This potential is generated by a density distribution of the form

$$\rho(r) = \frac{1}{4\pi G}\frac{1}{r^2}\frac{d}{dr}\left(r^2\frac{d\Phi}{dr}\right) = M\left[\frac{3(b+a)a^2 - r^2(b+3a)}{4\pi(b+a)^3 a^3}\right], \tag{2.37}$$

where $a = \sqrt{b^2 + r^2}$, which has the limiting forms

$$\rho(0) = \frac{3M}{16\pi Gb^3}, \qquad \rho(r) \simeq \frac{bM}{2\pi r^4} \quad (r \gg b). \tag{2.38}$$

To integrate the equations of motion in this potential, it is convenient to define a variable s by

$$s \equiv -\frac{GM}{b\Phi} = 1 + \sqrt{\frac{r^2}{b^2} + 1}, \qquad \frac{r^2}{b^2} = s^2\left(1 - \frac{2}{s}\right), \qquad s \geq 2. \tag{2.39}$$

We can now express the integrals that determine the trajectory of the particle in terms of s. Simple analysis gives the radial period as

$$T_r = \frac{2b}{\sqrt{-2E}}\int_{s_1}^{s_2}\frac{(s-1)\,ds}{\sqrt{(s_2-s)(s-s_1)}} = \frac{2\pi b}{\sqrt{-2E}}\left[\frac{1}{2}(s_1 + s_2) - 1\right], \tag{2.40}$$

where s_1 and s_2 corresponds to the turning points r_1 and r_2 and we have assumed that $E < 0$. These roots are determined by

$$2Es^2 - 2\left(2E - \frac{GM}{b}\right)s - \frac{4GM}{b} - \frac{L^2}{b^2} = 0, \tag{2.41}$$

from which it follows that the sum of the roots is given by $s_1 + s_2 = 2[1 - (GM/2Eb)]$; hence the radial period is

$$T_r = \frac{2\pi GM}{(-2E)^{3/2}}.$$ (2.42)

This result is independent of b and is, of course, valid even in the Keplerian limit. (The fact that the radial period is independent of the angular momentum lends the name *isochrone* to this potential.) The change in the azimuthal angle per cycle is given by

$$\Delta\psi = 2L \int_{s_1}^{s_2} \frac{dt}{r^2} = \frac{2L}{b\sqrt{-2E}} \int_{s_1}^{s_2} \frac{(s-1)\,ds}{s(s-2)\sqrt{(s_2-s)(s-s_1)}}$$

$$= \pi\left(1 + \frac{L}{\sqrt{L^2 + 4GMb}}\right).$$ (2.43)

In the limit of $b \to 0$, $\Delta\psi \to 2\pi$, which is the Keplerian result. On the other hand, as $b \to \infty$, $\Delta\psi \to \pi$, which is characteristic of orbits in the harmonic potential. In general, we have $\pi < \Delta\psi < 2\pi$. Because galaxies are more extended than point masses, a typical star in a galaxy completes one radial oscillation in a shorter time than is required for one complete azimuthal cycle around the galactic centre.

2.3.2 Rotation Curves of Disk Galaxies

We next consider issues related to orbits in axisymmetric systems like disk galaxies. There are several interesting features that emerge in this study, and we begin with the question of determining the circular velocity of a particle orbiting in the plane of a disk galaxy. We saw in Chap. 1, Section 1.5, that the surface-density profile of disks is fairly well described by the exponential form $\Sigma(R) = \Sigma_0 \exp(-R/R_d)$. We shall now determine the gravitational potential that is due to such a disk and the properties of stellar orbits.

The easiest way to determine the potential will be to begin by noting that, in axially symmetric cases, the functions

$$\Phi_{\pm}(R, z) = \exp(\pm kz)\, J_0(kR)$$ (2.44)

(where J_0 is the Bessel function) are solutions to the Laplace equation. Consider now the function defined by

$$\Phi_k(R, z) = \exp(-k|z|)\, J_0(kR),$$ (2.45)

which has the discontinuity in the derivative at $z = 0$ given by

$$\lim_{z \to 0+}\left(\frac{\partial \Phi_k}{\partial z}\right) = -k J_0(kR), \qquad \lim_{z \to 0-}\left(\frac{\partial \Phi_k}{\partial z}\right) = +k J_0(kR).$$ (2.46)

It follows from Gauss theorem that Φ_k is the gravitational potential that is due to the surface-density distribution,

$$\Sigma_k(R) = -\frac{k}{2\pi G} J_0(kR), \qquad (2.47)$$

located at $z = 0$. Superposing the density distribution by using a weightage function $S(k)$, we find that the surface-density profile,

$$\Sigma(R) = \int_0^\infty S(k)\Sigma_k(R)\,dk = -\frac{1}{2\pi G}\int_0^\infty S(k)J_0(kR)\,k\,dk, \qquad (2.48)$$

will generate the gravitational potential

$$\Phi(R, z) = \int_0^\infty S(k)\Phi_k(R, z)\,dk = \int_0^\infty S(k)J_0(kR)e^{-k|z|}\,dk. \qquad (2.49)$$

This analysis is sufficiently general and could have been done with a wide class of radial functions. The advantage of choosing $J_0(kR)$ lies in the fact that Eq. (2.48) can be inverted to give

$$S(k) = -2\pi G \int_0^\infty J_0(kR)\Sigma(R)R\,dR. \qquad (2.50)$$

Using this in Eq. (2.49), we can express the potential directly in terms of the density $\Sigma(R)$, thereby providing the complete solution to the problem. Given the form of the potential, we can also determine the rotation velocity in terms of $S(k)$:

$$v_c^2(R) = R\left(\frac{\partial \Phi}{\partial R}\right)_{z=0} = -R\int_0^\infty S(k)J_1(kR)k\,dk. \qquad (2.51)$$

For the case of exponential disks, we can evaluate the integral in Eq. (2.50) to obtain $S(k)$ as

$$S(k) = -\frac{2\pi G \Sigma_0 R_d^2}{[1 + (kR_d)^2]^{3/2}}. \qquad (2.52)$$

The potential is now given by Eq. (2.49):

$$\Phi(R, z) = -2\pi G\Sigma_0 R_d^2 \int_0^\infty \frac{J_0(kR)e^{-k|z|}}{[1 + (kR_d)^2]^{3/2}}\,dk, \qquad (2.53)$$

which unfortunately cannot be expressed in simple closed form for arbitrary z. In the $z = 0$ plane, however, the result can be expressed in terms of modified Bessel functions by

$$\Phi(R, 0) = -\pi G\Sigma_0 R\left[I_0(y)K_1(y) - I_1(y)K_0(y)\right], \qquad y \equiv \frac{R}{2R_d}, \qquad (2.54)$$

with the corresponding circular speed

$$v_c^2(R) = R\left(\frac{\partial \Phi}{\partial R}\right) = 4\pi G\Sigma_0 R_d y^2 [I_0(y)K_0(y) - I_1(y)K_1(y)]. \quad (2.55)$$

Figure 2.2 shows $v_c^2(R)/4\pi G\Sigma_0 R_d$ as a function of (R/R_d). A reasonable approximation to v_c for $R \lesssim 4R_d$ is given by

$$v_c(R) \simeq 0.876 \left(\frac{GM}{R_d}\right)^{1/2} \left(\frac{\bar{y}^{1.3}}{1 + \bar{y}^{2.3}}\right)^{1/2}, \quad (2.56)$$

where $\bar{y} = 0.533(R/R_d)$ and M is the total mass of the disk. This reaches a maximum around $(R/R_d)_{\text{max}} \simeq 2.3$ and falls as $v_c \propto R^{-1/2}$ for $(R/R_d) \gtrsim 7$.

Exercise 2.5
Embedding a disk in a halo: Consider a system with the density distribution $\rho(r, \theta) = \rho_0 S(\theta)\,(r_0/r)^2$ that is axisymmetric and falls as r^{-2}. Let $S(\theta)$ be normalised such that $\int_0^\pi S(\theta)\sin\theta\,d\theta = 2$. In general, we would have expected the radial and angular components of the force,

$$F_r = -\frac{\partial \Phi}{\partial r}, \qquad F_\theta = -\frac{1}{r}\frac{\partial \Phi}{\partial \theta}, \quad (2.57)$$

to depend on θ as well as on r. (a) Show that $(\partial F_r/\partial \theta) = 0$, i.e., the radial force is

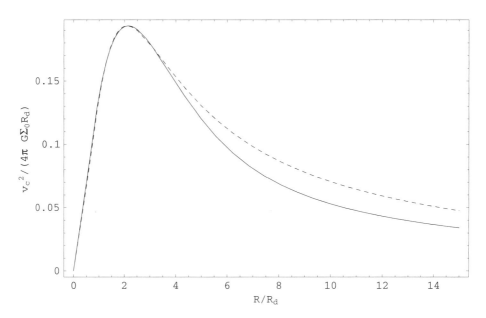

Fig. 2.2. Circular velocity of rotation as a function of radius in an exponential disk based on Eq. (2.55). The rotational velocity starts decreasing within a few factors of R_d. The dashed curve is the analytic fit given by relation (2.56).

independent of the latitude. Hence show that $F_r(r, \theta) = F_r(r) = -4\pi G\rho_0 r_0^2/r \equiv -v_0^2/r$.
(b) Show that the potential must have the form

$$\Phi(r, \theta) = v_0^2[\ln(r/r_0) + P(\theta)]. \tag{2.58}$$

Find the equation connecting $P(\theta)$ and $S(\theta)$. (c) What is the potential if $S(\theta) = a\delta_{\text{Dirac}}(\theta - \pi/2) + b$, which corresponds to the disk embedded in a spherical halo?

Exercise 2.6
Density from rotation curves: Obtain an expression for $\Sigma(R)$ in terms of $v_c^2(R)$. [Hint: Starting with the result in Eq. (2.50), use the integral expression for $\Sigma(R)$ in terms of $S_0(k)$.]

Exercise 2.7
Asymptotic form for potentials: Consider an axisymmetric body with density distribution $\rho(R, z)$ that is symmetric about the z axis and vanishes for $r > r_{\text{max}}$. Show that, for distances large compared with r_{max}, the gravitational potential is given by

$$\Phi(R, z) \simeq -\frac{GM}{r} - \frac{G\,(R^2 - 2z^2)}{4\,r^5} \int \rho(R', z')(R'^2 - 2z'^2)\,d^3\mathbf{r}'. \tag{2.59}$$

Apply this formula for the exponential disk with $\Sigma(R) = \Sigma_0 \exp(-R/R_d)$ and show that the potential is given by

$$\Phi(R, z) \simeq -\frac{GM}{r}\left[1 + \frac{3R_d^2(R^2 - 2z^2)}{2r^4}\right], \tag{2.60}$$

where M is the mass of the disk.

The potential for the exponential disk is of some practical significance in modelling the rotation curves of galaxies. The asymptotic behavior of Bessel functions in Eq. (2.55) (see Fig. 2.2) shows that the rotation curve that is due to the exponential disk alone will decrease with radius as $R \to \infty$. Because the observed rotation curve is either flat or even gently rising with radius, it is usual to assume that the density distribution has another component that is possibly due to a spherically distributed halo. The density profile of the halo needs to be obtained by a comparison of the observed rotation curve with the one that is due to the exponential disk. To illustrate the uncertainties involved in this modelling, let us consider a simple halo profile of the form

$$\rho_h(r) = \frac{\rho_h(0)}{1 + (r/a)^\gamma}. \tag{2.61}$$

This halo produces a rotational velocity given by

$$v_h^2(r) = \frac{GM_h(r)}{r} = \frac{4\pi G}{r} \int_0^r dx\, x^2 \rho_h(x). \tag{2.62}$$

The observed rotation curve in the plane of the galaxy is given by the addition

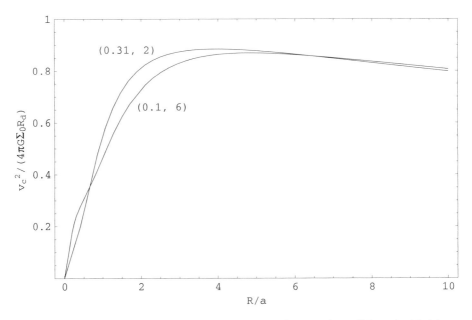

Fig. 2.3. Circular velocity as a function of radius for stars in a disk embedded in a halo. The curves have very different parameter values but still produce similar rotational velocities.

of the contributions in the quadrature: $v_c^2 = v_{\text{disk}}^2 + v_h^2$. Figure 2.3 illustrates this degeneracy in a simple situation. The y axis gives v_c^2 in units of $4\pi G\Sigma_0 R_d$ as a function of R/a. The curves are parameterised by the two ratios $(q_1, q_2) \equiv [\Sigma_0 R_d^2/\rho_h(0)a^3, a/R_d]$ in which the first parameter q_1 essentially gives the ratio of the disk mass to the halo mass and the second parameter q_2 gives the scale length of halo to the disk. The two curves are for $(q_1, q_2) = (0.31, 2)$ and $(q_1, q_2) = (0.1, 6)$ with $\gamma = 2.4$. One of the curves has a halo mass that is approximately three times the disk mass whereas the other one has a halo that is 10 times as massive as the disk. It is clear from the figure that the rotation curve at large r is essentially the same, indicating that we cannot distinguish between these two models, for example, from the asymptotic form of the rotation curves.

The difficulty described above persists in even more sophisticated modelling. Assuming a constant (M/L) ratio for the disk, and a disk scale length of ~ 3 kpc, good fits to observed rotation curves can be obtained for a wide class of model parameters. The model with maximum disk mass has $(M/L) \simeq 3.6$, consistent with the result for reasonable stellar population. However, this does not, in general, determine all the halo properties. For example, we can obtain equally good fits by varying in a correlated fashion γ and a of Eq. (2.61) in the range $1.9 < \gamma < 2.9$ and 7 kpc $< a < 12$ kpc. The core radius of the halo for this profile [usually defined to be $r_{\text{core}} = a(2\sqrt{2} - 1)^{1/\gamma}$], however, does not vary much and is given by $r_{core} = 12.5 \pm 1.5$ kpc. Thus rotation curves, by themselves, do not fix the disk and halo properties uniquely.

2.3.3 Epicyclic Approximation in Axisymmetric Potentials

A convenient coordinate system in the case of an axisymmetric galaxy is the cylindrical one with (R, ϕ) denoting the polar coordinates of a $z = $ constant plane, with z being measured along the axis of symmetry. The gravitational potential is $\Phi = \Phi(R, z)$ and will be independent of the angular coordinate ϕ.

If a star is confined to move in a $z = $ constant plane, then it cannot distinguish an axially symmetric potential from a spherically symmetric one, and the problem reduces to the one discussed in Subsection 2.3.1. Even when the motion of the star takes it out of the meridinal plane, it is possible to reduce the problem to a two-dimensional one along the following lines. The equation of motion $\ddot{\mathbf{r}} = -\nabla\Phi$ can be written in component form in the cylindrical coordinates as follows:

$$\frac{d}{dt}(R^2\dot{\phi}) = 0, \tag{2.63}$$

$$\ddot{R} - R\dot{\phi}^2 = -\frac{\partial\Phi}{\partial R}, \qquad \ddot{z} = -\frac{\partial\Phi}{\partial z}. \tag{2.64}$$

Equation (2.63) represents the conservation of the z component of angular momentum (per unit mass) $L_z = R^2\dot{\phi}$. Using this, the Eqs. (2.64) can be reduced to that of a two-dimensional problem with an effective potential Φ_{eff} in the form

$$\ddot{R} = -\frac{\partial\Phi_{eff}}{\partial R}, \qquad \ddot{z} = -\frac{\partial\Phi_{eff}}{\partial R}, \qquad \Phi_{eff} \equiv \Phi(R, z) + \frac{L_z^2}{2R^2}. \tag{2.65}$$

The dynamics can be determined once $\Phi(R, z)$ is given. The general behaviour of motion in such potentials has been discussed using surfaces of section in Vol. I, Chap. 2, Section 2.8.

From a practical point of view, it is interesting to study the behaviour of orbits near the minimum of the effective potential without assuming a specific form for $\Phi(R, z)$. The condition for the minimum, $\nabla\Phi_{eff} = 0$, has a fairly simple interpretation. The vanishing of the gradient with respect to z is satisfied everywhere in the equatorial plane $z = 0$ whereas the condition $(\partial\Phi_{eff}/\partial R) = 0$ is satisfied at a radius R_g, where

$$\left(\frac{\partial\Phi}{\partial R}\right)_{(R_g, z=0)} = \frac{L_z^2}{R_g^3} = R_g\dot{\phi}^2. \tag{2.66}$$

This is a condition for the existence of a circular orbit with angular speed $\dot{\phi}$. Thus the minimum of Φ_{eff} occurs at the radius of the circular orbit corresponding to the angular momentum L_z. To study the deviation from the circular motion, it is convenient to define a coordinate $x \equiv (R - R_g)$ and expand the potential in a

Taylor series around $(x, z) = (0, 0)$. To the lowest order, this will give a harmonic potential of the form

$$\Phi_{\text{eff}} \approx \frac{1}{2}(\kappa^2 x^2 + \nu^2 z^2) + \mathcal{O}(xz^2) + \text{constant}, \tag{2.67}$$

with the two frequencies (κ, ν) defined by the relations

$$\kappa^2 \equiv \left(\frac{\partial^2 \Phi_{\text{eff}}}{\partial R^2}\right)_{(R_g,0)} = \left(\frac{\partial^2 \Phi}{\partial R^2}\right)_{(R_g,0)} + \frac{3L_z^2}{R_g^4}, \quad \nu^2 = \left(\frac{\partial^2 \Phi}{\partial R^2}\right)_{(R_g,0)}. \tag{2.68}$$

Because the circular frequency at a radius R is given by

$$\Omega^2(R) = \frac{1}{R}\left(\frac{\partial \Phi}{\partial R}\right)_{(R,0)} = \frac{L_z^2}{R^4}, \tag{2.69}$$

we can also write

$$\kappa^2 = \left(R\frac{d\Omega^2}{dR} + 4\Omega^2\right)_{R=R_g}. \tag{2.70}$$

More generally, we can use this relation to define the function $\kappa(R)$ at any radius R. Usually Ω is nearly constant near the centre of a galaxy and decreases at larger radius. In this case we have $\Omega \leq \kappa \leq 2\Omega$.

Conventionally, we characterise the angular velocity $\Omega(R)$ in terms of two observationally determined quantities, called *Oort constants*. If the circular speed at the radius R is denoted by $v_c(R)$, then the Oort constants are defined by

$$A \equiv \frac{1}{2}\left(\frac{v_c}{R} - \frac{dv_c}{dR}\right), \quad B \equiv -\frac{1}{2}\left(\frac{v_c}{R} + \frac{dv_c}{dR}\right). \tag{2.71}$$

From these definitions, it follows that $\Omega(R) = (A - B)$ and

$$A \equiv \frac{1}{2}\left(\frac{v_c}{R} - \frac{dv_c}{dR}\right) = -\frac{1}{2}\left(R\frac{d\Omega}{dR}\right), \quad B = -\left(\frac{1}{2}R\frac{d\Omega}{dR} + \Omega\right). \tag{2.72}$$

For example, in our galaxy, at the location of the Sun $R = R_0 \simeq 8.5$ kpc, $v_c \approx 220$ km s^{-1}, $A \approx (14.8 \pm 0.8)$ km s^{-1} kpc^{-1}, and $B \approx (-12.4 \pm 0.6)$ km s^{-1} kpc^{-1}.[2] These imply that $(v_c/R) = (A - B) \approx 26.5$ km s^{-1} kpc^{-1}, which is consistent (within errors) with the value obtained from $v_c = 220$ km s^{-1} and $R_0 = 8.5$ kpc. Combining these, we get the value of κ at the location of the Sun as $\kappa_0^2 = -4B(A - B) = -4B\Omega_0$; the numerical value is $\kappa_0 \approx (36 \pm 10)$ km s^{-1} kpc^{-1}. Thus

$$\frac{\kappa_0}{\Omega_0} = 2\sqrt{\frac{-B}{A - B}} = 1.3 \pm 0.2, \tag{2.73}$$

implying that the Sun makes ~ 1.3 oscillations in the radial direction by the time it completes one orbit around the galactic centre.

Given the harmonic approximation to the potential (2.67), it is easy to solve for the orbits. The x and z motions are oscillatory and are given by

$$z = Z\cos(\nu t + \zeta), \qquad x(t) = X\cos(\kappa t + \psi), \qquad (2.74)$$

where Z and X are constants. To determine the azimuthal motion, we use the constancy of L_z and write

$$\dot{\phi} = \frac{L_z}{R^2} = \frac{L_z}{R_g^2}\left(1 + \frac{x}{R_g}\right)^{-2} \simeq \Omega_g\left(1 - \frac{2x}{R_g}\right). \qquad (2.75)$$

Substituting for $x(t)$ from Eqs. (2.74) and integrating, we get

$$\phi = \Omega_g t + \phi_0 - \frac{2\Omega_g X}{\kappa R_g}\sin(\kappa t + \psi). \qquad (2.76)$$

The first two terms on the right-hand side represent a rotation with angular velocity Ω_g that we can eliminate by introducing a local set of Cartesian axes (x, y, z) in which x and z are – as already defined – along the radial and the symmetry axes and the y coordinate is in the azimuthal direction. From Eq. (2.76), we get

$$y = -\frac{2\Omega_g}{\kappa}X\sin(\kappa t + \psi) \equiv -Y\sin(\kappa t + \psi). \qquad (2.77)$$

Clearly the motion of the star in the $(x-y)$ plane is an ellipse with a ratio of the semiaxis of $(X/Y) = (\kappa/2\Omega_g)$. The centre of the ellipse rotates at the angular speed Ω_g around the galactic centre.

As an application of the preceding results (usually called the *guiding centre* or the *epicyclic approximation*) let us consider the question of determining the Oort constants from the observed motion of stars. In principle, the measurements of proper motions in the x and the y directions should yield the maximum values κX and κY, the ratio of which contains the necessary information. Such a direct measurement, of course, is not feasible because of the time scales involved. A more indirect way begins with the measurement of the radial velocity v_R and the differential angular azimuthal velocity $v_\phi(R_0) - v_c(R_0)$ for a group of stars as they pass near the Sun at the radius R_0. This quantity is given by

$$v_\phi(R_0) - v_c(R_0) = R_0(\dot{\phi} - \Omega_0) = R_0(\dot{\phi} - \Omega_g + \Omega_g - \Omega_0)$$

$$\simeq R_0\left[(\dot{\phi} - \Omega_g) - \left(\frac{d\Omega}{dR}\right)_{R_0}x\right] \simeq -R_0 x\left[\frac{2\Omega_g}{R_g} + \left(\frac{d\Omega}{dR}\right)_{R_0}\right], \qquad (2.78)$$

where we have used Eqs. (2.75) and (2.77). When (Ω_g/R_g) is replaced by (Ω_0/R_0) – which is permissible in the coefficient of the small quantity x –

and the definition of the Oort constant is used, it follows that

$$v_\phi - v_c(R_0) \simeq 2Bx. \tag{2.79}$$

Similarly, we have

$$\langle v_R^2 \rangle = \langle \kappa^2 x^2 \rangle \simeq -4B(A - B)\langle x^2 \rangle, \tag{2.80}$$

where the average is over many stars. This allows us to write

$$\frac{\langle [v_\phi - v_c(R_0)]^2 \rangle}{\langle v_R^2 \rangle} \simeq \frac{-B}{A - B} = \frac{\kappa_0^2}{4\Omega_0^2}, \tag{2.81}$$

where all the quantities on the left-hand side can be determined from observations.

Exercise 2.8

Determining a product: The line-of-sight velocity along the direction l can be expressed by Eq. (1.94) of Chap. 1. Show how observations of line-of-sight velocity can be used to determine the product AR_0. [Answer. Using the Taylor expansion

$$\Omega(R = R_0 \sin l) - \Omega(R_0) \simeq \left(\frac{d\Omega}{dR}\right)_{R_0} R_0(\sin l - 1), \tag{2.82}$$

which is valid near $l = \pi/2$, and expressing $(d\Omega/dR)$ in terms of A, we get $v(l) \approx 2AR_0(1 - \sin l)$ near $l = \pi/2$. Observations near $l = \pi/2$ will now allow the determination of the product AR_0 fairly accurately, even though neither A nor R_0 is known to the same precision.]

2.3.4 Planar Nonaxisymmetric Potentials

When even the assumption of axisymmetry is relaxed, the orbits becomes difficult to tackle analytically. Some progress can be achieved if the motion is confined to the meridinal plane $z = 0$, with the potential now depending explicitly on x and y. In general, most orbits in this potential possess an extra integral (in addition to energy). The majority of orbits can be classified either as *loop* orbits (which have a fixed sense of rotation and never take the star close to the centre) or *box* orbits (which have no fixed sense of rotation and allow the star to pass arbitrarily close to the centre). When the axial ratio of equipotential curves is close to unity, the phase space is filled mostly with loop orbits; as the axial ratio deviates away from unity, box orbits become more common.

These can be illustrated by the technique of surface of section described in Vol. I, Chap. 2, Section 2.8. For definiteness, we study the results of numerically integrating the equations of motion for the potential

$$\phi(x, y) = \frac{1}{2}v_0^2 \ln\left(R_c^2 + x^2 + \frac{y^2}{q^2}\right). \tag{2.83}$$

For a system with $N = 2$, the phase space is four dimensional and any point in the phase space has the coordinates (x, y, p_x, p_y). Conservation of energy restricts the motion to a three-dimensional surface defined by

$$\frac{1}{2}(\dot{x}^2 + \dot{y}^2) + \phi(x, y) = E = \text{constant}. \tag{2.84}$$

(We take the mass of the star to be unity and use the symbols $\dot{x}$ and $\dot{y}$ instead of p_x and p_y.) The intersection of this three-dimensional surface with the surface $y = 0$ defines the "surface of section" for the dynamical problem, which has the coordinates $(x, \dot{x})$. Given x, $\dot{x}$, and $y = 0$, $\dot{y}$ is determined as

$$\dot{y} = +\{2[E - \phi(x, 0)] - \dot{x}^2\}^{1/2}, \tag{2.85}$$

where we have chosen the positive square root by convention.

Given the energy E and suitable initial conditions, the equations of motion can be integrated to give the orbit of the particle. This orbit will repeatedly cut the surface of section at different values of x and $\dot{x}$. If there are no other integrals of motion, we expect these points to be scattered in a two-dimensional region bounded by the curve

$$\frac{1}{2}\dot{x}^2 + \phi(x, 0) \leq E. \tag{2.86}$$

In case the system has some other hidden integrals of motion, the orbit will be confined to a lower-dimensional surface in phase space and hence will form a smooth curve in the x–$\dot{x}$ plane.

The surface of section for the potential is shown in Fig. 2.4 with $q = 0.9$, $R_c = 0.14$, and the conserved energy $E = -0.337$. The following conclusions can be drawn from the figure.

(1) The intersection of the orbit with the x–$\dot{x}$ plane occurs on reasonably smooth curves rather than on the entire allowed region in the plane. This suggests that there could be some other hidden integrals of motion.

(2) The orbits can be classified as belonging to the two different classes, usually called box orbits or loop orbits. Orbits like 1 or 2 (top panel of Fig. 2.4) revolve around the points A or B, respectively, and have a definite sense of rotation (loop orbits). These orbits do not pass through the origin. In contrast, orbits like 3 enclose both A and B and have no definite sense of rotation (box orbits). These orbits do cross the origin. The behaviour of these orbits in physical space is shown in the lower panel of Fig. 2.4. The two figures on the bottom right show the box orbits that oscillate through the origin with a superposed rotational motion. The two figures on the bottom left show the loop orbits that are confined to an annular region.

(3) As the initial conditions are changed, loop orbits can become more and more eccentric and the box orbits can become less and less elongated.

(4) The preceding discussion was based on a potential with $q = 0.9$. If $q = 1$ the potential depends on only $r^2 = x^2 + y^2$ and possesses an extra conserved quantity, namely the z component of the angular momentum, $J_z = x\dot{y} - y\dot{x} = x\dot{y}$ on the surface of

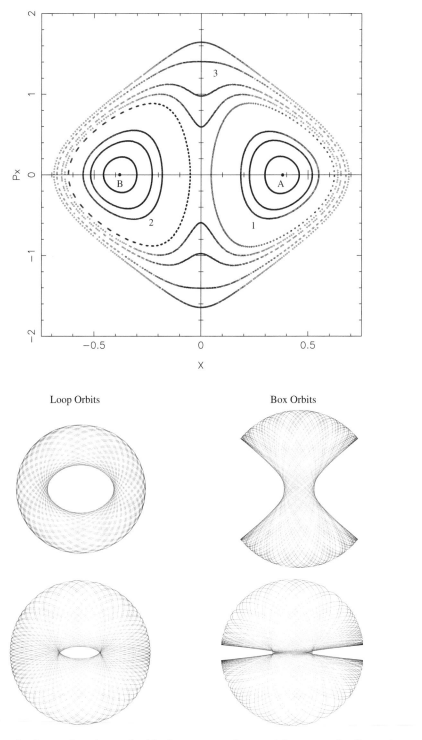

Fig. 2.4.   Surfaces of section and orbits in an external potential; see text for discussion.

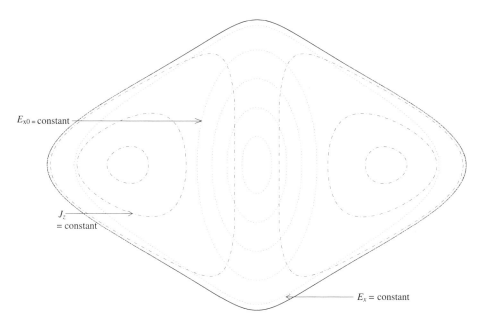

Fig. 2.5. Curves generated by the constancy of $x\dot{y}$ and E_x in the surface of section.

section with $y = 0$. In this case, all generic orbits are loop orbits. For $q \neq 1$, the expression for J_z becomes

$$(J_z)_{y=0} = x\dot{y} = x\left[2E - v_0^2 \ln\left(x^2 + R_c^2\right) - \dot{x}^2\right]^{1/2}. \tag{2.87}$$

Although this quantity is not strictly conserved when $q \neq 1$, it is useful to examine the nature of curves in the x–$\dot{x}$ plane, generated by the constancy of J_z. These are indicated in Fig. 2.5 by dotted–dashed curves. We see that these curves are similar to those corresponding to loop orbits. We may conclude that loop orbits owe their existence to an approximate conservation of a quantity like angular momentum.

(5) On the other extreme we may consider the motion in the potential when $r \ll R_c$, with $\phi(x, y)$ approximated as

$$\phi(x, y) \simeq \frac{v_0^2}{2R_c^2}\left(x^2 + \frac{y^2}{q^2}\right) + \text{constant}. \tag{2.88}$$

This is a double harmonic oscillator (discussed in Vol. I, Chap. 2, Section 2.5) which has an additional integral

$$E_{x0} = \frac{1}{2}\dot{x}^2 + \frac{v_0^2}{2R_c^2}x^2. \tag{2.89}$$

Constancy of E_{x0} will lead to the elliptical orbits shown in the middle of Fig. 2.5. We do not see evidence for such orbits in Fig. 2.4. However, if we generalize E_x to the form

$$E_x = \frac{1}{2}\dot{x}^2 + \phi(x, 0), \tag{2.90}$$

then we get better agreement with the orbits in Fig. 2.4. The outer dotted curve in Fig. 2.5

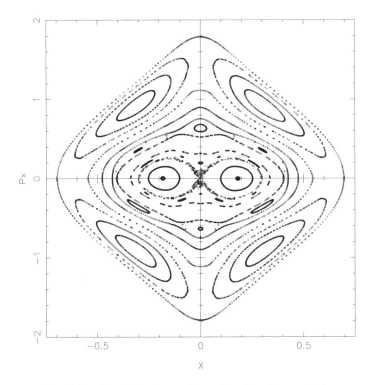

Fig. 2.6.   Islands in the surface of section for $q = 0.6$.

corresponds to $E_x =$ constant and is quite similar to the box orbits of Fig. 2.4. Thus box orbits owe their existence to an approximate conservation of the energy corresponding to x-axis motion.

(6) As q is decreased further, larger regions of phase space will be occupied by box orbits at the expense of loop orbits. When the potential is significantly nonaxisymmetric, several "islandlike" regions will form in the x–$\dot{x}$ plane with loop orbits around each of them. (See Fig. 2.6.) These results show that the stellar orbits in real potentials without any symmetries can be fairly complicated.

2.3.5 Potentials in the Rotating Frame

For most axisymmetric galaxies that have nonzero rotational motion about the z axis, the potential generated by the distribution of stars will be simplest in the corotating frame. Because such a frame is noninertial, the equations of motion pick up additional terms corresponding to Coriolis and centrifugal forces. If we take the galaxy as rotating with a bulk angular velocity $\Omega_b(> 0)$ about the z axis, then the equation of motion in the rotating coordinate system is given by (see Vol. I, Chap. 2, Section 2.3)

$$\ddot{\mathbf{r}} = -\nabla\Phi - 2(\mathbf{\Omega}_b \times \dot{\mathbf{r}}) - \mathbf{\Omega}_b \times (\mathbf{\Omega}_b \times \mathbf{r}). \tag{2.91}$$

Let (R, ϕ) be the polar coordinates in the frame that rotates with the potential so that the equations of motion, in component form, are

$$\ddot{R} - R\dot{\phi}^2 = -\frac{\partial \Phi}{\partial R} + 2R\dot{\phi}\Omega_b + \Omega_b^2 R, \tag{2.92}$$

$$R\ddot{\phi} + 2\dot{R}\dot{\phi} = -\frac{1}{R}\frac{\partial \Phi}{\partial \phi} - 2\dot{R}\Omega_b. \tag{2.93}$$

As for the potential, we take it to be

$$\Phi(R, \phi) = \Phi_0(R) + \Phi_1(R, \phi), \qquad \Phi_1(R, \phi) = \Phi_b(R)\cos(m\phi), \tag{2.94}$$

with $(\Phi_1/\Phi_0) \ll 1$. This form for Φ_1 is suggested by the fact that most galaxies have a barlike structure that rotates with a pattern speed Ω_b. The perturbation Φ_1, having an angular variation $\cos(m\phi)$ with $m = 2$, is the simplest form that takes this effect into account. To study the motion, we assume the trajectories to be

$$R(t) = R_0 + R_1(t), \qquad \phi(t) = \phi_0(t) + \phi_1(t), \tag{2.95}$$

substitute into the equations of motion, and equate the zeroth- and first-order terms. The zeroth-order term for the radial equation will determine the radius R_0 of the centrifugal equilibrium at which the angular velocity is $\Omega_0 = \Omega(R_0)$, where

$$\Omega(R) \equiv \pm\sqrt{\frac{1}{R}\frac{d\Phi_0}{dR}}. \tag{2.96}$$

With a suitable choice for the origin of time, the angular motion – at the lowest order – is given by a steady rotation, $\phi_0 = (\Omega_0 - \Omega_b)t$. In the first-order perturbation equations, we can replace ϕ with ϕ_0 when multiplied by other first-order terms. Elementary algebra now gives these equations as

$$\ddot{R}_1 + \left(\frac{d^2\Phi_0}{dR^2} - \Omega^2\right)_{R_0} R_1 - 2R_0\Omega_0\dot{\phi}_1 = -\left(\frac{d\Phi_b}{dR}\right)_{R_0} \cos[m(\Omega_0 - \Omega_b)t], \tag{2.97}$$

$$\ddot{\phi}_1 + 2\Omega_0\frac{\dot{R}_1}{R_0} = \frac{m\Phi_b(R_0)}{R_0^2}\sin[m(\Omega_0 - \Omega_b)t]. \tag{2.98}$$

Integrating Eq. (2.98), we get

$$\dot{\phi}_1 = -2\Omega_0\frac{R_1}{R_0} - \frac{\Phi_b(R_0)}{R_0^2(\Omega_0 - \Omega_b)}\cos[m(\Omega_0 - \Omega_b)t] + \text{constant}, \tag{2.99}$$

which can be used to eliminate $\dot{\phi}_1$ from the radial equation, leading to

$$\ddot{R}_1 + \kappa_0^2 R_1 = -\left[\frac{d\Phi_b}{dR} + \frac{2\Omega\Phi_b}{R(\Omega - \Omega_b)}\right]_{R_0} \cos[m(\Omega_0 - \Omega_b)t], \tag{2.100}$$

where

$$\kappa_0^2 \equiv \left(\frac{d^2\Phi_0}{dR^2} + 3\Omega^2 \right)_{R_0} = \left(R\frac{d\Omega^2}{dR} + 4\Omega^2 \right)_{R_0} \qquad (2.101)$$

is the square of the epicyclic frequency. [A constant has been ignored in arriving at Eq. (2.100) as it only leads to a shift in the origin of R_1.] Radial equation (2.100) is that of a forced harmonic oscillator with a driving frequency of $m(\Omega_0 - \Omega_b)$ and a natural frequency of κ_0. Its solution is given by

$$R_1(t) = C_1 \cos(\kappa_0 t + \psi) - \left[\frac{d\Phi_b}{dR} + \frac{2\Omega\Phi_b}{R(\Omega - \Omega_b)} \right]_{R_0} \frac{\cos[m(\Omega_0 - \Omega_b)t]}{\Delta},$$

$$(2.102)$$

where C_1 and ψ are constants and

$$\Delta \equiv \kappa_0^2 - m^2(\Omega_0 - \Omega_b)^2. \qquad (2.103)$$

Eliminating t between $R_1(t)$ and $\phi_0(t)$, we can write the trajectory in the form

$$R_1(\phi_0) = C_1 \cos\left[\frac{\kappa_0\phi_0}{(\Omega_0 - \Omega_b)} + \psi \right] + C_2 \cos(m\phi_0), \qquad (2.104)$$

with

$$C_2 \equiv -\frac{1}{\Delta}\left[\frac{d\Phi_b}{dR} + \frac{2\Omega\Phi_b}{R(\Omega - \Omega_b)} \right]_{R_0}. \qquad (2.105)$$

Several conclusions can be drawn from this solution. To begin with, note that $R_1(\phi_0)$ is periodic in ϕ_0 if and only if $C_1 = 0$. This will correspond to a closed-loop orbit. Further, the amplitude C_2 diverges (leading to a resonance condition) under two circumstances. The first case (known as *corotation resonance*) occurs when $\Omega_0 = \Omega_b$. The second corresponds to the condition $m(\Omega_0 - \Omega_b) = \pm\kappa_0$ and is known as *Lindblad resonance*. Radii at which these resonances occur are called *inner* and *outer Lindblad radii*. These resonances have a simple interpretation: For a star in circular orbit, there are two natural frequencies of oscillation. Along the radial direction it is κ_0 whereas along the azimuthal direction it is zero as the circular orbit is neutrally stable. The forcing frequency $m(\Omega_0 - \Omega_b)$, seen by the star, is at resonance with the two natural frequencies κ_0 and 0 at the Lindblad resonance and at the corotation resonance, respectively. In most galaxies, the bar structure corresponds to $m = 2$; Fig. 2.7 shows the behaviour of Ω, $\Omega + (1/2)\kappa$ and $\Omega - (1/2)\kappa$ for the isochrone potential. Note that $\Omega - (1/2)\kappa$ is approximately constant.

The perturbation analysis given above breaks down near the resonances, and we need to refine our approximation. We do this most easily by rewriting the perturbation equations assuming that R_1, $\dot{R}_1$ and $\dot{\phi}_1$ are small but not ϕ_1. We

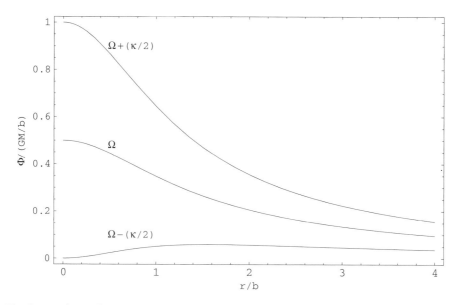

Fig. 2.7.   The variation of Ω, $\Omega + (1/2)\kappa$ and $\Omega - (1/2)\kappa$ for the isochrone potential.

shall illustrate this procedure for the resonance at the corotation radius. With the guiding centre located at the corotation radius $[\Omega(R_0) = \Omega_b; \phi_0 = (\pi/2)]$, the perturbation equations now become

$$\ddot{R}_1 + \left(\kappa_0^2 - 4\Omega_0^2\right)R_1 - 2R_0\Omega_0\dot{\phi}_1 = -\frac{\partial \Phi_1}{\partial R}, \qquad (2.106)$$

$$\ddot{\phi}_1 + 2\Omega_0\frac{\dot{R}_1}{R_0} = -\frac{1}{R_0^2}\frac{\partial \Phi_1}{\partial \phi}. \qquad (2.107)$$

Let us consider the relative strengths of various terms in this equation. If the nonaxisymmetric perturbation Φ_1 is of the order of a small parameter ϵ^2, then R_1 will be of the order of ϵ and ϕ_1 will be of the order of unity. The time derivatives of any quantity will be smaller by a factor ϵ compared with the original quantity. Therefore, in the radial equation, the three terms on the left-hand side are of the order of ϵ^3, ϵ, and ϵ, respectively, and the right-hand side is of the order of ϵ^2. Retaining only terms of the order of ϵ leads to the equation

$$\left(\kappa_0^2 - 4\Omega_0^2\right)R_1 - 2R_0\Omega_0\dot{\phi}_1 = 0. \qquad (2.108)$$

Using this relation to eliminate R_1 from angular equation (2.107) and substituting for the explicit form of Φ_1 with $m = 2$, we get the equation for angular part as

$$\ddot{\phi}_1 = -\frac{2\Phi_b}{R_0^2}\left(\frac{4\Omega_0^2 - \kappa_0^2}{\kappa_0^2}\right)\sin[2(\phi_0 + \phi_1)]. \qquad (2.109)$$

Taking $\Phi_b < 0$ and $\phi_0 = (\pi/2)$, we can recast this equation in the form

$$\frac{d^2\psi}{dt^2} = -p^2 \sin \psi, \tag{2.110}$$

where

$$\psi \equiv 2\phi_1, \qquad p^2 \equiv \frac{4}{R_0^2}|\Phi_b(R_0)|\frac{4\Omega_0^2 - \kappa_0^2}{\kappa_0^2}. \tag{2.111}$$

This is precisely the equation for an oscillating pendulum, and its solutions are known in terms of elliptic integrals. The singularity at the corotation radius has now disappeared because of the more careful analysis. The equilibrium point of the pendulum, $\phi_1 = 0$, is at the maximum of the potential Φ_1. If the integral of the motion $E_p = (1/2)\dot\psi^2 - p^2 \cos \psi$ is less than p^2, the star librates about the equilibrium point, whereas if $E_p > p^2$, the star circulates around the centre of the galaxy. The shape of the orbit is given by

$$R_1 = -\frac{2R_0\Omega_0\dot\phi_1}{4\Omega_0^2 - \kappa_0^2} = \pm\left(\frac{R_0\Omega_0}{4\Omega_0^2 - \kappa_0^2}\right)\sqrt{2[E_p + p^2 \cos(2\phi_1)]}, \tag{2.112}$$

which reduces to result (2.104) previously obtained in the limit of $E_p \gg p^2$. This result shows that there can be trapped orbits near the corotation radius. (It is possible to perform a similar analysis around the Lindblad radius, although it is mathematically more complicated.)

2.4 Application of the Jeans Equations

The discussion so far has concentrated on orbits of individual stars. We now take up the issue of studying the properties of a collection of stars described by a distribution function $f(\mathbf{x}, \mathbf{v}, t)$. It was mentioned in Vol. I, Chap. 10, Section 10.5, that one way of extracting useful information from the distribution function of stars in a galaxy will be to use the moments of the CBE and relate the different terms to the observed characteristics of a galaxy. It was shown there that the lowest-order moments lead to the equations

$$\frac{\partial\nu}{\partial t} + \frac{\partial(\nu\langle v_i\rangle)}{\partial x_i} = 0, \tag{2.113}$$

$$\nu\frac{\partial(\langle v_j\rangle)}{\partial t} + \nu\langle v_i\rangle\frac{\partial\langle v_j\rangle}{\partial x_i} = -\nu\frac{\partial\Phi}{\partial x_j} - \frac{\partial(\nu\sigma_{ij}^2)}{\partial x_i}, \tag{2.114}$$

where ν is the spatial density of stars, $\langle\mathbf{v}\rangle$ is the mean velocity, and σ_{ij}^2 is the velocity dispersion, defined through the relations

$$\nu \equiv \int f \, d^3\mathbf{v}, \qquad \langle v_i \rangle \equiv \frac{1}{\nu} \cdot \int f \, v_i \, d^3\mathbf{v},$$

$$\sigma_{ij}^2 \equiv \langle (v_i - \langle v_i \rangle)(v_j - \langle v_j \rangle) \rangle = \langle v_i v_j \rangle - \langle v_i \rangle \langle v_j \rangle.$$

(2.115)

Equation (2.113) expresses the conservation of mass, and Eq. (2.114) is the force equation in the stellar-dynamics context. The terms on the left-hand side of Eq. (2.114) give the acceleration; the first term on the right-hand side is the gravitational force and the second term arises from the velocity dispersion of particles. (These are called the *Jeans equations*.) If the system was collisional then this velocity dispersion could have been related to a pressure that in turn could be related to the density by an equation of state, thereby closing the system of equations. For a collisionless system, there is no simple way of closing the moment equations because there is no equation of state.

Because σ_{ij}^2 is a symmetric second-rank tensor, we can always choose the co-ordinate system such that it is diagonalised to the form $\sigma_{ij} = \text{dia}(\sigma_{11}, \sigma_{22}, \sigma_{33})$. An ellipsoid in this coordinate system with the principal axis along the coordinates and $(\sigma_{11}, \sigma_{22}, \sigma_{33})$ as its semiaxis lengths is called the *velocity ellipsoid*. Note that, in general, σ_{ij}^2 is a function of $\mathbf{x}$ and t; in steady state, the velocity ellipsoid will still depend on the location $\mathbf{x}$. The use of the above equations therefore relies heavily on the additional physical assumptions that we are prepared to make regarding the various averages that appear in these equations. We now consider several such applications of the Jeans equations.

2.4.1 *Asymmetric Drift*

Observations suggest that stellar populations in our galaxy show a circular velocity v_c that is different from the mean azimuthal velocity $\bar{v}_\phi$. We typically find that this difference is $v_a \equiv (v_c - \bar{v}_\phi) \approx \langle v_R^2 \rangle / D$, with $D \simeq 120$ km s^{-1}. This phenomenon, called *asymmetric drift*, can be understood in terms of the Jeans equations. To begin with, we write the Jeans equations in cylindrical polar coordinates by the standard procedure of multiplying the CBE by velocity components and integrating over the velocities. This gives

$$\frac{\partial (\nu \langle v_R \rangle)}{\partial t} + \frac{\partial \left(\nu \langle v_R^2 \rangle \right)}{\partial R} + \frac{\partial (\nu \langle v_R v_z \rangle)}{\partial z} + \nu \left(\frac{\langle v_R^2 \rangle - \langle v_\phi^2 \rangle}{R} + \frac{\partial \Phi}{\partial R} \right) = 0, \quad (2.116)$$

$$\frac{\partial (\nu \langle v_\phi \rangle)}{\partial t} + \frac{\partial (\nu \langle v_R v_\phi \rangle)}{\partial R} + \frac{\partial (\nu \langle v_\phi v_z \rangle)}{\partial z} + \frac{2\nu}{R} \langle v_\phi v_R \rangle = 0, \quad (2.117)$$

$$\frac{\partial (\nu \langle v_z \rangle)}{\partial t} + \frac{\partial (\nu \langle v_R v_z \rangle)}{\partial R} + \frac{\partial \left(\nu \langle v_z^2 \rangle \right)}{\partial z} + \frac{\nu \langle v_R v_\phi \rangle}{R} + \nu \frac{\partial \Phi}{\partial z} = 0. \quad (2.118)$$

Near the galactic equator ($z = 0$) we can take $(\partial v/\partial z) = 0$, and in steady state we can ignore the time derivatives in Eq. (2.116). Using the definitions $v_c^2 = R(\partial \Phi/\partial R)$, $v_a \equiv (v_c - \langle v_\phi \rangle)$, and defining the velocity dispersion $\sigma_\phi^2 = \langle v_\phi^2 \rangle - \langle v_\phi \rangle^2$, we find that Eq. (2.116) becomes

$$v_a(2v_c - v_a) \simeq 2v_c v_a \simeq \langle v_R^2 \rangle \left[\frac{\sigma_\phi^2}{\langle v_R^2 \rangle} - 1 - \frac{\partial \ln \left(v \langle v_R^2 \rangle \right)}{\partial \ln R} - \frac{R}{\langle v_R^2 \rangle} \frac{\partial (\langle v_R v_z \rangle)}{\partial z} \right],$$

(2.119)

where the first equality follows from the fact that we expect $v_a \ll v_c$.

Relation (2.119) is typical of the relations that we can obtain from the Jeans equations. To proceed further, we need to introduce a series of reasonable assumptions based on observational inputs. To begin with, let us consider the term $[\partial \ln(v \langle v_R^2 \rangle)/\partial \ln R]$. If the disk galaxy has an exponential profile, with a small thickness h_z in the vertical direction, we would expect $h_z \propto \sigma_{zz}^2/\Sigma$ with $\Sigma \propto \exp(-R/R_d)$. This suggests that $\sigma_{zz}^2 \propto \exp(-R/R_d)$. Further, if we take the velocity ellipsoid to be approximately constant in space so that $\sigma_{zz}^2 \propto \sigma_{RR}^2 \equiv \langle v_R^2 \rangle$, we can write

$$\frac{\partial \ln \left(v \langle v_R^2 \rangle \right)}{\partial \ln R} \simeq 2 \frac{\partial \ln v}{\partial \ln R}.$$

(2.120)

To estimate the other derivatives, we consider two extreme possibilities: (1) The principal axes of the velocity ellipsoid are in the direction of cylindrical coordinate axes or (2) the principal axes rotate in such a way as to retain alignment with the spherical coordinates centred on the galactic nucleus. In the first case $\langle v_R v_z \rangle$ is independent of z, and in the second case $\langle v_R v_z \rangle \simeq (\langle v_R^2 \rangle - \langle v_z^2 \rangle)(z/R)$. (This relation is most easily derived if v_R, v_z is expressed in terms of v_r, v_θ and the result is averaged with $\langle v_r v_\theta \rangle = 0$.) Taking the average of the derivatives in these two extreme cases and substituting in relation (2.119) we get the range of possibilities as given by

$$\frac{2v_c v_a}{\langle v_R^2 \rangle} \simeq \left[\frac{\sigma_\phi^2}{\langle v_R^2 \rangle} - \frac{3}{2} - 2 \frac{\partial \ln v}{\partial \ln R} + \frac{1}{2} \frac{\langle v_z^2 \rangle}{\langle v_R^2 \rangle} \pm \frac{1}{2} \left(\frac{\langle v_z^2 \rangle}{\langle v_R^2 \rangle} - 1 \right) \right].$$

(2.121)

From the analysis based on Oort constants [see relation (2.81)], we know that $\sigma_\phi^2 \simeq \langle v_z^2 \rangle \simeq 0.45 v_R^2$. Taking the disk of our galaxy to be exponential, with $v = v_0 \exp(-R/R_d)$, $(R/R_d) = 2.4$, and $v_c = 220$ km s^{-1}, we get the final result,

$$v_a \simeq \frac{\langle v_R^2 \rangle}{(110 \pm 7 \text{ km s}^{-1})},$$

(2.122)

which is in good agreement with observations.

2.4.2 Mass and Velocity Dispersion

Another application of the Jeans equations is in estimating the effects of aniso-tropic velocity dispersion on other derived quantities. Consider, for example, the problem of determining the mass $M(< R)$ contained within a projected radius R of a galaxy that contains a population of tracers (say, planetary nebu-las) distributed in a spherically symmetric manner with a volume number den-sity $n(r) \propto r^{-\gamma}$. Let us assume that the velocity dispersion is anisotropic with $\langle v_\theta^2 \rangle / \langle v_r^2 \rangle = 1 - \beta$; when $\beta < 0$ the system is dominated by tangential anisotropy whereas for $0 < \beta < 1$, the radial anisotropy dominates. (Spherical symmetry, of course, implies that $\langle v_\theta^2 \rangle = \langle v_\phi^2 \rangle$.) The Jeans equations now become

$$\frac{r}{n}\frac{d}{dr}\left(n\sigma_r^2\right) + 2\beta\sigma_r^2 = -r\frac{d\Phi}{dr} = -v_c^2, \qquad \sigma_r^2 = \langle v_r^2 \rangle, \qquad (2.123)$$

which can be expressed in the form

$$\left(\frac{d\ln n}{d\ln r} + \frac{d\ln \sigma_r^2}{d\ln r} + 2\beta\right)\sigma_r^2 = -v_c^2. \qquad (2.124)$$

If the circular velocity is approximately constant, then the right-hand side of Eq. (2.124) is independent of r; if we further assume that σ_r^2 is also independent of r and use $n \propto r^{-\gamma}$, then $\langle v_r^2 \rangle = v_c^2/(\gamma - 2\beta)$. The mass within a radius r is, of course, $M(< r) = v_c^2 r/G$ when v_c^2 is a constant. However, observations usually give the velocity dispersion $\langle v_l^2 \rangle$ along the line of sight rather than $\langle v_r^2 \rangle$. It is therefore necessary to relate these two quantities. If the radial distance to a point along the line of sight from the centre of the galaxy is r and the angle between the radial direction and the line of sight is $(\pi/2 - \theta)$, then the component v_z along the line of sight is

$$v_z = v_r \sin\theta + v_\theta \cos\theta. \qquad (2.125)$$

Squaring and averaging, using $\langle v_r v_\theta \rangle = 0$, $\langle v_\theta^2 \rangle = (1 - \beta)\langle v_r^2 \rangle$, we get

$$\langle v_z^2 \rangle = \langle v_r^2 \rangle(1 - \beta \cos^2\theta). \qquad (2.126)$$

The integral of this velocity dispersion along the line of sight is given by

$$\langle v_l^2 \rangle = \frac{\displaystyle\int_{-\infty}^{\infty} \langle v_z^2 \rangle n\, dz}{\displaystyle\int_{-\infty}^{\infty} n\, dz}. \qquad (2.127)$$

Using Eq. (2.126) and the relations $r = R \sec \theta$, $n(r) \propto R^{-\gamma} \cos^{\gamma} \theta$, and $z = R \tan \theta$, we can reduce this integral to the form

$$\langle v_l^2 \rangle = \langle v_r^2 \rangle \frac{\displaystyle\int_0^{\pi/2} (1 - \beta \cos^2 \theta)^{-1} \cos^{\gamma - 2} \theta \, d\theta}{\displaystyle\int_0^{\pi/2} \cos^{\gamma - 2} \theta \, d\theta}. \tag{2.128}$$

For $\gamma > 1$ this integral can be expressed in terms of beta functions $B(x, y)$, giving

$$\langle v_l^2 \rangle = \frac{v_c^2}{(\gamma - 2\beta)} \left(1 - \beta \frac{B\left[\frac{1}{2}, \frac{\gamma + 1}{2}\right]}{B\left[\frac{1}{2}, \frac{\gamma - 1}{2}\right]} \right), \quad (\gamma > 1). \tag{2.129}$$

The integral, however, simplifies for two special cases, $\gamma = 2$ and 3; we then get

$$\langle v_l^2 \rangle = \begin{cases} v_c^2 \dfrac{[1 - (\beta/2)]}{2(1 - \beta)} & (\gamma = 2) \\[2ex] \left(v_c^2 / 3 \right) & (\gamma = 3) \end{cases}. \tag{2.130}$$

We note that, if $\gamma = 3$, the result is independent of β, showing that the velocity anisotropy is not important. However, if $\gamma = 2$, we can only put an upper limit to $M(< r)$ from the observations of $\langle v_l^2 \rangle$.

Exercise 2.9
Velocity dispersion and mass estimate: Plot a graph of $\langle v_l^2 \rangle / v_c^2$ versus β for different values of γ. Explain why for $\gamma < 3$, $\langle v_l^2 \rangle / v_c^2$ increases as β increases, whereas for $\gamma > 3$ it decreases as β increases. Also explain why $\langle v_l^2 \rangle / v_c^2$ decreases with increasing γ for $\beta = 0$.

Exercise 2.10
Determining mass and velocity dispersion when $\beta = 0$: Let $I(R)$ and $\sigma_p^2(R)$ be the surface brightness and the line-of-sight velocity dispersion at the projected radius R. Show how these quantities can be used to determine $v(r)$ and $\langle v_r^2(r) \rangle$ if the system is spherically symmetric and $\beta = 0$. [Answer. When $\beta = 0$, Jeans equation (2.124) gives

$$M(r) = -\frac{r \langle v_r^2 \rangle}{G} \left(\frac{d \ln v}{d \ln r} + \frac{d \ln \langle v_r^2 \rangle}{d \ln r} \right). \tag{2.131}$$

Further, the projected quantities are related to actual three-dimensional variables by

$$I(R) = 2 \int_R^\infty \frac{v r \, dr}{\sqrt{r^2 - R^2}}, \qquad I(R) \sigma_p^2(R) = 2 \int_R^\infty \frac{v \langle v_r^2 \rangle r \, dr}{\sqrt{r^2 - R^2}}. \tag{2.132}$$

These equations, being Abel integral equations, can be inverted to provide

$$v(r) = -\frac{1}{\pi} \int_r^\infty \frac{dI}{dR} \frac{dR}{\sqrt{R^2 - r^2}}, \qquad v(r)\langle v_r^2 \rangle(r) = -\frac{1}{\pi} \int_r^\infty \frac{d(I\sigma_p^2)}{dR} \frac{dR}{\sqrt{R^2 - r^2}}.$$

(2.133)

The right-hand side is known from observations, allowing us to determine $v(r)$ and $\langle v_r^2(r) \rangle$. Once these are known, we can compute $M(r)$.]

2.4.3 Rotation of Elliptical Galaxies

The most natural explanation for the elliptical shape of some galaxies could be the flattening that is due to the systematic rotation of the components, much the same way as the ellipsoidal shape of Earth is due to the rotation. The Jeans equation gives a handle on analysing whether this is a feasible explanation.

To do this, we make the assumption that the velocity dispersion is isotropic, allowing us to write $\sigma_{ij}^2 = \sigma^2 \delta_{ij}$. The Jeans equation for an axisymmetric galaxy in steady state, with isotropic velocity dispersion and streaming motion in only the azimuthal direction, becomes

$$\frac{\partial(v\sigma^2)}{\partial R} - v\left(\frac{\bar{v}_\phi^2}{R} - \frac{\partial \Phi}{\partial R}\right) = 0,$$

(2.134)

$$\frac{\partial(v\sigma^2)}{\partial z} + v\frac{\partial \Phi}{\partial z} = 0.$$

(2.135)

Integrating Eq. (2.135) from $z = \infty$ to some finite value and assuming that the density vanishes at infinity, we can express the velocity dispersion in terms of the gravitational potential:

$$\sigma^2(R, z) = \frac{1}{v} \int_z^\infty v\frac{\partial \Phi}{\partial z} dz.$$

(2.136)

Using this in Eq. (2.134) we can find $\bar{v}_\phi$ as

$$\bar{v}_\phi^2(R, z) = R\frac{\partial \Phi}{\partial R} + \frac{R}{v}\frac{\partial}{\partial R} \int_z^\infty v\frac{\partial \Phi}{\partial z} dz.$$

(2.137)

These equations completely determine the velocity dispersion and the rotational streaming velocity in terms of the gravitational potential and density. Given any model for the density distribution (which will determine the gravitational potential), we can use these equations to relate the rotational velocity and the velocity dispersion.

To illustrate the idea, we take the gravitational potential Φ to be given by the logarithmic form of Eq. (2.27), discussed in Subsection 2.3.4:

$$\Phi = \Phi_L(R, z) \equiv \frac{1}{2} v_0^2 \ln\left(R_c^2 + R^2 + \frac{z^2}{q_\phi^2}\right).$$

(2.138)

The mass density $(4\pi G)^{-1}\nabla^2\Phi$ corresponding to the potential in Eq. (2.138) is given by

$$\rho_L(R, z) = \left(\frac{v_0^2}{4\pi G q_\Phi^2}\right)\frac{(2q_\Phi^2 + 1)R_c^2 + R^2 + 2(1 - \frac{1}{2}q_\Phi^{-2})z^2}{(R_c^2 + R^2 + z^2 q_\Phi^{-2})^2}. \quad (2.139)$$

It is, however, convenient to approximate this density distribution in a form similar to the potential by writing

$$\rho_L \simeq \nu_L(R, z) \equiv v_0^2 R_c^2\left(R_c^2 + R^2 + \frac{z^2}{q_\nu^2}\right)^{-1}. \quad (2.140)$$

This approximation is valid if $r \gg R_c$ and $1 \gg (1 - q_\Phi) \simeq 0.3(1 - q_\nu)$. The fractional error $\delta(R, z) = [\nu_L(R, z) - \rho_L(R, z)]/v_0^2$ for $q_\nu = 0.5$; $q_\Phi = 0.85$ is only $\sim 15\%$ at worst. The advantage of this approximation lies in the fact that the necessary integral in Eq. (2.137) can be performed in closed form. We get

$$\int_z^\infty \nu_L\frac{\partial\Phi_L}{\partial z}\,dz = \frac{\frac{1}{2}v_0 R_c^2 v_0^2}{(R^2 + R_c^2)[(q_\Phi/q_\nu)^2 - 1]}\ln\left[\frac{(R^2 + R_c^2)q_\Phi^2 + z^2}{(R^2 + R_c^2)q_\nu^2 + z^2}\right], \quad (2.141)$$

which determines σ^2 everywhere. Evaluating the expression at $z = 0$ and using it in Eq. (2.137), we get the rotational velocity on the equatorial plane as

$$\bar{v}_\Phi^2(R, 0) = \frac{v_0^2 R^2}{R^2 + R_c^2}\left[1 - \frac{2\ln(q_\Phi/q_\nu)}{(q_\Phi/q_\nu)^2 - 1}\right]. \quad (2.142)$$

If ϵ_ν and ϵ_Φ are the ellipticities of isodensity and isopotential surfaces, respectively, with $q_\nu = (1 - \epsilon_\nu)$, etc., and $\epsilon_\Phi \simeq 0.3\epsilon_\nu$, we can expand $\bar{v}_\Phi^2$ in the ellipticities to obtain

$$\bar{v}_\Phi^2(R, 0) = \frac{v_0^2 R^2}{R^2 + R_c^2}[(\epsilon_\nu - \epsilon_\Phi) + \cdots,]. \quad (2.143)$$

The circular speed v_c on the equatorial plane for a particle moving in this potential is given by

$$v_c = \frac{v_0 R}{\sqrt{R_c^2 + R^2}}. \quad (2.144)$$

A comparison of Eqs. (2.143) and (2.144) now shows that

$$(\bar{v}_\Phi/v_c) \approx (\epsilon_\nu - \epsilon_\Phi)^{1/2} = (0.7\epsilon_\nu)^{1/2}. \quad (2.145)$$

Because the right-hand side varies only as $\sqrt{\epsilon_\nu}$, even a small ellipticity requires appreciable rotation speed for its support. For example, if $\epsilon_\nu = 0.1$, we require

$\bar{v}_\phi \simeq 0.26 v_c$. For this potential, we can easily show from Eq. (2.141) that $\sigma \simeq$ $(v_c/1.4)$ so that we need $\bar{v}_\phi \simeq 0.36\sigma$ for rotational support. Giant ellipticals do not have such large rotations whereas dwarf spheroidals do. This indicates that the assumption of isotropic velocity dispersion is not contradicted by this analysis for low-luminosity ellipticals whereas giant ellipticals must have anisotropic velocity dispersion.

These results can be reinforced by an analysis based on virial theorem derived in Vol. I, Chap. 8, Section 8.2. There it was shown that the following result holds in general:

$$\frac{1}{2}\frac{d^2 I_{jk}}{dt^2} = 2T_{jk} + \Pi_{jk} + W_{jk}, \tag{2.146}$$

with

$$I_{jk} \equiv \int \rho x_j x_k \, d^3\mathbf{x}, \qquad T_{jk} \equiv \frac{1}{2}\int \rho \langle v_j \rangle \langle v_k \rangle \, d^3\mathbf{x}, \qquad \Pi_{jk} \equiv \int \rho \sigma_{jk}^2 \, d^3\mathbf{x}, \tag{2.147}$$

$$W_{jk} \equiv -\int \rho(\mathbf{x}) x_j \frac{\partial \Phi}{\partial x_k} \, d^3\mathbf{x} = -\frac{1}{2}G \int\int \rho(\mathbf{x})\rho(\mathbf{x}')\frac{(x_j' - x_j)(x_k' - x_k)}{|\mathbf{x}' - \mathbf{x}|^3} \, d^3\mathbf{x}\, d^3\mathbf{x}'. \tag{2.148}$$

Let us apply this relation to an axisymmetric system like an elliptical galaxy that rotates about the symmetry axis (z axis) and is observed edge-on. Taking the line of sight to the centre of the system as the x axis, it follows from symmetry that

$$W_{xx} = W_{yy}, \qquad W_{ij} = 0, \quad (i \neq j), \tag{2.149}$$

and similarly for Π_{ij} and T_{ij}. Hence the only two nontrivial virial equations in steady state are

$$2T_{xx} + \Pi_{xx} + W_{xx} = 0, \qquad 2T_{zz} + \Pi_{zz} + W_{zz} = 0. \tag{2.150}$$

Taking the ratio of these two, we get

$$\frac{2T_{xx} + \Pi_{xx}}{2T_{zz} + \Pi_{zz}} = \frac{W_{xx}}{W_{zz}}. \tag{2.151}$$

To proceed further, we need to make a series of physically motivated assumptions regarding the various terms. If the only streaming motion is rotation about the z axis, then $T_{zz} = 0$ and $2T_{xx} = (1/2)Mv_0^2$, where M is the mass of the system and v_0^2 is the mass-weighted mean-square speed of rotation. Similarly, $\Pi_{xx} = M\sigma_0^2$, where σ_0^2 is the mass-weighted mean-square random velocity along the line of sight. Further, we relate Π_{zz} to Π_{xx} by introducing a parameter δ that measures the aniostropy of the galaxy's velocity dispersion. We take

$$\Pi_{zz} = (1 - \delta)\,\Pi_{xx} = (1 - \delta)\,M\sigma_0^2. \tag{2.152}$$

Finally, we note that if the isodensity curves are similar concentric ellipsoids, then the ratios like (W_{xx}/W_{zz}) can depend on only the ellipticity ϵ of the surfaces. This result arises from the study of potentials generated by ellipsoidal figures and can be stated in the following form.[3] If the isodensity surfaces are triaxial ellipsoids on which the coordinates (x_1, x_2, x_3) are constrained to satisfy the condition that the combination

$$m^2 \equiv a_1^2 \sum_{i=1}^{3} \frac{x_i^2}{a_i^2} \tag{2.153}$$

is a constant, then we can take the density distribution to be a function of m^2 alone; that is, $\rho(\mathbf{x}) = \rho(m^2)$. It is also convenient to define an auxillary function

$$\psi(m) \equiv \int_0^{m^2} \rho(m^2) \, dm^2 \tag{2.154}$$

in terms of which the potential-energy tensor is given by

$$W_{jk} = -\pi^2 G \frac{a_2 a_3}{a_1^2} \left(\frac{a_j}{a_1}\right)^2 A_j \delta_{jk} \int_0^\infty [\psi(\infty) - \psi(m)]^2 \, dm. \tag{2.155}$$

The quantities A_j in this expression are functions of axial ratios like a_2/a_1, etc., and the integral is independent of a body's ellipticity and is the same for all components of W_{ij}. Hence, when we take a ratio between any two components of W_{ij}, the result can depend on only the ellipticity of the body. Putting all these together, we get the relation

$$\frac{v_0^2}{\sigma_0^2} = 2(1-\delta) \frac{W_{xx}}{W_{zz}} - 2 \equiv 2(1-\delta) f(\epsilon) - 2, \tag{2.156}$$

where $f(\epsilon) \equiv (W_{xx}/W_{zz})$ can be computed if the model for the galaxy is known. This equation shows that the quantity (v_0/σ_0) depends on only ϵ and δ. It is now possible to plot curves of constant δ in the $[v_0/\sigma_0, \epsilon]$ plane. Given the observed values of these quantities, we can infer the best possible value for σ for a given sample of galaxies. Such an analysis is complicated by the fact that we need to account for the effect of inclination in the sky plane, which is not easy to do for ellipticals. Nevertheless, studies show that most of the low-luminosity ellipticals are clustered around the $\delta = 0$ curve whereas the high-luminosity massive ellipticals do not show any systematic behaviour. This is an indication that giant ellipticals probably have significant and (random) anisotropy in the velocity dispersions whereas low-mass ellipticals are isotropic in the velocity space and are rotationally supported.

For the range of ellipticities encountered in the galaxies, $(W_{xx}/W_{zz}) \equiv f(\epsilon) \approx (a/b)^{0.89}$. If we assume that all the flattening is due to rotation and velocity anisotropy is negligible ($\delta \approx 0$), Eq. (2.156) gives

$$\frac{v_0}{\sigma_0} = \sqrt{2}\left[\left(\frac{a}{b}\right)^{0.89} - 1\right]^{1/2}.\tag{2.157}$$

For $(b/a) = 0.7$, this requires $(v_0/\sigma_0) \approx 0.8$. In the other extreme case, in which rotation is negligible and the flattening is entirely due to anisotropy, we can set $T_{xx} = T_{yy} = T_{zz} = 0$ in Eq. (2.151) and take $\Pi_{xx} = M\sigma_x^2$ and $\Pi_{zz} = M\sigma_z^2$. In this case, we need $\sigma_z/\sigma_x \approx (b/a)^{0.45} \approx 0.87$ if $(b/a) = 0.7$.

2.5 Stellar Dynamics at Galactic Cores

The dynamical issues discussed so far have dealt with general properties of the galaxies. However, in most galaxies, the central region has special characteristics and is of considerable importance. We have seen in Section 1.7 that the central region of most galaxies might contain a massive black hole. The question arises as to how the black hole will influence the stellar dynamics in the galactic core.

Direct observational evidence for black holes in the centres of galaxies comes from different kinds of observations in several systems. For example, Very Long Baseline Interferometer (VLBI) observations of megamasers in the nucleus of Seyfert galaxy NGC 4258 show individual knots of material revolving around a central source with near-Keplerian velocities. The mass of the compact central object seems to be approximately $3.6 \times 10^7\ M_\odot$, confined within an inner region of 13 pc. Similar conclusions are reached from the velocity field of matter close to the centre, measured through the optical emission lines. In M87, for example, HST observations of the [OII] λ 3727 line at three different positions separated by 0.2″ (with a resolution of 0.03″) allow the measurement of rotation curve up to 5 pc near the centre. The analysis of this data shows that the observed motion is consistent with the presence of a supermassive black hole with $M \simeq (3.2 \pm 0.9) \times 10^9\,M_\odot$. High-spatial-resolution HST studies of the central regions of other galaxies have led to similar conclusions. In NGC 6251, a giant E2 galaxy, there exists an extended nuclear disk, the dynamics of which is governed by a central compact object. The mass estimates vary in the range $(4\text{–}8) \times 10^8\ M_\odot$, depending on the model used for stellar mass density. Other galaxies, such as NGC 4261, NGC 4374, and NGC 7052, all indicate the existence of central black holes with masses in the range $(2\text{–}6) \times 10^8 M_\odot$. Finally, the central masses of AGN can also be measured by reverberation-mapping techniques (see Chap. 8). The central masses estimated from this procedure can be compared with those obtained from the standard photoionisation model of the BLR; they agree within the uncertainties of the measurements.

There are several evolutionary sequences in the core of the galaxy that can lead to the formation of a central black hole. In many of them the gravitational core collapse and dynamical friction (see Vol. I, Chap. 10, Section 10.7) play an important role. As an example of how such collisional dynamics operates, let us consider the nuclear region of the spiral galaxy M31 that is at a distance of $D = 700$ kpc. At a projected radius of 1 arcsec, stars in the nuclear region have a line-of-sight velocity dispersion of $\sigma = 150$ km s^{-1} and are also rotating about the nucleus at a similar speed of $v_\phi = 150$ km s^{-1}. The analysis based on Eq. (2.156) shows that $v_{rot} = \sigma$ will imply a specific value for eccentricity ϵ. Numerically, we get $\epsilon \approx 0.5$ for $\delta = 0$, implying that $(b/a) = \sqrt{1 - \epsilon^2} = 0.85$, which is fairly spherical. Jeans equation (2.134) now reduces to

$$\frac{\partial \phi}{\partial r} = \frac{v_\phi^2}{r} - \frac{1}{\nu} \frac{\partial(\nu\sigma^2)}{\partial r}. \tag{2.158}$$

Ignoring the variation of σ^2 with r and taking the density profile in the nuclear region to fall as r^{-2} [so that $(\partial \ln \nu / \partial \ln r) = -2$], we find that this reduces to

$$\frac{\partial \phi}{\partial r} = \frac{v_\phi^2}{r} + \frac{2\sigma^2}{r}. \tag{2.159}$$

A rough estimate of mass inside a radius r is given by $M(r) \approx (r^2/G)(\partial\phi/\partial r)$. In this case, we get, using the observed values for v_ϕ and σ,

$$M \simeq \frac{r}{G} \left(v_\phi^2 + 2\sigma^2 \right) \simeq 5 \times 10^7 M_\odot. \tag{2.160}$$

The luminosity in the core region of M31 is approximately $3 \times 10^6 L_\odot$ so that the mass-to-light ratio is ~ 17. The corresponding density profile can now be taken as

$$\rho(r) = 10^5 (M_\odot \text{ pc}^{-3}) \left(\frac{3.4 \text{ pc}}{r} \right)^2. \tag{2.161}$$

The gravitational two-body relaxation time in the core region is given by (see Vol. I, Chap. 10, Section 10.7)

$$t_R = 0.34 \frac{\sigma^3}{G^2 M \rho \ln \Lambda}, \tag{2.162}$$

where $\ln \Lambda \simeq \ln N \approx 14$, $\sigma \simeq 150$ km s^{-1}, and $M = 1 M_\odot$. This gives

$$t_R \simeq 10^{10} \left(\frac{4 \times 10^5 M_\odot \text{ pc}^{-3}}{\rho} \right) \text{ yr}. \tag{2.163}$$

If the age of galaxy is taken to be 10×10^9 years, then the two-body relaxation time is of the order of the age of the galaxy at $r \lesssim 1.7$ pc. The discussion in Vol. I, Chap. 10, Section 10.7, shows that such a system will undergo core collapse in a time scale that is typically 10^2 times the relaxation time. Because the relaxation time scales as $\rho^{-1} \propto r^2$, we would expect stars in a region smaller than $(1.7 \text{ pc}/\sqrt{10^2}) = 0.17$ pc to have undergone core collapse.

Another characteristic time scale at the core is that for a massive object with mass M – say a 10 $M_\odot$ blackhole – to sink to the centre. This time scale is given by (see Vol. I, Chap. 10, Section 10.7)

$$t_{\text{sink}} \simeq \frac{1}{3 \ln \Lambda} \frac{M(<r)}{M} t_{\text{orb}} \simeq \frac{1}{3 \ln \Lambda} \frac{v^2(r)r}{GM} \frac{2\pi r}{v} \simeq \frac{2\pi}{3 \ln \Lambda} \frac{r^2 v}{GM}; \quad (2.164)$$

that is,

$$r = \left(\frac{3 \ln \Lambda}{2\pi}\right)^{1/2} \sqrt{\frac{GM t_{\text{sink}}}{v}}. \quad (2.165)$$

For $t_{\text{sink}} \simeq 10^{10}$ yr, $v(r) = \sqrt{2}\sigma = 210$ km s^{-1}, $\ln \Lambda \approx 15$, and $M = 10\ M_\odot$, this gives a radius of $\sim$4.6 pc. These considerations show that the central regions of the nuclei are indeed dominated by gravitational dynamics.

It is interesting to ask what will be the effect of such a central black hole on a stellar system that interacts with it. The effect will depend on two parameters. The first is the ratio between the mass of the black hole, M_{BH}, and the amount of stellar mass, M_*, contained in the central region ($M_* \simeq \rho_0 r_0^3$, where r_0 is of the order of the core radius, say). The second is the ratio between the age of the system and the two-body relaxation time in the central region. Of these, it is usually safe to assume that $M_{\text{BH}} \ll M_*$. As regards the time scales, it is necessary to consider the two cases separately.

Let us begin with the situation in which $t \ll t_R$. If the black hole forms by slow accumulation of mass over a time scale that is much larger than the orbital time scale (R_0/σ) of the stars, then the evolution of the distribution function can be estimated in the adiabatic approximation. We have seen in Vol. I, Chap. 2, Section 2.6 that – in the adiabatic limit – the actions J_i remain constant. Hence the distribution function can be thought of as a given function of $\mathbf{J}(E, L)$, where $\mathbf{J}$ denotes the actions as functions of energy and angular momentum. If the initial distribution function is further assumed to be a Maxwellian, then we have

$$f(\mathbf{J}) = \frac{v_0}{(2\pi\sigma^2)^{3/2}} e^{-E_i(\mathbf{J})/\sigma^2}, \quad (2.166)$$

where $E_i(\mathbf{J})$ is the energy as a function of the actions in the initial potential. As an example of the application of this result, consider the density contributed by the stars that eventually become bound to the black hole. For these stars, $E_i \ll \sigma^2$, and we can take $f(J) \approx v_0/(2\pi\sigma^2)^{3/2}$. Then the density becomes

$$v(r) = 4\pi \int_0^{\sqrt{2GM_{\text{BH}}/r}} v^2 f(E, L)\, dv = \frac{4v_0}{3\sqrt{\pi}} \left(\frac{r_H}{r}\right)^{3/2}, \quad r_H \equiv \frac{GM_{\text{BH}}}{\sigma^2}, \quad (2.167)$$

where the parameter r_H represents the radius of influence of the hole. Note that the black hole has introduced a weak cusp ($v \propto r^{-3/2}$) in the density profile.

The situation in the case of $t \gg t_R$ is more complicated. A naive application of Maxwellian distribution in the energy will give a density of $\nu \propto$
$\exp[-\Phi(r)/\sigma^2] \propto \exp(r_H/r)$, which diverges exponentially near the origin. This
result, of course, is unrealistic because any star having a close encounter with
the black hole will be tidally shredded and swallowed by the black hole. If
these processes lead to a steady-state density distribution $\nu(r) = \nu_0(r_H/r)^s$,
then equilibrium would require an energy balance between the potential energy of the stars that are swallowed by the hole and the flux of positive energy
that must flow out through the cusp. Because $\nu^2 \propto r^{-1}$, the relaxation time
will vary as $t_R \propto (\nu^3/\nu) \propto r^{(s-3/2)}$. The number of stars $N(r)$ within a radius
r scales as $N(r) \propto r^{3-s}$ whereas the energy of a star at a given radius r scales
as $E(r) \propto r^{-1}$. Hence the rate of flow of energy through a radius r scales as
$[N(r)E(r)/t_R] \propto r^{(7/2-2s)}$. If this flow has to be independent of radius r, then
we must have $s = 7/4$, which is a slightly milder cusp compared with the one
previously obtained. Numerical simulations show that this result is indeed true.

Finally, let us consider the density of stars that are _not_ bound to the black
hole. Far away from the black hole, we may assume that the test particles have a
Maxwellian velocity distribution

$$f_0(v) = \frac{\nu_0}{(2\pi\sigma^2)^{3/2}} e^{-v^2/2\sigma^2} \tag{2.168}$$

with a velocity dispersion σ^2. For unbound particles the distribution function is
of the form $f(E) \propto \exp(-E/\sigma^2)$, with $E > 0$. The number of particles that are
not bound at any given radius r is given by

$$\nu(r) = \int_{v_{esc}(r)}^{\infty} 4\pi v^2 \, dv \, f(E), \tag{2.169}$$

where $v_{esc}(r) = (2GM_{BH}/r)^{1/2}$ is the escape velocity at the location r. It is
convenient to again define a scale length $r_H = (GM_{BH}/\sigma^2)$, in terms of which
the integral is easy to evaluate, and we get

$$\nu(r) = \nu_0 \left\{ 2\sqrt{\frac{r_H}{\pi r}} + e^{r_H/r} \left[1 - erf\left(\sqrt{\frac{r_H}{\pi r}}\right) \right] \right\}, \tag{2.170}$$

where

$$erf(x) = \frac{2}{\sqrt{\pi}} \int_0^x \exp(-t^2) \, dt \tag{2.171}$$

is the error function. Using the behaviour of $erf(x)$ near the origin, we easily see
that the second term in the braces in Eq. (2.170) also behaves as $r^{-1/2}$ near the
origin so that $\nu(r) \propto r^{-1/2}$ in the vicinity of the black hole.

2.6 Spiral Structure

We next consider the spiral pattern seen in the disk galaxies, which has attracted a considerable amount of theoretical attention. Visual appearance may suggest that such a pattern arises because the material making up the spiral arm winds up because of the differential rotation of the galaxy. As we shall subsequently see, this cannot be a possible explanation as this will lead to a very tightly wound spiral pattern in many cases. A more likely explanation is that a spiral structure is a quasi-stationary density wave in which a particular wave pattern stays in place even though the actual material forming the wave pattern does not reside in the location of the spiral. Such an idea is related to the stability of the differentially rotating disk, and we will see that the problem is closely connected with the resonances described earlier in Subsection 2.3.5.

Let us first show that the actual winding up of the material that is due to differential rotation, if that were the cause for the spiral pattern, is counterproductive. To see this, we consider the equation describing the location of a set of stars that were along a straight line $\phi = \phi_0$ at $t = 0$. At any given later time, the location of this strip of stars is given by $\phi(R, t) = \phi_0 + \Omega(R)t$ if they rotate differentially. The pitch angle i of the arm at any given radius is defined as the angle between the tangent to the arm and a circle of constant radius (see Fig. 2.8) so that $\cot i \equiv |R(\partial \phi / \partial R)|$. For our case, this gives $\cot i = Rt|(d\Omega/dR)|$. If ΔR is the separation between adjacent arms at a given angle ϕ, then we must have

$$2\pi = |\Omega(R + \Delta R) - \Omega(R)|t \qquad (2.172)$$

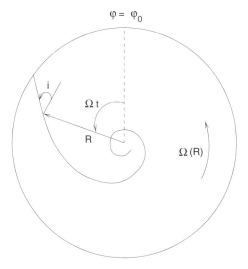

Fig. 2.8. The winding problem and the pitch angle.

so that, for $\Delta R \ll R$,

$$\Delta R = \frac{2\pi R}{\cot i}. \tag{2.173}$$

For a typical galaxy, with $\Omega R \simeq 220\,\mathrm{km\,s^{-1}}$, $R \simeq 10\,\mathrm{kpc}$, and $t \simeq 10^{10}\,\mathrm{yr}$, we get $i = 0.25°$ and $\Delta R \approx 0.28\,\mathrm{kpc}$. The actual observed separations are considerably larger; in other words, the spiral is not as tightly wound as suggested by this analysis.

It is possible, however, to come up with a density enhancement of stars along a spiral without leading to overwinding. To understand this idea in its simplest context, let us consider the orbit of stars in the epicycle approximation described in Section 2.3. If we concentrate on a specific guiding centre orbit at some radius R_0, it is easy to see that, in general, the orbit will not be closed in the inertial frame because the angular velocity $\Omega(R_0)$ and the epicyclic frequency $\kappa(R_0)$ will not be commensurate. On the other hand, if we view the orbit in a suitably chosen rotating frame (rotating with an angular pattern speed Ω_p), then the orbit can be closed provided that $\Omega' = (\Omega - \Omega_p)$ and κ are commensurate. We now note that, for a reasonable class of density profiles, the quantity $[\Omega(R) - \kappa(R)/2]$ is fairly constant at all R (see, e.g., Fig. 2.7). This suggests that if we take the pattern speed to be $\Omega_p = (\Omega - \kappa/2)$ so that $\Omega' = \kappa/2$, all the orbits will close. If we now arrange the closed orbits (which will in general be ellipses) with the position angle of the major axis increasing systematically, then stars will be concentrated along a spiral pattern (see Fig. 2.9). What is more, this pattern will persist in time as long as the mututal interaction of the stars is ignored. Note that stars are not winding up in this picture and the enhancement along the spiral is, in

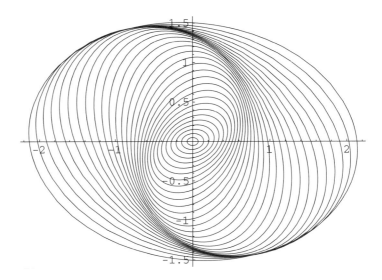

Fig. 2.9.   Kinematic spiral wave generated by nested ellipses.

some sense, kinematical. In the nonrotating, inertial frame, the spiral pattern will rotate at the pattern speed.

A more sophisticated and realistic approach to a spiral structure treats it as a density wave related to an instability in the rotating disk. To be rigorous, such an analysis needs to be carried out by treating the contents of the disk as a collisionless system of stars. It turns out, however, that somewhat similar results are obtained even if the material in the disk is treated as gaseous. Because the latter analysis is mathematically simpler, we shall describe it here and briefly mention the corresponding results for a stellar system.[3]

This analysis has three ingredients. First, we assume that there exists a surface-density wave capable of describing a spiral and we determine its gravitational potential by using Poisson's equation. Next, we determine how the motion of stars is affected by this gravitational potential. Finally, we match the modified orbital pattern of the stars to the initially assumed spiral-density wave, thereby obtaining a self-consistent model. The key assumption we shall use is that the response of the gravitational field to the perturbed density field is local and that the long-range effects are negligible. This approximation is reasonable, provided the spiral is not very loosely wound, and we need to confirm that this is indeed the case, at the end of the calculation, for consistency.

We begin by taking the surface density of a (zero-thickness) disk as the sum of two terms: $\Sigma_0(R)$ represents the unperturbed surface density and $\Sigma_1(R, \phi, t)$ represents the perturbation, where

$$\Sigma_1(R, \phi, t) = H(R, t)e^{i[m\phi + f(R,t)]}. \tag{2.174}$$

In this expression, $H(R, t)$ is a smooth function of radius that gives the amplitude of the spiral pattern. The location of m spiral arms ($m = 2, 3, \ldots,$) is defined by the relation

$$m\phi + f(R, t) = \text{ constant.} \tag{2.175}$$

If we introduce a radial wave number by

$$k(R, t) \equiv \frac{\partial f(R, t)}{\partial R}, \tag{2.176}$$

then the sign of k determines whether the arms are leading or trailing. If the galaxy rotates in the direction of increasing ϕ, then $k < 0$ implies leading arms, whereas $k > 0$ implies trailing arms. The pitch angle is now given by the relation $\cot i = |kR/m|$ and the condition for tight a spiral is that $\cot i \gg 1$. Observations show that the average pitch angle varies with the galaxy type and the median pitch angle is $\langle i \rangle \simeq 13°$, giving $\langle \cot i \rangle \approx 4$. This shows that the tight-coupling approximation is satisfied only marginally and hence the agreement of the results derived below to actual observed features is somewhat fortuitous.

We now need to determine the gravitational potential that is due to this surface-density distribution. In general this can be obtained only as an integral involving a two-dimensional Green's function to the Poisson equation. However, if the tight-winding approximation is true, then the sinusoidal oscillations of the surface density lead to a near-complete cancellation of the contribution from distant parts of the pattern to the local potential and the perturbed potential at a given location is essentially determined by the local density. In this case, we expand $f(R, t)$ in a Taylor series around R_0 to get

$$\Sigma_1(R, \phi, t) \simeq \Sigma_a e^{ik(R-R_0)}, \qquad \Sigma_a = H(R_0, t)e^{i[m\phi_0 + f(R_0, t)]}, \qquad k = k(R_0, t).$$
(2.177)

Because the radial dependence is also sinusoidal, the gravitational potential is trivial to determine in Fourier space. We get

$$\Phi_1(R, \phi, t) \simeq \Phi_a e^{ik(R-R_0)}, \qquad \Phi_a = -\frac{2\pi G \Sigma_a}{|k|}.$$
(2.178)

We are now free to set $R_0 = R$, thereby obtaining the gravitational potential at any location as

$$\Phi_1(R, \phi, t) = -\frac{2\pi G}{|k|} H(R, t)e^{i[m\phi + f(R, t)]} + \mathcal{O}(|kR|^{-1}).$$
(2.179)

We can obtain the corresponding inverse relation by differentiating relation (2.179) with respect to R and ignoring the slow variation of H:

$$\Sigma_1(R, \phi, t) = \frac{i \, \text{sign}(k)}{2\pi G} \frac{d}{dR} \Phi_1(R, \phi, t) + \mathcal{O}(|kR|^{-1}).$$
(2.180)

We next determine the response of the matter in the disk to this perturbed gravitational potential. We will assume that the matter is confined to the $z = 0$ plane and that the pressure p that acts on the disk is related to the density by an equation of state $p = K\Sigma^\gamma$. The corresponding specific enthalpy is given by $h = [\gamma/(\gamma - 1)]K\Sigma^{\gamma-1}$ (see Vol. I, Chap. 5, Section 5.6). The Euler equations, in component form, are given by

$$\frac{\partial v_R}{\partial t} + v_R \frac{\partial v_R}{\partial R} + \frac{v_\phi}{R} \frac{\partial v_R}{\partial \phi} - \frac{v_\phi^2}{R} = -\frac{\partial \Phi}{\partial R} - \frac{1}{\Sigma} \frac{\partial p}{\partial R},$$
(2.181)

$$\frac{\partial v_\phi}{\partial t} + v_R \frac{\partial v_\phi}{\partial R} + \frac{v_\phi}{R} \frac{\partial v_\phi}{\partial \phi} - \frac{v_\phi v_R}{R} = -\frac{1}{R} \frac{\partial \Phi}{\partial \phi} - \frac{1}{\Sigma R} \frac{\partial p}{\partial \phi}.$$
(2.182)

We can simplify the right-hand sides of these equations by using the specific enthalpy as we have

$$-\frac{\partial \Phi}{\partial R} - \frac{1}{\Sigma} \frac{\partial p}{\partial R} = -\frac{\partial \Phi}{\partial R} - \gamma K\Sigma^{\gamma-2} \frac{\partial \Sigma}{\partial R} = -\frac{\partial(\Phi + h)}{\partial R}.$$
(2.183)

We now solve Eqs. (2.181) and (2.182) in linear perturbation theory by taking the variables to be $v_R = v_{R1}$, $v_\phi = v_{\phi 0} + v_{\phi 1}$, $h = h_0 + h_1$, and $\Phi = \Phi_0 + \Phi_1$, where the subscript '1' denotes the first-order perturbations. In the lowest order, we assume the orbits to be circular with

$$\frac{v_{\phi 0}^2}{R} = \frac{d}{dR}(\Phi_0 + h_0) \simeq \frac{d\Phi_0}{dR} \tag{2.184}$$

where the second relation is valid because the interstellar sound speed ($v_s \approx$ 10 km s^{-1}) is negligible compared with the circular speed ($v_c \approx$ 220 km s^{-1}); the derivative of h_0 is smaller than the term retained by the factor $(v_s/v_c)^2$. The lowest-order perturbed equations now become

$$\frac{\partial v_{R1}}{\partial t} + \Omega \frac{\partial v_{R1}}{\partial \phi} - 2\Omega v_{\phi 1} = -\frac{\partial}{\partial R}(\Phi_1 + h_1), \tag{2.185}$$

$$\frac{\partial v_{\phi 1}}{\partial t} + \left[\frac{d(\Omega R)}{dR} + \Omega\right] v_{R1} + \Omega \frac{\partial v_{\phi 1}}{\partial \phi} = -\frac{1}{R}\frac{\partial}{\partial \phi}(\Phi_1 + h_1). \tag{2.186}$$

We can solve these equations by assuming that all the perturbed variables have dependences of the form $q_{a1}(R, \phi, t) = q_a(R)\exp[i(m\phi - \omega t)]$. Making these substitutions, we can solve for the velocity and enthalpy perturbations in terms of other quantities and obtain

$$v_{Ra} = -\frac{i}{\Delta}\left[(m\Omega - \omega)\frac{d}{dR}(\Phi_a + h_a) + \frac{2m\Omega}{R}(\Phi_a + h_a)\right], \tag{2.187}$$

$$v_{\phi a} = \frac{1}{\Delta}\left[-2B\frac{d}{dR}(\Phi_a + h_a) + \frac{m(m\Omega - \omega)}{R}(\Phi_a + h_a)\right], \tag{2.188}$$

$$h_a = \gamma K \Sigma_0^{\gamma-2}\Sigma_a \equiv \frac{v_s^2 \Sigma_a}{\Sigma_0}, \tag{2.189}$$

$$\Delta \equiv \kappa^2 - (m\Omega - \omega)^2, \tag{2.190}$$

with

$$\kappa^2(R) = R\frac{d\Omega^2}{dR} + 4\Omega^2 = -4B\Omega,$$

$$B(R) = -\frac{1}{2}\left[\frac{d(\Omega R)}{dR} + \Omega\right] = -\Omega - \frac{R}{2}\frac{d\Omega}{dR}. \tag{2.191}$$

So far we have considered only the Euler equations that govern the velocity response of the matter to the perturbed potential. Velocity perturbation induces a density perturbation through the continuity equation, which – to the lowest order – becomes

$$\frac{\partial \Sigma_1}{\partial t} + \frac{1}{R}\frac{\partial}{\partial R}(R\Sigma_0 v_{R1}) + \Omega\frac{\partial \Sigma_1}{\partial \phi} + \frac{\Sigma_0}{R}\frac{\partial v_{\phi 1}}{\partial \phi} = 0 \tag{2.192}$$

or, equivalently,

$$i(m\Omega - \omega)\Sigma_a + \frac{1}{R}\frac{d}{dR}(R\Sigma_0 v_{Ra}) + \frac{im\Sigma_0}{R}v_{\phi a} = 0. \tag{2.193}$$

Equations (2.187), (2.188), (2.189), and (2.193) provide four equations for the five variables (Σ_a, v_{Ra}, $v_{\phi a}$, h_a, Φ_a). The last relation needed to close the system is the Poisson equation connecting the potential to the density, which has already been solved in Eq. (2.179). This relation shows that Σ_a and – because of Eq. (2.189) – h_a contain exactly the same factor $\exp[i f(R)]$ as the potential Φ_a. Because this factor varies rapidly with R, we can drop terms involving the derivative with respect to R in Eqs. (2.187) and (2.188); we also write $d(\Phi_a + h_a)/dR \cong ik(\Phi_a + h_a)$ and $d(R\Sigma_0 v_{Ra})/dR \cong ik(R\Sigma_0 v_{Ra})$ as all these replacements involve only a fractional error of $\mathcal{O}(|kR|^{-1})$. This gives

$$v_{Ra} = \frac{(m\Omega - \omega)k(\Phi_a + h_a)}{\Delta}, \qquad v_{\phi a} = -\frac{2Bik(\Phi_a + h_a)}{\Delta}, \quad (2.194)$$

$$(m\Omega - \omega)\Sigma_a + k\Sigma_0 v_{Ra} = 0. \qquad (2.195)$$

Simple algebraic manipulation now leads to the condition

$$\Sigma_a \left(1 + \frac{k^2 v_s^2}{\Delta} - \frac{2\pi G \Sigma_0 |k|}{\Delta} \right) = 0, \qquad (2.196)$$

which provides the dispersion relation connecting ω and k:

$$(m\Omega - \omega)^2 = \kappa^2 - 2\pi G\Sigma|k| + k^2 v_s^2. \qquad (2.197)$$

We now consider the implications of this dispersion relation, beginning with the simple case of $m = 0$. This dispersion relation for $m = 0$ contains, as a further special case, the rigidly rotating fluid disk for which $\kappa = 2\Omega$. In this case, the dispersion relation becomes

$$\omega^2 = 4\Omega^2 - 2\pi G\Sigma_0|k| + k^2 v_s^2. \qquad (2.198)$$

The stability requires $\omega^2 \geq 0$, and – if there is no rotation, $\Omega = 0$ – then all the modes with $|k| < k_J \equiv (2\pi G\Sigma_0/v_s^2)$ are unstable. This is the standard Jeans instability (which has been discussed in Vol. I, Chap. 8, Section 8.5) but applied to the two-dimensional case. One important difference between Jeans instability in three-dimensional homogeneous media and a two-dimensional disk is the following. In the three-dimensional case, in the absence of pressure support ($v_s = 0$) and rotation ($\Omega = 0$), the instability grows as $\exp(\gamma t)$, where $\gamma^2 = 4\pi G\rho_0$, which is independent of the wavelength of the perturbation. In a two-dimensional disk, without pressure support and rotation, the corresponding growth rate is $\gamma^2 = 2\pi G\Sigma_0|k|$, which increases for small wavelengths.

This instability cannot be suppressed by rotation. In fact, when the sound speed is zero, all the modes with $|k| > (2\Omega^2/\pi G\Sigma_0)$ are unstable with a diverging growth rate at short wavelengths. It is interesting to observe that neither rotation nor pressure, by itself, can stabilise a sheet; a rotating sheet without pressure support is unstable at small wavelengths whereas a nonrotating sheet with pressure support is unstable at large wavelengths. Interestingly enough, the existence of *both* rotation and pressure can stabilise the sheet. When both effects are present, an examination of quadratic equation (2.198) shows that the sheet is stable at all wavelengths if the following condition is satisfied:

$$\frac{v_s \Omega}{G \Sigma_0} \geq \frac{\pi}{2} \cong 1.57. \tag{2.199}$$

Let us now return to the disk rotating differentially so that $\Omega(R)$ depends on the radius but still keep $m = 0$. The marginal stability determined by $\omega^2 = 0$ leads to the quadratic equation

$$\kappa^2 - 2\pi G \Sigma |k| + k^2 v_s^2 = 0. \tag{2.200}$$

It is easy to see that stability now requires that

$$Q \equiv \frac{v_s \kappa}{\pi G \Sigma} > 1, \tag{2.201}$$

which is similar to condition (2.199) previously obtained. The region of stability can be more conveniently expressed in terms of Q and a critical wavelength defined by $\lambda_{\text{crit}} \equiv (4\pi^2 G \Sigma / \kappa^2)$. The region of marginal stability has the equation $Q = 2[\zeta(1 - \zeta)]^{1/2}$, where $\zeta = (\lambda / \lambda_{\text{crit}})$ (see Fig. 2.10). In a solar neighbourhood, $\kappa \approx 32$ km s^{-1} kpc^{-1}, $\Sigma \approx 60\ M_\odot$ pc^{-2}, and $v_s = 30$ km s^{-1}, giving $Q \approx 1.2$. Observationally, therefore, we have no direct evidence for

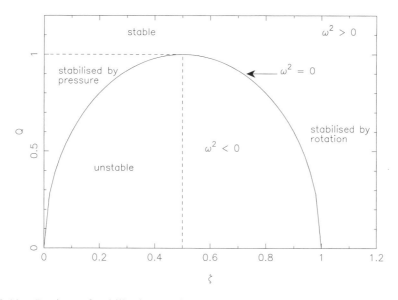

Fig. 2.10.   Regions of stability in an axisymmetric differentially rotating gaseous disk.

the occurrence of local instability in the solar neighbourhood, although the
uncertainties in the observations do not allow us to firmly rule out values of
Q equal to or marginally lower than unity.

   To analyse the case with $m \neq 0$, which can lead to spiral patterns with a pattern
speed of $\Omega_p \equiv (\omega/m)$, it is convenient to define a dimensionless frequency,
$s = [m(\Omega_p - \Omega)/\kappa]$, such that corotation resonance is at $s = 0$ and the Lindblad
resonances are at $s = \pm 1$. The actual radii at which these resonances occur, of
course, will depend on the functional form of $s(R)$. {For example, in a Mestel
disk, $s = (m/\sqrt{2})[(R/R_{CR}) - 1]$, where R_{CR} is the corotation radius and the
Lindblad resonances are at $(R/R_{CR}) = 1 \pm (\sqrt{2}/m)$.} The dispersion relation
provides a general curve connecting s to (k/k_{crit}) [where $k_{\mathrm{crit}} = (2\pi/\lambda_{\mathrm{crit}})$] for a
given value of Q:

$$\frac{|k|}{k_{\mathrm{crit}}} = \frac{2}{Q^2}\{1 \pm [1 - Q^2(1 - s^2)]^{1/2}\}. \tag{2.202}$$

The forms of this curve for different values of Q are shown in Fig. 2.11. It is
obvious that the dispersion relation is the same for leading $(k < 0)$ and trailing
$(k > 0)$ waves. Hence, for a given value of wave frequency ω and disk location
R, there can exist four kinds of waves: short trailing [ST; $k > 0$ with a plus sign
in Eq. (2.202)]; long trailing [LT; $k > 0$ with a minus sign in Eq. (2.202)]; short
leading [SL; $k < 0$ with a plus sign in Eq. (2.202)]; and long leading [LL; $k < 0$
with a minus sign in Eq. (2.202)]. All the long waves become infinitely long
at the outer and inner Lindblad resonances corresponding to $s = \pm 1$. At these
turning points, the character of the wave changes from sinusoidal to exponentially

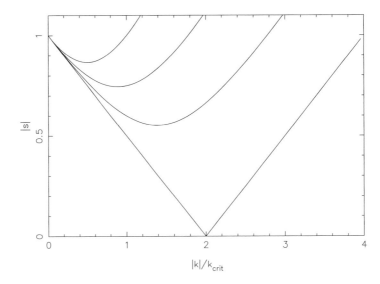

Fig. 2.11.   The dispersion relation in Eq. (2.202) is plotted for different values of Q.
The Q values (from bottom to top) are 1, 1.2, 1.5, and 2.

damped. Hence in galaxy models in which $\Omega \pm (\kappa/m)$ vary monotonically with R, long waves can propagate only in the region between the Lindblad resonances. When $Q > 1$, there is also a forbidden region around the corotation resonance in which the waves are exponentially damped. This is easy to see from the fact that the term within the square root in Eq. (2.202) vanishes when $s^2 = 1 - Q^{-2}$. For $Q \gg 1$, this condition is satisfied at the Lindblad resonances ($s = \pm 1$); if we lower Q throughout the disk, this point moves towards the corotation resonance ($s = 0$). For $Q \gg 1$, therefore, the long waves are excluded from the entire region between the Lindblad resonances, whereas for $Q \gtrsim 1$ they can propagate everywhere except in a narrow region surrounding the corotation region.

In summary, it may be stated that the following aspects of the dispersion relation are fairly generic: (1) In all disks, with $Q > 1$, there is a forbidden region around the corotation radius, and (2) between the corotation ($s = 0$) and the Lindblad resonances ($s = \pm 1$), there are two branches of the dispersion relation. The long wave begins at $k = 0$, $s = 1$, with $|k|$ increasing as s decreases; near $s = \pm 1$, the dispersion relation is independent of Q and varies as $|k| = k_{\rm crit}(1 - s^2)$. The short-wave branch has the limit $|k| \to (2\pi G\Sigma/v_s^2)$ as $|s| \to 1$ and $|k|$ decreases as s decreases.

The entire preceding discussion, including the stability analysis, used a gaseous disk rather than the distribution function for stars in the disk. It turns out, however, that the basic results are the same even for a collisionless set of particles. The stability criterion is modified only slightly and reads as

$$Q \equiv \frac{\sigma_R \kappa}{3.36\,G\,\Sigma} > 1 \qquad \text{(for stars)} \qquad (2.203)$$

for a stellar-dynamical disk, where σ_R is the velocity dispersion. The dispersion relation near Lindblad resonances is the same for both gaseous and stellar disks whereas the short-wave branch has $|k| \to \infty$ as $|s| \to 1$ in the stellar disk.

Given the dispersion relation for a given mode, we can construct propagating wave packets by adding different wave numbers with the propagation determined by the dispersion relation. For such waves, physical quantities like energy, angular momentum, etc., are propagated at the group velocity given by

$$v_g = \frac{\partial \omega(k, R)}{\partial k} = \text{sign}(k)\frac{|k|v_s^2 - \pi G\Sigma}{\omega - m\Omega}. \qquad (2.204)$$

In this equation, $v_s(R)$, $\Sigma(R)$, and $\Omega(R)$ are determined by the properties of the unperturbed disk and k is determined as a function of ω through the dispersion relation. As the wave propagates, k changes, leaving ω fixed. It is obvious that a wave packet localised around R will propagate outwards if $v_g(R) > 0$ and inwards if $v_g(R) < 0$. As an example, let us consider the wave propagation in a Mestel disk with $Q = 1.5 = \text{constant}$. Figure 2.12 shows the dispersion relation for $m = 2$ in the $R-k$ space. At point A, a packet of leading waves on the short

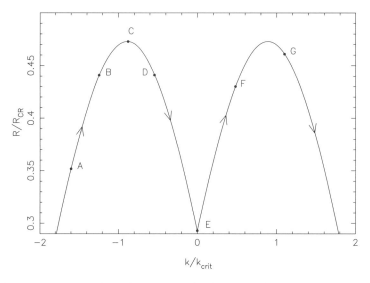

Fig. 2.12.   Propagation of a wave packet in a Mestel disk with $Q = 1.5$.

branch has positive group velocity and the wave packet propagates outwards
(towards increasing R) with $|k|$ decreasing. Eventually the wave reaches the
edge of forbidden region (at point C) where the group velocity changes sign,
and it starts propagating inwards as a leading wave on the long branch. This
change of direction is completely analogous to the reflection of wave packets by
potential barriers in quantum mechanics. Eventually the wave will reach the inner
Lindblad resonance (E) at which it reflects off the resonance and propagates out
as a trailing wave (F). (The WKB theory is suspect at the turning point; however,
a more rigorous analysis confirms this result.) It is reflected at the forbidden zone
again and propagates inwards as a ST branch wave.

   This entire process causes a tightly wound leading wave to first become a
loosely wound leading wave, then a loosely wound trailing wave, and finally a
trailing wave that becomes more and more tightly wound. This is reminiscent
of the winding up of material arms, and the rate of winding can be worked out
with the group velocity and dispersion relation. It is easy to show that the rate
of change of pitch angle i of a trailing-wave packet is given by

$$\frac{d}{dt}(\cot i) = \left.\frac{\partial \cot i}{\partial R}\right|_{\omega} v_g, \qquad (2.205)$$

which, for a Mestel disk with $\kappa \propto R^{-1}$, $\Omega \propto R^{-1}$, $\Sigma \propto R^{-1}$, and $v_s = \text{constant}$,
will give

$$\frac{d}{dt}(\cot i) = \frac{\omega}{m} = \Omega_p. \qquad (2.206)$$

In comparison, the material arms will wind up at a rate

$$\frac{d}{dt}(\cot i) = \left| R \frac{d\Omega}{dR} \right| = \Omega, \tag{2.207}$$

as $\Omega \propto R^{-1}$ for the Mestel disk. Clearly waves outside the corotation wind up faster whereas waves inside the corotation wind up slower.

The net result of the preceding analysis therefore is that – at the level of linear perturbation theory based on the WKB approximation – any tightly wound disturbance will wind up fast and disappear into the Lindblad resonance within a few rotation times. Hence linear analysis based on the WKB approximation *alone* cannot sustain a spiral pattern for a very long time. To understand how these patterns are sustained, we need to do a more careful analysis of what happens to the waves near the turning points. Such a study suggests that there are mechanisms that could lead to amplification of the wave, thereby possibly providing a consistent picture. We shall now briefly describe some of the ideas in this direction.

The travelling-wave packets previously described carry energy and angular momentum with a speed equal to the group velocity. Let us consider what happens to a LT wave carrying (-1) unit of angular momentum towards the corotation radius from the inner side. If $Q > 1$, then a ST wave carrying angular momentum ϵ will be transmitted and a wave with angular momentum $-(1 + \epsilon)$ will be refracted back. (Similar results hold for energy as well, and the process conserves energy and angular momentum at the reflection point.) Such a result – which is completely analogous to transmission and reflection of wave packets in quantum mechanics – shows that we can achieve a multiplicative factor of $(1 + \epsilon)$ in energy and angular momentum. If there is an inner Q barrier associated with, say, a galactic bulge, then the inwardly propagating ST wave will be reflected back into an outwardly propagating LT spiral wave. This wave will now undergo transmission and reflection at the corotation radius, with the transmitted ST wave carrying an angular momentum $\epsilon(1 + \epsilon)$ while the ST reflected wave carries an amount $(1 + \epsilon)^2$. This can now result in a feedback cycle, and the process can reinforce the original incident wave, provided an eigenvalue condition is satisfied. Thus we can arrive at a growing normal mode that amplifies by a factor $(1 + \epsilon)$ per feedback cycle (see Fig. 2.13).

There is another process that can lead to similar effects if the Q barrier rises sufficiently fast compared with the wavelength near the corrotation radius.[4] In that case, a SL wave incident upon the corotation radius from inside can lead not only to the refraction of a LL wave but also to the reflection of a ST wave. The latter (which arises through a nonadiabatic process and cannot be discussed in the WKB approximation) can lead to the transmission of $X(\gg 1)$ units of angular amplitude into the outer region and an amount $-(1 + X)$ units inwards in the

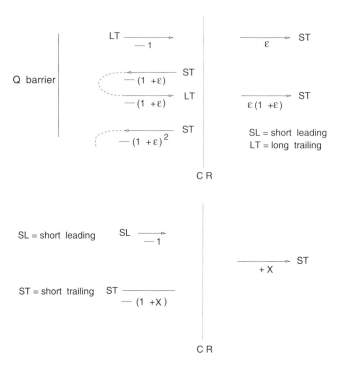

Fig. 2.13.   Feedback mechanisms capable of inducing spiral instability.

form of a ST wave. This process, called *swing amplification*, can often dominate the generation of a large-amplitude fluctuation. The physical reason for the swing amplification is a resonance between the epicyclic motion and the rotating spiral feature. To see this, we note that, for a material spiral arm with $\phi(R, t) = \phi_0 + \Omega(R)t$, the winding angle satisfies the relation $\cot i = -Rt(d\Omega/dR) \equiv 2At$. This will give the rate of change of pitch angle as

$$\frac{di}{dt} = \frac{2A}{1 + 4A^2t^2}.$$ (2.208)

For a tightly wound arm, (di/dt) is slow, but as the arm swings from leading to trailing branch, it reaches a maximum rotation rate of $2A$. This maximum rate is comparable with the average angular speed κ of stars around their epicycles; for example, $2A = \Omega$ and $\kappa = \sqrt{2}\Omega$ for a Mestel disk. Further, both the unwinding of the arm and the rotation of the stars around their epicycles are in the same sense (opposite to the direction of rotation) leading to a temporary resonance between the epicyclic motion and the rotating spiral feature. This enhances the effect of gravitational force from the spiral on the stellar orbit and could lead to rapid growth of the amplitude representing the spiral arm.

2.7 Warps

Disk galaxies, when observed edge on in the visible band, show a reasonably flat planar structure. The HI observations – which extend to much larger radii than the optical observations – however, often show noticeable warps at the edges. These are most easily seen in the external galaxies that are seen edge-on; for example, almost all the spiral galaxies in the local group are warped.

One possible explanation for the warps is the following. If the disk is embedded in a halo such that the symmetry axes of the disk and the halo are misaligned, then gaseous material in the galactic disk will settle at a configuration in which the net torque from the halo and the disk on each ring of material vanishes. At each radius, there is a unique plane (which in general is different at different radii) on which the net torque vanishes. This makes the gas reach a nonplanar configuration.

The determination of this surface is straightforward if both the disk and the halo are assumed to be axisymmetric and some simple gravitational potential is chosen for the halo. We choose the geometry such that the symmetry axis of the disk is along the z axis and the intersection of the equatorial planes of the disk and the halo is along the y axis. The inclination between the two symmetry axis will be taken to be v (see Fig. 2.14, left). We take the halo potential to be a simple quadratic in the inclined coordinates given by the form

$$\Phi_h = \frac{v_0^2}{2R_c^2}\left(x'^2 + y^2 + \frac{z'^2}{q^2}\right) + \text{constant}, \tag{2.209}$$

where q is a free parameter and $x' = x\cos v + z\sin v$ and $z' = z\cos v - x\sin v$. We now consider a ring of radius r_0 that has an inclination i and whose symmetry axis (like that of the halo) is in the x–z plane. Expressing Φ_h in terms of x, y, z and taking the gradients, we can calculate the components of the acceleration $\mathbf{g}_h$ that is due to the halo as a function of $\mathbf{x}$. We get

$$(g_{hx}, g_{hy}, g_{hz}) = \frac{v_0^2}{R_h^2}(x, y, z)$$

$$- \frac{v_0^2(q^{-2} - 1)(x\sin v - z\cos v)}{R_h^2}(\sin v, 0, \cos v). \tag{2.210}$$

This will lead to a torque on the material at the angular position between ϕ and $d\phi$ in this ring, given by

$$d\mathbf{N}_h = \frac{d\phi}{2\pi}\mathbf{x}(\phi) \times \mathbf{g}_h[\mathbf{x}(\phi)], \tag{2.211}$$

where

$$\mathbf{x}(\phi) = (r\cos\phi\cos i, r\sin\phi, r\cos\phi\sin i). \tag{2.212}$$

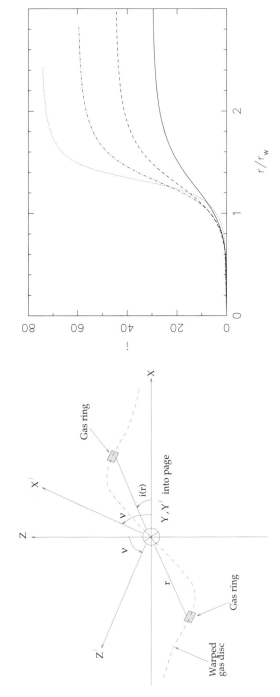

Fig. 2.14. The geometry determining the structure of the warp. The solid, dashed, dotted–dashed, and dotted curves in the right panel are for $\nu = 30°$, $45°$, $60°$, $75°$, respectively.

138

By symmetry only the y component of the torque will survive. Its total magnitude can be calculated by integration over all ϕ to give

$$N_{\text{halo}} = \frac{v_0^2 r_0^2}{4R_c^2} \left(\frac{1}{q^2} - 1\right) \sin 2(i - v). \tag{2.213}$$

We next need to estimate the force that is due to the disk. If the typical size of the optical disk is $r_{\max}$ then, at distances $r \gg r_{\max}$, the potential that is due to the disk can be expressed in the form (see exercise 2.7)

$$\Phi_D(R, z) = -\frac{GM_D}{\sqrt{R^2 + z^2}} - \frac{G}{4} \frac{(R^2 - 2z^2)}{(R^2 + z^2)^{5/2}} \int_0^\infty 2\pi R'^3 \Sigma(R') dR'. \tag{2.214}$$

Given this potential, we can compute the force and the torque that are due to the disk on the ring of material. This is also directed along the y axis and has a magnitude

$$N_{\text{disk}} = \frac{3G}{8r_0^2} \left[\int_0^\infty 2\pi R'^3 \Sigma(R') dR' \right] \sin 2i. \tag{2.215}$$

The equilibrium shape of the gas is decided by the surface on which $N_{\text{halo}} + N_{\text{disk}} = 0$. For an exponential disk with $\Sigma(R) = \Sigma_0 \exp(-R/R_d)$ and total mass $M_D = 2\pi \Sigma_0 R_d^2$, this equation leads to the condition

$$\frac{v_0^2 r^5}{9GM_D R_c^2 R_d^2} \left(\frac{1}{q^2} - 1\right) \equiv \left(\frac{r}{r_w}\right)^5 \left(\frac{1}{q^2} - 1\right) = \frac{\sin 2i}{\sin 2(v - i)}. \tag{2.216}$$

The right-hand side of Fig. 2.14 shows the shape of this surface for $v = 30°$, $45°$, $60°$, $75°$, $q = 0.9$. The x axis is in units of

$$r_w = \left(\frac{9GM_D R_c^2 R_d^2}{v_0^2}\right)^{1/5} q^{2/5} (1 - q^2)^{-1/5}, \tag{2.217}$$

which defines the characteristic scale for the warp. We note that for $r \ll r_w$, the equilibrium surface coincides with plane of the *disk*, whereas beyond $r = 2r_w$, it shifts to the equatorial plane of the *halo*.

Exercise 2.11
Filling in blanks: Prove Eqs. (2.213) and (2.215).

2.8 Chemical Evolution of Galaxies

The discussion so far treated galaxies as a chemically homogeneous medium interacting through different dynamical processes. Because stellar evolution leads to the synthesis of chemical elements that are distributed within the galaxy through supernova bursts, it is obvious that the chemical content of the galaxy will vary as stars evolve. This chemical evolution of a galaxy is a complex process because of the coupling between the stars and the ISM and the time dependence of the phenomena arising from stellar and galactic evolutions. At a fundamental level, this problem needs to be addressed in terms of local conservation laws for mass, momentum, and energy for the ISM and stars.[5] For example, the equation for mass conservation will read as

$$\frac{\partial \rho}{\partial t} + \nabla \cdot \rho \mathbf{v} = s_{mass}, \tag{2.218}$$

where the right-hand side takes into account conversion of interstellar gas into the stars as well as the addition of mass into the ISM from the stars through stellar winds and supernova explosions. We could write similar equations for conservation of energy and momentum. Given the details of stellar-evolution processes as well as the coupling between stars and the ISM, we can in principle solve these equations numerically, thereby obtaining the spatial and temporal variations of different properties (for example, the metallicity) in a galaxy. This procedure unfortunately is not practical because of mathematical complexity as well as the uncertainties in the input physics. Even if we succeeded in solving the detailed equations, it would be difficult to compare the results with observations at the same level of detail. Hence chemical evolution of the galaxy is usually studied in a coarse-grained manner in terms of parameters defined globally over a galaxy.

Some of the key limitations of this coarse-grained approach can be easily seen from the following argument. To ignore the spatial variations of physical processes (say, evolution of metallicity), we need to assume that homogenisation takes place at a time scale much shorter than the evolution time scale we are interested in. The relevant time scale to mix the constituents over 1 kpc, say, by physical processess operating in the ISM (with characteristic speeds of ~ 10 km s^{-1}) will be $\sim 10^8$ yr, which is comparable with the evolution time scale we are interested in. Similar estimates can be made for time scales for momentum and energy transfer as well. An object of mass m moving through interstellar gas of density ρ_{ISM} with a velocity v will accrete mass at the rate $\dot{M} = \rho_{ISM} v (\pi r_G^2)$, where $r_G = 2Gm/v^2$ is the radius over which the gravitational influence of the mass would be felt in the ISM. This gives

$$\frac{dM}{dt} = \rho_{ISM} \frac{4\pi (GM)^2}{v^3} \approx 5 \times 10^{11} n \text{ gm s}^{-1} \approx 8 \times 10^{-15} n \ M_\odot \text{ yr}^{-1}. \tag{2.219}$$

The characteristic time scale is given by $\tau_{acc} = m/\dot{M}$. This is very large ($\gtrsim 10^{14}$ yr) for a stellar-mass object moving through an ISM with $n \simeq 1$ cm^{-3}. However, in a giant molecular cloud (GMC), with $n = 10^4$ cm^{-3} and $v = 5 \times 10^4$ cm s^{-1}, we get $\tau_{acc} \simeq 10^6$ yr. Thus mass accretion could be important in GMCs where star formation takes place.

This accretion process leads to a momentum transfer that takes place at the same time scale τ_{acc}. In addition to this, stellar winds also act as a source of momentum in the ISM. The stellar winds exert a ram pressure of $P_w = \rho_w v_w^2$. Taking the mass-loss rate from a star as $(dm/dt) = 4\pi r^2 \rho_w v_w \simeq 10^{-6} \, M_\odot \, \mathrm{yr}^{-1}$ and $v_w \simeq 100 \, \mathrm{km \, s}^{-1}$, we get $P_w \simeq 5 \times 10^{26}/r^2 \, \mathrm{dyn \, cm}^{-2}$. The gas pressure of ISM is $P_{\mathrm{ISM}} \simeq 10^{-16} nT \, \mathrm{dyn \, cm}^{-2}$. This shows that $P_w = P_{\mathrm{ISM}}$ at a distance of $r \approx 7 \, \mathrm{pc}$ from the star. In other words, this effect can be important at the typical interstellar spacing of $r \simeq 1 \, \mathrm{pc}$, even in the vicinity of a star with a moderate mass-loss rate of $(dm/dt) \simeq 2 \times 10^{-8} \, M_\odot \, \mathrm{yr}^{-1}$. Hence we cannot ignore the input of momentum into the ISM in regions having stars with appreciable mass loss.

Although these results show that the coarse-grained approach has limitations, the absence of detailed knowledge forces us to model the average evolution of the galaxy by making several assumptions to simplify the analysis. We begin with a general description of this approach.

Integrating mass conservation law (2.218) over a given volume $\mathcal{V}$ will lead to an equation of the form

$$\frac{\partial G}{\partial t} + \nabla \cdot (G\mathbf{v}) = \int_{\mathcal{V}} s_{\mathrm{mass}} \, d^3x + I, \qquad I = -\int_{\Sigma} \rho \mathbf{v} \cdot d\Sigma, \quad (2.220)$$

where I denotes the net mass flowing through the external boundary Σ of the region $\mathcal{V}$. The first term on the right-hand side can be expressed in the form

$$\int_{\mathcal{V}} s_{\mathrm{mass}} \, d^3x = -F + E, \qquad (2.221)$$

where F is the rate of conversion of the ISM into stars and E is the rate of ejection of mass from the stars to the ISM. If we now further assume that the system is homogeneous, then we can ignore the spatial derivative on the left-hand side.

Although ignoring the spatial derivatives will make the equation depend only on time, it will remain in general nonlocal in time for the following reason. We note that the SFR F is given by

$$F = \int_0^\infty B(m, t) \, dm, \qquad (2.222)$$

where $B(m, t)$, usually called the *birthrate function*, is the rate at which gas is converted at time t into stars of mass m measured in units of solar mass; $m = M/M_\odot$. The ejection rate E can be expressed in the form

$$E(t) = \int_0^\infty \int_0^t B(m, t - a) R(a, m) \, dm \, da, \qquad (2.223)$$

where $R(a, m)$ is the rate at which a star of mass m and age a returns the mass to the ISM. It is obvious that the term $(E - F)$ is nonlocal in time.

These equations can be simplified if further assumptions are made with the inputs from stellar-evolution theory. Most of the stars return a negligible amount of mass to the ISM for most of their lifetime, the major exception being the massive stars with very short lifetimes. In their case, we can make an approximation that the mass ejection takes place at the end of their life at $a = \tau_m$. This will allow us to approximate the return function by $R(a, m) \approx \delta(a - \tau_m)R(m)$, so that

$$E(t) \approx \int_0^\infty B(m, t - \tau_m)R(m)\,dm. \qquad (2.224)$$

With this approximation, $E(t)$ and $F(t)$ can be determined from stellar evolution theory, allowing us to integrate the relevant equations.

It is, however, conventional to make another related approximation known as *instantaneous recyling*. If the rate of change of birthrate $B(m, t)$ is small compared with the lifetime of the stars, then we can write $B(m, t - \tau_m) \approx B(m, t)$. Further, we take the function $B(m, t)$ to be separable into the variables m and t in the form $B(m, t) = \Lambda(t)\xi(m)$, where $\xi(m)$ is the *initial mass function* (IMF). We now have $E(t) = f\Lambda(t)$, where f is a constant given by

$$f = \int_0^\infty R(m)\,\xi(m)\,dm. \qquad (2.225)$$

There is a considerable amount of uncertainty in the form of the IMF $\xi(m)$ (see Vol. II, Chap. 3, Section 3.6). One of the simplest characterisations of this function is in the form $\xi(m) = \xi_0 m^{-\alpha}$, with $\xi_0 \simeq 1.5 \times 10^{-2}\,\mathrm{pc}^{-3}$ and $\alpha \simeq 1.33$. Very often we use the normalised IMF $\phi(m)$ with

$$\int_{m_L}^{m_U} m\,\phi(m)\,dm = 1 \qquad (2.226)$$

with some lower and upper limits on the mass, $m_L \simeq 0.1$ and $m_U \simeq 100$. The sensitivity of the pure power-law functions to the limits of integration can be avoided if a suitable series of power laws is used in different ranges. One such fitting function has the form

$$m\phi(m) = Am^B, \qquad (2.227)$$

with the constants A and B having different values for different ranges as follows: $A = 0.93$, $B = 0.15$ for $0.1 \le m \le 0.5$; $A = 0.46$, $B = -0.85$ for $0.5 \le m \le 1$; $A = 0.46$, $B = -2.4$ for $1.0 \le m \le 3.16$; and $A = 0.21$, $B = -1.7$ for $3.16 \le m \le 100$. An approximate single power-law fit for this range is given by $m\phi(m) = 0.17m^{-1.35}$. Either of these will lead to 31% of the mass fraction in the interval $0.1 < m < 0.5$, another 31% in the range $0.5 < m < 1$, a mass fraction of 0.26 in the range $1 < m < 3.16$, and a mass fraction of 0.12 in the range $3.16 < m < 100$.

If the mass fraction locked up in stars and stellar remnants is α, then the return fraction is $R = 1 - \alpha$. The locked-up fraction can be computed as

$$\alpha = 1 - F_M(> m_\tau) + \int_{m_\tau}^{m_U} m_{\text{rem}}(m)\,\phi(m)\,dm, \qquad (2.228)$$

where $F_M(> m_\tau)$ is the fraction of stars with masses greater than a critical mass

m_τ, which we evaluate by equating the lifetime $\tau(m)$ to the age of the galaxy. In other words, $1 - F_M(> m_\tau)$ gives the fraction of mass that is locked up in long-lived stars and the last term of Eq. (2.228) gives the mass that is locked up in stellar remnants. The latter can be estimated as

$$m_{\mathrm{rem}} = \begin{cases} 0.11\,m + 0.45, & (m \le 6.8) \\ 1.5, & (m > 6.8) \end{cases}. \tag{2.229}$$

If we take $m_\tau = 1$, most of the IMFs will give $\alpha \simeq (0.7\text{–}0.8)$.

Given this background, an illustrative model for the chemical enrichment of a galaxy can be constructed along the following lines. Let us consider a region of galaxy containing a mass $M_g(t)$ in gas and $M_s(t)$ in stars, and let us assume that gas with metallicity Z_f is added or ejected from that region at a rate $\dot{M}_f$, where an overdot denotes the time derivative. Conservation of mass implies that

$$\dot{M}_g + \dot{M}_s = \dot{M}_f. \tag{2.230}$$

The conservation of heavy elements requires

$$d(ZM_g) = p\,dM_s - Z\,dM_s + Z_f\,dM_f. \tag{2.231}$$

The left-hand side is the change in the heavy elements in the gaseous phase; the first term on the right-hand side gives the nucleosynthesis yield that is due to processing in stars. We take this yield to be proportional to the processed stellar mass parameterised by a yield parameter p. The second term on the right-hand side is the amount of heavy elements locked in stars and the last term arises from sources or sinks present in the region. This equation assumes (as previously explained) that stellar processes are efficient in changing the heavy-element abundance instantaneously (i.e., at a time scale shorter than any other time scale in the problem) and that everything is spatially homogeneous. Dividing this equation by dt and using conservation of mass (2.230), we get

$$\dot{Z}M_g - p\dot{M}_s = (Z_f - Z)\,\dot{M}_f. \tag{2.232}$$

To proceed further, we need to make some specific assumptions regarding the sources and sinks. As the first example, let us consider a model in which the gas is blown out of the region at a rate that is a constant fraction ν of the SFR so that $\dot{M}_f = -\nu\dot{M}_s$. From mass conservation (2.230) we get $\dot{M}_g = -(1+\nu)\dot{M}_s$, which integrates to give

$$M_g(t) = M_g(0) - (1+\nu)M_s(t). \tag{2.233}$$

Using this in Eq. (2.232) with the assumption $Z_f = Z$, we get

$$Z = \frac{p}{1+\nu}\ln\frac{M_g(0)}{M_g(t)} = \frac{p}{1+\nu}\ln\left\{1 + (1+\nu)\left[\frac{1}{f_g(t)} - 1\right]\right\}, \tag{2.234}$$

where $f_g \equiv M_g/(M_g + M_s)$ is the fraction of gas.

The model shows that the metallicity increases as the gas is converted into stars and star formation effectively ceases when all the gas is exhausted, which occurs when $M_s = M_g(0)/(1 + \nu)$. Hence the mean stellar metallicity at late times is given by

$$\bar{Z}_s = \frac{1 + \nu}{M_g(0)} \int_0^{M_g(0)/(1+\nu)} Z \, dM_s = \frac{p}{1 + \nu}. \qquad (2.235)$$

The results from Eqs. (2.234) and (2.235) allow us to make some simple deductions that could be compared with observations. Note that the mean metallicity $\langle Z \rangle$ will approach the true yield p (if $\nu = 0$) or $p(1 + \nu)^{-1}$ (if $\nu \neq 0$) when the gas fraction goes to zero. It is also clear that the ratio of metallicities for two elements will be the same as the ratio of their yields: $Z_i/Z_j = p_i/p_j$. There have been several attempts to test these results against observations with a mixed level of success. Although there is some evidence suggesting that these predictions are true, there are also significant discrepancies in the case of some elements.

This model also predicts a logarithmic relation between the metallicity $Z(t)$ and the gas fraction $M_g(t)$, which can be tested if we can identify a population of objects with similar physical conditions but different amounts of gas fraction. This was attempted in a study of the oxygen abundance in a class of irregular and blue compact galaxies. The results show that the model is in broad agreement with the observations, although the error bars are fairly large. Similar results were obtained in the study of metallicities of F- and G-type stars as a function of their ages.

To get actual numbers, we need to estimate ν and p, which turns out to be fairly difficult. We can attempt to construct a simplified model for ν along the following lines. We have seen in Vol. I, Chap. 6, Section 6.12, that the cooling rate of a gas decreases with temperature in the range $(10^4 – 3 \times 10^6)$ K, indicating an instability. Hence we would expect the ISM of a galaxy to contain gas with temperatures $T < 10^4$ K and $T > 3 \times 10^6$ K. The one-dimensional velocity dispersion of the hotter component of gas will be $\sigma_h \approx 200$ km s^{-1}. For comparison, stars at the centres of dwarf spheroidals have a dispersion of $\sigma_* \approx (5\text{–}10)$ km s^{-1} whereas those at the centre of giant ellipticals have $\sigma_* \gtrsim 200$ km s^{-1}. Hence, in a dwarf spheroidal, the hot gas is likely to flow freely out of the system although in a giant elliptical we would expect the gravitational field to be able to confine the hot component of the ISM. To obtain a quantitative relationship between σ_* and Z in a giant elliptical, we can equate the rate at which ejected gas gains potential energy to the rate at which a supernova injects energy into the ISM. This would suggest that

$$\left| \Phi \frac{dM_g}{dt} \right| \simeq f \sigma_*^2 \nu \frac{dM_s}{dt} = E_{SN} \frac{dM_s}{dt}, \qquad (2.236)$$

where we have taken the gravitational potential to be $\Phi = -f\sigma_*^2$, with f denoting a numerical factor. Assuming that one core-collapse supernova injects $\sim 10^{51}$ ergs into the ISM and that one such event occurs for 200 $M_\odot$ of star formation, we get $E_{SN} \simeq 2.5 \times 10^{15}$ ergs gm^{-1}. This gives

$$\nu = \frac{0.6}{f} \left(\frac{200 \text{ km s}^{-1}}{\sigma_*} \right)^2, \qquad (2.237)$$

suggesting that ν will increase from a value of less than 1 for giant ellipticals to an order of a few for dwarf spheroidals. Combining Eq. (2.237) with Eq. (2.235) and using $\nu \propto \sigma_*^{-2}$, we would expect a relation of the form

$$\bar{Z}_s = \frac{p}{1 + (\sigma_0/\sigma_*)^2}, \qquad (2.238)$$

where σ_0 is a constant. Observationally, there are some data giving the abundance (based on MgII) for systems with different σ_* that seems to be in broad qualitative agreement with Eq. (2.238). There is no direct confirmation of flattening of $\bar{Z}$ for large σ_*, although this could depend on the relationship between the MgII line strength and total mean metallicity.

A similar analysis based on Eq. (2.235) can also be used to estimate the radial gradient of the metallicity in ellipticals. We have

$$\frac{d \ln \bar{Z}_s}{d \ln r} = -\frac{r}{1 + \nu} \frac{d\nu}{dr} = \frac{\nu}{1 + \nu} \frac{v_c^2}{\Phi} = \frac{E_{SN}}{E_{SN} - \Phi} \frac{v_c^2}{\Phi}, \qquad (2.239)$$

where we have used $\nu = E_{SN}/|\Phi(r)|$ and $v_c = \sqrt{rd\Phi/dr}$. For plausible values of Φ and E_{SN}, this gives a gradient of -0.2, which is in broad agreement with observation.

The second parameter that appears in the preceding model is the average yield p, which, in principle, could be estimated from stellar-evolution theory. From our discussion in Vol. 2, Chap. 3, Section 3.6, we know that stars of different initial masses produce different amounts of metals and lock up different amounts of mass in the final stage. To a very good approximation, the mass of heavy elements $M(Z)$ ejected during the evolution of a star of mass $M = mM_\odot$ is given by

$$M(Z) \approx M \left(0.4 - \frac{4}{m} \right) \qquad (10 \lesssim m \lesssim 25) \qquad (2.240)$$

for M in the range of 10–25 $M_\odot$. After approximately 5×10^9 yr, stars with mass $M < 1.2 \, M_\odot$ will still be in the main sequence; stars with mass $1.2 \, M_\odot < M < 7 \, M_\odot$ will leave white dwarf remnants of mass 0.6 $M_\odot$; and stars of mass $7 \, M_\odot < M < 25 \, M_\odot$ will leave neutron star remnants of mass 1.4 $M_\odot$. The extreme limits are more uncertain, but for simplicity we will assume that (1) no brown dwarfs are formed and (2) stars with $M > 25 \, M_\odot$ leave black hole remnants of mass $M/2$. Given a mass function of stars $\psi(m) \, dm$, we can then easily determine the total yield of heavy elements as

$$Y = \int_{10}^{\infty} dm\, \psi(m) M \left(0.4 - \frac{4}{m} \right). \tag{2.241}$$

The mass fraction α that is locked up in stars is given by Eq. (2.228). The parameter p is just the ratio $p = (Y/M_{\text{lock}})$. The actual numerical value, of course, depends sensitively on the asymptotic behaviour of $\psi(m)$. If we take the Miller–Scalo IMF with

$$\psi = Cm^{-1.9} 10^{-0.28(\log m)^2}, \tag{2.242}$$

we find that $Y \approx 0.161C$ and $M_{\text{lock}} \approx 2.67C$, giving $p = 0.06$. On the other hand, the Salpeter IMF gives $p \simeq 0.065/\alpha \simeq 0.08$. If we take $\nu \simeq 2$ from Eq. (2.237) then $p/(1 + \nu) \simeq 0.02$. Equation (2.234) suggests that $Z = Z_\odot \simeq 0.02$ when $M_g \simeq e^{-1} M_g(0) \simeq 0.37\, M_g(0)$.

The *simplest* model of chemical evolution, of course, will correspond to $\nu = 0$, which is usually called the *closed-box model*. Equation (2.234) with $\nu = 0$ shows that

$$f_s \equiv \frac{M_s(z)}{M} = 1 - \exp(-Z/p), \tag{2.243}$$

where $f_s \equiv M_s(Z)/M$ is the fraction of mass in stars when the metallicity is Z. It follows that

$$\frac{df_s}{d\log z} \propto z e^{-z}, \qquad z = \frac{Z}{p}, \tag{2.244}$$

implying that the number of stars with a given metallicity has a characteristic distribution that has a maximum at $z = 1$ or $Z = p$. We can test this result by studying the distribution of stars with different abundances for, say, [O/H]. Observations suggest that the theory fits the observations but with $p = 10^{-1.1} Z_\odot$. This modification can be easily incorporated into a model with $\nu \neq 0$, which allows for a mass loss at a rate proportional to the SFR. In such a case, the distribution in relation (2.244) will change to the form $ze^{-(1+\nu)z}$, which is equivalent to using an effective yield of $p/(1 + \nu)$.

Also note that, in the $\nu = 0$ model, the metallicity increases as the gas fraction decreases. Such a conclusion, of course, directly contradicts the existence of globular clusters that have very little gas but the stars have a metal-poor composition. Similarly dwarf spheroidal galaxies contain very little gas, although their stars have metal abundances which are 30–100 times lower than the galactic bulge. It is likely that, in both the cases, stars formed out of well-mixed initial gas and any residual gas were expelled from the system. This suggests that the flow of material in and out of the system needs to be taken into account for a more realistic result.

The most serious difficulty faced by the closed-box model is in the case of long-lived G-dwarf stars in a solar neighbourhood. Observations show that the fraction of low-metallicity G-dwarf stars is significantly lower than that predicted by this model. This problem, however, can be solved by a suitably modelled infall of gas (although it is not clear whether this is the best possible approach). One of the simplest models that solves the G-dwarf problem is based on a postulate that relates the gas mass to the star mass in the system by an (ad hoc) quadratic expression:

$$g(s) = \left(1 - \frac{s}{M}\right)\left(1 + s - \frac{s}{M}\right). \tag{2.245}$$

Here $g(s)$ is the gas mass, s is the mass in stars, and M is a constant parameterising the relation. All masses are measured in units of the initial mass, and M can be interpreted as the final mass in stars when $g = 0$. From Eq. (2.231) for the evolution of metallicity (in a slightly modified notation with $Z_f = 0$),

$$\frac{d}{ds}(Zg) = p - Z, \tag{2.246}$$

we can easily obtain the solution for $z = Z/p$ as

$$z(s) = \frac{Z}{p} = \left[\frac{M}{1 + s - (s/M)}\right]^2 \left[\ln\frac{1}{1 - s/M} - \frac{s}{M}\left(1 - \frac{1}{M}\right)\right]. \tag{2.247}$$

Correspondingly, the characteristic distribution in metallicity now becomes

$$\frac{ds}{d\ln z} = \frac{z[1 + s(1 - 1/M)]}{(1 - s/M)^{-1} - 2z(1 - 1/M)}. \tag{2.248}$$

Figure 2.15 shows this function for different values of M. Note that, compared with the original model of $M = 1$, this class of models with $M > 1$ allows for a relative enhancement of higher-metallicity stars with respect to G dwarfs.

Exercise 2.12
Chemical evolution with inflow: Another extreme form of evolution corresponds to assuming that the primordial gas flows into the region at a rate that is a constant fraction of the SFR. In this case $\dot{M}_f = +v\dot{M}_s$ and $Z_f = 0$. (a) Show that this implies that the amount of mass contained in stars with metallicity less than a particular value Z scales as

$$F(Z, Z_1) \equiv \frac{M_s(< Z)}{M_s(< Z_1)} = \frac{M_g(0) - M_g(< Z)}{M_g(0) - M_g(< Z_1)} = \frac{1 - q}{1 - q_1}, \qquad q = \frac{M_g(< Z, t)}{M_g(0)}. \tag{2.249}$$

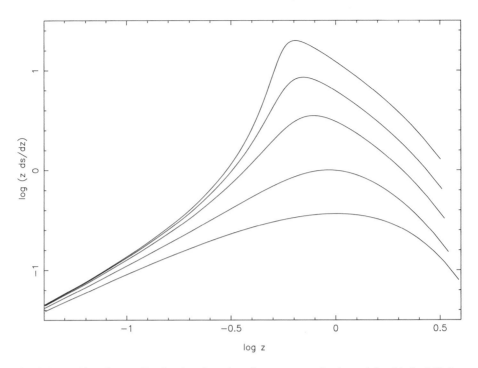

Fig. 2.15. Abundance distribution function for parameterised model with in-fall for
$M = 1, 2, 5, 10, 20$ from bottom to top.

Express this result in the form

$$F(Z, Z_1) = \frac{1}{1 - q_1} \left\{ 1 - \left[1 - \frac{Z}{Z_1} \left(1 - q_1^{\nu/(1-\nu)} \right) \right]^{(1-\nu)/\nu} \right\}. \qquad (2.250)$$

Also show that q_1 is related to the gas fraction by

$$q_1 = \left[1 + (1 - \nu) \left(\frac{1}{f_{g1}} - 1 \right) \right]^{-1}. \qquad (2.251)$$

(b) Observationally it is found that the F- and G-dwarf stars with $Z < (1/6)Z_1$ contribute
$\sim 2\%$; that is, $F[Z = (1/6)Z_1, Z_1] \simeq 0.02$. (In the solar neighbourhood we can take
$f_{g1} = 0.1$ for $Z_1 = 0.03 = 1.5 Z_\odot$.) Can ν be adjusted to get the correct abundance of
G dwarfs?

Finally, we describe a general result connecting the mass of the heavy elements
produced, the luminosity of the galaxy, and the index of the IMF. As we saw in
Chap. 1, Section 1.6, most of the present-day luminosity comes from stars whose
original masses were in a narrow range of values around the characteristic mass
$M_{\rm GB}$ of stars on the present giant branch, whereas the heavy elements generated
were all contributed by stars more massive than some critical mass M_Z. Assuming
that the massive stars reinject a fraction p of the original mass into the ISM in

the form of heavy elements, we find that the mass of heavy elements injected into the ISM will be

$$M_h = \int_{M_Z}^{\infty} pM \frac{dN}{dM} dM. \tag{2.252}$$

Taking a simple IMF with $(dN/dM) = K(M/M_\odot)^{-\alpha}$ for $M < M_{max}$ and 0 for $M > M_{max}$, we get

$$M_h = pKM_\odot^2 \int_{M_Z/M_\odot}^{M_{max}/M_\odot} m^{(1-\alpha)} dm = \frac{pKM_\odot^2}{\alpha - 2} \left[\left(\frac{M_\odot}{M_Z} \right)^{\alpha-2} - \left(\frac{M_\odot}{M_{max}} \right)^{\alpha-2} \right]. \tag{2.253}$$

Because $\alpha > 2$, we can ignore the second term if $M_{max} \gg M_Z$. Combining this with relation (1.55) of Chap. 1, Section 1.6, we get the mass of the heavy elements injected into the ISM per unit luminosity generated as

$$\frac{M_h}{L} = \frac{pM_\odot^2}{\alpha - 2} \left[\frac{25 \text{ Gyr}}{E_{GB}(M_{GB})} \right] \left(\frac{M_Z}{M_\odot} \right)^{2-\alpha} \left(\frac{M_{GB}}{M_\odot} \right)^{\alpha-3.5}. \tag{2.254}$$

Inserting the typical numbers $p \simeq 0.1$, $\alpha \simeq 2.4$, $M_{GB} \simeq 0.85 \, M_\odot$, and $E_{GB} \simeq 2.9 \times 10^{10} \, L_\odot$ yr, we get

$$\frac{M_h}{M_\odot} \simeq 0.26 \left(\frac{M_Z}{M_\odot} \right)^{2-\alpha} \frac{L}{L_\odot}. \tag{2.255}$$

The value for M_Z depends on the elements we are interested in. For carbon, nitrogen, and oxygen, $M_Z \simeq 2 \, M_\odot$, whereas for elements produced in a core-collapse supernova, $M_Z \simeq 8 \, M_\odot$. Because $(2 - \alpha) \approx -0.4$, the result is not very sensitive to the value of M_Z. We find that (M_h/L) varies in the range 0.11–0.20 $M_\odot/L_\odot$ for most elements.

2.9 Galaxy Interactions and Mergers

So far we have been treating galaxies as isolated noninteracting systems. It is, however, fairly easy to see that galaxy interactions cannot, in general, be ignored in the study of their dynamics. Even in the case of field galaxies, with a mean separation $l \simeq 1$ Mpc and relative velocities of $v \simeq 200$ km s^{-1}, the corresponding time scale is $t \simeq (l/v) \simeq 5 \times 10^9$ yr. This suggests that most galaxies would have undergone some form of gravitational interaction with another galaxy during the lifetime of the universe.[6] The conclusion is strengthened when we take into account the fact that galaxies are clustered and a fraction of them occur in high-density regions which will decrease l and increase v.

  The actual gravitational interaction of two galaxies is an enormously complicated affair that leads to a wide variety of effects. Even in the simplest contexts in which each galaxy is treated as being made up of point particles interacting only through gravity (thereby ignoring the entire gaseous phase of the ISM in the galaxy), there are several parameters that play a role in determining the outcome, and, in general, we must resort to numerical simulations to understand these processes. Given the density profile of the two galaxes, when they are far away from each other (thereby determining the individual masses, velocity dispersion, and gravitational potential energies) and the trajectories of the centres of mass of the galaxies under their mutual influence, we can identify a few cases in which analytic understanding is possible. We shall first describe these situations.

The first case corresponds to a situation in which the mass of one of the galaxies is significantly higher than that of the other. In this situation, the lighter galaxy will move in the gravitational potential of the heavier one and will, in general, lose energy and angular momentum because of dynamical friction, thereby merging with the heavier galaxy. This process can be effective not only in the case of galaxies but even for globular clusters orbiting in the halo of the galaxies. Mergers also occur between objects of comparable mass – usually called galactic cannibalism – but those cases are difficult to treat analytically.

The second situation in which some progress can be made corresponds to high-speed encounters that can be defined by the following criterion. If the colliding systems have masses M_1 and M_2, radii R_1 and R_2, impact paramter b, and relative velocity V, then the time scale for the encounter could be estimated as $t_{enc} \simeq \max(R_1, R_2, b)/V$. If the internal-velocity dispersions are σ_1 and σ_2, then the crossing times for stars in the galaxies will be $t_i = R_i/\sigma_i$ with $i = 1, 2$. If $t_i \gg t_{enc}$, we can assume that the stars have moved very little during the time of encounter. This translates into the condition $V \gg (\sigma_i/R_i) \max(R_1, R_2, b)$, which can be satisfied at high speeds.

It is straightforward to show that, when this condition is satisfied, we can also approximate the motion of the two centres of colliding galaxies as one of uniform velocity. Focussing the attention on one system (of mass M_1, which we will call the perturbed system), we can compute the change in the kinetic energy of the stars that is due to the passage of the second system of mass M_2 and trajectory $\mathbf{X}(t) \simeq (0, b, Vt)$. This was done in Vol. II, Chap. 10, Section 10.7, and we found that the change of velocity of the αth star is given by

$$\Delta \mathbf{v}_0 = - \int_{-\infty}^{\infty} \left[\nabla \Phi(\mathbf{r}_\alpha, t) - \frac{1}{M_1} \sum_\beta m_\beta \nabla \Phi(\mathbf{r}_\beta, t) \right] dt, \quad (2.256)$$

where $\Phi(\mathbf{r}, t)$ is the gravitational potential that is due to the perturber, which is treated as a known quantity. The potential energy of the stars does not change in this *impulse approximation*, whereas the kinetic energy of the perturbed system increases. The actual estimate of the change in the energy simplifies in two limiting cases. The first one was studied in Vol. II, Chap. 10, Section 10.7, under the tidal approximation in which it was assumed that R_i/b is a small quantity. In such distant encounters, the amount of energy transferred varies as b^{-4}. The other limiting case is a penetrating encounter corresponding to a head-on collision with $b \to 0$. A more general case can be treated (to a good approximation) as an interpolation between the two.

Immediately after the encounter, neither system is in virial equilibrium as their potential energies are unchanged but their kinetic energies have increased. As the systems relax back to virial equilibrium, the total kinetic energy actually *decreases*, as can be seen by the following argument. Originally, in virial equilibrium, the kinetic energy T_0 and the total energy E_0 are related by $T_0 = -E_0$. If the encounter increases the kinetic energy by δT, the final energy is $E_1 = E_0 + \delta T$, and the final kinetic energy in the final virialised configuration will be

$$T_1 = -E_1 = -(E_0 + \delta T) = T_0 - \delta T. \qquad (2.257)$$

Thus the final relaxation process has to decrease the kinetic energy from $T_0 + \delta T$ to $T_0 - \delta T$. This is usually accomplished by the system expanding after the encounter in some form, either by the ejection of more energetic components or by a gentler redistribution of mass. We will discuss the issue of mergers of systems with widely different masses in Subsection 2.9.1 and will study some of the effects of penetrating encounters in Subsection 2.9.2.

One of the key effects of interaction between galaxies seems to be an enhancement in the SFR – often in the form of star bursts. Because it is not possible to view a system both before and after interaction (except, of course, in simulations), our conclusions are necessarily based on statistical analysis of interacting and noninteracting samples of galaxies with respect to the SFR. Interacting systems are often seen to show evidence of young stars, an excess of type II supernova outbursts, evidence for line emission that indicates an overabundance of young OB stars, and other indications of enchanced SFR. Although the evidence is not completely conclusive, it is generally believed that the gravitational perturbation of one galaxy on another can induce instabilities in either or both of them involving density enhancements in gas and a consequent increase in SFR.

Mergers of galaxies also lead to striking visual signatures that include tails of material, polar rings, and shells of matter, or counterrotating stellar or gasesous subsystems. Detailed numerical simulations have been able to reproduce most of these features as a result of strong gravitational encounter between two galaxies, although the detailed modelling and statistical study do not seem to lead to any conclusions that are definitive and useful beyond the morphological aspects. In Subsections 2.9.1–2.9.3 we shall briefly describe the simple analytic results and some results from numerical simulations.

2.9.1 Galactic Cannibalism

One of the most dramatic effects of galaxy–galaxy interaction is a phenomena called *galactic cannibalism*, in which a large galaxy gobbles up smaller neighbours over a period of time. The chance for this phenomenon to occur over a Hubble time is fairly large and is enhanced by the fact that galaxies cluster around each other. The number density $n(r, L)dL$ of galaxies near another galaxy varies as (see Chap. 1, Section 1.13)

$$n(r, L)\delta L = \phi(L) \left[\left(\frac{r_0}{r} \right)^{1.8} + 1 \right] \delta L, \tag{2.258}$$

where $\phi(L)$ is the galaxy luminosity function and $r_0 \approx 5h^{-1}$ Mpc. If the orbiting galaxies slowly spiral towards the central galaxy, then the inward current that is due to dynamical friction will be $dJ(r, L) = 4\pi r^2 v_r n(r, L)dL$. The total number of galaxies eaten up by a large galaxy in one Hubble time $t_H \equiv H_0^{-1}$ is obtained by integration of $n(r, L)$ from $r = 0$ to a radius $r_i(t_H, L)$ at which $t_{\text{fric}} = t_H$. Taking $r \ll r_0$ and using the results of Section 2.5 (see Eq. (2.165)), we get

$$dN(L) = 4\pi \, dL \int_0^{r_i} n(r, L)r^2 \, dr = 9.5 \left(\frac{GQL_* \ln \Lambda}{H_0 r_0^2 v_c} \right)^{0.6} r_0^3 \phi(L) dL, \tag{2.259}$$

where we have taken the mass of the galaxy to be $M = QL$, with $Q \approx 12h(M_\odot/L_\odot)$ being the mass-to-light ratio of a typical galaxy. We can obtain the total amount of luminosity gained by a galaxy by integrating this expression with the Schecter luminosity function:

$$L_{\text{gain}} = \int_0^\infty L dN(L) \approx 9.5 \left(\frac{GQL_* \ln \Lambda}{H_0 r_0^2 v_c} \right)^{0.6} r_0^3 n_* L_* \int_0^\infty x^{1.6+\alpha} e^{-x} \, dx$$

$$= 9.5 \left(\frac{GQL_* \ln \Lambda}{H_0 r_0^2 v_c} \right)^{0.6} r_0^3 n_* L_* \Gamma(2.6 + \alpha). \tag{2.260}$$

For typical values of $v_c \approx 300$ km s^{-1}, $r_0 \approx 5h^{-1}$ Mpc, $\ln \Lambda = 3$, and the Schecter luminosity function, we find that L_{gain}/L_* is approximately 10%–15%. This shows that galactic cannibalism is a significant effect in increasing the intrinsic luminosity of the galaxies.

Although it is difficult to obtain direct observational signatures for determining whether a galaxy has undergone merging in its past, some insights can be gained from scaling arguments if we assume that the light distribution before and after merging satisfies some universal functional form: $\Sigma(r)/\Sigma_0 = F(r/r_0)$. For constant mass-to-light ratios, the total mass and binding energy of the galaxy will then scale as $M \propto \Sigma_0 r_0^2$, $E \simeq (GM^2/r_0) \propto \Sigma_0^2 r_0^3$. Mergers that conserve both mass and energy can be characterised by two extreme forms of behaviour. The first one corresponds to a massive galaxy's merging with several low mass galaxies with small binding energies, which will increase M at constant E, giving $\Sigma_0 \propto M^{-3}$. The second one involves the merging of nearly equal-mass galaxies in which case the binding energy increases in proportion to the mass, giving $\Sigma_0 \propto M^{-1}$. An appropriate projection of the fundamental plane of elliptical galaxies to get a relation between Σ_0 and M suggests that the actual scaling is roughly midway between the extreme models with $\Sigma_0 \propto M^{-2}$. Although this certainly does not *prove* the merger history of elliptical galaxies, it is consistent with the conjecture that elliptical galaxies have undergone a significant amount of mergers.

It is also possible to obtain simple scaling relations for the growth of total luminosity of a merging galaxy within a given aperture scale of observation r. If r_0 is small so that $r_0 \lesssim r$, we observe the full luminosity of the galaxy and $L \propto \Sigma_0 r_0^2 \propto M$ in both the extreme cases. However, at late times, r_0 is large so that $L \propto \Sigma_0 r^2$ (where r is the chosen scale of observation), which increases as M^{-1} or M^{-3} in the two extremes. Thus, for observations at fixed aperture, the magnitude of a cannibal galaxy can be very nearly constant for large parts of its history.

At a still smaller scale, mergers affect the globular clusters in a galaxy. The globular clusters that are distributed around the galaxy in an approximately spherical manner will experience dynamical friction when they move through the halo. If we take the density distribution in the halo to be $\rho(r) = (v_c^2/4\pi G r^2)$, then the general formula derived in Vol. II, Chap. 10, Section 10.5, will give the frictional force on a cluster of mass M moving with speed v_c at a radius r as

$$F = -\frac{4\pi(\ln\Lambda)G^2M^2\rho(r)}{v_c^2}\left[\mathrm{erf}(1) - \frac{2}{\sqrt{\pi}}e^{-1}\right] = -0.428(\ln\Lambda)\frac{GM^2}{r^2}.$$

(2.261)

(The Coloumb logarithm is $\ln\Lambda \simeq 10$ in this case.) Because the frictional force is tangential, it causes the cluster to lose the angular momentum (per unit mass) L at the rate

$$\dot{L} = \frac{Fr}{M}.$$

(2.262)

Assuming that the cluster continues to orbit at speed v_c as it spirals to the centre, we can take $\dot{L}$ to be equal to $v_c \dot{r}$. This leads to

$$r\frac{dr}{dt} = -0.428\frac{GM}{v_c}\ln\Lambda,$$

(2.263)

which can be easily integrated to find the time for the cluster to fall to the centre from some initial radius r_i; we get

$$t_{\mathrm{fric}} = \frac{1.17 r_i^2 v_c}{(\ln\Lambda)GM} = \frac{2.64 \times 10^{11}}{\ln\Lambda}\left(\frac{r_i}{2\ \mathrm{kpc}}\right)^2\left(\frac{v_c}{250\ \mathrm{km\ s^{-1}}}\right)\left(\frac{10^6\ M_\odot}{M}\right)\ \mathrm{yr}.$$

(2.264)

As an example, consider the globular clusters around M31 that have masses of approximately $M = 5 \times 10^6\ M_\odot$ and a circular speed of $v_c \approx 250\ \mathrm{km\ s^{-1}}$. Equation (2.264) shows that if $\ln\Lambda = 10$, massive clusters whose original orbits were at $r_i < 3$ kpc would have already spiralled into the centre of the galaxy. In other words, centres of galaxies like M31 could have accreted mass through the gobbling up of such clusters.

Another application of dynamical friction, in the context of interacting systems, can be illustrated by the Magellanic cloud, the centre of mass of which is orbitting our galaxy, at a distance of $r \approx 60$ kpc. Taking $M \approx 2 \times 10^{10} \, M_\odot$, we find that Eq. (2.164) gives

$$t_{\text{fric}} = \frac{1.0 \times 10^{10}}{\ln \Lambda} \left(\frac{r}{60 \text{ kpc}} \right)^2 \left(\frac{v_c}{220 \text{ km s}^{-1}} \right) \left(\frac{2 \times 10^{10} \, M_\odot}{M} \right) \text{ yr.} \quad (2.265)$$

For a Magellanic cloud, $\ln \Lambda \simeq 3$, and this relation suggests that the clouds will spiral to the centre of our galaxy in a fraction of Hubble time. There are two caveats to this conclusion that could provide an escape from this conclusion: (1) The argument assumes that the halo of our galaxy reaches up to a large Magellanic cloud (LMC); current observations neither confirm nor disprove this assumption strongly, but if the halo density falls more rapidly than r^{-2} near the location of the LMC, the time scale will increase. (2) The clouds were assumed to be on a circular orbit, which may not be necessarily true. If the orbit is eccentric then the time scale will depend on the location on the orbit and could be smaller or larger than our estimate, depending on whether the clouds are currently near the apocentre or the pericentre. Determining the orbital motion requires knowledge of the proper motions of the clouds, which are not well determined.

2.9.2 Galaxy Collisions

It was previously mentioned that the analysis of the energy transfer in the encounter between two galaxies simplifies in two limiting cases. In the first case, we assume that the effect of a perturber on a system can be calculated by studying the tidal effect in a Taylor series expansion in R/b, where R is the size of the system and b is the impact parameter. This analysis was performed in Vol. II, Chap. 10, Section 10.7, where it was shown that energy transfer varies as $\Delta E \propto b^{-4}$. In the case of galaxies, this acts as an effect that tidally transfers energy.

More dramatic effects occur in the case of penetrating encounters corresponding to $b = 0$. The previous approach is obviously invalid in this case, but we can obtain the necessary results along the following lines. Let us choose the coordinates such that the galaxy is centred at the origin ($R = 0, z = 0$) of the cylindrical coordinate system and the trajectory of the perturber is along the z axis given by [$R = 0, z = Z_p(t) = Vt$]. If the perturber can be treated as a spherical system with a potential $\Phi(r)$, the R component of the perturber's gravitational field at (R, z) is $(-R/r)(d\Phi/dr)$, where $r = \sqrt{(Z_p - z)^2 + R^2}$. This is the only component that is relevant because, by symmetry, the velocity increment Δv is generated towards only the perturber's line of motion. We may neglect any changes in the velocity parallel to the perturber's trajectory as equal and opposite impulses occur during the approaching and the receding phases of the perturber.

The magnitude of the velocity change now becomes

$$\Delta v_R = -\int_{-\infty}^{\infty} \frac{d\Phi}{dr} \frac{R}{r} \, dt = -\frac{2R}{V} \int_0^{\infty} \frac{d\Phi}{dr} \frac{dZ_p}{r}, \qquad (2.266)$$

corresponding to a net gain of energy:

$$\Delta E = \pi \int_0^{\infty} [\Delta v_R(R)]^2 \, \Sigma(R) R \, dR. \qquad (2.267)$$

For example, if both the galaxies are described by a Plummer model (see Subsection 2.2.1) with $\Phi = -GM/\sqrt{r^2 + a^2}$, then the net energy gained is $\Delta E = (G^2 M^3/3V^2 a^2)$. This complements the result for distant tidal encounters, $\Delta E = (4G^2 M_2^2 M_1 \langle r^2 \rangle/3b^4 V^2)$. In fact, any other encounter can be treated as a smooth interpolation between these two.

One of the results of a head-on impact of galaxies is the formation of ring-like structures seen in several disk galaxies. Consider two galaxies modelled as singular isothermal spheres with circular velocities v_c that collide with each other with a speed $V \gg v_c$. The change in the radial velocity of any given star located at a distance R from the line of motion is now given by Eq. (2.266) with $\Phi = v_c^2 \ln r$, so that

$$\Delta v_R = -2R \frac{v_c^2}{V} \int_0^{\infty} \frac{dZ_p}{R^2 + Z_p^2} = -\frac{\pi v_c^2}{V}. \qquad (2.268)$$

We now estimate the change in the trajectory of a star located in the stellar disk that is embedded in the target system. Taking the intruder's trajectory to be normal to the plane of the disk, we can calculate the change in trajectory by using the epicyclic approximation. Originally, the stars at R_0 execute small harmonic radial oscillations with frequency $\kappa(R_0) = \sqrt{2}(v_c/R_0)$. Taking the position of the star to be at $R(t, R_0)$ at time t, we must have $R(t, R_0) = R_0 - a \sin(\kappa t)$, where the amplitude a has to be fixed such that the velocity gained at $t = 0$ is given by expression (2.268). This gives

$$R(R_0, t) = R_0 \left[1 - \frac{\pi v_c}{\sqrt{2} V} \sin\left(\frac{\sqrt{2} v_c t}{R_0} \right) \right]. \qquad (2.269)$$

Figure 2.16 shows a plot of $R(R_0, t)$ in which the caustic structure and the enhancement of density along the lines of constant slope are apparent. This, in turn, suggests the propagation of a ring of material away from the galaxy at nearly constant speed.

If the unperturbed disk had a surface density $\Sigma_0(R_0)$, the new surface density is given by

$$\Sigma(R) = \Sigma_0(R_0) \left[\frac{R}{R_0} \left| \frac{\partial R}{\partial R_0} \right| \right]^{-1} = F(\tau) \Sigma_0(R_0), \qquad (2.270)$$

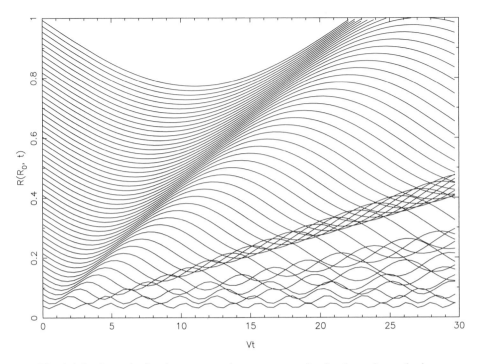

Fig. 2.16.   Perturbation in a penetrating encounter that leads to ring galaxies.

where

$$
\tau \equiv \frac{\sqrt{2}v_c t}{R_0}, \quad F(\tau) \equiv \left(1 - \frac{\pi v_c}{\sqrt{2}\,V}\sin\tau\right)^{-1}\left|1 - \frac{\pi v_c}{\sqrt{2}\,V}(\sin\tau - \tau\cos\tau)\right|^{-1}.
$$

$$(2.271)$$

We note that the new surface density is amplified by a factor $F(\tau)$ that depends on the ratio (t/R_0). This amplification has distinct maxima; if, for example, $(v_c/V) \approx 0.1$, then the maximum occurs at $\tau_m \approx 3.1$. At any given time t, the density is sharply amplified around the radius $R_m(t) = \sqrt{2}(v_c t/\tau_m)$, thereby producing a ringlike enhancement. Numerical simulations show that this result is fairly robust and is valid even when the perturber's trajectory is slightly inclined to the plane of the disk.

2.9.3 *Numerical Simulations*

The interaction and merger of galaxies have been traditionally studied by numerical simulations for several decades now. Although the quality of simulations and the level of details as regards input physics have increased significantly, clear quantitative results that allow us to classify the interactions in some systematic manner do not exist. We shall discuss some of the features of numerical simulations of galaxy encounters in this subsection.

If stars are treated as rigid hard spheres of small radius, interacting only through Newtonian gravity, then setting up a numerically feasible algorithm to study the encounter between two galaxies is relatively straightforward. However, the parameter space that needs to be explored in any such simulation is fairly large. To begin with, we need to specify the distribution function $f_i(\mathbf{x}, \mathbf{v}, 0)$, $i = 1, 2$, at $t = 0$ for the two galaxies. Choosing the coordinates and the Lorentz frame such that the centre of mass of the first galaxy is at rest at the origin at $t = 0$, we can describe the initial conditions for the centre of the second galaxy in terms of the position and velocity $(\mathbf{X_0}, \mathbf{V_0})$. Given these parameters, the integration of the equations of motion will allow the study of future evolution.

The simplest situation will correspond to spherically symmetric distributions of matter in the two galaxies, which are identical. In this case, there is no preferred orientation (for example, spin axis) for the individual galaxies, and we can choose the initial separation vector between the two centres of mass to be along the x axis; the initial velocity of the second galaxy can be chosen to lie in the x–y plane. The centre of masses will move broadly along a Kepler-like orbit, with tidal perturbations affecting the individual movement of the stars. If the two galaxies are gravitationally bound, then the transfer of energy contained in the centre-of-mass motion into the internal kinetic energy of the stars could eventually lead to a merger of the galaxies with the ejection of high-energy stars in the form of a halo. Even in this case, the situation can become complicated at late times, and a diverse class of results can be obtained if the radial distributions of matter in the two galaxies are not identical.

More parameters enter the discussion when the galaxies are not spherical. An extreme case will involve the interaction between two galaxies that are infinitesimally thin disks at $t = 0$. It is again convenient to keep the centre of mass of the first galaxy at the origin and choose a frame in which the first galaxy is at rest at $t = 0$. We can choose the x axis along the direction of the centre of mass of the second disk galaxy at $t = 0$. The second galaxy now has a distinct initial velocity as well as a direction of internal orbital angular momentum. One of these, say, the initial velocity vector, could be taken to lie in the x–y plane. This leaves two more angular coordinates that denote the direction of initial angular momentum of the second galaxy as arbitrary, and the results of the encounter, of course, will depend on this parameter.

Given the amount of arbitrariness in the parameters specifying the initial condition, it is not surprising that a wide variety of morphological features emerge in these simulations. Although these are visually striking and often bear a good resemblence to actual galactic systems observed in the sky, they are not very useful in the study of actual physical processes. In fact, most of the morphological features that emerge in these simulations are purely kinematic and can be reproduced in a fairly simple analysis that does not require genuine N-body simulations. We first describe this approach and some of its results in order to illustrate the importance of kinematics (or, more precisely, "reduced dynamics") in this particular context.

We model the encounter between two disk galaxies that were infinitesimally thin at $t = 0$ along the following lines: (1) All the mass in either of the galaxies will be taken to reside at the centres of the respective galaxies, with the masses being m_1 and m_2. (2) In each of the galaxies, we will distribute stars in circular Keplerian orbits at different radii in the range $r_{min} < r < r_{max}$, with the minimum and the maximum radii as well as the total number of stars being arbitrary for the two galaxies. Within each galaxy the initial distribution is uniform in the sense that the radii of the rings are uniformly spaced between r_{min} and r_{max} and the stars at any given radii are spaced uniformly on the circle. (3) The stars are treated as test particles and are affected by only the masses of the galaxies; that is, the stars do not interact with each other gravitationally. In such a situation, the dynamics can be separated into two parts. The two galaxies move in a Keplerian orbit under their mutual gravitational attraction. Any given star moves in the time-dependent potential generated by the two point masses, m_1 and m_2. The motion of each star is thus equivalent to a reduced version of the three-body problem. A simple numerical code that uses leapfrog integration can now generate wide varieties of morphology for different initial conditions.[7]

As the first example, let us consider the kinematic encounter between two spiral galaxies with the following initial conditions: (1) Both the galaxies are taken to be zero-thickness disks and coplanar with the stars rotating in circular Keplerian orbits around central masses. (2) The target galaxy has the mass of 5 units (one unit of mass corresponds to $2 \times 10^{10} M_\odot$ in these simulations; hence the actual mass is $10^{11} M_\odot$), and the second, the perturbing galaxy, has a mass that is one quarter of that of the target galaxy. (3) There are 1,000 stars distributed equally among 20 rings with radii varying between 10 and 30 units in the target galaxy; there are 250 stars distributed equally among 5 rings varying in radius between 2 and 7 units in the second galaxy. (The unit of distance is chosen such that 1 unit corresponds to 0.5 kpc.) The stars in both the galaxies are rotating in a counterclockwise direction. (4) The target galaxy is located at the origin with zero velocity at $t = 0$, and the perturber has the position $(30, -30, 0)$ and velocity $(0, 0.34, 0.34)$ in the chosen units. (The physical velocity corresponding to 1 unit is 400 km s^{-1}.) Figure 2.17 shows the configuration we obtain by integrating essentially a set of time-dependent three-body problems. The unit of time corresponds to 1.2×10^6 yr. (The units of time, length, and mass are chosen such that $G = 1$ in the numerical code.) It is clear that purely kinematic perturbation – which does not involve any self-gravity of the stars – can lead to striking spiral patterns similar to, for example, the ones seen in the whirpool galaxy.

In the preceding example, the angular momentum of the Kepler motion of the galaxies is in the same direction as the spin angular momentum of the individual stars in the galaxies (called *prograde motion*). This leads to a resonancelike

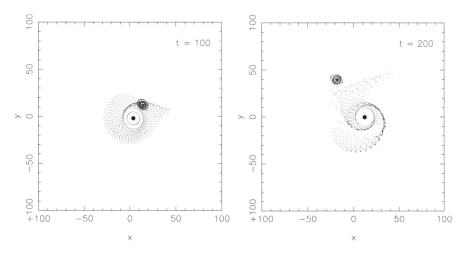

Fig. 2.17. Kinematic simulation of galaxy encounters that lead to a configuration similar to that of a whirpool galaxy.

condition, making the perturbing effect somewhat strong. To see this effect more clearly, we compare it with the retrograde initial condition in which the orbital angular momentum is in a direction opposite to the spin angular momentum. This effect is illustrated more dramatically in Fig. 2.18, in which the initial conditions differ only in the case of the position of the perturbing galaxy. In this case, we have taken the perturbing galaxy to have the same mass as that of the target galaxy

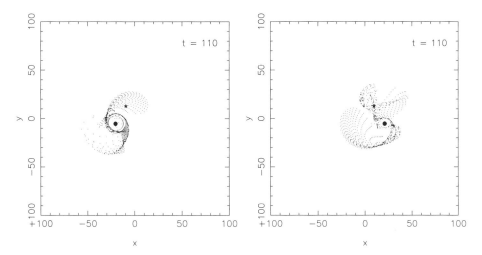

Fig. 2.18. Kinematic simulations of galactic encounters of two equally massive galaxies in retrograde (left) and prograde (right) motions. The stars in the perturber are not shown for simplicity. The retrograde initial conditions in the left frame lead to somewhat less disruption.

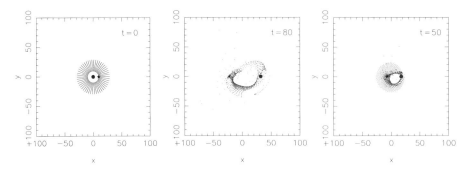

Fig. 2.19.   Kinematic simulation of a plunging galaxy in a near-head-on encounter.

(compared with the previous case shown in Fig. 2.17, in which the perturber had a mass that was 0.25 times that of the target).

Finally, Fig. 2.19 illustrates, through kinematic simulations, the effect that was previously discussed. In this case, the target galaxy has exactly the same parameters as those in the previous case whereas the perturber has the initial position of $(10, 0, 35)$ and initial velocity of $(0, 0, -1)$ so that it can plunge through the target galaxy. The masses of the galaxies are taken to be equal. It is clear from the figure that a ringlike structure can form and propagate outwards.

3

Friedmann Model of the Universe

3.1 Introduction

This chapter deals with the Friedmann model for the universe, which is used throughout the study of extragalactic astronomy. The basic framework needed in all the later chapters is also introduced. Some of the discussion requires concepts from the general theory of relativity developed in Vol. I, Chap. 11. The Latin indices go over $i, j = 0, 1, 2, 3$, and the Greek indices go over $\alpha, \beta = 1, 2, 3$. We shall use units with $c = 1$ in most sections.[1]

3.2 The Friedmann Model

To construct the simplest model of the universe, we begin by assuming that the geometrical properties of three-dimensional space are the same at all spatial locations and that these geometrical properties do not single out any special direction in space. Such a three-dimensional space is called *homogeneous* and *isotropic*.

The geometrical properties of the space are determined by the distribution of matter through Einstein's equations. It follows therefore that the matter distribution should also be homogeneous and isotropic. This is certainly not true in the observed universe, in which there exists a significant degree of inhomogeneity in the form of galaxies, clusters, etc. We assume that these inhomogeneities can be ignored and the matter distribution may be described by a smoothed-out average density in studying the large-scale dynamics of the universe. As we shall see in Chap. 5, there may be some justification in treating the matter distribution as homogeneous at scales larger than $\sim$100 Mpc.

Although this is a well-accepted procedure, the following subtlety must be stressed: Because Einstein's equations are nonlinear, there is no guarantee that the solution we obtain for a distribution of a source that is averaged over some region will be the same as that we obtain by first solving the Einstein's equations exactly and then averaging the solution.[2] In fact, it is possible to construct

161

counterexamples in which the operations of averaging and solving do not com-
mute. In spite of this mathematical difficulty, we continue to treat the geometrical
features of the universe as described by a solution to the Einstein's equation in
which the source is averaged over a sufficiently large scale.

The assumption of homogeneity and isotropy of the three-space singles out a
preferred class of observers, viz., those observers for whom the universe appears
isotropic. Another observer, who is moving with a uniform velocity with respect
to this fundamental class of observers, will find the universe to be anisotropic.
The description of physics will be simplest if we use the coordinate system
appropriate to this fundamental class of observers.

Let us determine the form of the space–time metric in such a coordinate
system (t, x^α) in which homogeneity and isotropy are self-evident. The general
space–time interval (ds^2) can be separated out as

$$ds^2 = g_{ik} \, dx^i \, dx^k = g_{00} \, dt^2 + 2g_{0\alpha} \, dt \, dx^\alpha + g_{\alpha\beta} \, dx^\alpha \, dx^\beta$$
$$\equiv g_{00} \, dt^2 + 2g_{0\alpha} \, dt \, dx^\alpha - \sigma_{\alpha\beta} \, dx^\alpha \, dx^\beta, \tag{3.1}$$

where $\sigma_{\alpha\beta}$ is a positive-definite spatial metric. (It is assumed that any index
that is repeated in a given term is summed over its range of values.) Isotropy of
space implies that $g_{0\alpha}$ must vanish; otherwise, they identify a particular direction
in space related to the three-vector v_α with components $g_{0\alpha}$. Further, in the
coordinate system determined by fundamental observers, we may use the proper
time of clocks carried by these observers to label the spacelike surfaces. This
choice for the time coordinate t implies that $g_{00} = 1$, bringing the space–time
interval to the form

$$ds^2 = dt^2 - \sigma_{\alpha\beta} \, dx^\alpha \, dx^\beta \equiv dt^2 - dl^2. \tag{3.2}$$

The problem now reduces to determining the three-metric $\sigma_{\alpha\beta}$ of a three-space
that, at any instant of time, is homogeneous and isotropic. This can be obtained
as follows.

Note that the assumption of isotropy – which implies spherical symmetry –
can be used to write the line interval in the form

$$dl^2 = a^2[\lambda^2(r) \, dr^2 + r^2(d\theta^2 + \sin^2\theta \, d\phi^2)], \tag{3.3}$$

where $a = a(t)$ can depend only on time. Computing the scalar curvature ^3R for
this three-dimensional space and using the formulas of Vol. 1, Chap. 11, Section
11.6, we find that

$$^3R = \frac{3}{2a^2r^3} \frac{d}{dr}\left[r^2\left(1 - \frac{1}{\lambda^2}\right)\right]. \tag{3.4}$$

Homogeneity implies that all geometrical properties are independent of r; hence

^3R must be a constant. Equating it to a constant and integrating the resulting equation, we get

$$r^2\left(1 - \frac{1}{\lambda^2}\right) = c_1 r^4 + c_2, \quad c_1, c_2 = \text{constants}. \tag{3.5}$$

To avoid any singularity at $r = 0$, we need $c_2 = 0$. Thus we get $\lambda^2 = \left(1 - c_1 r^2\right)^{-1}$. When $c_1 \neq 0$, we can rescale r and make $c_1 = 1$ or -1. This leads to the full space–time metric:

$$ds^2 = dt^2 - a^2(t)\left[\frac{dr^2}{1 - kr^2} + r^2(d\theta^2 + \sin^2\theta \, d\phi^2)\right]. \tag{3.6}$$

The prefactor a determines the overall scale of the spatial metric and, in general, can be a function of time: $a = a(t)$. This metric, called the Friedmann metric, describes a universe that is spatially homogeneous and isotropic at each instant of time. Note that the form of the metric has been determined entirely by symmetry considerations without any reference to the source T_{ik} or Einstein's equations. These geometrical considerations, however, will not allow us to determine the value of k and the form of the function $a(t)$ (called the *expansion factor*). They have to be determined with Einstein's equations once the matter distribution is specified. It is, however, possible to arrive at several general conclusions that are all independent of $a(t)$, which we shall now discuss.

Exercise 3.1
Filling in the details: Obtain relation (3.4).

The coordinates in Eq. (3.6) are, of course, chosen in such a way as to make the symmetries of the space–time self-evident. This coordinate system is called the *comoving* coordinate system. It is easy to show that world lines with $x^\alpha = $ constant are geodesics. To see this, consider a free material particle that is at rest at the origin of the comoving frame at some instant. No velocity can be induced on this particle by the gravitational field as no direction can be considered as special. Therefore the particle will continue to remain at the origin. Because spatial homogeneity allows us to choose any location as the origin, it follows that the world lines $x^\alpha = $ constant are geodesics. Observers following these world lines are called fundamental (or comoving) observers.

The spatial hypersurfaces of the Friedmann universe have positive, zero, and negative curvatures for $k = +1, 0$, and -1, respectively; the magnitude of the curvature, determined from Eq. (3.4), is ($6/a^2$) when k is nonzero. To study the geometrical properties of these spaces it is convenient to introduce a coordinate

χ, defined as

$$\chi = \int \frac{dr}{\sqrt{1 - kr^2}} = \begin{cases} \sin^{-1} r & \text{(for } k = 1) \\ r & \text{(for } k = 0) \\ \sinh^{-1} r & \text{(for } k = -1) \end{cases} . \tag{3.7}$$

In terms of (χ, θ, ϕ) the metric becomes

$$dl^2 = a^2 [d\chi^2 + S_k^2(\chi)(d\theta^2 + \sin^2 \theta d\phi^2)], \tag{3.8}$$

where

$$S_k(\chi) = \begin{cases} \sin \chi & \text{(for } k = +1) \\ \chi & \text{(for } k = 0) \\ \sinh \chi & \text{(for } k = -1) \end{cases} . \tag{3.9}$$

For $k = 0$, the space is the familiar, flat, Euclidian three-space; the homogeneity and the isotropy of this space are obvious. For $k = 1$, Eq. (3.8) represents a three-sphere of radius a embedded in an abstract flat four-dimensional Euclidian space. Such a three-sphere is defined by the relation

$$x_1^2 + x_2^2 + x_3^2 + x_4^2 = a^2, \tag{3.10}$$

where (x_1, x_2, x_3, x_4) are the Cartesian coordinates of some abstract four-dimensional space. We can introduce angular coordinates (χ, θ, ϕ) on the three-sphere by the relations

$$x_1 = a \cos \chi \sin \theta \sin \phi, \quad x_2 = a \cos \chi \sin \theta \cos \phi,$$
$$x_3 = a \cos \chi \cos \theta, \quad x_4 = a \sin \chi. \tag{3.11}$$

We can determine the metric on the three-sphere by expressing dx_i in terms of $d\chi, d\theta$, and $d\phi$ and substituting in the line element:

$$dL^2 = dx_1^2 + dx_2^2 + dx_3^2 + dx_4^2. \tag{3.12}$$

This leads to the metric

$$dL_{(3\text{-sphere})}^2 = a^2 [d\chi^2 + \sin^2 \chi (d\theta^2 + \sin^2 \theta \, d\phi^2)], \tag{3.13}$$

which is the same as Eq. (3.8) for $k = 1$.

The entire three-space of the $k = 1$ model is covered by the range of angles $[0 \leq \chi \leq \pi; 0 \leq \theta \leq \pi; 0 \leq \phi < 2\pi]$ and has a finite volume:

$$V = \int_0^{2\pi} d\phi \int_0^{\pi} d\theta \int_0^{\pi} d\chi \sqrt{g} = a^3 \int_0^{2\pi} d\phi \int_0^{\pi} \sin \theta \, d\theta \int_0^{\pi} \sin^2 \chi \, d\chi = 2\pi^2 a^3. \tag{3.14}$$

The surface area of a two-sphere, defined by $\chi = $ constant, is $S = 4\pi a^2 \sin^2 \chi$. As χ increases, S increases at first, reaches a maximum value of $4\pi a^2$ at $\chi = \pi/2$, and *decreases* thereafter. These are the properties of a three-space that is closed but has no boundaries.

In the case of $k = -1$, Eq. (3.8) represents the geometry of a hyperboloid embedded in an *abstract* four-dimensional space with Lorentzian signature. (This space should not be confused with the physical space–time.) Such a space is described by the line element

$$dL^2 = dx_1^2 + dx_2^2 + dx_3^2 - dx_4^2. \tag{3.15}$$

A three-dimensional hyperboloid, embedded in this space, is defined by the relation

$$x_4^2 - x_1^2 - x_2^2 - x_3^2 = a^2. \tag{3.16}$$

This three-space can be parameterised by the coordinates (χ, θ, ϕ) with

$$x_1 = a \sinh \chi \sin \theta \sin \phi, \quad x_2 = a \sinh \chi \sin \theta \cos \phi,$$
$$x_3 = a \sinh \chi \cos \theta, \quad x_4 = a \cosh \chi. \tag{3.17}$$

Expressing dx_i in terms of $d\chi$, $d\theta$, and $d\phi$ and substituting into Eq. (3.15), we can find the metric on the hyperboloid as

$$dL^2_{(\text{hyperboloid})} = a^2[d\chi^2 + \sinh^2 \chi (d\theta^2 + \sin^2 \theta \, d\phi^2)], \tag{3.18}$$

which is the same as Eq. (3.8) for $k = -1$.

To cover this three-space, we need the range of coordinates to be $[0 \le \chi \le \infty$; $0 \le \theta \le \pi$; $0 \le \phi < 2\pi]$. This space has infinite volume, just as the ordinary flat three-space does. The surface area of a two-sphere, defined by $\chi = \text{constant}$, is $S = 4\pi a^2 \sinh^2 \chi$. This expression increases monotonically with χ.

The notion of homogeneity (and isotropy) of these spaces can be expressed in the following manner: On the surface of a three-sphere, any point can be mapped onto any other point by a suitable rotation. Similarly, any point on the hyperboloid can be mapped on to any other point by a Lorentz-like transformation, which is a rotation by an imaginary angle (see Vol. 1, Chap. 3, Section 3.3). These mappings clearly leave the metric – and geometrical properties – of the three-space invariant. This shows that all points on these surfaces (and all directions) are physically equivalent.

Exercise 3.2
Properties of hyperboloid: (a) Although the three-sphere has a formal similarity to the two-sphere, there are certain significant differences. Show that the spatial sections of the $k = 1$ Friedmann model possess a translation symmetry that leaves *no* points fixed. (This will not be true in two dimensions; it is not possible to "comb" the two-sphere smoothly.) (b) Consider a three-dimensional space of velocities (v_x, v_y, v_z). Define a distance between any two nearby points A and B in this space to be the relative velocity of the observers having the velocities A and B. (The relative velocity has to be, of course,

found by use of special relativity.) Show that the metric in this space is

$$dV^2 = dv_r^2 + \sinh^2 v_r (dv_\theta^2 + \sin^2 \theta \, dv_\phi^2), \tag{3.19}$$

which has the form of a $(k = -1)$ Friedmann model.

Friedmann universes with $k = -1, 0$, and 1 are called *open, flat,* and *closed,* respectively. These terms refer to the topological nature of the three-space. The following point, however, should be noted: Our symmetry considerations (and Einstein's equations, which will be discussed later in Section 3.4) can determine the local geometry of only the space–time and not the global topology. Consider, for example, the $k = 0$ model that has infinite volume and a spatial topology of $\mathcal{R}^3$, if we allow the coordinates to take the full possible range of values: $-\infty < (x, y, z) < +\infty$. We could, however, identify the points with coordinates (x, y, z) and $(x + L, y + L, z + L)$, thereby changing the topology of this space to $\mathcal{S}^3$. Thus our considerations do not uniquely specify the topology of the space–time. The choices we have made for the three cases $k = 0, \pm 1$ discussed above should only be considered as the most natural choices for these models.

The full Friedmann metric in Eq. (3.6) can be expressed in either (t, r, θ, ϕ) coordinates or in (t, χ, θ, ϕ) coordinates. Sometimes it is convenient to use a different time coordinate, τ, related to t by $d\tau = a^{-1}(t) \, dt$. In the $(\tau, \chi, \theta, \phi)$ coordinates the Friedmann metric becomes

$$ds^2 = a^2(\tau)[d\tau^2 - d\chi^2 - S_k^2(\chi)(d\theta^2 + \sin^2 \theta \, d\phi^2)], \tag{3.20}$$

with $S_k^2(\chi)$ given by Eq. (3.9). In this form, all the time dependence is isolated into an overall multiplicative factor.

Exercise 3.3
Friedmann model in spherically symmetric coordinates: Express the Friedmann metric in a spherically symmetric coordinate system using a set of coordinates (T, R, θ, ϕ) in which the line interval must have the form

$$
\begin{aligned}
ds^2 &= a^2(\tau)[d\tau^2 - d\chi^2 - S_k^2(\chi)(d\theta^2 + \sin^2 \theta \, d\phi^2)] \\
&= e^\nu dT^2 - e^\lambda dR^2 - R^2(d\theta^2 + \sin^2 \theta \, d\phi^2),
\end{aligned}
\tag{3.21}
$$

where $\nu(T, R)$ and $\lambda(T, R)$ are functions of T and R. (a) Show that the transformation is of the form $R = ra(t)$, $T = F(q)$, where

$$q \equiv \int^r \frac{x \, dx}{1 - kx^2} + \int^t \frac{dy}{a(y)\dot{a}(y)} \tag{3.22}$$

and F is an *arbitrary* function.
(b) Determine the Newtonian limit of Friedmann metric in which the metric is given by

$$ds^2 \cong (1 + 2\phi_N) \, dT^2 - dR^2 - R^2(d\theta^2 + \sin^2 \theta \, d\phi^2), \tag{3.23}$$

where ϕ_N is the Newtonian gravitational potential. Show that

$$\phi_N(\mathbf{R}, t) = -\frac{1}{2}\frac{\ddot{a}}{a}R^2 \tag{3.24}$$

in the weak-gravity limit at scales $R^2 \ll (\ddot{a}/a)^{-1}$. Interpret this potential.

Exercise 3.4
Conformally flat form of the metric: The metric of the $k = 0$ Friedmann universe can be expressed in the form $g_{ik} = \Omega^2 \eta_{ik}$, where η_{ik} is the flat (Lorentzian) metric and $\Omega = \Omega(T)$. Even for the $k = \pm 1$ models, the metric can be reduced to the form $g_{ik} = \Omega^2 \eta_{ik}$, where $\Omega = \Omega(t, \mathbf{x})$ now depends on spatial coordinates as well. Construct this coordinate system.

3.3 Kinematics of the Friedmann Model

We now discuss several features related to the propagation of radiation and motion of material particles in the Friedmann universe. These results are independent of the explicit form of $a(t)$ and will be of use in the discussions that follow.

Because $a(t)$ multiplies the spatial coordinates, any proper distance $l(t)$ between spatial locations will change with time in proportion to $a(t)$:

$$l(t) = l_0 a(t) \propto a(t). \tag{3.25}$$

In particular, the proper separation between two observers, located at constant comoving coordinates, will change with time. Let the comoving separation between two such observers be δx, so that the proper separation is $\delta l = a(t)\delta x$. Each of the two observers will attribute to the other a velocity

$$\delta v = \frac{d}{dt}\delta l = \dot{a}\delta x = \left(\frac{\dot{a}}{a}\right)\delta l. \tag{3.26}$$

Consider now a narrow pencil of electromagnetic radiation that crosses these two comoving observers. The time for transit will be $\delta t = \delta l$. Let the frequency of the radiation measured by the first observer be ω. Because the first observer sees the second one as *receding* with velocity δv, the first observer will expect the second observer to measure a Doppler-shifted frequency $(\omega + \delta \omega)$, where

$$\frac{\delta \omega}{\omega} = -\frac{\delta v}{c} = -\delta v = -\frac{\dot{a}}{a}\delta l = -\frac{\dot{a}}{a}\delta t = -\frac{\delta a}{a}. \tag{3.27}$$

(Because the observers are separated by an infinitesimal distance of first order, δl, we can introduce a locally inertial frame that encompasses both the observers. The laws of special relativity can be applied in this frame.) This equation can be integrated to give

$$\omega(t)a(t) = \text{constant}. \tag{3.28}$$

In other words, the frequency of electromagnetic radiation changes because of expansion of the universe according to the law $\omega \propto a^{-1}$.

The preceding analysis did not use the wave nature of the electromagnetic radiation directly, although it is implicit in the Doppler-shift formula applied in the local frame. It is, of course, possible to obtain the same result directly from the solution to Maxwell's equations in Friedmann geometry. The dynamics of the electromagnetic field in curved space–time is described by the action

$$A_{\text{elec}} = \frac{1}{16\pi} \int F_{ik} F^{ik} \sqrt{-g}\, d^4x, \quad F_{ik} = \frac{\partial A_k}{\partial x^i} - \frac{\partial A_i}{\partial x^k} \tag{3.29}$$

(see Vol. I, Chap. 11, Section 11.5). Consider the transformations (called *conformal transformations*)

$$A_i \to A_i, \quad x^i \to x^i, \quad g_{ik} \to \Omega^2 g_{ik}, \quad g^{ik} \to \Omega^{-2} g^{ik} \tag{3.30}$$

where $\Omega(x^i)$ is an arbitrary nonsingular function of the space–time coordinates. Note that $(A^i)_{\text{new}} = g^{ik}_{\text{new}}\, (A_k)_{\text{new}} = \Omega^{-2} A^i_{\text{old}}$. Quite clearly,

$$F_{ik} F_{mn} g^{mi} g^{nk} \sqrt{-g} \to F_{ik} F_{mn} (\Omega^{-2} g^{mi})(\Omega^{-2} g^{nk})(\Omega^4 \sqrt{-g})$$

$$= F_{ik} F_{mn} g^{mi} g^{nk} \sqrt{-g}, \tag{3.31}$$

showing that A_{elec} is invariant under conformal transformations that, in turn, implies that Maxwell's equations and their solutions will be conformally invariant. The $k = 0$ Friedmann universe in the $(\tau, \chi, \theta, \phi)$ coordinate system is conformally flat with $g_{ik} = a^2(\tau) g^{\text{SR}}_{ik}$ where $g^{\text{SR}}_{ik} = \text{dia}(1, -1, -1, -1)$ is the standard metric of flat space–time. Because the electromagnetic field is conformally invariant, solutions to the wave equation for A_i in the metric g_{ik} are the same as the solutions in the flat space–time metric g^{SR}_{ik} and have the time dependence

$$A_i \propto \exp(-ik\tau) = \exp\left[-ik \int \frac{dt}{a(t)}\right]. \tag{3.32}$$

Because the time derivative of the phase of the wave defines the (instantaneous) frequency, we conclude that

$$\omega(t)a(t) = \text{constant}. \tag{3.33}$$

It is clear from the nature of the argument – and the form of the line element in Eq. (3.20) – that the result is also valid in $k \neq 0$ models. Even in this case, the *time* dependence of A_i is the same as that of Eq. (3.32).

The preceding result is of considerable importance. It shows that if the radiation is emitted by some source at time t_e and is observed at time $t = t_o$, then its wavelength will increase if $a(t_o) > a(t_e)$. In such an expanding phase of the universe, where $a(t)$ is an increasing function of time, we can associate a redshift

z with any time t in the past by the relation

$$1 + z(t) \equiv \frac{a(t_0)}{a(t)} \equiv \frac{a_0}{a(t)}, \qquad (3.34)$$

where t_0 is the value of time parameter at the current epoch and a_0 is the value of expansion factor at present. Thus we can use the variables t, a, or z interchangeably in a universe in which $a(t)$ is a monotonically increasing function of time.

For future use, we will derive the corresponding result in a slightly perturbed Friedmann model. Suppose the metric of the space–time is slightly different from that of a Friedmann model and is given by

$$ds^2 = dt^2 - a^2(t)[\delta_{\alpha\beta} - h_{\alpha\beta}(t, x)] dx^\alpha dx^\beta. \qquad (3.35)$$

A photon emitted by a distant source propagates through this space–time and is received by an observer at the present epoch. The proper distance between two observers, with a comoving separation $n^\alpha \delta x$, along the path of the photon is now given by

$$\delta l \cong a(t)\left[1 - \frac{1}{2} h_{\alpha\beta} n^\alpha n^\beta\right](\delta x) + \mathcal{O}(h^2), \qquad (3.36)$$

where we have retained only the terms that are linear in $h_{\alpha\beta}$. Using an argument similar to the one that led to Eq. (3.27), we find that

$$\frac{\delta\omega}{\omega} = -\frac{\delta l}{c}\frac{d}{dt} \ln[\delta l] = -\frac{\delta l}{c}\left[\frac{\dot{a}}{a} - \frac{1}{2}\dot{h}_{\alpha\beta} n^\alpha n^\beta\right] = -\frac{\delta a}{a} + \frac{1}{2}\dot{h}_{\alpha\beta} n^\alpha n^\beta \delta t \qquad (3.37)$$

or

$$\delta \ln(\omega a) = \frac{1}{2}\dot{h}_{\alpha\beta} n^\alpha n^\beta \delta t. \qquad (3.38)$$

Integrating both sides along the photon path from the moment of emission t_1 to the moment of absorption t_2, we find

$$\ln\left(\frac{\omega_2 a_2}{\omega_1 a_1}\right) = \frac{1}{2}\int_{t_1}^{t_2} \dot{h}_{\alpha\beta} n^\alpha n^\beta \, dt. \qquad (3.39)$$

This result will be useful in Chap. 6.

The expansion of the universe also affects the motion of material particles. Consider again two comoving observers separated by a proper distance δl. Let a material particle pass the first observer with velocity v. When it has crossed the proper distance δl (in a time interval δt), it passes the second observer whose velocity (relative to the first one) is

$$\delta u = \frac{\dot{a}}{a}\delta l = \frac{\dot{a}}{a}v dt = v\frac{\delta a}{a}. \qquad (3.40)$$

The second observer will attribute to the particle the velocity

$$v' = \frac{v - \delta u}{1 - v\delta u} = v - (1 - v^2)\,\delta u + \mathcal{O}[(\delta u)^2] = v - (1 - v^2)v\frac{\delta a}{a}. \quad (3.41)$$

This follows from the special relativistic formula for the addition of velocities, which is valid in an infinitesimal region around the first observer. Rewriting this equation as

$$\delta v = -v(1 - v^2)\frac{\delta a}{a} \quad (3.42)$$

and integrating, we get

$$p \equiv \frac{v}{\sqrt{1 - v^2}} = \frac{\text{constant}}{a}. \quad (3.43)$$

In other words, the magnitude of the three-momentum decreases as a^{-1} because of the expansion. If the particle is nonrelativistic, then $v \propto p$ and velocity itself decays as a^{-1}. This result can also, of course, be derived from the study of the geodesics in the Friedmann universe (see Exercise 3.6).

Exercise 3.5
Particle velocity: A material particle is released with some initial velocity in a Friedmann model. (a) Show that, in the $k = -1$ model, as $t \to \infty$, the velocity of the particle approaches that of some fundamental observer but the position of the particle is at constant proper distance from this observer. (b) What happens in the $k = 0$ and the $k = +1$ models?

Exercise 3.6
Geodesic equation: Derive the result $p(t) \propto a(t)^{-1}$ by studying the geodesic equation in the Friedmann universe. (a) Consider a particle travelling along $\theta = $ constant, $\phi = $ constant. Show that the zeroth component of the geodesic equation reads as

$$\frac{d^2t}{ds^2} + \frac{a\dot{a}}{1 - kr^2}\left(\frac{dr}{ds}\right)^2 = 0. \quad (3.44)$$

(b) Eliminate (dr/ds) between Eq. (3.44) and the first integral,

$$\left(\frac{dt}{ds}\right)^2 - \frac{a^2}{1 - kr^2}\left(\frac{dr}{ds}\right)^2 = 1, \quad (3.45)$$

to obtain

$$\frac{d^2t}{ds^2} + \frac{\dot{a}}{a}\left[\left(\frac{dt}{ds}\right)^2 - 1\right] = 0. \quad (3.46)$$

(c) Integrate this equation to obtain $a[(dt/ds)^2 - 1] = $ constant. If $p^a = (dx^a/ds)$ is the four-momentum of the particle, then the condition $p^a p_a = 1$ reduces to $[(dt/ds)^2 - \sigma_{\alpha\beta} \, p^\alpha p^\beta] = 1$. Show that $\sigma_{\alpha\beta} \, p^\alpha p^\beta \equiv |\mathbf{p}|^2 = (\text{constant}/a^2)$. (d) The geodesic for a particle, moving in a space–time with metric g_{ab}, can be obtained most efficiently from the Hamilton–Jacobi equation

$$g^{ab} \frac{\partial A}{\partial x^a} \frac{\partial A}{\partial x^b} = m^2, \tag{3.47}$$

where A is the action. Write this equation in the Friedmann metric and reduce the problem of determining the radial geodesics to quadrature.

One conclusion that can be reached immediately from Eq. (3.43) is the following: Consider a stream of particles propagating freely in the space–time. At some time t, a comoving observer finds dN particles in a proper volume dV, all having momentum in the range $(\mathbf{p}, \mathbf{p} + d^3\mathbf{p})$. The phase-space distribution function $f(\mathbf{x}, \mathbf{p}, t)$ for the particles is defined by the relation $dN = f \, dV d^3\mathbf{p}$. At a later instant $(t + \delta t)$, the proper volume occupied by these particles would have increased by a factor of $[a(t + \delta t)/a(t)]^3$ whereas the volume in the momentum space would be redshifted by $[a(t)/a(t + \delta t)]^3$, showing that the phase volume occupied by the particles does not change during the free propagation. Because the number of particles dN is also conserved, it follows that f is conserved along the streamline.

The preceding considerations have been purely classical. It is, however, sometimes convenient to think of electromagnetic radiation as consisting of photons with zero rest mass, for which $E = \hbar\omega = p$. The decay law $p \propto a^{-1}$ then implies the redshift for radiation that was derived earlier. A few other properties of radiation can also be derived easily by use of the photon concept. Let $f_\gamma(t, \mathbf{x}, \mathbf{p})$ be the phase-space density of photons that is conserved during the propagation. This conserved number of photons per unit phase-space volume, f_γ, can be expressed in the form

$$f_\gamma(t, \mathbf{x}, \mathbf{p}) = \frac{dN_\gamma}{d^3\mathbf{x} \, d^3\mathbf{p}} = \frac{dN_\gamma}{[c \, dt_e \, dA_e][\omega_e^2 \, d\omega_e \, d\Omega_e]} = \frac{dN_\gamma}{[c \, dt_r \, dA_r][\omega_r^2 \, d\omega_r \, d\Omega_r]}. \tag{3.48}$$

We have written the momentum-space volume as $d^3\mathbf{p} \propto p^2 \, dp \, d\Omega \propto \omega^2 \, d\omega \, d\Omega$, where $d\Omega$ is a solid angle around the direction of propagation and $d^3\mathbf{x} \propto c \, dt \, dA$, where dA is the area normal to the direction of propagation. The subscripts e and r represent the processes of emission and reception of photons, respectively. This shows that

$$\frac{dN}{dt \, dA \, d\omega \, d\Omega \, \omega^2} = \text{invariant.} \tag{3.49}$$

Because the energy is related to number of photons by $dE = \hbar\omega \, dN$ and the intensity is defined by $I \equiv (dE/dt\, dA\, d\omega\, d\Omega)$, it immediately follows that

$$I_{\text{rec}} = \left(\frac{\omega_{\text{rec}}}{\omega_{\text{em}}}\right)^3 I_{\text{em}}. \tag{3.50}$$

In other words, (I/ω^3) is invariant.

The quantity $I(\omega)$ will have units (for example) of ergs $\text{s}^{-1}\,\text{cm}^{-2}\,\text{Hz}^{-1}\,\text{sr}^{-1}$. The energy density $U(\omega)$ is defined as $U = (4\pi/c)I$ and will have the dimensions of ergs $\text{cm}^{-3}\,\text{Hz}^{-1}$. It is clear from the invariance of $[I/\omega^3]$ that I and U vary as a^{-3} in the expanding universe. More precisely, taking the redshift of the frequency into account, we obtain

$$I[\omega(1+z); z] = I[\omega; 0](1+z)^3. \tag{3.51}$$

The total flux of radiation, obtained by integration over all frequencies, varies as $(1+z)^4$:

$$F_{\text{total}} = \int_0^\infty I\, d\omega \propto (1+z)^4. \tag{3.52}$$

Radiation that has an intensity distribution of the form $I(\omega) = \omega^3 G(\omega/T)$ will retain the spectral shape during the expansion, with the parameter T varying with expansion as $T \propto a^{-1}$. The Planck spectrum has this form in which T corresponds to the temperature; it follows that the temperature of the radiation, which has Planck spectrum, decreases with expansion as $T(t) \propto a(t)^{-1}$.

Cosmological observations are mostly based on electromagnetic radiation that is received from faraway sources. Let an observer, located at $r = 0$, receive at time $t = t_0$ radiation from a source located at $r = r_{\text{em}}$. This radiation must have been emitted at some earlier time t_e such that the events (t_e, r_{em}) and $(t_0, 0)$ are connected by a null geodesic. Taking the propagation of the ray to be along $\theta =$ constant and $\phi =$ constant, we can write the equation for the null geodesic as

$$0 = ds^2 = dt^2 - a^2(t)\frac{dr^2}{1 - kr^2}. \tag{3.53}$$

Integrating this, we can find the relation between r_{em} and t_e:

$$\int_{t_e}^{t_0} \frac{dt}{a(t)} = \int_0^{r_{\text{em}}} \frac{dr}{(1 - kr^2)^{1/2}}. \tag{3.54}$$

The left-hand side is a definite function of time in a given cosmological model. Because the redshift z is also a unique function of time, we can express the left-hand side as a function of z, and hence r_{em} can be expressed as a function of z. This function $r_{\text{em}}(z)$ is of considerable use in observational astronomy, as it relates the radial distance of an object (from which the light is received) to the redshift at which the light is emitted. We shall see later in Section 3.4 that Einstein's equations allow us to determine $(\dot{a}/a)$ more easily than $a(t)$ itself. It is therefore convenient to define a quantity (called the *Hubble radius*) by

$$d_H(t) = d_H(z) \equiv \left(\frac{\dot{a}}{a}\right)^{-1}, \tag{3.55}$$

where the first equality shows that d_H can also be thought of as a function of the redshift z. This quantity allows us to convert integrals over dt to integrals over dz by using

$$dt = \left(\frac{dt}{da}\right)\left(\frac{da}{dz}\right) dz = -d_H(z)\left(\frac{dz}{1+z}\right). \tag{3.56}$$

It is now possible to write Eq. (3.54) as

$$\frac{1}{a_0}\int_0^z d_H(z)\,dz = S_k^{-1}(r_{em}), \tag{3.57}$$

where $S_k^{-1}(x) = [\sinh^{-1}(x), x, \sin^{-1}(x)]$ for $k = -1, 0, +1$, respectively. Therefore $r_{em}(z)$ can be written as

$$r_{em}(z) = S_k(\alpha), \quad \alpha \equiv \frac{1}{a_0}\int_0^z d_H(z)\,dz. \tag{3.58}$$

Two important observable quantities in which $r_{em}(z)$ plays a role are in (1) relating the luminosity of distant objects with the observed flux and (2) the measurement of angular sizes of distant objects, which we shall now discuss. Let $\mathcal{F}$ be the flux received from a source of luminosity $\mathcal{L}$ when the photons from the source reach us with a redshift z. The flux can be expressed as

$$\mathcal{F} = \frac{1}{(\text{area})}\left(\frac{dE_{rec}}{dt_{rec}}\right). \tag{3.59}$$

Using Eq. (3.50) and the fact that the proper area at $a = a_0$ of a sphere of comoving radius $r_{em}(z)$ is $4\pi a_0^2 r_{em}^2$, we get

$$\mathcal{F} = \frac{1}{4\pi a_0^2 r_{em}^2}\frac{1}{(1+z)^2}\left(\frac{dE_{em}}{dt_{em}}\right) = \frac{1}{4\pi a_0^2 r_{em}^2}\frac{1}{(1+z)^2}\mathcal{L}, \tag{3.60}$$

where $\mathcal{L} = (dE_{em}/dt_{em})$ is the luminosity of the source. Because the distance to an object of luminosity $\mathcal{L}$ and flux $\mathcal{F}$ can be expressed as $(\mathcal{L}/4\pi\mathcal{F})^{1/2}$ in flat Euclidian geometry, it is convenient to define distance $d_L(z)$, called the *luminosity distance*, by

$$d_L(z) \equiv \left(\frac{\mathcal{L}}{4\pi\mathcal{F}}\right)^{1/2} = a_0 r_{em}(z)(1+z) = a_0(1+z)S_k(\alpha). \tag{3.61}$$

The preceding expressions use the total luminosity and the total flux, which are quantities integrated over all frequencies. It is often necessary to work with luminosity L_ν and flux F_ν at some given frequency ν. To find the relation between these two quantities, we first note that – purely from dimensions – νF_ν and νL_ν should be related in the same way as L and F except for the redshift of the frequency. Hence $\nu_{ob} F(\nu_{ob}) = \nu_{em} L(\nu_{em})/(4\pi d_L^2)$ where $\nu_{em} = \nu_{ob}(1 + z)$. It follows that

$$F(\nu) = \frac{L[\nu(1 + z)](1 + z)}{4\pi d_L^2}. \tag{3.62}$$

Note that from Eqs. (3.58) and (3.61) we have

$$\alpha = \frac{1}{a_0} \int_0^z d_H(z)\, dz = S_k^{-1}\left[\frac{d_L(z)}{a_0(1 + z)}\right]. \tag{3.63}$$

Differentiating with respect to z gives

$$d_H(z) = \left[1 - \frac{k\, d_L^2(z)}{a_0^2(1 + z)^2}\right]^{-1/2} \frac{d}{dz}\left[\frac{d_L(z)}{1 + z}\right]. \tag{3.64}$$

Thus if $d_L(z)$ can be determined from observations, we can determine $d_H(z)$ and thus $(\dot{a}/a)$.

Another observable parameter for distant sources is the angular diameter. If D is the physical size of the object that subtends an angle δ to the observer, then, for small δ, we have $D = r_{em} a(t_e)\delta$. The "angular diameter distance" $d_A(z)$ for the source is defined by the relation $\delta = (D/d_A)$; so we find that

$$d_A(z) = r_{em} a(t_e) = a_0 r_{em}(t_e)(1 + z)^{-1}. \tag{3.65}$$

Quite clearly $d_L = (1 + z)^2\, d_A$.

The proper volume in the universe spanned by the region between comoving radii r_{em} and $r_{em} + dr_{em}$ can also be expressed in terms of $d_H(z)$ and $r_{em}(z)$. This infinitesimal proper volume element is given by

$$dV = a^3 \frac{dr_{em}}{\sqrt{1 - kr_{em}^2}} r_{em}^2 \sin\theta\, d\theta\, d\phi. \tag{3.66}$$

If we interpret this volume as the volume of the universe spanned along the backward light cone by the photons that are received today with redshifts between z and $z + dz$, then we can relate dr_{em} to dt_{em} by Eq. (3.54) and obtain

$$dV = a^2\, dt\left(r_{em}^2 \sin\theta\, d\theta\, d\phi\right) = \frac{a_0^2 r_{em}^2(z)\, d_H(z)}{(1 + z)^3} dz \sin\theta\, d\theta\, d\phi. \tag{3.67}$$

To proceed any further with any of these relations, such as Eqs. (3.61), (3.65), or (3.67), we need to know the form of $d_H(z)$, which in turn depends on the functional form of $a(t)$. If, however, we are interested in only small z [i.e., in a small time interval (t_0-t_e)] then we can Taylor expand $a(t)$ around t_0 and parameterise it by the coefficients of the Taylor expansion. It is conventional to write this expansion in the following form:

$$a(t) = a(t_0)\left[1 + \left(\frac{\dot{a}}{a}\right)_0 (t - t_0) + \frac{1}{2}\left(\frac{\ddot{a}}{a}\right)_0 (t - t_0)^2 + \cdots\right]$$

$$= a(t_0)\left[1 + H_0(t - t_0) - \frac{1}{2}q_0 H_0^2 (t - t_0)^2 + \cdots\right],$$

(3.68)

where

$$H_0 \equiv \left(\frac{\dot{a}}{a}\right)_{t=t_0}, \qquad q_0 \equiv -\left(\frac{\ddot{a}a}{\dot{a}^2}\right)_{t=t_0}.$$

(3.69)

Substituting this expansion into Eq. (3.54), expanding up to quadratic order in $(t - t_0)$ and r_{em}, and integrating, we get the relation between r_{em} and t_e:

$$r_{em} = \frac{1}{a_0}\left[(t_0 - t_e) + \frac{1}{2}H_0(t_e - t_0)^2 + \cdots\right].$$

(3.70)

Inverting Eq. (3.68), we can express $(t - t_0)$ in terms of $(1 + z) = (a_0/a)$:

$$(t - t_0) = -H_0^{-1}\left[z - \left(1 + \frac{q_0}{2}\right)z^2 + \cdots\right].$$

(3.71)

Finally, substituting Eq. (3.71) into Eq. (3.70), we can express r_{em} in terms of z:

$$a_0 r_{em} = H_0^{-1}\left[z - \frac{1}{2}(1 + q_0)z^2 + \cdots\right].$$

(3.72)

This shows that, to first order in z, the "redshift velocity" $v \equiv cz$ is proportional to the proper distance $a_0 r_{em}$. We can now use this relation to express d_L (and d_A) in terms of z; we find, for example,

$$d_L(z) = a_0 r_{em}(1 + z) = H_0^{-1}\left[z + \frac{1}{2}(1 - q_0)z^2 + \cdots\right].$$

(3.73)

We can determine the quantity d_L by measuring the flux $\mathcal{F}$ for a class of objects for which the intrinsic luminosity $\mathcal{L}$ is known. If we can also measure the redshift z for these objects, then a plot of d_L against z will allow us to determine the parameters H_0 and q_0. Observations suggest that

$$H_0 = 100h \text{ km s}^{-1} \text{ Mpc}^{-1}, \quad 0.5 \lesssim h \lesssim 1.$$

(3.74)

The quantity H_0, called the *Hubble constant*, which determines the rate at which the universe is expanding today, is of primary significance in cosmology. From H_0 we can construct the time scale $t_{univ} \equiv H_0^{-1} = 9.8 \times 10^9 h^{-1}$ yr and the length scale $l_{univ} \equiv ct_{univ} \cong 3000h^{-1}$ Mpc. These are the characteristic scales over which global effects of cosmological expansion will be important today.

3.4 Dynamics of the Friedmann Model

We now consider the evolution of the Friedmann model for different kinds of sources that populate the universe. The Friedmann metric contains a constant k and a function $a(t)$, both of which can be determined by Einstein's equations,

$$G_k^i = R_k^i - \frac{1}{2}\delta_k^i R = 8\pi G T_k^i, \qquad (3.75)$$

if the stress tensor for the source is specified. The assumption of isotropy implies that T_0^μ must be zero and that the spatial components T_β^α must have a diagonal form with $T_1^1 = T_2^2 = T_3^3$; homogeneity requires that all the components be independent of the spatial coordinate $\mathbf{x}$. It is conventional to write such a stress tensor as

$$T_k^i = \text{dia}[\rho(t), -p(t), -p(t), -p(t)]. \qquad (3.76)$$

The notation is suggested by the fact that if the source is an ideal fluid with energy density ρ and pressure p, then the stress tensor will have the preceding form (see Vol. I, Chap. 11, Section 11.7). Of course, there is no need for the source to be an ideal fluid, and hence this notation is only suggestive. In particular, there is no restriction that p should be positive. The tensor G_k^i on the left-hand side of Eq. (3.75) can be computed for the Friedmann metric in a straightforward manner. The nontrivial components are

$$G_0^0 = \frac{3}{a^2}(\dot{a}^2 + k), \qquad G_\nu^\mu = \frac{1}{a^2}(2a\ddot{a} + \dot{a}^2 + k)\delta_\nu^\mu. \qquad (3.77)$$

Thus Eq. (3.75) gives two independent equations:

$$\frac{\dot{a}^2 + k}{a^2} = \frac{8\pi G}{3}\rho, \qquad (3.78)$$

$$\frac{2\ddot{a}}{a} + \frac{\dot{a}^2 + k}{a^2} = -8\pi G p. \qquad (3.79)$$

These two equations, combined with the equation of state $p = p(\rho)$, completely determine the three functions $a(t)$, $\rho(t)$, and $p(t)$. We first discuss some general features of these equations.

From Eq. (3.78) it follows that

$$\frac{k}{a^2} = \frac{8\pi G}{3}\rho - \frac{\dot{a}^2}{a^2} = \frac{\dot{a}^2}{a^2}\left[\frac{\rho}{(3H^2/8\pi G)} - 1\right], \qquad H(t) \equiv \frac{\dot{a}}{a}, \qquad (3.80)$$

which suggests defining *critical density* $\rho_c(t)$ and a *density parameter* $\Omega(t)$ by

$$\rho_c(t) \equiv \frac{3H^2(t)}{8\pi G}, \qquad \Omega(t) \equiv \frac{\rho}{\rho_c}. \qquad (3.81)$$

These definitions are valid at any time t. Using these definitions and evaluating Eq. (3.80) at $t = t_0$, we get

$$\frac{k}{a_0^2} = H_0^2(\Omega - 1). \tag{3.82}$$

It is conventional to drop the subscript 0 in Ω_0 and use the notation $\Omega [\equiv \Omega(t_0)]$ when referring to the values at the current epoch. It is clear that $k = -1, 0$, or 1, depending on whether $\Omega < 1$, $\Omega = 1$, or $\Omega > 1$, respectively. Thus Ω determines the spatial geometry of the universe. Given H_0 and Ω and that $k \neq 0$, we can find a_0 from Eq. (3.82):

$$a_0 = H_0^{-1}(|\Omega - 1|)^{-1/2}. \tag{3.83}$$

The value of $\dot{a}_0$ can be fixed by

$$\dot{a}_0 = H_0 a_0 = (|\Omega - 1|)^{-1/2}. \tag{3.84}$$

Hence (H_0, Ω) are valid initial conditions for integrating Eqs. (3.78) and (3.79). [If $k = 0$, Einstein's equations allow the scaling $a \to \mu a$, making the normalisation of a arbitrary. It is conventional to take $a_0 = 1$ if $k = 0$; then the value of $(\dot{a}_0/a_0) = H_0 = \dot{a}_0$ determines $\dot{a}_0$.]

Substituting for $(\dot{a}^2 + k)/a^2$ into Eq. (3.79) from Eq. (3.78) and rearranging, we get

$$\frac{\ddot{a}}{a} = -\frac{4\pi G}{3}(\rho + 3p). \tag{3.85}$$

For normal matter, $(\rho + 3p) > 0$, implying that $\ddot{a} < 0$. The $a(t)$ curve (which has positive $\dot{a}$ at the present epoch t_0) must be convex; in other words, a will be smaller in the past and will become zero at sometime (in the past), say, at $t = t_{\text{sing}}$. It is also clear $(t_0 - t_{\text{sing}})$ must be less than value of the intercept $(\dot{a}/a)_0^{-1} = H_0^{-1}$. For convenience, we choose the time coordinate such that $t_{\text{sing}} = 0$, i.e., we take $a = 0$ at $t = 0$. In that case, the present "age" of the universe t_0 satisfies the inequality $t_0 < t_{\text{univ}}$, where

$$t_{\text{univ}} \equiv H_0^{-1} = 3.1 \times 10^{17} h^{-1} \text{ s} = 9.8 \times 10^9 h^{-1} \text{ yr}. \tag{3.86}$$

As a becomes smaller the components of the curvature tensor R^i_{klm} become larger, and when $a = 0$ these components diverge. Such a divergence (called *singularity*) is an artifact of our theory. When the radius of curvature of the space–time becomes comparable with the fundamental length $(G\hbar/c^3)^{1/2} \simeq 10^{-33}$ cm, constructed out of G, $\hbar$, and c, quantum effects of gravity will become important, rendering the classical Einstein's equations invalid. Therefore, in reality, t_0 is the time that has elapsed from the moment at which the classical equations became valid.

The quantities ρ and p are *defined* in Eq. (3.76) as the T_0^0 and T_1^1 (say) components of the stress tensor. The interpretation of p as "pressure" depends on treating the source as an ideal fluid. The source for a Friedmann model should *always* have the form in Eq. (3.76); but, as previously stated, if the source is not an ideal fluid then it is not possible to interpret the spatial components of T_k^i as pressure. It is therefore quite possible that the equation of state for matter at high energies does not obey the condition $(\rho + 3p) > 0$. The violation of this condition may occur at a later epoch (i.e., at a larger value of a) than the epoch at which the quantum gravitational effects become important. If this happens, then the "age of the universe" refers to the time interval since the breakdown of the condition $(\rho + 3p) > 0$.

We now consider the question of explicitly solving Einstein's equations. From Eq. (3.78), we see that $\rho a^3 = (3/8\pi G)a(\dot{a}^2 + k)$; differentiating this expression and using Eq. (3.79) we get

$$\frac{d}{dt}(\rho a^3) = -3a^2\dot{a}p = -p\frac{da^3}{dt} \tag{3.87}$$

or

$$\frac{d}{da}(\rho a^3) = -3a^2 p. \tag{3.88}$$

Given the equation of state $p = p(\rho)$, we can integrate Eq. (3.88) to obtain $\rho = \rho(a)$. Substituting this relation into Eq. (3.78), we can determine $a(t)$.

For an equation of state of the form $p = w\rho$, Eq. (3.88) gives $\rho \propto a^{-3(1+w)}$; in particular, for nonrelativistic matter ($w = 0$) and radiation ($w = 1/3$) we find $\rho_{NR} \propto a^{-3}$ and $\rho_R \propto a^{-4}$. If $w = -1$, then we find that $\rho = $ constant as the universe expands. In this case the pressure $p = -\rho$ is negative [because we must have $\rho > 0$ to maintain $(\dot{a}^2/a^2 > 0)$] and the negative pressure allows for the energy inside a volume to increase even when the volume expands. If $w = 1$, then $p = \rho$ and $\rho \propto a^{-6}$. This is called a *stiff* equation of state. In such a medium, the "speed of sound" $(\partial p/\partial \rho) = 1$ is the same as the speed of light.

The corresponding time evolution of $a(t)$ is easy to determine for $k = 0$. For $\rho \propto a^{-3(1+w)}$, the Friedmann equation becomes $(\dot{a}^2/a^2) \propto a^{-3(1+w)}$ or

$$\dot{a} \propto a^{-\frac{1}{2}(1+3w)}. \tag{3.89}$$

Integrating, we find

$$a(t) \propto t^{\frac{2}{3(1+w)}} \quad \text{(for } w \neq -1\text{)}$$
$$\propto \exp(\lambda t) \quad \text{(for } w = -1\text{)}, \tag{3.90}$$

where λ is a constant. For $w = 0$, $a \propto t^{2/3}$; for $w = 1/3$, $a \propto t^{1/2}$; and for $w = 1$, $a \propto t^{1/3}$.

Some particular values of w are of special importance. We have already seen above that nonrelativistic matter has $w = 0$, which is a good approximation for both dark matter and baryons in the universe at large scales; $w = (1/3)$ corresponds to all relativistic species, including radiation. The cases with $w = \pm 1$ arise naturally in the case of scalar fields, which we shall briefly mention. A scalar field ϕ with potential $V(\phi)$ can be described by a Lagrangian of the form

$$L = \frac{1}{2}\left(\frac{\partial \phi}{\partial x^i}\right)\left(\frac{\partial \phi}{\partial x_i}\right) - V(\phi). \tag{3.91}$$

The stress tensor for this scalar field is (see Vol. I, Chap. 11, Section 11.7)

$$T_{ik} = \left(\frac{\partial \phi}{\partial x^i}\right)\left(\frac{\partial \phi}{\partial x^k}\right) - g_{ik}L. \tag{3.92}$$

In a homogeneous universe, $\phi(t, \mathbf{x}) = \phi(t)$, and only the diagonal components of T^i_k remain nonzero:

$$T^0_0 = \frac{1}{2}\dot{\phi}^2 + V(\phi), \qquad T^1_1 = T^2_2 = T^3_3 = -\frac{1}{2}\dot{\phi}^2 + V(\phi), \tag{3.93}$$

giving

$$\rho = \frac{1}{2}\dot{\phi}^2 + V(\phi), \qquad p = \frac{1}{2}\dot{\phi}^2 - V(\phi). \tag{3.94}$$

When the kinetic energy ($\dot{\phi}^2/2$) of the field dominates over the potential energy $V(\phi)$, we get the equation of state $p = \rho$. If the potential $V(\phi)$ dominates over the kinetic energy ($\dot{\phi}^2/2$), then $p = -\rho$. Thus the scalar field can exhibit both these equations of state in appropriate ranges.

The equation of state $p = -\rho$ also arises if the Einstein equations are modified by adding a term $-\Lambda \delta^i_k$ is added on the left-hand side. For historical reasons the Λ introduced by such a modification is called a *cosmological constant*. Because this is completely equivalent to adding a term $\Lambda \delta^i_k$ on the right-hand side as a source, we shall take the point of view that any cosmological constant can be treated as a special kind of source. In that case, the contents of the universe can be taken to be (1) nonrelativistic matter, (2) relativistic matter, and (3) a cosmological constant that is also sometimes called the *vacuum energy density*. The quantities pertaining to the cosmological constant are denoted by the subscript V.

Exercise 3.7

de Sitter line element: (a) Solve Einstein's equations for $k = 1$ and $k = 0$ when the equation of state for matter is $p = -\rho$. (It is believed that matter in our universe is described by such an equation of state, during the very early stages of its evolution, for a short period of time.) Show that the line elements have the form

$$ds^2 = dt^2 - e^{2Ht}[dr^2 + r^2(d\theta^2 + \sin^2\theta \, d\phi^2)],$$

$$ds^2 = dT^2 - \cosh^2 HT \left[\frac{dR^2}{1 - R^2} + R^2(d\theta^2 + \sin^2\theta \, d\phi^2)\right].$$

(b) Because the source is the same, we expect the two line elements above to represent the same space–time. Prove that this is indeed the case by finding the coordinate transformation between (t, r) and (T, R).

Exercise 3.8
Thought provoker: The last problem illustrates a situation in which the specification of the source does not uniquely specify the value of k. (a) Investigate the geometric origin of this nonuniqueness. (b) Show that (fortunately!) there does not exist any other nonflat space–time that can be represented as Friedmann universes with different values of k.

We have seen above that the three energy densities vary as $\rho_{NR} \propto a^{-3}$, $\rho_R \propto a^{-4}$, and $\rho_V = $ constant, as the universe evolves. Hence the total energy density in the universe can be expressed as

$$\rho_{\text{total}}(a) = \rho_R(a) + \rho_{NR}(a) + \rho_V(a)$$
$$= \rho_c \left[\Omega_R \left(\frac{a_0}{a}\right)^4 + (\Omega_B + \Omega_{DM})\left(\frac{a_0}{a}\right)^3 + \Omega_V\right], \tag{3.95}$$

where ρ_c and various Ω refer to their values at $a = a_0$; Ω_B refers to the density parameter of baryons. Substituting this into Einstein's equation we get

$$\frac{\dot{a}^2}{a^2} + \frac{k}{a^2} = H_0^2 \left[\Omega_R \left(\frac{a_0}{a}\right)^4 + \Omega_{NR}\left(\frac{a_0}{a}\right)^3 + \Omega_V\right], \tag{3.96}$$

with $\Omega_{NR} = \Omega_B + \Omega_{DM}$. This equation can be cast in a more suggestive form. We write (k/a^2) as $(\Omega - 1)H_0^2(a_0/a)^2$ and move it to the right-hand side. Introducing a dimensionless time coordinate $\tau = H_0 t$ and writing $a = a_0 q(\tau)$, we find that our equation becomes

$$\frac{1}{2}\left(\frac{dq}{d\tau}\right)^2 + V(q) = E, \tag{3.97}$$

where

$$V(q) = -\frac{1}{2}\left(\frac{\Omega_R}{q^2} + \frac{\Omega_{NR}}{q} + \Omega_V q^2\right), \qquad E = \frac{1}{2}(1 - \Omega). \tag{3.98}$$

This equation has the structure of the first integral for the motion of a particle with energy E in a potential $V(q)$. For models with $\Omega = \Omega_{NR} + \Omega_V = 1$, we can take $E = 0$ so that $(dq/d\tau) = \sqrt{V(q)}$. Figure 3.1 shows the velocity $(dq/d\tau)$ as a function of the position $q = (1 + z)^{-1}$ for such models. Several features

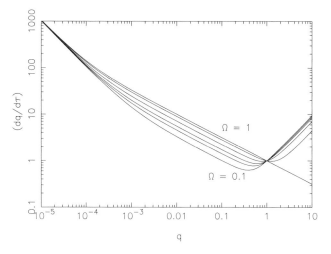

Fig. 3.1. The "velocity" of the universe $(dq/d\tau)$ plotted against the "position" $q = (1+z)^{-1}$ for different cosmological models with $\Omega_R = 2.56 \times 10^{-5}h^{-2}$, $h = 0.5$, and $\Omega_{NR} + \Omega_V = 1$. Curves are parameterised by the values of $\Omega_{NR} = 0.1, 0.2,$ 0.3, 0.5, 0.8, 1.0, going from bottom to top as indicated.

are clear from this figure. At high redshift (small q) the universe is radiation dominated and $\dot{q}$ is independent of the other cosmological parameters; hence all the curves asymptotically approach each other at the left end of the figure. At low redshifts, the presence of a cosmological constant makes a difference and in fact the velocity $\dot{q}$ changes from a decreasing function to an increasing function. In other words, the presence of a cosmological constant leads to an accelerating universe at low redshifts.

Note that, given the definition of the Hubble parameter $H(t)$, critical density $\rho_c(t)$, and density parameter $\Omega(t)$ at any given time t [see Eqs. (3.81)], it is easy to relate $\Omega(t)$ to $\Omega(t_0) \equiv \Omega$. The relation is given by

$$\Omega(t) - 1 = \frac{\Omega - 1}{1 - \Omega + \Omega_V a^2 + \Omega_{NR} a^{-1} + \Omega_R a^{-2}}. \tag{3.99}$$

Two important conclusions follow from this equation. First, for small a the right-hand side decreases rapidly and $\Omega(t) \approx 1$ in any model (other than the trivial case in which the only source is Ω_V). Hence in the study of the early phases of the universe we can approximate all models as those with $\Omega = 1$. Second, a given deviation of $\Omega(t_i)$ from unity at some very early epoch $t = t_i$ will get magnified during the evolution so that Ω will differ wildly from unity at late epochs. That is, we need to keep Ω very close to unity at an early epoch if we find that $\Omega = \mathcal{O}(1)$ at the present epoch. The only exception to this conclusion is in models in which Ω is strictly equal to unity at all times – which, of course, satisfies Eq. (3.99) trivially.

Let us now consider the solutions to Eq. (3.96) or – equivalently – to Eq. (3.97). To begin with, it is clear that, in the early evolution of the universe, the radiation term (Ω_R/q^2) in Eqs. (3.98) will dominate the dynamics. In this limit, the solution is given by

$$\left(\frac{a}{a_0}\right) \simeq \sqrt{2}\Omega_R^{1/4}(H_0 t)^{1/2}. \tag{3.100}$$

Thus for small a, we have $a \propto t^{1/2}$.

As the universe evolves, the matter density will catch up with radiation density and both a^{-4} and a^{-3} terms will be important. The equality of matter and radiation energies occurs at some time $t = t_{eq}$ in the past, corresponding to a value $a = a_{eq}$ and redshift $z = z_{eq}$. To determine the numerical values we need to know Ω_R. Assuming that most of the energy density is in the microwave background radiation at temperature T today, we get $\rho_R = (\pi^2/15)(k_B^4 T^4/c^3\hbar^3)$. Dividing this by $\rho_c \simeq 1.88 \times 10^{-29}h^2$ g cm^{-3} we can find Ω_R. Taking $T = 2.73$ K gives

$$\Omega_R h^2 = 2.56 \times 10^{-5}. \tag{3.101}$$

It follows that

$$(1 + z_{eq}) = \frac{a_0}{a_{eq}} = \frac{\Omega_{NR}}{\Omega_R} \simeq 3.9 \times 10^4(\Omega_{NR}h^2). \tag{3.102}$$

Because the temperature of the radiation grows as a^{-1}, the temperature of the universe at this epoch will be

$$T_{eq} = T_{now}(1 + z_{eq}) = 9.24(\Omega h^2)\,\text{eV}. \tag{3.103}$$

For $t \ll t_{eq}$ the energy density in the universe is dominated by radiation [with $p = (1/3)\rho$] whereas for $t \gg t_{eq}$ the energy density is dominated by matter (with $p \simeq 0$). When both radiation and matter terms are taken into consideration and other terms are ignored, Eq. (3.96) has the solution

$$H_{eq}t = \frac{2\sqrt{2}}{3}\left[\left(\frac{a}{a_{eq}} - 2\right)\left(\frac{a}{a_{eq}} + 1\right)^{1/2} + 2\right]. \tag{3.104}$$

This equation gives $a(t)$ in terms of the two (known) parameters:

$$H_{eq}^2 \equiv \frac{16\pi G}{3}\rho_{eq} \equiv 2H_0^2\Omega_R(1 + z_{eq})^4 = 2H_0^2\Omega_{NR}(1 + z_{eq})^3, \tag{3.105}$$

$$a_{eq} \equiv a_0(1 + z_{eq})^{-1} = H_0^{-1}|(\Omega - 1)|^{-1/2}(\Omega_R/\Omega_{NR}). \tag{3.106}$$

From Eq. (3.104) we can find the value of t_{eq}; setting $a = a_{eq}$ gives $H_{eq}t_{eq} \simeq$

0.552, or

$$t_{\text{eq}} = \frac{2\sqrt{2}}{3} H_{\text{eq}}^{-1}(2 - \sqrt{2}) \simeq 0.39 H_0^{-1} \Omega^{-\frac{1}{2}} (1 + z_{\text{eq}})^{-\frac{3}{2}} = 1.57 \times 10^{10} (\Omega h^2)^{-2} \text{ s.}$$
(3.107)

From Eq. (3.104), we can also find two limiting forms for $a(t)$ that are valid for $t \gg t_{\text{eq}}$ and $t \ll t_{\text{eq}}$:

$$\left(\frac{a}{a_{\text{eq}}}\right) = \begin{cases} (3/2\sqrt{2})^{2/3} (H_{\text{eq}}t)^{2/3} \\ (3/\sqrt{2})^{1/2} (H_{\text{eq}}t)^{1/2} \end{cases}.$$
(3.108)

Thus $a \propto t^{2/3}$ in the matter-dominated phase (when other contributions are negligible) and $a \propto t^{1/2}$ in the radiation-dominated phase.

At $z \ll z_{\text{eq}}$ we can ignore the radiation completely. The evolution now depends on the value of Ω_V and k. Let us take the case of $\Omega_V = 0$ first. Equations (3.98) can be integrated exactly when both the curvature and the matter terms are present, but $\Omega_V = 0$ and Ω_R is ignorable. The solution is

$$H_0 t = \frac{\Omega}{2(\Omega - 1)^{3/2}} \left[\cos^{-1}\left(\frac{\Omega z - \Omega + 2}{\Omega z + \Omega}\right) - \frac{2(\Omega - 1)^{1/2}(\Omega z + 1)^{1/2}}{\Omega(1 + z)} \right]$$
(3.109)

for $\Omega > 1$ and

$$H_0 t = \frac{\Omega}{2(1 - \Omega)^{3/2}} \left[\frac{2(1 - \Omega)^{1/2}(\Omega z + 1)^{1/2}}{\Omega(1 + z)} - \cosh^{-1}\left(\frac{\Omega z - \Omega + 2}{\Omega z + \Omega}\right) \right]$$
(3.110)

for $\Omega < 1$.

These expressions, together with the relation $a = a_0(1 + z)^{-1} = H_0^{-1}|(1 - \Omega)|^{-1/2}(1 + z)^{-1}$, completely determine $a(t)$. The energy density of matter varies as $\rho = \rho_c \Omega (1 + z)^3$ whereas the curvature term grows as $a^{-2} = H_0^2 |(\Omega - 1)|(1 + z)^2$.

If $\Omega < 1$, then the curvature term $(1 - \Omega)/2$ and the nonrelativistic matter term $\Omega_{\text{NR}}/2q$ in Eq. (3.97) will be equal at some $z = z_{\text{curv}}$, where

$$z_{\text{curv}} = \frac{1}{\Omega} - 2$$
(3.111)

(we have taken $\Omega \approx \Omega_{\text{NR}}$). It is possible for the curvature term to dominate over matter density at sufficiently small z, provided that $\Omega < 0.5$. During this phase, the effect of ρ is ignorable and the expansion factor $a(t)$ grows as t. For $z \gg z_{\text{curv}}$ we can ignore the curvature term.

Exercise 3.9

Nice features of conformal time: (a) Integrate the Friedmann equation for a $k = 0$ universe with matter and radiation by using the conformal time τ (defined through the relation $dt = a\,d\tau$). (b) Integrate the Friedmann equation for a matter-dominated universe with $\Omega = 1$ by using conformal time. Express $\Omega(t)$ and Ht in terms of τ. [Answers. (a) The Friedmann equation can now be reduced to the form

$$\left(\frac{da}{dq}\right)^2 = \Omega_{NR}a + \Omega_R, \quad q = H_0\tau, \tag{3.112}$$

which can be integrated to give

$$a = \sqrt{\Omega_R}(H_0\tau) + \frac{1}{4}\Omega_{NR}(H_0\tau)^2. \tag{3.113}$$

(b) The Friedmann equation in this case reduces to

$$d\tau = \frac{d\ln a}{(A/a - k)^{1/2}}, \quad A \equiv \frac{8\pi G\rho a^3}{3}. \tag{3.114}$$

Integrating, we can express a in terms of τ as

$$a = \frac{A}{k}\sin^2\left(k^{1/2}\tau/2\right). \tag{3.115}$$

Note that this is valid for $k = 0, 1$. (Similar results with hyperbolic functions exist for $k = -1$.) Because $dt = a\,d\tau$, integration gives

$$t = \frac{A}{2k^{3/2}}\left[k^{1/2}\tau - \sin\left(k^{1/2}\tau\right)\right]. \tag{3.116}$$

The Hubble parameter and the density parameter are given by

$$H(\tau) = \frac{c}{a}\left(\frac{A}{a} - k\right)^{1/2} = \frac{ck^{3/2}}{A}\frac{\cos\left(k^{1/2}\tau/2\right)}{\sin^3\left(k^{1/2}\tau/2\right)},$$

$$\Omega(\tau) = \frac{8\pi G\rho}{3H^2} = \frac{1}{1 - ka/A} = \frac{1}{\cos^2\left(k^{1/2}\tau/2\right)}. \tag{3.117}$$

The combination Ht is given by

$$Ht = \frac{\cos\left(k^{1/2}\tau/2\right)\left[k^{1/2}\tau - \sin\left(k^{1/2}\tau\right)\right]}{2\sin^3\left(k^{1/2}\tau/2\right)}. \tag{3.118}$$

Expressing the result in terms of Ω will lead to Eqs. (3.109) and (3.110).]

Finally let us consider the effect of Ω_V. If $\Omega_V > \Omega_{NR}$ then the Ω_V term will dominate over other terms at $z \ll z_V$, where $(1 + z_V) = (\Omega_V/\Omega_{NR})^{1/3}$. Keeping only the Ω_{NR} and Ω_V terms, we can integrate Eq. (3.96) (with $k = 0$) to determine $a(t)$. We get

$$\left(\frac{a}{a_0}\right) = \left(\frac{\Omega_{NR}}{\Omega_V}\right)^{1/3}\sinh^{2/3}\left(\frac{3}{2}\sqrt{\Omega_V}H_0t\right), \quad \Omega_{NR} + \Omega_V = 1. \tag{3.119}$$

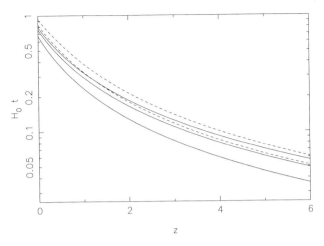

Fig. 3.2. The age of the universe at a redshift z for different values of Ω_{NR}. The solid curves are for a model with $\Omega_{tot} = \Omega_{NR} = (0.35, 0.5, 1)$ from top to bottom. The dashed curves are for a model with $\Omega_\Lambda + \Omega_{NR} = 1$, where $\Omega_{NR} = (0.35, 0.5, 1)$.

When $\sqrt{\Omega_V} H_0 t \ll 1$, this reduces to the matter-dominated evolution with $a^3 \propto t^2$; when $\sqrt{\Omega_V} H_0 t \gtrsim 1$, the growth is exponential with $a \propto \exp(\sqrt{\Omega_V} H_0 t)$. In this case, the age at any given redshift $t(z)$ is given by

$$t(z) = H_0^{-1} \frac{2}{3\sqrt{\Omega_V}} \sinh^{-1}\left[\left(\frac{\Omega_V}{\Omega_{NR}}\right)^{1/2} (1+z)^{-3/2}\right]. \tag{3.120}$$

Figure 3.2 gives $H_0 t$ as a function of z for different cosmologies parameterised by Ω. The three solid curves are for models with $\Omega_{tot} = \Omega_{NR} = (0.35, 0.5, 1)$ from top to bottom. The dashed curves are for models with $\Omega_\Lambda + \Omega_{NR} = 1$, where $\Omega_{NR} = (0.35, 0.5, 1)$. Obviously, models with cosmological constants have greater ages at a given redshift for a fixed value of Ω_{NR}.

We conclude this section with a comment on another important length scale in cosmology, viz., the *horizon size*. Suppose for a moment that the universe is described by the expansion factor $a(t) = a_0 t^n$, with $n < 1$ for *all* $t \geq 0$. Then, during the time interval $(0, t)$, a photon can travel a maximum coordinate distance of

$$\xi(t) = \int_0^t \frac{dx}{a(x)} = \frac{1}{a_0} \frac{t^{1-n}}{(1-n)}, \tag{3.121}$$

which corresponds to the proper distance,

$$h(t) = a(t)\xi(t) = (1-n)^{-1}t. \tag{3.122}$$

Numerically this quantity differs from the Hubble radius $(\dot{a}/a)^{-1} = n^{-1}t$ by only a constant factor of the order of unity. Conceptually, however, they are very different entities. To avoid any possible confusion between these two quantities, we emphasize the following fact: Note that $d_H(t)$ is a local quantity and its value at t is essentially decided by the behaviour of $a(t)$ near t; in contrast, the value of $h(t)$ depends on the entire past history of the universe. In fact, $h(t)$ depends very sensitively on the behaviour of $a(t)$ near $t = 0$ – something about which we know nothing. If, for example, $a(t) \propto t^m$ with $m \geq 1$ near $t = 0$, then $h(t)$ is infinite for all $t \geq 0$. Thus there can be several physical situations in which $h(t)$ and $d_H(t)$ differ widely; in such cases, we should examine each case carefully and decide which quantity is physically relevant.

Exercise 3.10

Horizon size: The maximum proper distance a photon can travel in the interval $(0, t)$ is given by the horizon size

$$h(t) = a(t) \int_0^t \frac{dx}{a(x)}.$$

Show that, for a matter-dominated universe,

$$h(z) = H_0^{-1}(1+z)^{-1}(\Omega - 1)^{-\frac{1}{2}} \cos^{-1}\left[1 - \frac{2(\Omega - 1)}{\Omega(1 + z)}\right] \quad \text{(for } \Omega > 1)$$

$$= 2H_0^{-1}(1+z)^{-\frac{3}{2}} \quad \text{(for } \Omega = 1)$$

$$= H_0^{-1}(1+z)^{-1}(1 - \Omega)^{-\frac{1}{2}} \cosh^{-1}\left[1 + \frac{2(1 - \Omega)}{\Omega(1 + z)}\right] \quad \text{(for } \Omega < 1).$$

Show also that $d_H \simeq 3H_0^{-1} \Omega_0^{-\frac{1}{2}} (1+z)^{-\frac{3}{2}}$ for $(1 + z) \gg \Omega^{-1}$.

Exercise 3.11

Loitering and other universes: Consider an Friedmann model with (1) only Ω_{NR} and Ω_V being nonzero, (2) $\Omega_{NR} \geq 0$, and (3) Ω_V being positive or negative. (a) Prove that if $\Omega_V < 0$ the universe always recollapses whereas if $\Omega_V > 0$ and $\Omega_{NR} < 1$ the universe expands forever. Also show that if $\Omega_{NR} > 1$ and

$$\Omega_V > 4 \Omega_{NR} \left\{ \cos\left[\frac{1}{3} \cos^{-1}\left(\Omega_{NR}^{-1} - 1\right) + \frac{4}{3}\pi\right] \right\}^3, \tag{3.123}$$

the universe does not re-collapse.

(b) Show that it is possible to adjust the values of Ω_V and Ω_{NR} such that the universe stays ("loiters") for a long time at a nearly constant scale factor $a = a_c$, say, at some $z > 0$. [Hint: In the model under consideration, the Hubble parameter can be expressed in the form

$$\frac{H^2(a)}{H_0^2} = \Omega_V\left(1 - a^{-2}\right) + \Omega_{NR}\left(a^{-3} - a^{-2}\right) + a^{-2}. \tag{3.124}$$

What happens when the right-hand side vanishes?]

3.5 Observational Tools in Friedmann Models

There are two Friedmann models that are of special interest in cosmology. The first one has $\Omega_V = 0$ and makes a transition from radiation-dominated to matter-dominated evolution at z_{eq}. At the late matter-dominated phase, it is essentially characterised by a single parameter, Ω_{NR}. The second is a $k = 0$ model, where both Ω_V and Ω_{NR} are nonzero. Because Ω_R is negligible, these models essentially have $\Omega_V = 1 - \Omega_{NR}$, and hence this model is also parameterised by just Ω_{NR}. We now discuss several features of these models that are observationally relevant.

The form of relations (3.110) and (3.119) at $z = 0$ and $a = a_0$ is of particular importance because it gives the age of the universe. The first relation is applicable for a universe that is dominated by nonrelativistic matter. However, in fact, it gives the age of the universe even in the presence of radiation because the radiation-dominated phase contributes a negligible amount of additional time to the evolution of the universe. We see that the age can be expressed in the form

$$t_{age} = t_0 = H_0^{-1} f(\Omega) = \frac{H_0^{-1}\Omega}{2(1-\Omega)^{3/2}} \left[\frac{2(1-\Omega)^{1/2}}{\Omega} - \cosh^{-1}\left(\frac{2-\Omega}{\Omega}\right) \right].$$

(3.125)

The top panel of Fig. 3.3 gives the curves of constant age in the $\Omega-H_0$ plane for open matter-dominated universes with $\Omega = \Omega_{NR}$. Given the ages of structures in the universe, this figure can provide important constraints on these parameters.

The age of a universe, with both Ω_V and Ω_{NR} making nonzero contributions (but with $\Omega_V + \Omega_{NR} = 1$), can again by setting $a = a_0$ in Eq. (3.119) and solving for t. This gives

$$t_{age} = t_0 = \frac{2}{3}\Omega_V^{-1/2} H_0^{-1} \ln\left(\frac{1+\Omega_V^{1/2}}{\Omega_{NR}^{1/2}}\right) \quad \text{(for } \Omega_{NR} + \Omega_V = 1, \ k = 0\text{)}.$$

(3.126)

The bottom panel of Fig. 3.3 gives the lines of constant age in the $\Omega_{NR}-h$ plane for models with $\Omega_{NR} + \Omega_V = 1$. It is obvious that, for a given Ω_{NR}, the age is higher for models with $\Omega_V \neq 0$. When both Ω_{NR} and Ω_V are present and are arbitrary, the age of the universe is determined by the integral

$$H_0 t_0 = \int_0^\infty \frac{dz}{(1+z)\sqrt{(1+z)^2(1+\Omega_{NR}z) - z(2+z)\Omega_V}}, \quad (3.127)$$

which cannot be expressed – in general – in terms of elementary functions. In the range of practical interest, with $0.1 \lesssim \Omega_{NR} \lesssim 1$, $|\Omega_V| \lesssim 1$, the result can, however, be approximated to a few percent accuracy by

$$H_0 t_0 \simeq \frac{2}{3}(0.7\Omega_{NR} + 0.3 - 0.3\Omega_V)^{-0.3}. \quad (3.128)$$

Let us next consider the explicit forms for different length scales obtained in

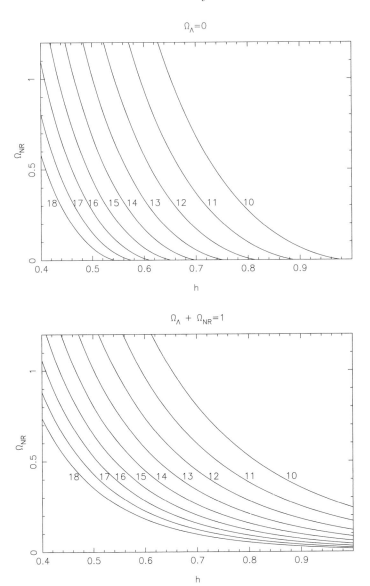

Fig. 3.3. Curves of constant age in the $\Omega_{\rm NR}$–h plane. The nine curves are for $t =$ 10, 11, 12, 13, 14, 15, 16, 17 and 18 Gyr, as shown. The top panel is for open models with $\Omega = \Omega_{\rm NR}$, and the bottom panel is for models with $\Omega_{\rm NR} + \Omega_V = 1$.

the last section in terms of $d_H(z)$ and $r_{\rm em}(z)$. Given the nature of the source, it is in principle possible to determine all the relevant parameters specifically. To begin with, it follows from of Eq. (3.96) that

$$d_H(z) = H_0^{-1}[\Omega_R(1+z)^4 + \Omega_{\rm NR}(1+z)^3 + (1-\Omega)(1+z)^2 + \Omega_V]^{-1/2}.$$

$$(3.129)$$

This has the limiting forms,

$$
d_H(z) \cong
\begin{cases}
H_0^{-1}\Omega_R^{-1/2}(1+z)^{-2} & (z \gg z_{eq}) \\
H_0^{-1}\Omega_{NR}^{-1/2}(1+z)^{-3/2} & (z_{eq} \gg z \gg z_{curv}, \Omega_V = 0) \\
H_0^{-1}(1+z)^{-1}(1+\Omega z)^{-1/2} & (z_{eq} \gg z, \Omega_{NR} \cong \Omega; \Omega_V = 0) \\
H_0^{-1}\Omega_{NR}^{-1/2}\left[(1+z)^3 + \Omega_{NR}^{-1} - 1\right]^{-1/2} & (z_{eq} \gg z, \Omega = \Omega_{NR} + \Omega_V = 1)
\end{cases}
\tag{3.130}
$$

during various epochs.

Although the integral in Eq. (3.57) cannot be done in closed form for the most general case, it is possible to obtain simple expressions for r_{em}, d_L, and d_A in the important case of a matter-dominated universe with only Ω_{NR} contributing to the energy density. In this case, we have the relations

$$
d_H(z) = H_0^{-1}(1+z)^{-1}(1+\Omega z)^{-1/2},
\tag{3.131}
$$

$$
r_{em}(z) = \frac{2\Omega z + 2(\Omega - 2)(\sqrt{\Omega z + 1} - 1)}{H_0 a_0 \Omega^2 (1+z)}.
\tag{3.132}
$$

Using these, we can express the angular diameter distance and the luminosity distance in the form

$$
d_A(z) = 2H_0^{-1}\Omega^{-2}[\Omega z + (\Omega - 2)(\sqrt{1+\Omega z} - 1)](1+z)^{-2},
\tag{3.133}
$$

$$
d_L(z) = 2H_0^{-1}\Omega^{-2}[\Omega z + (\Omega - 2)(\sqrt{1+\Omega z} - 1)].
\tag{3.134}
$$

Figure 3.4 gives the plot of $d_A(z)$; note that the angular diameter distance is not a monotonic function of z. Asymptotically, for large z, it has the limiting form,

$$
d_A(z) \cong 2(H_0\Omega)^{-1}z^{-1}.
\tag{3.135}
$$

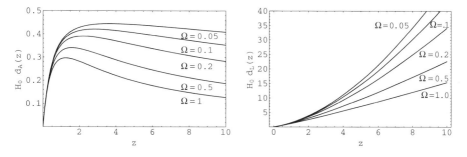

Fig. 3.4. The left panel gives the angular diameter distance in units of cH_0^{-1} as a function of redshift for open models with different values of Ω_{NR}. The right panel gives the luminosity distance in units of cH_0^{-1} as a function of redshift for open models with different values of Ω_{NR}.

It is sometimes convenient to define another distance scale in the Friedmann universe (sometimes called *proper motion distance* or *effective distance*) by the definition $D = \mathcal{R}\sin(r/\mathcal{R})$, where $\mathcal{R}$ is the radius of curvature at the present epoch for a closed model. (In the case of open models, we redefine D by taking $\mathcal{R} \to i\mathcal{R}$ and $\sin \to \sinh$.) In terms of D, the angular diameter distance and luminosity distance are

$$d_A = \frac{D}{1+z}, \qquad d_L = D(1+z). \qquad (3.136)$$

Also note that the bolometric flux of a source is given by

$$F_{\text{bol}} = \frac{L_{\text{bol}}}{4\pi D^2(1+z)^2}, \qquad (3.137)$$

and the flux density is

$$F_\nu(\nu_{\text{obs}}) = \frac{L_\nu(\nu_{\text{em}})}{4\pi D^2(1+z)}, \qquad \nu_{\text{obs}} = \frac{\nu_{\text{em}}}{1+z}. \qquad (3.138)$$

Very often we consider sources for which the luminosity can be approximated as a power law with $L_\nu \propto \nu^{-\alpha}$. In this case, the flux density becomes

$$F_\nu(\nu_{\text{obs}}) = \frac{L_\nu(\nu_{\text{obs}})}{4\pi D^2(1+z)^{1+\alpha}}. \qquad (3.139)$$

Finally, the difference between apparent and absolute magnitude for such a source (usually called the *distance modulus*) will be given by

$$m - M = 25 + 5\log[3000(1+z)H_0 D] - 5\log h + K, \qquad (3.140)$$

where the term K, which corrects for the fact that observed and emitted frequencies are different, is given by

$$K = -2.5\log\left[\frac{(1+z)L_\nu(\nu_{\text{em}})}{L_\nu(\nu_{\text{obs}})}\right]. \qquad (3.141)$$

This quantity is usually called the the *K correction* (see Chap. 8, Exercise 8.16 for a more detailed discussion).

Exercise 3.12
Angle subtended by horizon: Estimate the angle θ_h subtended by a region whose transverse length is equal to the horizon size $h(z)$ at redshift z. Assume that $z \gg 1$. Estimate θ_h for $z = 1300$ in different cosmological models. [Hint: This is most easily done by using the conformal time τ and the angular variable χ of Eq. (3.20). For $\Omega = 1$, matter-dominated universe, $\theta_h \approx z^{-1/2}$; for $z = 1300$, $\theta_h \approx 1.6°$. In the case of an open matter-dominated universe, $\theta_h \approx (\Omega_{\text{NR}}/z)^{1/2}$, which is smaller than the $\Omega = 1$ model by

the factor $\sqrt{\Omega_{NR}}$. In a model with $\Omega_{NR} + \Omega_V = 1$, the result is $\theta_h \approx (q/\sqrt{z})$ where

$$\frac{1}{q} = \int_0^1 \frac{dx}{\left[1 + \left(\frac{\Omega_V}{\Omega_{NR}}\right)x^6\right]^{1/2}}. \tag{3.142}$$

The integral is clearly less than unity, making $q > 1$; hence the angle is larger in this case. For $\Omega_{NR} = 0.35$, $\Omega_V = 0.65$, and $z = 1300$, the angle is $\theta = 1.78°$, which is nearly twice the corresponding value for an open model with $\Omega_{NR} = 0.35$.]

Exercise 3.13

Distance education: (a) Imagine moving a cluster with a physical size of 1 Mpc to larger and larger redshifts. How does the angle subtended by the cluster vary with z in a model with $\Omega_{NR} + \Omega_V = 1$? (b) A type Ia supernova observation can be used to fix the value of d_L at, say, $z = 0.5$. Given this value, we can constrain the cosmological model to lie on a line in the Ω_{NR}–Ω_V plane if we assume that there are no other sources of energy density. Determine the implicit equation to this line. (c) Consider a sphere of 5-Mpc radius placed at a redshift of $z = 1$. It subtends an angle $\Delta\theta$ in the transverse direction and occupies a redshift interval Δz along the line of sight. Find an expression for $(\Delta\theta/\Delta z)$ in a model with $\Omega_{NR} + \Omega_V = 1$. Compare the values of this ratio for $\Omega_V = 0$ and $\Omega_V = 0.35$.

Exercise 3.14

Luminosity function of gamma-ray bursts: Consider a class of sources that are detected in a survey with a limiting flux limit S_{lim}. For any given source in the catalogue, we define a variable $u \equiv (S/S_{lim})^{-3/2}$ so that $0 < u \le 1$. If the sources are distributed homogeneously in space, having Euclidean geometry, then the volume V within which a detected source could have been located scales as $V \propto r^3$, whereas $S \propto r^{-2}$, where r is the distance to the source.

(a) Argue that, if there are no correlations between the sources, then u is a random variable drawn from a uniform probability distribution $dF(u) = f(u)\,du$, with $f(u) = 1$. Hence argue that

$$\left\langle \frac{V}{V_{max}} \right\rangle \int_0^1 u f(u)\,du = 0.5. \tag{3.143}$$

(b) It is known that gamma-ray bursts have $\langle V/V_{max}\rangle = 0.33$. To model this, assume that the integral source counts are given by a function $N(S)$. Derive an expression for $f(u)$ in terms of $N(S)$, $N'(S)$, and S_{lim}. Next assume that the luminosity function of the gamma-ray bursts is given by $\phi(L, z)\,dL$. Obtain an expression for $N(S)$ in terms of $\phi(L, z)$. If the sources have a very sharp luminosity function, with $\phi(L, z) = n_0(1 + z)^\nu \delta(L - L_0)$, show that (in a flat universe)

$$F(u) = u \left\{ \frac{a[d_L(S)]}{a[d_L(S_{lim})]} \right\}^{3-\gamma} \tag{3.144}$$

where $a(d_L)$ is obtained by inverting the luminosity distance as a function of a with $a = (1 + z)^{-1}$ and $d_L(S) = (L/4\pi S)^{1/2}$. What is the dependence of a on u? Can cosmological effects or the evolution of number density account for $\langle u \rangle \simeq (1/3)$? What is the sign of γ if $\langle u \rangle < 0.5$?

(c) The luminosity function of gamma-ray bursts is quite uncertain. As the other extreme, assume a very broad luminosity function with $\phi(L, z) = (\phi_0/L_0)(1 + z)^\gamma$ $(L/L_0)^{-\alpha}$, where $1 < \alpha < 2$. Show that $F(u)$ is now independent of evolution and cosmology and depends on only α. What is α if $\langle u \rangle = 1/3$? [Answer: $\alpha = 1.75$.]

Exercise 3.15
Estimating accuracies: (a) Show that in the $\Omega = 1$ model, the apparent flux will differ from the z^{-2} dependence by 25% at $z = 0.5$. However, because galaxy luminosities evolve with time, we need to understand the galaxy evolution rather precisely to use this as a test for cosmological models. Estimate what the effective variation in Ω that will arise if we use the relation for $m-M$ without taking into account galaxy evolution. [Hint: We saw in Chap. 1 that the luminosities of the galaxies evolve approximately as $L \propto t^{-\alpha}$ [see relation (1.55) of Chap. 1]. Combine with $t \propto (1 + z)^{-3/2}$ to show that, for small z, we have $L \approx L_0[1 + \beta z + \mathcal{O}(z^2)]$ and estimate β. Use this to determine the effective change in Ω if the evolution is ignored.]

The geometrical features of the universe also play a role in determining the distribution of any population of objects in a given redshift range. To illustrate the concepts involved, let us consider a collection of objects (say, galaxies) with proper density $n(z)$ and typical proper radius r_G. Let us consider the probability of a given line of sight intersecting a galaxy with redshift in the range of $(z, z + dz)$. This is given by

$$dP = \pi r_G^2 n \, dl = \pi r_G^2 n(z) \frac{dl}{dz} \, dz, \qquad (3.145)$$

where dl is the line-of-sight distance travelled by the light ray during the redshift interval dz. The quantity dl can be written as

$$dl = dt = \left| \frac{da}{\dot{a}} \right| = \frac{dz}{(1 + z)} \left(\frac{a}{\dot{a}} \right) = \frac{dz}{(1 + z)} d_H(z). \qquad (3.146)$$

Given a cosmological model, we can determine the function $d_H(z)$ and thus the quantity $(dl/dz) = d_H/(1 + z)$. It follows that

$$\frac{dP}{dz} = \frac{\pi r_G^2 n(z) d_H(z)}{(1 + z)}. \qquad (3.147)$$

If the galaxies are neither created nor destroyed, $n(z) = n_0(1 + z)^3$, showing that

$$\frac{dP}{dz} = \left(\pi r_G^2 H_0^{-1} n_0 \right)[H_0 \, d_H(z)(1 + z)^2]. \qquad (3.148)$$

The total optical depth for intersection of objects up to some redshift z is given by the integrated probability

$$\tau(z) = \int_0^z dP = \int_0^z dx \, \frac{\pi r_G^2 n(x) d_H(x)}{(1 + x)}. \qquad (3.149)$$

For a $\Omega = \Omega_{NR} = 1$, the matter-dominated model, $d_H(z) = H_0^{-1}(1+z)^{-3/2}$; if galaxies are conserved, $n(z) = n_0(1+z)^3$. In this case, we get

$$\tau(z) = \frac{2}{3}\pi r_G^2 n_0 c H_0^{-1}(1+z)^{3/2}. \tag{3.150}$$

If $n_0 \approx 0.02\,h^3$ Mpc^{-3} and $r_G \approx 10\,h^{-1}$ kpc, we get $\tau \simeq 10^{-2}(1+z)^{3/2}$. At $z = 1$, the fraction of the sky covered by galaxies is $\tau(z=1) \simeq 0.04$. The optical depth reaches unity at a redshift of $z \simeq 20$.

As a second example, let us consider the expected number count of galaxies per unit of solid angle in the sky per unit of redshift interval. The area corresponding to a solid angle $d\Omega$ is $dA = a^2 r_{em}^2\,d\Omega = a_0^2 r_{em}^2(1+z)^{-2}\,d\Omega$. The distance along a line of sight corresponding to a redshift interval dz is given by Eq. (3.146). Combining these, we find that the proper volume element is

$$dV = \left[\frac{d_H(z)}{(1+z)}dz\right]\left[\frac{(a_0 r_{em})^2\,d\Omega}{(1+z)^2}\right] = \frac{a_0^2 r_{em}^2(z)\,d_H(z)}{(1+z)^3}dz\,d\Omega. \tag{3.151}$$

Thus the number count of galaxies per unit solid angle per redshift interval should vary as

$$\frac{dN}{d\Omega\,dz} = n(z)\frac{dV}{d\Omega\,dz} = \frac{n(z)a_0^2 r_{em}^2(z)\,d_H(z)}{(1+z)^3}. \tag{3.152}$$

To compute $(dN/d\Omega\,dz)$ explicitly, we need the function $r_{em}(z)$, which is given by Eq. (3.132) for a matter-dominated universe. In that case, we get

$$\frac{dN}{d\Omega\,dz} = \frac{4n(z)}{H_0^3(1+z)^6}\frac{[\Omega z + (\Omega - 2)(\sqrt{\Omega z + 1} - 1)]^2}{(\Omega z + 1)^{1/2}\Omega^4}. \tag{3.153}$$

The same procedure can be used for other cosmological models. For models with $\Omega_V \neq 0$, $r_{em}(z)$ needs to be determined by numerical integration. Figure 3.5 shows $(dN/d\Omega\,dz)$ as well as $(dN/d\Omega)$, which we obtain by integrating

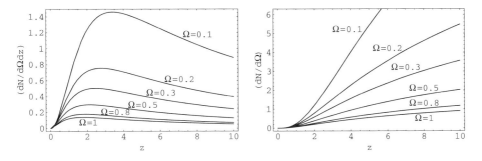

Fig. 3.5.   The left panel shows $(dN/d\Omega\,dz)$, and the right panel shows $(dN/d\Omega)$, which is obtained by integration of Eq. (3.152) in the range $(0, z)$; it is assumed that $n(z) = n_0(1+z)^3$. The y axis is in units of $n_0 H_0^{-3}$.

Eq. (3.152) in the range $(0, z)$ in the left and the right panels, respectively; it is assumed that $n(z) = n_0(1 + z)^3$. The y axis is in units of $n_0 H_0^{-3}$.

Finally, we discuss some of the radiative processes in the expanding universe and mention the key modifications that are required in comparison with the flat space–time situation. Consider some emission process that takes place at a redshift of z for which the emissivity (in units of ergs cm^{-3} s^{-1} Hz^{-1}) is $J_\omega(z)$. In an interval $(t, t + dt)$ this process will produce a spectral-energy density of radiation $dU_\omega(z) = J_\omega(z) dt$ (with units of ergs cm^{-3} Hz^{-1}). In the present epoch we will observe this radiation density as

$$dU_{\omega(1+z)^{-1}}(z = 0) = \frac{J_\omega(z) dt}{(1 + z)^3}, \tag{3.154}$$

where we have used the results $\omega \propto (1 + z)$ and $U_\omega \propto (1 + z)^3$, derived earlier [see the discussion following Eq. (3.50)]. If the sources are distributed in a redshift interval (z_1, z_2), then the total spectral intensity of radiation observed in the present epoch will be

$$U[\omega_0; z = 0] = \int_{z_1}^{z_2} \frac{J[\omega_0(1 + z); z]}{(1 + z)^3} \left| \frac{dt}{dz} \right| dz = \int_{z_1}^{z_2} \frac{J[\omega_0(1 + z); z]}{(1 + z)^4} d_H(z) dz. \tag{3.155}$$

The Ω dependence of the result arises through only the d_H factor. For a $\Omega_V = 0$, $\Omega = \Omega_{NR}$ model, we have, from relation (3.130),

$$d_H(z) = \frac{1}{H_0(1 + z)(1 + \Omega z)^{1/2}}, \tag{3.156}$$

leading to

$$U(\omega_0; z = 0) = \int_{z_1}^{z_2} \frac{J[\omega_0(1 + z); z]}{H_0(1 + z)^5(1 + \Omega z)^{1/2}} dz. \tag{3.157}$$

The corresponding intensity of radiation $I(\omega) = (c/4\pi)U(\omega)$ is

$$I(\omega_0; z = 0) = \frac{c}{4\pi H_0} \int_{z_1}^{z_2} \frac{J[\omega_0(1 + z); z] dz}{(1 + z)^5(1 + \Omega z)^{1/2}}. \tag{3.158}$$

(I has units of ergs cm^{-3} s^{-1} Hz^{-1} sr^{-1}.) This formula is often used to estimate the flux of radiation that is due to a class of sources whose number density is conserved [so that $n(z) = n_0(1 + z)^3$] and the individual luminosity is (dL/dv) per unit frequency range. In this case, $J = n(z)(dL/dv)$, and the formula becomes

$$I(\nu_{obs}; z = 0) = \frac{cn_0}{4\pi H_0} \int_{z_1}^{z_2} \frac{dL(\nu_{em})}{d\nu_{em}} \frac{dz}{(1 + z)^2(1 + \Omega z)^{1/2}}, \quad \nu_{em} = (1 + z)\nu_{obs}. \tag{3.159}$$

Some examples of the use of this formula are explored in Exercise 3.17.

Exercise 3.16

Emission processes: (a) As a first example of a radiative emission process in an expanding universe, consider the 21-cm emission from neutral hydrogen. Assume that the emissivity of the neutral hydrogen can be written in the form $J(\nu) = (3/4)An_H h\nu\delta(\nu - \nu_H)$, where $A = 2.85 \times 10^{-15}$ s^{-1} is the rate of spontaneous decay, $n_H = n_0(1 + z)^3[1 - x(z)]$ is the density of neutral hydrogen, $x(z)$ is the fraction of hydrogen gas that is in ionised form at redshift z, and n_0 is the present number density of hydrogen atoms plus ions. Show that we will observe the flux at all frequencies $\nu_0 < \nu_H$ and no flux at $\nu_0 > \nu_H$. Estimate the discontinuity in the flux at ν_H and show that it is given by

$$\Delta F = \frac{3ch}{16\pi H_0}(An_0)[1 - x(z = 0)] = \frac{3ch}{16\pi H_0}An_H(0), \qquad (3.160)$$

where $n_H(0)$ is the density of neutral atoms today. The observed value of ΔF can be used to estimate $n_H(0)$; when unobserved, bounds on ΔF can be converted into bounds on $n_H(0)$.

(b) As a second example, consider the Bremsstrahlung emission by the ionised hydrogen gas in the universe at temperatures higher than 10^6 K. If the temperature of the gas is $T_e(z)$ and the ionisation fraction is $x(z)$, find an expression for the total radiation flux at $z = 0$ in terms of the functions $x(z)$ and $T_e(z)$.

Exercise 3.17

Olber's paradox: Estimate the intensity $F(\nu_{obs})$ in a $\Omega = 1$ matter-dominated universe for a class of sources with $dL = L_0\nu^{-\alpha} d\nu$ in the range $\nu_{min} \le \nu \le \nu_{max}$. [Answer: Equation (3.159) now becomes

$$I(\nu_{obs}) = \frac{c}{4\pi}\frac{nL_0}{H_0}\int_{z_{min}}^{z_{max}}\nu_{em}^{-\alpha}\frac{dz}{(1+z)^{5/2}}, \qquad (3.161)$$

where

$$z_{min} = \max\left(0, \frac{\nu_{min}}{\nu_{obs}} - 1\right), \qquad z_{max} = \max\left(0, \frac{\nu_{max}}{\nu_{obs}} - 1\right). \qquad (3.162)$$

Using $\nu_{em} = (1 + z)\nu_{obs}$ and integrating, we get

$$\begin{aligned}
I(\nu_{obs}) &= \frac{ncL_0\nu_{obs}^{-\alpha}}{4\pi H_0\left(\frac{3}{2} + \alpha\right)}\left[1 - \left(\frac{\nu_{obs}}{\nu_{max}}\right)^{\frac{3}{2}+\alpha}\right] \\
&= \frac{nc}{4\pi H_0\left(\frac{3}{2} + \alpha\right)}\frac{dL(\nu_{obs})}{d\nu_{obs}}\left[1 - \left(\frac{\nu_{obs}}{\nu_{max}}\right)^{\frac{3}{2}+\alpha}\right].
\end{aligned} \qquad (3.163)$$

Because the total luminosity from a single object should be finite, we must have $\alpha > 1$, which allows us to take the limit $\nu_{max} \to \infty$. We then get

$$I(\nu_{obs}) = \frac{nc}{4\pi H_0\left(\frac{3}{2} + \alpha\right)}\frac{dL(\nu_{obs})}{d\nu_{obs}} \qquad (\nu_{max} \to \infty), \qquad (3.164)$$

which is the intensity of radiation that is due to a population of background sources at $z = 0$.]

To incorporate the effects of expansion on absorption, we need to modify the expression for the optical-depth property, which can be done as follows: A photon received with frequency ν_0 will have the frequency $\nu = \nu_0(1 + z)$ at the redshift z. During the interval $(t, t + dt)$, which corresponds to the redshift range $(z, z + dz)$, the optical depth increases by $d\tau_\nu = \sigma(\nu)n(z)c\,dt = c\sigma(\nu)n(z)|(dt/dz)|\,dz$. Integrating this expression, we find the total optical depth that is due to matter in the redshift range $(0, z)$ to be

$$\tau[\nu_0; z] = \int_0^z \sigma(\nu)n(z)c\left|\frac{dt}{dz}\right|dz = \frac{c}{H_0}\int_0^z \frac{\sigma[\nu_0(1+z)]n(z)\,dz}{(1+z)^2(1+\Omega z)^{1/2}}. \quad (3.165)$$

As an example of the use of this formula, consider the absorption of radiation by neutral hydrogen at the wavelength of 21 cm. The absorption cross section for this process is given by

$$\sigma(\nu) = \frac{3A}{16\pi}\left(\frac{h\nu}{2k_B T_{\text{sp}}}\right)\left(\frac{c}{\nu}\right)^2 \delta(\nu - \nu_H), \quad (3.166)$$

where T_{sp} is the spin temperature and the other symbols are defined in Exercise 3.16. Substituting the expression for $\sigma(\nu)$ into Eq. (3.165) we get

$$\tau[\nu_0; z] = \frac{3A}{32\pi}\left(\frac{h\nu_H}{kT_{\text{sp}}}\right)\left(\frac{c}{\nu_H}\right)^3 \frac{n_H(0)}{H_0}\frac{(1+z_c)^2}{(1+\Omega z_c)^{1/2}}[1 - x(z_c)] \quad \text{(for } z > z_c\text{)}, \quad (3.167)$$

where $(1 + z_c) = (\nu_H/\nu_0)$. Suppose we are receiving the radiation emitted by a class of sources at typical redshift of $z \simeq z_{\text{source}}$. Absorption can take place anywhere in the redshift range of $(0, z_{\text{source}})$; hence the intensity of the light we receive will be diminished at all frequencies in the band $[\nu_H(1 + z_{\text{source}})^{-1}, \nu_H]$. The discontinuity in τ at ν_H is

$$\Delta\tau = \frac{3A}{32\pi}\left(\frac{h\nu_H}{kT_{\text{sp}}}\right)\left(\frac{c}{\nu_H}\right)^3 \frac{n_H(0)}{H_0}[1 - x(0)], \quad (3.168)$$

which again allows us to determine (or put bounds on) the quantity $n_H(0)[1 - x(0)]$.

3.6 Gravitational Lensing

The deflection of light by a gravitational field, discussed in Volume I, Chap. 11, Subsection 11.9.2, suggests that images of distant sources will be affected by the intervening gravitational field. In particular, the deflection can lead to the formation of multiple images of the source, which is of astrophysical significance. We now develop the basic theory and discuss some simple applications.[3]

Consider a ray of light that travels from a source S to an observer O. The gravitational field along its trajectory will continuously deflect it. To simplify the calculation of this effect, we assume that all the deflection takes place when the light crosses the "deflector plane" L placed at some appropriate location between O and S.

It is then convenient to project all relevant quantities onto this two-dimensional plane, which is perpendicular to the line connecting the source and the observer. Let $\mathbf{s}$ and $\mathbf{i}$ denote the two-dimensional vectors that give the source and the image positions, respectively, projected on this two-dimensional plane, and let $\mathbf{d(i)}$ be the (vectorial) deflection produced by the lens. The geometry of gravitational lensing is shown in Fig. 3.6. The rays are assumed to propagate in straight lines to and from the lens plane and are deflected instantaneously in the lens plane. From the geometry of the diagram it follows that $\alpha D_{\mathrm{LS}} + \theta_s D_{\mathrm{OS}} = \theta_i D_{\mathrm{OS}}$. The projection of the lengths $\theta_s D_{\mathrm{OS}}$, αD_{LS}, and $\theta_i D_{\mathrm{OS}}$ onto the lens plane gives $\mathbf{s}$, $\mathbf{d}$, and $\mathbf{i}$. Therefore

$$D_{\mathrm{OS}}\,\mathbf{s} + D_{\mathrm{LS}}\,\mathbf{d} = D_{\mathrm{OS}}\,\mathbf{i} \tag{3.169}$$

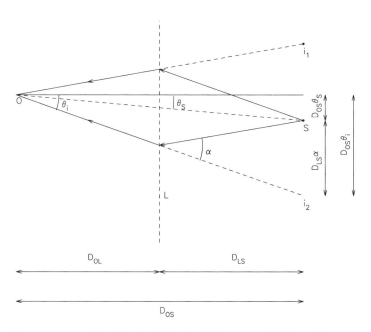

Fig. 3.6. The geometry of gravitational lensing. The source S is lensed by matter located at the plane L and produces two images i_1 and i_2 with respect to the observer O. We assume that all the deflection takes place at the plane of the lens.

or

$$s = i - \frac{D_{LS}}{D_{OS}} d(i). \tag{3.170}$$

To estimate the deflection $\mathbf{d}$, consider a bounded density distribution $\rho(\mathbf{x})$ that produces a gravitational potential $\phi(\mathbf{x})$. If a light ray is moving along the z axis, it will experience a transverse deflection by the amount

$$\frac{\mathbf{d}}{D_{OL}} = 2 \int_{-\infty}^{\infty} dz \left[\frac{^{(2)}\nabla \phi}{c^2} \right], \tag{3.171}$$

where $^{(2)}\nabla \equiv (\partial/\partial x, \partial/\partial y)$ is the gradient in the x–y plane and the factor 2 is due to the correction to Newtonian deflection that arises from general relativity, as obtained in Vol. I, Chap. 11, Section 11.9. Consider now the two-dimensional divergence $^{(2)}\nabla \cdot \mathbf{d}$. Using

$$^{(2)}\nabla^2 \phi = \nabla^2 \phi - \frac{\partial^2 \phi}{\partial z^2} = 4\pi G \rho(\mathbf{x}) - \frac{\partial^2 \phi}{\partial z^2}, \tag{3.172}$$

we get

$$^{(2)}\nabla \cdot \mathbf{d} = \frac{2 D_{OL}}{c^2} \int_{-\infty}^{\infty} dz \left[4\pi G \rho(x, y, z) - \frac{\partial^2 \phi}{\partial z^2} \right] = \frac{8\pi G D_{OL}}{c^2} \Sigma(x, y), \tag{3.173}$$

where $\Sigma(x, y)$ is the surface mass density corresponding to $\rho(\mathbf{x})$. (The $\partial^2 \phi/\partial z^2$ term vanishes on integration.) This two-dimensional Poisson equation has the solution

$$\mathbf{d}(i) = \frac{4 G D_{OL}}{c^2} \int d^2 x \, \Sigma(\mathbf{x}) \frac{(\mathbf{i} - \mathbf{x})}{|\mathbf{i} - \mathbf{x}|^2}, \tag{3.174}$$

which gives the deflection $\mathbf{d}$ in terms of the surface density $\Sigma(\mathbf{x})$. This equation, together with Eq. (3.170), gives the relation between source and image position.

A more straightforward way of obtaining this result is to note that the propagation of light rays obeys the Fermat principle. It is easy to verify that the total time delay along the path, defined by an image position $\mathbf{i}$ and a source position $\mathbf{s}$, is given by

$$t(\mathbf{i}) = t_{\text{geom}} + t_{\text{grav}} = \frac{1 + z_L}{c} \frac{D_{OL} D_{OS}}{D_{LS}} \left[\frac{1}{2} (\mathbf{i} - \mathbf{s})^2 - \psi(\mathbf{i}) \right], \tag{3.175}$$

where

$$\psi(\mathbf{i}) = \frac{4 G D_{OL} D_{LS}}{c^2 D_{OS}} \int d^2 x \, \Sigma(\mathbf{x}) \ln(|\mathbf{i} - \mathbf{x}|). \tag{3.176}$$

The first term in Eq. (3.175), a quadratic in $(\mathbf{i} - \mathbf{s})^2$, is purely geometrical and is due to the extra path length of the deflected light ray relative to the unperturbed trajectory. The second term is the time dilation introduced by the gravitational potential (see Vol. I, Chap. 11, Section 11.2)

$$\delta t = \int \frac{2}{c^3} |\phi| \, dl,$$

(3.177)

which is corrected by the extra factor $(1 + z_L)$ to take into account the cosmological expansion on the time scales. The Fermat principle requires that the trajectory of the light ray be given by the extremum of the function $t(\mathbf{i})$ for fixed $\mathbf{s}$. Ignoring overall multiplicative constants, we see that we can obtain the location of the images by calculating the extremum of the function

$$P(\mathbf{i}) = \frac{1}{2}(\mathbf{i} - \mathbf{s})^2 - \psi(\mathbf{i})$$

(3.178)

(treated as a function of $\mathbf{i}$ with fixed $\mathbf{s}$). The gradient of $\psi(\mathbf{i})$ with respect to $\mathbf{i}$ is

$$\nabla \psi = \frac{4G D_{OL} D_{LS}}{c^2 D_{OS}} \int d^2 x \, \Sigma(\mathbf{x}) \nabla \ln(|\mathbf{i} - \mathbf{x}|) = \frac{4G D_{OL} D_{LS}}{c^2 D_{OS}} \int d^2 x \, \Sigma(\mathbf{x}) \frac{(\mathbf{i} - \mathbf{x})}{|\mathbf{i} - \mathbf{x}|^2}.$$

(3.179)

It therefore follows that the equation $\nabla P = 0$ reduces to

$$\mathbf{i} = \mathbf{s} + \frac{4G D_{OL} D_{LS}}{c^2 D_{OS}} \int d^2 x \, \Sigma(\mathbf{x}) \frac{(\mathbf{i} - \mathbf{x})}{|\mathbf{i} - \mathbf{x}|^2},$$

(3.180)

which is exactly Eq. (3.170) combined with Eq. (3.174). Hence the extrema of the function $P(\mathbf{i})$ give the image positions. It is also obvious that $\psi(\mathbf{x})$ is proportional to the integral of the gravitational potential $\phi(\mathbf{x}, z)$ along the z axis and $\nabla^2_{(2)} \psi \propto \Sigma$. Equation (3.170) can now be written in component form as

$$s_a = i_a - \frac{\partial \psi}{\partial i_a} \quad (a = 1, 2).$$

(3.181)

The time-delay function $t(\mathbf{i})$ is of practical utility in a certain class of observations. The time difference between two stationary points of the function $t(\mathbf{i})$ will give the relative time delay in light propagation between the corresponding images. If the source shows detectable variability, it will be seen in both the images with a time delay determined by the height difference between the two stationary points of the time-delay surface $t(\mathbf{i})$ corresponding to the images.

For a smooth, spherically symmetric distribution of density (like that which is due to a galaxy) with a central concentration, the deflection will decrease with r far away from the lens. Near the origin of the lens, $\rho \approx$ constant and $\phi \propto r^2$, so the deflection will be linear in r. Hence the quantity $(D_{LS}/D_{OS})|\mathbf{d}|$ will vary with $|\mathbf{i}|$ roughly as shown in Fig. 3.7. We can determine the position of the images by finding the intersection of this curve with the line $y = (|\mathbf{i}| - |\mathbf{s}|)$. It is clear that,

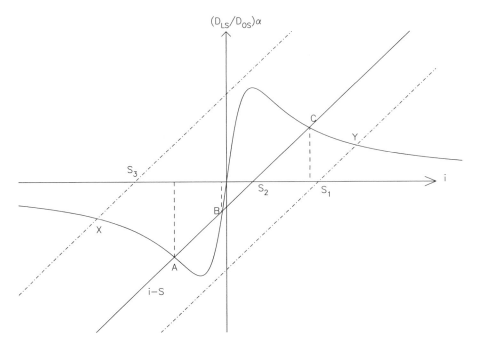

Fig. 3.7. Multiple images in the case of a spherical lens. When the source is near the centre of the lensing mass (at S_2, say), we obtain three images at the locations marked A, B, and C. If the source is farther away at S_1 or S_3, only one image (marked Y or X) is obtained.

when the source is far away from the centre of the lens, these curves intersect at only one point, giving rise to a single image. This is shown by the line through S_3 (or S_1) and the image is at X (or Y). When the source is closer to the centre, these curves intersect at three different points, giving rise to three images. This is shown by the line through S_2, and the images are at A, B, and C.

    This lensing action can also lead to magnification of the source, which can be calculated by considering the change in the image position $\delta\mathbf{i}$ for a small change in the source position $\delta\mathbf{s}$. The amplification will be the determinant of the matrix of the transformation between the two:

$$A = \det \left|\frac{\partial i_a}{\partial s_b}\right| = \left[\det\left|\frac{\partial s_b}{\partial i_a}\right|\right]^{-1}. \tag{3.182}$$

From Eq. (3.181) we get

$$\frac{\partial s_b}{\partial i_a} = \delta_{ab} - \frac{\partial^2 \psi}{\partial i_a \partial i_b} \equiv \delta_{ab} - \psi_{ab}. \tag{3.183}$$

The amplification as well as other properties of the lens can be conveniently studied in terms of the eigenvalues of the matrix ψ_{ab}. Because it follows from our definition [see Eqs. (3.170) and (3.181)] that $\nabla\psi = (D_{\mathrm{LS}}/D_{\mathrm{OS}})\mathbf{d}$, the two-dimensional Laplacian of ψ will be

$$^{(2)}\nabla^2\psi = \frac{8\pi G}{c^2}\frac{D_{\mathrm{OL}}D_{\mathrm{LS}}}{D_{\mathrm{LS}}}\Sigma \equiv 2\kappa. \qquad (3.184)$$

The last equality defines a quantity κ (called *convergence*) in terms of the surface density Σ and the distances involved in the lensing. We thus see that $\mathrm{Tr}\,(\psi_{ab}) = \nabla^2\psi = 2\kappa$. To characterise the matrix ψ_{ab} completely, we need to specify only two more components, which we take to be

$$\gamma_1(\mathbf{i}) = \frac{1}{2}(\psi_{11} - \psi_{22}) \equiv \gamma(\mathbf{i})\cos[2\phi(\mathbf{i})], \qquad (3.185)$$

$$\gamma_2(\mathbf{i}) = \psi_{12} = \psi_{21} \equiv \gamma(\mathbf{i})\sin[2\phi(\mathbf{i})]. \qquad (3.186)$$

With these definitions, we can write

$$\begin{aligned}
\psi_{ab} &= \begin{bmatrix} 1 - \kappa - \gamma_1 & -\gamma_2 \\ -\gamma_2 & 1 - \kappa + \gamma_1 \end{bmatrix} \\
&= (1 - \kappa)\begin{bmatrix} 1 & 0 \\ 0 & 1 \end{bmatrix} - \gamma\begin{bmatrix} \cos 2\phi & \sin 2\phi \\ \sin 2\phi & -\cos 2\phi \end{bmatrix}.
\end{aligned} \qquad (3.187)$$

The meanings of the different terms are now clear from the above decomposition. Convergence alone will cause isotropic focussing of light rays, thereby leading to isotropic magnification of a source. The shear terms, represented by γ_1 and γ_2, introduce anisotropic distortions in the image; the magnitude $\gamma = (\gamma_1^2 + \gamma_2^2)^{1/2}$ describes the total amount of shear, and the angle ϕ describes the relative orientation. A circular source of unit radius will be distorted to an elliptical image with major and minor axes $(1 - \kappa - \gamma)^{-1}$ and $(1 - \kappa + \gamma)^{-1}$, respectively. The magnification is now given by

$$A = \frac{1}{\det\psi} = \frac{1}{[(1 - \kappa)^2 - \gamma^2]}. \qquad (3.188)$$

There is a simple relationship between the amplification matrix ψ_{ab} and the geometrical structure of the time-delay surface $t(\mathbf{i})$ introduced in Eq. (3.175). In general, there are three kinds of stationary points on a two-dimensional surface defined by $t(\mathbf{i}) = \mathrm{constant}$. The nature of these stationary points will be related to the second derivatives of $t(\mathbf{i})$, which are proportional to the amplification matrix. If both the eigenvalues of the matrix

$$\mathcal{T} \equiv \frac{\partial^2 t(\mathbf{i})}{\partial i_a \partial i_b}, \qquad (3.189)$$

which describes the local curvature of the time-delay surface, are positive, the stationary point is a minimum and the images will have $\det\psi > 0$ and $\mathrm{Tr}\,\psi > 0$. If the eigenvalues of $\mathcal{T}$ have opposite signs then we have a saddle point with $\det$

$\psi < 0$; finally, if both the eigenvalues of $\mathcal{T}$ are negative, then det $\psi > 0$ and Tr $\psi < 0$. It is obvious that when we have a minimum or a maximum, the images have positive magnification and the saddle point leads to negative magnification, which is interpreted as a reversal in the parity of the image.

Because the curvature of the surface defined by $t(\mathbf{i})$ measures the reciprocal of magnification, it follows that the image is strongly magnified in any direction along which the curvature is small and vice versa. The lines in the image plane on which the curvature vanishes are called *critical lines*, and they correspond to infinite magnification (in the geometrical-optics limit, which of course breaks down around critical lines). The corresponding curves in the source plane are called *caustics*. We saw earlier that, when the separation between the lens and the source is small, three images are formed, whereas when the source is away from the lens, only one image is formed. Consider now the changes in the image configuration as we slowly move the source from a position close to the lens towards larger separations. As the source moves, two of the images have to approach each other, merge, and vanish. Because each of the images corresponds to a stationary point, it is clear that the curvature of the time-delay surface should vanish as the images approach each other. Thus the brightest image configurations are obtained when the pairs of images are close together on different sides of critical lines, just before merging. Critical lines and caustics provide a qualitative understanding of different image configurations and magnifications for any lensing geometry specified by $\psi(\mathbf{i})$.

The amplification introduced by a lens also influences the luminosity function of distant objects in the universe. If $n_0(> S, z)\,dz$ is the number density of a source population that has a flux greater than S in the redshift interval $(z, z + dz)$, then the amplification by gravitational lensing will change the number count to

$$n(>S, z) = \frac{1}{A} n_0 \left(> \frac{S}{A}, z \right). \tag{3.190}$$

It follows that lensing can either increase or decrease the local number counts, depending on the form of n_0.

We now consider some examples of the lensing that are due to different kinds of density distributions.

3.6.1 Constant Surface Density

For $\Sigma(x, y) = \Sigma_0$, Eq. (3.173) has the solution $\mathbf{d}(\mathbf{i}) = Q\mathbf{i}$, with $Q = 4\pi G D_{\mathrm{OL}} \Sigma_0$. Using this in Eq. (3.170), we get

$$|\mathbf{s}| = |\mathbf{i}| \left(1 - \frac{\Sigma_0}{\Sigma_{\mathrm{crit}}} \right), \qquad \Sigma_{\mathrm{crit}} \equiv \frac{c^2 D_{\mathrm{OS}}}{4\pi G D_{\mathrm{OL}} D_{\mathrm{LS}}}. \tag{3.191}$$

The corresponding magnification is

$$A = \left(1 - \frac{\Sigma_0}{\Sigma_{crit}}\right)^{-2}. \tag{3.192}$$

Note that when $\Sigma_0 = \Sigma_{crit}$ the sheet focusses the beam onto the observer, leading to infinite amplification. Lower surface densities cannot focus the beams whereas higher surface densities focus the beams before they reach the observer (so that the beams are again diverging when they reaches the observer). In a cosmological context, we may take $D_{OL} \approx D_{OS} \approx D_{LS} \approx H_0^{-1} \approx 3000$ Mpc; then the critical surface density is

$$\Sigma_{crit} \simeq \frac{c^2}{4\pi G D_{OS}} \approx 1 \text{ gm cm}^{-2}. \tag{3.193}$$

For comparison, the surface density of a typical galaxy is

$$\Sigma_{gal} \approx \frac{M_g}{R_g^2} \approx 0.3 \text{ gm cm}^{-2} \left(\frac{M_g}{10^{11} M_\odot}\right)\left(\frac{R_g}{10 \text{ kpc}}\right)^{-2}, \tag{3.194}$$

which is on the borderline for producing multiple images. For clusters, the corresponding value is $\Sigma_{clus} \approx 0.03$ if $R \approx 3$ Mpc and $M \approx 10^{15} M_\odot$, which may not be very effective in producing multiple images.

Exercise 3.18
Thought provoker: The critical density for multiple imaging by cosmological sources is of the order of $(c H_0/G) \approx \mathcal{O}(1)$ gm cm^{-2}. Is it a coincidence that this surface density is a laboratory-size quantity or does it have any deep significance?

3.6.2 Point Mass

For a point mass, $\Sigma(x, y) = M\delta(x)\delta(y)$; Eq. (3.174) then gives the deflection

$$\mathbf{d(i)} = \frac{4GM D_{OL}}{c^2}\left(\frac{\mathbf{i}}{i^2}\right). \tag{3.195}$$

Substituting into Eq. (3.170), we find

$$\mathbf{s} = \mathbf{i}\left(1 - \frac{L^2}{i^2}\right), \qquad L^2 = \frac{4GM D_{LS} D_{OL}}{c^2 D_{OS}}. \tag{3.196}$$

This result shows that $\mathbf{s}$ and $\mathbf{i}$ are in the same direction and their magnitudes are related by $s = i - (L^2/i)$. Solving this quadratic equation, we find that the two solutions for the image position i are

$$i_\pm = \frac{1}{2}(s \pm \sqrt{s^2 + 4L^2}). \tag{3.197}$$

Thus any source is imaged twice by a point-mass lens in which the two images are on either side of the source, with one image inside a circular ring of radius L (called an *Einstein ring*) and the other outside. As the source moves away from the lens (i.e., as s increases), one of the images approaches the lens and becomes fainter while the other approaches the true position of the source and tends towards a magnification of unity. The amplification of the two images is now given by

$$A_{\pm} = \left[1 - \left(\frac{L}{i_{\pm}}\right)^4\right]^{-1} = \frac{u^2 + 2}{2u\sqrt{u^2 + 4}} \pm \frac{1}{2}, \qquad (3.198)$$

where $u = s/L$. It is clear that the magnification of the image inside the Einstein ring is negative. The total magnification of the flux obtained when the absolute value of the two magnifications is added is given by

$$A_{\text{tot}} = |A_+| + |A_-| = \frac{u^2 + 2}{u\sqrt{u^2 + 4}}. \qquad (3.199)$$

In the cosmological context, all relevant distances can be taken to be $D \approx 3000$ Mpc. Then the angular size of this Einstein ring will be

$$\theta_L \equiv \frac{L}{D} = \left(\frac{4GM}{c^2 D}\right)^{1/2} \approx 2 \, \text{arcsec} \left(\frac{M}{10^{12} \, M_\odot}\right)^{1/2} \left(\frac{D}{3000 \, \text{Mpc}}\right)^{-1/2}. \qquad (3.200)$$

We can also think of θ_L as the typical angular separation between images produced by a lens of mass M located at cosmological distances. For a cluster $M \approx 10^{15} \, M_\odot$ and $\theta_L \approx 1$ arcmin.

An important application of the lensing by point mass is in a phenomenon called *microlensing*. If, for example, a compact object in our galaxy moves across a distant star in, say, a LMC, it will lead to a magnification of the light from the star under optimal conditions. The peak magnification at $u = 1$ is $A_{\text{tot}} = 1.34$ from Eq. (3.199), which corresponds to a brightening by 0.32 magnitude – and hence it should be easily detectable. The probability for lensing, of course, is low, and hence it is necessary to monitor a large number of stars for a long period of time in order to produce significant effects. It is also necessary to distinguish the variation in the light curve of the star caused by microlensing from the intrinsic variability of the star. By and large, this is feasible because microlensing produces light curves that are symmetric with respect to the peak – unlike those produced by intrinsic variability. The typical time scale for a microlensing event is

$$t = \frac{D_{\text{OL}}\theta_L}{v} = 0.214 \, \text{yr} \left(\frac{M}{M_\odot}\right)^{1/2} \left(\frac{D_{\text{OL}}}{10 \, \text{kpc}}\right)^{1/2} \left(\frac{D_{\text{LS}}}{D_{\text{OS}}}\right)^{1/2} \left(\frac{v}{200 \, \text{km s}^{-1}}\right)^{-1}. \qquad (3.201)$$

In the case of lenses located in our galactic halo and sources in the LMC, we need to sample the light curve between ~ 1 h and 1 yr to detect lenses with the mass range of 10^{-6}–10^2 $M_\odot$. The amplification as a function of time (with time measured from the point of closest approach) will vary as

$$A_{\text{tot}}(t) = \frac{2 + x^2}{x\sqrt{4 + x^2}}, \qquad x(t) = \sqrt{b^2 + a^2 t^2}, \qquad (3.202)$$

where b characterises the minimum-impact parameter and a is related to the characteristic inverse time scale for microlensing given by Eq. (3.201). Both these quantities can be treated as free parameters that are to be determined by fitting the observed light curve. The parameter a, of course, is the more important one as it contains information about the lens properties. In terms of the variables $q = (2/b^2)$ and $\tau = (at/b)$, we can write A_{tot} as

$$A_{\text{tot}}(\tau) = \frac{q + 1 + \tau^2}{\left[(1 + \tau^2)^{1/2}(1 + 2q + \tau^2)^{1/2}\right]}. \qquad (3.203)$$

Figure 3.8 plots $A_{\text{tot}}(\tau)$ for $q = 0.5, 1, 2$, and 10 for these values.

The probability for microlensing is usually given in terms of an optical depth that can be calculated as follows: If the number density of lenses is $n(D_{\text{OL}})$, and if we take the typical cross section of each lens to be $\pi\theta_L^2$, then the total optical depth along a line of sight will be

$$\tau = \frac{1}{\Delta\Omega} \int dV\, n(D_{\text{OL}})\pi\theta_L^2, \qquad (3.204)$$

where $dV = D_{\text{OL}}^2\, dD_{\text{OL}}\Delta\Omega$ is the volume element corresponding to the solid

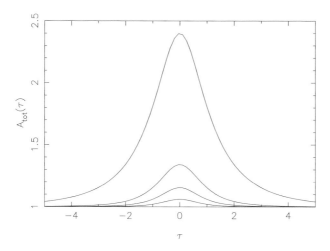

Fig. 3.8. The total amplification of a background source by a moving microlens for different values of the impact parameter b measured in units of L [see Eq. (3.196)]. The curves (from bottom to top) are for $q = (2/b^2) = 0.5, 1, 2, 10$.

angle $\Delta\Omega$. This gives

$$\tau = \int_0^{D_{OS}} \frac{4\pi G\rho}{c^2} \frac{D_{OL} D_{LS}}{D_{OS}} \, dD_{OL} = \frac{4\pi G D_{OS}^2}{c^2} \int_0^1 dx \rho(x) x(1-x), \quad (3.205)$$

where $x = D_{OL}/D_{OS}$ and ρ is the mass density of the lenses. In arriving at this expression, we have assumed that the lensing takes place in the local neighbourhood so that cosmological effects can be ignored, giving $D_{LS} = D_{OS} - D_{OL}$. If we take ρ to be constant along the line of sight, then we get

$$\tau = \frac{2\pi}{3} \frac{G\rho}{c^2} D_{OS}^2, \quad (3.206)$$

which depends on only the mass density of the lenses and not on the actual masses.

Given the optical depth, it is possible to make a rough estimate of the event rate along the following lines. If the source and the lens move with respect to each other with a speed u and impact parameter b, then the net area swept out in time t that is due to the motion is $2but$. If we assume that the fraction of the sky covered by the Einstein rings of the lenses is equal to the optical depth τ, then the number of lenses in 1 sr of the sky is $\tau/(\pi\theta_E^2)$, where θ_E is the angle subtended by the Einstein ring radius L. If we measure all lengths in units of L, then the number of lenses is τ/π per unit area. Hence the microlensing event rate is approximately $\mathcal{R} = 2but/\pi$. We can express b in terms of the maximum amplification A_{max} obtained by using Eqs. (3.202) as

$$\frac{b}{\sqrt{2}} = \left[(A_{max}^2 - 1) + A_{max}\sqrt{A_{max}^2 - 1} \right]^{-1/2}. \quad (3.207)$$

Hence, $\mathcal{R} \propto u\tau/A_{max}$ for large A_{max}, showing that the rate declines rather slowly with the peak amplification A_{max}.

3.6.3 Isothermal Sphere

For a source that is an isothermal sphere, $M(r) \cong (2\sigma^2/G)r$ at large r, where σ^2 is the velocity dispersion. The Newtonian potential for this mass distribution is

$$\phi(r) = \int^r \frac{GM(r)}{r^2} \, dr = 2\sigma^2 \ln r. \quad (3.208)$$

The deflection can be computed with Eq. (3.171):

$$\frac{\mathbf{d}}{D_{OL}} = \int_{-\infty}^{\infty} dz \left(\frac{4\sigma^2}{c^2} \right) {}^{(2)}\nabla \left(\ln \sqrt{x_\perp^2 + z^2} \right)$$

$$= \frac{4\sigma^2}{c^2} \int_{-\infty}^{\infty} dz \frac{\mathbf{x}_\perp}{(x_\perp^2 + z^2)} = 4\pi \left(\frac{\sigma}{c} \right)^2 \left(\frac{\mathbf{x}_\perp}{x_\perp} \right), \quad (3.209)$$

where $\mathbf{x}_\perp = (x, y)$. Note that the angular deflection is now a constant in magnitude and is given by

$$\alpha = 4\pi \left(\frac{\sigma}{c}\right)^2 \approx 2.6 \text{ arcsec} \left(\frac{\sigma}{300 \text{ km s}^{-1}}\right)^2. \tag{3.210}$$

Lensing equation (3.170) becomes

$$\mathbf{s} = \mathbf{i} - \frac{D_{\text{LS}} D_{\text{OL}}}{D_{\text{OS}}} \frac{\mathbf{i}}{|\mathbf{i}|} \alpha. \tag{3.211}$$

Taking the magnitudes, we find that

$$|\mathbf{i}| - |\mathbf{s}| = \frac{D_{\text{LS}} D_{\text{OL}}}{D_{\text{OS}}} \alpha. \tag{3.212}$$

Thus we get two images on the two sides of the source at $|\mathbf{s}| \pm (D_{\text{LS}} D_{\text{OL}}/D_{\text{OS}})\alpha$. The amplification is

$$A = \left(1 - \frac{D_{\text{LS}} D_{\text{OL}}}{D_{\text{OS}}} \frac{\alpha}{|\mathbf{i}|}\right)^{-1}. \tag{3.213}$$

In the cosmological context, we are often interested in the probability for different kinds of lensing phenomena to occur when the sources are objects such as, say, quasars located at large distances and the lenses are mass distributions such as galaxies located somewhere along the line of sight. To illustrate the ideas involved, let us consider a simple situation in which galaxies – treated as isothermal spheres – act as lenses. Let the ray of light pass a lens located at a redshift z_L within an impact parameter d. If the lens is an isothermal sphere, it induces a bending by an angle $\alpha = 4\pi(\sigma/c)^2$, which is independent of the impact parameter. From Fig. 3.6, we have $(d/D_{\text{OL}}) = (\alpha D_{\text{LS}}/D_{\text{OS}})$ or

$$d = \alpha \frac{D_{\text{OL}} D_{\text{LS}}}{D_{\text{OS}}}. \tag{3.214}$$

The probability for multiple imaging is the same as the probability for the line of sight to pass within a comoving distance d from a lens at redshift z_L. We can compute this by using Eq. (3.147), taking the effective cross section for the process as $A = \pi a^2 d^2$. We find that

$$\frac{dP}{dz_L} = 16\pi^3 a_0^2 \left(\frac{\sigma}{c}\right)^4 \left(\frac{D_{\text{OL}} D_{\text{LS}}}{D_{\text{OS}}}\right)^2 \frac{n(z_L) d_H(z_L)}{(1 + z_L)^3}, \tag{3.215}$$

where we have used $a^2 = a_0^2(1 + z)^{-2}$. We can obtain the optical depth for gravitational lensing up to a redshift z_s by integrating this expression in an interval $(0, z_s)$. Taking $n(z_L) = n_0(1 + z_L)^3$, we get

$$\tau = 16\pi^3 \left(\frac{\sigma}{c}\right)^4 n_0 \int_0^{z_s} \left(\frac{D_{\text{OL}} D_{\text{LS}}}{D_{\text{OS}}}\right)^2 d_H(z_L) dz_L. \tag{3.216}$$

Given a cosmological model, expressions (3.215) and (3.216) can be evaluated explicitly. If $\Omega = 1$, then $D_{LS} = D_{OS} - D_{OL}$ and the expression $(D_{OL}D_{LS}/D_{OS})$ is maximum when $D_{OL} = D_{OS}/2$; that is, when the lens is midway between the observer and the source. Further, in this case, $d_H(z_L) dz_L = dr_{OL} = d D_{OL}$; so we get

$$\tau = 16\pi^3 \left(\frac{\sigma}{c}\right)^4 \frac{n_0}{D_{OS}^2} \int_0^{D_{OS}} D_{OL}^2 (D_{OS} - D_{OL})^2 \, dD_{OL} = \frac{8\pi^3}{15} \left(\frac{\sigma}{c}\right)^4 n_0 D_{OS}^3.$$
(3.217)

For the $\Omega = 1$ case, $d_H(z) = H_0^{-1}(1+z)^{-3/2}$ and

$$D_{OS} = \int_0^{z_s} dz \, d_H(z) = 2H_0^{-1}\left[1 - \frac{1}{\sqrt{1+z_s}}\right],$$
(3.218)

leading to

$$\tau = \frac{64\pi^3}{15} \left(\frac{\sigma}{c}\right)^4 n_0 H_0^{-3} \left[1 - \frac{1}{\sqrt{1+z_s}}\right]^3.$$
(3.219)

Given the distribution of n_0 and σ for the lenses, we can calculate τ. Taking $n_0 \approx 0.001 \, h^3 \, \mathrm{Mpc}^{-3}$ and $\sigma \approx 250 \, \mathrm{km \, s^{-1}}$ as the values corresponding to galaxies, we find that

$$\tau \simeq 10^{-3} \left(\frac{n_0}{0.001 \, h^3 \, \mathrm{Mpc}^{-3}}\right) \left(\frac{\sigma}{250 \, \mathrm{km \, s^{-1}}}\right)^4 \left[1 - \frac{1}{\sqrt{1+z}}\right]^3.$$
(3.220)

Exercise 3.19
Cross sections for a given amplification: We are often interested in the proper area around a given lens through which a light ray should pass in order to be amplified by a factor greater than A. Show that, for a point mass, this is given by

$$\sigma(> A) = \frac{8\pi GM}{c^2} \frac{D_{OL}D_{LS}}{D_{OS}} \frac{1}{(A^2 - 1) + A\sqrt{A^2 - 1}},$$
(3.221)

whereas in the case of isothermal sphere, the corresponding result is

$$\sigma(> A) = \left(\frac{4\pi\sigma_v^2}{c^2}\right)^2 \left(\frac{D_{OL}D_{LS}}{D_{OS}}\right)^2 \frac{4\pi}{A^2} \qquad (A > 2),$$

$$\sigma(> A) = \left(\frac{4\pi\sigma_v^2}{c^2}\right)^2 \left(\frac{D_{OL}D_{LS}}{D_{OS}}\right)^2 \frac{\pi}{(A-1)^2} \qquad (A < 2).$$
(3.222)

Note that both the probabilities have the factor $D_{OL}D_{LS}/D_{OS}$ that is maximised when the lenses are halfway between the source and the observer.

Exercise 3.20
Lensing and geometry of space–time: Let a source be at redshift $z = 2$. Find the redshift of the lens, located optimally, in (a) an $\Omega = 1$ matter-dominated universe, (b) a very

low-density matter-dominated universe in the limit of $\Omega \to 0$, (c) a vacuum-dominated universe with $\Omega_V = \Omega_{tot}$. [Answers: If the distance to the source is r_{em}, then an optimally located lens should be at $r_{em}/2$. Converting this condition into redshift, we get the required answer. (a) In this case, $d_H \propto (1 + z)^{-3/2}$, so that

$$r_{em}(z) \propto \int_0^z d_H(z)\, dz \propto \left(1 - \frac{1}{\sqrt{1 + z}}\right). \tag{3.223}$$

The lens redshift is determined by the equation

$$\left(1 - \frac{1}{\sqrt{1 + z_L}}\right) = \frac{1}{2}\left(1 - \frac{1}{\sqrt{1 + z}}\right). \tag{3.224}$$

For $z = 2$, this gives $z_L = 0.608$. (b) In this case, $a \propto t$, giving $d_H \propto (1 + z)^{-1}$ and $r_{em}(z) \propto \ln(1 + z)$. The equation to be solved is $(1 + z_L) = (1 + z)^{1/2}$, which gives $z_L = 0.732$ for $z = 2$. (c) Now d_H is a constant, giving $r_{em}(z) \propto z$ and $z_L = (1/2)z$. Note that the lens redshift is larger for a vacuum-dominated universe compared with that of the matter-dominated universe.]

Exercise 3.21
Lensing probabilities with Λ: Show that the probability of a source at redshift z_s being lensed in a $\Omega_V + \Omega_m = 1$ universe (relative to the corresponding probability in a $\Omega_m = 1$, $\Omega_V = 0$ model) is given by

$$P_{lens} = \frac{15}{4}\left[1 - \frac{1}{\sqrt{1 + z_s}}\right]^{-3} \int_1^{a_s} \frac{H_0}{H(a)}\left[\frac{d_A(0, a)\, d_A(a, a_s)}{d_A(0, a_s)}\right] da, \tag{3.225}$$

where $a_s = (1 + z_s)^{-1}$. Plot this probability as a function of Ω_V for $z_s = 2$ and explain how this result can be used to put a bound on Ω_V.

The explicit analytic results previously given were mostly for the case of spherically symmetric lenses. The situation becomes a lot more complicated when the lens is not spherically symmetric. In general, the image configuration for such a lens needs to be analysed by numerical methods. A wide variety of images can arise when a spherically symmetric, distant source is lensed, for example, by an elliptical mass distribution in the foreground. Depending on the position of the source relative to the caustics and the critical curves, different kinds of images can be formed. In particular, when the source is close to the cusp, an extended arclike image will be produced. The study of such arclike distortions of distant galactic images by a cluster of galaxies in the foreground can be a powerful diagnostic tool for the mass distribution of the cluster.

4

Thermal History of the Universe

4.1 Introduction

To understand the features of the universe today, it is necessary to grasp the past history of the universe. We now tackle this issue and describe the physical processes that occur in the earlier phase of the universe. Section 2 develops the basic thermodynamics needed to understand these processes. In Sections 4.3 and 4.4, we consider the possible existence of a relic background of massless or massive fermions (like the neutrinos) in our universe today. In Section 4.5 we discuss the primordial nucleosynthesis and its observational relevance; we study the decoupling of matter from radiation in Section 4.6. In the last section the very early universe and inflationary models are described.[1]

4.2 Distribution Functions in the Early Universe

The analysis in Chap. 3 showed that the universe was dominated by radiation at redshifts higher than $z_{eq} \simeq 3.9 \times 10^4 (\Omega_{NR} h^2)$. In the radiation-dominated phase, the temperature of the radiation will be higher than $T_{eq} \simeq 9.2\ (\Omega_{NR} h^2) \text{eV} \simeq 1.07 \times 10^5 (\Omega_{NR} h^2) \text{K}$ and will be increasing as $T \propto (1+z)$.

The contents of the universe, at these early epochs, will be in a form very different from that in the present-day universe. Atomic and nuclear structures have binding energies of the order of a few tens of electron-volts and mega-electron-volts, respectively. When the temperature of the universe was higher than these values, such systems could not have existed as bound objects. Further, when the temperature T of the universe becomes higher than the rest mass m of a charged particle (say, electron or muon) then the photon energy will be large enough to produce these particles and their antiparticles in large numbers. For example, when $T \gg m_{elec} \simeq 0.5\,\text{MeV} \simeq 5.8 \times 10^9$ K, there will be large number of positrons in the universe. The typical energy of these particles will be T, making them ultrarelativistic.

Thus, depending on the temperature T, the early universe would be populated by different kinds of elementary particles at different times. To work out the physical processes at some time t, we need to know the distribution function $f_A(\mathbf{x}, \mathbf{p}, t) \equiv f_A(\mathbf{p}, t)$ of these particles. Here $A = 1, 2, \ldots$, labels different species of particles, like electrons, muons, etc; the dependence of f_A on the space coordinates is ruled out because of the homogeneity of the universe.

To determine the form of $f_A(\mathbf{p}, t)$ we may reason as follows: The different species of particles will be interacting constantly through various forces, scattering off each other and exchanging energy and momentum. If the rate of these reactions $\Gamma(t)$ is much higher than the rate of expansion of the universe, $H(t) = (\dot{a}/a)^{-1}$, then these interactions can produce (and maintain) thermodynamic equilibrium among the interacting particles with some temperature $T(t)$. All these interactions that occur between the particles have a short range. (The Coulomb force between charged particles has a long range; but, in a plasma, the process of Debye shielding reduces this range, making it effectively a short-range force; see Vol. I, Chap. 9, Section 9.2.) Therefore we may assume that the role of these interactions is limited to providing a mechanism for thermalisation, and we ignore their effects in deciding the *form* of the distribution function (see Exercise 4.1). In that case, the particles may be treated as an *ideal* Bose or Fermi gas with the distribution function

$$f_A(\mathbf{p}, t)\, d^3\mathbf{p} = \frac{g_A}{(2\pi)^3} \{\exp[(E_{\mathbf{p}} - \mu_A)/T_A(t)] \pm 1\}^{-1} d^3\mathbf{p}, \qquad (4.1)$$

where g_A is the spin-degeneracy factor of the species, $\mu_A(T)$ is the chemical potential, $E(\mathbf{p}) = (\mathbf{p}^2 + m^2)^{1/2}$, and $T_A(t)$ is the temperature characterising this species at time t. (We use units with $c = \hbar = 1$.) The upper sign $(+1)$ corresponds to fermions and the lower sign (-1) to bosons.

Exercise 4.1

Ideal-gas approximation: Consider a plasma with electron density n_e and temperature T. Under what circumstance can we treat the electrons as an *ideal* gas? Are these conditions satisfied in the early universe?

At any instant in time, the universe will also contain a blackbody distribution of photons with some characteristic temperature $T_\gamma(t)$. If a particular species couples to the photon directly or indirectly and if the rate of these A–γ interactions is high enough (i.e., $\Gamma_{A\gamma} \gg H$), then these particles will have the same temperature as photons: $T_A = T_\gamma$. Because this is usually the case, we often refer to the photon temperature by the term *temperature of the universe*. Of course, any sets of particle species $A, B, C, \ldots$, that are interacting among themselves at a high enough rate will also have the same temperature $T_A = T_B = T_C, \ldots$.

As the universe evolves, the temperature $T(t)$ changes because of expansion in a time scale of the order of $H^{-1}(t) \equiv (\dot{a}/a)^{-1}$; the *rate* at which temperature is changing is given by $H(t)$. The rate of interaction (per particle) can be expressed as $\Gamma \equiv n\langle\sigma v\rangle$, where n is the number density of target particles, v is the relative velocity, and σ is the interaction cross section. Because σ is usually a function of energy, $\langle\sigma v\rangle$ denotes an average value for this combination. As long as $\Gamma \gg H$, the interactions can maintain equilibrium. In that case, f will evolve adiabatically, maintaining the form of the equilibrium distribution given by Eq. (4.1), with the temperature corresponding to the instantaneous value.

The assumption of spatial homogeneity has played a crucial role in the preceding description. This can be seen as follows: In the characteristic time scale $H^{-1}(t)$, over which the parameters in the universe change, the particles can travel only a maximum distance $cH^{-1}(t)$. Therefore, if two regions in the universe had different temperatures at some time t_1, particle interactions may not be able to bring them to the same temperature at a later time $t > t_1$. By imposing strict homogeneity for the entire universe *at all times*, we have bypassed this problem.

It could happen that, at some instant, the *total* interaction rate $\Gamma_A(t)$ of a species A (taking into account interactions among themselves as well as with all other species) falls below the expansion rate $H(t)$: $\Gamma_A(t) \lesssim H(t)$; but the interaction rate among all other species Γ_{other} could still be much higher than expansion rate, $\Gamma_{\text{other}} \gg H$. In such a situation, the distribution function of all species other than A will be still given by Eq. (4.1) with a common temperature T. The particle species A, however, would have completely decoupled; its distribution function will not in general be given by Eq. (4.1). Its form, however, can be ascertained by the following argument.

Once species A is completely decoupled, each of the A particles will be travelling along a geodesic in the space–time. We have seen in Vol. I, Chap. 8, Section 8.2, that f is conserved during such free propagation. This allows us to obtain the function f_{dec}, after the species has decoupled, from the known form of f_{equi} before decoupling. For simplicity, let us assume that the decoupling occurred instantaneously at some time $t = t_D$ when the temperature was T_D and the expansion factor was a_D. For $t < t_D$, the distribution function is given by Eq. (4.1). At some later time, $t > t_D$, let the distribution function be $f_{\text{dec}}(p, t)$. Because of the redshift in momentum, all particles with momentum p at time t must have had momentum $p[a(t)/a(t_D)]$ at $t = t_D$. Therefore

$$f_{\text{dec}}(p, t) = f_{\text{equi}}\left[p\frac{a(t)}{a(t_D)}, t_D\right] \quad \text{(for } t > t_D\text{),} \qquad (4.2)$$

where f_{equi} is the equilibrium distribution function of Eq. (4.1). Thus, as long as species A was in equilibrium at *some* time, we can determine its distribution function at all later times.

From distribution function (4.1), we can calculate the number density n, energy density ρ, and pressure p (see Vol. I, Chap. 5, Section 5.9). Suppressing the time dependence and the subscript A for simplicity, we have

$$n = \int f(\mathbf{k})\, d^3\mathbf{k} = \frac{g}{2\pi^2} \int_m^\infty \frac{(E^2 - m^2)^{1/2} E\, dE}{\exp[(E - \mu)/T] \pm 1}, \tag{4.3}$$

$$\rho = \int E f(\mathbf{k})\, d^3\mathbf{k} = \frac{g}{2\pi^2} \int_m^\infty \frac{(E^2 - m^2)^{1/2} E^2\, dE}{\exp[(E - \mu)/T] \pm 1}, \tag{4.4}$$

$$p = \frac{1}{3} \int d^3\mathbf{k} f(\mathbf{k}) k v(\mathbf{k}) = \int \frac{1}{3} \frac{|\mathbf{k}|^2}{E} f(\mathbf{k})\, d^3\mathbf{k} = \frac{g}{6\pi^2} \int_m^\infty \frac{(E^2 - m^2)^{3/2}\, dE}{\exp[(E - \mu)/T] \pm 1}. \tag{4.5}$$

We use the symbol k to denote the momentum when the pressure is denoted by letter p. For a collection of relativistic particles, the velocity is $v = (k/E)$, which is used in Eq. (4.5).

Differentiating Eq. (4.5) with respect to T and treating μ as some specified function of T, we get

$$\frac{dp}{dT} = \frac{4\pi}{3} \int_0^\infty \frac{k^4\, dk}{E} f^2 \left[\exp \frac{(E - \mu)}{T} \right] \left[\frac{E}{T^2} + \frac{d}{dT} \left(\frac{\mu}{T} \right) \right]. \tag{4.6}$$

Using the relation

$$\frac{df}{dk} = -\frac{k}{ET} f^2 \exp \frac{(E - \mu)}{T}, \tag{4.7}$$

we can rewrite (dp/dT) as

$$\frac{dp}{dT} = -\frac{4\pi}{3} \int_0^\infty dk (k^3 T) \left(\frac{df}{dk} \right) \left[\frac{E}{T^2} + \frac{d}{dT} \left(\frac{\mu}{T} \right) \right]. \tag{4.8}$$

Integrating by parts and using the definitions of ρ and p, we find

$$\frac{dp}{dT} = \frac{1}{T}(\rho + p) + nT \frac{d}{dT} \left(\frac{\mu}{T} \right). \tag{4.9}$$

From Friedmann equations we have the relation $d(\rho a^3) = -p\, d(a^3)$, which can be written as

$$\frac{d}{dT}[(\rho + p)a^3] = a^3 \frac{dp}{dT}. \tag{4.10}$$

Substituting for (dp/dT) from Eq. (4.9) and rearranging the terms, we finally get

$$d(sa^3) \equiv d \left[\frac{a^3}{T}(\rho + p - n\mu) \right] = \left(\frac{\mu}{T} \right) d(na^3). \tag{4.11}$$

In most cases of interest to us either (na^3) will be approximately constant *or* we will have $\mu \ll T$. The preceding relation shows that, in either case, the quantity (sa^3) will be conserved.

When $\mu \ll T$, the expression for s reduces to $T^{-1}(\rho + p)$. Expanding the quantity $T d[a^3 T^{-1}(\rho + p)]$ and using relation (4.9) in the form $(dp/dT) \simeq T^{-1}(p + \rho)$, which is valid for $\mu \ll T$, we get

$$T d(sa^3) = T d \left[\frac{(\rho + p)a^3}{T} \right] = d[(\rho + p)a^3] - (\rho + p)a^3 \frac{dT}{T} \tag{4.12}$$

$$\simeq d[(\rho + p)a^3] - a^3 \, dp = d(\rho a^3) + p d(a^3).$$

Comparing Eq. (4.12) with the familiar thermodynamic relation $T dS = dE + p dV$, we see that $s = T^{-1} (\rho + p)$ may be interpreted as the entropy density. Then Eq. (4.11) shows that entropy density $s \propto a^{-3}$ during expansion, provided that $\mu \ll T$. Note that s is an additive quantity.

These equations simplify in some limiting cases. When the particles are highly relativistic $(T \gg m)$ and nondegenerate $(T \gg \mu)$, we get

$$\rho \cong \frac{g}{2\pi^2} \int_0^\infty \frac{E^3 \, dE}{e^{E/T} \pm 1} = \begin{cases} g_B(\pi^2/30)T^4 & \text{(bosons)} \\ \frac{7}{8}g_F(\pi^2/30)T^4 & \text{(fermions)} \end{cases}. \tag{4.13}$$

[To relate the integrals for fermion and bosons, we can use the following trick. Let

$$I_n^\pm \equiv \int_0^\infty \frac{x^n \, dx}{e^x \pm 1}. \tag{4.14}$$

Then

$$I_n^- - I_n^+ = \int_0^\infty dx \, x^n \frac{2}{e^{2x} - 1} = \frac{1}{2^n} \int_0^\infty dy \frac{y^n}{e^y - 1} = 2^{-n} I_n^-, \tag{4.15}$$

giving

$$I_n^+ = I_n^- \left(1 - \frac{1}{2^n} \right). \tag{4.16}$$

This accounts for the $(7/8)$ factor when $n = 3$.] Thus the total energy density contributed by all the relativistic species together can be expressed as

$$\rho_{\text{total}} = \sum_{i=\text{boson}} g_i \left(\frac{\pi^2}{30} \right) T_i^4 + \sum_{i=\text{fermion}} \frac{7}{8} g_i \left(\frac{\pi^2}{30} \right) T_i^4 = g_{\text{total}} \left(\frac{\pi^2}{30} \right) T^4, \tag{4.17}$$

where

$$g_{\text{total}} \equiv \sum_{\text{boson}} g_B \left(\frac{T_B}{T} \right)^4 + \sum_{\text{fermion}} \frac{7}{8} g_F \left(\frac{T_F}{T} \right)^4. \tag{4.18}$$

In writing g_{total}, we have explicitly taken into account the possibility that even though each of the species may have a thermal distribution they may not all have the same temperature. If all species have the same temperature, then $g = g_{\text{boson}} + (7/8)g_{\text{fermion}}$.

The pressure that is due to relativistic species is $p \simeq (\rho/3) = g(\pi^2/90)T^4$; so the entropy density of the relativistic species of particles will be

$$s \simeq \frac{1}{T}(\rho + p) = \frac{2\pi^2}{45}qT^3, \tag{4.19}$$

with

$$q \equiv q_{\text{total}} = \sum_{\text{boson}} g_B \left(\frac{T_B}{T}\right)^3 + \frac{7}{8} \sum_{\text{fermion}} g_F \left(\frac{T_F}{T}\right)^3. \tag{4.20}$$

Clearly, $q_{\text{total}} = g_{\text{total}}$ if all the particles have the same temperature. Our previous analysis shows that the quantity $S = qT^3a^3$ is conserved during the expansion.

The number density of relativistic particles can be computed in the same way:

$$n \simeq \frac{g}{2\pi^2} \int_0^\infty \frac{E^2\,dE}{e^{E/T} \pm 1} = \begin{cases} [\zeta(3)/\pi^2]g_B T^3 & \text{(boson)} \\ \frac{3}{4}[\zeta(3)/\pi^2]g_F T^3 & \text{(fermion)} \end{cases}, \tag{4.21}$$

where $\zeta(3) \simeq 1.202$ is the Riemann zeta function of the order of 3. Combining this with relation (4.13), we find that the mean energy of the particles $\langle E \rangle \equiv (\rho/n)$ is $\sim 2.7T$ for bosons and $3.15T$ for fermions. Note that s is proportional to the number density of relativistic particles if all species have the same temperature; in fact, $s \simeq 1.8qn_\gamma$, where n_γ is the photon number density.

In the opposite limit of $T \ll m$, the exponential in Eq. (4.1) is large compared with unity. Then we get, for *both* bosons and fermions,

$$\begin{aligned} n &\simeq \frac{g}{2\pi^2} \int_0^\infty p^2\,dp \exp\left[-\frac{(m-\mu)}{T}\right] \exp\left(-\frac{p^2}{2mT}\right) \\ &= g\left(\frac{mT}{2\pi}\right)^{3/2} \exp\left[-\frac{1}{T}(m-\mu)\right]. \end{aligned} \tag{4.22}$$

In this limit $\rho \simeq nm$ and $p = nT \ll \rho$. A comparison of relations (4.21) and (4.22) shows that the number (and energy) density of nonrelativistic particles is exponentially damped by the factor $\exp[-(m/T)]$ with respect to that of the relativistic particles.

For $t < t_{\text{eq}}$, in the radiation-dominated phase, we may ignore the contribution of nonrelativistic particles to ρ. We have seen earlier that, during the radiation-dominated phase, $a(t) \propto t^{1/2}$; therefore

$$\left(\frac{\dot{a}}{a}\right)^2 = H^2(t) = \frac{1}{4t^2} = \frac{8\pi G}{3}\rho = \frac{8\pi G}{3}g\left(\frac{\pi^2}{30}\right)T^4. \tag{4.23}$$

It is convenient to express these results in terms of Planck energy $m_{Pl} = G^{-\frac{1}{2}} = 1.22 \times 10^{19}$ GeV:

$$H(T) \cong 1.66 g^{1/2} \left(\frac{T^2}{m_{Pl}}\right), \qquad t \cong 0.3 g^{-\frac{1}{2}} \left(\frac{m_{Pl}}{T^2}\right) \simeq 1s \left(\frac{T}{1\ MeV}\right)^{-2} g^{-\frac{1}{2}}.$$

(4.24)

The factor g in these expressions counts the degrees of freedom of those particles that are *still* relativistic at this temperature T. As the temperature decreases, more and more particles will become nonrelativistic and g and q will decrease; thus $g = g(T)$ and $q = q(T)$ are slowly varying, decreasing functions of T. In the currently popular models for particle interactions, $g \simeq 10^2$ at $T \gtrsim 1$ GeV, $g \simeq 10$ for $T \simeq (100\text{–}1)$MeV, and $g \simeq 3$ for $T < 0.1$ MeV.

The slow variation of $q(T)$ has one important consequence. The expression for the conserved entropy in the radiation-dominated phase is $s \propto q(T)\ T^3 a^3$ [see relation (4.19)]. Therefore temperature T decreases as a^{-1} only as long as q is constant. If the number of degrees of freedom changes, then T will decrease slightly more slowly than a^{-1}; the correct relation is $q^{1/3}(T)T \propto a^{-1}$.

Finally, let us consider the distribution function for a species that has already decoupled. Expression (4.2) simplifies considerably if the decoupling occurs either when the species is ultrarelativistic ($T_D \gg m$) or when it is nonrelativistic ($T_D \ll m$). In the first case,

$$f_{dec}(p) = f_{equi}\left[p\,\frac{a(t)}{a(t_D)}, T_D\right] \cong \frac{g}{(2\pi)^3}\left\{\exp\frac{1}{T_D}\left[p\frac{a(t)}{a(t_D)}\right] \pm 1\right\}^{-1}. \quad (4.25)$$

This has the same form as f_{equi} for a relativistic species with the temperature

$$T(t) = T_D[a(t_D)/a(t)], \qquad (4.26)$$

even though this species is not in thermodynamic equilibrium any longer. The "temperature" in this distribution function falls *strictly* as a^{-1}; the entropy of these particles, $S_A = (s_A a^3)$, is separately conserved. Note that for the species that are still in thermal equilibrium, $T \propto q^{-\frac{1}{3}}a^{-1}$ falls more slowly.

The number density of these decoupled particles will be given by relation (4.21):

$$n = g_{eff}\left[\frac{\zeta(3)}{\pi^2}\right]T_D^3\left(\frac{a_D}{a}\right)^3, \qquad (4.27)$$

where $g_{eff} = (3g/4)$ for fermions and $g_{eff} = g$ for bosons. (Here g refers to the spin-degeneracy factor of the particular species that has decoupled.) This number density will be comparable with the number density of photons at any given time. In particular, any such decoupled species will continue to exist in our universe today as a relic background, with number densities comparable with the number densities of photons.

The following point should be noted: Suppose that a species with mass m decouples at the temperature T_D with $T_D \gg m$. At the time of decoupling, most of these particles will be ultrarelativistic and their (mean) momentum $p(t_D)$ and energy $E(t_D) = [p^2(t_D) + m^2]^{1/2} \simeq p(t_D)$ will be of the order of T_D. Their distribution function at $t = t_D$ is well approximated by the f_{equi} of zero-mass particles. Decoupling "freezes" the distribution function in this form. At a later time $(t > t_D)$, the mean momentum of the particles will be redshifted to a value $p(t) = p(t_D) (a_D/a) \simeq T_D(a_D/a)$. For $t \gg t_D$, most of the particles will have momentum $p(t)$, which is much smaller than m. Thus the individual particles would have become nonrelativistic when the universe has expanded sufficiently, which will happen when the temperature of the universe drops below $T_{\text{nr}} \simeq m$, that is, when $(a/a_D) \gtrsim (T_D/m)$. The energy of each of these particles will now be $E(t) = [p^2(t) + m^2]^{1/2} \simeq m$. *But the distribution function (and the number density) of the particles will still be given by the (frozen-in) form, which corresponds to relativistic particles.* Thus, for $t \gg t_D$, the number density of these particles will be similar to those of relativistic species but the energy density will be $\rho_{\text{dec}} \simeq nm$.

Consider next the other extreme case, that of a species that decouples when most of the particles are already nonrelativistic: $(T_D \ll m)$. In this case,

$$f_{\text{dec}}(p) = f_{\text{equi}}\left(p\frac{a}{a_D}, T_D\right) \simeq \frac{g}{(2\pi)^3} \exp\left[-\frac{(m-\mu)}{T_D}\right] \exp\left[-\frac{p^2}{2m}\frac{1}{T_D}\left(\frac{a}{a_D}\right)^2\right]$$

$$\simeq \frac{g}{(2\pi)^3} e^{-m/T_D} \exp\left[-\frac{p^2}{2mT_D}\left(\frac{a}{a_D}\right)^2\right], \tag{4.28}$$

where we have further assumed that $\mu \ll T_D$. This distribution function has the same form as that of a nonrelativistic Maxwell–Boltzmann gas with a "temperature" $T(t) \equiv T_D(a_D/a)^2$, which decreases as the *square* of the expansion factor. The corresponding number density is given by relation (4.22):

$$n = g\left(\frac{mT_D}{2\pi}\right)^{3/2} \left(\frac{a_D}{a}\right)^3 \exp-\frac{1}{T_D}(m-\mu)$$

$$\simeq g\left(\frac{mT_D}{2\pi}\right)^{3/2} \left(\frac{a_D}{a}\right)^3 \exp-\left(\frac{m}{T_D}\right) \quad \text{(for } \mu \ll T_D\text{).} \tag{4.29}$$

As is to be expected, $n \propto a^{-3}$. The energy density of these particles will be $\rho \simeq nm$.

To any species of particle that is not being created or destroyed ($n \propto a^{-3}$), we can assign a conserved number $N \propto na^3$; because $a^3 \propto s^{-1}$, we can conveniently define this number to be $N \equiv (n/s)$. From expressions (4.19), (4.21), and (4.22) (valid for $\mu \ll T$) it follows that

$$
N = \begin{cases} [45\zeta(3)/2\pi^4](g_{\mathrm{eff}}/q) & \cong 0.28\,(g_{\mathrm{eff}}/q) \\ [45/2\pi^4](\pi/8)^{1/2}(g/q)\left(\frac{m}{T}\right)^{3/2} e^{-m/T} & \cong 0.15\,(g/q)\left(\frac{m}{T}\right)^{3/2} e^{-m/T} \end{cases} ,
$$

(4.30)

where $g_{\mathrm{eff}} = g$ for bosons and $g_{\mathrm{eff}} = (3g/4)$ for fermions.

The assumption that decoupled particles travel along geodesics is equivalent to ignoring "gravitational collisions" between these particles. In other words, the gravitational force on each relic particle is assumed to be entirely due to the gravitational field produced by the smooth distribution of matter. We have seen in Vol. I, Chap. 10, Section 10.7 that the time scale for gravitational collisions is N times the dynamical time scale. Because the number of relic elementary particles inside a Hubble radius is enormous, the neglect of gravitational collisions is perfectly justified.

4.3 Relic Background of Relativistic Particles

The discussion so far has been based on general principles. Given the mass spectrum and the interactions of the elementary particles, the formalism developed in Section 4.2 can be used to make concrete predictions.

To understand the processes that occurred in the early universe when the temperature was T, we need to know the physics of particle interactions at energies $E \simeq T$. Based on our current knowledge of the latter, the study of the early universe can be divided into the following three different phases:

(1) Our understanding of particle interactions is reasonably complete for energies below 1 GeV. Correspondingly, we should be able to follow the evolution of the universe from the temperature of $T \simeq 1$ GeV $\simeq 1.2 \times 10^{13}$ K downwards with reasonable accuracy.

(2) There are several theoretical models that attempt to describe the particle interactions in the energy range of 1–10^{16} GeV. These models are comparatively more speculative, with the uncertainties increasing with the energy. (The range 1–100 GeV is somewhat better understood because it is accessible in particle accelerators.) Given a specific particle-physics model, the evolution of the universe in the temperature range of 10^{16} GeV $\simeq 1.2 \times 10^{29}$ K to 1 GeV $\simeq 1.2 \times 10^{13}$ K can be worked out. Because the models are not unique, we cannot obtain unique *predictions*. However, it is often possible to work out *some* consequences of these models that can be tested by cosmological observations.

(3) The physics at energies above 10^{16} GeV is very uncertain. Quantum gravitational effects, about which we know very little, will be significant at energies $E \gtrsim m_{\mathrm{Pl}} \simeq$

1.22×10^{19} GeV. The very basis for our discussion, classical general relativity, breaks down at these energies.

The uncertainty in our knowledge of particle interactions at high energies prevents us from *predicting* a unique material composition or evolutionary history for our universe. To make any progress, it is necessary to make reasonable assumptions about the material content of the universe at some moment in time and work out the consequences. Because the physical processes are relatively well understood at $T \lesssim 10^{12}$ K, we will discuss this part of the evolution first. Some comments regarding very early universe and inflationary scenarios will be made in Section 4.8.

Let us begin by ascertaining the composition of the universe at $T \lesssim 10^{12}$ K. Because the rest mass of electron $m_e \simeq 0.5$ MeV $= 6 \times 10^9$ K, we expect a significant number density (i.e., a number density comparable with that of photons) of ultrarelativistic electrons (e) and positrons ($\bar{e}$) at $T > 6 \times 10^9$ K. The only other particle species that could be relativistic at the temperature of 10^{12} K is the neutrino (ν). It is known that there are three kinds ("flavors") of neutrinos, viz., the electron neutrino (ν_e), the muon neutrino (ν_μ), and the tau neutrino (ν_τ). The experimental results on the masses of these particles are as follows: $m[\nu_e] \ll 5 \times 10^{-3}$ eV; $m[\nu_\mu] \simeq 5 \times 10^{-3}$ eV; $m[\nu_\tau] \lesssim 0.05$ eV. (These bounds are based on a claim for laboratory detection of mass oscillations between the neutrinos and a fair amount of theoretical modelling. At present these results must be considered tentative; we shall say more about these issues in Chap. 10.) Because the temperature of the universe at present is $T_0 \simeq 3$ K $\simeq 10^{-4}$ eV, we shall assume, for the time being, that $m_\nu = 0$ for all three species. There could be interesting physical consequences if any of the neutrinos has nonzero mass; this possibility needs to be discussed separately and we will take it up in Section 4.4.

The neutrons (n) and protons (p) contained in the present-day universe must have existed at $T \simeq 10^{12}$ K as well, as these particles could not have been produced at $T < 10^{12}$ K. The ratio between the number density of baryons (n_B) and the number density of photons (n_γ) has remained approximately constant from $T \simeq 10^{12}$ K until today. This number in the present-day universe is $(n_B/n_\gamma)_0 \simeq (\rho_c \Omega_B / m_B n_\gamma)_0 \simeq 10^{-8}$ to 10^{-10}. The smallness of this (conserved) number shows that we may ignore the effect of n_B on the overall dynamics of the radiation-dominated universe. There could also be a tiny fraction of muon–antimuon pairs; we will ignore them for simplicity.

Because photons are not conserved, the chemical potential for photons is identically zero. The reaction $e\bar{e} \leftrightarrow \gamma\gamma$ maintains the equilibrium among $e, \bar{e}$, and γ at this temperature. The conservation of chemical potential in this reaction implies that $\mu_e + \bar{\mu}_e = 0$; that is, $\bar{\mu}_e = -\mu_e$. (We denote a particle–antiparticle pair by $A, \bar{A}$ and the corresponding chemical potentials by $\mu_A, \bar{\mu}_A$.) The excess of electrons over positrons will then be

$$n - \bar{n} = \frac{g}{2\pi^2} \int_m^\infty E(E^2 - m^2)^{1/2}$$

$$\times \left[\frac{1}{\exp \frac{1}{T}(E - \mu) + 1} - \frac{1}{\exp \frac{1}{T}(E + \mu) + 1} \right] dE \quad (4.31)$$

$$\cong \frac{gT^3}{6\pi^2} \left[\pi^2 \left(\frac{\mu}{T}\right) + \left(\frac{\mu}{T}\right)^3 \right] \quad (\text{for } T \gg m).$$

As the universe cools to temperature $T \ll m_e$, electrons and positrons will annihilate in pairs and only this small excess will survive.

The only other *charged* particle that will be present in the universe is the proton. Because our universe appears to be electrically neutral [the bound on the net number density of free charges being $(n_Q/s) \lesssim 10^{-27}$; see Exercise 4.2], the electron excess $(n - \bar{n})$ should be equal to the number density of protons n_p. Because $(n_p/n_\gamma) \simeq 10^{-8}$ it follows that $[(n - \bar{n})/n_\gamma] \simeq 10^{-8}$. Using relations (4.21) and (4.31), we can write

$$\frac{n - \bar{n}}{n_\gamma} \cong \left(\frac{g_e}{g_\gamma}\right) \frac{\pi^2}{6\zeta(3)} \left[\frac{\mu}{T} + \frac{1}{\pi^2} \left(\frac{\mu}{T}\right)^3\right] \cong 1.33 \left(\frac{\mu}{T}\right) \simeq 10^{-8}. \quad (4.32)$$

Clearly $(\mu/T) \ll 1$ and we can set $\mu \cong 0$ for both electrons and positrons.

Exercise 4.2
Electric charge in the universe: (a) Can a $k = +1$ Friedmann universe contain a net electric charge? What about a net baryon number or lepton number?

(b) Consider an open Friedmann model with net charge density n_Q (distributed homogeneously). Show that the constraint $\Omega_{NR}h^2 \lesssim 1$ implies that $(n_Q/s) \lesssim 10^{-27}$.

(c) A massless, degenerate ($|\mu_\nu| \gg T$) neutrino will contribute an energy density $\rho_\nu \cong (\mu_\nu^4/8\pi^2)$. Show that the constraint $\Omega_{NR}h^2 \lesssim 1$ implies the bounds

$$(\mu_\nu/T_\nu) \lesssim 53, \qquad \frac{(n_\nu - n_{\bar{\nu}})}{n_\gamma} \lesssim 3.7 \times 10^3.$$

Similarly, from the reaction $\nu\bar{\nu} \leftrightarrow e\bar{e}$, it follows that $\mu_\nu + \bar{\mu}_\nu = \mu_e + \bar{\mu}_e = 0$. The excess of neutrinos over antineutrinos will again be given by an expression similar to Eq. (4.31). Unfortunately, the value of $[(n_\nu - \bar{n}_\nu)/n_\gamma]$ for our universe is not known. If this number is large, then our universe will have a large lepton number (L), which is far in excess of the baryon number B. Because our universe does not seem to have large values for any quantum number, a large value for L will require a special choice of initial conditions. This suggests that L should be small. In that case, $\mu_\nu = -\bar{\mu}_\nu \cong 0$. We will make this assumption in what follows.

The preceding arguments show that, at $T \simeq 10^{12}$ K, the energy density of the universe is essentially contributed by e, $\bar{e}$, ν, $\bar{\nu}$, and photons. Because the interactions among them maintain the equilibrium, they all have the same temperature. Taking $g_B = g_\gamma = 2$, $g_e = \bar{g}_e = 2$, and $g_\nu = \bar{g}_\nu = 1$ and including three flavors of neutrinos, we find

$$g_{\text{total}} = g_B + \frac{7}{8} g_F = 2 + \frac{7}{8}(2 + 2 + 2 \times 3) = \frac{43}{4} = 10.75 . \quad (4.33)$$

The g values for electrons and positrons represent the two possible spin states for the massive, spin-1/2 fermions. Although photons have spin-1, they have only two accessible states (corresponding to two states of polarisation), giving $g_\gamma = 2$. Massless spin-1/2 fermions, like neutrinos, exist only in left-handed or right-handed states, making $g_\nu = 1$. From relations (4.24), we can find the precise time–temperature relationship for this phase of the evolution:

$$H(T) \cong 5.44 \left(\frac{T^2}{m_{\text{Pl}}} \right), \qquad t \cong 0.09 \left(\frac{m_{\text{Pl}}}{T^2} \right). \quad (4.34)$$

Because neutrinos have no electric charge, they have no direct coupling with photons. Their interaction with baryons can be ignored because of the low density of baryons. Therefore they are kept in equilibrium essentially through reactions such as $\nu\bar{\nu} \leftrightarrow e\bar{e}$, $\nu e \leftrightarrow \nu e$, etc. The cross section $\sigma(E)$ for these weak-interaction processes is of the order of $(\alpha^2 E^2 / m_x^4)$, where $\alpha \simeq 2.8 \times 10^{-2}$ is related to the gauge-coupling constant $g_s \simeq 0.6$ by $\alpha = (g^2/4\pi)$ and $m_x \simeq 50$ GeV is the mass of the gauge-vector boson mediating the weak interaction. Defining the "Fermi coupling constant" $G_F = (\alpha/m_x^2) \simeq 1.17 \times 10^{-5} (\text{GeV})^{-2} = (293\ \text{GeV})^{-2}$ and using the fact that $E \simeq T$, we can write $\sigma \simeq G_F^2 E^2 \simeq G_F^2 T^2$. Because the number density of interacting particles is $n \cong [\zeta(3)g/\pi^2]T^3 \cong 1.3\ T^3$ and $\langle v \rangle \simeq c = 1$, the rate of interactions is given by

$$\Gamma = n\sigma|v| \simeq 1.3\ G_F^2 T^5. \quad (4.35)$$

The rate of expansion, from relations (4.34) is $H \cong 5.4\ (T^2/m_{\text{Pl}})$. Therefore

$$\frac{\Gamma}{H} \simeq 0.24\ T^3 \left(\frac{m_{\text{Pl}}}{G_F^{-2}} \right) \simeq \left(\frac{T}{1.4\ \text{MeV}} \right)^3 = \left(\frac{T}{1.6 \times 10^{10}\ \text{K}} \right)^3 . \quad (4.36)$$

The interaction rate of neutrinos becomes lower than the expansion rate when the temperature drops below $T_D \simeq 1$ MeV. At lower temperatures, the neutrinos are completely decoupled from the rest of the matter.

Exercise 4.3

Weak-interaction cross section: (a) Assume that weak interactions are mediated by a vector boson with mass m_x. Let $\sigma(E)$ be a typical cross section for a $2 \leftrightarrow 2$ process. Argue that for $E \lesssim m_x$, $\sigma(E) \simeq (\alpha^2 E^2 / m_x^4)$. How does $\sigma(E)$ behave for $E \gg m_x$?

(b) Assume that the rate Γ for some reaction is proportional to T^n with $n > 3$. Let t_D be defined as the time of decoupling at which $\Gamma(t_D) = H(t_D)$ with $t_D \ll t_{eq}$. Show that the number of *further* interactions that take place is less than unity.

Because the neutrinos are assumed to be massless, they are clearly relativistic at the time of decoupling. (This conclusion will be true even if neutrinos have a mass m_ν with $m_\nu \ll T_D \simeq 1$ MeV.) Their distribution function at later times is given by Eq. (4.25) with $T_\nu \propto a^{-1}$. The present-day universe should contain a relic background of these neutrinos.

At the time of decoupling, the photons, neutrinos, and the rest of the matter had the same temperature. As long as the photon temperature decreases as a^{-1}, neutrinos and photons will continue to have the same temperature even though the neutrinos have decoupled. However, the photon temperature will decrease at a slightly lower rate if the g factor is changing. In that case, T_γ will become higher than T_ν as the universe cools. Such a change in the value of g occurs when the temperature of the universe falls below $T \simeq m_e$. The electron rest mass $m_e \simeq 0.5$ MeV corresponds to a temperature of 5×10^9 K. When the temperature of the universe becomes lower than this value, the mean energy of the photons will fall below the energy required for creating $e\bar{e}$ pairs. Thus the backward reaction in $e\bar{e} \leftrightarrow \gamma\gamma$ will be severely suppressed. The forward reaction will continue to occur, resulting in the disappearance of the $e\bar{e}$ pairs.

This process clearly changes the value of g. At $T_D > T \gtrsim m_e$, the νs have decoupled and their entropy is separately conserved; but the photons ($g = 2$) are in equilibrium with electrons ($g = 2$) and positrons ($g = 2$). This gives $g(\gamma, e, \bar{e}) = 2 + (7/8) \times 4 = (11/2)$. For $T \ll m_e$, when the $e\bar{e}$ annihilation is complete, the only relativistic species left in this set is the photon ($g = 2$). The conservation of $S = q(Ta)^3$, applied to particles that are in equilibrium with radiation, shows that the quantity $q(T_\gamma a)^3 = g(aT_\gamma)^3$ remains constant during expansion. (Because γ, e, and $\bar{e}$ all have the same temperature, $q = g$.) Because g decreases during the $e\bar{e}$ annihilation, the value of $(aT_\gamma)^3$ after the $e\bar{e}$ annihilation will be higher than its value before:

$$\left[(aT_\gamma)^3_{\text{after}} / (aT_\gamma)^3_{\text{before}} \right] = (g_{\text{before}} / g_{\text{after}}) = \frac{11}{4}. \qquad (4.37)$$

The neutrinos, because they are decoupled, do not participate in this process. They are characterised by a temperature $T_\nu(t)$ that falls *strictly* as a^{-1} and their entropy ($s_\nu a^3$) is separately conserved. Let $T_\nu = Ka^{-1}$; originally, before the $e\bar{e}$ annihilations began, the photons and neutrinos had the same temperature:

$(aT_\nu)_{\text{before}} = (aT_\gamma)_{\text{before}} = K$. It follows that

$$
\begin{aligned}
(aT_\gamma)_{\text{after}} &= \left(\frac{11}{4}\right)^{1/3} (aT_\gamma)_{\text{before}} = \left(\frac{11}{4}\right)^{1/3} (aT_\nu)_{\text{before}} \\
&= \left(\frac{11}{4}\right)^{1/3} (aT_\nu)_{\text{after}} \simeq 1.4(aT_\nu)_{\text{after}}.
\end{aligned}
\tag{4.38}
$$

The first equality follows from Eq. (4.37), the second from the fact that $T_\gamma = T_\nu$ at $T \gtrsim m_e$, and the third from the strict constancy of (aT_ν). Thus the $e\bar{e}$ annihilations increase the temperature of photons compared with that of neutrinos by a factor of $(11/4)^{1/3} \simeq 1.4$.

Although the preceding analysis gives the asymptotic temperature correctly, it is also of interest to determine the evolution of temperature during the $e^+ - e^-$ annihilation phase. This is most easily done if the total entropy $s(T)a^3$ is set to a constant. Differentiating this result logarithmically gives the relation

$$
\frac{1}{s}\frac{ds}{dT}\frac{dT}{da} = -\frac{3}{a}.
\tag{4.39}
$$

Given entropy s as a function of temperature T, this equation can be integrated to provide $T(a)$. The form of $s(T) = s_\gamma(T) + s_{\text{ele}}(T)$ can be determined from the general relation $s = T^{-1}(\rho + p)$ and Eqs. (4.4) and (4.5). This gives the total entropy density as

$$
s = \frac{4}{3}a_B T^3 \left[1 + \frac{45}{2(\pi k_B T)^4} \int_0^\infty \frac{k^2\,dk}{\epsilon}\frac{\epsilon^2 + k^2/3}{e^{\epsilon/k_B T} + 1}\right], \qquad a_B \equiv \frac{\pi^2}{15}\frac{k_B^4}{\hbar^3 c^3},
\tag{4.40}
$$

where $\epsilon(k)$ is the energy corresponding to momentum k. With some simple manipulation, Eq. (4.39) can be recast in the form $aTf(m_e/T) = \text{constant}$, where

$$
f^3(x) = 1 + \frac{15}{2\pi^4} \int_0^\infty \frac{y^2\,dy}{\sqrt{x^2 + y^2}}\frac{3x^2 + 4y^2}{(\exp\sqrt{x^2 + y^2} + 1)}.
\tag{4.41}
$$

To determine the expansion rate, that is, a as a function of t, we need to determine the total energy density that is dominated by those of radiation, particle pairs, and other relativistic components. If T_ν is the neutrino temperature, then this energy density can be written as $\rho = a_B T_\nu^4 \mathcal{R}$, where

$$
\mathcal{R} = \left(\frac{T}{T_\nu}\right)^4 + \frac{7}{8}N_\nu + \frac{15}{\pi^2}\frac{\hbar^3}{(k_B T_\nu)^4}\int \frac{4\,d^3k}{(2\pi\hbar)^3}\frac{\epsilon}{e^{\epsilon/k_B T} + 1}.
\tag{4.42}
$$

The first term is the photon energy density, the second term is the contribution from all species of weakly interacting relativistic matter written in terms of the effective number N_ν of neutrino families existing at that epoch, and the last term is the energy density in electron–positron pairs. The evolution of T_ν is determined by

$$\left(\frac{\dot{T}_v}{T_v}\right)^2 = \frac{8}{3}\pi Ga_B T_v^4 \mathcal{R} = 0.047\left(\frac{T_v}{10^{10}\text{ K}}\right)^4 \mathcal{R} \text{ s}^{-2}. \tag{4.43}$$

(An equivalent form of this relation is given in Exercise 4.4.)

It can be easily verified that the photons released by this process ($e\bar{e} \to 2\gamma$) become thermalised rapidly because of the scattering with charged particles. (We will study these scattering processes in detail later in Section 4.7.) The preceding analysis, of course, is based on this tacit assumption.

Exercise 4.4

Details of e^+e^- annihilation: (a) Show that, during the e^+e^- annihilation, the relation $aT = $ constant is modified to the form $aTf(m_e/T) = $ constant, where f is given by Eq. (4.41).

(b) Show that the energy density during e^+e^- annihilation is given by $\rho = a_B T^4 \times \epsilon(m_e/T)$, where

$$\epsilon(x) = 1 + \left(\frac{21}{8}\right)\left(\frac{4}{11}\right)^{4/3} f^4 + \frac{30}{\pi^4}\int_0^\infty \frac{y^2\sqrt{x^2+y^2}\,dy}{(\exp\sqrt{x^2+y^2}+1)}. \tag{4.44}$$

(c) Using these, estimate how long it takes for the ratio (T_γ/T_v) to become, say, 1.39.

After the e^+e^- annihilations, the g factor does not change. Both T_γ and T_v fall as a^{-1} and the ratio $T_v = (4/11)^{1/3} T_\gamma \cong 0.71\, T_\gamma$ is maintained until today. The relic v background today should have the distribution given by Eq. (4.25) with $(T_v)_{\text{now}} \cong 0.71 \times 2.7\text{ K} = 1.9\text{ K}$.

Thus the species of particles that remain relativistic today will be photons ($g_\gamma = 2$) with a temperature $T_\gamma \cong 2.7\text{ K}$ and three flavors of massless neutrinos and antineutrinos ($g_F = 3 + 3 = 6$) with a temperature $T_v = (4/11)^{1/3} T_\gamma$. From Eqs. (4.18) and (4.20) we find

$$g(\text{now}) = 2 + \frac{7}{8} \times 6 \times \left(\frac{4}{11}\right)^{4/3} \cong 3.36,$$

$$q(\text{now}) = 2 + \frac{7}{8} \times 6 \times \left(\frac{4}{11}\right) \cong 3.91. \tag{4.45}$$

The energy and entropy densities of these relativistic particles in the present-day universe are

$$\rho_R = \frac{\pi^2}{30}gT^4 = 8.09 \times 10^{-34}\text{ gm cm}^{-3}$$

$$s = \frac{2\pi^2}{45}qT^3 \cong 2.97 \times 10^3\text{ cm}^{-3}. \tag{4.46}$$

This ρ_R corresponds to $\Omega_R = 4.3 \times 10^{-5} h^{-2}$. Note that $\rho_R = (g_{\text{total}}/g_\gamma) \rho_\gamma \simeq$ $1.68\rho_\gamma$; similarly, $\Omega_R = 1.68\,\Omega_\gamma$.

The matter density today is $\rho_{\text{NR}} = 1.88 \times 10^{-29}\,\Omega h^2\,\text{gm cm}^{-3}$. The redshift z_{eq} at which matter and radiation had equal energy densities is determined by the relation $(1 + z_{\text{eq}}) = (\rho_{\text{NR}}/\rho_R)$. We calculated this quantity z_{eq} in Chap. 3, assuming that Ω_R is contributed by photons alone. The correct value, if there are three massless neutrino species, is

$$(1 + z_{\text{eq}}) = \left(\frac{\Omega_{\text{NR}}}{\Omega_R} \right) = 2.3 \times 10^4 (\Omega_{\text{NR}} h^2). \tag{4.47}$$

This corresponds to the temperature

$$T_{\text{eq}} = T_0 (1 + z_{\text{eq}}) = 5.5(\Omega_{\text{NR}} h^2)\text{eV} \tag{4.48}$$

and time

$$t_{\text{eq}} \simeq \frac{2}{3} H_0^{-1} \Omega_{\text{NR}}^{-1/2} (1 + z_{\text{eq}})^{-3/2} = 5.84 \times 10^{10} (\Omega_{\text{NR}} h^2)^{-2}\,\text{s}. \tag{4.49}$$

As explained in Chap. 3, we have defined t_{eq} by using the expression for $a(t)$ that is valid in the matter-dominated phase [using the more precise form of $a(t)$ will give t_{eq} (exact) $= 0.585 t_{\text{eq}} = 3.41 \times 10^{10} (\Omega_{\text{NR}} h^2)^{-2}$ s]. The size of the Hubble radius at t_{eq} is $d_H(t_{\text{eq}}) \simeq c t_{\text{eq}} \simeq 1.75 \times 10^{21} (\Omega_{\text{NR}} h^2)^{-2}$ cm. This corresponds to the length scale

$$\lambda_{\text{eq}} = d_H(t_{\text{eq}})(1 + z_{\text{eq}}) \simeq 13\,\text{Mpc}(\Omega_{\text{NR}} h^2)^{-1} \tag{4.50}$$

today.

The entropy density s was used earlier to define the conserved quantities $N_A = (n_A/s)$ for those species of particles with $n_A \propto a^{-3}$. Knowing the present value of s, we can explicitly compute this number. For example, the present baryon number density $n_B = 1.13 \times 10^{-5} (\Omega_B h^2)\text{cm}^{-3}$ corresponds to the ratio

$$\left(\frac{n_B}{s} \right) = \left(\frac{\rho_c \Omega_B}{s m_B} \right) = 3.81 \times 10^{-9} (\Omega_B h^2). \tag{4.51}$$

Often the number density of photons n_γ is used rather than s to define a baryon-to-photon ratio, $\eta \equiv (n_B/n_\gamma)$, for our universe. Because

$$n_\gamma = \left[\frac{2\zeta(3)}{\pi^2} \right] T_0^3 = \left[\frac{45\zeta(3)}{\pi^4 q} \right] s \simeq 0.142 s \simeq 422\,\text{cm}^{-3}, \tag{4.52}$$

this ratio is

$$\eta = \left(\frac{n_B}{n_\gamma} \right) = \left(\frac{n_B}{s} \right) \left(\frac{s}{n_\gamma} \right) \simeq 7 \left(\frac{n_B}{s} \right) = 2.67 \times 10^{-8} (\Omega_B h^2). \tag{4.53}$$

This value is extremely low; the corresponding number in stellar interiors, for example, is $\sim 10^2$.

As mentioned earlier, the dominant-matter density in the universe is *not* contributed by the baryons. It seems very likely that nonbaryonic dark matter contributes a fraction $\Omega_{DM} \simeq 0.2 - 1$ whereas baryons contribute only $\Omega_B \simeq 0.02h^{-2}$ or so (see Section 4.5). Hence $\Omega = \Omega_B + \Omega_{DM} \approx \Omega_{DM}$. When the universe becomes matter dominated at $t \simeq t_{eq}$, it will be essentially dominated by the nonbaryonic dark matter and *not* by the baryons. The baryonic energy density will dominate over radiation when $\rho_B = \Omega_B \rho_c (1 + z)^3$ becomes larger than $\rho_R = \Omega_R \rho_c (1 + z)^4$. This occurs at the redshift z_{RB}, where

$$(1 + z_{RB}) = \frac{\Omega_B}{\Omega_R} = 2.3 \times 10^4 (\Omega_B h^2). \tag{4.54}$$

For $\Omega_B h^2 \simeq 0.02$, this is $z_{RB} \simeq 460$.

In this section we have started with a particular composition for the universe at $T \simeq 10^{12}$ K and worked out the consequences. This initial composition contained a small fraction of protons and neutrons but no antiprotons or antineutrons. Similarly, there was a small excess of electrons over positrons, so that after the $e\bar{e}$ annihilations an excess of electrons survived. At temperatures higher than a few giga-electron-volts, the universe would have contained a large number of antiprotons, antineutrons, etc., as well, but there should have been a slight excess of protons over antiprotons. As the universe cools through the temperature $T \simeq m_{proton} \simeq 1$ GeV, the protons and the antiprotons will annihilate each other, leaving a small excess of protons. It is therefore clear that we have put in by hand an excess of baryons over antibaryons in the initial conditions, so as to correctly reproduce the present-day universe. A truly fundamental theory should explain how this baryon excess arises in the earlier phases of the universe. Such an explanation is indeed possible in some of the particle-physics models, although the results are still not conclusive.

4.4 Relic Background of Wimps

In the discussion in Section 4.3 we assumed that the neutrinos are massless. If they have nonzero rest mass, then the physical consequences will be quite different. These consequences can be predicted in a rather general manner for any weakly interacting massive particle (usually called a *wimp*), which could be a neutrino or some other particle.

Let us first consider the case of a wimp that decouples while it is still relativistic, i.e., $T_D \gg m$, where m is the mass of the particle and T_D is the decoupling temperature. Such a particle is characterised by the conserved quantity

$$N = 0.28 \left(\frac{g_{\text{eff}}}{q} \right)_{T=T_D} = 0.21 \left[\frac{g}{q(T_D)} \right], \tag{4.55}$$

where we have assumed the particle to be a fermion and set $g_{\text{eff}} = (3g/4)$ in

Eq. (4.30). Such fermions will have a number density

$$n_0 = N s_0 = 2.97 \times 10^3 N \text{ cm}^{-3} \simeq 619 g [q(T_D)]^{-1} \text{ cm}^{-3} \qquad (4.56)$$

in the present universe.

These particles would have become nonrelativistic at some temperature $T_{\text{NR}} \simeq m$ in the past as long as $m > T_0$, i.e., $m \gtrsim 1.7 \times 10^{-4}$ eV. (If this is not the case, the particles will be relativistic even today and will behave just like the massless neutrinos discussed in the last section.) The energy contributed by each particle today is $E \simeq m$, so the energy density of these particles today will be

$$\rho = n_0 m = 6.19 \times 10^3 g [q(T_D)]^{-1} \left(\frac{m}{10 \text{ eV}}\right) \text{eV cm}^{-3}, \qquad (4.57)$$

which corresponds to

$$(\Omega h^2)_{\text{wimp}} = 0.59 \left(\frac{m}{10 \text{ eV}}\right) \left[\frac{g}{q(T_D)}\right]. \qquad (4.58)$$

In Chap. 3, we obtained the constraint $\Omega h^2 \lesssim 1$ from the observations related to the age of the universe. Combining this constraint with Eq. (4.58), we get a bound on the mass m:

$$m \lesssim 17 q(T_D) g^{-1} \text{ eV}. \qquad (4.59)$$

This bound is valid for any wimp that decouples while being still relativistic. The precise value of the right-hand side depends on the value of q at the time of decoupling.

Neutrinos or other wimps with masses of less than ~ 1 MeV decouple at $T_D \simeq (1-3)$ MeV; at this temperature $q = 10.75$. Therefore, for a single massive species with $g = 2$, we get

$$\Omega_\nu h^2 = \left(\frac{m_\nu}{91.5 \text{ eV}}\right), \quad m_\nu \lesssim 91.5 \text{ eV}, \quad (T_D \simeq 1 \text{ MeV}). \qquad (4.60)$$

The value of q will be higher at higher T_D. At temperatures above 300 GeV, the standard model predicts that eight gluons, W, $\bar{W}$, Z, one Higgs doublet, and three generations of quarks and leptons will all be relativistic, making $q \cong 106.5$. If a wimp decouples at $T_D \gtrsim 300$ GeV, then the corresponding bound will be

$$(\Omega h^2)_{\text{wimp}} = \left(\frac{m}{910 \text{ eV}}\right), \quad m \lesssim 910 \text{ eV}, \quad (T_D \gtrsim 300 \text{ GeV}). \qquad (4.61)$$

However, it should be noted that decoupling at such high energies is possible only if the particles have nonstandard interactions.

Consider next the wimps that decouple when they are nonrelativistic ($m \gtrsim 3 T_D$). The value of N for these particles is given by

$$N = 0.145 \left[\frac{g_A}{q(T_D)}\right] \left(\frac{m}{T_D}\right)^{3/2} e^{-m/T_D}. \qquad (4.62)$$

Unlike the previous case, N now depends on m quite strongly. To obtain a numerical estimate, we have to determine T_D by the criterion $\Gamma = H$. The reactions that are capable of changing the number of the wimps of type A are of the form $A\bar{A} \leftrightarrow X\bar{X}$, where X is some generic species of particle. (It will be assumed that the Xs are in thermal equilibrium.) The average value of σv for such annihilation processes can be expressed in the form

$$\langle \sigma v \rangle \equiv \sigma_0 \left(\frac{T}{m}\right)^k. \tag{4.63}$$

The value of k depends on the details of the dominant annihilation process (s wave, p wave, ..., etc.); it is usually of the order of unity. The value of σ_0 depends on m and has a simple form in the two extreme cases of $m \ll m_Z$ and $m \gg m_Z$, where $m_Z \simeq 10^2$ GeV is the mass of the Z boson that mediates weak interaction. Let us consider the $m \ll m_Z$ case first. For wimps with $m < m_Z$, the cross section σ_0 can be expressed as

$$\sigma_0 \simeq \frac{c}{2\pi} G_F^2 m^2. \tag{4.64}$$

The value of the constant c depends on the type of the "type" of fermion. Fermions with spin-1/2 are classified as "Dirac" type or "Majorana" type. A Dirac-type fermion will be distinct from its antiparticle whereas a Majorana-type fermion will be its own antiparticle. For the Dirac-type fermions that we will consider, $c \simeq 5$. The reaction rate therefore is

$$\Gamma = n\langle \sigma v \rangle \simeq g_A \left(\frac{mT}{2\pi}\right)^{3/2} e^{-m/T} \sigma_0 \left(\frac{T}{m}\right)^k = \frac{\sigma_0 g_A}{(2\pi)^{3/2}} T^3 \left(\frac{m}{T}\right)^{3/2-k} e^{-m/T}. \tag{4.65}$$

The expansion rate is $H = 1.66 g^{1/2} (T^2/m_{\text{Pl}})$; thus the condition $(\Gamma/H) = 1$ becomes

$$\frac{\Gamma}{H} = 3.825 \times 10^{-2} \left(\frac{g_A}{q^{1/2}}\right) \left(\frac{m}{T}\right)^{1/2-k} e^{-m/T} (\sigma_0 m m_{\text{Pl}}) = 1. \tag{4.66}$$

Solving this equation for $\exp[-(m/T)]$ and substituting into Eq. (4.62), we find

$$N \simeq \frac{3.79}{q^{1/2}} \left(\frac{m}{T_D}\right)^{k+1} (\sigma_0 m m_{\text{Pl}})^{-1} \simeq 2.87 \times 10^{-9} q^{-1/2} \left(\frac{m}{T_D}\right)^{k+1} \left(\frac{m}{1\text{ GeV}}\right)^{-3}. \tag{4.67}$$

This corresponds to a number density

$$n_0 = N s_0 = 8.523 \times 10^{-6} q^{-1/2} \left(\frac{m}{T_D}\right)^{k+1} \left(\frac{m}{1\text{ GeV}}\right)^{-3} \tag{4.68}$$

and the density parameter

$$(\Omega h^2)_{\text{wimp}} = 0.81 q^{-1/2} \left(\frac{m}{T_D}\right)^{k+1} \left(\frac{m}{1\ \text{GeV}}\right)^{-2}. \tag{4.69}$$

To make numerical estimates, Eq. (4.66) has to be solved for T_D. Taking logarithms, we can write Eq. (4.66) as

$$\frac{m}{T_D} = 17.966 + \ln\left(\frac{g_A}{q^{1/2}}\right) + \left(\frac{1}{2} - k\right) \ln\left(\frac{m}{T_D}\right) + 3\ln\left(\frac{m}{1\ \text{GeV}}\right). \tag{4.70}$$

This condition will determine the value of T_D in terms of m. Because g is a slowly varying function of T, it is best to solve this equation iteratively. Consider, as a typical example, a wimp with $m \gtrsim 1$ GeV. To leading order $(m/T_D) \simeq 17.966$; assuming, for simplicity, that $k = 0$, we find that the $\ln(m/T_D)$ term corrects this to $(m/T_D) \simeq 19.41$, giving $T_D \simeq 52$ MeV $(m/1\ \text{GeV})$; at this temperature, $g \simeq 10^2$. This gives $\ln(q^{1/2}/g_A) \simeq \ln 5 \simeq 1.61$, correcting (m/T_D) further to 17.8. Thus, to this order of iteration,

$$\left(\frac{m}{T}\right) \simeq 17.8 + 3\ln\left(\frac{m}{1\ \text{GeV}}\right). \tag{4.71}$$

On substitution into relations (4.67) and (4.69), we find

$$N \simeq 5.11 \times 10^{-9} \left(\frac{m}{1\ \text{GeV}}\right)^{-3}, \qquad \Omega_A h^2 \simeq 1.44 \left(\frac{m}{1\ \text{GeV}}\right)^{-2}. \tag{4.72}$$

The fermion A and its antiparticle $\bar{A}$ together will provide twice this value to Ω, corresponding to $\Omega_{A\bar{A}} h^2 \simeq 2.88\ (m/1\ \text{GeV})^{-2}$. The constraint $\Omega h^2 \lesssim 1$ then gives the mass bound

$$m \gtrsim 2\ \text{GeV}. \tag{4.73}$$

Finally, consider relics with $m \gg 100$ GeV. In this mass range the annihilation cross section σ_0 begins to *decrease* as m^2. Because σ_0 was increasing as m^2 in the previous case with $m < 100$ GeV, we have to change the value of N by only a factor m^4 to obtain the correct result. Therefore, for $m > 100$ GeV, we will get $N \propto m$ and $\Omega_A h^2 \propto m^2$. Repeating the preceding analysis and substituting the numbers, we will find that

$$\Omega_A h^2 \simeq \left(\frac{m}{1\ \text{TeV}}\right)^2. \tag{4.74}$$

Thus particles with $m \simeq 1$ TeV can also provide $\Omega h^2 \simeq 1$.

This analysis reveals that wimps in three different mass ranges can contribute significantly to the density of the universe, leading to $\Omega h^2 \lesssim 1$. For $m \lesssim 10^2$ eV and $T_D \simeq (1-3)$MeV, the wimps decouple while they are relativistic. Their number density today is comparable with those of photons and for $m \simeq 10^2$ eV, $\Omega h^2 \simeq 1$. (Such relics are called *hot* relics.) If, on the other hand, $m \gtrsim 1$ GeV,

$T_D \simeq (m/19) \simeq 52$ MeV, and the particles decouple while they are nonrelativistic. (They are called *cold* relics.) The number density of cold relics is suppressed by $\exp(-m/T_D)$. For $m \approx 2$ GeV, we again get $\Omega h^2 \simeq 1$. Finally, the range $m \simeq 1$ TeV can also lead to $\Omega h^2 \simeq 1$; such a particle will also be, of course, a cold relic. It is interesting to note that purely cosmological considerations (viz., $\Omega h^2 \lesssim 1$) rule out the possible existence of stable, weakly interacting fermions in the mass range 100 eV $< m < 2$ GeV.

Observations suggest that the universe contains a significant amount of dark matter. Wimps, if they exist, can provide much of the mass in the form of dark matter. For most of the discussion in the future chapters, we shall assume that this is indeed the case. Depending on the nature of the wimp, the dark matter will also be called *hot* or *cold*.

Exercise 4.5
The importance of being weak: Repeat the analysis in the text for a cold relic, keeping the value of σ_0 arbitrary and taking $k = 0$. Show that

$$(\Omega_A h^2) \simeq \mathcal{O}(1) \left(\frac{\sigma_0}{10^{-37} \text{ cm}^2} \right).$$

This shows why the annihilation cross section should be that of a weak-interaction process if $\Omega_A h^2 \simeq 1$.

4.5 Synthesis of Light Nuclei

The binding energies of the first four light nuclei ^{2}H, ^{3}H, ^{3}He, and ^{4}He are 2.22, 6.92, 7.72, and 28.3 MeV, respectively. As the universe cools below these temperatures, we expect these bound structures to form. The abundance of light elements, which are synthesised in the early universe, can be used to obtain important constraints on the cosmological parameters.

Although the energy considerations suggest that these nuclei could be formed when the temperature of the universe is in the range of 1–30 MeV, the actual synthesis takes place at only a much lower temperature, $T_{\text{nuc}} = T_n \simeq 0.1$ MeV. The main reason for this delay is the "high entropy" of our universe, i.e., the high value for the photon-to-baryon ratio, η^{-1}. This fact can be understood as follows:

Let us assume, for a moment, that the nuclear (and other) reactions are fast enough to maintain thermal equilibrium among various species of particles and nuclei. In thermal equilibrium, the number density of a nuclear species $^A N_z$ with atomic mass A and charge Z will be

$$n_A = g_A \left(\frac{m_A T}{2\pi} \right)^{3/2} \exp\left[-\left(\frac{m_A - \mu_A}{T} \right) \right]. \tag{4.75}$$

In particular, the equilibrium number densities of protons and neutrons will be

$$n_p = 2\left(\frac{m_p T}{2\pi}\right)^{3/2} \exp-\left[\frac{1}{T}(m_p - \mu_p)\right]$$

$$\cong 2\left(\frac{m_B T}{2\pi}\right)^{3/2} \exp-\left[\frac{1}{T}(m_p - \mu_p)\right], \qquad (4.76)$$

$$n_n = 2\left(\frac{m_n T}{2\pi}\right)^{3/2} \exp-\left[\frac{1}{T}(m_n - \mu_n)\right]$$

$$\cong 2\left(\frac{m_B T}{2\pi}\right)^{3/2} \exp-\left[\frac{1}{T}(m_n - \mu_n)\right]. \qquad (4.77)$$

The mass difference between the proton and the neutron, $Q \equiv m_n - m_p = 1.293$ MeV, has to be retained in the exponent but can be ignored in the prefactor $m_A^{3/2}$; we have set, in the prefactor, $m_n \cong m_p \cong m_B$, an average value.

Because the chemical potential is conserved in the reactions, producing $^A N_z$ out of Z protons and $(A - Z)$ neutrons, μ_A for any species can be expressed in terms of μ_p and μ_n:

$$\mu_A = Z\mu_p + (A - Z)\mu_n. \qquad (4.78)$$

Writing

$$\exp\left[\frac{1}{T}(\mu_A - m_A)\right] = \exp\frac{1}{T}[Z\mu_p + (A - Z)\mu_n]\exp-(m_A/T)$$

$$\qquad (4.79)$$

$$= [\exp(\mu_p/T)]^Z[\exp(\mu_n/T)]^{(A-Z)}\exp-(m_A/T)$$

and substituting for $\exp(\mu_p/T)$ and $\exp(\mu_n/T)$ from Eqs. (4.76) and (4.77), we get

$$\exp\left[\frac{1}{T}(\mu_A - m_A)\right] = 2^{-A}n_p^Z n_n^{(A-Z)}\left(\frac{2\pi}{m_B T}\right)^{3A/2}$$

$$\times \exp\left[\frac{1}{T}(Zm_p + (A - Z)m_n - m_A)\right] \qquad (4.80)$$

$$= 2^{-A}n_p^Z n_n^{(A-Z)}\left(\frac{2\pi}{m_B T}\right)^{3A/2}\exp(B_A/T),$$

where $B_A \equiv Zm_p + (A - Z)m_n - m_Z$ is the binding energy of the nucleus. Therefore the number density in Eq. (4.75) becomes

$$n_A = g_A 2^{-A} A^{3/2}\left(\frac{2\pi}{m_B T}\right)^{3(A-1)/2} n_p^Z n_n^{A-Z}\exp(B_A/T). \qquad (4.81)$$

We define the "mass fraction" of the nucleus A by $X_A = (An_A/n_B)$, where n_B is the number density of baryons in the universe. Substituting for n_A, n_p, and n_n

in Eq. (4.81) by $n_A = n_B(A^{-1}X_A) = \eta n_\gamma (A^{-1}X_A)$, $n_p = n_B X_p = \eta n_\gamma X_p$, and $n_n = n_B X_n = \eta n_\gamma X_n$, where $\eta = 2.68 \times 10^{-8} \, (\Omega_B h^2)$ is the baryon-to-photon ratio and $n_\gamma = [2\zeta(3)/\pi^2] \, T_0^3$ is the number density of photons, we get

$$X_A = F(A)(T/m_B)^{3(A-1)/2} \eta^{A-1} X_p^Z X_n^{A-Z} \exp(B_A/T), \tag{4.82}$$

where

$$F(A) = g_A A^{5/2} [\zeta(3)^{A-1} \pi^{(1-A)/2} 2^{(3A-5)/2}]. \tag{4.83}$$

Equation (4.82) shows why the high entropy of the universe, i.e., the small value of η, hinders the formation of nuclei. To get $X_A \simeq 1$, it is *not* enough that the universe cools to the temperature $T \lesssim B_A$; it is necessary that it cool *still further* so as to offset the small value of the η^{A-1} factor. The temperature T_A at which the mass fraction of a particular species A will be of the order of unity $(X_A \simeq 1)$ is given by

$$T_A \simeq \frac{B_A/(A-1)}{\ln(\eta^{-1}) + 1.5 \ln(m_B/T)}. \tag{4.84}$$

This temperature will be much lower than B_A; for ^{2}H, ^{3}He, and ^{4}He the values of T_A are 0.07, 0.11, and 0.28 MeV, respectively. Comparison with the binding energy of these nuclei shows that these values are lower than B_A by a factor of ~ 10, at least. (We have seen the same effect earlier in the case of photoionisation and $e^+ e^-$ pair production; see Vol. I, Chap. 5, Section 5.12.)

Thus, even when the thermal equilibrium is maintained, a significant synthesis of nuclei can occur only at $T \lesssim 0.3$ MeV and not at higher temperatures. If such is the case, then we would expect significant production $(X_A \lesssim 1)$ of nuclear species A at temperatures $T \lesssim T_A$. It turns out, however, that the rate of nuclear reactions is *not* high enough to maintain thermal equilibrium among various species. We have to determine the temperatures up to which thermal equilibrium can be maintained and redo the calculations to find nonequilibrium mass fractions.

In particular, we used the equilibrium densities for n_p and n_n in the preceding analysis. In thermal equilibrium, the interconversion between n and p is possible through the weak-interaction processes $(\nu + n \leftrightarrow p + e)$ and $(\bar{e} + n \leftrightarrow p + \bar{\nu})$ and the "decay" $(n \leftrightarrow p + e + \bar{\nu})$. From the conservation of the chemical potential in these reactions, we find that $\mu_n + \mu_\nu = \mu_p + \mu_e$, giving $(\mu_n - \mu_p) = (\mu_e - \mu_\nu) \simeq 0$, as $\mu_e \simeq 0$, $\mu_\nu \simeq 0$. The equilibrium (n/p) ratio will therefore be

$$\left(\frac{n_n}{n_p}\right) = \frac{X_n}{X_p} = \exp(-Q/T), \tag{4.85}$$

where $Q = m_n - m_p$. This ratio will be maintained only as long as the n–p reactions are rapid enough. However, when this reaction rate Γ falls below the expansion rate $H \simeq 5.5(T^2/m_{\rm Pl})$ at some temperature T_D, say, the (n/p) ratio will be "frozen" at the value $\exp(-Q/T_D)$. The only process that can continue to change this ratio thereafter will be the beta decay, $n \to p + e + \bar{\nu}$, of the free neutron. The neutron decay will continue to decrease this ratio until all the neutrons are used up in forming bound nuclei.

To determine T_D we have to estimate the interaction rate Γ and use the condition $\Gamma = H$. The rate Γ of the n–p reactions can be calculated from the theory of weak interactions. The total reaction rate $\Gamma(n \to p)$ can be written as the sum of the rates for the reactions $n + \bar{e} \to p + \bar{\nu}$, $n + \nu \to p + e$, and $n \to p + e + \nu$; similarly, the total reaction rate for $\Gamma(p \to n)$ can be written as the sum of the rates for $p + e \to n + \nu$, $p + \nu \to n + \bar{e}$, and $p + e + \bar{\nu} \to n$. We can express all these rates as functions of temperature T by evaluating them numerically. We shall now briefly indicate how this is done.

We consider first the decay rate for a free neutron ($n \to p + e^- + \bar{\nu}$). The probability per unit time for this process can be expressed in the form

$$\lambda_d = \kappa \int \frac{V d^3 p_\nu}{(2\pi\hbar)^3} \frac{2V d^3 p_e}{(2\pi\hbar)^3} \delta(Q - \epsilon_e - p_\nu) = \frac{\kappa V^2}{2\pi^4\hbar^6} m_e^5 f; \quad f \simeq 1.715,$$

(4.86)

where κ represents the relevant matrix element that needs to be computed from weak-interaction theory and f is obtained by evaluation of the integral and substitution of the numerical values for m_e and Q. Computation from weak-interaction theory shows that $\kappa \propto G_V^2[1 + 3(G_A^2/G_V^2)]$, where G_A and G_V are the coupling constants for the axial-vector-current interaction and the vector-current interaction. If the vector current is assumed to be conserved, then $G_V = G_F \cos\theta_C$, where $\theta_C \simeq 13°$ is the so-called Cabibo angle (which describes the mixing of quark weak eigenstates into the mass eigenstates) and G_F is the Fermi coupling constant, which is fairly well determined. Unfortunately, the weak axial current is *not* conserved, and G_A for nucleons is affected by strong interactions as well. These are nonperturbative effects that can affect the ratio G_A/G_V and cannot be computed reliably from theory. Thus we actually need to use the observed lifetime of the free neutron to determine the quantity κ that appears in Eq. (4.86) and parameterise all other decay rates. Observationally, we have

$$\tau_N = 886.7 \pm 1.9 \text{ s}, \quad \lambda_d = (1.127 \pm 0.003) \times 10^{-3} \text{ s}^{-1}. \quad (4.87)$$

Let us next consider the probability per unit time for the reaction $p + e^- \to n + \nu$; this is given by[2]

$$\lambda_1 = \kappa \int \frac{2V d^3 p_e}{(2\pi\hbar)^3} \frac{1}{e^{\epsilon_e/k_B T} + 1} \frac{V d^3 p_\nu}{(2\pi\hbar)^3} \frac{\delta_D(\epsilon_e - p_\nu - Q)}{1 + e^{-p_\nu/k_B T_\nu}}. \quad (4.88)$$

The first factor is an integral over incident electron momentum in a thermal distribution. The second factor is the probability that the final neutrino state is empty, given by

$$1 - \frac{1}{e^{p_v/k_B T_v} + 1} = \frac{1}{1 + e^{-p_v/k_B T_v}}. \tag{4.89}$$

The Dirac delta function ensures conservation of energy. Normalising this rate by using free neutron decay we get

$$\frac{\lambda_1}{\lambda_d} = \frac{1}{m_e^5 f} \int_0^\infty \frac{p_e \epsilon_e p_v^2 \, dp_v}{\left(e^{\epsilon_e/k_B T} + 1\right)\left(1 + e^{-p_v/k_B T_v}\right)}, \qquad \epsilon_e = p_v + Q. \tag{4.90}$$

We can similarly obtain the rates for other processes. The reverse process $n + v \to p + e^-$ has the same rate as that of Eq. (4.90) with a change in the exponential factors:

$$\frac{\hat{\lambda}_1}{\lambda_d} = \frac{1}{m_e^5 f} \int_0^\infty \frac{p_e \epsilon_e p_v^2 \, dp_v}{\left(e^{p_v/k_B T_v} + 1\right)\left(1 + e^{-\epsilon_e/k_B T}\right)}, \qquad \epsilon_e = p_v + Q. \tag{4.91}$$

The rate for $p + \bar{v} \to n + e^+$ is

$$\frac{\lambda_2}{\lambda_d} = \frac{1}{m_e^5 f} \int_0^\infty \frac{p_e^2 p_v^2 \, dp_e}{\left(e^{p_v/k_B T_v} + 1\right)\left(1 + e^{-\epsilon_e/k_B T}\right)}, \qquad p_v = \epsilon_e + Q, \tag{4.92}$$

with the reverse process $n + e^+ \to p + \bar{v}$ at the rate

$$\frac{\hat{\lambda}_2}{\lambda_d} = \frac{1}{m_e^5 f} \int_0^\infty \frac{p_e^2 p_v^2 \, dp_e}{\left(e^{\epsilon_e/k_B T} + 1\right)\left(1 + e^{-p_v/k_B T_v}\right)}, \qquad p_v = \epsilon_e + Q. \tag{4.93}$$

The normalisation automatically takes into account the spin dependence of the matrix elements. Also note that by the time free neutron decay is important, the pairs are sufficiently cool for us to ignore the blocking of states by the pairs.

It is now possible to write an equation for the net rate of conversion of neutrons and protons as

$$\frac{dn_n}{dt} = \lambda n_p - \hat{\lambda} n_n, \tag{4.94}$$

where

$$\lambda = \lambda_1 + \lambda_2, \qquad \hat{\lambda} = \hat{\lambda}_1 + \hat{\lambda}_2 + \lambda_d. \tag{4.95}$$

The quantities λt and $\hat{\lambda} t$ are plotted in Fig. 4.1. Numerical integration of Eq. (4.94) (under the assumption that no other process is present) gives the neutron-abundance fraction $x(t) \equiv n_n/(n_n + n_p)$, shown in Fig. 4.2.

At high temperatures, $T \gg Q \simeq 1.3$ MeV, both the rates vary as T^5; in the range of $T \simeq (0.1-1)$MeV, the rate $\lambda(n \to p) \propto T^{4.42}$ whereas $\lambda(p \to n)$ decreases faster; for $T < 0.1$ MeV, $\lambda(n \to p) \simeq \tau_n^{-1}$ is essentially dominated by the neutron decay whereas $\lambda(p \to n)$ drops exponentially.

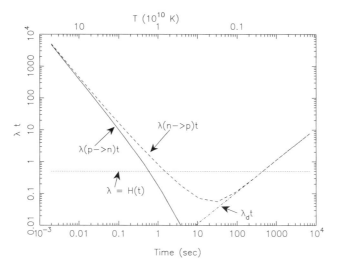

Fig. 4.1.   The conversion rate of neutrons through forward and return channels are plotted as functions of cosmic time t. The y axis is the dimensionless product of rate and cosmic time. The solid line gives $\lambda(p \rightarrow n)t$, and the dashed curve gives $\lambda(n \rightarrow p)t$. At late times, $\lambda(n \rightarrow p)t \approx (\lambda_d t)$ is dominated by neutron decay, as shown by the dotted–dashed line. The horizontal dotted line corresponds to $\lambda t = 0.5$ so that the intersection of the curves with the horizontal line occurs when the reaction rate λ is equal to the expansion rate $H(t) = 0.5/t$ of the radiation-dominated universe. The axis on top gives the corresponding values for the temperature.

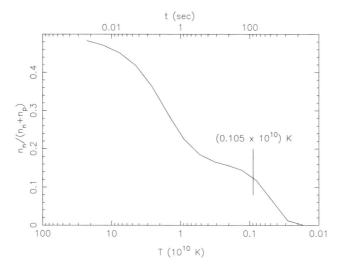

Fig. 4.2.   The neutron fraction as a function of cosmic temperature if the only processes operating are the weak decays that connect neutron and proton.

The reaction rates drop below the expansion rate around $T_D \simeq (0.8-0.7)$MeV. Thus thermal equilibrium between neutron and proton exists only for $T \gtrsim T_D \simeq 0.8$ MeV. At $T_D \simeq 0.7$ MeV, when the assumption of thermal equilibrium becomes invalid, the n/p ratio will be

$$\left(\frac{n}{p}\right) = \exp\left(-\frac{Q}{T_D}\right) \simeq \frac{1}{6}, \tag{4.96}$$

giving $X_n \simeq (1/7)$ and $X_p \simeq (6/7)$. Because T_A calculated in relation (4.84) is lower than T_D for all the light nuclei, none of the light nuclei will exist in significant quantities at this temperature T_D. For example, ^{2}H contributes only a mass fraction $X_2 \simeq 10^{-12}$ at this temperature.

The only way neutrons can survive is through the process of nuclear synthesis of light elements. As the temperature falls further to $T = T_{He} \simeq 0.28$ MeV, a significant amount of He could have been produced if the nuclear reaction rates were high enough. These reactions [D(D, n) ^{3}He(D, p) ^{4}He, D(D, p) ^{3}H(D, n) ^{4}He, D(D, γ) ^{4}He] are all based on D, ^{3}He, and ^{3}H, and do not occur rapidly enough because the mass fraction X_A of D, ^{3}He, and ^{3}H are still quite small (10^{-12}, 10^{-19}, and 5×10^{-19}, respectively) at $T \simeq 0.3$ MeV.

The reactions $n + p \rightleftharpoons d + \gamma$ will lead to an equilibrium-abundance ratio of D, given by

$$\frac{n_p n_n}{n_d n} = \frac{4}{3} \left(\frac{m_p m_n}{m_d}\right)^{3/2} \frac{(2\pi k_B T)^{3/2}}{(2\pi\hbar)^3 n} e^{-B/k_B T}$$

$$= \exp\left[25.82 - \ln \Omega_B h^2 T_{10}^{3/2} - \left(\frac{2.58}{T_{10}}\right)\right]. \tag{4.97}$$

The short vertical line in Fig. 4.2 indicates the temperature 0.105×10^{10} K at which the equilibrium deuterium abundance passes through unity for $\Omega_B h^2 = 0.013$. From Eq. (4.97) we see that the configuration swings sharply from free neutrons to deuterons, with most of the deuterium production occurring at this critical temperature. The rate for the reaction $n + p \rightleftharpoons d + \gamma$ is $\langle \sigma v \rangle \approx 4.6 \times 10^{-20}$ cm^3 s^{-1}, making $n\sigma vt \gg 1$ around this time, thereby cooking most of the free neutrons into deuterium. These abundances become nearly unity only at $T \lesssim 0.1$ MeV; therefore only at $T \lesssim 0.1$ MeV can these reactions be fast enough to produce an equilibrium abundance of ^{4}He. When the temperature becomes $T \lesssim 0.1$ MeV, the abundance of D and ^{3}H builds up, and these elements further react to form ^{4}He. A good fraction of D and ^{3}H is converted to ^{4}He. We can easily calculate the resultant abundance of ^{4}He by assuming that almost all neutrons end up in ^{4}He. Because each ^{4}He nucleus has two neutrons, $(n_n/2)$ He nuclei can be formed (per unit volume) if the number density of the neutrons is n_n. Thus the mass fraction of ^{4}He will be

$$Y = \frac{4(n_n/2)}{n_n + n_p} = \frac{2(n/p)}{1 + (n/p)} = 2x_c, \tag{4.98}$$

where $x_c = n/(n + p)$ is the neutron abundance at the time of the production of D. For $\Omega_B h^2 = 0.013$, $x_c \approx 0.11$, giving $Y \approx 0.22$. Increasing the baryon density to $\Omega_B h^2 = 1$ will make $Y \approx 0.25$. A reasonable fit to the neutron–proton ratio at the time of deuterium formation is given by

$$\frac{n_n}{n_p} \simeq 0.163(\Omega_B h^2)^{0.04} \left(\frac{N_\nu}{3}\right)^{0.2}, \tag{4.99}$$

where N_ν is the number of neutrino species. A more accurate fitting formula for the dependence of He abundance on various parameters is given by

$$Y = 0.226 + 0.025 \log \eta_{10} + 0.0075(g_* - 10.75) + 0.014\left[\tau_{1/2}(n) - 10.3 \min\right], \tag{4.100}$$

where η_{10} measures the baryon–photon ratio today by means of the relation

$$\Omega_B h^2 = 3.65 \times 10^{-3} \left(\frac{T_0}{2.73 \text{ K}}\right)^3 \eta_{10} \tag{4.101}$$

and g_* is the effective number of relativistic degrees of freedom that contribute to the energy density and $\tau_{1/2}(n)$ is the neutron half-life.[3] The best fits, with typical errors, to D, ^{3}He, and Li abundances calculated from the theory for the range $\eta = 10^{-10}$–10^{-9} are given by

$$Y_2 \equiv \left(\frac{D}{H}\right)_p = 3.6 \times 10^{-5 \pm 0.06} \left(\frac{\eta}{5 \times 10^{-10}}\right)^{-1.6},$$

$$Y_3 \equiv \left(\frac{^3\text{He}}{H}\right)_p = 1.2 \times 10^{-5 \pm 0.06} \left(\frac{\eta}{5 \times 10^{-10}}\right)^{-0.63}, \tag{4.102}$$

$$Y_7 \equiv \left(\frac{^7\text{Li}}{H}\right)_p = 1.2 \times 10^{-11 \pm 0.2} \left[\left(\frac{\eta}{5 \times 10^{-10}}\right)^{-2.38} + 21.7 \left(\frac{\eta}{5 \times 10^{-10}}\right)^{2.38}\right].$$

Qualitatively we can understand these results as follows. The n/p ratio at the time of "freeze-out" (t_D) was $\sim 1/6$; from t_D until the time of nucleosynthesis t_N, a certain fraction of neutrons would have decayed, lowering this ratio. Because the freeze-out occurred at $T_D \simeq 1$ MeV, $t_D \simeq 1$ s, and ^{4}He synthesis occurred at $T_N \simeq 0.1$ MeV, $t_N \simeq 3$ min, the decay factor will be $\exp(-t_N/\tau_n) \simeq 0.8$. Therefore the value of (n/p) at the time of ^{4}He synthesis will be $0.8 \times (1/6) \simeq (1/7)$. This gives $Y \simeq 0.25$.

As the reactions converting D and ^{3}H to ^{4}He proceed, the number densities of D and ^{3}H are depleted and the reaction rates – which are proportional to $\Gamma \propto X_A(\eta n_\gamma) \langle \sigma v \rangle$ – become small. These reactions soon freeze out, leaving a residual fraction of D and ^{3}H (a fraction of approximately 10^{-5}– 10^{-4}). Because $\Gamma \propto \eta$ it is clear that the fraction of (D, ^{3}H) left unreacted will decrease with η. In contrast, the ^{4}He synthesis – which is not limited by any reaction rate – is fairly independent of η and depends on only the n/p ratio at $T \simeq$ 0.1 MeV.

The production of still heavier elements – even those like ^{16}C, and ^{16}O that have higher binding energies than that of ^{4}He – is highly suppressed in the early universe. Two factors are responsible for this suppression: (1) Direct reactions between two He nuclei or between H and He will lead to nuclei with atomic masses of 8 or 5. Because there are no tightly bound isotopes with masses of 8 or 5, these reactions do not lead to any further synthesis. [The three-body-interaction, ^{4}He $+ ^4$He $+ ^4$He $\rightarrow ^{12}$C, is suppressed because of the low number density of ^{4}He nuclei; it is this ("triple-α") reaction that helps further synthesis in stellar interiors; see Vol. II, Chap. 2.] (2) For nuclear reactions to proceed, the participating nuclei must overcome their Coulomb repulsion. The probability of tunnelling through the Coulomb barrier is governed by the factor $F = \exp[-2A^{1/3} (Z_1 Z_2)^{2/3} (T/1 \text{ MeV})^{-1/3}]$, where $A^{-1} = A_1^{-1} + A_2^{-1}$. For heavier nuclei (with larger Z), this factor suppresses the reaction rate.

Small amount (approximately 10^{-10}–10^{-9} by mass) of ^{7}Li is produced by ^{4}He $(^3$H$, n)^7$Li or by ^{4}He $(^3$He$, \gamma)^7$Be followed by a decay of ^{7}Be to ^{7}Li. The first process dominates if $\eta \lesssim 3 \times 10^{-10}$ and the second process for $\eta \gtrsim 3 \times 10^{-10}$. In the second case, a small amount (10^{-11}) of ^{7}Be is left as a residue.

Given the various nuclear reaction rates, the primordial abundances of all the light elements can be computed by numerical integration of the relevant equations. The most uncertain input in these calculations is the neutron decay rate; the half-life of the neutron is known to only 2% accuracy. All the weak reaction rates $\Gamma \propto G_F^2 (1 + 3g_A^2)T^5$ are proportional to $T^5 \tau^{-1}$. An increase in τ decreases all Γs and makes the freeze-out [determined by $H(T_D) = \Gamma(T_D)$] occur at a higher temperature. Because $H \propto T^2$, $\Gamma \propto \tau^{-1} T^5$ we find that $T_D \propto \tau^{1/3}$. When T_D increases, so will the (n/p) value at freeze-out, resulting in a higher value of ^{4}He. The changes in the other nuclear abundances are not significant because they are within the present observational errors.

The two main *cosmological* parameters on which the results depend are the number of degrees of freedom, g (at $T \simeq 1$ MeV) and η. An increase in g increases $H(T) \propto g^{1/2} T^2$ and leads to a higher freeze-out temperature $T_D \propto g^{1/6}$ and a higher value of ^{4}He abundance. The dependence on η is more complicated. Because the mass fractions $X_A \propto \eta^{(A-1)}$, a larger value of η will allow the D, ^{3}He, and ^{3}H abundances to build up earlier and thus lead to an earlier formation of ^{4}He. Because the n/p ratio is higher at earlier times, this will result in more ^{4}He. However, the n/p ratio is varying only slowly with time at $T \simeq 0.1$ MeV, and hence this dependence is rather mild. The amount of residual D and ^{3}H, however, depends more strongly on η; it decreases as η^{-k} with $k \simeq 1.3$. The dominant channels for ^{7}Li production are different for $\eta \lesssim 3 \times 10^{-10}$ and for $\eta \gtrsim 3 \times 10^{-10}$. This fact leads to a trough in the abundance of ^{7}Li around $\eta \simeq 3 \times 10^{-10}$ where neither process is very efficient.

The calculated and observed values for several primordial abundances are shown in Fig. 4.3. The observations indicate, with reasonable certainty that (1) $(D/H) \gtrsim 1 \times 10^{-5}$, (2) $[(D + {}^3He)/H] \simeq (1-8) \times 10^{-5}$, (3) $({}^7Li/H) \sim 10^{-10}$, and (4) $0.236 < ({}^4He/H) < 0.254$. These observations are consistent with the predictions if $10.3 \text{ min} \lesssim \tau \lesssim 10.7 \text{ min}$, and

$$\eta = (3-10) \times 10^{-10}. \tag{4.103}$$

Because $\eta = 2.68 \times 10^{-8} \Omega_B h^2$, this leads to the following important conclusion:

$$0.011 \leq \Omega_B h^2 \leq 0.037. \tag{4.104}$$

When combined with the known bounds on h, $0.4 \leq h \leq 1$, the baryonic density of the universe can be constrained as

$$0.011 \leq \Omega_B \leq 0.23. \tag{4.105}$$

These are the most conservative bounds on Ω_B available today. It shows that if Ω_{total} has no contribution other than Ω_B then the universe must be open.

If some physical mechanism can be devised that will distribute neutrons and protons *differently* in space, then the results of the nucleosynthesis will vary from location to location. Such models involving inhomogeneous nucleosynthesis will change the Ω_B bounds. It turns out, however, that it is still not possible to produce viable models in which $\Omega_B = 1$.

Because the ${}^4He$ production depends on g, the observed value of ${}^4He$ restricts the number (N_ν) of light neutrinos (that is, neutrinos with $m_\nu \lesssim 1 \text{ MeV}$, which would have been relativistic at $T \simeq 1 \text{ MeV}$). The observed abundance is best explained by $N_\nu = 3$, is barely consistent with $N_\nu = 4$, and rules out $N_\nu > 4$. The laboratory bounds on the total number of particles including neutrinos, which are less massive than $(m_z/2) \simeq 46 \text{ GeV}$ and couple to the Z^0 boson, is $N_\nu = 2.79 \pm 0.63$. We determine this by measuring the decay width of Z^0; each particle with a mass of less than $m_z/2$ contributes $\sim 180 \text{ MeV}$ to this decay width. This bound is consistent with the cosmological observations.

It must be stressed that the observational determination of primordial abundances is not straightforward, and we often have to resort to a fair amount of processing of data to arrive at the final figure. We shall now briefly summarise these issues starting with the primordial D abundance, which is crucial in setting the bounds on the baryon fraction.

There are two advantages in using deuterium for the measurement of baryon density. The first one arises from the exponential sensitivity of the deuterium abundance on the baryon density, thereby allowing us to determine η to an accuracy of $\sim 5\%$ if the D abundance is known to $\sim 10\%$ accuracy. Second, deuterium can be destroyed in stellar evolution but not created; hence any observed deuterium abundance is a lower bound to the primordial abundance, thereby allowing us to put an upper bound to the baryonic density. High-redshift Lyman-α

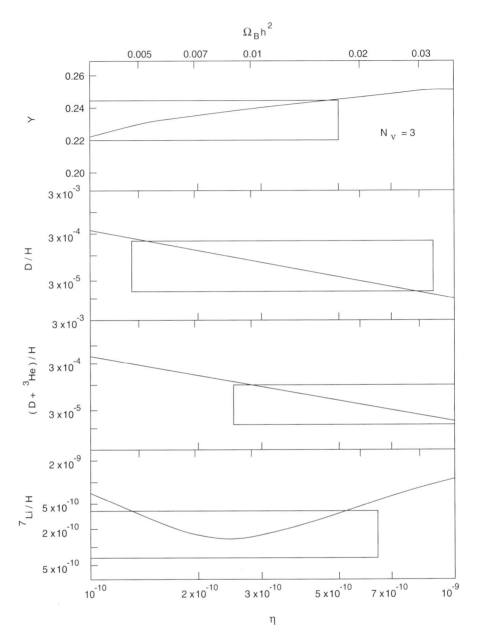

Fig. 4.3.   The theoretical calculations lead to abundances of light nuclei like the ones shown here. The precise values depend on several input parameters but the overall trend is as shown. The boxes are drawn based on the observed values of the light elements (marked in the *y* axis). The overlap of the allowed range permits only a narrow range for $\Omega_B h^2$ and excludes the value $\Omega_B = 1$.

absorbers allow, in principle, the determination of deuterium abundance in pure primordial form directly. In practice, we now have four such candidate measurements in which the results are inconclusive if not inconsistent. The systematic errors arise both from data reduction as well as from the fact that the D spectral lines need to be distinguished from the velocity-shifted hydrogen lines (called interlopers).

The absorption system towards $Q1937 - 1009$ at a redshift of $z = 3.572$ originally gave an abundance of $D/H = (2.3 \pm 0.3 \pm 0.3) \times 10^{-5}$, whereas a more precise reanalysis of a high-quality low-resolution spectrum has now led to a revised estimate of $(3.3 \pm 0.3) \times 10^{-3}$. The same data, when reanalysed with a different model for the velocity distribution of the absorbing gas, lead to the bound $3.5 \times 10^{-5} < D/H < 5.2 \times 10^{-5}$ at 95% confidence level.[4] Another absorption system at $z = 2.504$ towards $Q1009 + 2956$ gives an abundance ratio of $(4.0 \pm 0.7) \times 10^{-5}$. Reanalysis of these data along the lines previously mentioned leads to the range $(2.9–4.6) \times 10^{-5}$ at 95% confidence level. All these results are consistent with other measurements, which give values in the range $(2.9–4) \times 10^{-5}$.

There is, however, another low-redshift ($z = 0.701$) absorption system toward $Q1718 + 4807$ that gives a different value. The original analysis led to a very high deuterium abundance of $(20 \pm 5) \times 10^{-5}$ whereas reanalyses by two groups have led to $(4.1–4.7) \times 10^{-5}$ and $(8–57) \times 10^{-5}$. Only future analysis with better data can lead to an improvement in the situation. The lower value of the preceding two sets of abundances constrain η_{10} to be in the range 4.2–6.3 whereas the higher deuterium abundance will give η_{10} in the range 1.2–2.8. The groups that have analysed these data suggest an optimal value of $D/H = (3.4 \pm 0.5) \times 10^{-5}$ that corresponds to $\eta = (5.1 \pm 0.5) \times 10^{-10}$ that in turn translates to $\Omega_B h^2 = (0.019 \pm 0.0024)$.

Local observations of D abundance come from ISM data and solar system measurements. The best result from ISM currently available is $D/H = (1.60 \pm 0.09^{+0.05}_{-0.10}) \times 10^{-5}$. The solar abundance of D/H is inferred from the solar-wind measurements of ^3He as well as yields in meteorites. These measurements are capable of giving the fraction $(D + {}^3He)/H$, from which the $^3He/H$ ratio (obtained from meteoric measurements) needs to be subtracted to provide the D/H ratio. The best estimate available today gives $D/H = (2.6 \pm 0.6 \pm 1.4) \times 10^{-5}$.

Let us next consider He abundance. Because He is produced in the stars, we need to study systems of very low metallicity in order to extract the primordial abundance of He. These low-metallicity systems in turn are characterised by low abundances of elements like C, N, and O. The actual He abundance is best determined from the observations of HeII $\rightarrow$ HeI recombination lines in extragalactic HII regions. At present there exists a collection of over 70 such regions for which we have information about ^4He, O/H, and N/H. The data show a linear correlation between the mass fraction of He (Y) and the abundance ratio O/H or N/H. The intercept of this straight-line fit should give the value of Y at zero metallicity. Such an analysis gives $Y = (0.234 \pm 0.002 \pm 0.005)$. Although the analysis is straightforward in principle, it does have certain amount of underlying statistical uncertainty that is not captured in the error bars quoted above; some disparate results do appear in published literature regarding the abundance.

It may be noted that the slope of the Y versus the O/H line is also of interest in determining the chemical-evolution history of the galaxy. The abundance previously mentioned arises from the data having $\Delta Y / \Delta O \simeq (110 \pm 25)$. Chemical-evolution models typically give a much smaller value of ~ 20 and even models with outflow give only ~ 60. It is not clear whether this discrepancy has a bearing on the abundance estimate.

We next consider the Li abundance, which is usually determined from the atmospheres of old hot Pop II stars. For stars with surface temperatures higher than ~ 5500 K and metallicity lower than $(1/20)Z_\odot$, the effects of stellar convection can be ignored. When the ^{7}Li data from nearly 100 halo stars with [Fe/H] < -1.3 are plotted as functions of surface temperature, we see a plateau of constant abundance for $T > 5500$ K. In stars with lower temperatures, the surface abundance of Li could be depleted when the convection takes the Li through the hotter interior of the star where it could be destroyed. This is roughly what is seen in the data. The plateau value gives an abundance of Li/H as Li/$H = (1.6 \pm 0.1) \times 10^{-10}$.

The most significant model dependence arises from the conversion of B-V colour into temperature. Recent methods attempted to avoid this by using temperatures measured from Balmer lines or from IR flux. The average value obtained for the abundance is approximately $(1.6 \pm 0.36) \times 10^{-10}$, although there is another sample that gives $(1.73 \pm 0.21) \times 10^{-10}$. It has also been argued in the literature that both these values need to be corrected for depletion and that the primordial value is closer to $(2.3 \pm 0.5) \times 10^{-10}$. It must be stressed that some amount of stellar modelling is necessary to arrive at this result, and there could be an important source of systematic error if Li could be depleted compared with the initial abundance. Standard stellar models predict that the depletion of ^{7}Li would be accompanied by a more severe depletion of ^{6}Li. Recent observations of ^{6}Li in hot Pop II stars suggest that the depletion is probably not severe and the results are trustworthy.[5]

Exercise 4.6
Nonstandard nucleosynthesis: Determine whether the primordial nucleosynthesis will produce more (or less) He and D compared with the standard model if (a) there are many more neutrinos than antineutrinos (or photons) in the universe, (b) there are many more antineutrinos than neutrinos (or photons) in the universe, and (c) there is a significant contribution from gravitational radiation to the energy density of the early universe. [Answers: (a) If there are many more νs than $\bar{\nu}$s, then neutrinos will have a large negative chemical potential. Hence the reaction $\nu + n \rightarrow p + e$ will be driven to smaller value of

n/p, as a higher number of ν will convert more neutrons to protons. This results in smaller D and ^{4}He. The argument assumes that the total energy density in the neutrinos is not changed. (b) This will lead to a higher value of n/p through the reaction $\bar{\nu} + p \rightarrow n + e^+$. The result will be increased amounts of D and ^{4}He. (c) The existence of more energy density in any form will speed up the expansion, thereby causing the freeze-out of n/p to occur at higher temperatures and resulting in higher value of n/p. This will lead to more D and more ^{4}He.]

4.6 A Simplified Model for Primordial Nucleosynthesis

The results of primordial nucleosynthesis described in the last section are usually obtained by numerical work. These codes use numerous nuclear-reaction channels and simultaneously solve for the abundance of different elements. The typical differential equation describing the evolution of the abundance for an element has a form like

$$\dot{f}_i = J - \Gamma f_i, \tag{4.106}$$

where $J(t)$ and $\Gamma(t)$ are time-dependent source and sink terms that in turn depend on the abundances of other elements. Because the derivative term in this equation can be the difference between two large numbers and because there are two very different time scales, Γ^{-1} and H^{-1}, in the problem, numerical work requires high accuracy and a long computation time to reach a stable solution.

It is, however, possible to understand the dependence of the abundances on the cosmological parameters by use of the semianalytical model which we shall now describe.[6] The key ideas behind this approach are the following: (1) Although there are several nuclear reactions that contribute to each of the processes, few of them dominate the others in any given context. A careful study of the various reactions allows us to obtain fairly simple rate equations. (2) Even among various source and sink terms, the processes for which $\Gamma \gg H$ will quickly drive the solution to Eq. (4.106) to the time-*dependent* quasi-static solution $f_i(t) = J(t)/\Gamma(t)$. This fact can be used to determine the abundances of the "fast" elements that can be used in other reactions.

To do this effectively, we first need to understand the different stages through which nucleosynthesis proceeds, which we shall describe first. The first key event occurs when $T_W \approx 10^{10}$ K when the weak interactions freeze out. At this stage, the neutron abundance is $Y_n(T_W) \approx 0.16$. After weak interactions freeze out, neutron abundance changes only through the beta decay $n \rightarrow p + e + \bar{\nu}$, and hence we have

$$Y_n = 0.16 \exp(-t/\tau). \tag{4.107}$$

This process will continue until the neutrons are cooked into He. Let us therefore estimate the temperature T^α when the rate of formation of He dominates the rate of decay of free neutrons. During this epoch, the He production is governed by

$$\dot{Y}_\alpha = Y_d Y_T \sigma[dTn\alpha] + Y_p Y_T \sigma[pT\gamma\alpha] + Y_d Y_3 \sigma[d3\alpha p]$$
$$\cong Y_d Y_T \sigma[dTn\alpha] + Y_p Y_T \sigma[pT\gamma\alpha], \tag{4.108}$$

where the subscripts α, d, T, and 3 denote ^{4}He, D, tritium, and ^{3}He, respectively, and we have denoted the cross section of the nuclear reaction $a + b \to c + d$ by the symbol $\sigma[abcd]$. The two processes given on the right-hand side of Eq. (4.108) are the most dominant reactions out of a host of reactions that will lead to He formation. We now need to determine Y_d and Y_T, which are governed by

$$\dot{Y}_d = Y_n Y_p \sigma[npd\gamma] - Y_d Y_\gamma \sigma[d\gamma np], \tag{4.109}$$

$$\dot{Y}_T = Y_n Y_3 \sigma[n3pT] + Y_d Y_d \sigma[ddpT] - Y_d Y_T \sigma[dTn\alpha] - Y_p Y_T \sigma[pTn3]$$
$$- Y_p Y_T \sigma[pT\gamma\alpha] \cong Y_d Y_d \sigma[ddpT] - Y_d Y_T \sigma[dTn\alpha] - Y_p Y_T \sigma[pT\gamma\alpha]. \tag{4.110}$$

The crucial fact that we now use is that we can approximate the abundances of Y_d and Y_T by their equilibrium values (which we obtain by setting $\dot{Y}_d \cong 0$, $\dot{Y}_T \cong 0$) as these reactions are fast enough. This gives

$$Y_d = \frac{Y_n Y_p \sigma[npd\gamma]}{Y_\gamma \sigma[d\gamma np]}, \qquad Y_T = \frac{Y_d Y_d \sigma[ddpT]}{Y_d \sigma[dTn\alpha] + Y_p \sigma[pT\alpha\gamma]}. \tag{4.111}$$

Substituting these into Eq. (4.108) and using the known values for the reaction cross sections, we find that the He production is governed by

$$\dot{Y}_\alpha \cong \left(0.2 \times 10^{-12} \Omega_B h^2 Y_n Y_p e^{25.82/T_9}\right)^2 \sigma[ddpT]. \tag{4.112}$$

The abundance of neutrons from the time of freeze-out (when $T_f \approx 9.1 \times 10^9$ K and $Y_n = 0.16$) until nucleosynthesis is given by Eq. (4.107). The temperature at which $2\dot{Y}_\alpha = Y_n/\tau$ can be determined with Eq. (4.107) for Y_n and the reaction rate $\sigma[ddpT]$. In the relevant range of temperature, we can approximate the reaction rate by a simpler formula than the one given in Table 4.1. We take

$$\sigma[ddpT] \approx 7.5 \times 10^5 T_9^{7/3} \exp\left(-4.258/T_9^{1/3}\right) \Omega_B h^2 \ \text{s}^{-1} \tag{4.113}$$

and substitute into relation (4.112) to obtain

$$\frac{1}{\tau} = 6 \times 10^{-20} T_9^{7/3} (\Omega_B h^2)^3 \left(0.16 \, e^{-t/\tau}\right) Y_p^2 e^{51.64/T_9 - (4.258/T_9^{1/3})}, \tag{4.114}$$

Table 4.1. Some reaction rates

Reaction	Rate
$\sigma(np\,d\gamma)$	$4.40 \times 10^4 \left(1 - 0.860 T_9^{1/2} + 0.429 T_9\right)$
$\sigma(p\,d3\gamma)$	$2.24 \times 10^3 T_9^{-2/3} \exp\left(-3.720/T_9^{1/3}\right)$
	$\times \left(1 + 0.112 T_9^{1/3} + 3.38 T_9^{2/3} + 2.65 T_9\right)$
$\sigma(pT\gamma\alpha)$	$2.20 \times 10^4 T_9^{-2/3} \exp\left(-3.869/T_9^{1/3}\right)$
	$\times \left(1 + 0.108 T_9^{1/3} + 1.68 T_9^{2/3} + 1.26 T_9 + 0.551 T_9^{4/3} + 1.06 T_9^{5/3}\right)$
$\sigma(d\,dn3)$	$3.97 \times 10^8 T_9^{-2/3} \exp\left(-4.258/T_9^{1/3}\right)$
	$\times \left(1 + 0.098 T_9^{1/3} + 0.876 T_9^{2/3} + 0.600 T_9 - 0.041 T_9^{4/3} - 0.071 T_9^{5/3}\right)$
$\sigma(d\,dpT)$	$4.17 \times 10^8 T_9^{-2/3} \exp\left(-4.258/T_9^{1/3}\right)$
$\sigma(dTn\alpha)$	$8.09 \times 10^{10} T_9^{-2/3} \exp\left[-4.524/T_9^{1/3} - (T_9/0.12)^2\right]$
	$\times \left(1 + 0.092 T_9^{1/3} + 1.8 T_9^{2/3} + 1.16 T_9 + 10.5 T_9^{4/3} + 17.24 T_9^{5/3}\right)$
	$+ 8.73 \times 10^8 T_9^{-2/3} \exp(-0.523/T_9)$
$\sigma(d3p\alpha)$	$6.67 \times 10^{10} T_9^{-2/3} \exp\left[-7.181/T_9^{1/3} - (T_9/0.315)^2\right]$
	$\times \left(1 + 0.058 T_9^{1/3} - 1.14 T_9^{2/3} - 0.464 T_9 + 3.08 T_9^{4/3} + 3.18 T_9^{5/3}\right)$

which can be solved to give the relevant temperature as

$$T_9^\alpha = \frac{51.64}{39.6 - 3\log(\Omega_B h^2) - \left(0.196/T_9^2\right) + \left(4.258/T_9^{1/3}\right) - (7/3)\log(T_9)}$$

$$\cong \frac{51.64}{43.7 - 3\log(\Omega_B h^2)}, \tag{4.115}$$

where the second line gives a simpler fitting function valid in the relevant range. Substituting back into Eq. (4.107) and using the time–temperature relation $t(\text{sec}) = \theta T_9^{-2}$, where $\theta = 99.6$ for $T_9 \gtrsim 5$ and $\theta = 178$ for $T_9 \lesssim 1$, we get the neutron abundance at the temperature T_α as

$$Y_n^\alpha = 0.16 \exp\left[-0.197/\left(T_9^\alpha\right)^2\right] \tag{4.116}$$

and the corresponding He abundance as

$$X_4 = 2 Y_n\left(T_9^\alpha\right) \simeq 0.32 \exp\left\{-0.142\left[1 - \frac{3\log(\Omega_B h^2)}{44}\right]^2\right\}. \tag{4.117}$$

This is the simple analytic expression for the He abundance, which is accurate to ~5% in the relevant range. The agreement is remarkable considering the simplicity of the analysis.

Let us next consider the D abundance. For temperatures below T_α, the neutron abundance evolves as $\dot{Y}_n = -2\dot{Y}_\alpha$, where $\dot{Y}_\alpha$ is given by relation (4.112). As long as the number density of the neutrons is larger than that of the deuterons, we can determine the number of neutrons by this approach and solve for deuterons in quasi-static approximation. In this case, the two abundances are given by

$$Y_n = \frac{1}{(1/Y_n^\alpha) + (\Omega_B h^2)^3 e^{-46.92} T_9^5 \left[e^{(51.64/T_9)} - e^{(51.64/T_9^\alpha)} \right]}, \quad (4.118)$$

$$Y_d = 0.2 \times 10^{-12} \Omega_B h^2 Y_n Y_p T_9^{3/2} e^{25.82/T_9}. \quad (4.119)$$

The deuteron abundance given in Eq. (4.119) is essentially the thermal-equilibrium value. This becomes equal to the neutron abundance at the temperature T_d given by

$$T_9^d = \frac{25.82}{29.53 - \log(\Omega_B h^2) - (3/2)\log\left(T_9^d\right)}, \quad (4.120)$$

with $Y_d^i \equiv Y_d(T_d) \equiv Y_n(T_d)$. The rate of change of D now dominates over that of neutrons, and the conservation of neutrons requires that $\dot{Y}_d = -2\dot{Y}_\alpha$, in which the latter is determined by Eq. (4.108) having all the three terms present on the right-hand side. To solve this equation, we need to determine the quasi-static solutions for Y_T and Y_3 with the relevant equations for their production. In this particular range, these are given by the exact form of Eq. (4.110) and

$$\dot{Y}_3 = Y_d Y_p \sigma[pd3\gamma] + Y_T Y_p \sigma[pTn3] + Y_d Y_d \sigma[ddn3] \\ - Y_d Y_3 \sigma[d3p\alpha] - Y_n Y_3 \sigma[n3pT]. \quad (4.121)$$

Obtaining the quasi-static solutions and substituting into Eq. (4.108), we find that the D production is governed by

$$\dot{Y}_d = -2(Y_d)^2 \left(\sigma[ddpT] + \sigma[ddn3]\right) - 2Y_p Y_d \sigma[pd3\gamma]. \quad (4.122)$$

Initially, the Y_d^2 term dominates and the solution to this equation will be

$$Y_d^{\text{approx}} = \frac{1}{(1/Y_d^i) + \Omega_B h^2 e^{18.75} \left[e^{-4.258/(T_9^d)^{1/3}} - e^{-4.258/T_9^{1/3}} \right]}. \quad (4.123)$$

However, as the abundance Y_d begins to drop, the second term becomes important. To solve for this effect, we write $Y_d = Y_d^{\text{approx}} e^{-\delta}$ and obtain an equation for δ:

$$\dot{\delta}(t) = -f(t)(1 - e^{-\delta}) + g(t), \quad (4.124)$$

where

$$f(t) = 2Y_d^{\text{approx}} \left(\sigma[d\ dpT] + \sigma[d\ dn3] \right), \qquad g(t) = 2Y_p \sigma[p\ d3\gamma]. \quad (4.125)$$

This equation has an approximate solution that we obtain by retaining only $g(t)$ on the right-hand side. Retaining only the leading dependences, we get

$$\delta(T_9) \simeq 2.22 \times 10^3 \Omega_B h^2 \left[e^{-3.72/(T_9^d)^{1/3}} - e^{-3.72/T_9^{1/3}} \right]. \quad (4.126)$$

Substituting this result and Eq. (4.123) back into the definition of Y_d, we get the D abundance as a function of temperature to be

$$Y_d = \frac{\exp\left\{ -2.22 \times 10^3 \Omega_B h^2 \left[e^{-3.72/(T_9^d)^{1/3}} - e^{-3.72/T_9^{1/3}} \right] \right\}}{\left(1/Y_d^i \right) + \Omega_B h^2 e^{18.75} \left[e^{-4.258/(T_9^d)^{1/3}} - e^{-4.258/T_9^{1/3}} \right]}. \quad (4.127)$$

The asymptotic relic abundance of the D (when $T_9 \to 0$) is given by

$$Y_d^f = \frac{1}{\Omega_B h^2} \exp\left[-18.75 + \frac{4.258}{\left(T_9^d \right)^{1/3}} - 2.22 \times 10^3 \Omega_B h^2 e^{-3.72/(T_9^d)^{1/3}} \right],$$

$$(4.128)$$

where T_d is given by Eq. (4.127). Figure 4.4 shows Y_d^f as a function of η, related to $\Omega_B h^2$ by Eq. (4.101), in the relevant range; the exponential sensitivity is obvious. A similar analysis can be performed for all the other elements.

The analysis brings out clearly the following points: Once the rate of production of He dominates the rate of decay of free neutrons, He forms quite efficiently. This rate is proportional to the abundance of D, which is in thermal equilibrium

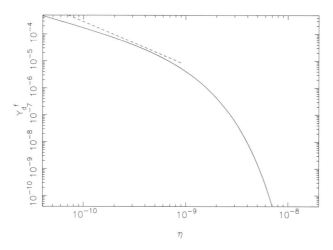

Fig. 4.4. Deuterium abundance estimated from the semianalytic model; the best-fit line, $2.2 \times 10^{-5}(\eta/5 \times 10^{-10})^{-1.6}$, from more sophisticated numerical work, is shown as a dashed line for comparison.

and hence varies as $\exp(-Q/T)$, where Q is the binding energy of D. However, the D abundance is also proportional to the baryon–photon ratio and rate of formation of tritium. Thus neutrons around temperature T_α have greater chance of being cooked into He than of decaying freely. This temperature is low compared with the binding energy of He because, for $T > T_\alpha$, there is not enough D and tritium to make α particles. Note that this result in *not* directly due to the low binding energy of D.

The strong dependence of D on the baryon density is also understandable in this approach. In Eq. (4.122), in which D abundance is determined, there are two sinks for D: (1) two D nuclei can form a tritium or ^{3}He or (2) D can be hit by a proton, leading to ^{3}He. The first term obviously scales as Y_d^2 whereas the second is proportional to Y_d. It is the second term that leads to an exponential depletion of D. In fact, our approximate solution for δ is essentially an integral over the $p + d \rightarrow$ ^{3}He $+ \gamma$ process and obviously this δ is proportional to the baryon density; this is the origin of the exponential dependence of Y_d on $\Omega_B h^2$. If we ignore the dependence of T_d on Ω_B and take for simplicity $T_d \approx 0.75 \times 10^9$ K and $h = 1$, then the logarithmic derivative of Y_d becomes

$$\frac{1}{Y_d} \frac{\partial Y_d}{\partial \Omega_B} = -\left(\frac{1}{\Omega_B} + 37.0 \right), \tag{4.129}$$

clearly showing the strong exponential dependence arising from the effect previously described.

4.7 Decoupling of Matter and Radiation

At temperatures below 0.1 MeV, the main constituents of the universe will be hydrogen nuclei, ^{4}He nuclei, electrons, photons, and decoupled neutrinos. Because $m_e \simeq 0.5$ MeV, the ions and the electrons may be considered nonrelativistic. These constituents interact among themselves and with the photons through various electromagnetic processes, like bremsstrahlung, Compton (and Thomson) scattering, recombination ($p + e \leftrightarrow H + \gamma$), and Coulomb scattering between charged particles. To decide whether these processes can maintain thermal equilibrium among the constituents, we should compare the various interaction rates with the expansion rate.

To begin with, the mean free time for Coulomb scattering among electrons is given by

$$t_{ee} \simeq (n_e \sigma v)^{-1} \propto n_e^{-1} \left(\frac{m}{T} \right)^{1/2} \left(\frac{e^2}{T} \right)^{-2} = n_e^{-1} \left(\frac{m}{T} \right)^{1/2} \left(\frac{e^2}{T} \right)^{-2} \left(\frac{1}{\pi \ln \Lambda} \right), \tag{4.130}$$

where $\ln \Lambda$ originates from distant collisions and is ~ 30 (see Vol. I, Chap. 9, Section 9.3). The n_e in this expression refers to the number density of *free* charged particles; because recombination changes this number, we write $n_e = x_e n_B$, where x_e is the fraction of charged particles that have *not* combined to form atoms. We will see later [see Eq. (4.193)] that x_e is nearly unity for $T \gtrsim 1$ eV and drops rapidly to a low value of $\sim 10^{-5}$ by $T \approx 0.1$ eV. Inserting the numerical values, we find

$$t_{ee} = 1.35 \text{ s}(T/1 \text{ eV})^{-\frac{3}{2}}(x_e \Omega_B h^2)^{-1}, \tag{4.131}$$

which is far smaller than the expansion time scale,

$$H^{-1}(T) = \begin{cases} 1.46 \times 10^{12} \text{ s}(T/1 \text{ eV})^{-2} & \text{(for } t < t_{\text{eq}}) \\ 1.13 \times 10^{12} \text{ s}(\Omega h^2)^{-\frac{1}{2}}(T/1 \text{ eV})^{-\frac{3}{2}} & \text{(for } t > t_{\text{eq}}) \end{cases} \tag{4.132}$$

at all times. (Even after x_e has fallen to a low final value of 10^{-5}, $t_{ee} \ll H^{-1}$.) A similar calculation will show that the ions and the electrons also interact at a sufficiently rapid rate. Thus these scatterings can easily maintain the thermal distribution for the matter.

The equality of temperature *between* matter and radiation has to be maintained by the interaction between photons and electrons. The simplest form of encounter between photons and electrons is the nonrelativistic, low-energy (Thomson) scattering with the mean free time

$$t_{\text{Th}} = \frac{1}{n_e \sigma_T c} = 6.14 \times 10^7 \text{ s } (T/1 \text{ eV})^{-3}(x_e \Omega_B h^2)^{-1} \tag{4.133}$$

where $\sigma_T = (8\pi/3)(e^2/m)^2$ is the Thomson scattering cross section. This process, however, cannot help in thermalisation because there is no energy exchange between the photon and electron in the Thomson scattering limit. Scattering with energy exchange occurs through Compton scattering. In addition to Compton scattering, bremsstrahlung and its inverse process (free–free absorption) can also help in maintaining thermal equilibrium between photons and matter. As we shall see, Compton scattering is the dominant mechanism at high temperatures.

The time scale for Compton scattering of photons can be estimated as follows: The mean free path for a photon between two collisions is $\lambda_\gamma = (\sigma_T n_e)^{-1}$, where σ_T is the Thomson scattering cross section. Thus, in travelling a distance l, the photon, performing a random walk, will undergo N collisions, where $N^{1/2}\lambda_\gamma = l$. The fractional frequency change in each Compton scattering is $(\delta \nu/\nu) \simeq (v/c)^2 \simeq (T/m)$. Treating this as another random walk in frequency space, we can determine the net change in the frequency after N collisions as $(\delta \nu/\nu) \simeq N^{1/2}(T/m) = (l/\lambda_\gamma)(T/m)$. Energy exchange by Compton scattering can be considered effective when $(\delta \nu/\nu) \simeq 1$, which gives the time scale

$$t_C \simeq \lambda_\gamma \left(\frac{m}{T}\right) = \frac{1}{n_e \sigma_T} \left(\frac{m}{T}\right). \tag{4.134}$$

(This is essentially the same result obtained in Vol. I, Chap. 6, Section 6.6.) Note that t_C is larger than the Thomson scattering time scale by the factor (m/T). This factor is much larger than unity in the temperature range we are interested in. Substituting the numbers, we get

$$t_C = 3.0 \times 10^{13} \text{ s} (\Omega_B h^2 x_e)^{-1} (T/1 \text{ eV})^{-4}. \tag{4.135}$$

Comparing with H^{-1}, we find that

$$\frac{t_C}{H^{-1}} = \left[\frac{T}{4.54 x_e^{-\frac{1}{2}} (\Omega_B h^2)^{-\frac{1}{2}} \text{ eV}} \right]^{-2}. \tag{4.136}$$

Thus this process is of importance in maintaining thermal equilibrium between matter and radiation for $T \gtrsim 5 \text{ eV}$, corresponding to $z > 1.7 \times 10^4$. The photons are strongly influenced by charged particles up to this redshift.

For the sake of comparison, let us consider the time scale for free–free absorption (see Vol. I, Chap. 6, Section 6.9). For a thermal plasma at temperature T, the time scale for free–free absorption $t_{ff}(\omega)$ at frequency ω is given by

$$t_{ff} \simeq \frac{3\sqrt{6\pi}}{4} \frac{(mT)^{1/2} m}{e^6 n_e^2} \frac{\omega^3}{(2\pi)^3} (1 - e^{-\omega/T})^{-1}. \tag{4.137}$$

For photons with frequency $\omega \simeq T$, we get

$$t_{ff} = 2 \times 10^{14} \text{ s} (\Omega_B h^2 x_e)^{-2} (T/1 \text{ eV})^{-\frac{5}{2}}, \tag{4.138}$$

which gives

$$\frac{t_{ff}}{H^{-1}} \cong (T/1.9 \times 10^4 \text{ eV})^{-\frac{1}{2}} (\Omega_B h^2)^{-2} \quad (\text{for } x_e \simeq 1). \tag{4.139}$$

Thus t_{ff} ceases to be effective at a very high temperature, $T \gtrsim 10^4 \text{ eV}$. In fact, even when $T \gtrsim 10^4 \text{ eV}$, free–free absorption is subdominant because Compton scattering is far more effective. The ratio between these two time scales will be

$$\frac{t_{ff}}{t_C} = \frac{1}{(\Omega_B h^2 x_e)} \left(\frac{T}{0.3 \text{ eV}} \right)^{3/2}. \tag{4.140}$$

Thus for all the way down to $T \simeq 0.3 \text{ eV}$, we have $t_C < t_{ff}$, and hence Compton scattering is more effective.

The time scale for photon production by free–free emission (bremsstrahlung) is comparable with t_{ff} for frequencies $\omega \simeq T$. So similar conclusions apply for this process as well. However, note that the effectiveness of these processes is strongly frequency dependent. A more careful calculation is needed if we are interested in the $\omega \ll T$ (or $\omega \gg T$) region of the spectrum; we will have occasion to comment about these effects in Chap. 6. Last, we should note that there exists another photon-production mechanism $(e + \gamma \rightarrow e + 2\gamma)$, called the *double-Compton process*, with a time scale of $t_{2C} \simeq 10^{20} \text{ s} (T/1 \text{ eV})^{-5} (\Omega_B h^2)^{-1}$

that dominates over the free–free processes at $T \gtrsim 1$ keV; this will be discussed in Chap. 6.

The preceding calculation, however, is somewhat oversimplified because expression (4.138) does not take into account the effect of Thomson scattering on the photons. If l_{ff} is the mean distance a typical photon travels before getting absorbed, it would have experienced $N = (l_{ff}/\lambda_\gamma)$ scatterings in this trip. Hence, it would have travelled a rms distance of $\lambda_{rms} = N^{1/2}\lambda_\gamma = (l_{ff}\lambda_\gamma)^{1/2}$ before being absorbed. This correction takes into account the fact that Thomson scatterings increase the effective path length for photon absorption. The corresponding values are

$$\bar{t} \equiv \frac{\lambda_{rms}}{c} = 1.1 \times 10^{11} \, \text{s} (T/1 \, \text{eV})^{-11/4} (\Omega_B h^2 x_e)^{-3/2},$$

$$\frac{\lambda_{rms}}{cH^{-1}} = \frac{\bar{t}}{H^{-1}} = (\Omega_B h^2 x_e)^{-3/2} (T/0.03 \, \text{eV})^{-3/4}, \tag{4.141}$$

$$\frac{\bar{t}}{t_C} = (\Omega_B h^2 x_e)^{-1/2} (T/88 \, \text{eV})^{5/4}. \tag{4.142}$$

The time scale $\bar{t}$ is still longer than t_C for $T \gtrsim 90$ eV, where Compton scattering dominates. At lower temperatures (90–1 eV), the (effective) free–free absorption can dominate over Compton scattering.

Another difference between Compton scattering and free–free transitions is that the former does not change the number of photons; thus this process alone can never lead to a Planck spectrum *if the system originally had the incorrect number of photons for a given total energy*. Because free–free absorption and emission can change the photon number, they can lead to true thermalisation. In this particular context, however, the system *does* have the correct number of photons because of the initial conditions.

The preceding analysis shows that there are several processes that can keep the radiation and matter tightly coupled until a temperature of approximately a few electron volts, provided that a sufficient number of free charged particles exist (i.e., $x_e \approx 1$).

A depletion of free charged particles occurs through the process of "re"combination, in which electrons and ions combine to form neutral, atomic systems. For the sake of simplicity, we consider only the formation of neutral H, which has a binding energy of 13.6 eV. As the universe cools to temperatures below this value, the formation of a bound atomic structure is *energetically* favoured. However, just as in the case of nucleosynthesis, the high entropy of the universe delays this process until the temperature drops to a much lower value, $T_{atom} \simeq 0.29$ eV. Below this temperature, most of the protons and the electrons combine together to form neutral atoms, leaving behind only a small fraction ($\sim 10^{-5}$) of free electrons (and protons). The disappearance of free charged particles reduces the scattering cross section between photons and charged particles, thereby increasing the mean free path (λ_γ) of the photons. When $\lambda_\gamma \gtrsim H^{-1}$, the photons decouple from the rest of the matter; this happens at $T = T_{dec} \simeq 0.26$ eV.

The temperature T_{atom} at which hydrogen atoms are formed can be computed with methods similar to those used in the last section if we make two simplifying assumptions: (1) The system is in thermodynamic equilibrium and (2) the recombination proceeds through the electron and the proton combining to form a hydrogen atom in the ground state. (Neither of these assumptions is quite correct; we will discuss the modifications that are required in a more realistic case in the end.) In thermodynamic equilibrium, the number densities of protons (n_p), electrons (n_e), and hydrogen atoms (n_H), are given by the usual formula,

$$n_i = g_i \left(\frac{m_i T}{2\pi} \right)^{3/2} \exp\left[\frac{1}{T}(\mu_i - m_i) \right], \qquad (4.143)$$

where $i = e, p$ or H. The equilibrium is maintained by the reaction $p + e \leftrightarrow n_H + \gamma$. Balance of the chemical potential in this reaction implies that $\mu_p + \mu_e = \mu_H$. Using this relation and expressing μ_p and μ_e in terms of n_p and n_e, we find that n_H becomes

$$n_H = \left(\frac{g_H}{g_p g_e} \right) n_p n_e \left(\frac{m_e T}{2\pi} \right)^{-\frac{3}{2}} \exp(B/T), \qquad (4.144)$$

where $B = m_p + m_e - m_H = 13.6$ eV is the binding energy and we have set $m_p \approx m_H$ in the prefactor, $m^{-\frac{3}{2}}$. Introducing the "fractional ionisation" x_i for each of the particle species and using the facts that $n_p = n_e$ and $n_p + n_H = n_B$, we find it follows that $x_p = x_e$ and $x_H = (n_H/n_B) = 1 - x_e$. Equation (4.144) can now be written in terms of x_e alone:

$$\frac{1 - x_e}{x_e^2} = \frac{4\sqrt{2}\zeta(3)}{\sqrt{\pi}} \eta \left(\frac{T}{m_e} \right)^{3/2} \exp(B/T) \cong 3.84\eta(T/m_e)^{3/2} \exp(B/T)$$

$$(4.145)$$

where $\eta = 2.68 \times 10^{-8}(\Omega_B h^2)$ is the baryon-to-photon ratio. We may define T_{atom} as the temperature at which 90% of the electrons, say, have combined with protons, i.e., when $x_e = 0.1$. This leads to the condition

$$(\Omega_B h^2)^{-1} \tau^{-\frac{3}{2}} \exp[-13.6\tau^{-1}] = 3.13 \times 10^{-18}, \qquad (4.146)$$

where $\tau = (T/1$ eV$)$. For a given value of $\Omega_B h^2$, this equation can be easily solved by iteration. Taking logarithms and iterating once, we find

$$\tau^{-1} \cong 3.084 - 0.0735 \ln(\Omega_B h^2), \qquad (4.147)$$

with the corresponding redshift $(1 + z) = (T/T_0)$ given by

$$(1 + z) = 1367[1 - 0.024 \ln(\Omega_B h^2)]^{-1}. \qquad (4.148)$$

For $\Omega_B h^2 = 1, 0.1$, and 0.01, we get $T_{\text{atom}} \cong 0.324, 0.307$, and 0.292 eV, respectively; these values correspond to the redshifts of 1367, 1296, and 1232. For the sake of definiteness, the mean values

$$(1 + z)_{\text{atom}} \equiv 1300, \quad T_{\text{atom}} = T_0(1 + z)_{\text{atom}} = 0.308 \text{ eV} \qquad (4.149)$$

will be adopted in future discussion as the epoch of the formation of atoms.

Because the preceding analysis was based on equilibrium densities, it is important to check that the rate of the reactions $p + e \leftrightarrow H + \gamma$ is fast enough to maintain equilibrium. The thermally averaged cross section for the process of recombination is given by

$$\langle \sigma v \rangle \cong 4.7 \times 10^{-24} \left(\frac{T}{1 \text{ eV}} \right)^{-\frac{1}{2}} \text{cm}^2. \qquad (4.150)$$

The cross section for interaction is proportional to the square of the de Broglie wavelength, $\lambda^2 \propto v^{-2}$; hence σv has $v^{-1} \propto T^{-1/2}$ dependence (see Vol. I, Chap. 6, Section 6.12). The reaction rate therefore will be

$$\Gamma = n_p \langle \sigma v \rangle = (x_e \eta n_\gamma) \langle \sigma v \rangle$$
$$= 2.374 \times 10^{-10} \text{ cm}^{-1} \tau^{7/4} e^{-(6.8/\tau)} (\Omega_B h^2)^{1/2}. \qquad (4.151)$$

In arriving at this expression, we have approximated Eq. (4.145) by

$$x_e \approx \left[\frac{\pi}{4\sqrt{2}\zeta(3)} \right]^{1/2} \eta^{-\frac{1}{2}} \left(\frac{T}{m_e} \right)^{-\frac{3}{4}} \exp \left(-\frac{6.8}{\tau} \right), \qquad (4.152)$$

which is valid for $x_e \ll 1$. The Γ in Eq. (4.151) has to be compared with the expansion rate H of the universe. Taking the universe to be matter dominated at this temperature, we have

$$H = 2.945 \times 10^{-23} \text{ cm}^{-1} (\Omega_{\text{NR}} h^2)^{1/2} \tau^{3/2}. \qquad (4.153)$$

Equating the two, we get

$$\tau^{-\frac{1}{4}} \exp(6.8/\tau) = 8.06 \times 10^{12} (\Omega_B / \Omega_{\text{NR}})^{1/2}. \qquad (4.154)$$

This equation can also be solved by taking logarithms and iterating the solution. We find that

$$\tau^{-1} \cong 4.316 - 0.074 \ln \left(\frac{\Omega_{\text{NR}}}{\Omega_B} \right), \quad (1 + z) = 977 \left[1 - 0.017 \ln \left(\frac{\Omega_{\text{NR}}}{\Omega_B} \right) \right]^{-1}. \qquad (4.155)$$

For $\Omega_{\text{NR}} \simeq 10 \Omega_B$, this gives $T_D \simeq 0.24$ eV. The fact that $T_D < T_{\text{atom}}$ justifies the assumption of thermal equilibrium used in the earlier calculation. There are, however, some other difficulties with this assumption, which will be discussed later. (See page 257).

When the reaction rate falls below the expansion rate, the formation of neutral atoms ceases. The remaining electrons and protons have negligible probability for combining with each other. The residual fraction can be estimated as the fraction present at $T = T_D$, i.e., $x_e(T_D)$. Combining Eq. (4.152) and Eq. (4.154), we find

$$x_e(T_D) \cong 7.4 \times 10^{-6} \left(\frac{T_D}{1 \text{ eV}}\right)^{-1} \left(\frac{\Omega_{NR}^{1/2}}{\Omega_B h}\right). \qquad (4.156)$$

For $T_D \simeq 0.24$ eV, this gives

$$x_e(T_D) \simeq 3 \times 10^{-5} \left(\frac{\Omega_{NR}^{1/2}}{\Omega_B h}\right). \qquad (4.157)$$

Thus a small fraction ($\sim 10^{-5}$) of electrons and protons will remain free in the universe.

The formation of atoms affects the photons, which were in thermal equilibrium with the rest of the matter through the various scattering processes previously described. It is easy to verify that the time scales for Compton scattering and free–free absorption become much larger than the expansion time scale when x_e drops to its residual value. The only scattering that is still somewhat operational is the Thomson scattering; this process merely changes the direction of the photon without any energy exchange. Its only effect is to make any given photon perform a random walk. When the number density of charged particles decreases, even this interaction rate Γ of the photons drops and, eventually, at some $T = T_{dec}$, becomes lower than the expansion rate. For $T < T_{dec}$, the photons are decoupled from the rest of the matter. The rate of Thomson scattering is given by

$$\Gamma = \sigma n_e = \sigma x_e n_B = \sigma x_e \eta n_\gamma$$
$$= 3.36 \times 10^{-11} (\Omega_B h^2)^{1/2} \tau^{9/4} \exp(-6.8/\tau) \text{cm}^{-1}. \qquad (4.158)$$

Comparing this with the expansion rate in Eq. (4.153), we get the condition

$$\tau^{-\frac{3}{4}} e^{6.8/\tau} = 1.14 \times 10^{12} (\Omega_B/\Omega_{NR})^{1/2}, \qquad (4.159)$$

where $\tau = (T_{dec}/1 \text{ eV})$. Solving this with one iteration, we get

$$\tau^{-1} \cong 3.927 + 0.0735 \ \ln(\Omega_B/\Omega_{NR}), \qquad (4.160)$$

which corresponds to the parameters

$$T_{dec} \cong 0.26 \text{ eV}, \quad (1 + z_{dec}) \cong 1100. \qquad (4.161)$$

For $T \lesssim 0.2$ eV, the neutral matter and photons evolve as uncoupled systems. The parameter T characterising the Planck spectrum continues to fall as a^{-1} because of the redshift of photons. The neutral matter behaves as a gaseous mixture of hydrogen and helium.

It should be stressed that three distinct events take place in the universe around $T \simeq (0.3\text{--}0.2)$eV: (1) Most of the protons and electrons combine to form a H atom; (2) the process of recombination stops, leaving a small fraction of free electrons and protons, when the interaction rate for $pe \leftrightarrow H\gamma$ drops below the expansion rate; and (3) the photon mean free path becomes larger than H^{-1}, decoupling radiation from matter. These events occur at almost the same epoch because $\eta \simeq 10^{-8}$ and $\Omega_{NR}h^2 \simeq 1, \Omega_B \lesssim 1$. For a different set of values for these parameters, these events could occur at different epochs.

After decoupling, the temperature of the neutral atoms falls faster than that of radiation. The decrease of matter temperature is governed by the equation

$$\frac{dT_m}{dt} + 2\frac{\dot{a}}{a}T_m = \frac{4\pi^2}{45}\sigma_T x_e \left(\frac{T^4}{m}\right)(T - T_m), \qquad (4.162)$$

where T is the radiation temperature (see Vol. I, Chap. 6, Section 6.7). The term $2(\dot{a}/a)T_m$ describes the cooling that is due to expansion, and the term on the right-hand side accounts for the energy transfer from radiation to matter. This process is now governed by the relaxation time for *matter*, which is given by

$$t_{\text{matter}} \simeq \frac{1}{\sigma_T x_e n_\gamma}\left(\frac{m}{T}\right). \qquad (4.163)$$

At high temperatures ($x_e \simeq 1$), $t_{\text{matter}} \simeq t_c \, (n_e/n_\gamma) \ll t_c$; then $T_m \approx T$ to a high degree of accuracy. As x_e becomes smaller, t_{matter} increases and the energy transfer from the radiation to matter becomes less and less effective. The adiabatic cooling term makes the matter temperature fall faster than the radiation temperature. The relaxation time t_{matter} becomes of the order of expansion time at $T \simeq 0.27$ eV ($z \simeq 1150$). At lower temperatures the matter temperature falls slightly slower than $T_m \propto a^{-2}$.

The small fraction of *ionised* matter ($n_e \simeq n_p \simeq 10^{-5}n_B$), however, continues to be affected by the photons. The *electron* mean free path, $\lambda_e = (\sigma_T n_\gamma)^{-1}$, that governs this process is much smaller than photon mean free path because $n_\gamma \gg n_e$. The corresponding time scale $t_{\text{elec}} = (n_e/n_\gamma)t_c$ will be

$$t_{\text{elec}} = 2.15 \times 10^6 \text{ s}(T/1 \text{ eV})^{-4}, \qquad (4.164)$$

which leads to the ratio

$$\frac{t_{\text{elec}}}{H^{-1}} = (\Omega_B h^2)^{\frac{1}{2}}(T/60 \text{ K})^{-\frac{5}{2}}. \qquad (4.165)$$

Thus the free electrons are tied to the radiation until a redshift of 20 or so. (In other words, a small number of electrons have many collisions with a *small* number of photons, although *most* of the photons are unaffected.) This process is therefore capable of maintaining the temperature of free electrons at the same value as T in the photon spectrum right up to $z \approx 20$. Of course, this interaction has very little effect on the photons because of the small number of charged particles present.

We now discuss the various approximations that have been made in the preceding calculation. To begin with, note that we assumed a recombination process that directly produces a H atom in the *ground* state. This will release a photon with energy of 13.6 eV in each recombination. If $n_\gamma(B)$ is the number density of photons in the background radiation with energy $B = 13.6$ eV, then

$$\frac{n_\gamma(B)}{n} \simeq \frac{16\pi}{n} T^3 \exp\left(-\frac{B}{T}\right) \simeq \frac{3 \times 10^7}{(\Omega_B h^2)} \exp\left(-\frac{13.6}{\tau}\right). \tag{4.166}$$

This ratio is unity at ~ 0.8 eV (i.e., at a redshift of $z \approx 3300$) and decreases rapidly at lower temperatures. Thus, at lower temperatures, the addition of 13.6 eV photons that are due to recombination significantly enhances the availability of ionising photons. These energetic photons have a high probability of ionising neutral atoms formed a little earlier. (That is, the "backward" reaction $H + \gamma \to p + e$ is enhanced.) Hence this process is not very effective in producing a *net* number of neutral atoms.

The dominant process that actually operates is the one in which recombination proceeds through an excited state: $(e + p \to H^\star + \gamma; H^\star \to H + \gamma_2)$. This will produce two photons, each of which has *lesser* energy than the ionisation potential of the H atom. The $2P$ and the $2S$ levels provide the most rapid route for recombination; the decay from the $2P$ state produces a single photon whereas the decay from the $2S$ state is through two photons. Because the reverse process does not occur at the same rate, this is nonequilibrium recombination.

Because of the above complication, the recombination proceeds at a slower rate compared with that predicted by Saha's equation. The actual fractional ionisation is higher than the value predicted by Saha's equation at temperatures below ~ 1300. For example, at $z = 1300$, these values differ by a factor of 3; at $z \simeq 900$, they differ by a factor of 200. The values of T_{atom}, T_{dec}, etc., however, do not change significantly. We now discuss some simple features of a more rigorous analysis.

We begin with the equation that determines the variation of radiation intensity in an expanding universe:

$$\frac{\partial}{\partial t}\left(\frac{vn_v}{n}\right) = \frac{v}{a}\frac{da}{dt}\frac{\partial}{\partial v}\left(\frac{vn_v}{n}\right) + \frac{vJ_v}{n}. \tag{4.167}$$

The first term on the right-hand side is due to the cosmological redshift that moves the energy density to the redder side and the second term is due to the net emission processes; n_v is the number of photons per unit volume per unit frequency interval, n is the number density of baryons and J_v is the net rate of production of photons per unit volume and frequency interval. The product vn_v/n does not vary because of expansion or redshift in the bandwidth and hence is a convenient parameter to work with. The net photon-production rate can be expressed in the form

$$J_\nu = \sigma_\nu c \left[\frac{8\pi\nu^3}{c^3} \frac{h^3}{(2\pi m_e k_B T)^{3/2}} e^{-(h\nu - B_1)/k_B T} n_e^2 - n_\nu n_{1s} \right], \quad (4.168)$$

where σ_ν is the photoionisation cross section, B_1 is the binding energy of the ground state ($n = 1$) of the hydrogen and atom, and n_e and n_{1s} are the number densities of electrons and H atoms, respectively, in the ground state. Once the recombination is appreciable, the mean free time for an ionising photon is significantly shorter than the expansion time, showing that the ionising part of the spectrum (that is, for $h\nu \geq B_1$) is in thermal equilibrium with matter. In this range, the right-hand side of Eq. (4.168) is nearly zero, allowing us to write

$$n_\nu = \frac{n_I h}{k_B T} e^{-(h\nu - B_1)/k_B T} \quad \text{(for } h\nu \geq B_1\text{),} \quad (4.169)$$

where n_I is the total number density of ionising photons. Substituting this result into Eq. (4.167), dividing by ν, and integrating the result over frequencies $h\nu \geq B_1$, we get the equation for the number density of the ionising photons:

$$\frac{d}{dt}\left(\frac{n_I}{n}\right) = -\frac{n_I}{n} \frac{B_1}{k_B T} \frac{1}{a} \frac{da}{dt} + \frac{J}{n}, \quad J = \alpha_{1s} n_e^2 - \sigma_I n_I c n_{1s}, \quad (4.170)$$

where α_{1s} is the recombination coefficient for direct transition to the ground state. The characteristic time scale governing the rate on the left-hand side of this equation is the the reciprocal of the cosmic time t^{-1} whereas the coefficient of n_I/n in the first term in the right-hand side, $(B_1/kT)(1/t) \approx 40/t$, is larger. To the first approximation we can ignore the left-hand side and solve the equation to determine n_I as

$$n_I \cong \alpha_{1s} n_e^2 \left(\sigma_I n_{1s} c + \frac{B_1}{k_B T} \frac{1}{a} \frac{da}{dt} \right)^{-1}. \quad (4.171)$$

Substituting into the expression for J in Eq. (4.170) we find that the net rate of creation of photons,

$$J \cong \alpha_{1s} n_e^2 \frac{B_1}{k_B T} \frac{1}{a} \frac{da}{dt} \left(\sigma_I n_{1s} c + \frac{B_1}{k_B T} \frac{1}{a} \frac{da}{dt} \right)^{-1}, \quad (4.172)$$

is suppressed by the factor $(\sigma_I c n t)(kT/B_1) \approx 10^7$. This is the quantitative reason for ignoring direct recombination to the ground state.

We next consider the recombination rates through other routes. Of the two routes for the production of atomic hydrogen, the first one – namely, decay from the metastable ^2S level – occurs at the rate $\Lambda = 8.23 \text{ s}^{-1}$. The second one, by means of Lyman-α photons, is a resonance transition and is described by a cross section

$$\sigma = \frac{3\lambda_\alpha^2}{8\pi} \frac{\gamma^2}{(\omega - \omega_\alpha)^2 + \gamma^2/4},$$ (4.173)

where γ is the total width of the line (see Vol. I, Chap. 7, Section 7.2). Because the natural width ($\gamma_{nat} = 6.28 \times 10^8 \text{ s}^{-1}$) is much smaller than the thermal width ($\delta\omega$), where

$$\frac{\delta\omega}{\omega_\alpha} = \frac{\nu}{c} = \left(\frac{3k_B T}{m_p c^2}\right)^{1/2} = 3 \times 10^{-5}$$ (4.174)

at $T = T_{dec}$, we need to consider only the latter. In that case, the cross section at resonance is given by

$$\bar{\sigma} \simeq \frac{3\lambda_\alpha^2}{8\pi} \left(\frac{\gamma^2}{\omega_\alpha^2}\right) \left(\frac{m_p c^2}{3k_B T}\right) = 3 \times 10^{-17} \text{ cm}^2.$$ (4.175)

The corresponding mean free time, $(\bar{\sigma} n_{dec} c)^{-1} \simeq 30 \Omega_B^{-1} h^{-2}$ s, is significantly shorter than the expansion time scale, implying that the population of $1S$ and $1P$ levels will be in thermal equilibrium.

The role of these Lyman photons is somewhat more complicated. If these photons can quickly encounter another hydrogen atom in a suitable excited state, then they will cause ionisation, thereby annulling the effect. A net effect is produced only because these photons are redshifted by the cosmic expansion to lower frequencies. Because photons are constantly fed into the Lyman-α line, the spectral distribution around $\nu \approx \nu_\alpha$ will be different at a frequency ν_- slightly below the line and at ν_+ slightly above the line. To determine the relative values, we integrate Eq. (4.167) across the line and obtain

$$\frac{d}{dt}\left(\frac{n_\alpha}{n}\right) = \frac{\nu_a}{n}\frac{1}{a}\frac{da}{dt}\left[n_{(\nu+)} - n_{(\nu-)}\right] + \frac{R}{n},$$ (4.176)

where R is the net rate of production of resonance-line photons per unit volume. In this equation, we note that $n_\alpha \approx n_- \delta\nu$, so that it can be treated as a first-order differential equation for n_-. However, the first term on the right-hand side is larger than that on the left-hand side by a factor of $(\nu/\delta\nu) \approx 10^5$, allowing us to ignore the left-hand side and solve the equation algebraically. This gives

$$N_\alpha = N_+ + \frac{R\lambda_\alpha^3 a}{8\pi \dot{a}},$$ (4.177)

where $N_\nu \equiv (c^3/8\pi \nu^2)n_\nu$ is the occupation number of the photons.

Let us next consider the distribution of hydrogen atoms in different energy levels. For all excited states we can take the relative population to be thermal with

$$n_{nl} = n_{2s}(2l + 1)e^{-(B_2 - B_n)/k_B T}.$$ (4.178)

To the same level of accuracy, we can also replace the occupation number of photons N_+ with its thermal equilibrium value, obtaining

$$N_\alpha = e^{-(B_2 - B_n)/k_B T} + \frac{R\lambda_\alpha^3 a}{8\pi \dot{a}}. \tag{4.179}$$

To determine N_α we need to eliminate R, which can be done as follows: We note that each net recombination results in the emission of either a Lyman-α photon or a two-photon decay. In such a case, the equation describing the rate of production of neutral H atoms can be written in the form

$$\alpha_{ex} n_e^2 - \beta_{ex} n_{2s} = R + \Lambda \left(n_{2s} - n_{1s} e^{-h\nu_\alpha/k_B T}\right). \tag{4.180}$$

We shall now describe each term in this equation.

The first term on the left-hand side is the rate of recombination to the excited levels (ignoring the ground state) with $\alpha_{ex} = \langle \sigma v \rangle$ giving the recombination rate (see Vol. I, Chap. 6, Section 6.12). The second term is the rate of ionisation from the excited states, which is dominated by the number of atoms in the $2S$ state in thermal equilibrium. The coefficient β_{ex} can be related to α_{ex} by the principle of detailed balance:

$$\beta_{ex} = \alpha_{ex} \frac{(2\pi m_e k_B T)^{3/2}}{(2\pi \hbar)^3} e^{-B_2/k_B T}, \tag{4.181}$$

where $B_2 = 3.4$ eV is the energy of the $n = 2$ level. On the right-hand side of Eq. (4.180), the first term, R, is the net rate of production of H atoms per unit volume by Lyman-α emission. The second term is the net rate of two-photon emission from the $2S$ state. This is given by the rate of the emission term (Λn_{2s}) minus the absorption rate for the two-photon process. The relative factor is again fixed by the principle of detailed balance.

To proceed further, we use the fact that thermal equilibrium across the line requires that $(n_{2s}/n_{1s}) = N_\alpha$. Using this relation in Eq. (4.180) and eliminating R between Eqs. (4.180) and (4.179), we can determine N_α as

$$N_\alpha = e^{-(B_2 - B_n)/k_B T} \frac{\left[1 + K \left(\alpha_{ex} n_e^2 e^{-(B_2 - B_n)/k_B T} + \Lambda_{2s,1s} n_{1s}\right)\right]}{\left[1 + K \left(\beta_{ex} + \Lambda_{2s,1s}\right) n_{1s}\right]}, \tag{4.182}$$

where

$$K \equiv \frac{\lambda_\alpha^3}{8\pi} \left(\frac{\dot{a}}{a}\right)^{-1}. \tag{4.183}$$

We can now write the net rate of production of hydrogen atoms in terms of the left-hand side of Eq. (4.180) with the form for N_α substituted. Direct algebra now gives

$$-\frac{d}{dt}\left(\frac{n_e}{n}\right) = \left(\frac{\alpha_{ex} n_e^2}{n} - \frac{\beta_{ex} n_{1s}}{n} e^{-h\nu_\alpha/k_B T}\right) C, \tag{4.184}$$

where

$$C = \frac{1 + K\Lambda n_{1s}}{1 + K(\Lambda + \beta_{ex})n_{1s}}, \qquad K \equiv \frac{\lambda_\alpha^3}{8\pi}\left(\frac{\dot{a}}{a}\right)^{-1}, \qquad (4.185)$$

and n gives the total baryon number density in both free protons and hydrogen atoms. Note that the ratio n_e/n is unaffected by the global expansion of the universe and any change in this quantity represents the net result of local physical processes. If we take $C \approx 1$, then this equation gives the recombination rate (leading to the decrease of free electron density) that is due to the $2S \to 1S$ transition. The existence of Lyman-α resonance photons introduces the β_{ex} term in the denominator of C and reduces this rate. The ratio of formation of hydrogen atoms by Lyman-α transitions to $2S \to 1S$ transitions is given by

$$\frac{R}{\Lambda\left(n_{2s} - n_{1s}e^{-h\nu_\alpha/k_B T}\right)} = \frac{1}{K\Lambda n_{1s}} \simeq \frac{0.01}{(1-x)}\left(\frac{\Omega_{NR}^{1/2}}{\Omega_B h}\right) \qquad (4.186)$$

at $z = z_{dec}$. When this ratio is much smaller than unity, most of the hydrogen is produced by two-photon decay and $C \cong \Lambda(\Lambda + \beta_{ex})^{-1}$. We have seen in Vol. I, Chap. 6, Section 6.12 that the recombination rate α_{ex} to excited states is approximated by the formula

$$\alpha_{ex} = 2.6 \times 10^{-13} T_4^{-0.8} \text{ cm}^3 \text{ s}^{-1}. \qquad (4.187)$$

From this we can easily estimate the ratio

$$\frac{\beta_{ex}}{\Lambda} = 8 \times 10^7 \, T_4^{0.7} \, e^{-3.95/T_4}. \qquad (4.188)$$

This factor is large at t_{dec} but drops rapidly, reaching unity around $T \approx 2300$ K at $z \approx 850$. After this epoch, $C \approx 1$.

It is possible to understand the basic physics of recombination by solving Eq. (4.184) in the approximation in which only the two-photon process is dominant and $C \approx \Lambda/(\Lambda + \beta_{ex})$. In this limit, we can also ignore the second term on the right-hand side of Eq. (4.184) and write it in the form

$$\frac{dx_e}{dt} \simeq -\frac{\Lambda\alpha_{ex}}{(\Lambda + \beta_{ex})}nx_e^2 \simeq \left[\frac{\alpha_{ex}n}{1 + (\beta_{ex}/\Lambda)}\right]x_e^2, \qquad x_e = \frac{n_e}{n}. \qquad (4.189)$$

Taking $n = n_0(1 + z)^3 = (\Omega_B\rho_c/m_p)(1 + z)^3$ and using Eqs. (4.187) and (4.188), we find that the expression within the square brackets is only a function of z (equivalently, t), so that this equation is trivial to integrate. Changing the derivative from t to z by using $\dot{z} \approx H_0\Omega_{NR}^{1/2}(1 + z)^{5/2}$, we find that this equation becomes

$$\frac{1}{x_e^2}\frac{dx_e}{dz} = -\frac{\mu}{[1 + (\beta_{ex}/\Lambda)]}(1 + z)^{-0.3} \equiv -\mu f(z), \qquad \mu = 634\left(\frac{\Omega_B h}{\Omega_{NR}^{1/2}}\right),$$

(4.190)

where we obtain the numerical value for the parameter μ by using the known numerical values for all other quantities. We integrate this equation with the boundary condition that at some very large value of the redshift, z_c, we have $x_e(z_c) = 1$. Then the solution is given by

$$x_e(z) = \left[1 + \mu \int_z^{z_c} f(z)\,dz\right]^{-1},$$

(4.191)

with

$$f(z) = \left\{10^{-2.1}\left(\frac{1+z}{10^3}\right)^{0.3} + 2.6 \times 10^5 \left(\frac{1+z}{10^3}\right)\right.$$

$$\left. \times \exp[-1.45 \times 10^4/(1+z)]\right\}^{-1}.$$

(4.192)

The function $x_e(z)$ is plotted as a function of redshift in Fig. 4.5 for the parameter values ($\Omega_{NR}h^2 = 0.1$, $\Omega_B h^2 = 0.1$), ($\Omega_{NR}h^2 = 1$, $\Omega_B h^2 = 0.1$), and ($\Omega_{NR}h^2 = 1$, $\Omega_B h^2 = 0.01$). We can obtain the asymptotic value of residual ionisation $x_e(0)$ by setting $z = 0$ in Eq. (4.191). For reasonable values of Ω_B and Ω_{NR}, the second term within the square brackets of Eq. (4.191) is significantly larger than unity, allowing us to write

$$x_e(0) \simeq \frac{1}{\mu}\left[\int_0^{z_c} f(z)\,dz\right]^{-1} \simeq 10^{-5}\frac{(\Omega_{NR}h^2)^{1/2}}{\Omega_B h^2},$$

(4.193)

which is comparable with the result obtained in relation (4.157). The second equality is obtained by numerical integration; the result is insensitive to the value of z_c as long as $z_c \gg 10^4$. The scaling is obvious from the fact that asymptotically $C \approx 1$, which allows the parameters in Eq. (4.184) to be rescaled, showing that $x \propto (\Omega_{NR}^{1/2}/\Omega_B h)$.

On the other hand, in the redshift range of $800 < z < 1200$, the fractional ionisation varies rapidly and is given (approximately) by the formula,

$$x_e = 2.4 \times 10^{-3}\frac{(\Omega_{NR}h^2)^{1/2}}{(\Omega_B h^2)}\left(\frac{z}{1000}\right)^{12.75}.$$

(4.194)

[This is obtained by fitting a curve to the numerical solution, as shown in Fig. 4.5.]

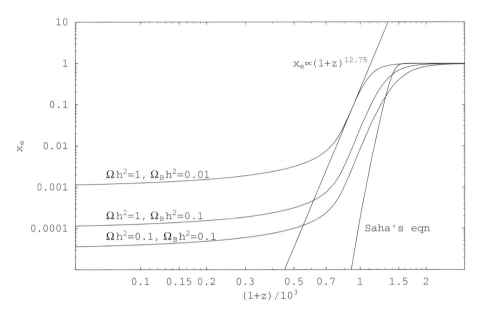

Fig. 4.5.   The ionisation fraction as a function of $1 + z$ in different contexts. The y axis is the ionisation fraction x_e, and the x axis is $(1 + z)$ normalised to unity at $(1 + z) = 10^3$. The three curves are for $(\Omega_{\mathrm{NR}}h^2 = 1, \Omega_B h^2 = 0.01)$, $(\Omega_{\mathrm{NR}}h^2 = 1, \Omega_B h^2 = 0.1)$, and $(\Omega_{\mathrm{NR}}h^2 = 0.1, \Omega_B h^2 = 0.1)$ from top to bottom at the left end. Of the two slanted vertical lines, the one on the left is the fit given in Eq. (4.194), corresponding to $x_e \propto (1 + z)^{12.75}$, and the one on the right is the ionisation fraction as predicted by Saha's ionisation equation (4.145).

Using this expression, we can compute the optical depth for photons as

$$\tau = \int_0^t n(t)x_e(t)\sigma_T \, dt = \int_o^z n(z)x_e(z)\sigma_T \left(\frac{dt}{dz}\right) dz \simeq 0.37 \left(\frac{z}{1000}\right)^{14.25},$$

(4.195)

where we have used the relation $H_0 \, dt \cong -\Omega_{\mathrm{NR}}^{-1/2} \, dz$, which is valid for $z \gg 1$. This optical depth is unity at $z_{\mathrm{dec}} = 1072$. Our approximate calculation earlier gave a value of 1100, which is quite close to the exact value.

From the optical depth, we can also compute the probability that the photon was last scattered in the interval $(z, z + dz)$. This is given by $(\exp -\tau) \, (d\tau/dz)$, which can be expressed as

$$P(z) = e^{-\tau} \frac{d\tau}{dz} = 5.26 \times 10^{-3} \left(\frac{z}{1000}\right)^{13.25} \exp\left[-0.37 \left(\frac{z}{1000}\right)^{14.25}\right].$$

(4.196)

This $P(z)$ has a sharp maximum at $z \simeq 1067$ and a width of $\Delta z \cong 80$. It is therefore reasonable to assume that decoupling occurred at $z \simeq 1070$ in an interval of $\Delta z \simeq 80$. We shall see in Chap. 6 that the finite thickness of the surface of the last scattering has important observational consequences.

The photons emitted during recombination cannot be thermalised effectively and thus distort the Planck spectrum. The distortions are on the high-frequency part ($\nu/T > 30$) of the spectrum, which corresponds to $\nu \gtrsim 1.5 \times 10^{12}$ Hz ($\lambda \lesssim 0.02$ cm) today. (The Planck spectrum has only 10^{-10} of all photons in the range $\nu/T > 30$.) Unfortunately, galactic nuclei and dust emit very strongly in the region $\lambda < 3 \times 10^{-2}$ cm, completely swamping this primordial signal.

The photons that decouple from matter at this epoch ($z \approx 10^3$) propagate freely and are present at $z = 0$ as thermal background radiation, usually called *cosmic microwave background radiation* (CMBR). Being a relic of an epoch at which the universe was nearly 1000 times smaller in size, this radiation contains very valuable information about the conditions that prevail at this epoch. We shall discuss the physics of CMBR in Chap. 6.

Exercise 4.7

Molecular H in the early universe: The residual free electrons can serve as catalysts for the formation of molecular H through the reactions (a) $H + e \leftrightarrow H^- + \gamma$; $H^- + H \leftrightarrow H_2 + e$ and (b) $H + p \leftrightarrow H_2^+ + \gamma$; $H_2^+ + H \leftrightarrow H_2 + p$. Because the binding energy of protons in H_2^+, $B_+ = 2.65$ eV, is considerably larger than the binding energy $B_- = 0.75$ eV of the electron in H^-, the second reaction produces molecular H at a higher redshift. Using equilibrium abundance ratios, show that most of the H_2 form at $z \approx 200$.

4.8 Very Early Universe and Cosmological Scalar Fields

The discussion in the previous sections dealt with the history of the universe during the epochs in which only the physical processes that take place are those tested in the laboratory. Such an approach has the advantage of being reasonably accurate because the input physics is well understood. However, it is unsatisfactory to the extent that it does not address the question of initial conditions that allow the universe to evolve in a particular way. There have been several attempts in the past to understand the very high-energy phase of the universe when the temperatures are higher than, say, $\sim 10^{14}$ GeV. If a fairly compelling model for the high-energy interaction of particles is available, then such a model could be used to study the very early phases of the universe with the hope of making predictions that can be tested by cosmology. Unfortunately, our theoretical understanding of high-energy interactions is not at a level to provide such a model. Hence it is difficult to make any firm statement about the very early evolution of the universe, and any conclusions drawn regarding this phase – based on our current understanding – should be considered as a mathematical exercise of a highly speculative nature.[7]

In spite of this serious weakness, it is worthwhile to try and identify some generic features that a wide variety of models of the early universe has attempted to incorporate. There are three such features that have emerged out of these studies: (1) The dynamics of a universe, dominated by a scalar field having a potential-energy term, seems to be sufficiently different from the evolutionary history discussed in the previous sections. We will spend sometime learning the interesting features of such an evolution. (2) A phase of very rapid expansion (usually called *inflation*) during the high-energy phase of the universe seems to have some nontrivial effects on the rest of the evolution of the universe. Although none of these effects have been conclusively demonstrated, they are worth discussing because of the attention they have received in the past two decades. (3) We shall see in Chap. 5 that the observed structures in the universe could have arisen out of gravitational instability of regions containing small density fluctuations during the early phase of the universe. Some of the models for the early universe are capable of producing such density inhomogeneities.

As additional motivation for discussing such speculative physics, we will begin by pointing out certain features of the conventional cosmological model that may be thought of as unsatisfactory. These features go under the names *horizon problem* and *flatness problem*, which we shall now discuss.

It seems reasonable to assume that the present features of the universe were determined by the initial conditions at some epoch $t = t_i$ when the temperature of the universe was very high, say $T \simeq 10^{14}$ GeV (which corresponds to the epoch at which most "grand unified theories" had a significant influence on the evolution of the universe). Let the proper distance a light signal could have travelled during the time interval $(0, t)$, assuming the universe was radiation dominated, be $R_H(t)$. Physical processes could have made the universe homogeneous over a sphere of radius $R_H(t_i)$ at the initial epoch $t = t_i$. If this sphere could expand to encompass the observed universe, then we will have a simple explanation for the observed homogeneity of the universe. We will estimate $R_H(t)$ and show that this is not possible in the conventional scenarios in which the universe is either radiation dominated or matter dominated.

Consider the radiation-dominated phase of the universe with the expansion law $a(t) \propto t^{1/2}$. The proper distance that a light signal could have travelled in the time interval $(0, t)$ is

$$R_H(t) \equiv a(t) \int_0^t \frac{dx}{a(x)} = 2t. \tag{4.197}$$

Therefore causal communication between two observers, O and O', can exist only if they are within a distance $2R_H(t) = 4t$. This boundary is called the *parti-cle horizon*. Two observers, O and O', separated by a proper distance larger than $2R_H(t)$ at an epoch t could have never influenced each other. Hence there is no a priori reason to expect points O and O' to have similar physical environments.

If the present features of the universe were essentially determined at some early epoch – say, at $t = t_i$ when the temperature of the universe was $T \simeq 10^{14}$ GeV – then we would expect a sphere, which had a radius $2R_H(t_i)$ at that epoch, to have expanded to encompass the currently observed universe. This will provide a natural explanation for the observed homogeneity of our universe. From the initial epoch $T \simeq 10^{14}$ GeV to the present epoch, with $T_0 = 2.75$ K $\approx 2.4 \times 10^{-4}$ eV, the universe has expanded by a factor of $(T/T_0) \simeq 4 \times 10^{26}$. However, when $T \simeq 10^{14}$ GeV, the time was $t \simeq 10^{-35}$ s and $2R_H \simeq 6 \times 10^{-25}$ cm. Thus the primordial sphere of homogeneity would have expanded to a size of only approximately 2.4×10^2 cm by today, a value that is far short of the size of the present universe.

The situation is worse as regards the isotropy of microwave background radiation. From the time t_i till the epoch of decoupling t_{dec}, the universe would have expanded by only a factor of $(T/T_{dec}) \simeq 4 \times 10^{23}$. The primordial sphere of homogeneity would have expanded to a size of only 0.24 cm at $t = t_{dec}$. We saw in Exercise 3.12 that a $1°$ scale in the sky probes a length scale of ~ 100 Mpc at $t = t_{dec}$. Two patches of the sky separated by, say, $10°$ could not have been in causal contact at the time of decoupling if the above analysis is correct. Nevertheless, such patches of sky show identical temperatures for the CMBR.

To increase 0.24 cm to ~ 100 Mpc, we require an additional expansion by a factor of $\sim 10^{27}$. This is easily achieved by a faster – say, exponential – growth of $a(t)$, known as inflation. One simple way of realising such an inflationary growth is in a model for the universe in which the universe was radiation dominated up to, say, $t = t_i$, but expanded exponentially in the interval $t_i < t < t_f$:

$$a(t) = a_i \exp H(t - t_i), \qquad t_i \leq t \leq t_f. \qquad (4.198)$$

For $t > t_f$, the evolution is again radiation dominated $[a(t) \propto t^{1/2}]$ until $t = t_{eq} \cong 5.8 \times 10^{10}(\Omega_{NR}h^2)^{-2}$ s. The evolution becomes matter dominated for $t_{eq} < t < t_{now} = t_0$. Typical values for $t_i, t_f,$ and H may be taken to be

$$t_0 \approx 10^{-35} \text{ s}; \qquad H \approx 10^{10} \text{ GeV}; \qquad t_f \approx 70\,H^{-1}, \qquad (4.199)$$

which give an overall "inflation" by a factor of approximately $A \equiv \exp N \cong \exp(70) \approx 2.5 \times 10^{30}$ to the scale factor during the period $t_i < t < t_f$. (In this section we use the symbol H to denote the Hubble constant *during* the inflationary phase.) At $t = t_i$, the temperature of the universe is $\sim 10^{14}$ GeV. During this exponential inflation, the temperature drops drastically; however, the matter is expected to be reheated to the initial temperature of $\sim 10^{14}$ GeV at $t \approx t_f$ by various high-energy processes. Thus inflation effectively changes the value of $S = T(t)a(t)$ by a factor of $A = \exp(70) \approx 10^{30}$. Note that this quantity S is conserved during the noninflationary phases of the expansion. The effect of inflation on our analysis can be easily estimated as follows: The coordinate size of the region in the last scattering surface (LSS) from where we receive signals today is

$$l(t_0, t_{dec}) = \int_{t_{dec}}^{t_0} \frac{dt}{a(t)} \simeq \frac{3}{a_{dec}} (t_{dec}^{2/3} t_0^{1/3}), \tag{4.200}$$

and the coordinate size of the horizon at $t = t_{dec}$ will now be

$$l(t_{dec}, 0) = \int_0^{t_{dec}} \frac{dt}{a(t)} \simeq \frac{4t_i}{a_{dec}} \left(\frac{t_{dec}}{t_f} \right)^{1/2} A. \tag{4.201}$$

[We have used the facts that $t_0 \gg t_{dec}$, $A \gg 1$, $t_i \simeq H^{-1}$, and $a_{dec} = a_i A(t_{dec}/t_f)^{1/2}$.] The ratio

$$R = \frac{l(t_{dec}, 0)}{l(t_0, t_{dec})} \cong 2 A \frac{t_i}{(t_f t_{dec})^{1/2}} \left[\frac{2}{3} \left(\frac{t_{dec}}{t_0} \right)^{1/3} \right] \approx 4 \times 10^4 \left(\frac{A}{10^{30}} \right), \tag{4.202}$$

is much larger than unity for $A \simeq 10^{30}$. Thus all the signals we receive today are from a causally connected domain in the LSS. Note that, in the absence of inflation, $l(t_{dec}, 0) = (2t_{dec}/a_{dec})$ so that $R = (2/3)(t_{dec}/t_0)^{1/3} \ll 1$. This value is amplified by a large factor in the course of inflation.

Such a rapid expansion also tackles another peculiar feature of the universe. We have seen earlier that the quantity $\Omega(t) - 1$ changes rapidly during the expansion of the universe if it is not identically zero [see Eq. (3.99)]. We can write, at any time t, $\Omega(t) - 1 = (k/\dot{a}^2)$. Assuming that $k \neq 0$, a comparison with the present epoch $t = t_0$ gives

$$\Omega(t) - 1 = \frac{\dot{a}_o^2}{\dot{a}^2} (\Omega_0 - 1). \tag{4.203}$$

Expressed as a function of temperature in the radiation-dominated era, this relation becomes

$$\Omega(T) - 1 \cong 4 \times 10^{-15}(\Omega_0 - 1) \left(\frac{T}{1 \text{ MeV}} \right)^{-2}. \tag{4.204}$$

The present astronomical observations imply that Ω_0 is between 0.1 and 5, say, with liberal allowance given for systematic errors. Thus $|\Omega_0 - 1|$ is of the order of unity. However, relation (4.204) shows that at earlier epochs $|\Omega_{NR} - 1|$ is far less than unity. If we assume that the initial conditions for the universe were "set" at Planck time, when $T \simeq 10^{19}$ GeV, then the a and $\dot{a}$ terms should have been matched at that epoch to an accuracy of $|(\Omega - 1)| \simeq 10^{-60}$ so as to result in the present values for a and $\dot{a}$. Had this fine-tuning not been resorted to, the universe would have contracted back to $a = 0$ (for $k = 1$) or diffused out to $a = \infty$ (for $k = -1$) *long before* the present epoch. In the absence of any physical mechanism, this fine-tuning has to be imposed in an ad hoc manner at some early epoch. (This is called the *flatness problem*.)

Inflation of $a(t)$ by a factor A decreases the value of the k/a^2 term by a factor of $A^{-2} \cong 10^{-60}$. Thus we can start with moderate values of k/a^2 before inflation and bring them down to a very small value at $t \gtrsim 10^{-33}$ s. This solves the flatness problem, interpreted as the smallness of k/a^2. Note that no classical process can change a $k \neq 1$ universe to a $k = 0$ universe, as it involves a change in topological properties. What inflation does is to decrease the value of k/a^2 so much that, for all practical purposes, we can ignore the dynamical effect of the curvature term, thereby having the same effect as setting $k = 0$. This has the consequence that Ω at present must be unity to a high degree of accuracy.

Exercise 4.8
Pseudosolution to a pseudoproblem?: (a) Show that inflation can keep the observed universe homogeneous only for a time $t < t_{\text{crit}}$, where

$$t_{\text{crit}} \simeq \frac{t_i^2}{t_f} \exp 2 \, H(t_f - t_i),$$

and t_i and t_f are the times at that the inflation started and ended. (b) Consider a $k = 1$ universe that went through an inflationary phase. Because such a model is expected to recollapse eventually, observations will reveal that $\Omega \neq 1$ at some stage, say, at $t > T$. Estimate T.

Having discussed the advantages of the rapid exponential growth, let us consider physical models that can produce such a growth. Because most of these models use one or more scalar fields as the source of expansion, we begin by recalling the energy density and pressure provided by a scalar field $\phi(t)$ that depends only on time. These are given by

$$\rho_\phi = \frac{1}{2}\dot{\phi}^2 + V(\phi), \qquad P_\phi = \frac{1}{2}\dot{\phi}^2 - V(\phi), \qquad (4.205)$$

where $V(\phi)$ is the potential for the scalar field (see Vol. I, Chap. 11, Section 11.7). Einstein's equation (for the $k = 0$ Friedmann model) with the above source can be written in the form

$$\frac{\dot{a}^2}{a^2} = H^2(t) = \frac{1}{3M_{\text{Pl}}^2}\left[V(\phi) + \frac{1}{2}\dot{\phi}^2\right], \qquad (4.206)$$

where $M_{\text{Pl}} \equiv (8\pi G)^{-1/2} \approx 2.4 \times 10^{18}$ GeV is a convenient energy scale for high-energy physics. The equation of motion for the scalar field reduces to (see Vol. I, Chap. 11, Section 11.7)

$$\ddot{\phi} + 3H\dot{\phi} = -\frac{dV}{d\phi}. \qquad (4.207)$$

In principle, Eqs. (4.206) and (4.207) can be integrated to give $a(t)$ and $\phi(t)$ once the form of $V(\phi)$ is given. The actual nature of solutions, of course, will depend critically on the form of $V(\phi)$ as well as on the initial conditions. Among these solutions, there exists a subset in which $a(t)$ is a rapidly growing function of t, either exponentially or as a power law $a(t) \propto t^p$ with an arbitrarily large value of p. We shall now provide a general procedure for determining a form of $V(\phi)$ such that any reasonable evolution characterised by a function $a(t)$ can be reproduced by use of a scalar field.

Let us consider a scalar field with an energy density and pressure given by Eqs. (4.205) with $w(t) \equiv P_\phi(t)/\rho_\phi(t)$. For a given $w(t)$, we can integrate the equation of motion $d(\rho a^3) = -w\rho\, d(a^3)$ to obtain

$$\frac{\dot{\rho}_\phi}{\rho_\phi} = -3\,H(1+w), \qquad \rho_\phi(t) \propto \exp\left[-3\int dt(1+w)H\right]. \quad (4.208)$$

On the other hand, in a universe with $k = 0$ with the scalar field acting as a source, we have the relation $\rho_\phi \propto H^2$ so that $(\dot{\rho}_\phi/\rho_\phi) = 2(\dot{H}/H)$. Combining the two relations we get

$$1 + w(t) = -\frac{2}{3}\frac{\dot{H}}{H^2}. \quad (4.209)$$

From the definition of w and Eq. (4.205), it follows that $\dot{\phi}^2/2V = (1+w) \times (1-w)^{-1} \equiv f(t)$, say. Writing this as $\dot{\phi}^2 = 2fV$, differentiating with respect to time, and using Eq. (4.207), we find that

$$\frac{\dot{V}}{V} = -\frac{\dot{f} + 6Hf}{1+f}. \quad (4.210)$$

Integrating this Equation and using the definition of $f(t)$ and Eq. (4.209) we get

$$V(t) = \frac{3H^2}{8\pi G}\left[1 + \frac{\dot{H}}{3H^2}\right]. \quad (4.211)$$

Substituting back in the relation $\dot{\phi}^2 = 2fV$, we can determine $\phi(t)$ as

$$\phi(t) = \int dt\left[-\frac{\dot{H}}{4\pi G}\right]^{1/2}. \quad (4.212)$$

Equations (4.211) and (4.212) completely solve the problem of finding a potential $V(\phi)$ that will lead to a given $a(t)$. In fact, the same method works even when matter other than a scalar field with some known energy density $\rho_k(t)$ is present in the universe. In this case, Eqs. (4.211) and (4.212) generalise to

$$V(t) = \frac{1}{16\pi G} H(1 - Q) \left[6H + \frac{2\dot{H}}{H} - \frac{\dot{Q}}{1 - Q} \right],$$ (4.213)

$$\phi(t) = \int dt \left[\frac{H(1 - Q)}{8\pi G} \right]^{1/2} \left[\frac{\dot{Q}}{1 - Q} - \frac{2\dot{H}}{H} \right]^{1/2},$$ (4.214)

where $Q(t) \equiv [8\pi G \rho_k(t)/3 H^2(t)]$.

As an example of using Eqs. (4.211) and (4.212), let us consider a universe in which $a(t) = a_0 t^p$. Elementary algebra now gives the potential to be of the form

$$V(\phi) = V_0 \exp\left(-\sqrt{\frac{2}{p}} \frac{\phi}{M_{\mathrm{Pl}}} \right),$$ (4.215)

where V_0 and p are constants and $M_{\mathrm{Pl}}^{-2} = 8\pi G$. The corresponding evolution of $\phi(t)$ is given by

$$a(t) = a_0 t^p, \qquad \frac{\phi(t)}{M_{\mathrm{Pl}}} = \sqrt{2p} \ln \left[\sqrt{\frac{V_0}{p(3p - 1)}} \frac{t}{M_{\mathrm{Pl}}} \right].$$ (4.216)

As the second example, consider an evolution of the form

$$a(t) \propto \exp(At^f), \qquad f = \frac{\beta}{4 + \beta}, \qquad 0 < f < 1, \qquad A > 0.$$ (4.217)

In this case, we can determine the potential to be

$$V(\phi) \propto \left(\frac{\phi}{M_{\mathrm{Pl}}} \right)^{-\beta} \left(1 - \frac{\beta^2}{6} \frac{M_{\mathrm{Pl}}^2}{\phi^2} \right),$$ (4.218)

where β is a constant.

Both these examples show that it is indeed possible to have a rapid growth of $a(t)$ for a suitable choice for the potential. [In fact, we have proved a stronger result. The preceding analysis shows that cosmologies with a scalar field are trivial in the sense that, for any sensible $a(t)$ and any given source of matter $\rho_k(t) < \rho_c$, it is possible to invent a $V(\phi)$ such that we have a $k = 0$ universe with the given $a(t)$. This fact is often not appreciated in the literature.] We next enquire about the general conditions that $V(\phi)$ should satisfy in order to lead to this behaviour.

One possible way of achieving this is through potentials that allow what is known as a *slow rollover*. Such potentials have a gently decreasing form for $V(\phi)$ for a range of values for ϕ during which $\phi(t)$ evolves very slowly. Assuming a sufficiently slow evolution of $\phi(t)$, we can ignore the $\ddot{\phi}$ term in Eq. (4.207). Similarly, we can ignore the kinetic-energy term $\dot{\phi}^2$ in comparison with the potential energy $V(\phi)$ in Eq. (4.206). In this limit, Eqs. (4.206) and (4.207) become

$$H^2 \simeq \frac{V(\phi)}{3M_{\text{Pl}}^2}, \qquad 3\,H\dot{\phi} \simeq -V'(\phi). \tag{4.219}$$

Let us assume that the slow rollover approximation is valid until a time $t = t_{\text{end}}$. The amount of inflation can be characterised by the ratio $a(t_{\text{end}})/a(t)$; a large value for this number indicates that the universe expands by a large factor from the time t until the time t_{end}. Because this is a large number in most models, it is conventional to work with the quantity $N(t) \equiv \ln[a(t_{\text{end}})/a(t)]$. From relations (4.219) it is easy to obtain an expression for N in terms of the potential $V(\phi)$:

$$N \equiv \ln \frac{a(t_{\text{end}})}{a(t)} = \int_t^{t_{\text{end}}} H\,dt \simeq \frac{1}{M_{\text{Pl}}^2} \int_{\phi_{\text{end}}}^{\phi} \frac{V}{V'}\,d\phi. \tag{4.220}$$

This provides a general procedure for generating and quantifying the rapid growth of $a(t)$ arising from a given potential. The validity of a slow rollover requires that the following two parameters be sufficiently small:

$$\epsilon(\phi) = \frac{M_{\text{Pl}}^2}{2}\left(\frac{V'}{V}\right)^2, \qquad \eta(\phi) = M_{\text{Pl}}^2\,\frac{V''}{V}. \tag{4.221}$$

Of these, ϵ has to be small in order to ignore the kinetic energy in comparison with the potential energy, whereas η has to be small in order to ignore the acceleration of the scalar field. The end point for inflation can be taken to be the epoch at which ϵ becomes comparable with unity.

Exercise 4.9
Hamilton–Jacobi formulation of inflation: Consider an inflationary model with a scalar field for which $\dot{\phi} > 0$. Given the functions $H(t)$ and $\phi(t)$, it is possible to define another function $H(\phi)$. Show that $H(\phi)$ satisfies

$$[H'(\phi)]^2 - \frac{3}{2M_{\text{Pl}}^2}H^2(\phi) = -\frac{1}{2M_{\text{Pl}}^4}V(\phi). \tag{4.222}$$

Show further that the slow-rollover phase requires the conditions $\epsilon_H \ll 1, \eta_H \ll 1$, where $\epsilon_H = -(d \ln H/d \ln a)$ and $\eta_H = -(d \ln H'/d \ln a)$. Describe how this formalism allows us to determine the form of $V(\phi)$ for a given cosmological example. [Hint: Differentiate Eq. (4.206) with respect to t and substitute into Eq. (4.207). Dividing by $\dot{\phi}$ will give $\dot{\phi}$ in terms of $H(\phi)$. Substituting in the Friedmann equation gives the necessary result.]

Exercise 4.10
Inflation with $V(\phi) = \lambda\phi^4$: It is possible to satisfy the conditions of slow rollover even in fairly simple potentials. As an example, let us consider the case of $V(\phi) = \lambda\phi^4$, with $\phi(t)$

evolving from some initial value ϕ_i at $a = a_i$ towards $\phi = 0$. Show that the slow-rollover solutions are given by

$$\phi = \phi_i \exp\left[-\sqrt{\frac{32\lambda M_{\text{Pl}}^2}{6}}(t - t_i)\right], \tag{4.223}$$

$$a = a_i \exp\left(\frac{\phi_i^2}{8M_{\text{Pl}}^2}\left\{1 - \exp\left[-\sqrt{\frac{64\lambda M_{\text{Pl}}^2}{3}}(t - t_i)\right]\right\}\right). \tag{4.224}$$

Determine when the inflation ends and the value of N in such a model.

5

Structure Formation

5.1 Introduction

The formalism developed in Chaps. 3 and 4 needs to be modified to take into account the inhomogeneities present in our universe, and we shall try to reconstruct the observed universe in the following manner: We assume that, at some time in the past, there were small deviations from homogeneity in our universe. These deviations can then grow because of gravitational instability over a period of time and, eventually, form galaxies, clusters etc. The first task is to develop the mathematical machinery capable of describing the growth of structures.[1]

In a universe made of nonbaryonic dark matter, baryons, and radiation, we need to discuss each component separately. The simplest of the three components is the dark matter, which is affected only by gravity and is collisionless. In the fluid limit, we can ignore the velocity dispersion of the dark-matter particles, and there is no effective pressure term in the equations. If the velocity dispersion is important, there will arise an effect called *free streaming*, which we will discuss in Section 5.6. The physics of radiation is complicated by the fact that a photon can traverse a distance of the order of $d_H \approx ct$ within a cosmic time t; hence any perturbation in δ_R at length scales $\lambda \lesssim ct$ will be wiped out by the free propagation of photons. At $\lambda > ct$, radiation can be treated as a fluid influenced only by the gravitational perturbations. The baryons are the most difficult to handle. They are strongly coupled to the photons until $z = z_{\text{dec}}$ and the photon–baryon system acts as a single fluid for $z > z_{\text{dec}}$; for lower values of z, the evolution of baryons is complicated by several heating and cooling processes. We shall concentrate on dark matter in this chapter and take up radiation in Chap. 6 and baryons in Chap. 7.

It turns out that the equations describing the evolution of dark-matter perturbations are fairly complicated and exact solutions to these equations are impossible to obtain. To make progress we need to resort to one of the following three possible approaches:

(1) When the density contrast $\delta = (\rho - \bar{\rho})/\bar{\rho}$ is small, it is possible to linearise the equations in δ and obtain solutions that describe the growth of linear perturbations. Because the density contrasts are expected to be small in the early universe this approximation will be valid for considerable period of time.

(2) When the density contrast is comparable with unity, the relevant equations can be solved only if specific assumptions (e.g., spherical symmetry) are made regarding the solutions. The usefulness of such solutions, of course, depends on the validity of the underlying assumptions.

(3) In the nonlinear stage ($\delta \gg 1$) the equations have to be integrated numerically. Such an integration is time consuming, and it is usually not easy to obtain physical insight about the evolution even when the numerical integration is performed successfully. To tackle this difficulty we often introduce some simplifying ansatz while integrating the equations of motion. By comparing the results of exact numerical integration with the simplified picture, it is possible to gain some physical insight about the dynamics.

The linear evolution described in (1) will be discussed in Sections 5.2–5.6, and the nonlinear evolution [(2) and (3)] will be discussed in Sections 5.9–5.12. It should be noted that this approach *assumes* the existence of small inhomogeneities at some initial time. To be considered complete, the cosmological model should also *produce* these initial inhomogeneities by some viable physical mechanism. Some attempts in this direction will be discussed in Section 5.7.

5.2 Growth of Inhomogeneities

We begin by describing an important conceptual issue before proceeding to the development of the technical formalism. Consider a perturbation of the metric $g_{ik}(x)$ and the source T_{ik} into the form $(g_{ik} + \delta g_{ik})$ and $(T_{ik} + \delta T_{ik})$, where the set (g_{ik}, T_{ik}) corresponds to the background Friedmann universe and the set $(\delta g_{ik}, \delta T_{ik})$ denotes the perturbation. Assuming the latter to be "small," we can linearise Einstein's equations to obtain a second-order differential equation of the form

$$\hat{\mathcal{L}}(g_{ik})\delta g_{ik} = \delta T_{ik}, \tag{5.1}$$

where $\hat{\mathcal{L}}$ is a linear differential operator depending on the background space–time. Because this is a linear equation, it is convenient to expand the solution in terms of some suitably chosen mode functions. For the sake of simplicity, let us consider the spatially flat ($\Omega = 1$) universe. The mode functions will then be plane waves, and by Fourier transforming the variables, we can obtain a set of separate equations $\hat{\mathcal{L}}_{(k)}\delta g_{(k)} = \delta T_{(k)}$ for each mode, labelled by a wave vector **k**. Solving this set of equations, we can determine the evolution of each mode separately. (A similar procedure, of course, works for the case with $\Omega \neq 1$. In this case, the mode functions will be more complicated than the plane waves; but, with a suitable choice of orthonormal functions, we can obtain a similar set of equations.)

There is, however, one major conceptual difficulty in carrying out this program. In general relativity, the form (and numerical value) of the metric coefficients g_{ik} (or the stress-tensor components T_{ik}) can be changed by a relabelling of coordinates $x^i \to x^{i'}$. By such a trivial change we can make a small δT_{ik} large or even generate a component that was originally absent. Thus the perturbations may grow at different rates – or even decay – when we relabel coordinates. It is necessary to tackle this ambiguity before we can talk meaningfully about the growth of inhomogeneities.

There is a simple way of handling this problem for modes that have proper wavelengths much smaller than the Hubble radius. The general relativistic effects that are due to the curvature of space–time will be negligible at length scales far smaller than the Hubble radius. In such regions, we shall see that there exists a natural choice of coordinates in which Newtonian gravity is applicable. All physical quantities can be unambiguously defined in this context.

Although such a Newtonian analysis provides valuable insight into the behaviour of inhomogeneities, it suffers from the following difficulty: As we shall see, the proper wavelength of any mode will be bigger than the Hubble radius at sufficiently early epochs. Thus the early evolution of *any* mode needs to be tackled by general relativity and the coordinate ambiguities have to be settled.

There are two different ways of handling such difficulties in general relativity. The first method is to resolve the problem by force: We may choose a particular coordinate system and compute everything in that coordinate system. If the coordinate system is physically well motivated, then the quantities computed in that system can be interpreted easily; for example, we will treat δT_0^0 as the perturbed mass (energy) density even though it is coordinate dependent. The difficulty with this method is that we cannot fix the gauge *completely* by simple physical arguments; the residual gauge ambiguities do create some problems.

The second approach is to construct quantities – linear combinations of various perturbed physical variables – that are scalars under coordinate transformations. Einstein's equations are then rewritten as equations for these gauge-invariant quantities. This approach, of course, is manifestly gauge invariant from start to finish. However, it is more complicated than the first one; besides, the gauge-invariant objects do not, in general, possess any simple interpretation. We shall therefore be mainly concerned with the first approach.

Because the gauge ambiguity is a purely general relativistic effect, we begin by determining when such effects are significant. The effects that are due to the curvature of space–time will be important at length scales bigger than (or comparable with) the Hubble radius, defined as $d_H(t) \equiv (\dot{a}/a)^{-1}$. Writing the

Friedmann equation as

$$\frac{\dot{a}^2}{a^2} = H_0^2\left[\Omega_R\left(\frac{a_0}{a}\right)^4 + \Omega_{NR}\left(\frac{a_0}{a}\right)^3 + \Omega_V + (1-\Omega)\left(\frac{a_0}{a}\right)^2\right], \quad (5.2)$$

where Ω_R, Ω_{NR}, Ω_V, and Ω represent the density parameters for relativistic matter [with $p_R = (1/3)\rho_R$; $\rho_R \propto a^{-4}$], nonrelativistic matter (with $p_{NR} = 0$; $\rho_{NR} \propto a^{-3}$), cosmological constants ($p_V = -\rho_V$; $\rho_V = $ constant), and total energy density ($\Omega = \Omega_R + \Omega_{NR} + \Omega_V$), respectively, it follows that

$$d_H(z) = H_0^{-1}[\Omega_R(1+z)^4 + \Omega_{NR}(1+z)^3 + (1-\Omega)(1+z)^2 + \Omega_V]^{-1/2}.$$

$$(5.3)$$

This has the limiting forms

$$d_H(z) \cong \begin{cases} H_0^{-1}\Omega_R^{-1/2}(1+z)^{-2} & (z \gg z_{eq}) \\ H_0^{-1}\Omega_{NR}^{-1/2}(1+z)^{-3/2} & (z_{eq} \gg z \gg z_{curv}; \Omega_V = 0) \end{cases} \quad (5.4)$$

during the radiation-dominated and matter-dominated epochs, where

$$(1+z_{eq}) \equiv \frac{\Omega_{NR}}{\Omega_R}, \quad (1+z_{curv}) \equiv \frac{1}{\Omega_{NR}} - 1. \quad (5.5)$$

(The universe is radiation dominated for $z \gg z_{eq}$ and makes the transition to matter-dominated phase at $z \simeq z_{eq}$. It becomes "curvature dominated" sometime in the past, for $z \lesssim z_{curv}$, if $\Omega_{NR} < 0.5$. We have set $\Omega_V = 0$ for simplicity.) The physical wavelength λ_0, characterising a perturbation of size λ_0 today, will evolve as $\lambda(z) = \lambda_0(1+z)^{-1}$. Because d_H increases faster with redshift, [as $(1+z)^{-3/2}$ in the matter-dominated phase and as $(1+z)^{-2}$ in the radiation-dominated phase] $\lambda(z) > d_H(z)$ at sufficiently large redshifts. For a given λ_0 we can assign a particular redshift z_{enter} at which $\lambda(z_{enter}) = d_H(z_{enter})$. For $z > z_{enter}$, the proper wavelength is bigger than the Hubble radius and general relativistic effects are important, whereas for $z < z_{enter}$ we have $\lambda < d_H$, and we can ignore the effects of general relativity. It is conventional to say that the scale λ_0 "enters the Hubble radius" at the epoch z_{enter}.

The exact relation between λ_0 and z_{enter} differs in the case of radiation-dominated and matter-dominated phases because $d_H(z)$ has different scalings in these two cases. Using relation (5.4) we easily verify that (1) a scale

$$\lambda_{eq} \cong \left(\frac{H_0^{-1}}{\sqrt{2}}\right)\left(\frac{\Omega_R^{1/2}}{\Omega_{NR}}\right) \cong 14 \text{ Mpc}(\Omega_{NR}h^2)^{-1} \quad (5.6)$$

enters the Hubble radius at $z = z_{eq}$. (2) Scales with $\lambda > \lambda_{eq}$ enter the Hubble radius in the matter-dominated epoch with

$$z_{enter} \simeq 900(\Omega_{NR}h^2)^{-1}\left(\frac{\lambda_0}{100 \text{ Mpc}}\right)^{-2}. \quad (5.7)$$

(3) Scales with $\lambda < \lambda_{eq}$ enter the Hubble radius in the radiation-dominated epoch with

$$z_{enter} \simeq 4.55 \times 10^5 \left(\frac{\lambda_0}{1\,\text{Mpc}} \right)^{-1}. \tag{5.8}$$

We can characterise the wavelength λ_0 of the perturbation more meaningfully as follows: As the universe expands, the wavelength λ grows as $\lambda(t) = \lambda_0[a(t)/a_0]$ and the density of nonrelativistic matter decreases as $\rho(t) = \rho_0[a_0/a(t)]^3$. Hence the mass of nonrelativistic matter $M(\lambda_0)$ contained inside a sphere of radius $\lambda/2$ remains constant at

$$M = \frac{4\pi}{3}\rho(t)\left[\frac{\lambda(t)}{2}\right]^3 = \frac{4\pi}{3}\rho_0\left(\frac{\lambda_0}{2}\right)^3 = 1.45 \times 10^{11}\, M_\odot(\Omega_{NR}h^2)\left(\frac{\lambda_0}{1\,\text{Mpc}}\right)^3. \tag{5.9}$$

This relation shows that a comoving scale $\lambda_0 \approx 1$ Mpc contains a typical galaxy mass and $\lambda_0 \approx 10$ Mpc contains a typical cluster mass. From relation (5.8), we see that all these – astrophysically interesting – scales enter the Hubble radius in the radiation-dominated epoch.

This feature suggests the following strategy for studying the gravitational clustering. At $z \gg z_{enter}$ (for any given λ_0), the perturbations need to be studied by use of general relativistic, linear perturbation theory. For $z \ll z_{enter}$, general relativistic effects are ignorable and the problem of gravitational clustering can be studied with Newtonian gravity in proper coordinates. Observations indicate that the perturbations are only of the order of 10^{-4}–10^{-5} at $z \simeq z_{enter}$ for all λ_0. Hence the nonlinear epochs of gravitational clustering occur only in the regime of Newtonian gravity. In fact the *only* role of general relativity in this formalism is to evolve the initial perturbations up to $z \lesssim z_{enter}$, after which Newtonian gravity can take over. Also note that, in the nonrelativistic regime ($z \lesssim z_{enter}; \lambda \lesssim d_H$), there exists a natural choice of coordinates in which Newtonian gravity is applicable. Hence all the physical quantities can be unambiguously defined in this context.

5.3 Linear Growth in the General Relativistic Regime

Let us start by analysing the growth of the perturbations when the proper wavelength of the mode is larger than the Hubble radius. Because $\lambda \gg d_H$ we cannot use Newtonian perturbation theory. Nevertheless, it is easy to determine the evolution of the density perturbation by the following argument.

Consider a spherical region of radius $\lambda(\gg d_H)$ containing energy density $\rho_1 = \rho_b + \delta\rho$, embedded in a $k = 0$ Friedmann universe of background density ρ_b. It follows from spherical symmetry that the inner region is not affected by the matter outside; hence the inner region evolves as a $k \neq 0$ Friedmann universe.

Therefore we can write, for the two regions,

$$H^2 = \frac{8\pi G}{3}\rho_b, \qquad H^2 + \frac{k}{a^2} = \frac{8\pi G}{3}(\rho_b + \delta\rho). \tag{5.10}$$

The change of density from ρ_b to $\rho_b + \delta\rho$ is accommodated by the addition of a spatial curvature term (k/a^2). If this condition is to be maintained at all times, we must have

$$\frac{8\pi G}{3}\delta\rho = \frac{k}{a^2}, \tag{5.11}$$

or

$$\frac{\delta\rho}{\rho_b} = \frac{3}{8\pi G(\rho_b a^2)}. \tag{5.12}$$

If $(\delta\rho/\rho_b)$ is small, $a(t)$ on the right-hand side will differ only slightly from the expansion factor of the unperturbed universe. This allows us to determine how $(\delta\rho/\rho_b)$ scales with a for $\lambda > d_H$. Because $\rho_b \propto a^{-4}$ in the radiation-dominated phase ($t < t_{eq}$) and $\rho_b \propto a^{-3}$ in the matter-dominated phase ($t > t_{eq}$) we get

$$\left(\frac{\delta\rho}{\rho}\right) \propto \begin{cases} a^2 & (\text{for } t < t_{eq}) \\ a & (\text{for } t > t_{eq}) \end{cases}. \tag{5.13}$$

Exercise 5.1
Perturbation growth in terms of conformal time: Show that, within the approximation made above, $\delta \propto \tau^2$ both in the radiation- and matter-dominated epochs where τ is the conformal time coordinate defined through $d\tau = dt/a$.

Thus the amplitude of the mode with $\lambda > d_H$ always grows; as a^2 in the radiation-dominated phase and as a in the matter-dominated phase. Because no microscopic processes can operate at scales bigger than d_H all components of density (dark matter, baryons, photons) grow in the same manner as $\delta \propto (\rho_b a^2)^{-1}$ when $\lambda > d_H$.

A more formal way of obtaining this result is as follows: We first recall that there is an *exact* equation in general relativity connecting the geodesic acceleration $\mathbf{g}$ with the density and pressure (see Vol. I, Chap. 11, Section 11.6):

$$\nabla \cdot \mathbf{g} = -4\pi G(\rho + 3p). \tag{5.14}$$

Perturbing this equation, in a medium with the equation of state $p = w\rho$, we get

$$\nabla_r \cdot [\delta\mathbf{g}] = -4\pi G(\delta\rho + 3\delta p) = -4\pi G\rho_b(1 + 3w)\delta = a^{-1}\nabla_x \cdot [\delta\mathbf{g}], \tag{5.15}$$

where $\delta = (\delta\rho/\rho)$ is the density contrast. Let us produce $\delta\mathbf{g}$ by perturbing the proper coordinate $\mathbf{r} = a(t)\mathbf{x}$ to $\mathbf{r} + \mathbf{l} = a(t)\mathbf{x}[1 + \epsilon]$ such that $\mathbf{l} \cong a\mathbf{x}\epsilon$. The corresponding perturbed acceleration is given by $\delta\mathbf{g} = \mathbf{x}[a\ddot{\epsilon} + 2\dot{a}\dot{\epsilon}]$. Taking the divergence of this $\delta\mathbf{g}$ with respect to $\mathbf{x}$ and using Eq. (5.15) we get

$$\nabla_{\mathbf{x}} \cdot [\delta\mathbf{g}] = 3[a\ddot{\epsilon} + 2\dot{a}\dot{\epsilon}] = -4\pi G\rho_b a(1+3w)\,\delta. \tag{5.16}$$

(This approach works because, at scales $\lambda \gg d_H$, we are not concerned with the spatial gradients and can work in terms of a density contrast that depends only on time. Obviously, this method – as well as the final equation we derive – will not be valid in a more general context in which spatial gradients and peculiar velocities are important.) This perturbation also changes the proper volume by an amount

$$(\delta V/V) = (3l/r) = 3\epsilon. \tag{5.17}$$

If we now consider a *metric* perturbation of the form $g_{ik} \to g_{ik} + h_{ik}$, the change in the proper volume that is due to the change in $\sqrt{-g}$ is

$$(\delta V/V) = -(h/2), \tag{5.18}$$

where h is the trace of h_{ik}. Comparison of the expressions for $(\delta V/V)$ suggests that, as far as the dynamics is concerned, the equation satisfied by 3ϵ and that satisfied by $-(h/2)$ will be identical. Substituting $\epsilon = (-h/6)$ into Eq. (5.16), we get

$$\ddot{h} + 2\left(\frac{\dot{a}}{a}\right)\dot{h} = 8\pi G\rho_b(1+3w)\,\delta. \tag{5.19}$$

(A more formal approach – which uses the full machinery of general relativity – leads to the same equation.) We next note that $\dot{\delta}$ and $\dot{h}$ can be related through conservation of mass. From the equation $d(\rho V) = -p\,dV$ we obtain

$$\delta = \frac{\delta\rho}{\rho} = -(1+w)\frac{\delta V}{V} = -3(1+w)\epsilon, \tag{5.20}$$

giving

$$\dot{\delta} = -3\dot{\epsilon}(1+w) = +(1+w)\frac{\dot{h}}{2}. \tag{5.21}$$

Combining Eqs. (5.19) and (5.21), we find the equation satisfied by δ to be

$$\ddot{\delta} + 2\frac{\dot{a}}{a}\dot{\delta} = 4\pi G\rho_b(1+w)(1+3w)\,\delta. \tag{5.22}$$

This is the equation that is satisfied by the density contrast in a medium with equation of state $p = w\rho$.

To solve this equation, we need the background solution that determines $a(t)$ and $\rho_b(t)$. When the background matter is described by the equation of state $p = w\rho$, the background density evolves as $\rho_b \propto a^{-3(1+w)}$. In that case, the Friedmann equation (with $\Omega = 1$, which is a valid approximation at large redshifts) leads to

$$a(t) \propto t^{[2/3(1+w)]}; \qquad \rho_b = \frac{1}{6\pi G(1+w)^2 t^2}, \tag{5.23}$$

provided that $w \neq -1$. When $w = -1$, $a(t) \propto \exp(\mu t)$ with a constant μ. We consider the $w \neq -1$ case first. Substituting the solution for $a(t)$ and $\rho_b(t)$ into Eq. (5.22), we get

$$\ddot{\delta} + \frac{4}{3(1+w)} \frac{\dot{\delta}}{t} = \frac{2(1+3w)}{3(1+w)} \frac{\delta}{t^2}. \tag{5.24}$$

This equation is homogeneous in t and hence admits power-law solutions. Using an ansatz $\delta \propto t^n$ and solving the quadratic equation for n, we find the two linearly independent solutions (δ_g, δ_d) to be

$$\delta_g \propto t^n, \qquad \delta_d \propto \frac{1}{t}, \qquad n = \frac{2(1+3w)}{3(1+w)}. \tag{5.25}$$

In the case of $w = -1$, $a(t) \propto \exp(\mu t)$, and the equation for δ reduces to

$$\ddot{\delta} + 2\lambda\dot{\delta} = 0. \tag{5.26}$$

This has the solution $\delta_g \propto \exp(-2\mu t) \propto a^{-2}$. All the preceding solutions can be expressed in a unified manner. By direct substitution it can be verified that δ_g in all the above cases can be expressed as

$$\delta_g \propto \frac{1}{\rho_b a^2}, \tag{5.27}$$

which is exactly the result obtained originally in Eq. (5.12). This allows us to evolve the perturbation from an initial epoch until $z = z_{\text{enter}}$, after which Newtonian theory can take over.

5.4 Gauge Dependence of Perturbations: An Illustration

Although the result in Section 5.3 is quite useful, it must be stressed that it is gauge dependent as we are dealing with modes which are bigger than the Hubble radius. To illustrate how gauge dependence can completely alter the apparent evolutionary behaviour of the density contrast, we now consider Einstein's equation for a different class of perturbations. The results obtained below will be of use in the study of CMBR anisotropies in Chap. 6.

The most general perturbation to the Friedmann metric can be expressed in the form

$$ds^2 = a^2(\tau)\{(1+2\psi)d\tau^2 - w_\alpha d\tau dx^\alpha - [(1-2\phi)\gamma_{\alpha\beta} + 2h_{\alpha\beta}]dx^\alpha dx^\beta\}. \tag{5.28}$$

In this expression, $\psi(\tau, \mathbf{x})$ and $\phi(\tau, \mathbf{x})$ are scalar fields, $\omega_\alpha(\tau, \mathbf{x})$ is a three-vector field, and $h_{\alpha\beta}(\tau, \mathbf{x})$ is a second-rank tensor field. (Note that Greek indices like α or β vary over 1, 2, 3 whereas Latin indices vary over 0, 1, 2, 3.) The three-metric $\gamma_{\alpha\beta}$ describes the spatial geometry of the unperturbed Friedmann universe and is used to raise and lower all the three-dimensional tensorial quantities. We are using the conformal time coordinate as it turns out to be more convenient. It is obvious from Eq. (5.28) that the trace of $h_{\alpha\beta}$ can be incorporated into ϕ. Therefore we can take $h_{\alpha\beta}$ to be a traceless tensor (that is, $h_{\alpha\beta}\gamma^{\alpha\beta} = 0$) with five independent components. Together with three components of ω_α and two scalars ψ and ϕ, we have 10 independent degrees of freedom, which is precisely what an arbitrary metric perturbation should have. [There is, however, still some ambiguity in the definition of vector and tensor parts of the perturbations in the preceding equations. This is because any symmetric second-rank tensor in three-dimensions can be separated into parts that arise from (1) a vector h_α, (2) a scalar h, and (3) a transverse, traceless tensor $h_{\alpha\beta}^T$. Such a decomposition can be expressed mathematically in the form

$$h_{\alpha\beta} = h_{\alpha\beta}^T + \left(\partial_\alpha\partial_\beta - \frac{1}{3}\gamma_{\alpha\beta}\nabla^2\right)h + (h_{\alpha,\beta} + h_{\beta,\alpha}). \tag{5.29}$$

We will not worry about this aspect as the vector perturbations turn out to be irrelevant and the transverse traceless tensor part decouples from the rest.]

The definition of perturbed metric given by Eq. (5.28), of course, is valid in an arbitrary coordinate system. By choosing a specific coordinate system, it is possible to bring 4 of the 10 components of perturbations into preassigned form. This can, of course, be done in infinite number of ways; however, two particular choices turn out to be very convenient. The first choice, usually called the *synchronous gauge*, uses the coordinate freedom to set $\psi = 0$ and $\omega_\alpha = 0$. If we transform back to the Friedmann time coordinate $dt = a\,d\tau$ and absorb the scalar ϕ into the trace of $h_{\alpha\beta}$, then the perturbed metric in the synchronous gauge will be

$$ds^2 = dt^2 - a^2(t)(\gamma_{\alpha\beta} + h_{\alpha\beta})\,dx^\alpha\,dx^\beta. \tag{5.30}$$

Note that the vector part of h_α could be thought of as vanishing in this gauge, allowing us to deal with only the trace $h = \gamma^{\alpha\beta}h_{\alpha\beta}$ (which includes the original ϕ) and the transverse traceless part of $h_{\alpha\beta}$ (which is a genuine tensor mode).

We obtain another popular coordinate choice, called the *Poisson gauge*, by setting $\omega_\alpha = 0$ and $h_{\alpha\beta}^\alpha = 0$. In this case, we have two scalars ψ and ϕ and a transverse traceless tensor. (Once again, we can take the vector part to be zero.) It turns out that, if the perturbed stress tensor δT_{ik} has no anisotropic stress components, then perturbed Einstein's equations allow us to set $\psi = \phi$. Thus, as before, we have to deal with one scalar degree of freedom and a transverse traceless tensor field.

These two gauges have very different properties and it is convenient to use them in different contexts. The synchronous gauge uses the time coordinate that is comoving with the material particles, and we could choose the trajectories of the dark-matter particles such that they have negligible peculiar velocities in this gauge. However, this gauge does not have a simple Newtonian limit (because $g_{00} = 1$), and hence it is difficult to think in terms of Newtonian gravitational potential in this gauge. This gauge is therefore useful for studying the growth of density perturbations at scales bigger than the Hubble radius. The Poisson gauge, on the other hand, has a simple Newtonian limit, and the variable ϕ can be identified with Newtonian potential in this limit. In this gauge, the observers will detect a velocity field of particles falling into the potential well, and it is convenient to use this gauge to study perturbations much smaller than the Hubble radius. This is broadly what we will do in the later sections after briefly describing the mathematical structure of both the gauges.[2]

5.4.1 Synchronous Gauge

In the synchronous gauge, using the conformal time coordinate for convenience, we can take the metric to be

$$ds^2 = a^2(\tau)[d\tau^2 - (\delta_{\alpha\beta} + h_{\alpha\beta})\,dx^\alpha\,dx^\beta]. \tag{5.31}$$

As previously described, the metric can be decomposed into a trace and a traceless part, and we are essentially interested in the scalar parts of $h_{\alpha\beta}$. This is most easily identified in Fourier space, where we write $h_{\alpha\beta}$ in terms of two fields, $h(\mathbf{k}, \tau)$ and $\eta(\mathbf{k}, \tau)$, in Fourier space as

$$h_{\alpha\beta}(\mathbf{x}, \tau) = \int d^3k e^{i\mathbf{k}\cdot\mathbf{x}} \left[\hat{\mathbf{k}}_\alpha \hat{\mathbf{k}}_\beta h(\mathbf{k}, \tau) + \left(\hat{\mathbf{k}}_\alpha \hat{\mathbf{k}}_\beta - \frac{1}{3}\delta_{\alpha\beta} \right)\eta(\mathbf{x}, \tau) \right] \tag{5.32}$$

where the caret denotes the unit vector. We expand all the perturbed quantities like (1) the perturbed energy density δ, (2) the perturbed velocity v^i, and (3) the perturbed pressure δp, as well as h and η in Fourier modes, and we concentrate on the Fourier space amplitudes for any given wave vector $\mathbf{k}$. The perturbed Einstein equation then becomes an ordinary differential equation involving only the time derivative for the amplitudes in Fourier space. To derive these perturbation equations, it is convenient to begin with the equation $T^{ik}_{;k} = 0$ and use the variable $\theta = \partial_\alpha v^\alpha$. Direct calculation then gives

$$\dot{\delta} = -(1+w)\left(\theta + \frac{\dot{h}}{2}\right) - 3\frac{\dot{a}}{a}\left(\frac{\delta p}{\delta\rho} - w\right)\delta, \tag{5.33}$$

$$\dot{\theta} = -\frac{\dot{a}}{a}(1-3w)\theta - \frac{\dot{w}}{1+w}\theta + \frac{\delta p}{\delta\rho}\frac{1}{1+w}k^2\delta, \tag{5.34}$$

where the overdots denote derivatives with respect to conformal time τ, $w = p/\rho$, and $(\delta p/\delta\rho) = (\dot{p}/\dot{\rho})$. Note that p and ρ denote the *total* background pressure and density, respectively, and hence both w and $c_s^2 = (\delta p/\delta\rho)$ can, in general, be functions of τ. (We have dropped the subscript 'b' on ρ_b etc. to simplify the notation.) In writing these equations, we have assumed that the perturbations do not involve any anisotropic stress that has nonzero values for the off-diagonal spatial components of the energy-momentum tensor. These are two equations for the three variables δ, θ, and h. As the third equation, we may take any one of the Einstein equations or a combination of them. One particular choice that does not involve η is given by

$$\ddot{h} + \frac{\dot{a}}{a}\dot{h} = -8\pi Ga^2\rho\left(1 + 3c_s^2\right)\delta, \qquad c_s^2 \equiv \frac{\delta p}{\delta\rho}. \tag{5.35}$$

Equations (5.33)–(5.35) form a complete set that determines the evolution of δ, θ, and h in synchronous gauge.

As an example of the use of this gauge, let us consider modes that are much bigger than the Hubble radius so that we can set $k \simeq 0$ in these equations. Further, if there is only one component as the source, then we can take the equation of state $w = c_s^2 =$ constant. In this limit, $\theta \simeq 0$, and Eq. (5.33) gives

$$\delta = -\frac{1}{2}(1 + w)h. \tag{5.36}$$

Substituting this into Eq. (5.35), we get

$$\ddot{\delta} + \frac{\dot{a}}{a}\dot{\delta} = 4\pi G\rho a^2(1 + 3w)(1 + w)\delta. \tag{5.37}$$

Converting the derivatives in this equation – which are with respect to the conformal time τ – to derivatives with respect to t by using $(d/d\tau) = a(d/dt)$, we find that this equation is identical to Eq. (5.22). We thus conclude that the density contrast grows as $(1/\rho a^2)$ for modes with $\lambda \gg d_H$ in the synchronous gauge.

5.4.2 Poisson Gauge

Let us next consider the equations for perturbed quantities in the Poisson gauge with $\omega_\alpha = 0$ and $\psi = \phi$. Taking the background metric to be spatially flat and using the conformal time, we find that the metric now becomes

$$ds^2 = a^2(\tau)\{(1 + 2\phi)\,d\tau^2 - [(1 - 2\phi)\delta_{\alpha\beta} + 2h_{\alpha\beta}]\,dx^\alpha\,dx^\beta\}. \tag{5.38}$$

In this case, the conservation law $T^i_{k;i} = 0$ leads to the equations

$$\dot{\rho} + 3\left(\frac{\dot{a}}{a} - \dot{\phi}\right)(\rho + p) + \partial_\alpha[(\rho + p)v^\alpha] = 0, \tag{5.39}$$

$$\partial_\tau[(\rho + p)v_\alpha] + 4\frac{\dot{a}}{a}(\rho + p)\,v_\alpha + \partial_\alpha p + (\rho + p)\phi_{,\alpha} = 0. \tag{5.40}$$

If we relate the perturbed pressure δp and the perturbed density $\delta \rho$ by $\delta p = c_s^2 \delta \rho$ and take $p = w\rho$ for the background, the perturbed density and velocity will obey the equations

$$\dot{\delta} + 3\left(c_s^2 - w\right)\frac{\dot{a}}{a}\delta + (1 + w)(\partial_\alpha v^\alpha - 3\dot{\phi}) = 0, \tag{5.41}$$

$$\frac{1}{\rho a^4}\partial_\tau[a^4\rho(1 + w)v_\alpha] + c_s^2\partial_\alpha\delta + (1 + w)\phi_{,\alpha} = 0. \tag{5.42}$$

Using the fact that $\dot{a}^2 = (8\pi G/3)\rho a^4$ and using the Friedmann equation, we can simplify Eq. (5.42) to

$$\dot{\mathbf{v}} + (1 - 3w)\frac{\dot{a}}{a}\mathbf{v} = -\frac{c_s^2}{1 + w}\nabla\delta - \nabla\phi. \tag{5.43}$$

The Einstein equations reduce to the following set of equations for these variables:

$$\nabla^2\phi - 3\frac{\dot{a}}{a}\left(\dot{\phi} + \frac{\dot{a}}{a}\phi\right) = 4\pi Ga^2\rho\delta, \tag{5.44}$$

$$\dot{\phi}_{,\alpha} + \frac{\dot{a}}{a}\phi_{,\alpha} = -4\pi Ga^2\rho(1 + w)v_\alpha, \tag{5.45}$$

$$\ddot{\phi} - \nabla^2\phi + 6\frac{\dot{a}}{a}\dot{\phi} + 2\left[\frac{\ddot{a}}{a} + \frac{\dot{a}^2}{a^2}\right]\phi = -4\pi Ga^2\rho\left(1 - c_s^2\right)\delta, \tag{5.46}$$

$$\left(\partial_\tau^2 - \nabla^2 + 2\frac{\dot{a}}{a}\partial_\tau\right)h_{\alpha\beta} = 0. \tag{5.47}$$

The overdot denotes a derivative with respect to conformal time τ, and the gradient symbols are spatial derivatives evaluated with the unperturbed three-metric. Of these, Eq. (5.47) represents the propagation of gravitational waves in the Friedmann background and decouples from the rest. Among Eqs. (5.41), (5.42), (5.44), (5.45), and (5.46), only three are independent as the Einstein equations imply the conservation law $T^i_{k;i} = 0$. It is convenient to take these to be Eqs. (5.41), (5.42), and (5.44).

Let us consider the solutions to these equations in the case of a single-component universe with $w = c_s^2 = $ constant. Then Eqs. (5.41) and (5.42) become

$$\dot{\delta} + (1 + w)(\nabla \cdot \mathbf{v} - 3\dot{\phi}) = 0, \tag{5.48}$$

$$\frac{1}{\rho a^4}\partial_\tau[a^4\rho(1 + w)\mathbf{v}] + w\nabla\delta + (1 + w)\nabla\phi = 0. \tag{5.49}$$

Although the general solution to these equations are difficult to obtain (and not very illuminating), it is easy to determine the nature of solutions in the limit of long wavelengths ($\lambda \gg d_H$) and short wavelengths ($\lambda \ll d_H$). In the first case, spatial gradients may be ignored relative to time derivatives and terms involving $(\dot{a}/a)$. Then these equations can be solved to give

$$\delta = 3(1+w)\phi, \qquad v \propto (\rho a^4)^{-1}. \tag{5.50}$$

In the same limit, Poisson equation (5.44) becomes

$$-3\frac{\dot{a}}{a}\left(\dot{\phi}+\frac{\dot{a}}{a}\phi\right) \cong 4\pi G\rho a^2\delta. \tag{5.51}$$

Converting the derivatives with respect to τ to derivatives with respect to a and using $\delta = 3(1+w)\phi$ and the Friedmann equation in conformal time, $(\dot{a}/a)^2 = (8\pi G/3)\rho a^2$, we can reduce this equation to the simple form

$$\frac{d}{da}(\phi a) = -\frac{3(1+w)}{2}\phi. \tag{5.52}$$

This integrates to give

$$\phi = \phi_i\left(\frac{a}{a_i}\right)^{-\frac{1}{2}(5+3w)} \qquad (a > a_i), \tag{5.53}$$

with the initial condition $\phi = \phi_i$ at $a = a_i$. This result shows that, in the long-wavelength limit, ϕ (and hence δ) has no growing solution.

If we do not assume that $w = c_s^2$ and keep them independent, the relevant equations become, in the long-wavelength limit,

$$\dot{\delta} + 3\left(c_s^2 - w\right)\frac{\dot{a}}{a}\delta - 3(1+w)\dot{\phi} = 0, \tag{5.54}$$

$$\partial_\tau[a^4\rho(1+w)\mathbf{v}] = 0, \tag{5.55}$$

$$\dot{\phi} + \frac{\dot{a}}{a}\phi = -\frac{1}{2}\frac{\dot{a}}{a}\delta. \tag{5.56}$$

(The last equation is obtained after using the Friedmann equation to simplify the time derivatives of a.) We can eliminate δ to obtain the closed equation for the potential ϕ as

$$\ddot{\phi} + 3\left(1+c_s^2\right)\frac{\dot{a}}{a}\dot{\phi} + 3\left(c_s^2 - w\right)\left(\frac{\dot{a}}{a}\right)^2\phi = 0. \tag{5.57}$$

It is obvious that there are no driving terms in this equation, and hence there is no growing mode for ϕ at long wavelengths even now. The same conclusion holds for δ as well. This is quite different from the result we obtained in the synchronous gauge in which $\delta \propto (\rho a^2)^{-1}$ at large scales, illustrating the fact that the actual results obtained for perturbed density, etc., depend on the gauge and hence care should be exercised in interpreting them.

No such difficulty arises at small scales. In this case, the time-derivative terms can be dropped in Eq. (5.44), leading to the simple Poisson equation

$\nabla^2 \phi = 4\pi G a^2 \rho \delta$. In Eq. (5.41), the third term is dominated by the spatial derivative of velocity rather than the time derivative of the potential and we get

$$\dot{\delta} + (1 + w)\nabla \cdot \mathbf{v} + 3(c_s^2 - w)\frac{\dot{a}}{a}\delta = 0. \tag{5.58}$$

This equation, along with Eq. (5.43), determines the evolution now. In the
case of a single-component universe, we have $w = c_s^2$; taking the divergence of
Eq. (5.43) and using Eq. (5.58), we get the equation satisfied by δ as

$$\ddot{\delta} - (3w - 1)\frac{\dot{a}}{a}\dot{\delta} - w\nabla^2\delta = 4\pi G\rho(1 + w)a^2\delta. \tag{5.59}$$

Converting back to the cosmic time $dt = a\,d\tau$, we find that this will become

$$\ddot{\delta} + (2 - 3w)\frac{\dot{a}}{a}\dot{\delta} - \frac{w}{a^2}\nabla^2\delta = 4\pi G\rho(1 + w)\delta. \tag{5.60}$$

This agrees with the corresponding result in synchronous gauge for nonrelativistic matter with $w = 0$.

The situation becomes more complicated in the case of a multicomponent fluid,
for which w and c_s^2 will, in general, depend on time. We shall now comment on
some of the technical issues that arise in the presence of photons, baryons, and
dark matter.

All these components contribute energy densities to the background universe
and drive the expansion. For $z > z_{eq}$ the dominant contribution comes from radiation whereas for $z < z_{eq}$ it comes from dark matter, with baryons being subdominant throughout. When the smooth energy density in any of these components are
perturbed, a density contrast $\delta_A \equiv (\delta\rho_A/\rho_A)$ arises in that component, where the
subscripts $A = R$, DM, and B will indicate radiation, dark matter, and baryons,
respectively. The equation for the evolution of δ_A for each A will contain terms
that drive the instability and the terms that suppresses the growth of inhomogeneity. In all the cases, the *perturbed* part of the gravitational potential drives
the inhomogeneity whereas the expansion of the universe works against the development of inhomogeneity. The expansion is contributed by the total, smooth
energy density and the driving term is contributed by the total perturbation in
the energy density. For example, in a universe dominated by dark matter and
radiation, the perturbed energy density is given by

$$\delta \equiv \frac{\delta\rho_{total}}{\rho_{total}} = \frac{\rho_R\delta_R + \rho_{DM}\delta_{DM}}{\rho_R + \rho_{DM}} = \frac{\delta_R + y\delta_{DM}}{1 + y}, \tag{5.61}$$

where

$$y = \frac{\rho_{DM}}{\rho_R} = \frac{a}{a_{eq}}. \tag{5.62}$$

It is this total perturbation δ that leads to a perturbed gravitational potential that
drives the instability. To describe the two-component (R and DM) universe, we

need one more variable. The second convenient variable is the perturbation in the entropy per particle $s \propto (T_R^3 / \rho_{DM})$. Taking a logarithmic derivative, we find that

$$\sigma \equiv \left(\frac{\delta s}{s} \right) = \frac{3 \delta T_R}{T_R} - \frac{\delta \rho_{DM}}{\rho_{DM}} = \frac{3}{4} \delta_R - \delta_{DM}. \tag{5.63}$$

In arriving at the last equality, we have used the fact that $\rho_R \propto T_R^4$, giving $\delta_R = 4 \delta T_R / T_R$. In general, gravity couples the growth rate of δ_R and δ_{DM}, and hence the equations for the growth of either one component will contain the other term. It is, however, more convenient to write all the equations in terms of σ and δ instead of δ_R and δ_{DM}. At very large scales ($\lambda \gg ct$) microscopic physics is unimportant and gravity does not distinguish between radiation and matter. If a perturbation is set up with $\sigma = 0$ and $\delta \neq 0$, this condition will be maintained at sufficiently large scales. Such perturbations are usually called *isoentropic*, as they correspond to the entropy per particle being kept constant. On the other hand, it is also possible to consider perturbations for which $\delta = 0$ but $\sigma \neq 0$ initially. When δ vanishes, so does the perturbed gravitational potential and the additional space–time curvature induced by it. Such perturbations are hence called *isocurvature perturbations*. The preceding discussion shows that it is convenient to characterise the perturbation by decomposing them into an isoentropic part and an isocurvature part, with the general perturbation being a linear combination of both. Evolution can, of course, generate one of these components from the other, but this coupling is easier to handle with the variables σ and δ than with δ_R and δ_{DM}.

In the case of a universe with *both* dark matter and radiation, it is better to recast our original equations in a simpler and more meaningful form without making any assumptions regarding the form of $a(\tau)$ or the kind of matter that is perturbed. We also set $h_{\alpha\beta} = 0$ because we are not interested in the tensor modes. We first rewrite the time–time component of Einstein equation (5.44) as

$$4\pi G a^2 \delta \rho = \nabla^2 \phi - 3 \frac{\dot{a}^2}{a^2} \frac{d}{da}(a\phi). \tag{5.64}$$

We interpret this equation as giving the perturbation in the *total* density contrast if the effective Newtonian potential ϕ is known. The term involving the Laplacian is an obvious extension of Newtonian gravity, whereas the time-dependent term arises because of purely general relativistic effects. The space–time component of the Einstein equation can be rewritten as

$$4\pi G a^2 \nabla_\alpha [(\rho + p)v^\alpha] = -\left(\frac{\dot{a}}{a} \right) \frac{d}{da} \nabla^2 (a\phi), \tag{5.65}$$

which can be integrated for an irrotational velocity field, giving

$$4\pi G a^2 (\rho + p)v^\alpha = -\left(\frac{\dot{a}}{a} \right) \frac{d}{da} \nabla^\alpha (a\phi), \tag{5.66}$$

which is equivalent to Eq. (5.45). This equation gives the perturbed velocity field once the perturbed potential ϕ is known. Equation (5.65) shows that the momentum density in the universe can also be a source of curvature; however, this is typically smaller than the effects produced by $\delta\rho$ by a factor of v/c. Hence the correction to the Laplacian in Eq. (5.64) is significant only at scales $\lambda \gtrsim d_H$.

Equations (5.64) and (5.66) give the perturbations in the energy density and velocity in terms of the perturbations in the gravitational potential. To form a closed system, we need the equation that gives the evolution of ϕ. This is given by Eq. (5.46), which we can simplify by eliminating the $\nabla^2\phi$ term by using Eq. (5.44). This leads to the result

$$\ddot{\phi} + 3\frac{\dot{a}}{a}\dot{\phi} + \left[2\frac{d}{d\tau}\left(\frac{\dot{a}}{a}\right) + \left(\frac{\dot{a}}{a}\right)^2\right]\phi = 4\pi Ga^2\delta p, \qquad (5.67)$$

where $\delta p = p - \bar{p}$ is the perturbation in the pressure. This equation, however, is not very useful as δp is not a simple quantity to handle. It is, however, possible to derive a simpler equation by using the definition of entropy perturbation.

To do this, we first define an analogue of sound speed by the relation $c_s^2 = (\dot{p}/\dot{\rho})$, where $\rho = (\rho_R + \rho_M + \rho_V)$ and $p = (1/3)\rho_R - \rho_V$. (We use the subscript M to denote quantities pertaining to nonrelativistic matter made of baryons and dark matter.) We get

$$c_s^2 = \frac{(4/3)\rho_R}{4\rho_R + 3\rho_M} = \frac{1}{3}\left(1 + \frac{3}{4}\frac{\rho_M}{\rho_R}\right)^{-1} = \frac{1}{3}\left(1 + \frac{3}{4}y\right)^{-1}, \qquad (5.68)$$

where

$$y \equiv \frac{\rho_M}{\rho_R} = \frac{a}{a_{eq}} \approx x^2 + 2x, \qquad x \equiv \left(\frac{\Omega_M}{4a_{eq}}\right)^{1/2} H_0\tau. \qquad (5.69)$$

The expression for y in terms of τ is obtained from the solution to the Friedmann equation that is valid at high redshifts in which vacuum energy density can be ignored and $\Omega \approx 1$ (see Exercise 3.9.). With this definition, it is easy to relate δp and $\delta\rho$ to the entropy perturbation σ. We have

$$\delta p - c_s^2\delta\rho = \frac{1}{3}\delta\rho_R - c_s^2(\delta\rho_R + \delta\rho_M) = \rho_M c_s^2\sigma, \qquad \sigma \equiv \frac{3}{4}\delta_R - \delta_M. \qquad (5.70)$$

This equation is exact for linear perturbations and does not assume that the two components are tightly coupled. Because we expect σ to be a slowly varying quantity at very large scales and to vanish for isentropic perturbations, expression (5.70) suggests that we rewrite our equations by using $(\delta p - c_s^2\delta\rho)$ as the source. We do this easily by combining Eqs. (5.64) and (5.67). We get

$$\ddot{\phi} + 3\left(1 + c_s^2\right)\frac{\dot{a}}{a}\dot{\phi} + \left[2\frac{d}{d\tau}\left(\frac{\dot{a}}{a}\right) + \left(1 + 3c_s^2\right)\left(\frac{\dot{a}}{a}\right)^2\right]\phi - c_s^2\nabla^2\phi$$

$$= 4\pi Ga^2\rho_M c_s^2\sigma. \qquad (5.71)$$

This equation shows how the curvature perturbations characterised by ϕ can be generated from entropy perturbations characterised by σ. [The term in the square brackets can be simplified by Friedmann equations to give the third term on the left-hand side of Eq. (5.57).] To complete the analysis, we also need an equation for the evolution of σ. However, this is unnecessary for the study of modes bigger than the Hubble radius, as we shall see.

At scales bigger than the Hubble radius, we can use two simplifying assumptions. First, the spatial-derivative term $c_s^2 \nabla^2 \phi$ can be ignored because the pressure gradients are not important at large scales. Second, we can take the entropy perturbation to be frozen in time and treat σ as a constant; $\sigma = \sigma_i$. Changing the variables from τ to y, the differential equation for ϕ now becomes, after some algebra,

$$y\phi'' + \frac{y\phi'}{2(1+y)} + 3\left(1+c_s^2\right)\phi' + \frac{3c_s^2\phi}{4(1+y)} = \frac{3c_s^2\sigma}{2(1+y)}, \qquad (5.72)$$

where the prime denotes a derivative with respect to y and c_s^2 is given by Eq. (5.68). This equation is easy to solve in principle. The two homogeneous solutions, corresponding to $\sigma = 0$, are given by

$$f_1(y) = 1 + \frac{2}{9y} - \frac{8}{9y^2} - \frac{16}{9y^3}, \qquad f_2(y) = \frac{\sqrt{1+y}}{y^3}. \qquad (5.73)$$

Any particular solution can be easily obtained by a suitable linear superposition. In this limit, ignoring the Laplacian, we can determine the density contrast from Eq. (5.64) to be

$$\delta \equiv \frac{\delta\rho}{\rho} = -2\frac{d(y\phi)}{dy}. \qquad (5.74)$$

We can obtain the perturbations in matter and radiation individually by solving the two equations

$$\sigma = \frac{3}{4}\delta_R - \delta_M, \qquad \delta = \frac{\delta_R + y\delta_M}{1+y}. \qquad (5.75)$$

For future reference, we note that the perturbations in radiation and the velocity field are given by the expressions

$$\delta_R = \frac{-2(1+y)\,d(y\phi)/dy + y\sigma}{1+(3/4)y}, \qquad (5.76)$$

$$v_\alpha = -\frac{3c_s^2}{2(\dot{a}/a)}(1+y)\nabla_\alpha\frac{d(y\phi)}{dy}. \qquad (5.77)$$

The fundamental solutions f_1 and f_2 in Eqs. (5.73) are singular at $y = 0$. For a physically meaningful initial condition with a finite perturbation in the potential at the initial epoch, it is necessary to take a suitable linear combination of f_1 and

f_2. It is easy to see that the combination

$$f_3(y) \equiv \frac{9}{16} f_1 + f_2 \tag{5.78}$$

has a finite limit as $y \to 0$ with $f_3 \to (5/8) + \mathcal{O}(y)$ as $y \to 0$. Normalising this function to some initial potential ϕ_i, we find that the solution to isoentropic perturbations ($\sigma = 0$) with an initial gravitational potential ϕ_i is given by

$$\phi(\tau, \mathbf{x}) = \left[\frac{9}{10} f_1 + \frac{8}{5} f_2 \right] \phi_i(\mathbf{x}). \tag{5.79}$$

As the universe evolves, y increases but ϕ changes very little. For $y \gg 1$, ϕ tends to the asymptotic value $\phi \approx (9/10)\phi_i$. If we had retained a finite initial entropy perturbation, then the asymptotic limit would change to

$$\phi(\tau, \mathbf{x}) \approx \frac{9}{10} \phi_i(\mathbf{x}) + \frac{1}{5} \sigma_i(\mathbf{x}) \quad \text{(for } y \gg 1). \tag{5.80}$$

A comparison of the relative terms shows that entropy perturbation is fairly inefficient in producing a corresponding density perturbations.

In this gauge, the potential ϕ is effectively frozen at scales bigger than the Hubble radius and, to the extent that Eq. (5.74) is applicable, so is δ. The density contrast δ changes under the gauge transformation, thereby modifying the relation between δ and ϕ. This fact shows that one has to be careful in dealing with the nature of growth of density contrast at scales bigger than the Hubble radius.

5.5 Gravitational Clustering in the Newtonian Limit

Once the mode enters the Hubble radius, dark-matter perturbations can be treated by Newtonian theory of gravitational clustering. Although $\delta_\lambda \ll 1$ at $z \lesssim z_{\text{enter}}$, we shall develop the full formalism of Newtonian gravity at one go rather than do the linear perturbation theory separately.

The evolution of dark-matter perturbations (in Newtonian limit) can be analysed at three different levels. (1) To begin with, we can study the evolution of individual trajectories $\mathbf{x}_i(t)$ of each dark-matter particle in an expanding universe. This provides the most detailed level of description and is often used in simulations. (2) At the next level, we can try to describe the dark matter by a distribution function $f(\mathbf{x}, \mathbf{p}, t)$ such that a small element in phase volume $d^3\mathbf{x} d^3\mathbf{p}$ contains $dN = f d^3\mathbf{x} d^3\mathbf{p}$ particles. At this level we ignore the individual trajectories but content ourselves with an average description. For dark matter, the gravitational time scale for collisions is enormous and the description in terms of the CBE is quite adequate. (3) At the third level, we treat the system as made of pressureless dust and describe the dark matter by using a smoothed density $\rho(t, \mathbf{x})$ and mean velocity $\mathbf{v}(t, \mathbf{x})$. The key difference between (2) and (3) is the following: The distribution function f allows for particles with different velocities to exist at some given location $\mathbf{x}$; in other words, there exists a velocity dispersion at each location $\mathbf{x}$. In the pressureless dust limit, we ignore the velocity dispersion and assume that at any given point there is a single velocity of flow $\mathbf{v}(t, \mathbf{x})$. We now study each of these descriptions, starting with (1).

In any region that is small compared with d_H, we can set up an unambiguous coordinate system in which the *proper* coordinate of a particle $\mathbf{r}(t) = a(t)\mathbf{x}(t)$ satisfies the Newtonian equation $\ddot{\mathbf{r}} = -\nabla_{\mathbf{r}}\Phi$, where Φ is the gravitational potential. Expanding $\ddot{\mathbf{r}}$ and writing $\Phi = \Phi_{\mathrm{FRW}} + \phi$, where Φ_{FRW} is due to the smooth background density and ϕ is due to the perturbations, we get

$$\ddot{a}\mathbf{x} + 2\dot{a}\dot{\mathbf{x}} + a\ddot{\mathbf{x}} = -\nabla_{\mathbf{r}}\Phi_{\mathrm{FRW}} - \nabla_{\mathbf{r}}\phi = -\nabla_{\mathbf{r}}\Phi_{\mathrm{FRW}} - a^{-1}\nabla_{\mathbf{x}}\phi. \qquad (5.81)$$

The first terms on both sides of the equation ($\ddot{a}\mathbf{x}$ and $-\nabla_{\mathbf{r}}\Phi_{\mathrm{FRW}}$) should match as they refer to the global expansion of the background FRW universe. Equating them individually gives the results

$$\ddot{\mathbf{x}} + 2\frac{\dot{a}}{a}\dot{\mathbf{x}} = -\frac{1}{a^2}\nabla_x\phi, \qquad \Phi_{\mathrm{FRW}} = -\frac{1}{2}\frac{\ddot{a}}{a}r^2 = -\frac{2\pi G}{3}(\rho_b + 3p_b)r^2, \quad (5.82)$$

where ϕ is generated by the perturbed Newtonian mass density through

$$\nabla_x^2\phi = 4\pi Ga^2(\delta\rho) = 4\pi G\rho_b a^2\delta. \qquad (5.83)$$

If $\mathbf{x}_i(t)$ is the trajectory of the ith particle, then equations for Newtonian gravitational clustering can be summarised as

$$\dot{\mathbf{x}}_i + \frac{2\dot{a}}{a}\dot{\mathbf{x}}_i = -\frac{1}{a^2}\nabla_x\phi, \qquad \nabla_x^2\phi = 4\pi Ga^2\rho_b\delta, \qquad (5.84)$$

where ρ_b is the smooth background density of matter. It should be stressed that, in the nonrelativistic limit, the perturbed potential ϕ satisfies the usual Poisson equation.

Exercise 5.2

Trajectories in a perturbed metric: Consider the trajectories of particles in a slightly perturbed Friedmann universe with the metric

$$ds^2 = a^2(\tau)[(1 + 2\phi)d\tau^2 - (1 - 2\phi)(dx^2 + dy^2 + dz^2)], \qquad (5.85)$$

where ϕ is the perturbed gravitational potential arising from the perturbed energy density. (a) Show that, in the Newtonian limit, this metric leads to an effective Newtonian potential $\phi_N = -(1/2)(\ddot{a}/a)R^2 + \phi$. (b) Derive the equations of motion for a particle in this metric correct to linear order in ϕ and show that

$$\dot{\mathbf{x}} = \frac{\mathbf{p}}{\epsilon}\left[1 + \phi + \phi\left(\frac{2 + p^2/a^2}{1 + p^2/a^2}\right)\right], \qquad (5.86)$$

$$\dot{\mathbf{p}} = -\epsilon\nabla\phi - \epsilon\left(\frac{p^2/a^2}{1 + p^2/a^2}\right)\nabla\phi, \qquad \epsilon = a\sqrt{1 + \frac{p^2}{a^2}}. \qquad (5.87)$$

(c) Obtain the extreme relativistic and Newtonian limit of the equation of motion and show that in the extreme relativistic limit we get

$$\dot{\mathbf{x}} \cong (1 + 2\phi)\frac{\mathbf{p}}{|\mathbf{p}|}, \qquad \dot{\mathbf{p}} \cong -2|\mathbf{p}|\nabla\phi, \tag{5.88}$$

whereas in the nonrelativistic limit we have

$$\dot{\mathbf{x}} = \frac{\mathbf{p}}{a}, \qquad \dot{\mathbf{p}} = -a\nabla\phi. \tag{5.89}$$

Because the density contrast can be expressed in terms of the trajectories of the particles, it should be possible to write a differential equation for $\delta(t, \mathbf{x})$ based on the equations for the trajectories $\mathbf{x}_i(t)$ previously derived. It is, however, somewhat easier to write an equation for $\delta_{\mathbf{k}}(t)$ that is the spatial Fourier transform of $\delta(t, \mathbf{x})$. To do this, we begin with the fact that the density $\rho(\mathbf{x}, t)$ that is due to a set of point particles, each of mass m, is given by

$$\rho(\mathbf{x}, t) = \frac{m}{a^3(t)} \sum_i \delta_D[\mathbf{x} - \mathbf{x}_i(t)], \tag{5.90}$$

where $\mathbf{x}_i(t)$ is the trajectory of the ith particle. To verify the a^{-3} normalisation, we can calculate the average of $\rho(\mathbf{x}, t)$ over a large volume V. We get

$$\rho_b(t) \equiv \int \frac{d^3\mathbf{x}}{V} \rho(\mathbf{x}, t) = \frac{m}{a^3(t)} \left(\frac{N}{V}\right) = \frac{M}{a^3 V} = \frac{\rho_0}{a^3}, \tag{5.91}$$

where N is the total number of particles inside the volume V and $M = Nm$ is the mass contributed by them. Clearly $\rho_b \propto a^{-3}$, as it should. The density contrast $\delta(\mathbf{x}, t)$ is related to $\rho(\mathbf{x}, t)$ by

$$1 + \delta(\mathbf{x}, t) \equiv \frac{\rho(\mathbf{x}, t)}{\rho_b} = \frac{V}{N} \sum_i \delta_D[\mathbf{x} - \mathbf{x}_i(t)] = \int d\mathbf{q}\,\delta_D[\mathbf{x} - \mathbf{x}_T(t, \mathbf{q})].$$

$$\tag{5.92}$$

In arriving at the last equality we have taken the continuum limit by replacing (1) $\mathbf{x}_i(t)$ with the trajectory $\mathbf{x}_T(t, \mathbf{q})$, where $\mathbf{q}$ stands for a set of quantities like the initial velocity, the initial position, etc., of a particle and uniquely labels it; and (2) (V/N) with $d^3\mathbf{q}$ as both represent volume per particle. Fourier transforming both sides, we get

$$\delta_{\mathbf{k}}(t) \equiv \int d^3\mathbf{x}\, e^{i\mathbf{k}\cdot\mathbf{x}}\delta(\mathbf{x}, t) = \int d\mathbf{q}\, \exp[-i\mathbf{k}\cdot\mathbf{x}_T(t, \mathbf{q})] - (2\pi)^3\delta_D(\mathbf{k}). \tag{5.93}$$

Differentiating this expression and using equation of motion (5.84) for the trajectories give, after straightforward algebra, the equation

$$\ddot{\delta}_{\mathbf{k}} + 2\frac{\dot{a}}{a}\dot{\delta}_{\mathbf{k}} = 4\pi G\rho_b\delta_{\mathbf{k}} + A_{\mathbf{k}} - B_{\mathbf{k}}, \tag{5.94}$$

with

$$A_{\mathbf{k}} = 4\pi G \rho_b \int \frac{d^3\mathbf{k}'}{(2\pi)^3} \delta_{\mathbf{k}'} \delta_{\mathbf{k}-\mathbf{k}'} \left[\frac{\mathbf{k} \cdot \mathbf{k}'}{k'^2} \right], \tag{5.95}$$

$$B_{\mathbf{k}} = \int d\mathbf{q} (\mathbf{k} \cdot \dot{\mathbf{x}}_T)^2 \exp[-i\mathbf{k} \cdot \mathbf{x}_T(t, \mathbf{q})]. \tag{5.96}$$

Equation (5.96) is exact but involves $\dot{\mathbf{x}}_T(t, \mathbf{q})$ on the right-hand side and hence cannot be considered as closed.

The structure of Eqs. (5.94) and (5.96) can be simplified somewhat if we use the perturbed gravitational potential (in Fourier space) $\phi_{\mathbf{k}}$ related to $\delta_{\mathbf{k}}$ by

$$\delta_{\mathbf{k}} = -\frac{k^2 \phi_{\mathbf{k}}}{4\pi G \rho_b a^2} = -\left(\frac{k^2 a}{4\pi G \rho_0} \right) \phi_{\mathbf{k}} = -\left(\frac{2}{3H_0^2} \right) k^2 a \phi_{\mathbf{k}} \tag{5.97}$$

and write the integrand for $A_{\mathbf{k}}$ in the symmetrised form as

$$\delta_{\mathbf{k}'} \delta_{\mathbf{k}-\mathbf{k}'} \left[\frac{\mathbf{k} \cdot \mathbf{k}'}{k'^2} \right] = \frac{1}{2} \delta_{\mathbf{k}'} \delta_{\mathbf{k}-\mathbf{k}'} \left[\frac{\mathbf{k} \cdot \mathbf{k}'}{k'^2} + \frac{\mathbf{k} \cdot (\mathbf{k} - \mathbf{k}')}{|\mathbf{k} - \mathbf{k}'|^2} \right]$$

$$= \frac{1}{2} \left(\frac{\delta_{\mathbf{k}'}'}{k'^2} \right) \left(\frac{\delta_{\mathbf{k}-\mathbf{k}'}}{|\mathbf{k} - \mathbf{k}'|^2} \right) [(\mathbf{k} - \mathbf{k}')^2 \mathbf{k} \cdot \mathbf{k}' + k'^2 (k^2 - \mathbf{k} \cdot \mathbf{k}')]$$

$$= \frac{1}{2} \left(\frac{2a}{3H_0^2} \right)^2 \phi_{\mathbf{k}'} \phi_{\mathbf{k}-\mathbf{k}'} [k^2 (\mathbf{k} \cdot \mathbf{k}' + k'^2) - 2(\mathbf{k} \cdot \mathbf{k}')^2]. \tag{5.98}$$

In terms of $\phi_{\mathbf{k}}$, Eq. (5.94) becomes, for an $\Omega = 1$ universe,

$$\ddot{\phi}_{\mathbf{k}} + 4\frac{\dot{a}}{a} \dot{\phi}_{\mathbf{k}} = -\frac{1}{2a^2} \int \frac{d^3\mathbf{k}'}{(2\pi)^3} \phi_{\mathbf{k}'} \phi_{\mathbf{k}-\mathbf{k}'} \left[\mathbf{k}' \cdot (\mathbf{k} + \mathbf{k}') - 2\left(\frac{\mathbf{k} \cdot \mathbf{k}'}{k} \right)^2 \right]$$

$$+ \left(\frac{3H_0^2}{2} \right) \int \frac{d\mathbf{q}}{a} \left(\frac{\mathbf{k} \cdot \dot{\mathbf{x}}}{k} \right)^2 e^{i\mathbf{k} \cdot \mathbf{x}}, \tag{5.99}$$

where $\mathbf{x} = \mathbf{x}_T(t, \mathbf{q})$. We shall see later in Section 5.12 how this helps us to understand power transfer in gravitational clustering.

5.6 Linear Perturbations in the Newtonian Limit

If the density contrasts are small and linear perturbation theory is to be valid, we should be able to ignore the terms $A_{\mathbf{k}}$ and $B_{\mathbf{k}}$ in Eq. (5.94). Then the linear perturbation theory in Newtonian limit is governed by the equation

$$\ddot{\delta}_{\mathbf{k}} + 2\frac{\dot{a}}{a} \dot{\delta}_{\mathbf{k}} = 4\pi G \rho_b \delta_{\mathbf{k}}. \tag{5.100}$$

From the structure of Eq. (5.94) it is clear that we will obtain the linear equation if $A_{\mathbf{k}} \ll 4\pi G \rho_b \delta_{\mathbf{k}}$ and $B_{\mathbf{k}} \ll 4\pi G \rho_b \delta_{\mathbf{k}}$. A *necessary* condition for this is $\delta_{\mathbf{k}} \ll 1$, but this is *not* a sufficient condition – a fact often ignored or incorrectly treated in literature. For example, if, at $t = t_0$, we have $\delta_{\mathbf{k}} \to 0$ for certain range of $\mathbf{k}$ but it is nonzero for some other range of $\mathbf{k}$, then $A_{\mathbf{k}}$ can be larger than $4\pi G \rho_b \delta_{\mathbf{k}}$ and the growth of perturbations around $\mathbf{k}$ will be entirely determined by nonlinear effects. [The approximation in Eq. (5.100) also ignores the velocity dispersion; see Exercise 5.7.] For the present, we shall assume that $A_{\mathbf{k}}$ and $B_{\mathbf{k}}$ are ignorable and study the resulting system of equations.

Exercise 5.3

Choice of time: Use a time coordinate $b(t)$ rather than t in Eq. (5.84) and choose $b(t)$ to be a solution of Eq. (5.100). Introduce a new "velocity" $\mathbf{w} \equiv (d\mathbf{x}/db)$ and a new potential

$$\psi = \frac{2}{3H_0^2 a_0^3} \left(\frac{a}{b}\right) \phi = \frac{1}{4\pi G \rho_{bm} a^2} \frac{\phi}{b}. \tag{5.101}$$

Show that Eq. (5.84) now becomes

$$\frac{d\mathbf{w}}{db} = -\frac{3A}{2b}(\mathbf{w} + \nabla\psi), \qquad \nabla^2 \psi = \left(\frac{\delta}{b}\right), \qquad A = \left(\frac{\rho_{bm}}{\rho_c}\right)\left(\frac{\dot{a}b}{a\dot{b}}\right)^2. \tag{5.102}$$

At $z \lesssim z_{\text{enter}}$, the perturbation can be treated as linear ($\delta \ll 1$) and Newtonian ($\lambda \ll d_H$). In this case, the equations to be solved are (5.100) and

$$\frac{\dot{a}^2}{a^2} + \frac{k}{a^2} = \frac{8\pi G}{3}(\rho_R + \rho_{\text{DM}} + \rho_V). \tag{5.103}$$

We will also assume that the dark matter is made of collisionless matter and is perturbed whereas the energy densities of radiation and the cosmological constant are left unperturbed, making the right-hand side of Eq. (5.100) $\rho_b \delta_k = \rho_{\text{DM}} \delta_k$. Changing the variable from t to a in Eq. (5.100) and using Eq. (5.103), we find that the perturbation equation becomes, after straightforward algebra,

$$2a^2 \left[\rho_R + \rho_{\text{DM}} + \rho_V - \frac{3k}{8\pi G a^2}\right]\frac{d^2\delta}{da^2}$$
$$+ a\left[2\rho_R + 3\rho_{\text{DM}} + 6\rho_V - 4\left(\frac{3k}{8\pi G a^2}\right)\right]\frac{d\delta}{da} = 3\rho_{\text{DM}}\delta. \tag{5.104}$$

By introducing the variable $q \equiv (a/a_0) = (1+z)^{-1}$ and by writing $\rho_i = \Omega_i \rho_c$ for the ith species and $k = -(8\pi G/3)\rho_c a_0^2(1 - \Omega)$, we can recast the equation in the form

$$2q\left[\Omega_V q^4 + (1 - \Omega)q^2 + \Omega_{\text{DM}}q + \Omega_R\right]\delta''$$
$$+ \left[6\Omega_V q^4 + 4(1 - \Omega)q^2 + 3\Omega_{\text{DM}}q + 2\Omega_R\right]\delta' = 3\Omega_{\text{DM}}\delta, \tag{5.105}$$

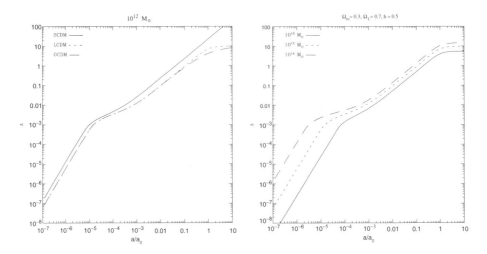

Fig. 5.1.   The growth of pertubations in different models. The left frame shows the evolution of perturbation containing a mass of $10^{12}\ M_\odot$ in three different cosmological models, marked SCDM ($\Omega_{\text{total}} = \Omega_{\text{NR}} = 1$), OCDM ($\Omega_{\text{total}} = \Omega_{\text{NR}} = 0.3$), and LCDM ($\Omega_V = 0.7$, $\Omega_{\text{NR}} = 0.3$). The right frame compares the evolution of three different mass scales in the LCDM model. A value of $\delta_{\text{enter}} \approx 10^{-3}$ has been taken for all the models.

where the prime denotes derivatives with respect to q. This equation is in a form convenient for numerical integration from $q = q_{\text{enter}} = (1 + z_{\text{enter}})^{-1}$ to $q = 1$.

The exact solution to Eq. (5.105) cannot be given in terms of elementary functions, but it is straightforward to integrate this equation numerically. The results of the numerical integration for different models are shown in Fig. 5.1. The left frame shows the evolution of perturbation containing a mass of $10^{12}\ M_\odot$ in three different cosmological models, marked SCDM ($\Omega_{\text{total}} = \Omega_{\text{NR}} = 1$), OCDM ($\Omega_{\text{total}} = \Omega_{\text{NR}} = 0.3$) and LCDM ($\Omega_V = 0.7$, $\Omega_{\text{NR}} = 0.3$). [The acronym CDM stands for cold dark matter; 'S' is for 'standard', 'O' is for 'open' and 'L' is 'lambda'.] The right frame compares the evolution of three different mass scales in the LCDM model. In all the models, the perturbation grows as a^2 before it enters the Hubble radius. We have also taken the value for $\delta_{\text{enter}} \approx 10^{-3}$ for all the models. During the epoch $a_{\text{enter}} < a < a_{\text{eq}}$, there is very little growth in the perturbation, as seen by the rather flat portion of the curves near the middle. For $a \gtrsim a_{\text{eq}}$, the perturbation grows as a. Such a growth continues unhindered in the SCDM model. In the OCDM and LCDM models, the universe becomes dominated by curvature or the cosmological constant, respectively, near $a \lesssim 1$. After this occurs, the growth of perturbation becomes stagnated at all scales. Note that the suppression of growth during $a_{\text{enter}} < a < a_{\text{eq}}$ has the effect of reducing the difference between δ at various scales by bringing the curves closer together in the right frame.

It is possible to obtain some insight into the form of the solution described above if the different epochs are considered separately. Let us first consider the epoch $1 \ll z \lesssim z_{enter}$ when we can take $\Omega_V = 0$ and $\Omega = 1$, reducing Eq. (5.105) to

$$2q(\Omega_{DM}q + \Omega_R)\delta'' + (3\Omega_{DM}q + 2\Omega_R)\delta' = 3\Omega_{DM}\delta. \qquad (5.106)$$

Dividing thoughout by Ω_R and changing the independent variable to

$$y \equiv q\left(\frac{\Omega_{DM}}{\Omega_R}\right) = \frac{a}{a_0(\Omega_R/\Omega_{DM})} = \frac{a}{a_{eq}}, \qquad (5.107)$$

we get

$$2y(1+y)\frac{d^2\delta_{DM}}{dy^2} + (2+3y)\frac{d\delta_{DM}}{dy} = 3\delta_{DM}, \quad y = \frac{a}{a_{eq}}. \qquad (5.108)$$

One solution to this equation can be written by inspection:

$$\delta_{DM} = 1 + \frac{3}{2}y. \qquad (5.109)$$

In other words $\delta_{DM} \approx$ constant for $a \ll a_{eq}$ (no growth in the radiation-dominated phase) and $\delta_{DM} \propto a$ for $a \gg a_{eq}$ (growth proportional to a in the matter-dominated phase).

We now have to find the second solution Δ, which is most easily obtained by the Wronskian condition $(Q'/Q) = -[(2+3y)/2y(1+y)]$, where $Q = \delta_{DM}\Delta' - \delta'_{DM}\Delta$. Writing the second solution as $\Delta = f(y)\delta_{DM}(y)$ and substituting into this equation, we find

$$\frac{f''}{f'} = -\frac{2\delta'_{DM}}{\delta_{DM}} - \frac{2+3y}{2y(1+y)}, \qquad (5.110)$$

which can be integrated to give

$$f = -\int \frac{dy}{y(1+3y/2)^2(1+y)^{1/2}}. \qquad (5.111)$$

The integral is straightforward, and the second solution is

$$\Delta = f\delta_{DM} = \left(1 + \frac{3y}{2}\right)\ln\left[\frac{(1+y)^{1/2}+1}{(1+y)^{1/2}-1}\right] - 3(1+y)^{1/2}. \qquad (5.112)$$

Thus the general solution to the perturbation equation for a mode that is inside the Hubble radius is the linear superposition $\delta = A\delta_{DM} + B\Delta$ with the asymptotic forms:

$$\delta_{gen}(y) = A\delta_{DM}(y) + B\Delta(y) = \begin{cases} A + B\ln(4/y) & (y \ll 1) \\ (3/2)Ay + (4/5)By^{(-3/2)} & (y \gg 1) \end{cases}. \qquad (5.113)$$

This result shows that dark-matter perturbations can grow only logarithmically during the epoch $a_{enter} < a < a_{eq}$. During this phase the universe is dominated by radiation that is unperturbed. Hence the damping term that is due to expansion $(2\dot{a}/a)\dot{\delta}$ in Eq. (5.100) dominates over the gravitational potential term on the right-hand side and restricts the growth of perturbations. In the matter-dominated phase with $a \gg a_{\rm eq}$, the perturbations grow as a. This result, combined with that of Section 5.2, shows that, in the matter-dominated phase, *all the modes* (i.e., modes that are inside or outside the Hubble radius) grow in proportion to the expansion factor.

Combining the preceding result with that of Section 5.3, we can determine the evolution of density perturbations in dark matter during all relevant epochs. The general solution after the mode has entered the Hubble radius is given by Eq. (5.113). The constants A and B in this solution have to be fixed by matching this solution to the growing solution, which was valid when the mode was bigger than the Hubble radius. Because the latter solution is given by $\delta(y) = y^2$ in the radiation-dominated phase, the matching conditions become

$$
\begin{aligned}
y_{\rm enter}^2 &= [A\delta_{\rm DM}(y) + B\Delta(y)]_{y=y_{\rm enter}}, \\
2y_{\rm enter} &= [A\delta_{\rm DM}'(y) + B\Delta'(y)]_{y=y_{\rm enter}}.
\end{aligned}
\tag{5.114}
$$

This determines the constants A and B in terms of $y_{\rm enter} = (a_{\rm enter}/a_{\rm eq})$, which, in turn, depends on the wavelength of the mode through $a_{\rm enter}$.

As an example, we consider a mode for which $y_{\rm enter} \ll 1$. The second solution has the asymptotic form $\Delta(y) \simeq \ln(4/y)$ for $y \ll 1$. Using this and matching the solution at $y = y_{\rm enter}$ we get the properly matched mode, inside the Hubble radius, as

$$
\delta(y) = y_{\rm enter}^2 \left[1 + 2\ln\left(\frac{4}{y_{\rm enter}}\right) \right] \left(1 + \frac{3y}{2} \right) - 2y_{\rm enter}^2 \ln\left(\frac{4}{y}\right).
\tag{5.115}
$$

During the radiation-dominated phase – that is, until $a \lesssim a_{\rm eq}$, $y \lesssim 1$ – this mode can grow by a factor

$$
\begin{aligned}
\frac{\delta(y \simeq 1)}{\delta(y_{\rm enter})} &= \frac{1}{y_{\rm enter}^2}\delta(y \simeq 1) \cong 5\ln\left(\frac{1}{y_{\rm enter}}\right) \\
&= 5\ln\left(\frac{a_{\rm eq}}{a_{\rm enter}}\right) = \frac{5}{2}\ln\left(\frac{t_{\rm eq}}{t_{\rm enter}}\right).
\end{aligned}
\tag{5.116}
$$

Because the time $t_{\rm enter}$ for a mode with wavelength λ is fixed by the condition $\lambda a_{\rm enter} \propto \lambda t_{\rm enter}^{1/2} \simeq d_H(t_{\rm enter}) \propto t_{\rm enter}$, it follows that $\lambda \propto t_{\rm enter}^{1/2}$. Hence,

$$
\frac{\delta_{\rm final}}{\delta_{\rm enter}} \cong 5\ln\left(\frac{\lambda_{\rm eq}}{\lambda}\right) \cong \frac{5}{3}\ln\left(\frac{M_{\rm eq}}{M}\right)
\tag{5.117}
$$

for a mode with wavelength $\lambda \ll \lambda_{\rm eq}$. [Here, M is the mass contained in a sphere of radius $(\lambda/2)$; see Eq. (5.9).] The growth in the radiation-dominated phase therefore is logarithmic. Notice that the matching procedure has brought in an amplification factor *that depends on the wavelength*.

In the preceding discussion, we have assumed that $\Omega = 1$, which is a valid assumption in the early phases of the universe. However, during the later stages of evolution in a matter-dominated phase, we have to take into account the actual value of Ω and solve Eq. (5.100). This can be done along the following lines.

Let $\rho(t)$ be a solution to the background Friedmann model dominated by pressureless dust. Consider now the function $\rho_1(t) \equiv \rho(t + \epsilon)$, where ϵ is some constant. Because the Friedmann equations contain t only through the derivative, $\rho_1(t)$ is also a valid solution. If we now take ϵ to be small, then $[\rho_1(t) - \rho(t)]$ will be a small perturbation to the density. The corresponding density contrast is

$$\delta(t) = \frac{\rho_1(t) - \rho(t)}{\rho(t)} = \frac{\rho(t + \epsilon) - \rho(t)}{\rho(t)} \cong \epsilon \frac{d \ln \rho}{dt} = -3\epsilon H(t), \qquad (5.118)$$

where the last relation follows from the fact that $\rho \propto a^{-3}$ and $H(t) \equiv (\dot{a}/a)$. Because ϵ is a constant, it follows that $H(t)$ is a solution to be the perturbation equation.

This curious fact, of course, can be verified directly: From the equations describing the Friedmann model, it follows that $\dot{H} + H^2 = (-4\pi G\rho/3)$. Differentiating this relation and using $\dot{\rho} = -3H\rho$, we immediately get $\ddot{H} + 2H\dot{H} - 4\pi G\rho H = 0$. Thus H satisfies the same equation as δ. Because the cosmological constant contributes only a constant ρ and p to Friedmann equation, the derivation goes through even for a universe with both Ω_{NR} and Ω_V nonzero.

Because $\dot{H} = -H^2 - (4\pi G\rho/3)$, we know that $\dot{H} < 0$; that is, H is a decreasing function of time and the solution $\delta = H \equiv \delta_d$ is a decaying mode. We can find the growing solution ($\delta \equiv \delta_g$) again by using the fact that, for any two linearly independent solutions of Eq. (5.100), the Wronskian ($\dot{\delta}_g\delta_d - \dot{\delta}_d\delta_g$) has a value of a^{-2}. This implies that

$$\delta_g = \delta_d \int \frac{dt}{a^2\delta_d^2} = H(t) \int \frac{dt}{a^2 H^2(t)} = H(a) \int \frac{da}{(Ha)^3}. \qquad (5.119)$$

Thus we see that the $H(t)$ of the background space–time allows us to completely determine the evolution of density contrast. It is convenient for numerical work to rewrite the growing mode in the form

$$\delta_g(a) = \frac{5}{2}\Omega_{NR} H(a) \int_0^a \frac{dx}{x^3 H^3(x)}, \qquad (5.120)$$

with

$$H(x) = [\Omega_{NR}x^{-3} + \Omega_V + (1 - \Omega_{NR} - \Omega_V)x^{-2}]^{1/2}. \qquad (5.121)$$

We have taken advantage of overall multiplicative freedom in δ_g to normalise $a = 1$ and $H = 1$ at the present epoch. For small values of a, the integral gives

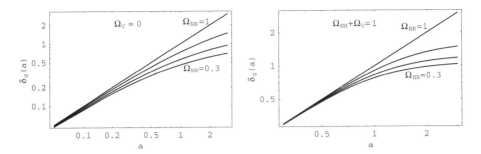

Fig. 5.2. The growth of linear density contrast as a function of the expansion factor $a(t)$ for two classes of cosmological models. The left panel is for open models with $\Omega_{NR} = 0.3, 0.4, 0.6,$ and 1 from bottom to top; the right panel is for models with $\Omega_{NR} + \Omega_V = 1$ for the same values of Ω_{NR} from bottom to top. The x axis is scaled to $a = 1$ at the present epoch and the y-axis scale is arbitrary.

$\delta_g(a) \approx a$. Figure 5.2 shows this functional form of $\delta_g(a)$ for open models (left panel) and models with $\Omega_{NR} + \Omega_V = 1$ (right panel). It is obvious that the growth is suppressed with respect to the $\Omega_{NR} = 1$ models in both the cases with the effect increasing with decreasing Ω_{NR}. Between the two cases, there is less growth in open models compared with models with a cosmological constant.

It is more convenient to express this result in terms of the redshift z. For a universe with $\Omega_V = 0$ and $\Omega_{NR} = \Omega_{tot}$, we have the relations

$$a(z) = a_0(1 + z)^{-1}, \qquad H(z) = H_0(1 + z)(1 + \Omega_{NR}z)^{1/2}, \qquad (5.122)$$

$$H_0 \, dt = -(1 + z)^{-2}(1 + \Omega_{NR}z)^{-\frac{1}{2}} \, dz. \qquad (5.123)$$

Taking $\delta_d = H(z)$, we get

$$\delta_g = \delta_d(z) \int a^{-2}\delta_d^{-2}(z) \left(\frac{dt}{dz}\right) dz$$

$$= (a_0 H_0)^{-2}(1 + z)(1 + \Omega_{NR}z)^{1/2} \int_z^\infty dx(1 + x)^{-2}(1 + \Omega_{NR}x)^{-\frac{3}{2}}. \quad (5.124)$$

This integral can be expressed in terms of elementary functions:

$$\delta_g = \frac{1 + 2\Omega_{NR} + 3\Omega_{NR}z}{(1 - \Omega_{NR})^2}$$

$$- \frac{3}{2} \frac{\Omega_{NR}(1 + z)(1 + \Omega_{NR}z)^{1/2}}{(1 - \Omega_{NR})^{5/2}} \ln\left[\frac{(1 + \Omega_{NR}z)^{1/2} + (1 - \Omega_{NR})^{1/2}}{(1 + \Omega_{NR}z)^{1/2} - (1 - \Omega_{NR})^{1/2}}\right].$$

$$(5.125)$$

Thus $\delta_g(z)$ for an arbitrary Ω_{NR} can be given in closed form. The solution in Eq. (5.125) is not normalised in any manner; normalisation can be achieved by multiplying δ_g by some constant, depending on the context.

For large z (i.e., early times), $\delta_g \propto z^{-1}$. This is to be expected because, for large z, the curvature term can be ignored and the Friedmann universe can be approximated as an $\Omega_{NR} = 1$ model. [The large z expansion of the logarithm in Eq. (5.125) has to be taken up to $O(z^{-5/2})$ to get the correct result; it is easier to obtain the asymptotic form directly from the integral in Eq. (5.124)]. For $\Omega_{NR} \ll 1$, we can see that $\delta_g \simeq$ constant for $z \ll \Omega_{NR}^{-1}$. This is the curvature-dominated phase, in which the growth of perturbations is halted by rapid expansion.

Exercise 5.4

An alternative derivation: A more formal way of deriving Eq. (5.119) is by a method known as the variation of parameters. Let $a(C, k, t)$ be a solution to the Friedmann equations in which C is the origin for time and k is the curvature parameter. Argue that, in a universe with pressureless fluid and cosmological constant, the time evolution of the density contrast can be expressed in the form

$$\delta = -3\frac{1}{a}\frac{\partial a}{\partial \alpha}\delta\alpha, \tag{5.126}$$

where α is any parameter characterising the solution $a(\alpha, t)$. Hence show that the growing and decaying solutions for the linear perturbation equation can be expressed in the form

$$\delta_1(t) = -\frac{3X^{1/2}}{2a}\int^a \frac{da}{X^{3/2}}, \qquad \delta_2(t) = 3\frac{X^{1/2}}{2a}, \tag{5.127}$$

where

$$X = \dot{a}^2 = \frac{8}{3}\pi G\rho_b a^2 + k + \frac{1}{3}\Lambda a^2. \tag{5.128}$$

Demonstrate that this approach leads to the same result as that given in the text. Explain why this method works in the presence of the cosmological constant but not in the presence of, say, radiation.

Exercise 5.5

Growth in the presence of a smooth background: Consider a universe containing two kinds of dark-matter particles (and nothing else) conventionally called hot dark matter (HDM) and cold dark matter (CDM) such that $\Omega_{CDM} + \Omega_{HDM} = 1$. The energy density of HDM is unperturbed and that of CDM is perturbed. Show that the growth law now becomes $\delta_{CDM} \propto t^\alpha$, where $6\alpha + 1 = (25 - 24\Omega_{HDM})^{1/2}$. [Hint: We need to solve the equation $\ddot{\delta} + 2(\dot{a}/a)\dot{\delta} = 4\pi G\rho\delta$. On the right-hand side, we need to keep only the perturbed energy density, and hence the right-hand side is $2(1 - \Omega_{HDM})/3t^2$. Using $a \propto t^{2/3}$ and looking for power-law solutions, we get the quadratic equation $\alpha(\alpha - 1) + 4\alpha/3 = 2(1 - \Omega_{HDM})/3$, which leads to the result cited in the question.]

The situation is more complicated if $\Omega_V \neq 0$. Consider, for example, the case with $\Omega_V + \Omega_{NR} = 1$. To find the growing mode, we now have to evaluate the integral in Eq. (5.120) with $H(x) = [\Omega_{NR}x^{-3} + \Omega_V]^{1/2}$. Unfortunately, the integral can be expressed in terms of only elliptic functions; however, an accurate fit to the solution can be provided in terms of simple functional form as follows:

$$\delta(z, \Omega_V) = \delta(z, \Omega_V = 0) \left[\frac{g(\Omega)}{\Omega} \right], \tag{5.129}$$

where

$$g(\Omega) = \frac{5}{2}\Omega \left(\frac{1}{70} + \frac{209\Omega}{140} - \frac{\Omega^2}{140} + \Omega^{4/7} \right)^{-1},$$

$$\Omega(z) = \Omega_{NR} \frac{(1+z)^3}{1 - \Omega_{NR} + (1+z)^3 \Omega_{NR}}. \tag{5.130}$$

Comparing the fitting function in Eq. (5.130) with actual numerical integration for different values of Ω_{NR} shows that the fit is moderately accurate; errors can be $\sim 30\%$. Incidentally, even in the case of an open FRW model (with $\Omega_{NR} < 1$, $\Omega_V = 0$) when it is possible to integrate the perturbation equation [see Eq. (5.125)] it is often more convenient to use an approximation like the one given above. In this case, $g(\Omega)$ is given by

$$g(\Omega) = \frac{5}{2}\Omega \left(1 + \frac{\Omega}{2} + \Omega^{4/7} \right)^{-1}. \tag{5.131}$$

We next consider the evolution of perturbations in the gravitational potential and peculiar velocity. Because the solution for δ we have obtained holds for all $\mathbf{k}$, the $\delta(\mathbf{x}, t)$ in real space also grows as some function $b(t)$, preserving the spatial form; hence we can write $\delta(\mathbf{x}, t) = b(t)q(\mathbf{x})$. Because Eq. (5.102) gives $\nabla^2\psi = (\delta/b)$, it follows that $\psi(\mathbf{x}, t) = \psi(\mathbf{x})$ is a constant in time.

As regards the velocity perturbations, we note that the first equation in (5.102) becomes

$$\frac{\partial w^\alpha}{\partial b} \cong -\frac{3A}{2b} [\partial^\alpha \psi + w^\alpha] \tag{5.132}$$

in linear order. Because $\partial^\alpha \psi$ is independent of t (or b), this equation has the solution

$$\mathbf{w}(\mathbf{x}) = -\nabla\psi(\mathbf{x}), \tag{5.133}$$

for which both sides vanish identically. Thus, in the linear approximation, the scaled gravitational potential $\psi(t, \mathbf{x}) = \psi(\mathbf{x})$ completely specifies the solution:

$$\delta(t, \mathbf{x}) = b(t)\nabla^2\psi(\mathbf{x}), \qquad \mathbf{w}(t, \mathbf{x}) = \mathbf{w}(\mathbf{x}) = -\nabla\psi(\mathbf{x}). \tag{5.134}$$

The true gravitational potential is $\phi = (4\pi G\rho_b a^2 b)\psi$ and the true peculiar velocity is $\mathbf{v} = a\dot{b}\mathbf{w}$. In terms of ϕ and $\mathbf{v}$, Eq. (5.133) becomes

$$\frac{\mathbf{v}}{a\dot{b}} = -\frac{1}{(4\pi G\rho_b a^2 b)}\nabla\phi \qquad (5.135)$$

or

$$\mathbf{v} = -\left(\frac{\dot{b}}{4\pi G\rho_b ba}\right)\nabla\phi = \left(\frac{\dot{b}}{4\pi G\rho_b b}\right)\mathbf{g} = \frac{2}{3H\Omega}\left[\frac{\dot{b}a}{b\dot{a}}\right]\mathbf{g}, \qquad (5.136)$$

where $\mathbf{g} = -a^{-1}\nabla\phi$, $H = (\dot{a}/a)$ and $\Omega = [\rho_b(t)/\rho_c(t)]$. Writing $f \equiv (\dot{b}a/b\dot{a})$, we get

$$\mathbf{v} = \frac{2f}{3H_b\Omega}\mathbf{g} = -\frac{2f}{3H_b\Omega}\frac{1}{a}\nabla\phi, \qquad (5.137)$$

where ϕ is the Newtonian potential generated by the excess density $\rho_b\delta$. Thus the peculiar velocity $\mathbf{v}$ is proportional to the peculiar acceleration $\mathbf{g}$ with the coefficient of proportionality $(2f/3H\Omega)$, where

$$f(a) = \frac{a}{\delta}\frac{d\delta}{da} = -\frac{(1+z)}{\delta}\frac{d\delta}{dz}. \qquad (5.138)$$

The peculiar velocities and accelerations observed in our universe at the present epoch are related by the value of f at $z = 0$, which – in turn – will depend only on Ω. Although $f(\Omega, z = 0)$ can be calculated exactly from Eq. (5.119), its functional form is not convenient for further manipulations. It turns out, however, that $f(\Omega, z = 0)$ is very well approximated by the power law $f(\Omega) \approx \Omega^{0.6}$, which is often used, instead of the exact form, for estimates. With this approximation,

$$\mathbf{v} \simeq \frac{2}{3H\Omega}\Omega^{0.6}\mathbf{g}, \qquad (5.139)$$

where all the quantities are evaluated at present and $\mathbf{g}$ is the peculiar acceleration generated by the density contrast.

Exercise 5.6

Velocity and light: Show that the peculiar-velocity field generated by a given distribution of matter with density contrast δ can be expressed in the form

$$\mathbf{v}(\mathbf{x}) = a\frac{fH}{4\pi}\int\frac{\mathbf{y}-\mathbf{x}}{|\mathbf{y}-\mathbf{x}|^3}\delta(\mathbf{y})\,d^3y. \qquad (5.140)$$

How does the intensity of radiation from a distribution of luminous matter distributed with a density contrast δ_{lum} vary? Explain how the measurement of a peculiar-velocity field, as well as the distribution of luminous matter, can provide a handle on the relative distribution of dark and luminous matter in the universe.

Exercise 5.7
Vlasov equation and the linear limit: Consider the situation in which the system is described by a distribution function $f(\mathbf{x}, \mathbf{p}, t)$, defined in such a way that the number of particles dN in a phase volume $d^3x d^3p$ is $dN = f(\mathbf{x}, \mathbf{p}, t) d^3x d^3p$.
    (a) Introduce a time coordinate η such that $d\eta = (dt/a^2)$ and show that the Vlasov equation becomes

$$\frac{\partial f}{\partial \eta} + \frac{\mathbf{p}}{m} \frac{\partial f}{\partial \mathbf{x}} = ma^2(\eta)\nabla\phi\frac{\partial f}{\partial \mathbf{p}}. \tag{5.141}$$

    (b) Substitute $f(\mathbf{x}, \mathbf{p}, \eta) = f_0(\mathbf{p}, \eta) + f_1(\mathbf{x}, \mathbf{p}, \eta)$ into Eq. (5.141) and linearise the equation in f_1. Define the variables in Fourier space by

$$f_1(\mathbf{x}, \mathbf{p}, \eta) = \int \frac{d^3k}{(2\pi)^3} f_{\mathbf{k}}(\mathbf{p}, \eta)e^{i\mathbf{k}\cdot\mathbf{x}}, \qquad \phi(\eta, \mathbf{x}) = \int \frac{d^3k}{(2\pi)^3} \phi_{\mathbf{k}}(\eta)e^{i\mathbf{k}\cdot\mathbf{x}} \tag{5.142}$$

and show that the equation satisfied by $f_{\mathbf{k}}(\mathbf{p}, \eta)$ is

$$\frac{\partial f_{\mathbf{k}}}{\partial \eta} + \left(\frac{i\mathbf{k}\cdot\mathbf{p}}{m}\right) f_{\mathbf{k}} = ma^2(\eta)\left(i\mathbf{k}\cdot\frac{\partial f_0}{\partial \mathbf{p}}\right)\phi_{\mathbf{k}}(\eta). \tag{5.143}$$

    (c) Consider next the equation satisfied by δ_k. Show that δ_k satisfies the integral equation

$$\delta_k(\eta) = \frac{3}{2}\frac{H_0^2}{k}\int_{\eta_i}^{\eta} ds\, a(s)\delta_k(s)J_k(\eta - s), \tag{5.144}$$

with

$$J_k(\eta - s) \equiv \frac{4\pi m}{n_0}\int_0^{\infty} dp\, p f_0(p)\sin\left[\frac{pk(\eta - s)}{m}\right]. \tag{5.145}$$

    (d) Note that this equation is still linear in δ but is nonlocal. Earlier, we obtained a general equation governing the evolution of δ, including the nonlinear terms [see Eq. (5.94)]. If $A_{\mathbf{k}}$ and $B_{\mathbf{k}}$ are ignored completely in Eq. (5.94), then we obtain a *differential* equation for δ_k that is, of course, different from *integral* equation (5.144), even though both pertain to the linear evolution of δ. Explain why.
    (e) In the limit of long wavelengths (that is, when $k \to 0$), we do expect Eq. (5.144) to describe the same evolution as Eq. (5.100) for pressure-free matter. Show that in this limit $\delta_k(\eta)$ satisfies the equation

$$\delta_k'' - \frac{3}{2}H_0^2 a(\eta)\delta_k \cong 0, \tag{5.146}$$

which is same as Eq. (5.100) for $\rho_b \propto a^{-3}$ in terms of the new time coordinate η.
    (f) As an example, consider the situation with distribution function

$$f_0(\mathbf{p}) \propto (\alpha^2 + p^2)^{-2}. \tag{5.147}$$

Show that, in this case, Eq. (5.144) is equivalent to

$$\ddot{\delta}_k + \frac{2\alpha k}{m}\dot{\delta}_k + \frac{\alpha^2 k^2}{m^2}\delta_k = -k^2 a^2 \phi_k. \tag{5.148}$$

Converting back to t with $(d/d\eta) = a^2(d/dt)$ and using Poisson equation to relate ϕ_k to δ_k, we get

$$\frac{d^2\delta_k}{dt^2} + 2\left(\frac{\dot{a}}{a} + \frac{\alpha k}{a^2}\right)\frac{d\delta_k}{dt} + \left(\frac{\alpha^2 k^2}{a^4} - 4\pi G\rho_b\right)\delta_k = 0. \tag{5.149}$$

Interpret the terms in this equation.

Exercise 5.8
Free streaming and trajectories: The proper distance travelled by a particle in the time interval $(0, t)$ can be written as

$$l_{FS}(t) = a(t)\int_0^t \frac{v(t')}{a(t')}\,dt'. \tag{5.150}$$

Consider a particle that is relativistic for $t < t_{NR}(< t_{eq})$ and nonrelativistic for $t > t_{NR}$. Show that

$$\frac{l_{FS}(t)}{a(t)} \cong \begin{cases} (2t_{NR}/a_{NR}^2)a = (2t/a) & (t < t_{NR}) \\ (2t_{NR}/a_{NR})[1 + \ln(a/a_{NR})] & (t_{NR} < t < t_{eq}) \\ (2t_{NR}/a_{NR})[(5/2) + \ln(a_{eq}/a_{NR})] & (t_{eq} \ll t) \end{cases} \tag{5.151}$$

and

$$\lambda_{FS} \equiv l_{FS}(t_0) \cong \left(\frac{a_0}{a_{NR}}\right)(2t_{NR})\left(\frac{5}{2} + \ln\frac{a_{eq}}{a_{NR}}\right). \tag{5.152}$$

[Numerically, this corresponds to,

$$\lambda_{FS} \approx 28\,\text{Mpc}\left(\frac{m_\nu}{30\,\text{eV}}\right)^{-1} \tag{5.153}$$

if the dark matter is made of massive neutrinos of mass m_ν. This length scale contains a mass of

$$M_{FS} \approx 4 \times 10^{15}\,M_\odot\left(\frac{m_\nu}{30\,\text{eV}}\right)^{-2}, \tag{5.154}$$

which corresponds to large clusters. Thus, if dark matter is made of massive neutrinos, it will have very little power at scales with $\lambda < \lambda_{FS}$.]

Exercise 5.9
Fluid limit: The equations for a fluid can be obtained from the moments of the Vlasov equation. It is, however, more illuminating to derive these equations directly from the equations describing the pressureless fluid in the proper coordinates. Show that, if the

equations are

$$\frac{\partial \delta}{\partial b} + \partial_\alpha [u^\alpha (1 + \delta)] = 0, \qquad \nabla^2 \psi = \left(\frac{\delta}{b}\right), \tag{5.155}$$

$$\frac{\partial u^\alpha}{\partial b} + u^\beta \partial_\beta u^\alpha = -\frac{3A}{2b} [\partial^\alpha \psi + u^\alpha], \qquad A = \left(\frac{\rho_b}{\rho_c}\right)\left(\frac{\dot{a}b}{a\dot{b}}\right)^2, \tag{5.156}$$

the quantity $u^\alpha = u^\alpha(\mathbf{x}, b)$ is the velocity of the fluid at $(\mathbf{x}, b)$. Because we have ignored the velocity dispersion, u^α satisfies the same equation as the velocity w^α of individual particles [Eq. (5.102)].

Exercise 5.10
Equation for δ in the fluid limit: Decompose the derivative of the velocity $\partial_\alpha u_\beta$ into shear $\sigma_{\alpha\beta}$, rotation Ω^γ, and expansion θ by writing

$$\partial_\alpha u_\beta = \sigma_{\alpha\beta} + \epsilon_{\alpha\beta\gamma} \Omega^\gamma + \frac{1}{3} \delta_{\alpha\beta} \theta. \tag{5.157}$$

Obtain an equation for $\delta(b, \mathbf{x})$ in terms of $\Omega^2 = \Omega^\alpha \Omega_\alpha$ and $\sigma^2 = \sigma^{\alpha\beta} \sigma_{\alpha\beta}$. [Hint: From Eqs. (5.155) we find that

$$\frac{d\delta}{db} \equiv \frac{\partial \delta}{\partial b} + u^\alpha \partial_\alpha \delta = -(1 + \delta)\theta. \tag{5.158}$$

Taking the derivative of Eqs. (5.156) with respect to x^α we get

$$\frac{d\theta}{db} + \frac{3A}{2b}\theta + \frac{1}{3}\theta^2 + \sigma^2 - 2\Omega^2 = -\frac{3A}{2b^2}\delta, \tag{5.159}$$

where $\sigma^2 \equiv \sigma^{ab} \sigma_{ab}$ and $\Omega^2 \equiv \Omega^\alpha \Omega_\alpha$. Combining Eqs. (5.159) and (5.158) show that δ satisfies the equation

$$\frac{d^2\delta}{db^2} + \frac{3A}{2b}\frac{d\delta}{db} - \frac{3A}{2b^2}\delta(1 + \delta) = \frac{4}{3}\frac{1}{(1 + \delta)}\left(\frac{d\delta}{db}\right)^2 + (1 + \delta)(\sigma^2 - 2\Omega^2). \tag{5.160}$$

From the last term on the right-hand side we see that shear contributes positively to $(d^2\delta/db^2)$ whereas rotation Ω^2 contributes negatively. Thus shear helps the growth of inhomogenities whereas rotation works against it.]

5.7 Origin of Density Perturbations

The formalism developed in the last section can be used to evolve the initial perturbations as long as $\delta \ll 1$. We still need to generate some initial perturbation before this procedure is effective. We now consider some of the issues related to the generation of these perturbations.

We saw earlier that the physical wavelength of any mode characterising the density perturbation will be larger than the Hubble radius at sufficiently early epochs. The proper wavelength $\lambda(M)$ of a perturbation containing mass M will be bigger than the Hubble radius at all redshifts $z > z_{enter}(M)$, where,

$$z_{enter}(M) \approx \begin{cases} 1.41 \times 10^5 (\Omega h^2)^{\frac{1}{3}} (M/10^{12} \, M_\odot)^{-1/3}, & M < M_{eq} \approx 3.2 \times 10^{14} \, M_\odot (\Omega h^2)^{-2} \\ 1.10 \times 10^6 (\Omega h^2)^{-\frac{1}{3}} (M/10^{12} \, M_\odot)^{-2/3}, & M > M_{eq} \approx 3.2 \times 10^{14} \, M_\odot (\Omega h^2)^{-2} \end{cases}.$$

$$(5.161)$$

Note that a galactic mass perturbation was bigger than the Hubble radius for redshifts larger than a moderate value of $\sim 10^6$.

This result leads to a major difficulty in conventional cosmology. Normal physical processes can act coherently only over length scales smaller than the Hubble radius. Thus any physical process leading to density perturbations at some early epoch, $t = t_i$, could have operated only at scales smaller than $d_H(t_i)$. However, most of the relevant astrophysical scales (corresponding to clusters, groups, galaxies, etc.) were much bigger than $d_H(t)$ at early epochs. Thus, if we want the seed perturbations to have originated in the early universe, then it is difficult to understand how any physical process could have contributed to it. To tackle this difficulty, we must arrange matters such that, at sufficiently small t, $\lambda(t) < d_H(t)$. If this can be done, then the physical processes can lead to an initial density perturbation.

We first show that $d_H(t)$ will rise faster than $a(t)$ as $t \to 0$ if the source of expansion of the universe has positive energy density and positive pressure. Thus the preceding problem *cannot* be tackled in any Friedmann model with $\rho > 0$ and $p > 0$. To see this, we begin by noting that, as $t \to 0$, we can ignore the k/a^2 term in the Friedmann equations and write

$$d_H^2 = \frac{a^2}{\dot{a}^2} = \frac{3}{8\pi G \rho}.$$

$$(5.162)$$

Therefore

$$\frac{d \ln d_H}{dt} = -\frac{1}{2} \frac{d \ln \rho}{dt} = \frac{3}{2} \left(1 + \frac{p}{\rho}\right) \frac{d \ln a}{dt},$$

$$(5.163)$$

where we have used the relation $d(\rho a^3) = -p \, da^3$. The proper wavelengths scale as $\lambda \propto a(t)$, giving $d(\ln \lambda) = d(\ln a)$. Combining with Eq. (5.163), we get

$$\frac{d \ln d_H}{d \ln \lambda} = \frac{3}{2} \left(1 + \frac{p}{\rho}\right).$$

$$(5.164)$$

For $(p/\rho) > 0$, the right-hand side is greater than unity. Therefore d_H grows as λ^n with $n > 1$ during the early epochs of evolution. [Of course, n itself could vary with time if (p/ρ) is not a constant; but we will always have the condition $n > 1$.] It follows that $(d_H/\lambda) < 1$ for sufficiently small a, proving that the Hubble radius will be smaller than the wavelength of the perturbation at *sufficiently* early epochs (as $a \to 0$), provided that $(p/\rho) > 0$. Hence, in order to generate density perturbation from a region that is smaller than the Hubble radius, it is necessary to have models with ρ or p negative. However, to satisfy the first Friedmann equation with $k = 0$, we need $\rho > 0$; thus the only models in which we can have $d_H > \lambda$ for small t are the ones for which $p < 0$.

One simple way of achieving this is to make $a(t)$ increase rapidly (for example, exponentially) with t for a brief period of time. Such a rapid growth is called *inflation*, and in conventional models of inflation, the energy density during the inflationary phase is provided by a scalar field with a potential $V(\Phi)$. We saw in Chap. 4 that when the potential energy dominates over the kinetic energy, such a scalar field can act as an ideal fluid with the equation of state $p = -\rho$. Because the field is assumed to be inherently quantum mechanical, it will have characteristic quantum fluctuations. It is possible for these quantum fluctuations to eventually manifest as classical density perturbations. We now work out the basic mechanism behind this phenomenon.

Let us assume that, during the inflationary phase, the universe can be described as a Friedmann model with small inhomogeneities. The source of expansion can be assumed to be some classical scalar field $\Phi(t, \mathbf{x})$ that can be split as $\phi_0(t) + f(t, \mathbf{x})$, where ϕ_0 is the average homogeneous part of the field that drives the inflation and $f(t, \mathbf{x})$ is the space-dependent fluctuating part. Because the energy density that is due to a scalar field is $\rho \cong (1/2)\dot{\phi}^2 + V_0$, where V_0 is a constant, we get,

$$\delta\rho(t, \mathbf{x}) = \rho(\mathbf{x}, t) - \bar{\rho}(t) \cong \dot{\phi}_0(t)\dot{f}(t, \mathbf{x}), \tag{5.165}$$

where we have assumed that $f \ll \phi_0$. Fourier transforming this equation, we find that

$$\delta\rho(\mathbf{k}, t) \cong \dot{\phi}_0(t)\dot{Q}_\mathbf{k}(t), \tag{5.166}$$

with

$$f(t, \mathbf{x}) \equiv \int \frac{d^3\mathbf{k}}{(2\pi)^3} Q_\mathbf{k}(t)e^{i\mathbf{k}\cdot\mathbf{x}}. \tag{5.167}$$

Because the average energy density during inflation is dominated by the constant term V_0, the density contrast will be

$$\delta(\mathbf{k}, t) \cong \frac{\delta\rho}{V_0} = \frac{\dot{\phi}_0(t)\dot{Q}_\mathbf{k}(t)}{V_0}. \tag{5.168}$$

The mean value ϕ_0 and the fluctuating field $f(t, \mathbf{x})$ appearing in this relation are supposed to be some *classical* objects *mimicking* the quantum fluctuations. It is not easy to devise and justify such quantities. What is usually done is to choose some convenient quantum-mechanical measure for fluctuations and *define* ϕ_0 and $Q_\mathbf{k}$ in terms of this quantity.

In quantum theory, the field $\hat{\phi}(t, \mathbf{x})$ and its Fourier coefficients $\hat{q}_{\mathbf{k}}(t)$ will become operators related by

$$\hat{\phi}(t, \mathbf{x}) = \int \frac{d^3\mathbf{k}}{(2\pi)^3} \hat{q}_{\mathbf{k}}(t) e^{i\mathbf{k}\cdot\mathbf{x}}. \tag{5.169}$$

We can specify the quantum state of the field by giving the quantum state $\psi_{\mathbf{k}}(q_{\mathbf{k}}, t)$ of each of the modes $\hat{q}_{\mathbf{k}}$. We can think of $q_{\mathbf{k}}$ as coordinates of a particle and $\psi_{\mathbf{k}}(q_{\mathbf{k}}, t)$ as the wave function describing this particle. The fluctuations in $q_{\mathbf{k}}$ can be characterised by the dispersion

$$\sigma_{\mathbf{k}}^2(t) = \langle\psi|\hat{q}_{\mathbf{k}}^2(t)|\psi\rangle - \langle\psi|\hat{q}_{\mathbf{k}}(t)|\psi\rangle^2 = \langle\psi|\hat{q}_{\mathbf{k}}^2(t)|\psi\rangle \tag{5.170}$$

in this quantum state. [The mean value of the scalar field operator $\langle\hat{\phi}(t, \mathbf{x})\rangle$ is zero in the inflationary phase. Therefore we have set $\langle\hat{q}_{\mathbf{k}}\rangle$ to zero in Eq. (5.170). Note that we are interested in only the $\mathbf{k} \neq 0$ modes.] Expressing $\hat{q}_{\mathbf{k}}$ in terms of $\hat{\phi}(t, \mathbf{x})$, we easily see that

$$\sigma_{\mathbf{k}}^2(t) = \int d^3\mathbf{x} \, \langle\psi|\hat{\phi}(t, \mathbf{x}+\mathbf{y})\hat{\phi}(t, \mathbf{y})|\psi\rangle \, e^{i\mathbf{k}\cdot\mathbf{x}}. \tag{5.171}$$

In other words, the "power spectrum" of fluctuations $\sigma_{\mathbf{k}}^2$ is related to the Fourier transform of the two-point correlation function of the scalar field. Because $\sigma_{\mathbf{k}}^2(t)$ appears to be a good measure of quantum fluctuations, we may attempt to *define* $Q_{\mathbf{k}}(t)$ as

$$Q_{\mathbf{k}}(t) = \sigma_{\mathbf{k}}(t). \tag{5.172}$$

This is equivalent to *defining* the fluctuating classical field $f(t, \mathbf{x})$ as

$$f(t, \mathbf{x}) \equiv \int \frac{d^3\mathbf{k}}{(2\pi)^3} \sigma_{\mathbf{k}}(t) e^{i\mathbf{k}\cdot\mathbf{x}}, \tag{5.173}$$

leading to the result

$$\delta(\mathbf{k}, t) = \frac{\dot{\phi}_0(t)}{V_0} \dot{\sigma}_{\mathbf{k}}(t). \tag{5.174}$$

It should be stressed that the expression chosen for $Q_{\mathbf{k}}$ and ϕ_0 are only two out of the many possible choices available. Such an ambiguity cannot be avoided when semiclassical expressions have to be computed from quantum-mechanical operators.

During the inflationary phase the proper wavelengths of perturbations increase exponentially while the Hubble radius remains constant. It follows that a mode will exit the Hubble radius at some time during the inflation. After the inflationary phase ends, the universe becomes radiation dominated and the Hubble radius grows faster than the wavelength of the mode. Eventually the mode will reenter the Hubble radius at some epoch. We now show that, at the time of reentry, the amplitude of the density perturbation will be of the order of

$$\delta(\mathbf{k}, t_{\text{enter}}) \cong \left(\frac{\dot{\sigma}_{\mathbf{k}}}{\dot{\phi}_0} \right)_{t=t_{\text{exit}}}.$$ (5.175)

The right-hand side can be evaluated from the study of the quantum fields in the inflationary phase. This equation provides the spectrum of perturbation at the time when the mode enters the Hubble radius.

Let us suppose that the particular mode (corresponding to the galactic scale, say) we are interested in leaves the Hubble radius at $t = t_1$ during the inflationary phase, with the amplitude $\epsilon(t_1) \equiv \delta[\mathbf{k}, t_1(\mathbf{k})]$. We have suppressed the $\mathbf{k}$ dependence in ϵ to simplify the notation. Let $t = t_f$ be the instant at which the universe makes a transition from the inflationary phase to the radiation dominated phase. During the time $t_1 < t < t_f$, the mode is outside the Hubble radius in an inflating universe. We derived a growth law in Section 5.3 for modes bigger than the Hubble radius:

$$\left(\frac{\delta\rho}{\rho} \right) \propto \frac{1}{\rho_{\text{bg}} a^2}.$$ (5.176)

In the present context, $\rho_{\text{bg}} = \text{constant}$ and we get

$$\epsilon(t) \propto \exp(-2Ht).$$ (5.177)

This result can also be expressed as

$$\frac{\epsilon(t_f)}{\epsilon(t_1)} = \left[\frac{H(t_1)a(t_1)}{H(t_f)a(t_f)} \right]^2,$$ (5.178)

as $H(t)$ is a constant during this interval. If the transition from the inflationary phase to the radiation-dominated phase is approximated as instantaneous, then the radiation density ρ_R and the fluctuations in the radiation density $\delta\rho_R$ originate directly from the corresponding terms in the inflationary phase: $\rho_R \simeq \dot{\phi}^2$ and $\delta\rho$ (radiation phase) $\approx \delta\rho$ (inflationary phase). Further, because the background energy density ρ in the inflationary (de Sitter) phase is dominated by V_0, we can write

$$\left(\frac{\delta\rho}{\rho} \right)_{\text{rad}} \simeq \left(\frac{\delta\rho}{\rho} \right)_{\text{de Sitter}} \cdot \left(\frac{\rho_{\text{de Sitter}}}{\rho_{\text{rad}}} \right) \simeq \left(\frac{\delta\rho}{\rho} \right)_{\text{de Sitter}} \cdot \left(\frac{V_0}{\dot{\phi}^2} \right).$$ (5.179)

During the time $t_f < t < t_{\text{enter}}$, the mode is bigger than the Hubble radius and is evolving in the radiation-dominated universe. We know from Section 5.3 that such modes grow as $\delta \propto a^2 \propto t$, while the quantity $a(t)H(t) \propto t^{-1/2}$. Hence, at $t = t_{\text{enter}} \equiv t_2$,

$$
\begin{aligned}
\epsilon(t_2) &= \left(\frac{V_0}{\dot{\phi}^2}\right)\left(\frac{\delta\rho}{\rho}\right)_{\text{de Sitter}} \cdot \left[\frac{H(t_f)a(t_f)}{H(t_2)a(t_2)}\right]^2 \\
&= \epsilon(t_1)\left(\frac{V_0}{\dot{\phi}^2}\right)\left[\frac{H(t_1)a(t_1)}{H(t_f)a(t_f)}\right]^2\left[\frac{H(t_f)a(t_f)}{H(t_2)a(t_2)}\right]^2 \\
&= \epsilon(t_1)\left(\frac{V_0}{\dot{\phi}^2}\right)\left[\frac{H(t_1)a(t_1)}{H(t_2)a(t_2)}\right]^2.
\end{aligned}
\tag{5.180}
$$

However, for a mode that is leaving the Hubble radius at t_1 and entering it again at t_2, $H(t_1)a(t_1) = H(t_2)a(t_2)$. Therefore

$$
\epsilon(t_2) = \epsilon(t_1)\left(\frac{V_0}{\dot{\phi}^2}\right).
\tag{5.181}
$$

On using Eq. (5.174), we get

$$
\delta(\mathbf{k}, t_{\text{enter}}) \cong \left(\frac{\dot{\sigma}_{\mathbf{k}}}{\dot{\phi}_0}\right)_{t=t_{\text{exit}}}.
\tag{5.182}
$$

A more exact matching of energy densities at $t = t_f$ gives an additional factor of $(4/3)$.

We now need to estimate the quantity $\dot{\sigma}_{\mathbf{k}}$ to complete the calculation. This can be achieved as follows. During the inflationary phase, the potential $V(\phi)$ remains approximately constant. Hence for the study of quantum fluctuations in this field, we may ignore the potential and treat ϕ as a massless scalar field. By introducing the Fourier transform $q_{\mathbf{k}}(t)$ of the fluctuating part of the field $\phi(t, \mathbf{x})$ through the relation

$$
\phi(t, \mathbf{x}) = \int \frac{d^3\mathbf{k}}{(2\pi)^3} q_{\mathbf{k}}(t)e^{i\mathbf{k}\cdot\mathbf{x}},
\tag{5.183}
$$

we can decompose the action for the scalar field into that of a bunch of harmonic oscillators:

$$
\begin{aligned}
\mathcal{A} &= \frac{1}{2}\int d^3\mathbf{x}\, dt\, \sqrt{-g}(\phi^i\phi_i) = \frac{1}{2}\int dt\, d^3\mathbf{x}\, a^3\left(\dot{\phi}^2 - \frac{|\nabla\phi|^2}{a^2}\right) \\
&= \frac{1}{2}\int \frac{d^3\mathbf{k}}{(2\pi)^3}\int dt\, a^3\left(|\dot{q}_{\mathbf{k}}|^2 - \frac{k^2}{a^2}|q_{\mathbf{k}}|^2\right).
\end{aligned}
\tag{5.184}
$$

To quantise this field, we have to quantise each independent oscillator $q_{\mathbf{k}}$. For the kth oscillator, we have the Schrodinger equation:

$$
i\frac{\partial\psi_{\mathbf{k}}}{\partial t} = -\frac{1}{2a^3}\frac{\partial^2\psi_{\mathbf{k}}}{\partial q_{\mathbf{k}}^2} + \frac{1}{2}a^3\omega_{\mathbf{k}}^2 q_{\mathbf{k}}^2\psi_{\mathbf{k}}, \qquad \omega = \frac{|\mathbf{k}|}{a}.
\tag{5.185}
$$

The quantum state of the field can be expressed by a wave function ψ, which is the product of ψ_k for all k. This Schrodinger equation can be solved by the ansatz:

$$\psi_k = A_k(t) \exp\{-B_k(t)[q_k - f_k(t)]^2\}. \tag{5.186}$$

Substituting this ansatz into the Schrodinger equation and equating the coefficients of various powers of q_k, we get three equations

$$i\dot{B} = \frac{2B^2}{a^3} - \frac{1}{2}ak^2, \tag{5.187}$$

$$i(\dot{B}f + B\dot{f}) = \frac{2B^2}{a^3}f, \tag{5.188}$$

$$i\frac{\dot{A}}{A} = i\dot{B}f^2 + 2iBf\dot{f} + \frac{B}{a^3} - \frac{2B^2f^2}{a^3}, \tag{5.189}$$

where we have suppressed the index k for simplicity. We can transform these equations to a simpler form by introducing a variable Q defined by the relation: $B = -(i/2)a^3(\dot{Q}/Q)$. Simple algebra then shows that

$$f = (\text{const})(a^3\dot{Q})^{-1}, \qquad A = (\text{const})Q^{-1/2}\exp\left(-\frac{i}{2}\int k^2af^2\,dt\right), \tag{5.190}$$

and Q satisfies the linear equation

$$\frac{1}{a^3}\frac{d}{dt}\left(a^3\frac{dQ}{dt}\right) + \frac{k^2}{a^2}Q = 0. \tag{5.191}$$

From Eq. (5.186) we also see that

$$|\psi_k|^2 = N_k \exp\left[-\frac{(q_k - R_k)^2}{2\sigma_k^2}\right], \tag{5.192}$$

with

$$\sigma_k^2 = \frac{1}{2}(B_k + B_k^*)^{-1}, \qquad R_k = \frac{B_k f_k + B_k^* f_k^*}{B_k + B_k^*}. \tag{5.193}$$

We can further simplify Eqs. (5.193) by noting that Eq. (5.191) implies the relation

$$\frac{d}{dt}[a^3(Q_k^*\dot{Q}_k - \dot{Q}_k^*Q_k)] = 0, \tag{5.194}$$

giving $Q_k^*\dot{Q}_k - \dot{Q}_k^*Q_k = i(\text{constant})a^{-3}$. Using this result in Eqs. (5.193), we get

$$\sigma_k^2 = (\text{const})|Q_k|^2, \qquad R_k = \text{Re}\left[(\text{const})\sigma_k^2 B_k f_k\right]. \tag{5.195}$$

Thus all relevant quantities can be expressed in terms of Q_k.

To solve for $q_{\mathbf{k}}(t)$, it is convenient to transform the independent variable from t to another time coordinate τ with $d\tau = [dt/a(t)]$. Then

$$t = -H^{-1}\ln|(1 - H\tau)|; \quad a(\tau) = (1 - H\tau)^{-1}, \qquad (5.196)$$

and the equation for Q becomes

$$\frac{d^2 Q}{d\tau^2} - \frac{2H}{H\tau - 1}\frac{dQ}{d\tau} + k^2 Q = 0. \qquad (5.197)$$

Changing the variable to $x = kH^{-1}(1 - H\tau) = kH^{-1}e^{-Ht}$, we can write Eq. (5.197) as

$$\frac{d^2 Q}{dx^2} - \frac{2}{x}\frac{dQ}{dx} + Q = 0. \qquad (5.198)$$

The general solution to this equation is

$$Q(x) = a(1 + ix)e^{-ix} + b(1 - ix)e^{ix}, \qquad (5.199)$$

where a and b are two constants that depend on the initial conditions imposed on $Q(x)$. Because the quantum state of the oscillator $\psi_{\mathbf{k}}(q_{\mathbf{k}})$ is completely determined by the function $Q_{\mathbf{k}}(t)$, we now have a *set* of quantum states, parameterised by the constants $a_{\mathbf{k}}$ and $b_{\mathbf{k}}$. Expectation values of physical variables will, of course, depend on the quantum state in which they are evaluated. Therefore, to proceed further, we need to make a specific choice for the quantum state of the field.

Because the minimum amount of fluctuations will arise from the "ground state" of the system, it seems natural to choose the quantum state as the ground state of the system. It is, however, not easy to define a ground state in the expanding background. One way of doing this is the following: We know that when $a = 1$ and $H = 0$, we should recover the ordinary flat-space field theory. As $H \to 0$, Eq. (5.199) becomes

$$Q_{\text{flat}}(\tau, \mathbf{k}) = \lim_{H \to 0}\left[a\left(1 + \frac{ik}{H} - ik\tau\right)e^{i(k\tau - kH^{-1})} + b\left(1 - \frac{ik}{H} + ik\tau\right)e^{-i(k\tau - kH^{-1})}\right],$$

$$= \alpha e^{ik\tau} + \beta e^{-ik\tau}, \qquad (5.200)$$

provided we keep $\alpha = ika H^{-1}\exp(-ik/H)$ and $\beta = -ikb H^{-1}\exp(ik/H)$ as *finite* constants when the limit is taken. However, to obtain the "standard vacuum" state in the flat-space–time, each oscillator must be described by the wavefunction

$$\psi = N\exp\left(-\frac{i}{2}k\tau\right)\exp\left(-\frac{1}{2}kq^2\right) = (\text{const})Q^{-1/2}\exp\left(+\frac{i}{2}\frac{\dot{Q}}{Q}q^2\right). \qquad (5.201)$$

Therefore we must satisfy three conditions: (1) $(\dot{Q}/Q) = ik$, (2) $\beta = 0$, and

(3) α is independent of k. This implies that our solution for Q_k must have the form

$$Q_k = a(1 + ix)e^{-ix} = \alpha H \, e^{ik/H}(ik)^{-1}(1 + ix)e^{-ix}. \tag{5.202}$$

(Note that $\tau = t$ in the limit of $H \to 0$.) We can now compute all the physical quantities: Direct calculation gives

$$B_k = \frac{k^3 H^{-2}}{2(1 + k^2/H^2 a^2)}\left[1 - \frac{iHa}{k}\right], \qquad \sigma_k^2 = \frac{1}{2}(B + B^*)^{-1} = \frac{H^2}{2k^3} + \frac{1}{2ka^2}. \tag{5.203}$$

To compute $\delta(k, t_{\text{enter}})$ we need the value of $\dot{\sigma}_k$ and $\dot{\phi}_0$ at $t = t_{\text{exit}}$. From the expression for $\sigma_k(t)$, we have

$$|\dot{\sigma}_k| = \frac{H}{\sqrt{2}k^{3/2}}\frac{1}{2}\left(1 + \frac{k^2}{H^2 G^2}\right)^{-1/2}\frac{2k^2 \dot{a}}{H^2 a^3} = \frac{H^2}{\sqrt{2}k^{3/2}}\left(1 + \frac{k^2}{H^2 a^2}\right)^{-1/2}\left(\frac{k^2}{H^2 a^2}\right) \tag{5.204}$$

at any time t. At $t = t_{\text{exit}}$, $(k/Ha) = 2\pi$. So, at $t = t_{\text{exit}}$, we find

$$|\dot{\sigma}_k| = \frac{4\pi^2}{\sqrt{2}(1 + 4\pi^2)^{1/2}}\frac{H^2}{k^{3/2}} \simeq 4\left(\frac{H^2}{k^{3/2}}\right) \qquad (\text{at } t = t_{\text{exit}}). \tag{5.205}$$

We saw earlier in relation (5.182) that this spectrum of perturbation at the time of reentry has the form $\delta_{\mathbf{k}}(t_{\text{enter}}) \simeq (\dot{\sigma}_{\mathbf{k}}/\dot{\phi}_0)$. The $\mathbf{k}$ dependence of $\delta_{\mathbf{k}}$ arises solely from the $\mathbf{k}$ dependence of $\dot{\sigma}_{\mathbf{k}}$. The preceding analysis shows that $k^3|\delta_{\mathbf{k}}(t_{\text{enter}})|^2$ is independent of $\mathbf{k}$.

To determine the amplitude of the perturbations, we also need to calculate $\dot{\phi}_0$ when the mode leaves the Hubble radius. We can determine this quantity by solving the equation of motion for the scalar field,

$$\ddot{\phi} + 3\frac{\dot{a}}{a}\dot{\phi} + V'(\phi) = 0, \tag{5.206}$$

with a given $V(\phi)$. To make a simple estimate, we assume that $V(\phi)$ can be approximated as

$$V(\phi) \approx V_0 - \frac{\lambda}{4}\phi^4 \tag{5.207}$$

and that we can ignore the $\ddot{\phi}$ term in the equation of motion. For a potential with $V(\phi) \approx (V_0 - \lambda\phi^4/4)$, we have $V'(\phi) \approx -\lambda\phi^3$. During inflation, let us assume that ϕ varies from ϕ_i to ϕ_f. In the "slow-rollover" phase, $\dot{\phi} \approx -V'(\phi)/3H$, giving

$$\delta_k = \frac{\dot{\sigma}_k}{\dot{\phi}} = \frac{4H^2}{k^{3/2}}\left[\frac{3H}{-V'(\phi)}\right] = \frac{12H^3}{k^{3/2}[-V'(\phi)]}. \tag{5.208}$$

The expression on the right-hand side is to be evaluated at a time $t = t_{\text{exit}}$ when the mode leaves the Hubble radius. During the time $t_{\text{exit}} < t < t_f$ the universe inflates by the factor $\exp N$, with

$$N = \int_{t_{\text{exit}}}^{t_f} \frac{\dot{a}}{a} dt \cong 8\pi G \int_{\phi}^{\phi_f} \frac{V_0}{[-V']} d\phi = \frac{3H^2}{2\lambda} \left(\frac{1}{\phi^2} - \frac{1}{\phi_f^2} \right) \approx \frac{3H^2}{2\lambda\phi^2}, \tag{5.209}$$

as we may take $\phi \ll \phi_f$. Using this, we get

$$-V'(\phi) = \lambda\phi^3 = \left(\frac{3}{2} \right)^{3/2} \frac{H^3}{\lambda^{1/2} N^{3/2}}. \tag{5.210}$$

Using this in Eq. (5.208) we get

$$k^{3/2}\delta_k = 12H^3 \left(\frac{2}{3} \right)^{3/2} \frac{\lambda^{1/2} N^{3/2}}{H^3} \approx 12 \left(\frac{2}{3} \right)^{3/2} \lambda^{1/2} N^{3/2}. \tag{5.211}$$

To have sufficient inflation, we need $N \approx 50$. Then $k^{3/2}\delta_k \approx 2.3 \times 10^3\lambda^{1/2}$, which is far too large for $\lambda \approx 0.1$–1. To get $k^{3/2}\delta_k \approx 10^{-4}$ we need $\lambda \approx 10^{-15}$.

This has been the most serious difficulty faced by all the realistic inflationary models: They all produce too large an inhomogeneity. The qualitative reason for this result can be found from relation (5.182). To obtain a *slow* rollover and sufficient inflation, we need to keep $\dot{\phi}_0$ small; this leads to an increase in the value of δ. The difficulty could have been avoided if it were possible to keep σ_k arbitrarily small; unfortunately the inflationary phase induces a fluctuation of approximately $H/2\pi$ on any quantum field because of field theoretical reasons. This lower bound prevents us from getting acceptable values for δ unless we fine-tune the dimensionless parameters of $V(\phi)$. Several "solutions" have been suggested in the literature to overcome this difficulty, but none of them appear to be very compelling.

The same mechanism that produces the density inhomogenieties will also produce gravitational-wave perturbations as both the scalar field and the gravitational-wave modes satisfy the same equations. The perturbation equations describing gravitational waves, Eq. (5.47), can be used to describe the generation of gravitational waves in inflationary models. If we take each of the two transverse traceless modes of the gravitational wave such that $h_{ik} = \sqrt{16\pi G}\phi\, e_{ik}$, where e_{ik} is the unit tensor describing the polarisation of the waves and $\phi(t, \mathbf{x})$ is the scalar function that satisfies the wave equation in the Friedmann universe, an analysis similar to the one done previously will show that these perturbations will also have a scale-invariant power spectrum and a rms amplitude of approximately $(H/10^{19}$ GeV$)$. The power spectrum will be

$$P_{\text{grav}}(k) \cong \frac{k^3 |h_k|^2}{2\pi^2} = \frac{4}{\pi} \left(\frac{H}{m_P} \right)^2, \tag{5.212}$$

where h_k is a typical Fourier amplitude at wave vector k and $m_P^2 = (\hbar c / G)$ is the Planck mass. When this mode enters the Hubble radius, it will carry the energy density

$$\frac{k}{\rho} \frac{d\rho_{\text{grav}}}{dk} \simeq \frac{4}{3\pi} \left(\frac{H}{m_P} \right)^2 \quad (\text{at } t = t_{\text{enter}}), \tag{5.213}$$

where ρ is the energy density of the background. Once the mode has entered the Hubble radius, its energy density decreases as a^{-4}. If $\Omega_{\text{grav}}(k) = [k(d\rho_{\text{grav}}/dk)/\rho_c]$ is evaluated today, then

$$\Omega_{\text{grav}}(k)h^2 \simeq 10^{-5} \left(\frac{M}{m_P} \right)^4, \tag{5.214}$$

where M is the energy scale at which inflation took place. Such perturbations can induce a quadrapole anisotropy in the MBR background (see Chap. 6). The present value of the quadrapole anisotropy ($\lesssim 10^{-5}$) suggest that $H < 10^{15}$ GeV. The value of Ω_{grav} can be also restricted by the timing measurements of the millisecond pulsar (see Vol. II, Chap. 6, Section 6.6); the present bound is $\Omega_{\text{grav}}(\lambda \sim 1 \text{ pc}) \lesssim 3 \times 10^{-7}$.

The results obtained above can be easily generalised for a wide class of inflationary models characterised by two small parameters, ϵ and η, defined by [see Eqs. (4.221) of Chap. 4]

$$\epsilon = \frac{m_P^2}{16\pi} \left(\frac{V'}{V} \right)^2, \qquad \eta = \frac{m_P^2}{8\pi} \left(\frac{V''}{V} \right). \tag{5.215}$$

For these models Eq. (5.211) generalises to $k^{3/2}\delta_k \propto k^{(1/2)(n-1)}$ and the power spectrum for the gravitational-wave modes can be expressed in the form $P_{\text{grav}}(k) \propto k^q$. [The quantity $(n-1)$ is called the tilt of the power spectrum.] It is fairly easy to obtain an order-of-magnitude estimate of the amplitudes of the scalar and the tensor perturbations as well as the *tilt* for the a general inflationary model. We note that within factors of the order of unity, we would have expected

$$\Delta_{\text{tensor}}^2 \sim k^3 |h_k|^2 \sim \left(\frac{H^2}{m_P^2} \right) \sim \left(\frac{V}{m_P^4} \right), \tag{5.216}$$

which follows from relation (5.212) with $H^2 \propto V m_P^{-2}$ and

$$\Delta_{\text{scalar}}^2 \sim k^3 |\delta_k|^2 \sim \frac{H^6}{(V')^2} \sim \left(\frac{V^3}{m_P^6 V'^2} \right), \tag{5.217}$$

where $m_P^2 = G^{-1}$ [see Eq. (5.208)]. This shows that $(\Delta_T / \Delta_S)^2 \simeq m_P^2 (V'/V)^2 \simeq 16\pi\epsilon$. A more precise analysis gives the ratio as 12.4ϵ. To determine the tilt of the spectrum, we need to compute the quantity $(1 - n) = -d \ln \Delta_S^2 / d \log k$, with the relevant quantities being evaluated when the mode leaves the Hubble radius, the condition for which is $(a/k) = H^{-1}$. Because H is nearly constant during inflation, we can replace $d/d \ln k$ with $d/d \ln a$ and obtain, in the slow-rollover approximation,

$$\frac{d}{d \ln k} = a \frac{d}{da} = \frac{\dot{\phi}}{H} \frac{d}{d\phi} = -\frac{m_P^2}{8\pi} \frac{V'}{V} \frac{d}{d\phi}. \tag{5.218}$$

Taking the fluctuation at the time of crossing the Hubble radius as

$$\Delta_S^2 = \frac{H^4}{(2\pi \dot{\phi})^2} = \frac{128\pi}{3} \left(\frac{V^3}{m_P^6 V'^2} \right) \tag{5.219}$$

and evaluating the derivative of $\ln \Delta_S^2$ with respect to $\ln k$, we get $(1-n) = 6\epsilon - 2\eta$, where $\epsilon \equiv (m_P^2/16\pi)(V'/V)^2 \ll 1$ and $\eta \equiv (m_P^2/8\pi)(V''/V) \ll 1$. A similar analysis gives $q = -2\epsilon$. In the slow-rollover approximation, $n \approx 1$ and $q \approx 0$.

This analysis provides the lowest-order corrections to the spectral index in terms of the parameters of the inflationary model ϵ and δ. In principle, detection of both the perturbations (Δ_S^2, Δ_T^2) separately will allow us to determine ϵ and δ and thus have a direct handle on the nature of inflationary potential. For generic inflationary potentials, $|\eta| \approx |\epsilon|$ so that $(1-n) \approx 4\epsilon$. Combined with the relation $(\Delta_T/\Delta_S)^2 \approx 12.4\epsilon$, it follows that $(\Delta_T/\Delta_S)^2 \approx \mathcal{O}(3)(1-n)$, which is a relation connecting three potentially observable quantities. This could serve as a consistency check on inflationary predictions in the future when high-quality observations are available.

Exercise 5.11
Tilt in an explicit model: Work out the expression for tilt in the case of inflation in a power-law potential with $V(\phi) \propto \phi^\alpha$ assuming that the inflation lasts for $\sim 60e$ folding times. [Answer: $(1-n) = (2+\alpha)/120$; $n = 0.97$ for $\alpha = 2$ and $n = 0.95$ for $\alpha = 4$.]

5.8 Transfer Functions and Statistical Indicators

In the past few sections, we have discussed the generation and the growth of linear perturbations in relativistic theory (which is required when the wavelength is larger than Hubble radius) and in Newtonian theory (which is valid when the wavelength is much smaller than the Hubble radius). Given any initial density contrast, we are now in a position to evolve it forward in time until $\delta \approx 1$, at which time the linear perturbation theory will break down. We now discuss the form of the final perturbations, given its initial value within the linear regime.

As long as $\delta(t, \mathbf{x}) \ll 1$, then we can describe the evolution of $\delta(t, \mathbf{x})$ by linear perturbation theory, in which each mode $\delta_{\mathbf{k}}(t)$ will evolve independently. We can therefore write

$$\delta_{\mathbf{k}}(t) = T_{\mathbf{k}}(t, t_i)\delta_{\mathbf{k}}(t_i), \tag{5.220}$$

where $T_k(t, t_i)$ depends on only the dynamics and not on the initial conditions. We now determine the form of $T_k(t, t_i)$.

Let $\delta_\lambda(t_i)$ denote the amplitude of the dark-matter perturbation corresponding to some wavelength λ at the initial instant t_i. To each λ, we can associate a wave number $k \propto \lambda^{-1}$ and a mass $M \propto \lambda^3$; accordingly, we may label the perturbation as $\delta_M(t)$ or $\delta_k(t)$, as well, with the scalings $M \sim \lambda^3$, $k \sim \lambda^{-1}$. We are interested in the value of $\delta_\lambda(t)$ at some $t \gtrsim t_{\text{dec}}$.

To begin with, consider the modes that enter the Hubble radius in the radiation-dominated phase; their growth is suppressed in the radiation-dominated phase by the rapid expansion of the universe; therefore they do not grow significantly until $t = t_{\text{eq}}$, giving $\delta_\lambda(t_{\text{eq}}) = L\delta_\lambda(t_{\text{enter}})$, where $L \simeq 5 \ln(\lambda_{\text{eq}}/\lambda)$ is a logarithmic factor determined in relation (5.117). After matter begins to dominate, the amplitude of these modes grows in proportion to the scale factor a. Thus

$$\delta_M(t) = L\delta_M(t_{\text{enter}}) \left(\frac{a}{a_{\text{eq}}} \right) \quad \text{(for } M < M_{\text{eq}}). \tag{5.221}$$

Consider next the modes with $\lambda_{\text{eq}} < \lambda < \lambda_H$, where $\lambda_H \equiv H^{-1}(t)$ is the Hubble radius at the time t when we are studying the spectrum. These modes enter the Hubble radius in the matter-dominated phase and grow proportional to a afterwards. Therefore

$$\delta_M(t) = \delta_M(t_{\text{enter}}) \left(\frac{a}{a_{\text{enter}}} \right) \quad \text{(for } M_{\text{eq}} < M < M_H), \tag{5.222}$$

which may be rewritten as

$$\delta_M(t) = \delta_M(t_{\text{enter}}) \left(\frac{a_{\text{eq}}}{a_{\text{enter}}} \right) \left(\frac{a}{a_{\text{eq}}} \right). \tag{5.223}$$

However, note that, because t_{enter} is fixed by the condition $\lambda a_{\text{enter}} \propto t_{\text{enter}} \propto \lambda t_{\text{enter}}^{2/3}$, we have $t_{\text{enter}} \propto \lambda^3$. Further, $(a_{\text{eq}}/a_{\text{enter}}) = (t_{\text{eq}}/t_{\text{enter}})^{2/3}$, giving

$$\left(\frac{a_{\text{eq}}}{a_{\text{enter}}} \right) = \left(\frac{\lambda_{\text{eq}}}{\lambda} \right)^2 = \left(\frac{M_{\text{eq}}}{M} \right)^{2/3}. \tag{5.224}$$

Substituting Eq. (5.224) into Eq. (5.223), we get

$$\delta_M(t) = \delta_M(t_{\text{enter}}) \left(\frac{\lambda_{\text{eq}}}{\lambda} \right)^2 \left(\frac{a}{a_{\text{eq}}} \right) = \delta_M(t_{\text{enter}}) \left(\frac{M_{\text{eq}}}{M} \right)^{2/3} \left(\frac{a}{a_{\text{eq}}} \right). \tag{5.225}$$

Comparing Eqs. (5.225) and (5.221) we see that the mode that enters the Hubble radius after t_{eq} has its amplitude decreased by a factor of $L^{-1}M^{-2/3}$, compared with its original value.

Finally, consider the modes with $\lambda > \lambda_H$ that are still outside the Hubble radius at t and will enter the Hubble radius at some *future* time $t_{\text{enter}} > t$. During the time interval (t, t_{enter}), they will grow by a factor of (a_{enter}/a). Thus

$$\delta_\lambda(t_{\text{enter}}) = \delta_\lambda(t) \left(\frac{a_{\text{enter}}}{a}\right) \tag{5.226}$$

or

$$\delta_\lambda(t) = \delta_\lambda(t_{\text{enter}}) \left(\frac{a}{a_{\text{enter}}}\right) = \delta_M(t_{\text{enter}}) \left(\frac{M_{\text{eq}}}{M}\right)^{2/3} \left(\frac{a}{a_{\text{eq}}}\right) \quad (\lambda > \lambda_H). \tag{5.227}$$

(The last equality follows from the previous analysis.) Thus the behaviour of the modes is the same for the cases $\lambda_{\text{eq}} < \lambda < \lambda_H$ and $\lambda_H < \lambda$, i.e., for all wavelengths $\lambda > \lambda_{\text{eq}}$. Combining all these pieces of information, we can state the final result as follows:

$$\delta_\lambda(t) = \begin{cases} L\delta_\lambda(t_{\text{enter}})(a/a_{\text{eq}}) & (\lambda < \lambda_{\text{eq}}) \\ \delta_\lambda(t_{\text{enter}})(a/a_{\text{eq}})(\lambda_{\text{eq}}/\lambda)^2 & (\lambda_{\text{eq}} < \lambda) \end{cases} \tag{5.228}$$

or, equivalently,

$$\delta_M(t) = \begin{cases} L\delta_M(t_{\text{enter}})(a/a_{\text{eq}}) & (M < M_{\text{eq}}) \\ \delta_M(t_{\text{enter}})(a/a_{\text{eq}})(M_{\text{eq}}/M)^{2/3} & (M_{\text{eq}} < M) \end{cases}. \tag{5.229}$$

Thus the amplitude at late times is completely fixed by the amplitude of the modes when they enter the Hubble radius.

In this approach, to determine $\delta(\mathbf{x}, t)$ or $\delta_{\mathbf{k}}(t)$ at time t, we need to know its exact space dependence (or $\mathbf{k}$ dependence) at some initial instant $t = t_i$ [e.g., to determine $\delta(t, \mathbf{x})$, we need to know $\delta(t_i, \mathbf{x})$]. Often we are not interested in the *exact* form of $\delta(t, \mathbf{x})$ but only in its "statistical properties" in the following sense: We may assume that, for sufficiently small t_i, each Fourier mode $\delta_{\mathbf{k}}(t_i)$ was a Gaussian random variable with

$$\langle \delta_{\mathbf{k}}(t_i)\delta_{\mathbf{p}}^*(t_i)\rangle = (2\pi)^3 P(\mathbf{k}, t_i)\delta_D(\mathbf{k} - \mathbf{p}), \tag{5.230}$$

where $P(\mathbf{k}, t_i)$ is the power spectrum of $\delta(t_i, \mathbf{x})$ and $\langle \cdots \rangle$ denotes an ensemble average. Then,

$$\langle \delta_{\mathbf{k}}(t)\delta_{\mathbf{p}}^*(t)\rangle = T_{\mathbf{k}}(t, t_i)T_{\mathbf{p}}^*(t, t_i)\langle \delta_{\mathbf{k}}(t_i)\delta_{\mathbf{p}}^*(t_i)\rangle$$

$$= (2\pi)^3 |T_{\mathbf{k}}(t, t_i)|^2 P(\mathbf{k}, t_i)\delta_D(\mathbf{k} - \mathbf{p}) \tag{5.231}$$

and the statistical nature of $\delta_{\mathbf{k}}$ is preserved by evolution with the power spectrum evolving as

$$P(\mathbf{k}, t) = |T_{\mathbf{k}}(t, t_i)|^2 P(\mathbf{k}, t_i). \tag{5.232}$$

For any random field we can define a power spectrum and study its evolution along the lines described below. In the case of a *Gaussian* random field with zero mean, the power spectrum contains the complete information; in other cases the power spectrum will provide only partial information. This is the key difference between Gaussian and other statistics. Some theories of structure formation describing the origin of initial perturbations – like inflation, described in Section 5.7 – *predict* the statistics of the perturbations to be Gaussian. Because this seems to be fairly natural, we shall confine ourselves to this case in our discussion. It should be stressed that, as far as linear evolution of perturbations are concerned, the statistics of the perturbations are maintained.

Writing $\delta_{\mathbf{k}} = r_{\mathbf{k}} \exp i\phi_{\mathbf{k}}$, we can express the same probability distribution as a probability distribution for $r_{\mathbf{k}}$ and $\phi_{\mathbf{k}}$:

$$
\begin{aligned}
g_{\mathbf{k}}(r_{\mathbf{k}}, \phi_{\mathbf{k}}; t)\, dr_{\mathbf{k}}\, d\phi_{\mathbf{k}} &= \frac{1}{2\pi \mu_k^2} \exp\left(-\frac{1}{2} \frac{r_{\mathbf{k}}^2}{\mu_k^2} \right) \cdot r_{\mathbf{k}}\, dr_{\mathbf{k}}\, d\phi_{\mathbf{k}} \\
&= \frac{2(r_{\mathbf{k}}\, dr_{\mathbf{k}})}{\sigma_k^2} \left(\frac{d\phi_{\mathbf{k}}}{2\pi} \right) \exp\left(-\frac{r_{\mathbf{k}}^2}{\sigma_k^2} \right), \qquad \sigma_k^2 = 2\mu_k^2.
\end{aligned}
$$
(5.233)

This shows that the $\phi_{\mathbf{k}}$ are distributed uniformly in the allowed range of $(0, 2\pi)$ whereas $r_{\mathbf{k}}^2$ obeys an exponential distribution with $\langle r_{\mathbf{k}}^2 \rangle^{1/2} = \sigma_k = \sqrt{2}\mu_k$. Thus the phases of the $\delta_{\mathbf{k}}$ are in a completely random manner in the interval $(0, 2\pi)$.

Our analysis of $\delta_{\mathbf{k}}$ can be used to determine the growth of $P(k)$ as well. In practice, a more relevant quantity characterising the density inhomogeneity is $\Delta_k^2 \equiv [k^3 P(k)/2\pi^2]$, where $P(k) = |\delta_k|^2$ is the power spectrum. Physically, Δ_k^2 represent the power in each logarithmic interval of k. From Eq. (5.228) we find that the quantity behaves like

$$
\Delta_k^2 = \begin{cases} L^2(k)\Delta_k^2(t_{\text{enter}})(a/a_{\text{eq}})^2 & (\text{for } k_{\text{eq}} < k) \\ \Delta_k^2(t_{\text{enter}})(a/a_{\text{eq}})^2(k/k_{\text{eq}})^4 & (\text{for } k < k_{\text{eq}}) \end{cases}.
$$
(5.234)

Let us next determine $\Delta_k^2(t_{\text{enter}})$ if the initial power spectrum, when the mode was much larger than the Hubble radius, was a power law with $\Delta_k^2 \propto k^3 P(k) \propto k^{n+3}$. This mode was growing as a^2 while it was bigger than the Hubble radius (in the radiation-dominated phase). Hence $\Delta_k^2(t_{\text{enter}}) \propto a_{\text{enter}}^4 k^{n+3}$. In the radiation-dominated phase, we can relate a_{enter} to λ by noting that $\lambda a_{\text{enter}} \propto t_{\text{enter}} \propto a_{\text{enter}}^2$; so $\lambda \propto a_{\text{enter}} \propto k^{-1}$. Therefore

$$
\Delta_k^2(t_{\text{enter}}) \propto a_{\text{enter}}^4 k^{n+3} \propto k^{n-1}.
$$
(5.235)

Using this in Eq. (5.234) we find that

$$
\Delta_k^2 = \begin{cases} L^2(k)k^{n-1}(a/a_{\text{eq}})^2 & (\text{for } k_{\text{eq}} < k) \\ k^{n+3}(a/a_{\text{eq}})^2 & (\text{for } k < k_{\text{eq}}) \end{cases}.
$$
(5.236)

This is the shape of the power spectrum for $a > a_{\text{eq}}$. It retains its initial primordial shape ($\Delta_k^2 \propto k^{n+3}$) at very large scales ($k < k_{\text{eq}}$ or $\lambda > \lambda_{\text{eq}}$). At smaller scales, its amplitude is essentially reduced by four powers of k (from k^{n+3} to k^{n-1}). This arises because the small-wavelength modes enter the Hubble radius earlier on and their growth is suppressed more severely during the phase $a_{\text{enter}} < a < a_{\text{eq}}$.

Note that the index $n = 1$ (which is indeed generated by the inflationary models discussed in Section 5.7) is special. In this case, $\Delta_k^2(t_{\text{enter}})$ is independent of k and all the scales enter the Hubble radius with the same amplitude. The preceding analysis suggests that, if $n = 1$, then all scales in the range $k_{\text{eq}} < k$ will have nearly the same power except for the weak, logarithmic dependence through $L^2(k)$. Small scales will have slightly more power than the large scales because of this factor.

There is another – completely different – reason why the $n = 1$ spectrum is special. If $P(k) \propto k^n$, the power spectrum for gravitational potential $P_\varphi(k) \propto [P(k)/k^4]$ varies as $P_\varphi(k) \propto k^{n-4}$. The power per logarithmic band *in the gravitational potential* varies as $\Delta_\varphi^2 \equiv (k^3 P_\varphi(k)/2\pi^2) \propto k^{n-1}$. For $n = 1$, this is independent of k and each logarithmic interval in k space contributes the same amount of power to the gravitational potential. Hence *any* fundamental physical process that is scale invariant will generate a spectrum with $n = 1$. Thus observational verification of the index to $n = 1$ *only* verifies the fact that the fundamental process that led to the primordial fluctuations is scale invariant. The inflationary model, described in Section 5.7, has this property.

The transfer functions obtained by the simple reasoning above are approximate, and for more precise work it is necessary to integrate the equations of linear perturbation theory from some initial epoch until the present moment. Because this equation depends on $\dot{a}/a$ as well as on ρ_b, the results will depend crucially on the nature of dark matter as well as on the background cosmology. The general solution to this equation is, unfortunately, not expressible in closed form, and hence it is conventional to use a numerically fitted function for studies involving transfer functions. For the sake of completeness, we describe some of the commonly used transfer functions.

The simplest model for dark matter is based on CDM particles with a density parameter Ω_{NR} and a background cosmology with or without a cosmological constant.[3] In the presence of a cosmological constant we will assume that $\Omega_{\text{NR}} + \Omega_V = 1$. (This class of models is therefore parameterised by a single parameter Ω_{NR} and includes – as a special case – the pure CDM universe with $\Omega_{\text{NR}} = 1$.) In such a universe the transfer function is given by

$$T(q) = \frac{\ln(1 + 2.34q)}{2.34q} [1 + 3.89q + (16.1q)^2 + (5.46q)^3 + (6.71q)^4]^{-1/4},$$

(5.237)

where $q = (k/\Gamma h)$ Mpc^{-1} and

$$\Gamma = \Omega_{\text{NR}} h \exp(-\Omega_B - \sqrt{2h}\Omega_B/\Omega_{\text{NR}}).$$

(5.238)

These are, of course, empirical fitting functions but they are in reasonable agreement with the simulations.

In a model with both HDM and CDM, the transfer function becomes modified because of the existence of a smooth background. Fitting functions to describe the evolution in such mixed-dark-matter (MDM) models exist in the literature, and one reasonably accurate fitting function (that relates CDM and MDM models at a specific redshift) is given by

$$T_{\mathrm{MDM}}(k, z) = T_{\mathrm{CDM}}(k)\, D(k, z), \tag{5.239}$$

with

$$D(q, z) = \left[\frac{1 + (Aq)^2 + a_{\mathrm{eq}}(1 + z)(1 - \Omega_\nu)^{1/\beta}(Bq)^4}{1 + (Bq)^2 - (Bq)^3 + (Bq)^4} \right]^\beta, \tag{5.240}$$

where

$$\beta = \frac{5}{4}\left(1 - \sqrt{1 - 24\Omega_\nu/25}\right),$$

$$A = 17.266\frac{(1 + 10.912\Omega_\nu)\sqrt{\Omega_\nu(1 - 0.9465\Omega_\nu)}}{1 + (9.259\Omega_\nu)^2},$$

$$B = 2.6823\frac{1.1435}{\Omega_\nu + 0.1435},$$

$$a_{\mathrm{eq}} = \frac{4.212 \times 10^{-5}}{h^2}, \qquad q = \frac{k}{h^2}e^{2\Omega_B}. \tag{5.241}$$

A quantity closely related to the power spectrum is the two-point correlation function, defined as

$$\xi(\mathbf{x}) = \langle \delta(\mathbf{x} + \mathbf{y})\delta(\mathbf{y}) \rangle = \int \frac{d^3k}{(2\pi)^3} \frac{d^3p}{(2\pi)^3} \langle \delta_{\mathbf{k}}\delta_{\mathbf{p}}^* \rangle e^{i\mathbf{k}\cdot(\mathbf{x}+\mathbf{y})} e^{-i\mathbf{p}\cdot\mathbf{y}}, \tag{5.242}$$

where $\langle \cdots \rangle$ is the ensemble average. Using

$$\langle \delta_{\mathbf{k}}\delta_{\mathbf{p}}^* \rangle = (2\pi)^3 P(\mathbf{k})\delta_D(\mathbf{k} - \mathbf{p}), \tag{5.243}$$

we get

$$\xi(\mathbf{x}) = \int \frac{d^3k}{(2\pi)^3} P(\mathbf{k})e^{i\mathbf{k}\cdot\mathbf{x}}, \tag{5.244}$$

that is, the correlation function is the Fourier transform of the power spectrum. A simple property of $\xi(\mathbf{x})$ can be derived directly from its definition. Consider the integral

$$I = \int d^3x\, \xi(\mathbf{x}) = \int d^3x\, \langle \delta(\mathbf{y} + \mathbf{x})\delta(\mathbf{y}) \rangle. \tag{5.245}$$

Because the processes of statistical averaging and spatial integration can be interchanged, this integral can be written as

$$I = \left\langle \delta(\mathbf{y}) \int d^3x\, \delta(\mathbf{y} + \mathbf{x}) \right\rangle = 0, \tag{5.246}$$

where we have used the fact that the integral of $\delta(\mathbf{x})$ over all space vanishes. Because $\xi(\mathbf{x}) = \xi(|\mathbf{x}|) \equiv \xi(x)$, in an isotropic universe,

$$I = \int_0^\infty 4\pi x^2 \xi(x) \, dx = 0, \tag{5.247}$$

which implies that $\xi(x)$ should change sign at some scale $x = x_0$.

A physical interpretation for the correlation function can be given along the following lines: Suppose that $\delta(\mathbf{x})$ is such that there are regions with a typical size L on which $\delta > 0$ and there are similar regions with $\delta < 0$. If $|\mathbf{x}| \ll L$, then $(\mathbf{x} + \mathbf{y})$ and $(\mathbf{y})$ in Eq. (5.242) will both lie in an overdense or underdense region most of the time; in either case, the product $\delta(\mathbf{x} + \mathbf{y})\delta(\mathbf{y})$ will be positive, making $\xi(\mathbf{x})$ positive. On the other hand, if $|\mathbf{x}| \gtrsim L$, then $(\mathbf{x} + \mathbf{y})$ and $(\mathbf{y})$ are likely to be in regions with $\delta > 0$ and $\delta < 0$, making $\xi(\mathbf{x})$ negative. Therefore $\xi(\mathbf{x})$ will be positive for small $|\mathbf{x}|$ and will become negative for $x \gtrsim L$. Thus the first zero of $\xi(x)$ will give the typical size of overdense and underdense regions.

Finally, we mention a few other related measures of inhomogeneity. Given a variable $\delta(\mathbf{x})$, we can smooth it over some scale by using window functions $W(\mathbf{x})$ of suitable radius and shape [we suppress the t dependence in the notation, writing $\delta(\mathbf{x}, t)$ as $\delta(\mathbf{x})$]. Let the smoothed function be

$$\delta_W(\mathbf{x}) \equiv \int \delta(\mathbf{x} + \mathbf{y}) W(\mathbf{y}) \, d^3 \mathbf{y}. \tag{5.248}$$

Fourier transforming $\delta_W(\mathbf{x})$, we find that

$$\delta_W(\mathbf{x}) = \int \frac{d^3 \mathbf{k}}{(2\pi)^3} \delta_{\mathbf{k}} W_{\mathbf{k}}^* e^{i\mathbf{k}\cdot\mathbf{x}} \equiv \int \frac{d^3 \mathbf{k}}{(2\pi)^3} Q_{\mathbf{k}}. \tag{5.249}$$

If $\delta_{\mathbf{k}}$ is a Gaussian random variable, then $Q_{\mathbf{k}}$ is also a Gaussian random variable. Clearly $\delta_W(\mathbf{x})$ – which we obtain by adding several Gaussian random variables $Q_{\mathbf{k}}$ – is also a Gaussian random variable. Therefore, to find the probability distribution of $\delta_W(\mathbf{x})$, we need to know only the mean and the variance of $\delta_W(\mathbf{x})$. These are

$$\langle \delta_W(\mathbf{x}) \rangle = \int \frac{d^3 \mathbf{k}}{(2\pi)^3} \langle \delta_{\mathbf{k}} \rangle W_{\mathbf{k}}^* e^{i\mathbf{k}\cdot\mathbf{x}} = 0,$$

$$\langle \delta_W^2(\mathbf{x}) \rangle = \int \frac{d^3 \mathbf{k}}{(2\pi)^3} P(\mathbf{k}) |W_{\mathbf{k}}|^2 \equiv \mu^2. \tag{5.250}$$

Hence the probability of δ_W having a value q at any location is given by

$$\mathcal{P}(q) = \frac{1}{(2\pi\mu^2)^{1/2}} \exp\left(-\frac{q^2}{2\mu^2}\right). \tag{5.251}$$

Note that this is independent of $\mathbf{x}$, as expected.

A more interesting construct is based on the following question: What is the probability that the value of δ_W at two points $\mathbf{x_1}$ and $\mathbf{x_2}$ are q_1 and q_2? Once we choose $(\mathbf{x_1}, \mathbf{x_2})$, $\delta_W(\mathbf{x_1})$ and $\delta_W(\mathbf{x_2})$ are *correlated* Gaussians, with $\langle \delta_W(\mathbf{x_1})\delta_W(\mathbf{x_2}) \rangle = \xi_R(\mathbf{r})$, where $\mathbf{r} = \mathbf{x_1} - \mathbf{x_2}$. The simultaneous probability distribution for $\delta_W(\mathbf{x_1}) = q_1$ and $\delta_W(\mathbf{x_2}) = q_2$ for two correlated Gaussians is given by

$$P[q_1, q_2] = \frac{1}{2\pi\mu^2} \left(\frac{1}{1 - A^2} \right)^{1/2} \exp -Q[q_1, q_2], \qquad (5.252)$$

where

$$Q[q_1, q_2] = \frac{1}{2} \left(\frac{1}{1 - A^2} \right) \frac{1}{\mu^2} \left[q_1^2 + q_2^2 - 2Aq_1q_2 \right], \qquad (5.253)$$

with $A \equiv [\xi_R(r)/\mu]$. (We easily verify this by computing $\langle q_1 \rangle$, $\langle q_2 \rangle$ and $\langle q_i q_j \rangle$ explicitly.) We can now ask: What is the probability that both q_1 and q_2 are high-density peaks? Such a question is particularly relevant because we may expect high-density regions to be the locations of galaxy formation in the universe. Then the correlation function of the galaxies could be the correlation between the *high-density* peaks of the underlying Gaussian random field. This is easily computed as

$$P_2[q_1 > \nu\mu, q_2 > \nu\mu] = \int_{\nu\mu}^{\infty} dq_1 \int_{\nu\mu}^{\infty} dq_2 P[q_1, q_2] \equiv P_1^2(q > \nu\mu)[1 + \xi_\nu(r)],$$

$$(5.254)$$

where $\xi_\nu(r)$ denotes the correlation function for regions with a density that is ν times higher than the variance of the field. Explicit computation now gives

$$P_2 \propto \int_{\nu}^{\infty} dt_1 \int_{\nu}^{\infty} dt_2 \exp\left[-\frac{1}{2} \frac{1}{1 - A^2} \left(t_1^2 + t_2^2 - 2At_1t_2 \right) \right]. \qquad (5.255)$$

This result can be expressed in terms of an error function. An interesting special case in which this expression can be approximated occurs when $A \ll 1$ and $\nu \gg 1$ (although $A\nu^2$ is arbitrary). Then we get

$$P_2 \cong \frac{1}{2\pi} e^{-\nu^2} \exp(A\nu^2) \cong P_1^2(q > \nu\mu)\exp(A\nu^2), \qquad (5.256)$$

so that

$$\xi_\nu(r) = \exp(A\nu^2) - 1 = \exp\left[\frac{\nu^2}{\mu^2}\xi_R(r) \right] - 1. \qquad (5.257)$$

In other words, the correlation function of high-density peaks of a Gaussian random field can be significantly higher than the correlation function of the underlying field. If we further assume that $A \ll 1$, $\nu \gg 1$, and $A\nu^2 \ll 1$, then

$$\xi_\nu(r) \cong \nu^2 \frac{\xi_R(r)}{\xi_R(0)} = \left(\frac{\nu}{\mu}\right)^2 \xi_R(r). \qquad (5.258)$$

In this limit, $\xi_\nu(r)$, $\propto \xi_R(r)$, with the correlation increasing as ν^2. If we assume that (1) galaxies preferentially form at the peaks of the dark-matter density field and (2) the conditions $A \ll 1$, $\nu \gg 1$, and $A\nu^2 \ll 1$ are satisfied, then the correlation function of galaxies will be a constant multiple of the correlation function of the underlying dark-matter distribution. The ratio $b(r) \equiv [\xi_{\text{galaxy}}(r)/\xi_{\text{DM}}(r)]^{1/2}$ – called *bias* – will then be a useful quantity for describing the statistical distribution of galaxies. However, as we shall see in Chap. 7, galaxy formation involves several complex processes and cannot be described by a constant ratio between the two correlation functions. Hence the bias parameter b (which will, in general, be a function of scale r) can, at best, serve as a convenient shorthand notation describing the ratio between the two correlation functions.

In order to avoid possible misunderstanding, we stress the following fact: The correlation function for baryonic structures, like galaxies, will definitely be different – in general – from the correlation function for the underlying dark-matter distribution. It is also true that, at high redshifts, higher peaks of the density field will collapse and form structures before the lower peaks collapse. Hence there *will* exist a nontrivial bias between baryonic and dark-matter distributions in the universe. However, there is no reason to believe that such a bias is independent of (1) the scale at which the correlations are measured or (2) the epoch at which they are measured, or even (3) the sample (ellipticals, spirals, etc.) by which they are measured. It is in this sense that the bias b does not add to our knowledge, except in the form of a convenient notation.

A specific, simple example of the window function arises in the following context. Consider the mass contained within a sphere of radius R centred at some point $\mathbf{x}$ in the universe. As we change $\mathbf{x}$, keeping R constant, the mass enclosed by the sphere will vary randomly around a mean value $M_0 = (4\pi/3)\rho_{\text{bg}} R^3$, where ρ_{bg} is the matter density of the background universe. The mean-square fluctuation in this mass $\langle(\delta M/M)_R^2\rangle$ is a good measure of the inhomogeneities present in the universe at the scale R. In this case, the window function is $W(\mathbf{y}) = 1$ for $|\mathbf{y}| \le R$ and zero otherwise. The variance in Eqs. (5.250) becomes

$$\sigma_{\text{sph}}^2(R) = \langle\delta_W^2\rangle = \int \frac{d^3k}{(2\pi)^3} P(k) W_{\text{sph}}(k)$$
$$= \int_0^\infty \frac{dk}{k}\left(\frac{k^3 P}{2\pi^2}\right)\left[\frac{3(\sin kR - kR\cos kR)}{k^3 R^3}\right]^2. \qquad (5.259)$$

This will be a useful statistic in many contexts.

Another quantity that we will use extensively in later sections is the average value of the correlation function within a sphere of radius r, defined as

$$\bar{\xi} = \frac{3}{r^3} \int_0^r \xi(x) x^2 \, dx. \tag{5.260}$$

Using

$$\xi(\mathbf{x}) \equiv \int \frac{d^3 \mathbf{k}}{(2\pi)^3} P(\mathbf{k}) e^{i\mathbf{k}\cdot\mathbf{x}} = \int_0^\infty \frac{dk}{k} \left[\frac{k^3 P(k)}{2\pi^2} \right] \left(\frac{\sin kx}{kx} \right) \tag{5.261}$$

and Eq. (5.260), we find that

$$
\begin{aligned}
\bar{\xi}(r) &= \frac{3}{r^3} \int_0^\infty \frac{dk}{k^2} \left(\frac{k^3 P}{2\pi^2} \right) \int_0^r dx (x \sin kx) \\
&= \frac{3}{2\pi^2 r^3} \int_0^\infty \frac{dk}{k} P(k)(\sin kr - kr \cos kr).
\end{aligned}
\tag{5.262}
$$

A simple computation relates $\sigma_{\text{sph}}^2(R)$ to $\xi(x)$ and $\bar{\xi}(x)$. We can show that

$$\sigma_{\text{sph}}^2(R) = \frac{3}{R^3} \int_0^{2R} x^2 \, dx \, \xi(x) \left(1 - \frac{x}{2R} \right)^2 \left(1 + \frac{x}{4R} \right), \tag{5.263}$$

$$\sigma_{\text{sph}}^2(R) = \frac{3}{2} \int_0^{2R} \frac{dx}{(2R)} \bar{\xi}(x) \left(\frac{x}{R} \right)^3 \left[1 - \left(\frac{x}{2R} \right)^2 \right]. \tag{5.264}$$

Note that σ_{sph}^2 at R is determined entirely by $\xi(x)$ [or $\bar{\xi}(x)$] in the range $0 \le x \le 2R$.

The Gaussian nature of δ_k cannot be maintained if the evolution couples the modes for different values of $\mathbf{k}$. Equation (5.94), which describes the evolution of $\delta_\mathbf{k}(t)$, shows that the modes do mix with each other as time goes on. Thus, in general, the Gaussian nature of $\delta_\mathbf{k}$ cannot be maintained in the nonlinear epochs.

Exercise 5.12
Truncated power laws: Consider a power spectrum of the form $\Delta^2 = (k/k_0)^{n+3}$ $\times \exp(-k/k_c)$. Show that the corresponding correlation function is given by

$$\xi(r) = \frac{(k_c/k)^{n+3}}{y(1+y^2)^{1+n/2}} \Gamma(2+n) \sin\left[(2+n) \arctan y \right], \quad y = k_c r. \tag{5.265}$$

Exercise 5.13
Lower-dimensional power spectra: Consider a Gaussian random field in three dimensions that has a power spectrum $\Delta_{3D}^2(k)$. By considering the values of the field along a randomly chosen straight line or on a randomly chosen plane in three dimensions, we can

construct a corresponding Gaussian random field in one dimension and two dimensions, respectively. Prove that

$$\Delta_{1D}^2 = k \int_k^\infty \Delta_{3D}^2(y) y^{-2}\, dy, \qquad \Delta_{2D}^2 = k^2 \int_k^\infty \Delta_{3D}^2(y) \frac{y^{-2}\, dy}{\sqrt{y^2 - k^2}}. \tag{5.266}$$

(Hint: Use the fact that the Gaussian random field is isotropic, which implies that the correlation function should not change under dimensional reduction.)

Exercise 5.14
Peaks in Gaussian random fields: Consider a Gaussian random field in one dimension with the Fourier coefficients δ_k. Define the parameters

$$\sigma_m^2 \equiv \left(\frac{1}{2\pi}\right) \int_{-\infty}^\infty |\delta_k|^2 k^{2m}\, dk \tag{5.267}$$

that measure the moments in k space. Prove that the number density of points along one dimension at which the density field is stationary [that is, $\delta' = 0$] and has a height $v\sigma_0$ is given by

$$dN = \frac{e^{-Q/2}}{(2\pi)^{3/2}(1-\gamma^2)^{1/2} R_*} |x|\, dx\, dv, \qquad Q = \frac{(v - \gamma x)^2}{1 - \gamma^2} + x^2, \qquad x \equiv -\frac{\delta''}{\sigma_2}, \tag{5.268}$$

where

$$v \equiv \frac{\delta}{\sigma_0}, \qquad \gamma \equiv \frac{\sigma_1^2}{\sigma_0 \sigma_2}, \qquad R_* \equiv \sqrt{n}\frac{\sigma_1}{\sigma_2}. \tag{5.269}$$

The peaks in the Gaussian random field will correspond to $x > 0$. Show that the integral peak density is

$$P(> v) = \frac{1}{2}\left(\mathrm{erfc}\left[\frac{v}{\sqrt{2(1 - \gamma^2)}}\right] + \gamma e^{-v^2/2}\left\{1 + \mathrm{erf}\left[\frac{\gamma v}{\sqrt{2(1 - \gamma^2)}}\right]\right\}\right) \tag{5.270}$$

and the total peak density is $N_{\mathrm{pk}} = (1/2\pi R_*)$. [Hint: If $\delta(x)$ is a Gaussian random field, so are δ', δ'', ..., etc. The joint probability distribution will then be given by

$$p(\delta, \delta', \delta'', \dots,) = \frac{|\mathbf{M}|^{1/2}}{(2\pi)^{m/2}} \exp\left(-\frac{1}{2}\mathbf{V}^\mathsf{T}\mathbf{M}\mathbf{V}\right), \tag{5.271}$$

where $\mathbf{V} = (\delta, \delta', \delta'', \dots,)$ and $\mathbf{M}$ is the covariance matrix. Consider the case of having only the field, its first derivative, and its second derivative. Express the covariance matrix in terms of σ_m^2. The stationary points will correspond to $\delta' = 0$, which needs to be expressed properly in terms of the volume element in the space made of the components of the vector $\mathbf{V}$. The rest of the results follow from straightforward integration.]

5.9 Zeldovich Approximation

We next consider the evolution of perturbations in the nonlinear epochs. This is an intrinsically complex problem, and the only exact procedure for studying it involves setting up large-scale numerical simulations. Unfortunately, numerical simulations tend to obscure the basic physics contained in the equations and essentially acts as a "black box." Hence it is worthwhile to analyse the nonlinear regime by use of some simple analytic approximations in order to obtain insights into the problem. In Sections 5.9–5.11 we shall describe a series of such approximations with an increasing degree of complexity. The first one – called the Zeldovich approximation – is fairly simple and leads to an idea of the kind of structures that generically form in the universe. This approximation, however, is not of much use for more detailed work. The second approximation described in Sections 5.10 and 5.11 is more useful and allows the modelling of the universe based on the evolution of the initially overdense region. Finally in Section 5.12 we discuss a fairly sophisticated approach involving nonlinear scaling relations that are present in the dynamics of gravitational clustering. In between the discussion of these approximations, we also describe some useful procedures which can be adopted to answer questions that are directly relevant to structure formation.

We can obtain a useful insight into the nature of linear perturbation theory (as well as nonlinear clustering) by examining the nature of particle trajectories that lead to the growth of the density contrast $\delta_L(a) \propto a$. To determine the particle trajectories corresponding to the linear limit, let us start by writing the trajectories in the form

$$\mathbf{x}_T(a, \mathbf{q}) = \mathbf{q} + \mathbf{L}(a, \mathbf{q}), \tag{5.272}$$

where $\mathbf{q}$ is the Lagrangian coordinate (indicating the original position of the particle) and $\mathbf{L}(a, \mathbf{q})$ is the displacement. The corresponding Fourier transform of the density contrast is given by the general expression

$$\delta(a, \mathbf{k}) = \int d^3\mathbf{q} \, e^{-i\mathbf{k}\cdot\mathbf{q} - i\mathbf{k}\cdot\mathbf{L}(a,\mathbf{q})} - (2\pi)^3 \delta_{\text{Dirac}}[\mathbf{k}]. \tag{5.273}$$

In the linear regime, we expect most of the particles to have moved very little, and hence we can expand the integrand in Eq. (5.273) in a Taylor series in $(\mathbf{k} \cdot \mathbf{L})$. This gives, to the lowest order,

$$\delta(a, \mathbf{k}) \cong -\int d^3\mathbf{q} \, e^{-i\mathbf{k}\cdot\mathbf{q}}[i\mathbf{k} \cdot \mathbf{L}(a, \mathbf{q})] = -\int d^3\mathbf{q} \, e^{-i\mathbf{k}\cdot\mathbf{q}} \left(\nabla_q \cdot \mathbf{L}\right), \tag{5.274}$$

showing that $\delta(a, \mathbf{k})$ is the Fourier transform of $-\nabla_q \cdot \mathbf{L}(a, \mathbf{q})$. This allows us to identify $\nabla \cdot \mathbf{L}(a, \mathbf{q})$ with the original density contrast in real space $-\delta(a, \mathbf{q})$. Using the Poisson equation (for $\Omega = 1$, which is assumed for simplicity), we can write $\delta(a, \mathbf{q})$ as a divergence; that is,

$$\nabla \cdot \mathbf{L}(a, \mathbf{q}) = -\delta(a, \mathbf{q}) = -\frac{2}{3}H_0^{-2}a\nabla \cdot (\nabla\phi), \qquad (5.275)$$

which, in turn shows that *a consistent set* of displacements that will lead to $\delta(a) \propto a$ is given by

$$\mathbf{L}(a, \mathbf{q}) = -(\nabla\psi)a \equiv a\mathbf{u}(\mathbf{q}), \qquad \psi \equiv (2/3)H_0^{-2}\phi. \qquad (5.276)$$

The trajectories in this limit are therefore linear in a:

$$\mathbf{x}_T(a, \mathbf{q}) = \mathbf{q} + a\mathbf{u}(\mathbf{q}). \qquad (5.277)$$

A useful approximation to describe the quasi-linear stages of clustering is obtained by use of the trajectory in Eq. (5.277) as an ansatz valid *even at quasi-linear epochs*. In this approximation, called the Zeldovich approximation, the proper Eulerian position $\mathbf{r}$ of a particle is related to its Lagrangian position $\mathbf{q}$ by

$$\mathbf{r}(t) \equiv a(t)\mathbf{x}(t) = a(t)[\mathbf{q} + a(t)\mathbf{u}(\mathbf{q})], \qquad (5.278)$$

where $\mathbf{x}(t)$ is the comoving Eulerian coordinate. This relation in Eq. (5.277) gives the comoving position ($\mathbf{x}$) and proper position ($\mathbf{r}$) of a particle at time t, given that at some time in the past it had the comoving position $\mathbf{q}$. If the initial, unperturbed density is $\bar{\rho}$ (which is independent of $\mathbf{q}$), then the conservation of mass implies that the perturbed density will be

$$\rho(\mathbf{r}, t)d^3\mathbf{r} = \bar{\rho}d^3\mathbf{q}. \qquad (5.279)$$

Therefore

$$\rho(\mathbf{r}, t) = \bar{\rho}\left[\det\left(\frac{\partial q_i}{\partial r_j}\right)\right]^{-1} = \frac{\bar{\rho}/a^3}{\det(\partial x_j/\partial q_i)} = \frac{\rho_b(t)}{\det[\delta_{ij} + a(t)(\partial u_j/\partial q_i)]}, \qquad (5.280)$$

where we have set $\rho_b(t) = [\bar{\rho}/a^3(t)]$. Because $\mathbf{u}(\mathbf{q})$ is a gradient of a scalar function, the Jacobian in the denominator of Eq. (5.280) is the determinant of a real symmetric matrix. This matrix can be diagonalised at every point $\mathbf{q}$ to yield a set of eigenvalues and principal axes as functions of $\mathbf{q}$. If the eigenvalues of $(\partial u_j/\partial q_i)$ are $[-\lambda_1(\mathbf{q}), -\lambda_2(\mathbf{q}), -\lambda_3(\mathbf{q})]$ then the perturbed density is given by

$$\rho(\mathbf{r}, t) = \frac{\rho_b(t)}{[1 - a(t)\lambda_1(\mathbf{q})][1 - a(t)\lambda_2(\mathbf{q})][1 - a(t)\lambda_3(\mathbf{q})]}, \qquad (5.281)$$

where $\mathbf{q}$ can be expressed as a function of $\mathbf{r}$ by the solution of (5.278). This expression describes the effect of deformation of an infinitesimal, cubical volume (with the faces of the cube determined by the eigenvectors corresponding to λ_n) and the consequent change in the density. A positive λ denotes collapse and negative λ signals expansion.

In a overdense region, the density will become infinite if one of the terms in brackets in the denominator of (5.281) becomes zero. In the generic case, these eigenvalues will be different from each other so that we can take, say, $\lambda_1 \geq \lambda_2 \geq \lambda_3$. At any particular value of $\mathbf{q}$ the density will diverge for the first time when $[1 - a(t)\lambda_1] = 0$; at this instant the material contained in a cube in the $\mathbf{q}$ space gets compressed to a sheet in the $\mathbf{r}$ space along the principal axis corresponding to λ_1. Thus sheetlike structures, or "pancakes," will be the first nonlinear structures to form when gravitational instability amplifies density perturbations.

The trajectories in the Zeldovich approximation, given by Eq. (5.277) can be used in Eq. (5.99) to provide a *closed*-integral equation for $\phi_\mathbf{k}$. In this case,

$$
\mathbf{x}_T(\mathbf{q}, a) = \mathbf{q} + a\nabla\psi, \qquad \dot{\mathbf{x}}_T = \left(\frac{2a}{3t}\right)\nabla\psi, \qquad \psi = \frac{2}{3H_0^2}\varphi, \qquad (5.282)
$$

and, to the same order of accuracy, $B_\mathbf{k}$ in Eq. (5.96) becomes

$$
\int d^3\mathbf{q}\,(\mathbf{k} \cdot \dot{\mathbf{x}}_T)^2\, e^{-i\mathbf{k}\cdot(\mathbf{q}+\mathbf{L})} \cong \int d^3\mathbf{q}(\mathbf{k} \cdot \dot{\mathbf{x}}_T)^2 e^{-i\mathbf{k}\cdot\mathbf{q}}. \qquad (5.283)
$$

Substituting these expressions into Eq. (5.99), we find that the gravitational potential is described by the closed-integral equation:

$$
\ddot{\phi}_\mathbf{k} + 4\frac{\dot{a}}{a}\dot{\phi}_\mathbf{k} = -\frac{1}{3a^2}\int \frac{d^3\mathbf{p}}{(2\pi)^3}\phi_{\frac{1}{2}\mathbf{k}+\mathbf{p}}\phi_{\frac{1}{2}\mathbf{k}-\mathbf{p}}\mathcal{G}(\mathbf{k}, \mathbf{p}),
$$

$$
\mathcal{G}(\mathbf{k}, \mathbf{p}) = \frac{7}{8}k^2 + \frac{3}{2}p^2 - 5\left(\frac{\mathbf{k}\cdot\mathbf{p}}{k}\right)^2. \qquad (5.284)
$$

This equation provides a powerful method for analysing nonlinear clustering as estimating $(A_\mathbf{k} - B_\mathbf{k})$ by the Zeldovich approximation has a very large domain of applicability.

It is also possible to determine the power spectrum corresponding to these trajectories by using our general formula[4]

$$
P(\mathbf{k}, a) = |\delta(\mathbf{k}, a)|^2 = \int d^3q d^3q' e^{-i\mathbf{k}\cdot(\mathbf{q}-\mathbf{q}')}\langle e^{-i\mathbf{k}\cdot[\mathbf{L}(a,\mathbf{q})-\mathbf{L}(a,\mathbf{q}')]}\rangle. \qquad (5.285)
$$

The ensemble averaging can be performed with the general result for Gaussian random fields:

$$
\langle e^{i\mathbf{k}\cdot\mathbf{V}}\rangle = \exp[-k_i k_j \sigma^{ij}(V)/2], \qquad (5.286)
$$

where σ^{ij} is the covariance matrix for the components V^a of a Gaussian random field. This quantity can be expressed in terms of the power spectrum $P_L(k)$ in the linear theory, and a straightforward analysis gives

$$
P(k, a) = \int_0^\infty 2\pi q^2\, dq \int_{-1}^{+1} d\mu\, e^{ikq\mu} \exp -k^2[F(q) + \mu^2 q F'(q)], \qquad (5.287)
$$

where

$$F(q) = \frac{a^2}{2\pi^2} \int_0^\infty dk \, P_L(k) \frac{j_1(kq)}{kq}. \tag{5.288}$$

The integrals, unfortunately, need to be evaluated numerically except in the case of $n = -2$. In this case, we get

$$\Delta^2(k, a) \equiv \frac{k^3 P}{2\pi^2} = \frac{16}{\pi} \frac{a^2 k}{[1 + (2a^2 k)^2]^2} \left\{ 1 + \frac{3\pi}{4} \frac{a^2 k}{[1 + (2a^2 k)^2]^{1/2}} \right\}, \tag{5.289}$$

which shows that $\Delta^2 \propto a^2$ for small a but decays as a^{-2} at late times because of the dispersion of particles. Clearly, the Zeldovich approximation breaks down beyond a particular epoch and is of limited validity.

Exercise 5.15
Power spectrum in the Zeldovich approximation: Obtain the results in Eqs. (5.287) and (5.289).

5.10 Spherical Approximation

In the nonlinear regime – when $\delta \gtrsim 1$ – it is not possible to solve Eq. (5.94) exactly. Some progress, however, can be made if we assume that the trajectories are homogeneous, i.e., $\mathbf{x}(t, \mathbf{q}) = f(t)\mathbf{q}$, where $f(t)$ is to be determined. In this case, the density contrast is

$$\delta_{\mathbf{k}}(t) = \int d^3 q \, e^{-if(t)\mathbf{k}\cdot\mathbf{q}} - (2\pi)^3 \delta_D(\mathbf{k})$$

$$= (2\pi)^3 \delta_D(\mathbf{k})[f^{-3} - 1] \equiv (2\pi)^3 \delta_D(\mathbf{k})\delta(t), \tag{5.290}$$

where we have defined $\delta(t) \equiv [f^{-3}(t) - 1]$ as the amplitude of the density contrast for the $\mathbf{k} = 0$ mode. It is now straightforward to compute A and B in Eq. (5.94). We have

$$A = 4\pi G\rho_b \delta^2(t)[(2\pi)^3 \delta_D(\mathbf{k})], \tag{5.291}$$

$$B = \int d^3 q (k^a q_a)^2 \dot{f}^2 e^{-if(k_a q^a)} = -\dot{f}^2 \frac{\partial^2}{\partial f^2} [(2\pi)^3 \delta_D(f\mathbf{k})]$$

$$= -\frac{4}{3} \frac{\dot{\delta}^2}{(1+\delta)} [(2\pi)^3 \delta_D(\mathbf{k})], \tag{5.292}$$

so that Eq. (5.94) becomes

$$\ddot{\delta} + 2\frac{\dot{a}}{a}\dot{\delta} = 4\pi G\rho_b(1+\delta)\delta + \frac{4}{3} \frac{\dot{\delta}^2}{(1+\delta)}. \tag{5.293}$$

To understand what this equation means, let us consider, at some initial epoch t_i, a spherical region of the universe that has a slight constant overdensity compared with the background. As the universe expands, the overdense region will expand more slowly compared with the background, will reach a maximum radius, contract, and virialise to form a bound nonlinear system. Such a model is called a *spherical top hat*. For this spherical region of radius $R(t)$ containing dustlike matter of mass M (in addition to other forms of energy densities), the density contrast for dust will be given by

$$1 + \delta = \frac{\rho}{\rho_b} = \frac{3M}{4\pi R^3(t)} \frac{1}{\rho_b(t)} = \frac{2GM}{\Omega_{NR} H_0^2 a_0^3} \left[\frac{a(t)}{R(t)} \right]^3 \equiv \mu \frac{a^3}{R^3}. \tag{5.294}$$

[Note that, with this definition, $f \propto (R/a)$.] Using this in Eq. (5.293) we can obtain an equation for $R(t)$ from the equation for δ; straightforward algebra now gives

$$\ddot{R} = -\frac{GM}{R^2} - \frac{4\pi G}{3}(\rho + 3p)_{rest} R. \tag{5.295}$$

This equation could have been written "by inspection" with the relations

$$\ddot{R} = -\nabla\phi_{tot}, \qquad \phi_{tot} = \phi_{FRW} + \delta\phi = -(\ddot{a}/2a)R^2 - G\delta M/R. \tag{5.296}$$

Note that this equation is valid for perturbed "dustlike" matter in *any* background space–time with density ρ_{rest} and pressure p_{rest} contributed by the rest of the matter. Our homogeneous trajectories $x(q, t) = f(t)q$ actually describe the spherical top-hat model.

This model is particularly simple for the $\Omega = 1$, matter-dominated universe, in which $\rho_{rest} = p_{rest} = 0$, and we have to solve the equation

$$\frac{d^2 R}{dt^2} = -\frac{GM}{R^2}. \tag{5.297}$$

This is the equation for the radial Kepler problem and can be solved by standard techniques. The final result for the evolution of a spherical overdense region can be summarised by the following relations:

$$R(t) = \frac{R_i}{2\delta_i}(1 - \cos\theta) = \frac{3x}{10\delta_0}(1 - \cos\theta), \tag{5.298}$$

$$t = \frac{3t_i}{4\delta_i^{3/2}}(\theta - \sin\theta) = \left(\frac{3}{5}\right)^{3/2} \frac{3t_0}{4\delta_0^{3/2}}(\theta - \sin\theta), \tag{5.299}$$

$$\rho(t) = \rho_b(t)\frac{9(\theta - \sin\theta)^2}{2(1 - \cos\theta)^3}. \tag{5.300}$$

The density can be expressed in terms of the redshift by the relation $(t/t_i)^{2/3} =$

$(1 + z_i)(1 + z)^{-1}$. This gives

$$(1 + z) = \left(\frac{4}{3}\right)^{2/3} \frac{\delta_i(1 + z_i)}{(\theta - \sin\theta)^{2/3}} = \left(\frac{5}{3}\right)\left(\frac{4}{3}\right)^{2/3} \frac{\delta_0}{(\theta - \sin\theta)^{2/3}}, \quad (5.301)$$

$$\delta = \frac{9}{2} \frac{(\theta - \sin\theta)^2}{(1 - \cos\theta)^3} - 1. \quad (5.302)$$

Given an initial density contrast δ_i at redshift z_i, these equations define (implicitly) the function $\delta(z)$ for $z > z_i$. Equation (5.301) defines θ in terms of z (implicitly); Eq. (5.302) gives the density contrast at that $\theta(z)$.

For comparison, note that linear evolution gives the density contrast δ_L, where

$$\delta_L = \frac{\bar{\rho}_L}{\rho_b} - 1 = \frac{3}{5} \frac{\delta_i(1 + z_i)}{1 + z} = \frac{3}{5}\left(\frac{3}{4}\right)^{2/3} (\theta - \sin\theta)^{2/3}. \quad (5.303)$$

We can estimate the accuracy of the linear theory by comparing $\delta(z)$ and $\delta_L(z)$. To begin with, for $z \gg 1$, we have $\theta \ll 1$ and we get $\delta(z) \simeq \delta_L(z)$. When $\theta = (\pi/2)$, $\delta_L = (3/5)(3/4)^{2/3}(\pi/2 - 1)^{2/3} = 0.341$ and $\delta = (9/2)(\pi/2 - 1)^2 - 1 = 0.466$; thus the actual density contrast is $\sim$40% higher. When $\theta = (2\pi/3)$, $\delta_L = 0.568$ and $\delta = 1.01 \simeq 1$. If we interpret $\delta = 1$ as the transition point to nonlinearity, then such a transition occurs at $\theta = (2\pi/3)$, $\delta_L \simeq 0.57$. From Eq. (5.301), we see that this occurs at the redshift $(1 + z_{nl}) = 1.06\delta_i(1 + z_i) = (\delta_0/0.57)$.

The spherical region reaches the maximum radius of expansion at $\theta = \pi$. From our equations, we find that the redshift z_m, the proper radius of the shell r_m, and the average density contrast δ_m at "turnaround" are

$$(1 + z_m) = \frac{\delta_i(1 + z_i)}{\pi^{2/3}(3/4)^{2/3}} = 0.57(1 + z_i)\delta_i$$

$$= \frac{5}{3} \frac{\delta_0}{(3\pi/4)^{2/3}} \simeq \frac{\delta_0}{1.062}, \quad (5.304)$$

$$r_m = \frac{3x}{5\delta_0}, \quad \left(\frac{\bar{\rho}}{\rho_b}\right)_m = 1 + \bar{\delta}_m = \frac{9\pi^2}{16} \approx 5.6.$$

The first equality of Eq. (5.304) gives the redshift at turnaround for a region parameterised by the (hypothetical) linear density contrast δ_0 extrapolated to the present epoch. If, for example, $\delta_i \simeq 10^{-3}$ at $z_i \simeq 10^4$, such a perturbation would have turned around at $(1 + z_m) \simeq 5.7$ or when $z_m \simeq 4.7$. The second equality gives the maximum radius reached by the perturbation. The third equality shows that the region under consideration is nearly six times denser than the background universe at turnaround. This corresponds to a density contrast of $\delta_m \approx 4.6$, which is definitely in the nonlinear regime. The linear evolution gives $\delta_L = 1.063$ at $\theta = \pi$.

After the spherical overdense region turns around it will continue to contract. Equation (5.300) suggests that at $\theta = 2\pi$ all the mass will collapse to a point.

However, long before this happens, the approximation that matter is distributed in spherical shells and that random velocities of the particles are small [implicit in the assumption of homogeneous trajectories $\mathbf{x} = f(t)\mathbf{q}$] will break down. The collisionless (dark-matter) component will relax to a configuration with radius r_{vir}, velocity dispersion v, and density ρ_{coll}. After virialisation of the collapsed shell, the potential energy U and the kinetic energy K will be related by $|U| = 2K$ so that the total energy $\mathcal{E} = U + K = -K$. At $t = t_m$ all the energy was in the form of potential energy. For a spherically symmetric system with constant density, $\mathcal{E} \approx -3GM^2/5r_m$. The "virial velocity" v and the "virial radius" r_{vir} for the collapsing mass can be estimated by

$$K \equiv \frac{Mv^2}{2} = -\mathcal{E} = \frac{3GM^2}{5r_m}, \qquad |U| = \frac{3GM^2}{5r_{vir}} = 2K = Mv^2. \qquad (5.305)$$

We get

$$v = (6GM/5r_m)^{1/2}, \qquad r_{vir} = r_m/2. \qquad (5.306)$$

The time taken for the fluctuation to reach virial equilibrium, t_{coll}, is essentially the time corresponding to $\theta = 2\pi$. From Eq. (5.301), we find that the redshift at collapse, z_{coll}, is

$$(1 + z_{coll}) = \frac{\delta_i(1 + z_i)}{(2\pi)^{2/3}(3/4)^{2/3}} = 0.36\delta_i(1 + z_i) = 0.63(1 + z_m) = \frac{\delta_0}{1.686}. \qquad (5.307)$$

The density of the collapsed object can also be determined fairly easily. Because $r_{vir} = (r_m/2)$, the mean density of the collapsed object is $\rho_{coll} = 8\rho_m$, where ρ_m is the density of the object at turnaround.

We have $\rho_m \cong 5.6\rho_b(t_m)$ and $\rho_b(t_m) = (1 + z_m)^3 (1 + z_{coll})^{-3}\rho_b(t_{coll})$. Combining these relations, we get

$$\rho_{coll} \simeq 2^3\rho_m \simeq 44.8\rho_b(t_m) \simeq 170\rho_b(t_{coll}) \simeq 170\rho_0(1 + z_{coll})^3, \qquad (5.308)$$

where ρ_0 is the present cosmological density. This result determines ρ_{coll} in terms of the redshift of formation of a bound object. Once the system has been virialised, its density and size does not change. Because $\rho_b \propto a^{-3}$, the density contrast δ increases as a^3 for $t > t_{coll}$. The evolution is described schematically in Fig. 5.3.

Exercise 5.16

Approximating a sphere: In the preceding discussion we could obtain δ in terms of δ_L only in parametric form through the angle θ. Try an approximate fit of the form $1 + \delta = [1 - (\delta_L/n)]^{-n}$, with n as a free parameter. Determine n so that this relation fits the spherical evolution as accurately as posiible. Comment on the result.

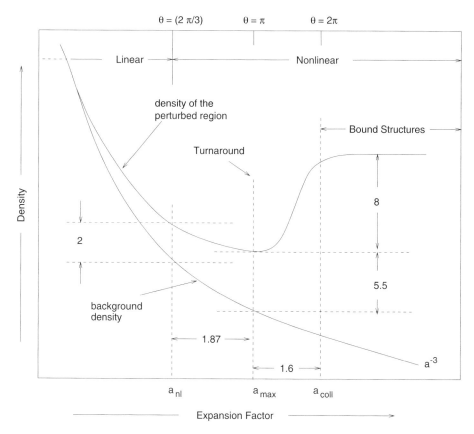

Fig. 5.3. Evolution of an overdense region in a spherical top-hat approximation.

Exercise 5.17
Spherical top hat in an $\Omega \neq 1$ universe: Generalise the discussion in the text for a spherical top-hat model in an underdense universe with $\Omega < 1$. Show that the results for the density contrast at virialisation can be well approximated by the formula $(1 + \delta) \approx 170\,\Omega^{-0.7}$.

This approach can be easily generalised to describe the situation in which the initial density profile is given by $\rho(r_i)$. Given an initial density profile $\rho_i(r)$, we can calculate the mass $M(r_i)$ and energy $E(r_i)$ of each shell labelled by the initial radius r_i. In spherically symmetric evolution, M and E are conserved and each shell will be described by Eq. (5.297). Assuming that the average density contrast $\bar{\delta}_i(r_i)$ decreases with r_i, the shells will never cross during the evolution. Each shell will evolve in accordance with Eqs. (5.298) and (5.299) with δ_i replaced with the mean initial density contrast $\bar{\delta}_i(r_i)$ characterising the shell of initial radius r_i. Equation (5.300) gives the mean density inside each of the shells from which the density profile can be computed at any given instant.

A simple example for this case corresponds to a scale-invariant situation in which $E(M)$ is a power law. If the energy of a shell containing mass M is taken to be

$$E(M) = E_0 \left(\frac{M}{M_0} \right)^{2/3-\epsilon} < 0, \tag{5.309}$$

then the turnaround radius and turnaround time are given by

$$r_m(M) = -\frac{GM}{E(M)} = -\frac{GM_0}{E_0} \left(\frac{M}{M_0} \right)^{\frac{1}{3}+\epsilon}, \tag{5.310}$$

$$t_m(M) = \frac{\pi}{2} \left(\frac{r_m^3}{2GM} \right)^{1/2} = \frac{\pi GM}{(-E_0/2)^{3/2}} \left(\frac{M}{M_0} \right)^{3\epsilon/2}. \tag{5.311}$$

To avoid shell crossing, we must have $\epsilon > 0$ so that outer shells with more mass turn around at later times. In such a case, the inner shells expand, turn around, collapse and virialise first, and the virialisation proceeds progressively to the outer shells. We assume that each virialised shell settles down to a final radius that is a fixed fraction of the maximum radius. Then the density in the virialised part will scale as (M/r^3), where M is the mass contained inside a shell whose turnaround radius is r. Using Eq. (5.310) to relate the turnaround radius and mass, we find that

$$\rho(r) \propto \frac{M(r_m = r)}{r^3} \propto r^{3/(1+3\epsilon)} r^{-3} \propto r^{-9\epsilon/(1+3\epsilon)}. \tag{5.312}$$

Two special cases of this scaling relation are worth mentioning: (1) If the energy of each shell is dominated by a central mass m located at the origin, then $E \propto Gm/r \propto M^{-1/3}$. In that case, $\epsilon = 1$ and the density profile of the virialised region falls as $r^{-9/4}$. The situation corresponds to an accretion onto a massive object. (2) If $\epsilon = 2/3$, then the binding energy E is the same for all shells. Then we get $\rho \propto r^{-2}$, which corresponds to an isothermal sphere.

Exercise 5.18
Generalisation of a spherical model: Generalise the model discussed in the text to the case with trajectories $x^a(t, \mathbf{q}) = f^{ab}(t)q_b$.

[Hint: In this case, it is convenient to decompose the derivative of the velocity $\partial_a u_b = \dot{f}_{ab}$ into shear σ_{ab}, rotation Ω^c, and expansion θ by writing

$$\dot{f}_{ab} = \sigma_{ab} + \epsilon_{abc}\Omega^c + \frac{1}{3}\delta_{ab}\theta, \tag{5.313}$$

where σ_{ab} is the symmetric traceless part of f_{ab}; $\epsilon_{abc}\Omega^c$ is the antisymmetric part and $(1/3)\delta_{ab}\theta$ is the trace. Then Eq. (5.293) gets generalised to

$$\ddot{\delta} + 2\frac{\dot{a}}{a}\dot{\delta} = 4\pi G\rho_b(1+\delta)\delta + \frac{4}{3}\frac{\dot{\delta}^2}{(1+\delta)} + \dot{a}^2(1+\delta)(\sigma^2 - 2\Omega^2), \qquad (5.314)$$

where $\sigma^2 \equiv \sigma^{ab}\sigma_{ab}$ and $\Omega^2 \equiv \Omega^i\Omega_i$. From the last term on the right-hand side we see that shear contributes positively to $\ddot{\delta}$ whereas rotation Ω^2 contributes negatively. Thus shear helps the growth of inhomogenities and rotation works against it. To see this explicitly, we again introduce a function $R(t)$ by the definition

$$1 + \delta = \frac{9GMt^2}{2R^3} \equiv \mu\frac{a^3}{R^3}, \qquad (5.315)$$

where M and μ are constants. Using this relation between δ and $R(t)$, we can convert Eq. (5.314) into the following equation for $R(t)$:

$$\ddot{R} = -\frac{GM}{R^2} - \frac{1}{3}\dot{a}^2(\sigma^2 - 2\Omega^2)R, \qquad (5.316)$$

where the first term represents the gravitational attraction that is due to the mass inside a sphere of radius R and the second gives the effect of the shear and angular momentum.]

5.11 Scaling Laws

Before describing more sophisticated analytic approximations to gravitational clustering, we briefly discuss some simple scaling laws that can be obtained from our knowledge of linear evolution. These scaling laws are sufficiently powerful to allow reasonable predictions regarding the growth of structures in the universe and hence are useful for quick diagnostics of a given model. We confine our attention to the scaling relations for a power-law spectrum for which $|\delta_k|^2 \propto k^n$ and $\sigma^2(R) \propto R^{-(n+3)} \propto M^{-(n+3)/3}$.

Let us begin by asking what restrictions can be put on the index n. The integrand defining σ^2 in Eq. (5.259) behaves as $k^2|\delta_k|^2$ near $k = 0$. (Note that $W_k \simeq 1$ for small k in any window function.) Hence the finiteness of σ^2 will require the condition $n > -3$. The behaviour of the integrand for large values of k depends on the window function W_k. If we take the window function to be a Gaussian, then the convergence is ensured for all n. This might suggest that n can be made as large as we want, that is, we can keep the power at small k (i.e., large wavelengths) to be as small as we desire. This result, however, is not quite true for the following reason: As the system evolves, small-scale nonlinearities will develop in the system that can actually affect the large scales. If the large scales have too little power intrinsically (i.e., if n is large), then the long-wavelength power will soon be dominated by the "tail" of the short-wavelength power arising from the nonlinear clustering. This occurs because, in Eq. (5.94), the nonlinear terms $A_\mathbf{k}$ and $B_\mathbf{k}$ can dominate over $4\pi G\rho_b\delta_\mathbf{k}$ at long wavelengths (as $\mathbf{k} \to 0$). Thus there will be an *effective* upper bound on n.

The actual value of this upper bound depends, to some extent, on the details of the small-scale physics. It is, however, possible to argue that the *natural* value for this bound is $n = 4$. The argument runs as follows: Let us suppose that a large number of particles, each of mass m, are distributed carefully in space in such a way that there is very little power at large wavelengths, that is, $|\delta_k|^2 \propto k^n$ with $n \gg 4$ for small k. As time goes on, the particles influence each other gravitationally and will start clustering. The density $\rho(\mathbf{x}, t)$ that is due to the particles in some region will be

$$\rho(\mathbf{x}, t) = \sum_i m\delta[\mathbf{x} - \mathbf{x}_i(t)], \tag{5.317}$$

where $\mathbf{x}_i(t)$ is the position of the ith particle at time t and the summation is over all the particles in some specified region. The density contrast in the Fourier space will be

$$\delta_{\mathbf{k}}(t) = \frac{1}{N} \sum_i \{\exp[i\mathbf{k} \cdot \mathbf{x}_i(t)] - 1\}, \tag{5.318}$$

where N is the total number of particles in the region. For small enough $|\mathbf{k}|$, we can expand the right-hand side in a Taylor series, obtaining

$$\delta_{\mathbf{k}}(t) = i\mathbf{k} \cdot \left[\frac{1}{N} \sum_i \mathbf{x}_i(t) \right] - \frac{1}{2} k^a k^b \left[\frac{1}{N} \sum_i x_i^a(t) x_i^b(t) \right] + \cdots. \tag{5.319}$$

If the motion of the particles is such that the centre of mass of each of the sub-regions under consideration do not change, then $\sum \mathbf{x}_i$ will vanish; under this (reasonable) condition, the lowest-order contribution will be from the second term, which is quadratic in k. Isotropy requires $|\delta_k|^2$ to depend on only $k = |\mathbf{k}|$. Hence we must have $|\delta_{\mathbf{k}}|^2 \propto k^4$ for small k. Note that this result follows, essentially, from three assumptions: The small-scale graininess of the system, conservation of mass, and the conservation of momentum in local subregions. This will lead to a long-wavelength tail with $|\delta_k|^2 \propto k^4$, which corresponds to $n = 4$. The corresponding power spectrum for gravitational potential, $P_\varphi(k) \propto k^{-4}|\delta_k|^2$, is a constant. Thus, for all practical purposes, $-3 < n < 4$. The value $n = 4$ corresponds to $\sigma_M^2(R) \propto R^{-7} \propto M^{-7/3}$. For comparison, note that purely Poisson fluctuations will correspond to $(\delta M/M)^2 \propto (1/M)$, i.e., $\sigma_M^2(R) \propto M^{-1} \propto R^{-3}$ with an index of $n = 0$.

A more formal way of obtaining the k^4 tail is to solve Eq. (5.284) for long wavelengths, i.e., near $\mathbf{k} = 0$. Writing $\phi_{\mathbf{k}} = \phi_{\mathbf{k}}^{(1)} + \phi_{\mathbf{k}}^{(2)} + \cdots + $ where $\phi_{\mathbf{k}}^{(1)} = \phi_{\mathbf{k}}^{(L)}$ is the time-*independent* gravitational potential in the linear theory and $\phi_{\mathbf{k}}^{(2)}$ is the next order correction, we get from Eq. (5.284) the equation correct to the lowest order:

$$\ddot{\phi}_{\mathbf{k}}^{(2)} + 4\frac{\dot{a}}{a}\dot{\phi}_{\mathbf{k}}^{(2)} \cong -\frac{1}{3a^2} \int \frac{d^3\mathbf{p}}{(2\pi)^3} \phi_{\frac{1}{2}\mathbf{k}+\mathbf{p}}^L \phi_{\frac{1}{2}\mathbf{k}-\mathbf{p}}^L \mathcal{G}(\mathbf{k}, \mathbf{p}). \tag{5.320}$$

Writing $\phi_{\mathbf{k}}^{(2)} = aC_{\mathbf{k}}$, we can determine $C_{\mathbf{k}}$ from the above equation. Plugging it back, we find the lowest order correction to be,

$$\phi_{\mathbf{k}}^{(2)} \cong -\left(\frac{2a}{21H_0^2}\right) \int \frac{d^3\mathbf{p}}{(2\pi)^3} \phi_{\frac{1}{2}\mathbf{k}+\mathbf{p}}^L \phi_{\frac{1}{2}\mathbf{k}-\mathbf{p}}^L \mathcal{G}(\mathbf{k},\mathbf{p}). \qquad (5.321)$$

Near $\mathbf{k} \simeq 0$, we have

$$\phi_{\mathbf{k}\simeq 0}^{(2)} \cong -\frac{2a}{21H_0^2} \int \frac{d^3\mathbf{p}}{(2\pi)^3} |\phi_{\mathbf{p}}^L|^2 \left[\frac{7}{8}k^2 + \frac{3}{2}p^2 - \frac{5(\mathbf{k}\cdot\mathbf{p})^2}{k^2}\right]$$

$$= \frac{a}{126\pi^2 H_0^2} \int_0^\infty dp\, p^4 |\phi_{\mathbf{p}}^L|^2, \qquad (5.322)$$

which is independent of $\mathbf{k}$ to the lowest order. Correspondingly, the power spectrum for density $P_\delta(k) \propto k^4 P_\varphi(k) \propto k^4$ in this order.

The generation of a long-wavelength k^4 tail is easily seen in simulations if we start with a power spectrum that is sharply peaked in $|\mathbf{k}|$. Figure 5.4 shows the results of such a simulation in which the y axis is $[\Delta(k)/a(t)]$. In linear theory,

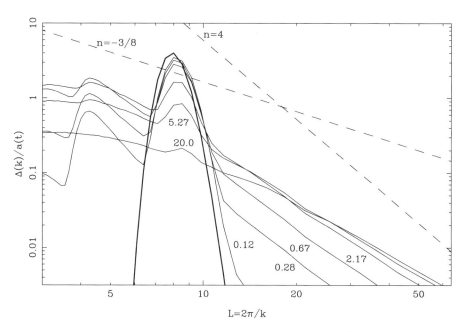

Fig. 5.4.   The transfer of power to long wavelengths forming a k^4 tail is illustrated by simulation results. Power is injected in the form of a narrow peak at $L = 8$, and the growth of power over and above the linear growth is shown in the figure. Note that the y axis is (Δ/a) so that there will be no change of shape under linear evolution with $\Delta \propto a$. As time goes on, a k^4 tail is generated, which itself evolves according to a nonlinear scaling relation, which will be discussed later.

$\Delta \propto a$, and this quantity should not change. The curves labelled by $a = 0.12$ to $a = 20.0$ show the effects of nonlinear evolution, especially the development of the k^4 tail.

Some more properties of the power spectra with different values of n can be obtained if the nonlinear effects are taken into account. We know that, in the matter-dominated phase, linear perturbations grow as $\delta_k(t) \propto a(t) \propto t^{2/3}$. Hence $\sigma^2(R) \propto t^{4/3} R^{-(3+n)}$. We may assume that the perturbations at some scale R become nonlinear when $\sigma(R) \simeq 1$. It follows that the time t_R at which a scale R becomes nonlinear satisfies the relation

$$t_R \propto R^{3(n+3)/4} \propto M^{(n+3)/4}. \tag{5.323}$$

For $n > -3$, the time scale t_R is an increasing function of M; small scales become nonlinear at earlier times. The proper size L of the region that becomes nonlinear is

$$L \propto Ra(t_R) \propto Rt_R^{2/3} \propto R^{(5+n)/2} \propto M^{(5+n)/6}. \tag{5.324}$$

Further, the objects that are formed at $t = t_R$ will have density ρ, which is larger than the background density $\bar{\rho}$ of the universe at t_R by a constant factor. Because $\bar{\rho} \propto t^{-2}$, we get

$$\rho \propto t_R^{-2} \propto R^{-3(3+n)/2} \propto M^{-(3+n)/2}. \tag{5.325}$$

Combining relations (5.324) and (5.325), we get $\rho \propto L^{-\beta}$, with

$$\beta = \frac{3(3+n)}{(5+n)}. \tag{5.326}$$

In the nonlinear case, we may interpret the correlation function ξ as $\xi(L) \propto \rho(L)$; this would imply that $\xi(x) \propto x^{-\beta}$. (We shall see later that such behaviour is to be expected on more general grounds.) The gravitational potential that is due to these bodies is

$$\phi \simeq G\rho(L)L^2 \propto L^{(1-n)/(5+n)} \propto M^{(1-n)/6}. \tag{5.327}$$

The same scaling, of course, can be obtained from $\phi \propto (M/L)$. This result shows that the binding energy of the structures increases with M for $n < 1$. In that case, the substructures will be rapidly erased as larger and larger structures become nonlinear. For $n = 1$, the gravitational potential is independent of the scale and $\rho \propto L^{-2}$.

5.12 Nonlinear Scaling Relations

Given an initial density contrast, we can trivially obtain the density contrast at any later epoch in the *linear* theory. If there is a procedure for relating the nonlinear density contrast and linear density contrast (even approximately) then we can make considerable progress in understanding nonlinear clustering. It is actually possible to make one such ansatz along the following lines.[5]

Let $v_{rel}(a, x)$ denote the relative pair velocities of particles separated by a distance x, at an epoch a, averaged over the entire universe. This relative velocity is a measure of gravitational clustering at the scale x at the epoch a. Let $h(a, x) \equiv -[v_{rel}(a, x)/\dot{a}x]$ denote the ratio between the relative pair velocity and the Hubble velocity at the same scale. In the extreme nonlinear limit ($\bar{\xi} \gg 1$), bound structures do not expand with Hubble flow. To maintain a stable structure, the relative pair velocity $v_{rel}(a, x)$ of particles separated by x should balance the Hubble velocity $Hr = \dot{a}x$; hence, $v_{rel} = -\dot{a}x$ or $h(a, x) \cong 1$.

The behaviour of $h(a, x)$ for $\bar{\xi} \ll 1$ is more complicated and can be derived as follows: Let the peculiar-velocity field be $\mathbf{v}(\mathbf{x})$ [we suppress the a dependence as we will be working at constant a]. The mean relative velocity at a separation $\mathbf{r} = (\mathbf{x} - \mathbf{y})$ is given by

$$
\begin{aligned}
\mathbf{v}_{rel}(\mathbf{r}) &\equiv \langle [\mathbf{v}(\mathbf{x}) - \mathbf{v}(\mathbf{y})][1 + \delta(\mathbf{x})][1 + \delta(\mathbf{y})] \rangle \\
&\cong \langle [\mathbf{v}(\mathbf{x}) - \mathbf{v}(\mathbf{y})]\delta(\mathbf{x}) \rangle + \langle [\mathbf{v}(\mathbf{x}) - \mathbf{v}(\mathbf{y})]\delta(\mathbf{y}) \rangle
\end{aligned}
\tag{5.328}
$$

to lowest order because the δ^2 term is higher order and $\langle \mathbf{v}(\mathbf{x}) - \mathbf{v}(\mathbf{y}) \rangle = 0$. Denoting $[\mathbf{v}(\mathbf{x}) - \mathbf{v}(\mathbf{y})]$ by $\mathbf{v}_{xy}$ and writing $\mathbf{x} = \mathbf{y} + \mathbf{r}$, we find that the radial component of the relative velocity is

$$
\mathbf{v}_{xy} \cdot \mathbf{r} = \int \mathbf{v}(\mathbf{k}) \cdot \mathbf{r} \left[e^{i\mathbf{k}\cdot(\mathbf{r}+\mathbf{y})} - e^{i\mathbf{k}\cdot\mathbf{y}} \right] \frac{d^3k}{(2\pi)^3},
\tag{5.329}
$$

where $\mathbf{v}(\mathbf{k})$ is the Fourier transform of $\mathbf{v}(\mathbf{x})$. This quantity is related to $\delta_{\mathbf{k}}$ by

$$
\mathbf{v}(\mathbf{k}) = i H a \left(\frac{\delta_{\mathbf{k}}}{k^2} \right) \mathbf{k}.
\tag{5.330}
$$

[This equation is the same as $\mathbf{u} = -\nabla\psi$, used in Eq. (5.276), expressed in Fourier space.] Using this in Eq. (5.329) and writing $\delta(\mathbf{x})$, $\delta(\mathbf{y})$ in Fourier space, we find that

$$
\begin{aligned}
&\mathbf{v}_{xy} \cdot \mathbf{r}[\delta(\mathbf{x}) + \delta(\mathbf{y})] \\
&= i H a \int \frac{d^3k}{(2\pi)^3} \int \frac{d^3p}{(2\pi)^3} \left(\frac{\mathbf{k}\cdot\mathbf{r}}{k^2} \right) \delta_{\mathbf{k}}\delta_{\mathbf{p}}^* e^{i(\mathbf{k}-\mathbf{p})\cdot\mathbf{y}} [e^{i\mathbf{k}\cdot\mathbf{r}} - 1][e^{-i\mathbf{p}\cdot\mathbf{r}} + 1].
\end{aligned}
\tag{5.331}
$$

We average this expression by using $\langle \delta_{\mathbf{k}} \delta_{\mathbf{p}}^* \rangle = (2\pi)^3 \delta_D(\mathbf{k} - \mathbf{p}) P(k)$ to obtain

$$\mathbf{v}_{\mathrm{rel}} \cdot \mathbf{r} \equiv \langle \mathbf{v}_{xy} \cdot \mathbf{r}[\delta(\mathbf{x}) + \delta(\mathbf{y})] \rangle$$

$$= iHa \int \frac{d^3k}{(2\pi)^3} \frac{P(k)}{k^2} (\mathbf{k} \cdot \mathbf{r}) [e^{i\mathbf{k} \cdot \mathbf{r}} - e^{-i\mathbf{k} \cdot \mathbf{r}}]$$

$$= -2Ha \int \frac{d^3k}{(2\pi)^3} \frac{P(k)}{k^2} (\mathbf{k} \cdot \mathbf{r}) \sin(\mathbf{k} \cdot \mathbf{r}). \tag{5.332}$$

From the symmetries in the problem, it is clear that $\mathbf{v}_{\mathrm{rel}}(\mathbf{r})$ is in the direction of $\mathbf{r}$. Therefore $\mathbf{v}_{\mathrm{rel}} \cdot \mathbf{r} = v_{\mathrm{rel}} r$. The angular integrations are straightforward and give

$$r v_{\mathrm{rel}} = \langle \mathbf{v}_{xy} \cdot \mathbf{r}[\delta(\mathbf{x}) + \delta(\mathbf{y})] \rangle = \frac{Ha}{r\pi^2} \int_0^\infty \frac{dk}{k} P(k)[kr \cos kr - \sin kr]. \tag{5.333}$$

Using expression (5.262) for $\bar{\xi}(r)$, we can write this as

$$r v_{\mathrm{rel}}(r) = -\frac{2}{3}(Har^2)\bar{\xi}. \tag{5.334}$$

Dividing by r and noting that $Hr_{\mathrm{prop}} = Har$, we get

$$h = -\frac{v_{\mathrm{rel}}(r)}{Hr_{\mathrm{prop}}} = -\frac{v_{\mathrm{rel}}(r)}{aHr} = \frac{2}{3}\bar{\xi}. \tag{5.335}$$

We thus reach the important conclusion that $h(a, x)$ depends on (a, x) only through $\bar{\xi}(a, x)$ in the linear limit, whereas $h \cong -1$ is the nonlinear limit. This suggests that the ansatz h depends on a and x only through some measure of the density contrast at the epoch a at the scale x. As a measure of the density contrast we use $\bar{\xi}(a, x)$ itself as the result in Eq. (5.335) clearly singles it out. In other words, we assume that $h(a, x) = h[\bar{\xi}(a, x)]$.

We now obtain an equation connecting h and $\bar{\xi}$, solving which we can relate $\bar{\xi}$ and $\bar{\xi}_L$. The mean number of neighbours within a distance x of any given particle is

$$N(x, t) = (na^3) \int_o^x 4\pi y^2 \, dy [1 + \xi(y, t)] \tag{5.336}$$

when n is the comoving number density. Hence the conservation law for pairs implies that

$$\frac{\partial \xi}{\partial t} + \frac{1}{ax^2} \frac{\partial}{\partial x}[x^2(1 + \xi)v] = 0, \tag{5.337}$$

where $v(t, x)$ denotes the mean relative velocity of pairs at separation x and epoch t (we have dropped the subscript 'rel' for simplicity). Using

$$(1 + \bar{\xi}) = \frac{1}{3x^2} \frac{\partial}{\partial x}[x^3(1 + \bar{\xi})] \tag{5.338}$$

in Eq. (5.337), we get

$$\frac{1}{3x^2}\frac{\partial}{\partial x}\left[x^3\frac{\partial}{\partial t}(1+\bar{\xi})\right] = -\frac{1}{ax^2}\frac{\partial}{\partial x}\left\{\frac{v}{3}\frac{\partial}{\partial x}[x^2(1+\bar{\xi})]\right\}. \tag{5.339}$$

Integrating, we find

$$x^3\frac{\partial}{\partial t}(1+\bar{\xi}) = -\frac{v}{a}\frac{\partial}{\partial x}[x^3(1+\bar{\xi})]. \tag{5.340}$$

(The integration would allow the addition of an arbitrary function of t on the right-hand side. We have set this function to zero so as to reproduce the correct limiting behaviour.) It is now convenient to change the variables from t to a, thereby getting an equation for $\bar{\xi}$:

$$a\frac{\partial}{\partial a}[1+\bar{\xi}(a,x)] = \left(\frac{v}{-\dot{a}x}\right)\frac{1}{x^2}\frac{\partial}{\partial x}\{x^3[1+\bar{\xi}(a,x)]\} \tag{5.341}$$

or, defining $h(a,x) = -(v/\dot{a}x)$,

$$\left(\frac{\partial}{\partial \ln a} - h\frac{\partial}{\partial \ln x}\right)(1+\bar{\xi}) = 3h(1+\bar{\xi}). \tag{5.342}$$

This equation shows that the behaviour of $\bar{\xi}(a,x)$ is essentially decided by h, the dimensionless ratio between the mean relative velocity v and the Hubble velocity $\dot{a}x = (\dot{a}/a)x_{\text{prop}}$, both evaluated at scale x. We now assume that

$$h(x,a) = h[\bar{\xi}(x,a)]. \tag{5.343}$$

This assumption, of course, is consistent with the extreme linear limit $h = (2/3)\bar{\xi}$ and the extreme nonlinear limit $h = 1$. When $h(x,a) = h[\bar{\xi}(x,a)]$, it is possible to find a solution to Eq. (5.343) that reduces to the form $\bar{\xi} \propto a^2$ for $\bar{\xi} \ll 1$ as follows: Let $A = \ln a$, $X = \ln x$, and $D(X,A) = (1+\bar{\xi})$. We define curves ("characteristics") in the X, A, D space that satisfy

$$\frac{dX}{dA}\bigg|_c = -h[D[X,A]], \tag{5.344}$$

i.e., the tangent to the curve at any point (X, A, D) is constrained by the value of h at that point. Along this curve, the left-hand side of Eq. (5.342) is a total derivative that allows us to write it as

$$\left[\frac{\partial D}{\partial A} - h(D)\frac{\partial D}{\partial X}\right]_c = \left(\frac{\partial D}{\partial A} + \frac{\partial D}{\partial X}\frac{dX}{dA}\right)_c \equiv \frac{dD}{dA}\bigg|_c = 3hD. \tag{5.345}$$

This determines the variation of D along the curve. Integrating, we get

$$\exp\left[\frac{1}{3}\int\frac{dD}{Dh(D)}\right] = \exp(A+c) \propto a. \tag{5.346}$$

Squaring and determining the constant from the initial conditions at a_0 in the linear regime, we get

$$\exp\left[\frac{2}{3}\int_{\bar{\xi}(a_0,l)}^{\bar{\xi}(x)}\frac{d\bar{\xi}}{h(\bar{\xi})(1+\bar{\xi})}\right] = \frac{a^2}{a_0^2} = \frac{\bar{\xi}_L(a,l)}{\bar{\xi}_L(a_0,l)}. \tag{5.347}$$

We now need to relate the scales x and l. We can write Eq. (5.344), by using Eq. (5.345), as

$$\frac{dX}{dA} = -h = \frac{1}{3D}\frac{dD}{dA}, \tag{5.348}$$

giving

$$3X + \ln D = \ln[x^3(1+\bar{\xi})] = \text{constant}. \tag{5.349}$$

Using the initial condition in the linear regime, we get

$$x^3(1+\bar{\xi}) = l^3. \tag{5.350}$$

This shows that $\bar{\xi}_L$ should be evaluated at $l = x(1+\bar{\xi})^{1/3}$. It can be checked directly that Eqs. (5.350) and (5.347) satisfy Eq. (5.342). The final result can therefore be summarised by the equation [equivalent to Eqs. (5.347) and (5.350)]

$$\bar{\xi}_L(a,l) = \exp\left[\frac{2}{3}\int^{\bar{\xi}(a,x)}\frac{d\mu}{h(\mu)(1+\mu)}\right], \qquad l = x[1+\bar{\xi}(a,x)]^{1/3}. \tag{5.351}$$

Given the function $h(\bar{\xi})$, this relates $\bar{\xi}_L$ and $\bar{\xi}$ or – equivalently – gives the mapping $\bar{\xi}(a,x) = U[\bar{\xi}_L(a,l)]$ between the nonlinear and the linear correlation functions evaluated at different scales x and l. The lower limit of the integral is chosen to give $\ln\bar{\xi}$ for small values of $\bar{\xi}$ in the linear regime. It may be mentioned that Eq. (5.348) and its integral in Eq. (5.350) are independent of the ansatz $h(a,x) = h[\bar{\xi}(a,x)]$.

The following points need to be stressed regarding this result: (1) Among all statistical indicators, it is *only* $\bar{\xi}$ that obeys a nonlinear scaling relation (NSR) of the form $\bar{\xi}_{\text{NL}}(a,x) = U\left[\bar{\xi}_L(a,l)\right]$. Attempts to write similar relations for ξ or $P(k)$ have no fundamental justification. (2) The nonlocality of the relation represents the transfer of power in gravitational clustering and cannot be ignored – or approximated by a local relation between $\bar{\xi}_{NL}(a,x)$ and $\bar{\xi}_L(a,x)$.

Given the form of $h(\bar{\xi})$, Eq. (5.351) determines the relation $\bar{\xi} = U[\bar{\xi}_L]$. It is, however, easier to determine the form of U directly along the following lines: In the linear regime $(\bar{\xi} \ll 1, \bar{\xi}_L \ll 1)$ we clearly have $U(\bar{\xi}_L) \simeq \bar{\xi}_L$. To determine its form in the quasi-linear regime, consider a region surrounding a density peak in the linear stage around which we expect the clustering to take place. From the definition of $\bar{\xi}$ it follows that the density profile around this peak can be described by

$$\rho(x) \approx \rho_{bg}[1 + \xi(x)]. \tag{5.352}$$

Hence the initial mean density contrast scales with the initial shell radius l as $\bar{\delta}_i(l) \propto \bar{\xi}_L(l)$ in the initial epoch, when linear theory is valid. This shell will expand to a maximum radius of $x_{\max} \propto l/\bar{\delta}_i \propto l/\bar{\xi}_L(l)$. In scale-invariant radial-collapse models, each shell may be approximated as contributing with an effective scale that is proportional to $x_{\max}$. Taking the final effective radius x as proportional to $x_{\max}$, we find that the final mean correlation function will be

$$\bar{\xi}_{QL}(x) \propto \rho \propto \frac{M}{x^3} \propto \frac{l^3}{[l^3/\bar{\xi}_L(l)]^3} \propto \bar{\xi}_L(l)^3, \tag{5.353}$$

that is, the final correlation function in the quasi-linear regime, $\bar{\xi}_{QL}$, at x is the cube of initial correlation function at l, where $l^3 \propto x^3 \bar{\xi}_L^3 \propto x^3 \bar{\xi}_{QL}(x)$.

Note that we did not assume that the initial power spectrum is a power law to get this result. In the case in which the initial power spectrum *is* a power law, with $\bar{\xi}_L \propto x^{-(n+3)}$, then we immediately find that

$$\bar{\xi}_{QL} \propto x^{-3(n+3)/(n+4)}. \tag{5.354}$$

[If the correlation function in linear theory has the power-law form $\bar{\xi}_L \propto x^{-\alpha}$, then the process previously described changes the index from α to $3\alpha/(1 + \alpha)$.] For the power-law case, the same result can be obtained by more explicit means. For example, in power-law models the energy of spherical shell with mean density $\bar{\delta}(x_i) \propto x_i^{-b}$ will scale with its radius as $E \propto G\delta M(x_i)/x_i \propto G\delta x_i^2 \propto x_i^{2-b}$. Because $M \propto x_i^3$, it follows that the maximum radius reached by the shell scales as $x_{\max} \propto (M/E) \propto x_i^{1+b}$. Taking the effective radius as $x = x_{\text{eff}} \propto x_i^{1+b}$, we find that the final density scales as

$$\rho \propto \frac{M}{x^3} \propto \frac{x_i^3}{x_i^{3(1+b)}} \propto x_i^{-3b} \propto x^{-3b/(1+b)}. \tag{5.355}$$

In this quasi-linear regime, $\bar{\xi}$ will scale like the density, and we get $\bar{\xi}_{QL} \propto x^{-3b/(1+b)}$. We can relate the index b to n by assuming that the evolution starts at a moment when linear theory is valid. Because the gravitational potential energy (or the kinetic energy) scales as $E \propto x_i^{-(n+1)}$ in the linear theory, it follows that $b = n + 3$. This leads to the correlation function in the quasi-linear regime, given by relation (5.354).

If $\Omega = 1$ and the initial spectrum is a power law, then there is no intrinsic scale in the problem. It follows that the evolution has to be self-similar and $\bar{\xi}$ can depend on only the combination $q = xa^{-2/(n+3)}$. This allows us to determine the a dependence of $\bar{\xi}_{QL}$ by substituting q for x in relation (5.354). We find

$$\bar{\xi}_{QL}(a, x) \propto a^{6/(n+4)} x^{-3(n+3)/(n+4)}. \tag{5.356}$$

We know that, in the linear regime, $\bar{\xi} = \bar{\xi}_L \propto a^2$. Relation (5.356) shows that, in the quasi-linear regime, $\bar{\xi} = \bar{\xi}_{QL} \propto a^{6/(n+4)}$. Spectra with $n < -1$ grow faster than a^2, spectra with $n > -1$ grow slower than a^2, and the $n = -1$ spectrum grows as a^2. Direct algebra shows that

$$\bar{\xi}_{QL}(a, x) \propto [\bar{\xi}_L(a, l)]^3, \tag{5.357}$$

reconfirming the local dependence on a and the nonlocal dependence on spatial coordinates. This result has no trace of the original assumptions (spherical evolution, scale-invariant spectrum, etc.) left in it and hence we would strongly suspect that it will have a far more general validity.

Let us now proceed to the fully nonlinear regime. If we ignore the effect of mergers, then it seems reasonable that virialised systems should maintain their densities and sizes in proper coordinates, i.e., the clustering should be "stable." This would require the correlation function to have the form $\bar{\xi}_{NL}(a, x) = a^3 F(ax)$. (The factor a^3 arises from the decrease in background density.) From our previous analysis we expect this to be a function of $\bar{\xi}_L(a, l)$ where $l^3 \approx x^3 \bar{\xi}_{NL}(a, x)$. Let us write this relation as

$$\bar{\xi}_{NL}(a, x) = a^3 F(ax) = U[\bar{\xi}_L(a, l)], \tag{5.358}$$

where $U[z]$ is an unknown function of its argument that needs to be determined. Because the linear correlation function evolves as a^2 we know that we can write $\bar{\xi}_L(a, l) = a^2 Q[l^3]$, where Q is some known function of its argument. (We are using l^3 rather than l in defining this function just for future convenience of notation.) In our case $l^3 = x^3 \bar{\xi}_{NL}(a, x) = (ax)^3 F(ax) = r^3 F(r)$, where we have changed variables from (a, x) to (a, r) with $r = ax$. Equation (5.358) now is

$$a^3 F(r) = U[\bar{\xi}_L(a, l)] = U[a^2 Q[l^3]] = U[a^2 Q[r^3 F(r)]]. \tag{5.359}$$

Consider this relation as a function of a at constant r. Clearly we need to satisfy $U[c_1 a^2] = c_2 a^3$, where c_1 and c_2 are constants. Hence we must have

$$U[z] \propto z^{3/2}. \tag{5.360}$$

Thus in the extreme nonlinear end we should have

$$\bar{\xi}_{NL}(a, x) \propto [\bar{\xi}_L(a, l)]^{3/2}. \tag{5.361}$$

[Another way to derive this result is to note that if $\bar{\xi} = a^3 F(ax)$, then $h = 1$. Integrating Eq. (5.351) with appropriate boundary condition leads to relation (5.361).] Once again we did not need to invoke the assumption that the spectrum is a power law. *If* it is a power law, then we get

$$\bar{\xi}_{NL}(a, x) \propto a^{(3-\gamma)} x^{-\gamma}, \qquad \gamma = \frac{3(n+3)}{(n+5)}. \tag{5.362}$$

This result is based on the assumption of "stable clustering"; it can be directly verified that the right-hand side of this equation can be expressed in terms of q alone, as we would have expected.

Putting all our results together, we find that the nonlinear mean correlation function can be expressed in terms of the linear mean correlation function by the relation

$$\bar{\xi}(a, x) = \begin{cases} \bar{\xi}_L(a, l) & \text{(for } \bar{\xi}_L < 1, \, \bar{\xi} < 1) \\ \bar{\xi}_L(a, l)^3 & \text{(for } 1 < \bar{\xi}_L < 5.85, \, 1 < \bar{\xi} < 200). \\ 14.14\bar{\xi}_L(a, l)^{3/2} & \text{(for } 5.85 < \bar{\xi}_L, \, 200 < \bar{\xi}) \end{cases} \quad (5.363)$$

The numerical coefficients have been determined by continuity arguments. We have assumed the linear result to be valid up to $\bar{\xi} = 1$ and the virialisation to occur at $\bar{\xi} \approx 200$, which is the result arising from the spherical model. The *exact* values of the numerical coefficients can be obtained only from simulations.

The true test of such a model, of course, is N-body simulations and, remarkably enough, simulations are very well represented by relations of form of Eq. (5.363). The simulation data for CDM, for example, is well fitted by

$$\bar{\xi}(a, x) = \begin{cases} \bar{\xi}_L(a, l) & \text{(for } \bar{\xi}_L < 1.2, \, \bar{\xi} < 1.2) \\ \bar{\xi}_L(a, l)^3 & \text{(for } 1 < \bar{\xi}_L < 5, \, 1 < \bar{\xi} < 125), \\ 11.7\bar{\xi}_L(a, l)^{3/2} & \text{(for } 5 < \bar{\xi}_L, \, 125 < \bar{\xi}) \end{cases} \quad (5.364)$$

which is fairly close to the theoretical prediction.

A comparison of relations (5.363) and (5.364) shows that the physical processes that operate at different scales are well represented by our model. In other words, the processes described in the quasi-linear and nonlinear regimes for an *individual* lump still models the *average* behaviour of the universe in a statistical sense. It must be emphasised that the key point is the "flow of information" from l to x, which is an exact result. Only when the results of the specific model are recast in terms of suitably chosen variables do we get a relation that is of general validity. It would have been, for example, incorrect to use a spherical model to obtain a relation between linear and nonlinear densities at the same location or to model the function h. Figure 5.5 shows the evolution of $\bar{\xi}$ for a cosmological model calculated with the nonlinear scaling relations previously described above. The y axis plots the ratio of $\bar{\xi}^{1/2}$ to the linear growth factor $D(z)$ so that the linear evolution will not change the figure. It is obvious that nonlinear evolution increases $\bar{\xi}$ at small scales whereas the large scales remain linear.

It may be noted that, to obtain the result in the nonlinear regime, we need to invoke the assumption of stable clustering, which has not been deduced from any fundamental considerations. In the case in which the mergers of structures are

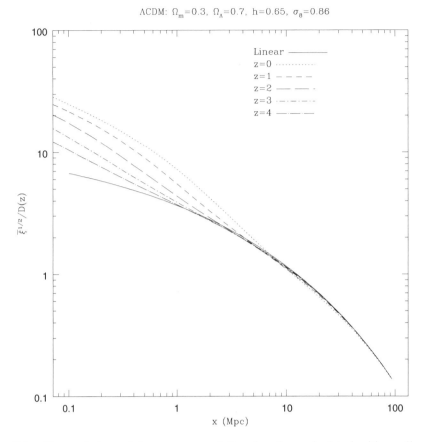

ΛCDM: $\Omega_m=0.3$, $\Omega_\Lambda=0.7$, h=0.65, $\sigma_8=0.86$

Fig. 5.5. The evolution of the mean correlation function calculated with nonlinear scaling relations. The y axis is rescaled by the linear growth factor $D(z)$ so that linear evolution cannot modify the curve. It is clear that the correlation function increases because of nonlinear effects at small scales whereas the large scales remain linear.

important, we would consider this assumption to be suspect. We can, however, generalise the above argument in the following manner: If the virialised systems have reached stationarity in the statistical sense, the function h – which is the ratio between two velocities – should reach some constant value. In that case, we can integrate Eq. (5.351) and obtain the result $\bar{\xi}_{NL} = a^{3h} F(a^h x)$, where h now denotes the asymptotic value. A similar argument will now show that

$$\bar{\xi}_{NL}(a, x) \propto [\bar{\xi}_L(a, l)]^{3h/2} \qquad (5.365)$$

in the general case. For the power-law spectra, we would get

$$\bar{\xi}(a, x) \propto a^{(3-\gamma)h} x^{-\gamma}, \qquad \gamma = \frac{3h(n+3)}{2 + h(n+3)}. \qquad (5.366)$$

Simulations are not accurate enough to fix the value of h; in particular, the asymptotic value of h could depend on n within the accuracy of the simulations.

It may be possible to determine this dependence by modelling mergers in some simplified form.

If $h = 1$ asymptotically, the correlation function in the extreme nonlinear end depends on the linear index n. We may feel that the physics at the highly nonlinear end should be independent of the linear spectral index n. This will be the case if the asymptotic value of h satisfies the scaling

$$h = \frac{3c}{n + 3} \tag{5.367}$$

in the nonlinear end with some constant c. Only high-resolution numerical simulations can test this conjecture that $h(n + 3) = \text{constant}$.

We can obtain similar relations between $\xi(a, x)$ and $\xi_L(a, l)$ in two dimensions as well by repeating the preceding analysis. In two dimensions the scaling relations turn out to be

$$\bar{\xi}(a, x) \propto \begin{cases} \bar{\xi}_L(a, l) & \text{(linear)} \\ \bar{\xi}_L(a, l)^2 & \text{(quasi-linear)}, \\ \bar{\xi}_L(a, l)^h & \text{(nonlinear)} \end{cases} \tag{5.368}$$

where h again denotes the asymptotic value. For a power-law spectrum the

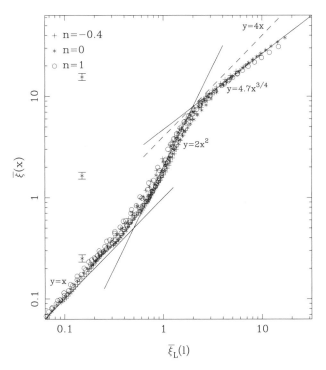

Fig. 5.6. The comparison between theory and simulations in two-dimensional gravitational clustering.

nonlinear correction function will be $\bar{\xi}_{NL}(a, x) = a^{(2-\gamma)h} x^{-\gamma}$, with

$$\gamma = \frac{2h(n+2)}{2+h(n+2)}. \tag{5.369}$$

If we generalise the concept of stable clustering to a mean constancy of h in the nonlinear epoch, then the correlation function will behave as $\bar{\xi}_{NL}(a, x) = a^{2h} F(a^h x)$. In this case, if the spectrum is a power law then the nonlinear and linear indices are related by Eq. (5.369). All the features discussed in the case of three dimensions are present here as well. For example, if the asymptotic value of h scales with n such that $h(n+2) = $ constant then the nonlinear index will be independent of the linear index.[6] Figure 5.6 shows the results of numerical simulation in two dimensions, which suggests that $h = 3/4$ asymptotically.

6

Cosmic Microwave Background Radiation

6.1 Introduction

Cosmic microwave background radiation (CMBR) is a relic from the redshift $z \approx 10^3$, beyond which the universe is optically thick in most of the wave bands. This radiation therefore carries vital information about the state of the universe at an epoch that is probably as early as we could probe by direct electromagnetic measurements. A considerable amount of theoretical and observational progress has been achieved in this topic in the past decade, and future observations of CMBR hold the promise for allowing us to determine the parameters of the universe with unprecedented accuracy.

The temperature anisotropies in CMBR and related issues are discussed in this chapter.[1] Anisotropies that are due to peculiar velocities and fluctuations in the gravitational potential are derived in Section 6.3 and discussed in detail in Sections 6.4 and 6.5. The damping of anisotropies and the distortions that arise because of the astrophysical processes are studied in Section 6.7.

6.2 Processes Leading to Distortions in CMBR

We have seen in Chap. 4 that the photons in the universe decoupled from matter at a redshift of $\sim 10^3$. These photons have been propagating freely in space–time since then and can be detected today. In an *ideal* Friedmann universe, a *comoving* observer will see these photons as a blackbody spectrum at some temperature T_0. The deviations in the metric from that of the Friedmann universe, the motion of the observer with respect to the comoving frame, and the astrophysical processes that take place along the trajectory of the photon can all lead to potentially observable effects in this radiation. These effects can be of two kinds: (1) The spectrum may not be strictly blackbody, that is, the radiation intensity $I(\omega)$ at different frequencies, coming to us from a *particular* direction in the sky, may not correspond to a Planck spectrum with a single temperature T_0. (2) The spectrum may be Planckian in any given direction; but the temperature of the radiation may be different in *different* directions; that is, $T_0 = T_0(\theta, \phi)$, where (θ, ϕ) are the angular coordinates on the sky.

No deviation of the first kind has been detected in CMBR, and there exist tight observational bounds on the spectral deviations of CMBR from the ideal blackbody form. This negative result is useful in establishing constraints on the models describing structure formation, especially those aspects that are related to the IGM. On the other hand, CMBR does show the distortions of the second kind at different angular scales. Direct measurements of these distortions, which are currently available at different angular scales, turn out to be more restrictive than the bounds on spectral deviations. Because of this, we concentrate on angular anisotropies in this chapter.

Angular anisotropies, leading to the directional dependence $T = T(\theta, \phi)$ of the temperature, arise primarily for the following reasons: (1) If the observer is moving with respect to the comoving frame, then the photons reaching the observer from different directions with respect to the direction of motion will be redshifted (or blueshifted) by different amounts. This will lead to a dependence of T on $\cos \theta$, where θ is the angle between the direction of motion and the direction of the photon. (2) If the matter that scattered the radiation in our direction had a peculiar velocity (with respect to the comoving frame) when the scattering occurred, then an effect similar to that of (1) can occur. Because the peculiar velocity of matter will be different at different locations in the LSS, corresponding to different directions in the sky today, this will lead to an anisotropy. (3) If the local gravitational potential on the LSS was different at different locations, then the photons will be climbing out of different gravitational potential wells and hence will experience different amounts of redshift. This will lead to an angular dependence for the observed temperature. (4) The energy density of radiation in the LSS can have an intrinsic inhomogeneity $\delta_R = (\delta \rho_R / \rho_R)$. This will appear as a temperature fluctuation $\delta T / T$. (5) Processes that take place along the path of the photon reaching us can also produce $\delta T / T$. For example, consider a photon that travels through a large-mass concentration that is collapsing. Because the depth of the gravitational potential well of a collapsing structure is increasing with time, the blueshift suffered by the photon as it travels towards the centre of the condensation will be lower than the redshift that occurs while it is emerging from the centre. Similarly, photons travelling through hot gas in a cluster will be affected by the Compton scattering with the charged particles in the gas. These two processes can cause temperature distortions in specific directions. (6) Last, we must also consider effects that can *wipe out* any $\delta T / T$ produced by some of the processes previously mentioned. Such an effect can arise primarily for two reasons: First, it should be noted that the decoupling was not an instantaneous event but took an interval of $\Delta z \simeq 80$ in the redshift space. This allows the photon to perform a random walk during the interval $(z_{\rm dec}, z_{\rm dec} + \Delta z)$, which can wipe out the original temperature fluctuations at small scales. Second, if the matter in the universe was reionised at any redshift $z_{\rm ion}$ (which is lower than

$z_{\text{dec}} \simeq 10^3$), then the photons would have interacted with those charged particles at $z \lesssim z_{\text{ion}}$. Thus the "last" scattering was not at $z \simeq z_{\text{dec}}$ but at $z \simeq z_{\text{ion}}$, and some of the anisotropies induced at $z \simeq z_{\text{dec}}$ would be wiped out.

The physical processes described above operate at different characteristic length scales. For example, the effect in (3) will be governed by a (proper) length scale L over which the gravitational potential ϕ varies on the $z \simeq 10^3$ surface. Such a proper length L will subtend an angle $\theta = [L/d_A(z)]$ in the sky, where $d_A(z)$ is the angular diameter distance discussed in Chap. 3. In particular, the angle subtended by the region that has a size equal to that of the Hubble radius at z_{dec} turns out to be $\sim 1°$ (see Chap. 3, Section 3.5). Therefore angular separation of more than $1°$ in the sky would correspond to regions that were bigger than the Hubble radius at the time of decoupling. This fact makes the angle of $\sim 1°$ a natural dividing line between what may be called *small angles* $(\theta < 1°)$ and *large angles* $(\theta > 1°)$. Because most of the nongravitational processes operate at proper lengths smaller than Hubble radius, they predominantly produce anisotropies at small angles. On the other hand, fluctuations in a gravitational field can cause anisotropies at large angles. A linear scale of 100 Mpc today has evolved from a region of size $100(1 + z_{\text{dec}})^{-1}$ at recombination. It turns out that this region subtends $\sim \Omega h$ deg in the sky (see Chap. 3, Section 3.5); a scale of 1000 Mpc corresponds to only $\sim 10\Omega h$ deg. Therefore observations of $\delta T/T$ at these angular scales probe the density fluctuations at length scales much larger than those available for astronomical investigations today, say, in the study of galaxy–galaxy clustering.

Exercise 6.1
Thought provoker: Think of a photon travelling radially through a spherically symmetric mass distribution that is collapsing inwards. As the photon moves from the edge to the centre, it gains gravitational potential energy, and on its outward motion from centre to the edge, it loses gravitational energy. Because the mass distribution would have collapsed a little further during the light transit times, the potential well would have become deeper when the photon was climbing out, thereby making it lose more energy than it gains. This could be a simple picture of the *Rees–Sciama* effect. Where did the energy of the photon disappear?

Exercise 6.2
Some uses of CMBR I: A cosmic-ray proton of energy 10^{20} eV will see – in its rest frame – microwave photons as having an energy of ~ 100 MeV. Such an energetic photon, hitting a proton at rest, is almost at the threshold for producing a pion. (a) How would a terrestrial observer view this phenomenon? (b) What is the mean free path for a proton with energy higher than 10^{20} eV?

Exercise 6.3
Some uses of CMBR II: It is possible to produce $e\bar{e}$ pairs in the energetic collisions of γ rays. The cross section σ for this process reaches a value of 10^{-25} cm^2 at the threshold

of the production. Consider a cosmic-ray photon of energy E (as measured by the terrestrial observer) interacting with microwave photons. (a) Show that $e\bar{e}$ pairs should be produced by cosmic-ray photons with $E \gtrsim 2.5 \times 10^{14}$ eV. (b) Show that the mean free path for this process (i.e., the mean distance a cosmic-ray photon will travel before producing $e\bar{e}$) is $\sim 10^{22}$ cm (this is smaller than the size of the galaxy). (c) What will be the fate of these $e\bar{e}$ pairs?

6.3 Angular Pattern of CMBR Anisotropies

A physical process characterised by a length scale L at $z = z_{\text{dec}}$ will subtend an angle $\theta(l) = [l/d_A(z_{\text{dec}})]$ in the sky today, where $d_A(z)$ is the angular diameter distance. We saw in Chap. 3, Section 3.6, that $d_A = a_0 r_1(z)(1+z)^{-1}$, where $r_1(z)$ is given by Eq. (3.132) in a model with $\Omega_V = 0$, $\Omega_{\text{NR}} = \Omega$. For large z, this reduces to $r_1(z) \approx 2(H_0 a_0 \Omega)^{-1}$, so that, for $z \gg 1$, the angular diameter distance becomes $d_A(z) \cong 2H_0^{-1}(\Omega z)^{-1}$, giving

$$\theta(L) \cong \left(\frac{\Omega}{2}\right)\left(\frac{Lz}{H_0^{-1}}\right) = 34.4''(\Omega h)\left(\frac{\lambda_0}{1\,\text{Mpc}}\right). \tag{6.1}$$

[As always, we quote the numerical values with the length scales by extrapolating them to the present epoch. Thus $\lambda_0 = L(1 + z_{\text{dec}}) \simeq Lz_{\text{dec}}$ is the proper length *today* that would have been L at the redshift of z_{dec}. We will not bother to indicate this fact with a subscript 0 when no confusion is likely to arise.] In particular, consider the angle subtended by the region that has a size equal to that of the Hubble radius at z_{dec}; that is, we take L to be $d_H(z_{\text{dec}}) = H^{-1}[z = z_{\text{dec}}] = H_0^{-1}(\Omega z_{\text{dec}})^{-1/2} z_{\text{dec}}^{-1}$ so that

$$\lambda = d_H(z_{\text{dec}})[1 + z_{\text{dec}}] \cong d_H(z_{\text{dec}})z_{\text{dec}} \cong H_0^{-1}(\Omega z_{\text{dec}})^{-1/2}. \tag{6.2}$$

Then

$$\theta_H \equiv \theta(d_H) \cong 0.87° \,\Omega^{1/2}\left(\frac{z_{\text{dec}}}{1100}\right)^{-\frac{1}{2}} \simeq 1°. \tag{6.3}$$

Therefore angular separation of more than $1°$ in the sky would correspond to regions that were bigger than the Hubble radius at the time of decoupling. (For comparison, note that the scale λ_{eq} subtends an angle of $\sim 0.13° \, h^{-1}$.)

Let us now estimate the magnitude of the anisotropy $\Delta T/T$ that is due to processes (1), (2), and (3) mentioned in the last section. We begin with the intrinsic anisotropies in the radiation field. We first consider large angular scales, $\theta > \theta_H$.

If baryons and photons are tightly coupled, then we expect the anisotropy in the energy density of the radiation, δ_R, to be comparable to δ_B. The exact relation between these two depends on the initial process that generated the fluctuations. Most popular models generate fluctuations that are "isentropic" in the sense that the entropy per baryon is unaffected by the fluctuation (see Chap. 5). Because the entropy is mostly contributed by the radiation, isentropic fluctuations are characterised by

$$0 = \delta \left[\frac{s_R}{n_B} \right] = \frac{s_R}{n_B} \left[\frac{\delta s_R}{s_R} - \frac{\delta \rho_B}{\rho_B} \right]. \tag{6.4}$$

With $s_R \propto T^3$ and $\rho_R \propto T^4$, it follows that $(\delta s_R/s_R) = (3/4)\delta_R$. Therefore, for adiabatic fluctuations, we will have $(3/4)\delta_R = \delta_B$ or

$$\left(\frac{\Delta T}{T} \right)_{\text{int}} = \frac{1}{3} \delta_B(t_{\text{dec}}). \tag{6.5}$$

We can relate $\delta_B(t_{\text{dec}})$ to $\delta_{\text{DM}}(t_{\text{dec}})$ in the following manner: Because baryons were tightly coupled to the photons, for $t_{\text{ent}} < t < t_{\text{dec}}$ there is no growth in this period, i.e., $\delta_B(t_{\text{dec}}) = \delta_B(t_{\text{enter}}) = \delta_{\text{DM}}(t_{\text{enter}})$. However, the dark-matter perturbations were growing by very little for $a_{\text{ent}} < a < a_{\text{eq}}$ but will grow as $\delta \propto a$ during the phase $a_{\text{eq}} < a < a_{\text{dec}}$. Therefore $\delta_{\text{DM}}(t_{\text{dec}}) = (a_{\text{dec}}/a_{\text{eq}})\delta_{\text{DM}}(t_{\text{enter}})$. It follows that $\delta_B(t_{\text{dec}}) \simeq (a_{\text{dec}}/a_{\text{eq}}) \, \delta_{\text{DM}}(t_{\text{dec}}) \simeq (\delta_{\text{DM}}/20 \, \Omega h^2)$. We can thus relate the intrinsic anisotropy in the radiation field to the density contrast of the *dark-matter* perturbation at decoupling:

$$\left(\frac{\Delta T}{T} \right)_{\text{int}} \cong \frac{1}{60 \, \Omega h^2} \delta_{\text{DM}}(t_{\text{dec}}). \tag{6.6}$$

To determine the angular scale over which this effect is significant, we proceed as follows: Each Fourier mode of $\delta_{\text{DM}}(k)$, labelled by a wave vector k, will correspond to a wavelength $\lambda \propto k^{-1}$ and will contribute at an angle θ given by relation (6.1). The mean-square fluctuation in temperature will therefore scale with the angle θ as

$$\left(\frac{\Delta T}{T} \right)_{\text{int}}^2 \propto d^3k \, P(k) \propto \frac{dk}{k} P(k) k^3 \propto \lambda^{-(3+n)} \propto \theta^{-(3+n)} \tag{6.7}$$

for $\theta > \theta_H$. [If $n = 1$, then $(\Delta T/T)_{\text{in}} \propto \theta^{-2}$.]

To relate the theoretical predictions to observations, it is usual to expand the temperature anisotropies in the sky in terms of the spherical harmonics. The temperature anisotropy in the sky will provide $\Delta = \Delta T/T$ as a function of two angles, θ and ψ. Just as it is convenient to use the Fourier modes for discussing spatially varying anisotropies, it is convenient to use spherical harmonics to describe angular anisotropies. Accordingly we expand the temperature anisotropy distribution in the sky in spherical harmonics:

$$\Delta(\theta, \psi) \equiv \frac{\Delta T}{T}(\theta, \psi) = \sum_{l,m}^{\infty} a_{lm} Y_{lm}(\theta, \psi). \tag{6.8}$$

All the information is now contained in the angular coefficients a_{lm}.

If **n** and **m** are two directions in the sky with an angle α between them, the two-point correlation function of the temperature fluctuations in the sky can be defined as

$$\mathcal{C}(\alpha) = \langle S(\mathbf{n})S(\mathbf{m})\rangle = \sum\sum\langle a_{lm}a^*_{l'm'}\rangle Y_{lm}(\mathbf{n})Y^*_{l'm'}(\mathbf{m}). \tag{6.9}$$

Because the sources of temperature fluctuations are related linearly to the density inhomogeneities, the coefficients a_{lm} will be random fields with some power spectrum. In that case $\langle a_{lm}a^*_{l'm'}\rangle$ will be nonzero only if $l = l'$ and $m = m'$. Writing

$$\langle a_{lm}a^*_{l'm'}\rangle = C_l\delta_{ll'}\delta_{mm'} \tag{6.10}$$

and using the addition theorem of spherical harmonics, we find that

$$\mathcal{C}(\alpha) = \sum_l \frac{(2l+1)}{4\pi}C_l P_l(\cos\alpha), \tag{6.11}$$

with $C_l = \langle|a_{lm}|^2\rangle$. In this approach, the pattern of anisotropy is contained in the variation of C_l with l. Roughly speaking, $l \propto \theta^{-1}$ and we can think of the (θ, l) pair as an analogue of $(\mathbf{x}, \mathbf{k})$ variables in three dimensions. C_l is similar to the power spectrum $P(\mathbf{k})$.

Exercise 6.4
Thought provoker: Explain what is wrong with the following argument: "The recombination of protons and electrons forming the hydrogen atom occurs at some constant temperature T related to the binding energy of the hydrogen atom. Hence the CMBR photons come from an essentially constant-temperature surface, and there should be no intrinsic temperature variations."

Consider next the contribution to $\Delta T/T$ from the velocity of the scatterers in the last scattering surface (LSS). If a particular photon that we receive was last scattered by a charged particle moving with a velocity v, then the photon will suffer a Doppler shift of the order of v/c. This will contribute $(\Delta T/T)^2 \simeq (v^2/c^2)$. At large scales (for $\lambda \gtrsim d_H$), the velocity of the baryons is essentially due to the gravitational force exerted by the dark-matter potential wells and $v \propto \nabla\phi$. By an analysis similar to the above, we conclude that

$$\left(\frac{\Delta T}{T}\right)^2_{\text{Dopp}} \propto d^3k\,k^2\frac{P}{k^4} \propto \frac{dk}{k}kP(k) \propto \lambda^{-(1+n)} \propto \theta^{-(1+n)} \tag{6.12}$$

for $\theta > \theta_H$. [If $n = 1$, then $(\Delta T/T)_{\text{Dopp}} \propto \theta^{-1}$.]

Finally, let us consider the contribution to $\Delta T/T$ from the gravitational potential along the path of the photons. When the photons travel through a region of gravitational potential ϕ, they undergo a redshift of the order of ϕ/c^2. In the $\Omega = 1$, matter-dominated universe, the perturbed gravitational potential is independent of time, i.e., $\phi(t, \mathbf{x}) = \phi(\mathbf{x})$. We may therefore estimate the temperature anisotropies that are due to the gravitational potential as $(\Delta T/T) \simeq (\phi/c^2)$ where ϕ is the gravitational potential *on* the LSS. This contribution to temperature anisotropy is usually called the *Sachs–Wolfe effect*. Its angular dependence will be

$$
\left(\frac{\Delta T}{T}\right)^2_{\text{SW}} \propto d^3k \frac{P(k)}{k^4} \propto \frac{dk}{k^2} P(k) \propto \begin{cases} \lambda^{5-n} & (\text{for } \lambda < \lambda_{\text{eq}}) \\ \lambda^{1-n} & (\text{for } \lambda > d_H) \end{cases}
$$

$$
\propto \begin{cases} \theta^4 & (\text{for } \theta < \theta_{\text{eq}}) \\ \text{constant} & (\text{for } \theta > \theta_H) \end{cases}.
$$

(6.13)

(The second line assumes that $n = 1$.)

Let us next consider the anisotropies at small angular scales $\theta < \theta_H$. The contribution $(\Delta T/T)^2_{\text{SW}}$ varies as θ^4 and hence is not very significant at small θ. As regards $(\Delta T/T)^2_{\text{int}}$ and $(\Delta T/T)^2_{\text{Dopp}}$, we should note the following: At small scales, the evolution of δ_B is governed by the pressure support in the baryonic fluid and the density contrast δ_B, and velocity v will oscillate as an acoustic wave with a wavelength $\lambda \simeq \lambda_J \simeq \lambda_H$. Hence $\Delta T/T$ at small angular scales will have crests and troughs separated at wave numbers $\Delta k \simeq (1/\lambda_J) \simeq 0.01h \text{ Mpc}^{-1}$. This corresponds to an angular separation of $\Delta\theta \simeq 1.0°$ in the sky. The amplitudes of all three contributions will be comparable around $\theta \approx \theta_{\text{eq}}$. At smaller θ, the potential contribution is ignorable and the other two contributions will be comparable in magnitude but (approximately) opposite in phase.

There is, however, another effect that needs to be taken into account at small angular scales. The above analysis assumes that decoupling took place at a sharp value of redshift z_{dec}. In reality, the LSS has a finite thickness of $\Delta z \approx 80$ (see Chap. 4, Section 4.7). Any photon detected today has a probability $\mathcal{P}(z)\,dz$ of having been last scattered in the redshift interval $(z, z + dz)$. The observed $\Delta T/T$ has to be computed as

$$
\left(\frac{\Delta T}{T}\right)_{\text{obs}} = \int dz \left\{ \begin{matrix} (\Delta T/T) \text{ if the last} \\ \text{scattering was at } z \end{matrix} \right\} \times \mathcal{P}(z). \quad (6.14)
$$

We saw in Chap. 4, Section 4.7, that $\mathcal{P}(z)$ is very well approximated by a Gaussian peaked at $z = z_{\text{dec}}$ with a width of $\Delta z \approx 80$. This width corresponds to a line-of-sight (comoving) distance of

$$
\Delta l = c\left(\frac{dt}{dz}\right)\Delta z(1 + z_{\text{dec}}) \approx H_0^{-1}\frac{\Delta z}{\Omega^{1/2}z_{\text{dec}}^{3/2}} \approx 8(\Omega h^2)^{-1/2} \text{ Mpc}. \quad (6.15)
$$

Hence the anisotropies at length scales smaller than Δl will be suppressed because of the finite thickness of the LSS. This length scale corresponds to an angular scale of $\theta_{\Delta z} \simeq 3.8' \, \Omega^{1/2}$.

Another effect that operates at small scales is the photon diffusion. As recombination proceeds, the diffusion length approaches infinity and hence can be important.[2] The actual length scales below which this effect can wipe out fluctuations is somewhat model dependent but could be $l_{\text{diff}} \simeq 35$ Mpc, corresponding to $\theta_{\text{diff}} \simeq 24'$. This can be seen as follows: Consider a time interval Δt in which a photon suffers $N = [\Delta t / l(t)]$ collisions. Between successive collisions it travels a proper distance $l(t)$ or – equivalently – a coordinate distance $[l(t)/a(t)]$. Because of this random walk, it acquires a mean-square coordinate displacement:

$$(\Delta x)^2 = N \left(\frac{l}{a}\right)^2 = \frac{\Delta t}{l(t)} \frac{l^2}{a^2} = \frac{\Delta t}{a^2} l(t). \tag{6.16}$$

The total mean-square coordinate distance travelled by a typical photon until the time of decoupling is

$$x^2 \equiv \int_0^{t_{\text{dec}}} \frac{dt}{a^2(t)} l(t) = \frac{3}{5} \frac{t_{\text{dec}} l(t_{\text{dec}})}{a^2(t_{\text{dec}})}, \tag{6.17}$$

where we have used the facts that $l(t) = [\sigma n_e(t)]^{-1} \simeq 10^{29}$ cm $(1+z)^{-3} \times (\Omega_B h^2)^{-1}$ and that $a \propto t^{2/3}$. This corresponds to the proper distance

$$l_{\text{diff}} = a(t_{\text{dec}})x = \left[\frac{3}{5} t_{\text{dec}} l(t_{\text{dec}})\right]^{1/2} \simeq 35 \text{ Mpc} \left(\frac{\Omega_B h^2}{0.02}\right)^{-1/2} \left(\Omega h_{50}^2\right)^{-1/4}. \tag{6.18}$$

The effect of this diffusion is equivalent to multiplying each mode $(\Delta T/T)_k$ in the Fourier space by the Gaussian $\exp[-k^2(\Delta l)^2/2] = \exp[-(k/k_T)^2]$, with $k_T \simeq 0.02h$ Mpc^{-1}.

When this effect is taken into account, the final pattern of anisotropies will look like the one in Fig. 6.1. The contribution at large angular scales that is due to intrinsic anisotropy, the Doppler effect, and the Sachs–Wolfe effect are shown by the dotted curve, the dotted–dashed curve, and the dashed curve, respectively. The dashed–triple-dotted curve indicates the Gaussian cutoff that is due to the photon diffusion. The thick curve clearly shows the acoustic peaks and the small-scale suppression.

At large scales, the anisotropies are dominated by the Sachs–Wolfe effect and are independent of the angular scale. As we move to smaller scales, other contributions add up, making $(\Delta T/T)$ increase. At scales $\theta \lesssim \theta_{\text{eq}}$, the oscillations in δ_B and v make their presence felt, and we see characteristic peaks called *acoustic peaks*. At still smaller scales, the effects that are due to the finite thickness of the LSS drastically reduce the amplitude of fluctuations.

The fact that several different processes contribute to the structure of angular anisotropies makes CMBR a valuable tool for extracting cosmological information. To begin with, the anisotropy at very large scales directly probes modes that are bigger than the Hubble radius at the time of decoupling and allows us to directly determine the primordial spectrum. As we have seen, the effect of initial

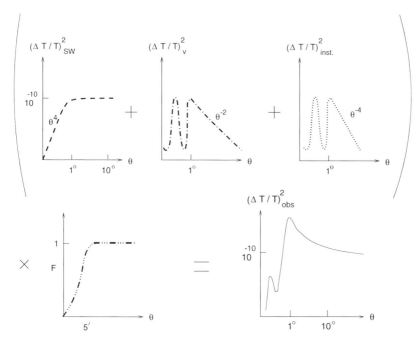

Fig. 6.1.   Schematic diagram showing the expected anisotropies in the CMBR temperature at different angular scales.

fluctuations in the scalar gravitational potential leads to a temperature anisotropy with the angular dependence $\Delta T/T \propto \lambda^{1-n} \propto \theta^{1-n}$. In particular, if $\Delta T/T$ is independent of the angular scale at large angles, then the spectrum is scale invariant with $n = 1$ in an $\Omega = \Omega_{NR} = 1$ universe. The behaviour is different if there is a cosmological constant (or if $\Omega < 1$), but in a predictable way (see Section 6.6). Thus, in general, if the angular dependence of the spectrum at very large scales is known, we can work backwards and determine the initial power spectrum. [In generic inflationary models, there will also be some amount of gravitational waves that is generated primordially (see Chap. 5, Section 5.7). In some models they could also contribute a significant fraction of the anisotropy, which we will discuss in Section 6.6. Even in this case, the anisotropy pattern at large scales is predictable and can be used to extract useful information about the initial power spectrum.]

The behaviour of $\Delta T/T$ at large angles also depends on cosmological parameters such as h and Ω, but this dependence is not very strong. The structure of acoustic peaks at small scales, however, provides a much more reliable procedure for estimating the cosmological parameters. To illustrate this point, let us consider the location of the first acoustic peak. The wavelength of the acoustic oscillations, which lead to the first acoustic peak, will occur at a scale comparable with the Hubble radius at t_{dec}. However, the angle subtended by the Hubble radius at $z = z_{dec}$, corresponding to a length scale ct_{dec}, given by Eq. (6.3), depends directly on $\Omega_0 = \Omega_{NR}$; the angle increases with increasing Ω_0 and vice versa. Figure 6.2

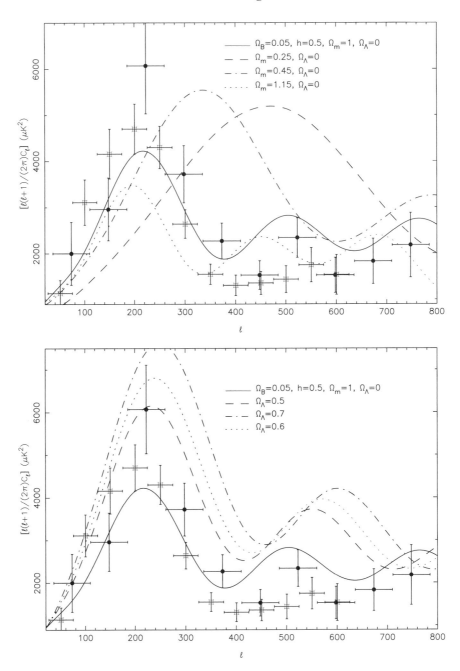

Fig. 6.2.   The variation of the anisotropy pattern with $\Omega_m = \Omega_{NR}$ and Ω_V. The top frame shows the $\Delta T/T$ in universes with $\Omega_{NR} = 0.25, 0.45, 1.0$, and 1.15, $\Omega_V = 0$, with the first acoustic peak moving from right to left. The y axis is essentially a measure of $(\Delta T/T)^2$, and the x axis is a measure of $1/\theta$. The bottom frame is for universes with $\Omega_{NR} + \Omega_V = 1$. The curves are for $\Omega_V = 0.0, 0.5, 0.6$, and 0.7. (Figure courtesy of S. Sethi.)

shows the variation in the structure of acoustic peaks when Ω_0 is changed, keeping $\Omega_V = 0.3$ (The curves are based on the rigorous theory that is described in Section 6.5). The four curves are for $\Omega_0 = \Omega_{NR} = 0.25, 0.45, 1.0$, and 1.15, with the first acoustic peak moving from right to left. C_l in the y axis is essentially a measure of $(\Delta T/T)^2$ at an angular scale θ related to the l in the x axis, which is a measure of $1/\theta$. Clearly θ_{peak} increases with Ω_0. The data points on the figures are from the first results of MAXIMA and BOOMERANG experiments (see Section 6.6) and are included to give a feel for the error bars in current observations. It is obvious that the overall geometry of the universe can be easily fixed by the study of CMBR anisotropy.

The situation is slightly more complicated in models with $\Omega_{NR} + \Omega_V = 1$. Here the relevant angle is

$$\theta \approx \frac{ct_{\text{dec}}}{d_A} \approx \frac{(\Omega_{NR}h^2)^{-1/2}}{h^{-1}f(\Omega_{NR}, \Omega_V)}. \tag{6.19}$$

At high redshifts, the angular diameter distance will increase with Ω_V (see Exercise 3.12). Thus the angle subtended by the acoustic peak will decrease with an increasing cosmological constant and vice versa. We thus conclude that the location of the acoustic peak will depend monotonically on Ω_V as well. This is illustrated in the bottom frame of Fig. 6.2, which gives the anisotropy pattern for universes with $\Omega_{NR} + \Omega_V = 1$. The curves are for $\Omega_V = 0.0, 0.5, 0.6$, and 0.7. The variation is substantially less in this figure than in the top frame of Fig. 6.2; the location of the peak is very sensitive to the geometry of the universe determined predominantly by the total value of Ω and somewhat less dependent on the composition when $\Omega_{\text{tot}} = 1$.

The situation is summarised in the top frame of Fig. 6.4, which shows the anisotropies for two cosmological models with $\Omega_{NR} = 0.35$ but with different Ω_{total}. One model has $\Omega_{NR} = \Omega_{\text{tot}} = 0.35$, and the other one has $\Omega_{NR} + \Omega_V = \Omega_{\text{tot}} = 1$. It is clear that the location of the peak is different in the two cases and – as previously described – this is essentially due to differing values of Ω_{tot}.

The heights of acoustic peaks also contain important information. In particular, the height of the first acoustic peak relative to the second one depends sensitively on Ω_B. This is because the damping of the anisotropies arises from the finite thickness of the surface of recombination, which, in turn, depends on the strength of the coupling between photons and baryons. Increasing the amount of baryons increases this coupling and thus increases the effect of damping on the second peak compared to first. Figure 6.3 shows this effect in a universe with $\Omega_{NR} = 1, \Omega_V = 0$ for $\Omega_B = 0.015, 0.025, 0.05$, and 0.08, with the height of the peak increasing with Ω_B.

The structure of the anisotropies also depend on other parameters, notably the Hubble constant. However, this dependence is somewhat more complicated and not monotonic. For example, the bottom frame of Fig. 6.4 shows the variation of

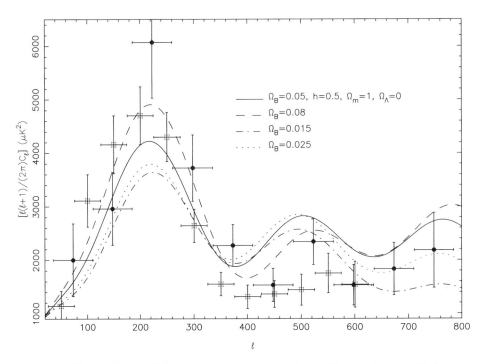

Fig. 6.3.   The variation of the anisotropy pattern with Ω_B for a universe with $\Omega_m = \Omega_{NR} = 1$ and $\Omega_V = 0$. The figure shows $\Delta T/T$ for universes with $\Omega_B = 0.015, 0.025,$ 0.05, and 0.08, with the first acoustic peak increasing in height with increasing Ω_B. The y axis is essentially a measure of $\Delta T/T$ and the x axis is a measure of $1/\theta$. (Figure courtesy of S. Sethi.)

the height of the first peak as a function of the Hubble constant and Ω_B. For any given value of h, the variation is monotonic and the height increases with Ω_B. However, for a given value of Ω_B, the variation is not monotonic with h – the height first decreases and then increases with h because of several conflicting factors coming into play.

It is also important to stress that not all cosmological parameters can be measured *independently* with CMBR data alone. For example, different models with the same values for $(\Omega_{DM} + \Omega_V)$ and $\Omega_B h^2$ will give anisotropies that are fairly indistinguishable. The structure of the peaks will be almost identical in these models. This shows that although CMBR anisotropies can, for example, determine the total energy density $(\Omega_{DM} + \Omega_V)$, we will need some other independent cosmological observations to determine the individual components.

6.4 CMBR Anisotropies: Simplified Derivation

We now proceed with the derivation of the anisotropies at different angular scales expected in CMBR. At large scales, we need to deal with only gravitational

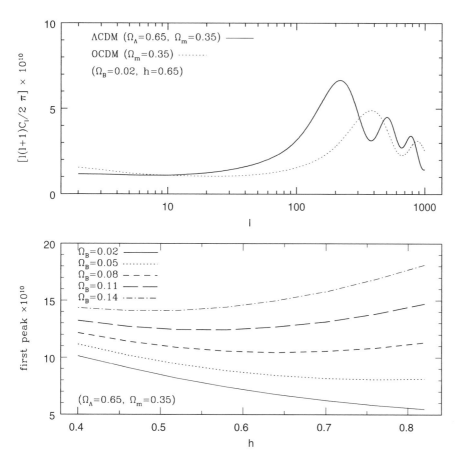

Fig. 6.4. The top frame shows the anisotropies for two cosmological models with $\Omega_m = 0.35$ but with different Ω_{total}. One model has $\Omega_m = \Omega_{tot} = 0.35$, and the other one has $\Omega_m + \Omega_V = \Omega_{tot} = 1$. It is clear that the location of the peak is different in the two cases. The bottom frame shows the variation of the height of the first peak as a function of the Hubble constant and Ω_B. For any given value of h, the variation is monotonic and the height increases with Ω_B. However, for a given value of Ω_B, the variation is not monotonic with h – the height first decreases and then increases with h. (Figure courtesy of S. Sethi.)

effects, but some complications arise because of the gauge dependence of the description. At small scales, we can ignore the issues related to gauge invariance but we must contend with microscopic physical processes that strongly couple baryons and photons. To handle these difficulties, we shall discuss the theory behind CMBR anisotropies on two levels. In this section we provide a simplified derivation that ignores issues related to gauge dependence. In Section 6.5 we will discuss a somewhat more sophisticated approach that can handle the gauge dependence correctly.

Because the temperature anisotropy δT is proportional to the frequency shift $\Delta \omega$, the result of Chap. 3, Section 3.3, can be written as

$$\ln \left(\frac{T_{\text{ob}} a_{\text{ob}}}{T_{\text{em}} a_{\text{em}}} \right) = \frac{1}{2} \int_{t_{\text{em}}}^{t_{\text{ob}}} \dot{h}_{\alpha\beta} \left[t, x^\mu(t) \right] n^\alpha n^\beta \, dt. \tag{6.20}$$

To avoid possible misunderstanding, we stress the following point: We obtain the quantity $\dot{h}_{\alpha\beta} \left[t, x^\mu(t) \right]$ by taking the partial derivative of $h_{\alpha\beta}(t, \mathbf{x})$ with respect to t (at constant x^i), *followed* by the replacement of $\mathbf{x}$ with the function $\mathbf{x}(t) = \mathbf{n}\tau(t)$. We cannot therefore write the result of the integration as $[h_{\alpha\beta}^{(2)} - h_{\alpha\beta}^{(1)}]$. The path of the photon can be expressed as follows. Consider a photon that was emitted at $t = t_e$ and reaches an observer at $t = t_0$ travelling along a null geodesic. Let the *unit* vector pointing along this null ray *from* the observer *to* the source be $\mathbf{n}$. Assuming the observer is at the origin, the location of the photon at some time t will be at

$$\mathbf{x} = \mathbf{n}\tau(t), \qquad \tau(t) = \int_{t}^{t_0} \frac{dt'}{a(t')}. \tag{6.21}$$

This is the path of propagation in the unperturbed metric; as we shall see, we do not need the perturbed trajectory.

In the absence of perturbations, $T \propto a^{-1}$; blackbody radiation at an epoch when the temperature of the universe was T_{em} would have appeared as blackbody radiation with temperature $(T_{\text{em}} a_{\text{em}} / a_{\text{ob}}) \equiv T_0$ today. The deviation ΔT between the observed temperature T_{ob} and the expected temperature T_0 will be a small quantity caused by $h_{\alpha\beta}$. Therefore we can replace the left-hand side of Eq. (6.20) with

$$\ln \left(\frac{T_0 + \delta T_{\text{ob}}}{T_0} \right) \simeq \left(\frac{\delta T}{T} \right)_{\text{obs}}. \tag{6.22}$$

Further note that, on the right-hand side of Eq. (6.20), $h_{\alpha\beta}$ is already a first-order quantity. Therefore in evaluating the integral we need only the trajectory $\mathbf{x}(t)$ of the photon to zeroth-order accuracy; i.e., we need only the trajectory of Eq. (6.21) in the unperturbed metric. Therefore we can write the final answer as

$$\left(\frac{\delta T}{T_0} \right)_{\text{obs}} = \frac{1}{2} n^\alpha n^\beta \int_{t_e}^{t_0} dt \left\{ \frac{\partial}{\partial t} h_{\alpha\beta}[t, \mathbf{n}\tau(t)] \right\}. \tag{6.23}$$

It is possible to relate the derivative of the metric perturbation $\dot{h}_{\alpha\beta}$ to the perturbed gravitational potential by a simple trick. We know that

$$\dot{\delta} = \frac{\partial}{\partial t} [b(t) f(\mathbf{x})] = \frac{\dot{b}}{b} \delta(t, \mathbf{x}), \tag{6.24}$$

where $b(t)$ is the linear density growth factor in the universe. Further, the peculiar-velocity field is related to the gravitational potential by [see Chap. 5, Section 5.6]

$$v_\alpha = -\frac{(\dot{b}/b)}{4\pi G\rho_B a}\frac{\partial\phi}{\partial x^\alpha}. \tag{6.25}$$

Consider now two particles with comoving separation, $\delta x^\beta = n^\beta \delta x$. The relative velocity of these two particles is given by

$$\delta v = n^\alpha \delta v_\alpha = -\frac{(\dot{b}/b)}{4\pi G\rho_B a}\frac{\partial^2\phi}{\partial x^\alpha \partial x^\beta}n^\alpha n^\beta \delta x. \tag{6.26}$$

However, we have seen in Chap. 3, Section 3.3, that the same quantity can be expressed as

$$\delta v = -\frac{1}{2}\dot{h}_{\alpha\beta}n^\alpha n^\beta \delta l = -\frac{1}{2}\dot{h}_{\alpha\beta}n^\alpha n^\beta a\delta x. \tag{6.27}$$

Equating the two expressions for δv, we find

$$\dot{h}_{\alpha\beta} = \frac{1}{2\pi G\rho_B a^2}\left(\frac{\dot{b}}{b}\right)\frac{\partial^2\phi}{\partial x^\alpha \partial x^\beta} \equiv \dot{b}\frac{\partial^2 K(\mathbf{x})}{\partial x^\alpha \partial x^\beta}, \tag{6.28}$$

where the last equality defines the function $K(\mathbf{x})$ as

$$K(\mathbf{x}) = \frac{\phi}{2\pi G\rho_B a^2 b} = -\frac{1}{2\pi}\int \frac{\delta(\mathbf{x}',t)}{b(t)}\frac{d^3 x'}{|\mathbf{x}-\mathbf{x}'|}, \tag{6.29}$$

which is proportional to the gravitational potential generated by the inhomogeneity. Note that, because $\delta(t,\mathbf{x}) \propto b(t)$, the right-hand side is actually independent of time. Putting all these results together, we can express the temperature anisotropy in the form

$$\frac{\delta T}{T} = \frac{1}{2}n^\alpha n^\beta \int_{t_e}^{t_0} dt\,\frac{db}{dt}\frac{\partial^2 K}{\partial x^\alpha \partial x^\beta}. \tag{6.30}$$

The integrand in Eq. (6.23) involves this quantity evaluated along the photon path $\mathbf{x} = \mathbf{n}\tau(t)$. For any function $f(x^\alpha)$ evaluated along a path $x^\alpha = n^\alpha\tau(t)$, $(df/d\tau) = (\partial f/\partial x^\alpha)n^\alpha$. Therefore, along the path, we can write

$$n^\alpha n^\beta \frac{\partial^2 K}{\partial x^\alpha \partial x^\beta} = \left(\frac{d^2 K}{d\tau^2}\right), \tag{6.31}$$

where $K = K(x^\alpha = n^\alpha\tau) = K(\tau)$ is now treated as a function of τ. Thus

$$\frac{\delta T}{T} = \frac{1}{2}\int_{t_e}^{t_0} dt\,\frac{db}{dt}\frac{d^2 K}{d\tau^2}, \qquad d\tau = -\frac{dt}{a}, \tag{6.32}$$

which, on integrating by parts twice, will give

$$\frac{\delta T}{T} = -\frac{a_0}{2}\left(\frac{db}{dt}\right)_0 \frac{\partial K}{\partial x^\alpha} n^\alpha - \left[\frac{a}{2}\frac{d}{dt}\left(a\frac{db}{dt}\right)K\right]_{t_e}^{t_0} + \frac{1}{2}\int_{t_e}^{t_0} dt\, K \frac{d}{dt} a \frac{d}{dt} a \frac{db}{dt}.$$
(6.33)

We can express the first term in this equation by using Eq. (6.29) in a more meaningful form as

$$\left(\frac{\delta T}{T}\right)_{1\text{term}} = \frac{a_0}{4\pi}\left(\frac{\dot{b}}{b}\right)_0 n^\alpha \int d^3 x' \frac{(\mathbf{x}' - \mathbf{x})^\alpha}{|\mathbf{x}' - \mathbf{x}|^3}\, \delta(\mathbf{x}', t_0).$$
(6.34)

This is essentially the dipole anisotropy caused by our peculiar motion through the MBR. The second term in Eq. (6.33) has contributions at the upper limit ($t = t_0$) and at the lower limit ($t = t_e$); the contribution from the present epoch ($t = t_0$) is a constant and can be ignored. As regards the contribution at $t = t_e$, we note that, because this is evaluated at $z \gg 1$, we can approximate the evolution of the universe as that of an $\Omega = 1$ model with $b \propto a \propto t^{2/3}$. Then the contribution that is due to the second term becomes

$$\left(\frac{\delta T}{T}\right)_{2\text{term}} = -\frac{1}{18\pi}\frac{b}{b_0}\frac{a^2}{t^2}\int \frac{d^3 x'}{|\mathbf{x}' - \mathbf{x}|}\, \delta(\mathbf{x}', t_0),$$
(6.35)

where $b_0 = b(t_0)$ and we have taken $b(t) \simeq a(t)/a_0$ at high redshifts. Using

$$t = \frac{2}{3 H_0 \Omega^{1/2}}\left(\frac{a}{a_0}\right)^{3/2},$$
(6.36)

we get the answer as

$$\left(\frac{\delta T}{T}\right)_{2\text{term}} = \frac{\phi_0}{3b_0}, \qquad \phi_0(xn^\alpha) = -G\rho_0 \int d^3 r' \frac{\delta(\mathbf{r}', t_0)}{|\mathbf{r}' - r\hat{\mathbf{n}}|}.$$
(6.37)

This is essentially the gravitational potential that is due to the density fluctuation extrapolated to the present epoch in linear perturbation theory.

Finally, the third term in Eq. (6.33) gives the effect of the time variation on the gravitational potential along the path of radiation. A region with gravitational potential ϕ, when averaged over a proper distance l, will contribute an amount

$$\frac{\delta T}{T} \simeq l \frac{\phi}{G\rho_B a^2 b}\frac{a^2 b}{t^3} \simeq \phi\frac{l}{t},$$
(6.38)

where the first factor is the path length, the second factor is essentially K, and the last factor is the order of magnitude of the time derivative. This is subdominant to the second term by a factor of l/t at low redshifts. As to be expected, this term is identically zero in an $\Omega = 1$ universe in which $b(t) = a(t) \propto t^{2/3}$.

In fact, the analysis simplifies considerably for an $\Omega = 1$ universe, which is worth describing. Here we start with the relation

$$n^\alpha n^\beta \dot{h}_{\alpha\beta} = \frac{\dot{a}}{Q} n^\alpha n^\beta \frac{\partial^2 \phi}{\partial x^\alpha \partial x^\beta},$$ (6.39)

where $Q \equiv 2\pi G\rho_B a^3$ is a constant. From Eq. (6.21), it follows that $dt = -a\, d\tau$, with the minus sign arising from the lower limit of the integral. When the variable is converted from t to τ, Eq. (6.23) becomes

$$\frac{\delta T}{T_0} = \frac{1}{2Q} \int dt\, \dot{a} \left(\frac{d^2\phi}{d\tau^2}\right) = \frac{1}{2Q} \int d\tau \left(\frac{da}{d\tau}\right) \left(\frac{d^2\phi}{d\tau^2}\right).$$ (6.40)

Integrating by parts, we get

$$\frac{\delta T}{T_0} = \left[\frac{1}{2Q} \left(\frac{da}{d\tau}\right) \left(\frac{d\phi}{d\tau}\right)\right]_{\tau_1}^{\tau_1} - \frac{1}{2Q} \int d\tau \frac{d^2 a}{d\tau^2} \left(\frac{d\phi}{d\tau}\right).$$ (6.41)

In the matter-dominated case, $a \propto t^{2/3}$, so that $\tau \propto t^{1/3}$ and $a \propto \tau^2$. Therefore, $(d^2 a/d\tau^2)$ in the second term is a constant. Further note that, in the expression

$$\frac{1}{2Q} \left(\frac{d^2 a}{d\tau^2}\right) \left(\frac{d\phi}{d\tau}\right) d\tau = \frac{1}{(4\pi G\rho_B a^3)} \left[a \frac{d}{dt}(a\dot{a})\right] \dot{\phi}\, dt,$$ (6.42)

a appears three times in both the numerator and the denominator. Thus the proportionality constant in $a \propto t^{2/3}$ relation is irrelevant. Taking just $a = t^{2/3}$, we find the right-hand side of Eq. (6.42) to be

$$\frac{\dot{\phi}\, dt}{(4\pi G\rho_B t^2)} \left[t^{2/3} \frac{d}{dt} \left(t^{2/3} \frac{2}{3} t^{-1/3}\right)\right] = \frac{2}{9} \frac{\dot{\phi}\, dt}{(4\pi G\rho_B t^2)} = \frac{1}{3} \dot{\phi}\, dt.$$ (6.43)

In arriving at the last step, we have used the relation $6\pi G\rho_B t^2 = 1$ for the matter-dominated phase. Consider now the first term in Eq. (6.41), which can be written as

$$-\frac{1}{(4\pi G\rho_B a^3)}(a\dot{a}) \left(\frac{d\phi}{d\tau}\right) = -\frac{(\dot{a}/a)}{(4\pi G\rho_B)} \frac{1}{a} n^\alpha \left(\frac{\partial\phi}{\partial x^\alpha}\right) = \mathbf{v} \cdot \mathbf{n}$$ (6.44)

if we use Eq. (6.25). Substituting this result and Eq. (6.43) into Eq. (6.41), we find that the temperature fluctuation becomes

$$\frac{\delta T}{T_0} = \mathbf{n} \cdot (\mathbf{v}_{ob} - \mathbf{v}_{em}) - \frac{1}{3} \int \dot{\phi}\, dt = \mathbf{n} \cdot (\mathbf{v}_{ob} - \mathbf{v}_{em}) - \frac{1}{3} [\phi(0) - \phi(\mathbf{x}_{em})].$$ (6.45)

Each of these terms has a simple interpretation: $(\mathbf{n} \cdot \mathbf{v}_{ob})$ is the Doppler shift that is due to the motion of the observer with respect to the comoving frame; $(-\mathbf{n} \cdot \mathbf{v}_{em})$ is the corresponding Doppler effect in the emitting surface; $(1/3)[\phi(0) - \phi(\mathbf{x}_{em})]$ is caused by the local variations in the gravitational potential at the source and the observer. The term $(1/3)\phi(0)$ merely adds a constant to $\delta T/T_0$ and will not introduce any directional dependence in $\delta T/T_0$. The other three terms, of course, depend on **n** and hence vary from direction to direction, causing angular anisotropies in the sky.

The derivation did not take into account the effect of any primordial gravitational waves that could be present in the universe. We saw in Chap. 5, Section 5.7, that inflationary models are capable of generating both tensor and scalar perturbations generically. The effect of tensor modes can be easily taken into account by our general formula (6.20), with $h_{\alpha\beta}$ contributed by the gravitational waves. In most inflationary models, the contribution that is due to these waves is important only at large angles and could be a fraction of the contribution from the scalar Sachs–Wolfe effect.

Although the result in Eq. (6.45) is strikingly simple, it hides some subtleties: (1) We shall see in the next section that the factor $1/3$ in the gravitational potential term arises *after* part of the intrinsic anisotropy in the radiation field cancels with the full gravitational potential contribution. (2) At smaller scales, it is possible to use this expression with an addition (and careful interpretation) of the intrinsic anisotropy. (3) Finally, note that the results are valid for only isoentropic fluctuations.

Exercise 6.5

Some uses of CMBR III: In Chap. 5, we decomposed the velocity field u_i of matter into a "rotation" term (ω_{ij}), a "shear" term (σ_{ij}), and the Hubble flow ($H\delta_{ij}$). What bounds can be put on (ω/H) and (σ/H) from the observed isotropy of MBR?

Exercise 6.6

Being unconventional: Consider the possibility that MBR originates from a collection of discrete sources and is not a primordial relic. (a) Suppose MBR is due to the integrated emission from N sources per steradian, each of flux density S that existed at some particular epoch in the past. Let the beam angle of the telescope be Ω. Show that the lack of statistical fluctuations in MBR will imply that $N \gtrsim 10^{12}$ sr^{-1}. Is this an acceptable number? (b) Suppose MBR is due to the emission from a uniform population of sources with luminosity P and extending up to cH_0^{-1}, with $N(\geq S) \propto S^{-3/2}$ for $S \geq S_{\min}$. Show that $N(\geq S_{\min}) \simeq (27\,\Omega)^{-1}(\Delta I/I)^{-3}$. Is this bound acceptable?

6.5 CMBR Anisotropies: A More Rigorous Derivation

We shall now provide a more general and rigorous derivation of the temperature anisotropies, before we discuss the comparison of theory with observations. We will work in the Poisson gauge described in Chap. 5, Subsection 5.4.2, and use the conformal time $d\tau = dt/a$ and the pseudoangular variable χ as our time and radial coordinates. The specific intensity I_ν measured by an observer at $(\tau_o, \chi_o = 0)$ in a particular direction **n** in the sky at frequency ν will contain complete information about the background radiation. However, at the frequencies we are interested in, we can relate the intensity to a brightness temperature through $I_\nu = B_\nu[T(x^\alpha, \tau, \nu, n^\alpha)]$ at every space–time event along which the radiation is propagating. Because the intensity distribution is expected to be a Planckian with redshifted temperature T_0/a plus small perturbations, it is convenient to write $T = (T_0/a)(1 + \Delta)$, where Δ are the fractional perturbations in the temperature that we need to determine. For the primary anisotropies we are considering, both T and Δ are independent of ν, and the latter is of the form $\Delta = \Delta(x^\alpha, \tau, n^\alpha)$. The distribution function for photons is then

$$f(x^\alpha, \tau, E, n^\alpha) = \frac{I_\nu}{2\pi \nu^3} = f_P\left(\frac{aE}{1 + \Delta}\right), \qquad f_P(\epsilon) \equiv 2\left[\exp(\epsilon/T_0) - 1\right]^{-1},$$

(6.46)

which satisfies the equation

$$\frac{df}{d\tau} = \left(\frac{aE}{1 + \Delta}\right) f_P'\left(\frac{aE}{1 + \Delta}\right)\left[\frac{d\ln(aE)}{d\tau} - \frac{d\Delta}{d\tau}\right] = \left(\frac{df}{d\tau}\right)_c, \quad (6.47)$$

where the right-hand side describes the collisional and scattering terms and the left-hand side is the standard propagation term for a collisionless system. (We are using units with $\hbar = c = k_B = 1$, measuring both frequency and temperature in energy units.) The time derivative of Δ is given by

$$\frac{d\Delta}{d\tau} = \frac{\partial\Delta}{\partial\tau} + \frac{\partial\Delta}{\partial x^\alpha}\frac{dx^\alpha}{d\tau} + \frac{\partial\Delta}{\partial E}\frac{dE}{d\tau} + \frac{\partial\Delta}{\partial n^\alpha}\frac{dn^\alpha}{d\tau} \cong \partial_\tau\Delta + n^\alpha\partial_\alpha\Delta, \quad (6.48)$$

where we have used the facts that $(\partial\Delta/\partial E) = 0$ and $dx^\alpha/d\tau \approx n^\alpha$; $dn^\alpha/d\tau = 0$ to the lowest order in perturbation. The rate of change of photon energy can be determined from the geodesic equation in the perturbed metric,

$$ds^2 = a^2(\tau)[(1 + 2\phi)\,d\tau^2 - (1 - 2\psi)\,dl^2]. \quad (6.49)$$

To the lowest order in the perturbed metric coefficients, the geodesic equation gives

$$\frac{d\ln(aE)}{d\tau} = -n^\alpha\partial_\alpha\phi + \partial_\tau\psi. \quad (6.50)$$

The factor a and the first term on the right-hand side are due to the redshift arising from the g_{00} term; the second term on the right-hand side is a purely relativistic effect arising from spatial curvature. Using Eq. (6.50) in Eq. (6.47) we can write the equation for the evolution of Δ:

$$\frac{d\Delta}{d\tau} = \partial_\tau\Delta + n^\alpha\partial_\alpha\Delta = -n^\alpha\partial_\alpha\phi + \partial_\tau\psi + \left(\frac{d\Delta}{d\tau}\right)_c, \quad (6.51)$$

which shows that brightness Δ can change because of gravitational redshift, a time variation of the spatial curvature, or radiative processes described by the collisional term, which – in turn – can be expressed as

$$\left(\frac{d\Delta}{d\tau}\right)_c = an_e\sigma_T\left(-\Delta + \frac{1}{4}\delta_R + v_e^\alpha n_\alpha + \frac{1}{2}\Pi^{\alpha\beta}n_\alpha n_\beta\right) \qquad (6.52)$$

(see Vol. I, Chap. 6, Section 6.8). The factor outside the parentheses is the Thomson scattering rate with (unperturbed) electron density n_e. The first two terms describe the scattering in terms of absorption and reemission of radiation. When the scattering is assumed to be isotropic in the electron rest frame, the absorption gives the $-\Delta$ term and the emission will give a term that is the angular average of Δ – which is $(1/4)\delta_R$; the factor of 4 in the second term arises while relating the temperature to energy density of photons. The third term is due the to the first-order Doppler effect from the velocity field of electrons. The last term is due to anisotropic processes, which we shall ignore. (It arises essentially because the polarisation that is due to scattering couples to the angular term in the Thomson scattering.) For perfect blackbody radiation (with no anisotropies or inhomogeneities), the scattering by an electron gas at rest (however inhomogeneous it may be) has no effect. Scattering produces an anisotropy only if there are some fluctuations in the radiation energy density or when the electron gas is moving. Both these effects are taken into account in Eq. (6.52).

The general solution to Eq. (6.51), correct to linear order in the perturbations, is easy to obtain:

$$\Delta_0(n^\alpha) = \int_0^{\tau_0} d\chi\, e^{-q_T(\chi)}\left[-n^\alpha\partial_\alpha\phi + \partial_\tau\psi + an_e\sigma_T\left(\frac{1}{4}\delta_R + v_e^\alpha n_\alpha\right)\right]_{\mathrm{ret}},$$
$$(6.53)$$

where

$$q_T(\chi) \equiv \int_0^\chi d\chi\, (an_e\sigma_T)_{\mathrm{ret}} \qquad (6.54)$$

and the subscript "ret" means that the quantity is evaluated at $\tau = \tau_0 - \chi$. We ignore the $\Pi^{\alpha\beta}$ term and take the integral along the backward light cone. To interpret this equation, it is convenient to define an *integrated visibility function*

$$\zeta(\tau) \equiv \exp\left[-\int_\tau^{\tau_0} d\tau'\, a(\tau')n_e(\tau')\sigma_T\right] = \exp{-q_T\left[(\tau_0 - \tau)\right]}. \quad (6.55)$$

This is effectively a step function rising rapidly from zero to unity at the time of emission $\tau = \tau_e$, corresponding to the surface of last scattering. Its derivative, $d\zeta/d\tau = an_e\sigma_T\exp(-q_T)$, is essentially a Dirac delta function that appears in Eq. (6.53). Converting the spatial-derivative terms to total time derivatives by

$-n^\alpha \partial_\alpha = d/d\tau + \partial_\tau$ and integrating them by parts, we can write Eq. (6.53) as

$$\Delta_0(n^\alpha) = \int_0^{\tau_0} d\chi \, \frac{d\zeta(\tau_0 - \chi)}{d\tau} \left(\phi + \frac{1}{4}\delta_R + v_e^\alpha n_\alpha \right)$$

$$+ \int_0^{\tau_0} d\chi \, \zeta(\tau_0 - \chi) \partial_\tau (\phi + \psi)_{\text{ret}}, \tag{6.56}$$

where the monopole term $\phi(0)$ is ignored as it is unobservable. The first integral essentially is contributed in the LSS. Because $d\zeta/d\tau$ is like a delta function and the relevant terms are due to gravitational potential, intrinsic anisotropy, and the Doppler effect. The second term arises from the time dependence of the gravitational potential from the LSS until today. This term (called the *integrated Sachs–Wolfe effect*) is irrelevant in $\Omega = 1$ matter-dominated models but can be significant in other cosmologies – like low-Ω models or models that become Λ dominated at late times.

If we assume that the universe recombined instantaneously, we can approximate $d\zeta/d\tau$ as a Dirac delta function. Then the anisotropies are given by

$$\Delta_0(n^\alpha) = \Delta_e + \phi_e - \phi_0 + \int_0^{\chi_e} d\chi \, \partial_\tau (\phi + \psi)_{\text{ret}}, \tag{6.57}$$

with

$$\Delta_e = \left(\frac{1}{4}\delta_R + v_e^\alpha n_\alpha \right)_e. \tag{6.58}$$

In this limit, tight coupling between the photons and the electrons leads to $v_e \approx v_R$; further, the $\Pi_{\alpha\beta}$ term actually vanishes because multiple scattering before recombination damps out the polarisation. Any polarisation-dependent term in $(\Delta T/T)$ has to arise because of the finite thickness of the LSS (which is a reason why they are small). The first term in Eq. (6.58) gives the intrinsic anisotropy, and the second term is the contribution from Doppler effect.

The explicit form of the anisotropies can be determined from the solutions to the perturbation equations that determine ϕ, $v_i = v_R$, and δ_R. The solution to the perturbation equation simplifies tremendously at large scales, where microscopic processes like the pressure gradient can be ignored. We consider this case first.

The solution for this case was found in Chap. 5, Section 5.4, and it depends on whether the perturbations are isentropic or isocurvature. We found that, for a general situation containing both isentropic and isocurvature modes (with $\phi = \psi$), the gravitational potential perturbations can be expressed in the form [see Eq. (5.80)]

$$\phi(\tau, \mathbf{x}) \approx \frac{9}{10}\phi_i(\mathbf{x}) + \frac{1}{5}\sigma_i(\mathbf{x}), \tag{6.59}$$

and the velocity and the density perturbations of the radiation field are expressible in the form

$$v_\alpha = -\frac{3c_s^2}{2(\dot{a}/a)}(1+y)\nabla_\alpha \frac{d(y\phi)}{dy}, \qquad \delta_R = \frac{-2(1+y)d(y\phi)/dy + y\sigma}{1+(3/4)y},$$

$$(6.60)$$

where $y = a/a_{\rm eq}$. From these we get, for $y \gg 1$,

$$\Delta_e = \begin{cases} -(2/3)\phi_e - (2/3\tau_e)n^\alpha\nabla_\alpha\phi_e & \text{(isentropic)} \\ \phi_e - (2/3\tau_e)n^\alpha\nabla_\alpha\phi_e & \text{(isocurvature)} \end{cases}.$$

$$(6.61)$$

It is clear that the velocity term becomes small for wavelengths much larger than the Hubble radius τ_e^{-1} at recombination. The gravitational potential contributions are also quite different for the isoentropic and isocurvature cases. Combining all the factors, we get the large-scale anisotropy along a given direction in the sky specified by either a unit vector or two angles θ and ψ (we shall use the notation ψ to denote the polar angle in order to distinguish it from the gravitational potential ϕ):

$$\Delta(\theta, \psi) = \left(C + \frac{2}{3\tau_e}\frac{\partial}{\partial\chi}\right)\phi(\chi_e, \theta, \psi, \tau_e) + 2\int_0^{\chi_e} d\chi\,\partial_\tau\phi(\chi, \theta, \psi, \tau_0 - \chi),$$

$$(6.62)$$

with $C = 1/3$ for isoentropic perturbations and $C = 2$ for isocurvature fluctuations. This derivation shows that the origin of the factor $1/3$ multiplying the gravitational potential is specific to isoentropic perturbations and arises because a term $(2/3)\phi$ [the first term in Eq. (6.61)] is subtracted from the original term ϕ when the intrinsic anisotropies are taken into account. The contribution of isocurvature potential is nearly six times larger than for isoentropic fluctuation in the gravitational potential term. Thus, for a given gravitational potential fluctuation, essentially determined by dark-matter fluctuation, we will obtain much more anisotropy from isocurvature fluctuation.

To relate to the observations, we have to expand Δ in spherical harmonics as in Eq. (6.8) and evaluate the C_l that appear in Eq. (6.11). All relevant quantities can be expressed in terms of C_l. For example, the mean-square fluctuation in the temperature is

$$\left(\frac{\Delta T}{T}\right)_{\rm rms}^2 = \mathcal{C}(0) = \frac{1}{4\pi}\sum_{l=2}^{\infty}(2l+1)C_l,$$

$$(6.63)$$

where we have omitted the dipole contribution corresponding to $l = 1$. This term will be swamped by the much larger effect arising from our motion through CMBR, and hence the leading term in this sum arises from the quadrapole ($l = 2$), which has the magnitude

$$\left(\frac{\Delta T}{T}\right)_Q^2 = \frac{5}{4\pi}C_2.$$

$$(6.64)$$

To evaluate the C_l we may proceed as follows. Expanding the anisotropy in terms of plane waves, we can write Eq. (6.62) in Fourier space as

$$\Delta(\mathbf{n}) = \int d^3k \left[e^{-i\chi_e \mathbf{k}\cdot\mathbf{n}} \left(C - \frac{2i\mathbf{k}\cdot\mathbf{n}}{3\tau_e} \right) \phi(\mathbf{k}, \tau_e) \right.$$
$$\left. + 2 \int_0^{\chi_e} d\chi\, e^{-i\chi_e \mathbf{k}\cdot\mathbf{n}} \partial_\tau \phi(\mathbf{k}, \tau_0 - \chi) \right] \qquad (6.65)$$

and use the spherical-wave expansion for the plane wave:

$$e^{ix\mu} = \sum_{l=0}^{\infty} i^l (2l + 1)\, j_l(x)\, P_l(\mu). \qquad (6.66)$$

This gives, after some algebra,

$$\Delta(\mathbf{n}) = \int d^3k \sum_{l=0}^{\infty} (-i)^l (2l + 1)\, \Delta_l(\mathbf{k}, \tau_0) P_l(\mathbf{k}\cdot\mathbf{n}), \qquad (6.67)$$

where

$$\Delta_l(\mathbf{k}, \tau_0) = \left[C j_l(k\chi_e) + \frac{2}{3\tau_e} j_l'(k\chi_e) \right] \phi(\mathbf{k}, \tau_e)$$
$$+ 2 \int_0^{\chi_e} d\chi\, j_l(k\chi)\, \partial_\tau \phi(\mathbf{k}, \tau_0 - \chi). \qquad (6.68)$$

[The derivative term of j_l arises from the trick used to introduce the factor $(-i\mathbf{k}\cdot\mathbf{n})$ in the Doppler term by differentiating the exponential.] Using now the addition theorem for spherical harmonics,

$$P_l(\mathbf{n_1}\cdot\mathbf{n_2}) = \frac{4\pi}{2l + 1} \sum_{m=-l}^{l} Y_{lm}(\mathbf{n_1})\, Y_{lm}^*(\mathbf{n_2}), \qquad (6.69)$$

we get an explicit formula for a_{lm} in terms of Δ_l:

$$a_{lm} = (-i)^l\, 4\pi \int d^3k\, Y_{lm}^*(\hat{\mathbf{k}})\, \Delta_l(\mathbf{k}, \tau_0). \qquad (6.70)$$

Let us now consider the contribution to C_l arising from the gravitational potential fluctuations $\Delta(\mathbf{n}) = (1/3)\phi(\tau_e, -\chi_e\mathbf{n})$ in an isoentropic case having the initial power spectrum $P(k) = Ak^n$. Using the result

$$\int_0^{\infty} x^\nu j_l(x)\, j_{l'}(x)\, dx = \frac{\pi 2^{\nu-2} \Gamma(1 - \nu) \Gamma\left(\frac{l+l'+\nu+1}{2}\right)}{\Gamma\left(\frac{l-l'-\nu+2}{2}\right) \Gamma\left(\frac{l'-l-\nu+2}{2}\right) \Gamma\left(\frac{l+l'-\nu+3}{2}\right)}, \qquad (6.71)$$

we can easily to show that

$$C_l = \frac{AH_0^{n+3}}{16} \frac{\Gamma(3-n)}{\Gamma^2[(4-n)/2]} \frac{\Gamma[(2l+n-1)/2]}{\Gamma[(2l+5-n)/2]}. \tag{6.72}$$

For a scale-invariant spectrum with $n = 1$, this reduces to

$$C_2 = \frac{AH_0^4}{24\pi}, \qquad C_l = \frac{6C_2}{l(l+1)}. \tag{6.73}$$

Obviously the observations of C_l from small l will help in constraining the primordial spectral index. We will discuss this further in Section 6.6.

The tensor modes that are due to primordial gravitational waves, if present, will also make a contribution to the anisotropy at large angular scales, which can be computed along the line mentioned at the end of Section 6.4. The results, of course, depend on the detailed model for the inflation that generates the perturbations. If $P_{grav} \propto k^{n_{grav}}$, then $l(l+1)C_l = \mathcal{O}(1)P_{grav}(aH)$ for $l \lesssim 10$ and rapidly goes to zero for higher l. In general, the results of Chap. 5, Section 5.7, shows that the ratio of the contribution from gravitational waves to the contribution from the scalar potential at large angular scales is

$$r \equiv \frac{C_l(\text{grav})}{C_l(\text{scalar})} \simeq 12.4\,\epsilon \tag{6.74}$$

where $\epsilon = (1/16\pi G)(V'/V)^2 \ll 1$ is the slow-rollover parameter. This shows that, for a generic inflationary model, the tensor contribution is subdominant.

We next consider the anisotropies at small angular scales, corresponding to large l. The general solution obtained in Eq. (6.53) continues to be valid, although determining the growth of perturbation in the presence of microscopic physical processes becomes a more difficult problem. For any accurate estimate, it is necessary to integrate the relevant equations numerically. We shall, however, provide a somewhat simplified picture of the features that could be expected at small scales.

To do this, we begin by writing the equations in the Fourier space obeyed by the baryon–photon fluid for a mode labelled by wave vector k. The continuity equations for the photons and the baryons are given by

$$\left(\frac{3}{4}\right)\dot{\delta}_R = -kv_R - 3\dot{\psi}, \qquad \dot{\delta}_B = -kv_B - 3\dot{\psi}. \tag{6.75}$$

In these equations, the second term, $3\dot{\psi}$, arises because the spatial volume gets stretched by the factor $\sqrt{^3g} \approx (1 - 3\psi)$, which needs to be accounted for while conserving the number of particles. The factor $(3/4)$ converts the energy density of photons to number density. The Euler equations for photons and baryons are

$$\dot{v}_R = k(\Theta + \phi) - \dot{q}_T(v_R - v_B), \tag{6.76}$$

$$\dot{v}_B = -\frac{\dot{a}}{a}v_B + k\phi + \frac{\dot{q}_T(v_R - v_B)}{R}, \tag{6.77}$$

where $\Theta = \delta_R/4$. In the case of photons, cosmological expansion – which causes the redshift of temperature – is already taken into account, whereas for baryons this appears explicitly as the first term on the right-hand side of Eq. (6.77). The terms proportional to $k\phi$ arise because of the standard effect of transforming the gravitational potential gradient $\nabla\phi$ into Fourier space. For photons, stress gradients in the fluid pressure proportional to $k\delta p_R/(p_R + \rho_R) = k\Theta$ also provide a force term. Finally, the term proportional to $v_R - v_B$ is the drag between photons and baryons that are coupled by Thompson scattering. The quantity $\dot{q}_T = n_e\sigma_T a$ is the differential Thompson optical depth. The drags on the baryons and photons are different because of the difference in energy densities and the factor

$$R \equiv \frac{p_B + \rho_B}{p_R + \rho_R} \simeq \frac{3\rho_B}{4\rho_R} \approx \left(\frac{450}{1+z}\right)\left(\frac{\Omega_B h^2}{0.015}\right) \tag{6.78}$$

accounts for this. The overdots represent derivative with respect to conformal time τ, with $d\tau = dt/a$. If the scattering rate $\dot{q}_T^{-1}$ is rapid compared with k^{-1}, the photon–baryon system behaves as a perfect fluid. Then, to the lowest order in $k/\dot{q}_T$, Eqs. (6.75)–(6.77) can be combined to give

$$\ddot{\delta}_R + \frac{\dot{R}}{(1+R)}\dot{\delta}_R + k^2 c_s^2 \delta_R = F, \tag{6.79}$$

where the forcing function F is

$$F = 4\left[\ddot{\phi} + \frac{\dot{R}}{(1+R)}\dot{\phi} - \frac{1}{3}k^2\phi\right] \tag{6.80}$$

and the speed of sound is $c_s = [3(1+R)]^{-1/2}$.

To get a feel for the physics described by this equation, let us consider a simple case in which the gravitational potential and the speed of sound are independent of time. In that case, we need only the initial conditions to solve this harmonic-oscillator equation. We note that we can take $\dot{\delta}_R(0) = 0$ although $\delta_R(0)$ cannot be set to vanish. This is because of Eqs. (6.60), which give $\Theta = -2\phi/3$ in the matter-dominated era. [This initial condition corresponds to isoentropic perturbations whereas isocurvature perturbations would have had $\delta_R(0) = 0$; see the discussion after Eq. (6.61).] We shall concentrate on the isoentropic perturbations in what follows. Using this result to set the initial condition to be $\Theta(0) = -2\phi/3$, we find that our solution for the intrinsic temperature anisotropy $\Delta_{T_0} = (\Delta T/T)$ becomes

$$\Delta_{T_0} + \phi_0 = \frac{\phi_0}{3}(1 + 3R)\cos(kc_s\tau_e) - \phi_0 R, \quad v_R = -\phi_0(1 + 3R)c_s\sin(kc_s\tau_e), \tag{6.81}$$

where $\tau_e \approx 190(\Omega_m h^2)^{-1/2}$ Mpc is the comoving time at the surface of recombination. The preceding solution is to be computed *on* the LSS and the result is to be treated as a function of k, which in turn will translate to an angle in the sky. When the Fourier mode dependence $\exp(ikx)$ is incorporated, the solution becomes

$$\Delta_T(\hat{\mathbf{n}}) = e^{ikD_{\mathrm{LSS}}\cos\theta}\, S, \tag{6.82}$$

$$S = \phi_0 \frac{(1+3R)}{3}\left[\cos(kc_s\tau_e) - \frac{3R}{(1+3R)} - i\sqrt{\frac{3}{1+R}}\,\cos\theta\,\sin(kc_s\tau_e)\right], \tag{6.83}$$

where D_{LSS} is the distance to the LSS (which is $\tau_0 - \tau_e$ in a flat universe). This solution has the intrinsic anisotropy as well as the velocity dependent term, as in Eq. (6.58).

The contribution of any plane wave to the angular harmonic of the order of l can now be obtained by direct integration, which is most easily performed by taking the wave vector to be aligned along z axis. Then

$$a_{lm} = \int d\Omega\, Y^*_{lm}(\hat{\mathbf{n}})\,\Delta_T(\hat{\mathbf{n}})$$

$$= \delta_{m0}\phi_0\left\{\left[\frac{(1+3R)}{3}\cos(kc_s\tau_e) - R\right]j_l(kD_{\mathrm{LSS}})\right.$$

$$\left. + (1+3R)c_s\sin(kc_s\tau_e)j_l'(kD_{\mathrm{LSS}})\right\} \tag{6.84}$$

$$\equiv \phi_0[S^m\, j_l(kD_{\mathrm{LSS}}) + S^v\, j_l'(kD_{\mathrm{LSS}})].$$

The superscripts m and v denote the monopole and the velocity contributions, respectively. To obtain the C_l we need to add these contributions in quadrature and integrate over all $\mathbf{k}$. The variance $\langle\phi_0^2\rangle$ will give the power spectrum of the initial gravitational potential, allowing us to write

$$C_l = \int d^3\mathbf{k}\, P_\phi(k)[S^m\, j_l(kD_{\mathrm{LSS}}) + S^v\, j_l'(kD_{\mathrm{LSS}})]^2. \tag{6.85}$$

Given any initial power spectrum, this integral can be evaluated to find the C_l for the theory. It is, however, possible to make an estimate of the integral along the following lines. The Bessel functions for each value of l have a sharp maximum followed by several oscillatory peaks of rapidly decreasing height. Hence the peak contribution to the integral comes from wave numbers at which $kD_{\mathrm{LSS}} \approx l$. Defining a wave number $k^* = l/D_{\mathrm{LSS}}$, we can approximate the contribution from the first (monopole) term as

$$C_l^m \approx \left[\frac{(1+3R)}{3}\cos(k^*c_s\tau_e) - R\right]^2\int d^3\mathbf{k}\, P_\phi\, j_l^2(kD_{\mathrm{LSS}}). \tag{6.86}$$

Similarly, the second term gives

$$C_l^v \approx \frac{(1+3R)^2}{3}c_s^2\sin^2(k^*c_s\tau_e)\int d^3\mathbf{k}\, P_\phi\, j_l'^2(kD_{\mathrm{LSS}}). \tag{6.87}$$

In the case of a scale-invariant power spectrum with $P_\phi = Ak^{-3}$, the integral over the Bessel function gives

$$\int \frac{dk}{k} j_l^2(kD_{\text{LSS}}) \propto \frac{1}{l(l+1)} \tag{6.88}$$

and the expression for C_l becomes

$$l(l+1)C_l = A\left\{\left[\frac{(1+3R)}{3}\cos(k^*c_s\tau_e) - R\right]^2 + \frac{(1+3R)^2}{3}c_s^2 \sin^2(k^*c_s\tau_e)\right\}. \tag{6.89}$$

This expression contains the basic features of the angular anisotropy at different scales. To begin with, note that if $R = 0$ and $c_s = 1/\sqrt{3}$ we get $l(l+1)C_l = A/9$, giving a constant contribution. This is precisely what happens in the absence of baryon–photon coupling. The baryon drag is represented by the nonzero R terms and the C_l have a series of peaks at $k^*c_s\tau_e = \pi, 2\pi, 3\pi, \ldots$, corresponding to $l_{\text{peak}} = n\pi D_{\text{LSS}}/c_s\tau_e$. The monopole term has two contributions with opposite signs so that the peaks at which $\cos(k^*c_s\tau_e)$ is positive will be lower than those at which this term is negative. The difference in the heights is $[(1 + 2R)^2 - 1]$, which will be bigger for larger photon–baryon ratios.

The analysis so far did not incorporate the photon diffusion or the finite thickness of the LSS. This, is however, easily taken into account because we can approximate the scattering region to be a Gaussian in redshift space with some width. In the Fourier space, photon diffusion will exponentially suppress modes with wave vectors larger than the reciprocal of the diffusion length. Thus we only need to modify the Fourier components by $\exp(-k^2L^2)$, where L is the diffusion length scale.

For a more accurate solution, we can use the WKB approximation to solve Eq. (6.79). In the WKB limit, the solution to Eq. (6.79) is given by

$$[1 + R(\tau)]^{1/4}\,\Theta(\tau) = \Theta(0)\cos[kr_s(\tau)] + \frac{\sqrt{3}}{k}\left[\dot\Theta(0) + \frac{1}{4}\dot R(0)\Theta(0)\right]\sin[kr_s(\tau)]$$

$$+ \frac{\sqrt{3}}{k}\int_0^\tau [1 + R(\tau')]^{3/4}\sin[kr_s(\tau) - kr_s(\tau')]\,F(\tau')\,d\tau', \tag{6.90}$$

where $r_s(\tau)$ is the sound horizon defined by

$$r_s(\tau) = \int_0^\tau c_s(\tau')\,d\tau' = \frac{c}{\sqrt{3}H_0\Omega_m^{1/2}}\int_0^{a_R} \frac{da}{(a + a_{\text{eq}})^{1/2}}\frac{1}{R^{1/2}}$$

$$= \frac{2}{3}\frac{1}{k_{\text{eq}}}\sqrt{\frac{6}{R_{\text{eq}}}}\ln\left[\frac{\sqrt{1+R} + \sqrt{R + R_{\text{eq}}}}{1 + \sqrt{R_{\text{eq}}}}\right]. \tag{6.91}$$

Here R_{eq} is the value of R at $z = z_{eq} \approx 2.4 \times 10^4 \, \Omega h^2$ [see Eq. (6.78)] and $k_{eq} \approx (14 \text{ Mpc})^{-1} \Omega h^2$. This solution exhibits oscillations, with the peaks occurring at the l values of the order of

$$l_m = \mathcal{O}(1)m \left[\frac{\pi d_A(z_R)}{r_s} \right], \tag{6.92}$$

where, for a general cosmology,

$$d_A = \frac{c}{H_0 |\Omega_k|^{1/2}} \sin_k \left(|\Omega_k|^{1/2} x \right), \tag{6.93}$$

$$x \approx \int_{a_R}^{1} \frac{da}{(\Omega_m a + \Omega_k a^2 + \Omega_V a^4)^{1/2}}, \tag{6.94}$$

with $\Omega_k = 1 - \Omega_V - \Omega_m$ and $\sin_k = \sinh$ when Ω_k is greater than zero.

The preceding analysis reproduces all the essential results of CMBR anisotropy at different angular scales. In general, there are three characteristic length scales in the problem. The first one leads to an angular harmonic value $l_A = (\pi d_A / r_s)$, where r_s, given by Eq. (6.91), is the sound horizon at the LSS and

$$d_A \approx \frac{2[1 + \ln(1 - \Omega_V)^{0.085}]^{1+1.14(1+w)}}{\sqrt{\Omega_m H_0^2 \Omega_t^{(1-\Omega_V)^{-0.76}}}} \equiv \frac{2}{\sqrt{\Omega_m H_0^2}} d \tag{6.95}$$

is a fitting function for the angular diameter distance in a universe with matter and a comological constant having $\Omega_t = \Omega_m + \Omega_V$. (The second equality defines the dimensionless factor d.) This harmonic l_A can be expressed numerically in the form[4]

$$l_A \approx 172d \left(\frac{z_e}{10^3} \right)^{1/2} \left(\frac{1}{\sqrt{R_e}} \ln \frac{\sqrt{1 + R_e} + \sqrt{R_e + r_e R_e}}{1 + \sqrt{r_e R_e}} \right)^{-1}, \tag{6.96}$$

with

$$r_e \equiv \frac{\rho_R(z_e)}{\rho_m(z_e)} = 0.042(\Omega_m h^2)^{-1} \left(\frac{z_e}{10^3} \right), \quad R_e \equiv \frac{3\rho_B(z_e)}{4\rho_R(z_e)} = 30(\Omega_B h^2) \left(\frac{z_e}{10^3} \right)^{-1}, \tag{6.97}$$

where the subscript e denotes evaluation at the LSS. The second length scale is the size of the Hubble radius at $z = z_{eq}$, which corresponds to the harmonic

$$l_{eq} \equiv \left(2\Omega_m H_0^2 z_{eq} \right)^{1/2} d_A \approx 438d(\Omega_m h^2)^{1/2}. \tag{6.98}$$

The last length scale corresponds to the damping that is due to photon diffusion and the finite thickness of the scattering surface, which can be characterised by the harmonic

$$l_D \equiv k_D d_A \approx \frac{2240d}{[(1 + r_e)^{1/2} - r_e^{1/2}]^{1/2}} \left(\frac{z_e}{10^3} \right)^{5/4} (\Omega_B h^2)^{0.24} (\Omega_m h^2)^{-0.11}. \tag{6.99}$$

The pattern of the angular anisotropies depends on cosmological parameters essentially through these harmonics. Numerical work shows that these harmonics scale in the following manner[5]:

$$l_A \propto h^{-0.34} \, \Omega_B^{0.07} \Omega_m^{-0.15} \Omega_t^{-1.4}, \tag{6.100}$$

$$l_{eq} \propto h \, \Omega_m^{0.59} \Omega_t^{-1.4}, \tag{6.101}$$

$$l_D \propto h^{-0.02} \, \Omega_B^{0.20} \Omega_m^{-0.12} \Omega_t^{-1.4}. \tag{6.102}$$

Most of the effects seen in Figs. 6.2 and 6.3 can be understood in terms of variations of these scales.

Exercise 6.7

The photon–baryon coupling: The damping of the perturbations in the photon–baryon system (discussed in the text) can be derived more rigorously in the following manner: Consider the distribution function $f_t(x^a, p^b)$ for the photons. Using $t, x^j, |\mathbf{p}| = \epsilon$, and the direction cosines γ^α of the vector $\mathbf{p}$ as the independent variables, we can write the evolution equation for f_t as

$$\frac{\partial f_t}{\partial t} + \frac{\partial f_t}{\partial x^\alpha} \dot{x}^\alpha + \frac{\partial f_t}{\partial \epsilon} \dot{\epsilon} + \frac{\partial f_t}{\partial \gamma^\alpha} \dot{\gamma}^\alpha = c_1 - c_2, \tag{6.103}$$

where c_1 is the amount of photons scattered into the beam of radiation travelling in a particular direction and c_2 is the amount of photons scattered out of this beam.

(a) Separate f_t as $(f_0 + f)$, where f_0 is the homogeneous, isotropic background and f is a perturbation. Show that in the *rest frame* of the matter, f changes between two infinitesimally separated points by the amount

$$df' = \delta t' \sigma n_e(g'(\epsilon') - f'), \tag{6.104}$$

where the prime denotes quantities measured in the rest frame of matter, ϵ' is the energy of the photon, and

$$g'(\epsilon') = \int \frac{d\Omega}{4\pi} f'(\epsilon', \gamma^{\alpha'}). \tag{6.105}$$

(b) Prove that the quantities in the primed and unprimed coordinates are related by $\delta t' = (\epsilon'/\epsilon)\delta t$ and $\epsilon' \cong \epsilon(1 - \gamma_\alpha v^\alpha)$ to first order in v^α.

(c) Argue that, to the linear order in perturbation, $(\epsilon'/\epsilon)(g' - f') \approx (g' - f')$ and $(\partial f / \partial \gamma^\alpha)(\dot{\gamma}^\alpha) \approx 0$.

(d) Using the above facts, derive the equation

$$\frac{\partial I}{\partial t} + \frac{\gamma^\alpha}{a} \frac{\partial I}{\partial x^\alpha} \cong \sigma n_e(\delta_R + 4\gamma_\alpha v^\alpha - I), \tag{6.106}$$

where the "brightness function" I is defined through the relation

$$\int dp p^3 f_t = \frac{\rho_b}{4\pi} [1 + I(\gamma^\alpha, \mathbf{x}, t)] \tag{6.107}$$

and we have ignored the effects that are due to the perturbed gravitational field. Interpret the origin of each term in this equation.

(e) Fourier transform the variables, taking $\mathbf{k}$ along the z axis, to obtain

$$\dot{I} + \frac{ik\mu}{a} I = \sigma_T n_e [\delta_R + 4\mu v - I], \quad \mu = \hat{k}^\alpha \gamma_\alpha. \tag{6.108}$$

Also show that, to the same order of approximation, the equation of motion for matter is

$$\rho_m \left(\dot{v} + \frac{\dot{a}}{a} v \right) = \sigma_T n_e \rho_\gamma \left[P - \frac{4}{3} v \right], \quad P \equiv \frac{1}{2} \int_{-1}^{+1} d\mu \, I, \tag{6.109}$$

$$\dot{\delta}_m = -\frac{ik}{a} v. \tag{6.110}$$

(f) Because $t_c \equiv (\sigma_T n_e)^{-1}$ is much smaller than the expansion time scale, these equations can be solved by a systematic expansion in t_c. Writing

$$I = \delta_R + 4\mu v + t_c \left[\dot{I} + \frac{ik\mu}{a} I \right] \tag{6.111}$$

and iterating systematically, show that, to first order in t_c, we have the equations

$$\dot{\delta}_R \cong \frac{-ik}{a} \left[\frac{4}{3} v - t_c \left(\frac{4}{3} \dot{v} + \frac{ik}{3a} \delta_R \right) \right]$$

$$\rho_m \dot{v} \cong \rho_\gamma \left[-\frac{4}{3} \dot{v} - \frac{ik}{3a} \delta_R + t_c \left(\frac{4}{3} \ddot{v} + \frac{2}{3} \frac{ik}{a} \dot{\delta}_R - \frac{4}{5} \frac{k^2}{a^2} v \right) \right]. \tag{6.112}$$

These equations admit solutions with the time dependence $\exp(-\Gamma t)$, where

$$\Gamma = \frac{k^2 t_c}{6a^2} \left(1 - \frac{6}{5b} + \frac{1}{b^2} \right) \pm \frac{ik}{a\sqrt{3b}}, \quad b = 1 + \frac{3}{4} \frac{\rho_m}{\rho_\gamma}. \tag{6.113}$$

Demonstrate that the oscillations correspond to the acoustic vibrations with the correct adiabatic sound speed of the photon–baryon gas while the damping defines a critical wavelength λ_D, where

$$\lambda_D^2 = \frac{4\pi^2}{6} t_c t. \tag{6.114}$$

Compare this result with the one derived in the text.

(g) If the viscous drag is completely ignored, we get the *zeroth*-order approximation. Show that, in this limit,

$$\delta_R = \frac{4}{3} \delta_m. \tag{6.115}$$

Thus the coupling between radiation and matter will preserve the adiabatic nature of baryonic perturbations.

6.6 Comparison with Observations

Because the temperature anisotropies are approximately a few parts in 10^{-5}, observing them (as well as interpreting them) is a fairly involved process, and several special techniques and methodologies have been developed to deal with this problem. We now provide a overview of what current results imply for theoretical models.

6.6.1 Dipolar Anisotropy

Because the most dominant anisotropy in CMBR is dipolar, which could be interpreted as being due to the motion of our galaxy with respect to the cosmic rest frame, we begin with a discussion of this anisotropy and what it implies.

The dipole anisotropy of CMBR suggests that our Local Group is moving with respect to MBR at a speed of 627 ± 22 km s^{-1} in the direction $l = 276 \pm 3°$, $b = 30 \pm 3°$. To avoid possible misunderstanding, it is worthwhile to state clearly what this velocity refers to.[6] The quantity that is actually observed in a MBR experiment will be the velocity of Earth with respect to MBR. If the rotation and the revolution of Earth – which vary in a very short time scale, but are known quite precisely – are subtracted out, we obtain the "heliocentric" velocity, which is $\sim 370.6 \pm 0.4$ km s^{-1} in the direction of $l = 264.31 \pm 0.17°$, $b = 48.5 \pm 0.10°$. This is the velocity that corresponds to the MBR anisotropy that a hypothetical observer located on the Sun will detect. If we further correct for the motion of the Sun (which is due to the Sun's rotation in the Milky Way) and the motion of the Milky Way relative to galaxies in the Local Group, we obtain the velocity vector with respect to the centre of mass of the Local Group. The last two steps are achieved as follows.

A star in a *circular* orbit, at the location of the Sun, will be moving with a speed of 222 ± 5 km s^{-1} towards $l = 91.1 \pm 0.4°$, $b = 0$. The motion of such a hypothetical star is used to define a frame of reference, usually called the *local standard of rest* (LSR) (see Chap. 1, Section 1.9). The actual motion of the Sun deviates from the LSR somewhat; the velocity of the Sun *with respect to* the LSR is 20.0 ± 1.4 km s^{-1} in the direction of $l = 57 \pm 4°$, $b = 23 \pm 4°$. Because the (v_x, v_y, v_z) components of a vector $\mathbf{v}$ in the direction (l, b) – which we will denote as $v(l, b)$ collectively – are $v \cos b \cos l$, $v \cos b \sin l$, and $v \sin b$, we can easily compute the velocity of the Sun with respect to the centre of the Milky Way, $v_{\odot MW}$. We get

$$v_{\odot - MW} = (10, 235.4, 7.8) \text{km s}^{-1}. \tag{6.116}$$

To proceed further we need to know the motion of the centre of our galaxy with respect to the centre of mass of the Local Group. This value, however, is somewhat uncertain. As a first approximation, we may assume that the centre of the Milky Way is moving towards Andromeda, which is in the direction specified by the unit vector $(-0.476, 0.79, -0.41)$. The relative velocity of Andromeda towards the Milky Way is $\sim$119 km s^{-1}. If we assume that $m_{MW}\mathbf{v}_{MW} = -m_A\mathbf{v}_A$ and that $m_A \approx 2m_{MW}$, then we find that the centre of the Milky Way must be moving with the velocity $(-38, 62.7, -29)$km s^{-1} with respect to the centre of mass of the Local Group. Combining this value with the previous result, we find that the velocity of the Sun with respect to the centre of mass of the Local Group is $(-28, 298, -21)$km s^{-1}. Different measurements of the motion in the Local Group have led to different values for this number. A detailed analysis, taking into account the velocities of all the members of the Local Group, for example, gives the value

$$\mathbf{v}_{\odot-LG} = (-89.3, 292.6, -36.96)\text{km s}^{-1}. \tag{6.117}$$

To facilitate an easy comparison of the results, we usually use the "standard" result of $(0, 300, 0)$km s^{-1}. Given some specified value for $\mathbf{v}_{\odot-LG}$ and the velocity of the Sun with respect to MBR, we can compute the velocity of the centre of mass of the Local Group with respect to MBR. Taking $\mathbf{v}_{\odot-LG} = (0, 300, 0)$km s^{-1} gives

$$\mathbf{v}_{LG-MBR} = 627 \pm 22 \text{ km s}^{-1}(l = 276° \pm 3°, b = 30° \pm 3°). \tag{6.118}$$

On the other hand, if we take $\mathbf{v}_{\odot-LG} = (-89.3, 292.6, -37)$km s^{-1}, then we find that $\mathbf{v}_{LG-MBR} = 622 \pm 20$ km s^{-1} ($l = 277° \pm 2°, b = 30° \pm 2°$). Note that the final values that we obtain for this velocity are not too different.

Two immediate questions arise from the preceding observation: (1) What is the physical origin of this, rather large, velocity of the Local Group and (2) how far does it extend? In other words, what is the size of the region within which the motion of the Local Group is shared by other galaxies?

To answer the first question we may proceed as follows: Because peculiar velocities are generated by the mass concentration around us, it is natural to see whether our velocity points towards any of the known mass concentrations.[7] One of the largest clusters near us is the Virgo cluster, which is in the direction $(284°, 74°)$. The v_{MBR} can be resolved into three components in the following convenient manner. We first resolve the velocity vector into a component perpendicular to the supergalactic plane (v_z) and another in the supergalactic plane. The component in the supergalactic plane is further resolved into one along the direction of the Virgo cluster (v_v) and one in the perpendicular direction (v_s). These components have the values $v_z = -355 \pm 25$ km s^{-1}, $v_v = 418 \pm 25$ km s^{-1}, and $v_s = 277 \pm 25$ km s^{-1}. It is obvious that the Virgo cluster cannot be the sole source of this peculiar velocity; it accounts, at best, only for $\sim$46% of the total amplitude. Two other nearby clusters are Hydra and Centaurus, and it is certainly conceivable that some of the gravitational pull is exerted by these systems. To settle this issue, it is necessary to measure the effects of a deeper and more extensive set of galaxies.

In the linear theory, peculiar velocities are in the same direction as the peculiar acceleration, which we can compute by adding the contributions M_i/r_i^2 from each galaxy vectorially. Assuming that galaxies trace mass and that they have a constant M/L ratio, we can replace this sum with the addition of L_i/r_i^2 for each galaxy. Such calculations have been performed with optical and IR catalogues.[8] The "light dipoles" are in reasonable alignment ($7°$ to $15°$) with the MBR dipole. These results suggest that the mass distribution causing the acceleration lies outside the Local Group and extends to a sphere of $40h^{-1}$ Mpc or so.

If most of the accelerating mass is far away, then all the galaxies in our local region will feel typically the same acceleration and will move together as a unit. In other words, there could exist a "bulk flow" in our neighbourhood. This leads us to the second question, which is an intrinsically more difficult one. Although it is rather easy to measure *our* peculiar velocity with respect to MBR, it is not so easy to estimate the peculiar velocities of other galaxies with respect to MBR. What is available observationally are the redshifts z of the various galaxies around us. Redshift data will give an equivalent *radial* velocity (v) for each galaxy. However, this radial velocity will be partly due to Hubble flow and partly due to peculiar motions. To isolate the latter, we need an estimate of the distance r to the galaxy (which should not be, of course, based on the redshift). If this information is available, then we can compute the radial peculiar velocity, v_{pec}, by the relation

$$v_{\text{pec}} = v - Hr, \qquad (6.119)$$

thereby obtaining a detailed map of the radial-velocity field of the universe. There have been several attempts to map the peculiar-velocity field around us by use of different types of astronomical objects (spirals, ellipticals, ...,) and by use of different estimators for the distance. Because these distance estimators are rather crucial to our discussion, we first examine some of their properties.

There are two distance estimators that have been used extensively in these studies. These are based on the IR Tully–Fisher (IRTF) relation for spirals (see Chap. 1, Section 1.7) ($L \propto v^4$, where L is the absolute luminosity and v is the circular velocity) and the diameter–velocity dispersion ($D_n - \sigma$) relation (see Chap. 1, Section 1.8) for ellipticals. (This relation is $D_n^3 \propto \sigma^4$, where D_n is the angular diameter and σ is the central velocity dispersion.[9] The diameter D_n is defined by use of the isophote within which the mean surface brightness is 20.75 mag arcsec^{-2}. Normally, the "diameters" are defined either with a faint isophote at 25 mag arcsec^{-2} or by use of central brightness in the photographic plate. Both these extremes are avoided in the preceding definition for D_n.) These relations represent empirically determined correlations between a distance-dependent quantity (like flux or angular diameter) and another distance-independent quantity (circular velocity or central velocity dispersion).

A study of nearly 385 elliptical galaxies by use of the $D_n - \sigma$ relation suggested that the galaxies in the Centaurus cluster themselves are moving in the same direction as we are.[10] Detailed modelling of these galaxies within a velocity sphere of 6000 km s^{-1} shows that there is a bulk motion of ~600 km s^{-1} in the direction of $l = 312°$, $b = 6°$. This direction is still $\sim 40°$ away from the MBR dipole direction. We now examine this result in greater detail.

The first question to settle will be the estimate of the size of the region in which all the galaxies exhibit a common, shared flow. This can be decided by plotting the peculiar velocities of nearby galaxies, measured in the Local Group velocity frame, against the observed distance. If a galaxy completely shares the MBR motion of the Local Group, its peculiar velocity in the Local Group frame should be nearly zero. Such an analysis shows that galaxies within a distance of ~ 500 km s^{-1} (which is equivalent to a distance of $5h^{-1}$ Mpc) share the MBR motion of the Local Group.[11] The motion of the galaxies begin to deviate in a systemic manner around 600–750 km s^{-1}. This motion seems to be shared by a part of the Coma-Sculptor cloud. This cloud is in the form of a narrow, thin distribution fully contained in the plane of the Local Supercluster. The major axis of this cloud extends nearly 750 km s^{-1} in the direction $l = 134°$, $b = 62°$ and ~ 400 km s^{-1} in the opposite direction. The minor axis is along $l = 315°$, $b = 28°$. The v_z velocity of the Local Group is shared, at least in projection, by the whole cloud. There is some evidence that the galaxies at the edge of the cloud are moving towards the centre of the cloud.

The overall impression from this analysis is that there exists a net flow towards the Centaurus region along with a net inward motion on either side of the supergalactic plane from the Local Group. Analysing these data, one concludes that there are significant dipole and quadrupole components in the flow. The net dipole motion has an amplitude of 500 ± 50 km s^{-1} in the direction $l = 314 \pm 15°$, $b = 14 \pm 5°$. It must be emphasised that these peculiar velocities are measured with the $D_n - \sigma$ relation for estimating distances; any systematic error in this relation could affect the results drastically, even though such a possibility seems unlikely.

6.6.2 Anisotropies at Large Angular Scales

We next consider the anisotropies at large angular scales in an $\Omega = 1$ model (we shall discuss the modifications arising from different cosmological models later). At large angular scales, where the effects that are due to the gravitational potential dominates the temperature anisotropy, we have, from Eq. (6.45),

$$\left(\frac{\Delta T}{T}\right) = -\frac{1}{3}\phi(\mathbf{x}_{em}) = -\frac{1}{3}G\rho_b(t_0)\int d^3\mathbf{x}' \frac{\delta(\mathbf{x}', t_0)}{|\mathbf{x}_{em} - \mathbf{x}'|}$$

$$= -\frac{1}{2}H_0^2 \int \frac{d^3\mathbf{k}}{(2\pi)^3}\frac{\delta_k}{k^2}\exp(-i\mathbf{k}\cdot\mathbf{x}_{em}). \tag{6.120}$$

In arriving at the last expression, we have used $G\rho_b = (3H^2/8\pi)$ and introduced the Fourier components. Comparing Eqs. (6.120) and (6.8) and using the fact that $\mathbf{x}_{em} = 2H_0^{-1}\mathbf{n}$, we can express $\langle|a_{lm}|^2\rangle$ in terms of $|\delta_k|^2$. We get, after a straightforward calculation,

$$\langle |a_{lm}|^2 \rangle = C_l = \frac{H_0^4}{2\pi} \int_0^\infty dk \frac{|\delta_k|^2}{k^2} \left| j_l \left(2H_0^{-1}k\right) \right|^2, \tag{6.121}$$

where j_l is the spherical Bessel function of the order of l. This equation, along with Eq. (6.11), expresses the temperature correlation function in terms of the power spectrum of the theory. The mean-square fluctuation in the temperature is given by Eq. (6.63) and the leading term in this sum (which arises from the quadrupole) has the magnitude given by Eq. (6.64).

The preceding formulas, especially Eq. (6.63), are correct provided that the observations can probe all values of l with equal sensitivity. However, in practice, any instrument will have a finite angular resolution, θ_c. This implies that the response of the instrument will decrease significantly for modes with $l > l_c$, where $l_c \simeq \theta_c^{-1}$. This effect can be taken into account if a response profile is introduced for the detector. If we take the profile to be Gaussian, the modified formulas for the anisotropy will then be

$$\left(\frac{\Delta T}{T}\right)_{\text{rms}}^2 = \frac{1}{4\pi} \sum_{l=2}^\infty (2l+1)C_l \exp\left(-\frac{l^2\theta_c^2}{2}\right), \tag{6.122}$$

$$\left(\frac{\Delta T}{T}\right)_Q^2 = \frac{5}{4\pi} C_2 \exp\left(-2\theta_c^2\right). \tag{6.123}$$

For example, the very first detection of temperature anisotropies at large angular scales came from the Cosmic background explorer (COBE) experiment, for which $\theta_c \simeq 10°$. Assuming that the power spectrum at large scales has $n = 1$, that is, $P(k) = Ak$, and substituting the expressions for C_l from Eqs. (6.73) into Eqs. (6.122) and (6.123), we get

$$\left(\frac{\Delta T}{T}\right)_Q^2 \simeq \left(\frac{5}{96\pi^2}\right)\left(AH_0^4\right) = (5.28 \times 10^{-3})\left(AH_0^4\right), \tag{6.124}$$

$$\left(\frac{\Delta T}{T}\right)^2 = \frac{1}{4\pi} \sum_{l=2}^\infty (2l+1) \left[\frac{6C_2}{l(l+1)}\right] \exp\left(-\frac{1}{2}l^2\theta_c^2\right)$$

$$= \frac{C_2}{4\pi} \times 28.45 = 0.03\left(AH_0^4\right). \tag{6.125}$$

The quantity $\left(AH_0^4\right)$ is directly related to the fluctuations in the gravitational potential $\Phi^2 = (k^3|\phi_k|^2/2\pi^2)$ at large scales. Because $\phi_k = (4\pi G\rho_b)\,(\delta_k/k^2) = (3/2)H^2(\delta_k/k^2)$ we find that

$$\Phi^2(k) = \frac{k^3|\phi_k|^2}{2\pi^2} = \frac{9}{4}\left(\frac{H}{k}\right)^4 \left(\frac{k^3|\delta_k|^2}{2\pi^2}\right) = \frac{9}{8\pi^2}\left(AH_0^4\right) \tag{6.126}$$

if $|\delta_k|^2 \cong Ak$. Therefore $AH_0^4 = (8\pi^2/9)\Phi^2$, and we can reexpress relations (6.124) and (6.125) as

$$\left(\frac{\Delta T}{T}\right)_Q \cong 0.22\Phi, \qquad \left(\frac{\Delta T}{T}\right)_{rms} \cong 0.51\Phi. \qquad (6.127)$$

We are now in a position to compare the theoretical results at large angular scales with the COBE observations. To begin with, relations (6.127) predict that $(\Delta T_{rms}/\Delta T_Q) \cong 2.3$ if the spectrum has $n = 1$. The COBE results allow this ratio to fall between 1.43 and 3.94, with a mean value of 2.29. This is consistent with the assumption of $n = 1$, although the uncertainty is rather large.

The parameter Φ and the amplitude A can now be determined by comparing relations (6.127) with COBE result. Ideally, of course, both $(\Delta T/T)_{rms}$ and $(\Delta T/T)_Q$ should lead to the same value for Φ and A. However, because of instrumental and systematic errors, we get slightly different values that are – of course – consistent within error bars. For a scale-invariant spectrum in the $\Omega = 1$ model, the normalisation from COBE is

$$\Delta^2(k) = \frac{k^3|\delta_k|^2}{2\pi^2} = \frac{Ak^4}{2\pi^2} = \left(\frac{k}{0.0737h \text{ Mpc}^{-1}}\right)^4. \qquad (6.128)$$

This corresponds to $\Phi \cong 3.1 \times 10^{-5}$ and $A \cong (28.6h^{-1} \text{ Mpc})^4$.

For the sake of completeness, we shall mention the modifications to these results when some of the parameters are varied. Most of the results quoted below are obtained by numerical work and should be thought of as approximate fitting functions. To begin with, let us consider the effects of (1) the change in the value of the spectral index from unity and (2) the existence of tensor perturbations. When the spectrum is *not* scale invariant, it is necessary to choose a scale at which the spectrum is specified; we will take it to be the current Hubble radius. Because the parameters n, n_{grav}, and r – with r defined by Eq. (6.74) – are not strictly constant, it is also necessary to choose a point (usually called the *pivot point*) at which the predictions of models with different n, r will agree and evaluate n, n_{grav}, and r at this scale.[12] This can be done only after the models are fully worked out. The numerical results show that the pivot point can be taken to be $k_p \approx 7a_0 H_0$. With these choices, it is found that the dependence of Φ on n and r can be fitted by

$$\Phi(n, r) = \Phi(1, 0)\frac{\exp[1.01(n - 1)]}{\sqrt{1 + 0.75r}}. \qquad (6.129)$$

The denominator takes into account the effect of gravitational waves. Although r was defined in Eq. (6.74) as the ratio between tensor and scalar contributions, that definition assumed that (1) only the Sachs–Wolfe effect contributes to $(\Delta T/T)$ and (2) the universe is matter dominated at the LSS. Correcting for these assumptions lead to the coefficient 0.75 rather than unity for r.

The preceding result was for a matter-dominated universe with $\Omega = 1$. If we introduce a cosmological constant but keep $\Omega_{NR} + \Omega_V = 1$, then the fitting function becomes

$$\Phi(n, r, \Omega_{NR}) = \Phi(1, 0, 1)\frac{\exp[1.01(n-1)]}{\sqrt{1 + f(\Omega_{NR})r}}\frac{[1 - 0.18(1-n)\Omega_V - 0.03r\Omega_V]}{\Omega_{NR}^{0.8+0.05\ln\Omega_{NR}}},$$

(6.130)

where $f = 0.75 - 0.13\Omega_V^2$. These fitting functions are accurate to a few percent for the range $0.7 < n < 1.3, r < 2$, and $0.2 < \Omega_{NR} < 1$. Part of the effect arises because of differences in the growth rate of perturbations (from the case with $\Omega_V = 0$) and part of the effect is due to the relative weights of scalar and tensor components in the presence of a cosmological constant.

6.6.3 Anisotropies at Small Angular Scales

Let us next consider the results from smaller scales. The study of the CMBR spectra at smaller scales needs to be done by some variant of a best-fit procedure for all the data points. The value of such an analysis lies in the possibility of extracting cosmological parameters by comparing the results with other observations. Some attempts in this direction are shown in Figs. 6.5 and 6.6. The results

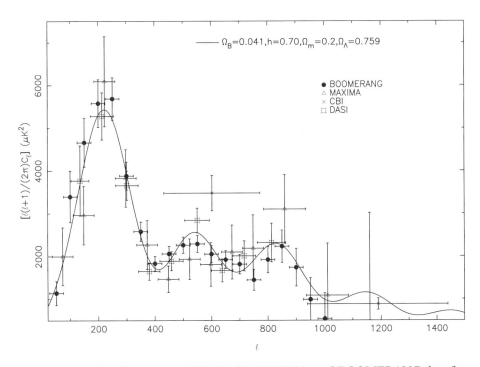

Fig. 6.5.   The best-fit curves to CBI, DASI, MAXIMA, and BOOMERANG data for a universe with $\Omega_{tot} = 1$ and $n = 1$. The best-fit model is plotted along with data from different experiments. The errors shown do not include calibration errors.

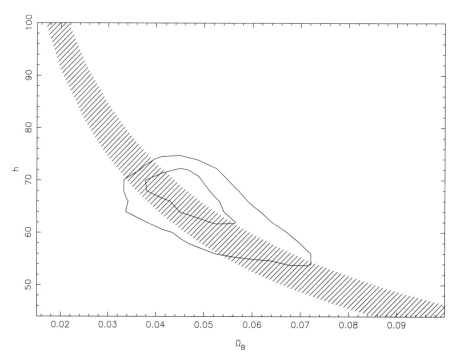

Fig. 6.6. The contours correspond to allowed 1σ and 2σ regions by the BOOMERANG experiment. The hatched region corresponds to the 95% region ($\simeq 2\sigma$) from primordial nucleosynthesis.

depend on the number of free parameters as well as on the assumptions used in the statistical analysis.[13] If we take $\Omega = 1$ and $n = 1$, then a best-fit curve for the currently available data looks like the one shown in Fig. 6.5. Depending on the prior assumptions, the best-fit parameters change somewhat but broadly favour models with $\Omega_V \approx 0.7$ and $\Omega_{NR} \approx 0.3$. It is possible to make the data fit $\Omega_{NR} = 1$ and $\Omega_V = 0$ within 2σ errors. However, the best-fit curve given in Fig. 6.5 is also consistent with other observations, like those based on supernova, which we will discuss in Chap. 10. In Fig. 6.6 the value of $\Omega_B h^2$ from CMBR data is compared with big bang nucleosynthesis constraints, and we can see that the results are consistent within error bars.

The following point should be noted regarding the statistical analysis of CMBR data: The actual $\Delta T / T$ projected on the sky is a Gaussian random field, provided that the initial perturbations obey Gaussian statistics. The definition of a_{lm} in terms of $\Delta T / T$ shows that the former is a sum over the latter with Y_{lm} acting as the weightage functions. Because Y_{lm} is real for $m = 0$ but complex for $m \neq 0$, it follows that a_{lm} for $m \neq 0$ are two-dimensional Gaussian random fields in the complex plane. In the case of a quadrapole, we are interested in the sum

$$S = \sum_{m=-2}^{m=2} |a_{lm}|^2. \qquad (6.131)$$

Each term in the sum is uncorrelated with the others, and hence the sum has a χ^2 distribution with 5 degrees of freedom. The observed quadrupole is one realisation of the distribution and hence is subject to the unavoidable fluctuations of the background distribution, usually called *cosmic variance*.

In describing the anisotropies, we have assumed that the matter recombined around $z \approx 10^3$ and has remained neutral since then. If this is the case, we would have expected to see some amount of neutral hydrogen in the IGM. (It will be quite difficult for the galaxy formation to be 100% efficient, thereby succeeding in making *all* the gas condense out as bound structures.) The light we receive from distant sources (such as quasars) has to pass through this neutral hydrogen before reaching us. We should therefore be able to see a characteristic signature of neutral hydrogen in the absorption spectra of quasars. No such signature that is due to smoothly distributed neutral hydrogen is seen. (This observation will be discussed in Chap. 9.) The absence of any detectable, smoothly distributed neutral hydrogen in the galactic medium suggests that there could have been a reionisation of the IGM at some redshift higher than $z \gtrsim 6$. Such a reionisation can make the CMBR photons interact with plasma once again and can lead to distortions in the spectrum and isotropy.

Let us suppose that the reionisation occurred at a redshift of z_{ion} and that the IGM remained fully ionised for all $z < z_{\mathrm{ion}}$. Then the optical depth that is due to Thomson scattering up to a redshift of z ($< z_{\mathrm{ion}}$) will be

$$
\begin{aligned}
\tau &= \int \sigma_T n_e \, dt = \left(\frac{\sigma_T \Omega_B \rho_c}{m_p} \right) \int_0^z H_0^{-1} \, dz (1+z)^3 \frac{1}{(1+z)^2 (1+\Omega_{\mathrm{NR}} z)^{1/2}} \\
&= \frac{\sigma_T H_0}{4\pi G m_p} \frac{\Omega_B}{\Omega_{\mathrm{NR}}^2} \left[2 - 3\Omega_{\mathrm{NR}} + (1+\Omega_{\mathrm{NR}} z)^{1/2} (\Omega_{\mathrm{NR}} z + 3\Omega_{\mathrm{NR}} - 2) \right],
\end{aligned}
$$

(6.132)

where we have assumed that $\Omega_V = 0$ and $\Omega_{\mathrm{tot}} = \Omega_{\mathrm{NR}}$. This expression has the limiting forms

$$
\begin{aligned}
\tau &= \frac{\sigma_T H_0}{4\pi G m_p} \frac{3}{2} (\Omega_B z) = 0.026 (\Omega_B z) & \text{(for } \Omega_{\mathrm{NR}} z \ll 1) \\
&= \frac{\sigma_T H_0}{4\pi G m_p} \left(\frac{\Omega_B}{\Omega_{\mathrm{NR}}^{1/2}} \right) z^{3/2} = 0.017 \left(\frac{\Omega_B}{\Omega_{\mathrm{NR}}^{1/2}} \right) z^{3/2} & \text{(for } \Omega_{\mathrm{NR}} z \gg 1).
\end{aligned}
$$

(6.133)

If $\tau \simeq 1$ for some $z \leq z_{\mathrm{ion}}$, then the photons will be significantly scattered by the reionised plasma. This will partially wipe out information about the $z = z_{\mathrm{dec}}$ surface. For universes with $\Omega_{\mathrm{NR}} \simeq 1$ and $\Omega_B \simeq 0.1$, $\tau = 1$ occurs near $z \simeq 70$ whereas for $\Omega_{\mathrm{NR}} \simeq \Omega_B \simeq 0.2$, $\tau = 1$ can occur at a much lower redshift: $z \simeq 25$; thus in standard dark-matter models reionisation has to occur at fairly high redshifts (greater than 70) to produce any effect. If such reionisation occurs at a redshift of z_{ion}, its primary effect will be to wipe out the original $\delta T / T$ at scales smaller than the Hubble radius $d_H(z_{\text{ion}})$ at z_{ion}. From Eq. (6.3), we see that d_H subtends the angle

$$\theta_{\text{ion}}(z_{\text{ion}}) \cong 3° \, \Omega_{\text{NR}}^{1/2} \left(\frac{100}{z_{\text{ion}}} \right)^{1/2} \tag{6.134}$$

in the sky. Because $\tau = 1$ corresponds to $z \cong 15.08 \, \Omega_{\text{NR}}^{1/3} (\Omega_B h)^{-2/3}$, the maximum value of θ_{ion} will be $\theta_m(z_m) \simeq 7.4° (\Omega_B \Omega_{\text{NR}} h)^{1/3}$. If such reionisation occurs, the trace of primordial anisotropies will remain only at larger angles. This case, however, is not very realistic as the current CMBR data suggest that $\tau \lesssim 0.1$.

Although reionisation wipes out the original anisotropies, the peculiar velocities at $z \simeq z_{\text{ion}}$ regenerate new anisotropies at small scales.[14] The actual value of these anisotropies is fairly model dependent. In the simplest models, these secondary anisotropies arise because of the coupling between the motion of the scatterers and the fluctuation in the electron density along the line of sight. This effect is dominant in the angular scales $1'$ to $10'$ and produces $\delta T / T$ of the order of 10^{-6}.

Although reionisation at $z \approx 100$ is a possibility, it may not be a *likely* possibility. To provide a physical mechanism for reionisation, it is necessary to form the first generation of structures at $z > z_{\text{ion}}$. In standard scenarios with adiabatic initial perturbations, this can happen only at small z_{ion}; in that case, the optical depth τ will be far less than unity and the effect will be insignificant. (It is comparatively easier to form structures at high redshifts in models with isocurvature perturbations.)

6.7 Spectral Distortions of CMBR

Coupling between radiation and matter can lead to observable distortions of CMBR under certain circumstances. For example, if a significant amount of energy is supplied to matter by some astrophysical process, the coupling between photons and charged particles will transfer part of this energy into the CMBR. Depending on the redshift at which this energy injection occurs, this process can distort the spectrum of CMBR. There are also regions in the universe – like the clusters of galaxies – that contain hot, ionised gas. When the MBR photons pass through these regions, they will be scattered by the electrons (which are at a much higher temperature) and gain energy. This will also distort the MBR spectrum in the vicinity of a cluster of galaxies. We now study these two effects.

6.7.1 Distortions Due to Global-Energy Injection

The spectral distortions that are due to energy injection depend crucially on the redshift at which the process occurs. When the matter is ionised, Compton scattering, double-Compton scattering ($\gamma e \rightarrow \gamma e + \gamma$), bremsstrahlung, and its inverse process (free–free absorption) will come into operation and the plasma will evolve towards a new equilibrium configuration. Whether such an equilibrium can be reached will depend on the various time scales in the problem.[15]

It is convenient to distinguish three different cases, depending on the redshift at which the energy was injected into the medium. If the energy is injected sufficiently early (say, at a redshift higher than a critical redshift z_P), then the radiation can relax to a thermal distribution independently of the amount of energy released. This necessarily requires processes that can change the total number of photons because thermal spectrum for radiation requires a prespecified ratio between the energy and the number density of the photons. Unless absorption and emission of photons occur at time scales shorter than the expansion time scale, it will not be possible to redistribute an arbitrary amount of energy into a thermal spectrum. We first determine this redshift z_P.

Two main processes that can change the photon number are the bremsstrahlung and double-Compton scattering. The rate of production of photons by bremsstrahlung can be expressed in the form (see Vol. I, Chap. 6, Section 6.9)

$$\left.\frac{\partial n}{\partial t}\right|_{\text{brem}} = \frac{1}{t_{\gamma e}}\left[\frac{Qg(x)}{e^x}\right]\frac{1}{x^3}[1 - n(e^x - 1)], \qquad (6.135)$$

where $x = (h\nu/k_B T)$, T denotes the radiation temperature, $t_{\gamma e}$ is the time scale for Thomson scattering for a hydrogen–helium plasma,

$$t_{\gamma e} = \frac{1}{(n_e \sigma_T)} \simeq 6.6 \times 10^{-11}\,\text{s}\left(\frac{T}{\text{MeV}}\right)^{-3}(\Omega_B h^2)^{-1}, \qquad (6.136)$$

and

$$Q = 2\sqrt{2\pi}\left(\frac{m_e}{T_e}\right)^{1/2}\alpha\,n_B\,T_e^{-3} \simeq 1.7 \times 10^{-10}\left(\frac{T}{\text{MeV}}\right)^{-1/2}\left(\frac{T}{T_e}\right)^{7/2}\Omega_B h^2, \qquad (6.137)$$

where T_e is the electron temperature and the Gaunt factor is given by

$$g(x) \simeq \ln\frac{2.25}{x} \qquad (\text{for } x \lesssim 1)$$
$$\simeq \frac{\ln 2.25}{\sqrt{x}} \qquad (\text{for } x \gtrsim 1). \qquad (6.138)$$

On the other hand, the rate of production of photons by double-Compton scattering ($e\gamma \rightarrow e\gamma\gamma'$) is given by

$$\frac{\partial n}{\partial t}\bigg|_{dc} = \frac{1}{t_{\gamma e}}\left[\frac{4\alpha}{3\pi}\left(\frac{T_e}{m_e}\right)^2 I(t)\right]\frac{1}{x^3}[1 - n(e^x - 1)],$$

$$I(t) = \int_0^\infty x^4(1 + n)n\, dx. \tag{6.139}$$

Both these rates vanish for a Planck spectrum of radiation with

$$n = n_P \equiv \frac{1}{(e^x - 1)}, \qquad T_e = T, \tag{6.140}$$

as they should. Here, and in what follows, we measure frequency, temperature, and mass in energy units, thereby taking $\hbar = 1, k_B = 1$, and $c = 1$.

In addition to these processes, which change the photon number, we also need to take into account the standard Compton scattering (see Vol. I, Chap. 6, Section 6.6) that changes the energy of photons but not their number. The evolution of the photon number under Compton scattering is determined by the Kompaneets equation

$$\frac{\partial n}{\partial t}\bigg|_{comp} = \frac{1}{t_{\gamma e}}\frac{T_e}{m_e}\frac{1}{x^2}\frac{\partial}{\partial x}\left[x^4\left(\frac{\partial n}{\partial x} + n + n^2\right)\right], \qquad x \equiv \frac{\omega}{T_e}. \tag{6.141}$$

This Compton scattering process – acting alone – will lead to a Bose–Einstein distribution for which the right-hand side of Eq. (6.141) vanishes (see Vol. I, Chap. 6, Section 6.6) in a characteristic time scale

$$t_{comp} = \left(\frac{m_e}{T_e}\right)t_{\gamma e} \simeq 3.4 \times 10^{-11}\, \text{s}\left(\frac{T}{\text{MeV}}\right)^{-4}\left(\frac{T}{T_e}\right)(\Omega_B h^2)^{-1}. \tag{6.142}$$

The full evolution equation for the photon number, taking into account all three processes, will then be

$$\frac{\partial n}{\partial t} = \frac{\partial n}{\partial t}\bigg|_{comp} + \frac{\partial n}{\partial t}\bigg|_{brem} + \frac{\partial n}{\partial t}\bigg|_{dc}. \tag{6.143}$$

When the system is away from equilibrium, the radiative processes create soft photons that are upscattered by the Comptonisation process. As the number density of photons approaches the density in a Planck spectrum around a given frequency, absorption processes become important. The spectrum relaxes to a Planckian shape, first at low frequencies at which new photons are absorbed effectively before they are scattered to higher frequencies. Slowly the process propagates upwards in frequency. A numerical integration of Eq. (6.143) shows that the typical frequency x_c below which Planck spectrum has been established is

$$x_{\text{brem}} \simeq \left(\frac{m_e}{8T_e} \, Q \, \ln \frac{2.25}{x_{\text{brem}}} \right)^{1/2} \ll 1 \tag{6.144}$$

if bremsstrahlung is the dominant process and

$$x_{\text{dc}} \simeq \left[\frac{\alpha}{6\pi} \, I(t) \, \frac{T_e}{m_e} \right]^{1/2} \tag{6.145}$$

if double-Compton scattering is dominant. In the first case, x_c is constant in time, whereas in the second case x_c moves up in frequency with time.

From Eqs. (6.135) and (6.139) we can work out the relevant time scales for the two processes. For bremsstrahlung the time scale for complete thermalisation is given by

$$t_{\text{brem}} \simeq \frac{2\zeta(3)}{Q \, \ln^2(2.25/x_{\text{brem}})} \, t_{\gamma e}$$

$$\simeq 9.4 \, \text{s} \left(\frac{T}{\text{MeV}} \right)^{-5/2} \left(\frac{T}{T_e} \right)^{7/2} (\Omega_B h^2)^{-1} \ln^{-2} \left[\frac{2.3 \times 10^6}{(\Omega_B h^2)^{1/2}} \left(\frac{T}{\text{MeV}} \right)^{3/4} \right],$$

$$\tag{6.146}$$

and for the double-Compton scattering, the corresponding time scale is

$$t_{\text{dc}} = \frac{\pi}{8\alpha} \left(\frac{m_e}{T_e} \right)^2 t_{\gamma e} \simeq 9.3 \times 10^{-10} \, \text{s} \left(\frac{T}{\text{MeV}} \right)^{-5} \left(\frac{T}{T_e} \right)^2 (\Omega_B h^2)^{-1}. \tag{6.147}$$

It should, however, be noted that bremsstrahlung increases the photon number linearly with time whereas double-Compton scattering – which requires some initial photons – increases the photon number nearly exponentially. Comparing the two, we find that Compton process dominates over bremsstrahlung if

$$\frac{n_R}{n_B} \gtrsim \left(\frac{m_e}{T_e} \right)^{5/2}, \tag{6.148}$$

which holds at sufficiently early stages.

To determine a critical redshift z_{crit} above which a given scattering process is operational, we need to solve the equation

$$\int_{t(z_{\text{crit}})}^{t_0} \frac{dt}{t_{\text{scat}}} = \int_0^{z_{\text{crit}}} \frac{dt}{dz} \frac{dz}{t_{\text{scat}}} = 1 \tag{6.149}$$

for the given process. Because the double-Compton process is dominant earlier on, we determine the redshift above which this process can thermalise an arbitrarily large amount of energy by evaluating the condition

$$\int_{t(z_P)}^{t_0} \frac{dt}{t_{dc}} = 1. \tag{6.150}$$

Numerical integration of the relevant parameter gives the redshift

$$z_P \simeq \frac{3.9 \times 10^6}{(\Omega_B h^2)^{1/3}}. \tag{6.151}$$

Thus an arbitrarily large injection of energy can be tolerated (without the thermal spectrum being affected at low redshifts), provided that such injection takes place at $z \gtrsim 10^7$. For consistency, we should check that bremsstrahlung cannot be more effective at lower redshifts. This is indeed true, and a corresponding calculation for bremsstrahlung gives a critical redshift of $\sim 10^9 (\Omega_B h^2)^{-4}$.

Let us next consider the injection of energy at lower redshifts. In this case, because neither bremsstrahlung nor double-Compton scattering is effective, it is *not* possible for thermalisation to take place. The best that can happen is for the spectrum to relax to a Bose–Einstein form with a nonzero chemical potential. (This was discusssed in Vol. I, Chap. 5, Section 5.14.) For this to occur, energy must be injected at redshifts higher than z_{BE} at which the Comptonisation parameter

$$y(z) \equiv \int_{t(z)}^{t_0} \frac{dt}{t_{comp}} \tag{6.152}$$

is unity. This is given by

$$z_{BE} \simeq 2.2 \times 10^4 \, (\Omega_B h^2)^{-1/2}. \tag{6.153}$$

In this case, some amount of thermalisation can still take place at *low* frequencies at which bremsstrahlung can be effective. This frequency dependence can be parameterised in terms of a chemical potential that itself is frequency dependent. Below the critical frequency x_{brem} given in relation (6.144) the chemical potential is effectively zero, whereas above this frequency the chemical potential is equal to the initial value determined by the amount of energy injection. An approximate fitting form for such a chemical potential is

$$\mu(x) \simeq \mu_i \exp\left(\frac{-2\sqrt{2}\, x_{brem}}{x}\right) \qquad \text{(for } x \ll 1\text{)}. \tag{6.154}$$

The equivalent brightness temperature of the spectrum, defined by (see Vol. I, Chap. 6, Section 6.2)

$$T_B = \frac{\omega}{\ln[1 + n^{-1}]}, \tag{6.155}$$

will now have a dip in the middle of the Rayleigh–Jeans region around $x \approx 2\sqrt{2}\, x_{\text{brem}}$.

Observations of CMBR puts a 95% confidence interval limit on the chemical potential at $\mu < 9 \times 10^{-5}$. To obtain a bound on energy injection from this bound on the chemical potential, we note that the energy and the number densities of a Bose–Einstein distribution can be expressed in the form

$$\rho_{\text{BE}} = \rho_R + \Delta\rho_R \equiv \frac{\pi^2}{15} T_e^4 \, f(\mu), \qquad n_{\text{BE}} = n_R + \Delta n_R \equiv \frac{2\zeta(3)}{\pi^2} T_e^3 \, \phi(\mu),$$

$$(6.156)$$

with

$$f(\mu) = \frac{15}{\pi^4} \int_0^\infty \frac{x^3}{(e^{x+\mu}-1)}\, dx \simeq \begin{cases} \dfrac{90}{\pi^4} e^{-\mu} & (\text{for } \mu \gg 1) \\[2mm] 1 - \dfrac{90\zeta(3)}{\pi^4}\mu & (\text{for } \mu \ll 1) \end{cases}, \quad (6.157)$$

$$\phi(\mu) = \frac{1}{2\zeta(3)} \int_0^\infty \frac{x^2}{(e^{x+\mu}-1)}\, dx \simeq \begin{cases} \dfrac{e^{-\mu}}{\zeta(3)} & (\text{for } \mu \gg 1) \\[2mm] 1 - \dfrac{\pi^2}{6\zeta(3)}\mu & (\text{for } \mu \ll 1) \end{cases}, \quad (6.158)$$

where $\Delta\rho_R$ is the energy injected into an original spectrum with energy density ρ_R, etc. If a negligible number of photons are added when the energy is injected into the system, then the electrons are initially heated to the temperature $T_e \approx T[\phi(\mu)]^{-1/3}$ and the initial chemical potential is given by

$$\frac{\Delta\rho_R}{\rho_R} = \frac{f(\mu)}{[\phi(\mu)]^{4/3}} - 1 \simeq \begin{cases} 1.18\, e^{\mu/3} - 1 & (\text{for } \mu \gg 1) \\ 0.714\, \mu & (\text{for } \mu \ll 1) \end{cases}. \quad (6.159)$$

The bound on the chemical potential now translates to the constraint

$$\frac{\Delta\rho_R}{\rho_R} < 6.4 \times 10^{-5} \qquad (z_P > z > z_{\text{BE}}). \quad (6.160)$$

Finally, let us consider an energy injection at $z \lesssim z_{\text{BE}}$ that cannot be thermalised and will effectively heat the electrons to a temperature $T_e > T$. The approximate form of the Kompaneets equation (discussed in Vol. I, Chap. 6, Section 6.7)

$$\frac{\partial n(x, \bar{y})}{\partial \bar{y}} = \frac{1}{x^2} \frac{\partial}{\partial x}\left(x^4 \frac{\partial n}{\partial x}\right), \quad x \equiv \frac{\omega}{T}, \quad \bar{y}(z) \equiv \int_{t(z)}^{t_0} \frac{dt}{t_{\text{comp}}}\left(1 - \frac{T}{T_e}\right) \quad (6.161)$$

is adequate to treat this case. (The variable $\bar{y}$ reduces to the Compton y parameter when $T \ll T_e$.) In this case, there is a change in the energy without any change in the photon number. Let us suppose that the initial spectrum was Planckian with $n(x, 0) = (\exp x - 1)^{-1}$. Because the photons are now being redistributed at different frequencies, with an increase in the net energy, we expect an overall shift in the spectrum to *higher* frequencies. This would imply that the effective temperature of the spectrum at the Rayleigh–Jeans end will be *lowered*. We can obtain the general solution by transforming the variables in Eqs. (6.161) and converting them into a diffusion equation. The solution in our case will be (see Vol. I, Chap. 6, Section 6.7)

$$n(x, \bar{y}) = \frac{1}{\sqrt{4\pi \bar{y}}} \int_0^\infty \frac{dz}{z} \exp\left\{-\frac{[\ln(x/z) + 3\bar{y}]^2}{4\bar{y}}\right\} \frac{1}{(e^z - 1)}, \quad (6.162)$$

which reduces to a simple form when the Comptonisation parameter is small:

$$n \simeq n + \Delta n = \frac{1}{e^x - 1}\left\{1 + \frac{\bar{y}xe^x}{e^x - 1}\left[\frac{x}{\tanh(x/2)} - 4\right]\right\}. \quad (6.163)$$

To the lowest order in the Compton parameter y, this leads to a fractional change in the radiation temperature by the amount

$$\frac{\delta T}{T_R} = -\frac{2k_B T_e}{m_e c^2} \tau = -2y \quad (6.164)$$

when $T_e \gg T$ so that $\bar{y} = y$. This result is valid in the Rayleigh–Jeans limit. The total energy content is related to the Comptonisation parameter by (see Vol. I, Chap. 6, Section 6.7)

$$\frac{\Delta \rho_R}{\rho_R} = e^{4y} - 1 \simeq 4y \quad \text{(for } y \ll 1). \quad (6.165)$$

The observations lead to a 95% confidence bound on y to be $y < 1.2 \times 10^{-5}$. The corresponding bound on the energy injection for all redshifts $z < z_{BE}$ will be

$$\frac{\Delta \rho_R}{\rho_R} < 3 \times 10^{-5}. \quad (6.166)$$

These considerations show that the thermal history of photons is very well constrained for a wide range of redshifts in the universe. In the regime $0 < z < z_{BE} \simeq 10^5$, the fractional energy injection has to be less than 3×10^{-5}. In the interval $z_{BE} \lesssim z \lesssim z_P \simeq 10^7$, the bound is slightly more relaxed and the fractional energy injection has to be less than 6.4×10^{-5}. There are effectively no constraints for energy injection at redshifts higher than z_P.

The following point, however, should be noted: The analysis given above assumes that an efficient transfer of energy from electrons to photons can take place in spite of the expansion of the universe. The cooling time for a plasma (through Compton scattering) is given by

$$t_{cool} \cong \frac{1}{\sigma_T n_\gamma}\left(\frac{m_e}{T}\right). \quad (6.167)$$

This time scale exceeds the Hubble time $(\dot{a}/a)^{-1}$ for $z < 8$. Thus, for $z_{ion} < 8$, only a small part of the energy injected into the matter will be transferred to radiation.

6.7.2 Sunyaev–Zeldovich Effect

Let us next consider the interaction of the MBR photons with the hot gas in a cluster of galaxies.[16] If the optical depth is τ, then a fraction τ of the MBR photons, seen in the direction of the cluster, may be assumed to have been scattered (at least) once by the hot electrons. This fraction gains in energy and is redistributed at the higher-frequency region of the spectrum. Because the total number of photons is conserved, this results in a lateral shift of the spectrum called the *Sunyaev–Zeldovich* (SZ) effect. The net effect is similar to that in Eq. (6.164). The temperature in the Rayleigh–Jeans region of the spectrum decreases by the amount calculated in relations (6.164) and (6.163).

In large clusters of galaxies, we can use the SZ effect along with x-ray emission by the hot electron gas to model several aspects of the clusters. If the bremsstrahlung emissivity at energy E from a hot gas of electrons at temperature T_e is taken to be $\Lambda(E, T_e)$, then the x-ray brightness at energy E that is due to an optically thin cluster at redshift z is given by the integral along the line of sight:

$$b_X(E) = \frac{1}{4\pi(1+z)^3} \int n_e(\mathbf{r})^2 \, \Lambda[E, T_e(\mathbf{r})] \, dl. \tag{6.168}$$

The other two parameters relevant for the electron gas in the cluster are the optical depth and the Compton y parameter:

$$\tau_e = \int n_e(\mathbf{r}) \, \sigma_T \, dl, \qquad y = \int n_e(\mathbf{r}) \, \sigma_T \, \frac{k_B T_e(\mathbf{r})}{m_e c^2} \, dl. \tag{6.169}$$

Although $b_X(E)$ can be obtained from x-ray measurements, there is no unique inversion of Eq. (6.168) to predict n_e and T_e. What is usually attempted is to choose different models for the cluster and check for consistency with respect to all observations.

To illustrate this procedure, let us consider a simple model of an isothermal cluster with a characteristic central electron density n_0 and a radial profile with a length scale r_0. The central x-ray brightness of the cluster will then scale as

$$I(0) \propto T^{-1/2} e^{-h\nu/k_B T} n_0^2 r_0.$$

Fitting this to the observed spectrum will allow us to determine T and $n_0^2 r_0$. If the SZ effect is observed, it will allow a determination of the combination $y \propto T n_0 r_0$. Knowing T and the observed value of y, we can determine $n_0 r_0$. Once both the combinations $n_0^2 r_0$ and $n_0 r_0$ are determined, we can solve for n_0 and r_0 individually, thereby determining the parameters that characterise the cluster. Further, because the angle subtended by the cluster $\theta \propto r_0/d_A(z)$ as well as its redshift could be independently determined, we can estimate the Hubble constant that sets the overall scale of d_A. We shall discuss more about this procedure in Chap. 10.

In addition, the motion of the cluster with a peculiar velocity v relative to CMBR will induce a Doppler shift, thereby causing an effective temperature change:

$$\frac{\Delta T_R}{T_R} = 2\tau \, \frac{v}{c}, \tag{6.170}$$

where τ is the optical depth for the scattering. This result is subdominant to the SZ effect in magnitude. However, in contrast to the SZ effect, this is independent of frequency and can be a source of noise in temperature anisotropy measurements.

Exercise 6.8

Molecular transitions and CMB: Consider a cloud containing H_2 molecules of density n and kinetic temperature T_K intermixed with another set of molecules X, with the latter ones modelled as a two-level system with a higher energy state $|u>$ and lower energy state $|l>$. The X molecules are excited both radiatively and by collisions with H_2, with the collisional deexcitation rate being σv. The frequency of transition between the two energy levels is v. The system is embedded in a region containing CMBR with temperature $T_R = 2.73$ K. (a) Solve the relevant rate equations and show that – in thermal equilibrium – the excitation temperature T_s of the molecular levels is

$$\exp\left(-\frac{hv}{k_B T_s}\right) \equiv \frac{n_u g_l}{n_l g_u} = \frac{\exp(-hv/k_B T_R) + \xi \exp(-hv/k_B T_K)}{1 + \xi}, \tag{6.171}$$

where

$$\xi = \frac{n\sigma v}{A_{ul}}[1 - \exp(-hv/k_B T_R)] \tag{6.172}$$

and A_{ul} is the Einstein coefficient. Show that at low densities $T_s \simeq T_R$ whereas at high densities $T_s \simeq T_K$. (b) Assume that $k_B T_K \gg hv$. Estimate the density n_c at which the excitation temperature will be twice the background temperature. Assuming that $\sigma \approx (3\,\text{Å})^2$ and $T_K \simeq 100$ K, estimate the critical density for the OH molecule ($v = 1.667$ GHz, $A_{ul} = 8 \times 10^{-11}$ s^{-1}), the NH_3 molecule ($v = 23.69$ GHz, $A_{ul} = 4 \times 10^{-8}$ s^{-1}) and HCN molecule ($v = 88.63$ GHz, $A_{ul} = 2 \times 10^{-5}$ s^{-1}). [Hint: The critical density at which the excitation temperature is twice the background temperature is given by

$$n_c = \frac{A_{ul}}{\sigma v} \frac{\exp(-hv/2k_B T_R)}{1 - \exp(-hv/k_B T_R)}. \tag{6.173}$$

This can be computed with the given numerical values.]

7

Formation of Baryonic Structures

7.1 Introduction

We now take up the study of galaxy formation, which requires understanding the growth of structure in the baryonic component of the universe. This chapter uses concepts from several earlier chapters, especially Chaps. 3–5, and will be needed for Chap. 9.

The study of galaxy formation from fundamental physical considerations is made difficult by a wide variety of physical processes that we need to take into account. To begin with, galaxies by themselves show a variety of morphological and physical properties, even at $z = 0$. The formation process should be such that, starting from relatively structureless density enhancements at high redshifts ($z \gtrsim 25$), one is capable of producing such a variety at $z = 0$. Second, the observational situation as regards galaxylike structures at high redshifts is still very unsatisfactory. The samples are small in number and often we have to decide how to make the correspondence between the sources seen at high redshifts and those seen at low redshifts. This issue is further complicated by the fact that a certain kind of population could have existed at a certain interval of time and could have vanished outside this epoch. Such a conjecture of invoking new populations is often convenient, but is not very satisfactory unless we could back it with some physical reasoning. Third, any scenario for galaxy formation should also naturally tie up with other high-redshift phenomena, notably the existence of AGN. Fourth, we must also ensure that the galaxy-formation scenario fits with the broader picture at different scales. For example, the pattern of density inhomogeneities at $z = 1000$ (and other cosmological parameters) inferred from CMBR anisotropy studies must be used as the initial condition for the growth of perturbations that lead to galaxy formation at lower redshifts. Fifth, it must be stressed that, to a great extent, the visible part of the galaxies is made of stars and the formation of a galaxy is essentially a study of cosmic star-formation history. As the stars are formed, they leak out radiation that should be detectable today at different wave bands; stars also produce metals, and the evolution of metallicity with redshift will be related to the history of star formation. It is important that galaxy-formation scenarios are consistent with the observations of EBL and the evolution of metallicities. Finally, galaxy surveys in the local universe allow us to determine statistical parameters like the two-point correlation function of galaxies, and any theory should be able to reproduce such statistical indicators. We will now attempt to address several of these questions, starting with the evolution of linear perturbations in the baryonic fluid.[1]

7.2 Linear Perturbations in Baryons

Baryons in the universe are tightly coupled to the photons until matter and radiation decouple at $z \simeq 10^3$. This tight coupling provides the baryonic fluid with a pressure that can resist the force of gravity. Hence small-scale perturbations in a baryonic fluid cannot grow in amplitude at $z > z_{\mathrm{dec}}$. However, at sufficiently large scales, we would expect gravity to overpower the pressure gradient, thereby allowing the perturbation to grow.

The analysis for baryonic perturbations proceeds in a manner similar to that in Section 5.6. In the proper coordinates $(t, \mathbf{r})$ with $\mathbf{r} = a(t)\mathbf{x}$, the equations describing an ideal self-gravitating baryonic fluid can be written as

$$
\frac{\partial \rho}{\partial t} + \nabla_r \cdot (\rho \mathbf{U}) = 0,
$$

$$
\frac{\partial \mathbf{U}}{\partial t} + (\mathbf{U} \cdot \nabla)\mathbf{U} = -\nabla \Psi_{\mathrm{tot}} - \frac{1}{\rho}\nabla p, \tag{7.1}
$$

$$
\nabla^2 \Psi_{\mathrm{tot}} = 4\pi G \rho_{\mathrm{tot}}.
$$

Comparing these equations with the basic equations in Section 5.6, we see that the *only* difference between the baryonic fluid and the dark matter treated as a fluid is the additional term $\nabla p / \rho$ in the Euler equation, which is due to the pressure support. We now introduce the perturbed potential ϕ, perturbed density contrast of baryons δ_B, perturbed pressure δp, and perturbed velocity $\mathbf{v}$, by using the equations

$$
\phi = \Psi - \phi_{\mathrm{FRW}}, \qquad \mathbf{v} = \mathbf{U} - H\mathbf{r},
$$

$$
\delta_B = \frac{(\rho - \bar{\rho}_B)}{\bar{\rho}_B}, \qquad \delta p = c_s^2(\delta\rho). \tag{7.2}
$$

We have assumed that the perturbed pressure δp and the perturbed density $\delta\rho$ are related by $\delta p = c_s^2 \delta\rho$, where c_s is the speed at which disturbances propagate in the fluid. We define the perturbed potential and velocity by subtracting out the corresponding quantities for the homogeneous Friedmann model. The perturbed density contrast is defined by the background density of *baryons*; the perturbed pressure is related to the perturbed density through c_s^2. When photons and baryons are tightly coupled, we would expect $c_s^2 \simeq (\partial p_{\rm rad}/\partial \rho_{\rm rad}) = (1/3)$. After the matter has decoupled from radiation, c_s^2 will be equal to the adiabatic sound speed in the gas.

We now linearise the equations in the perturbed variables and change from proper to comoving coordinates. The analysis proceeds exactly as in that Section 5.6, and we get

$$\ddot{\delta}_k^{(B)} + 2\frac{\dot{a}}{a}\dot{\delta}_k^{(B)} + \frac{k^2 c_s^2}{a^2}\delta_k^{(B)} = 4\pi G \bar{\rho}_B \delta_k^{(B)} + 4\pi G \bar{\rho}_{\rm DM}\delta_k^{\rm DM}. \qquad (7.3)$$

This differs from the equation for dark-matter perturbations in two respects: (1) The term $(k^2 c_s^2/a^2)$ arises because of the pressure gradient that was not present for the dark matter and (2) on the right-hand side we have now retained terms that arise from both baryonic and dark-matter perturbations. When dark-matter perturbations dominate over baryonic perturbations (which is often the case), the baryonic driving term can be ignored on the right-hand side. This equation can be rewritten in the form

$$\ddot{\delta}_k^{(B)} + 2\frac{\dot{a}}{a}\dot{\delta}_k^{(B)} + \left(\frac{k^2 c_s^2}{a^2} - 4\pi G \bar{\rho}_B\right)\delta_k^{(B)} = 4\pi G \bar{\rho}_{\rm DM}\delta_k^{\rm DM}, \qquad (7.4)$$

which is similar to that of a forced, damped, harmonic oscillator. The $\dot{a}/a$ term represents the damping that is due to expansion and the term on the right-hand side provides the driving force.

In the absence of dark matter, the nature of the solution is essentially determined by the sign of the third term on the left-hand side. If $k^2 > k_J^2$, where

$$k_J^2 = \frac{4\pi G \bar{\rho}_B}{c_s^2}, \qquad (7.5)$$

then the "frequency" of the oscillator is real and we get oscillatory solutions. The perturbation in the baryonic component oscillates as an acoustic vibration. On the other hand, if $k^2 < k_J^2$, then the "frequency" is imaginary and we get growing and decaying amplitudes. In such a case, the perturbations in the baryonic component can grow.

When both dark matter and baryons are present, then $\bar{\rho}_{\rm DM}\delta_{\rm DM}$ is usually dominant over $\bar{\rho}_B \delta_B$. The growth in $\delta_{\rm DM}$ can induce a corresponding growth in δ_B at large scales. At small scales, however, the $(k^2 c_s^2/a^2)$ term will reduce this effect significantly.

The condition for the growth of modes could also be expressed in terms of the physical wavelength, $\lambda_{\rm phy} \equiv (2\pi/k)a(t)$. Modes with $\lambda_{\rm phy} < \lambda_J$ do not grow whereas long-wavelength modes with $\lambda > \lambda_J$ can grow, where

$$\lambda_J = \left(\frac{\pi c_s^2}{G\rho}\right)^{1/2} \qquad (7.6)$$

is called Jeans length and k_J is the corresponding Jeans wave number. It is convenient to define a quantity called the *Jeans mass*, M_J, as the amount of baryonic mass contained within a sphere of radius $\lambda_J/2$. By definition, the Jeans mass scales as

$$M_J \propto \rho_B \lambda_J^3 \propto \rho_B \left(\frac{c_s^3}{\rho_B^{3/2}} \right) \propto \left(\frac{c_s^3}{\rho_B^{1/2}} \right). \tag{7.7}$$

During $a_{eq} < a < a_{dec}$, the baryons and the photons are tightly coupled and $c_s \simeq$ constant. Hence $M_J \propto \rho_B^{-1/2} \propto a^{3/2}$. For $a > a_{dec}$, $c_s \propto a^{-1}$ because the velocity dispersion decreases as the inverse of the expansion factor for an adiabatically expanding gas. Then $M_J \propto (a^{-3}/a^{-3/2}) \propto a^{-3/2}$. More importantly, after decoupling, the pressure support provided by the photons vanishes and the baryonic gas can resist gravity only by normal gas pressure. At $T = T_{dec}$, the radiation pressure is given by $p_{rad} \simeq n_\gamma k T_{dec}$ and the gas pressure is given by $p_{gas} \simeq n_B k T_{dec}$. Because $n_B \simeq 10^{-8} n_\gamma$, the pressure drops drastically at $t = t_{dec}$. In other words, the sound speed c_s (as well as the Jeans length and the Jeans mass) drops by orders of magnitude at $t = t_{dec}$. Putting in numbers, we can easily see that the Jeans mass before decoupling is

$$M_J^{(1)} = M_J(t < t_{dec}) \simeq 10^{16} \, M_\odot \left(\frac{\Omega_B}{\Omega} \right) (\Omega h^2)^{-1/2}, \tag{7.8}$$

and the Jeans mass after decoupling is

$$M_J^{(2)} = M_{JB}(t \gtrsim t_{dec}) \simeq 10^4 \, M_\odot \left(\frac{\Omega_B}{\Omega} \right) (\Omega h^2)^{-1/2}. \tag{7.9}$$

Perturbations carrying masses with $M > M_J^{(1)}$ will always be growing because the pressure support cannot prevent their growth. Perturbations containing masses in the range $M_J^{(2)} < M < M_J^{(1)}$ cannot grow until $a = a_{dec}$ because for these scales the pressure support can withstand gravity. They can, however, grow after decoupling because the Jeans mass would have now dropped to a low value.

   Consider now the modes with $\lambda_{phy} \ll \lambda_J$ for which the pressure term dominates over gravity. We may assume that, for $z \gtrsim 10^2$, the temperature of matter is approximately equal to that of radiation, which scales as a^{-1}; i.e., we can take $c_s^2 = v_0^2(a_0/a)$. When the pressure term dominates over gravity and $c_s^2 = (k_B T/m_p) = (k_B T_0/m_p)(1/a)$, baryonic perturbation equation (7.4) becomes

$$\ddot{\delta}_B + 2\frac{\dot{a}}{a}\dot{\delta}_B + \left(\frac{k_B T_0}{m_p} \right) \left(\frac{k^2}{a^3} \right) \delta_B \simeq 0. \tag{7.10}$$

Changing the independent variable from t to $x = (a/a_0)^{1/2}$, we get

$$\frac{d^2 \delta_B}{dx^2} + \frac{2}{x}\frac{d\delta_B}{dx} + \frac{\omega^2 \delta_B}{x^2} = 0, \quad \omega^2 \equiv \frac{4k^2}{H^2 a^3} \left(\frac{k_B T_0}{m_p} \right), \quad H(t) = \frac{\dot{a}}{a}, \tag{7.11}$$

or

$$\frac{1}{x^2}\frac{d}{dx}\left(x^2\frac{d\delta_B}{dx}\right) + \frac{\omega^2\delta_B}{x^2} = 0. \qquad (7.12)$$

Note that ω^2 is a constant because $H^2 \propto \rho \propto a^{-3}$. Further, $\omega \gg 1$ because we are considering modes for which $\lambda_{\rm phy} \ll d_H$. Equation (7.12) has the solution $\delta_B = x^n$ with $n \cong [(-1/2) \pm i\omega]$ for $\omega \gg 1$. Therefore, using $x \propto a^{1/2} \propto t^{1/3}$, we can write the density contrast as

$$\delta_B \cong t^{-1/6} \exp\left(\pm\frac{i\omega}{3}\ln t\right), \qquad (7.13)$$

which decays as $t^{-1/6}$.

This decay law has a simple interpretation. The oscillations in the baryon density can be thought of as acoustic vibrations in a medium. The frequency of these vibrations scales as $\nu = (v/\lambda) \propto a^{-1/2}a^{-1} \propto a^{-3/2} \propto t^{-1}$, whereas the energy of the vibrations in a volume V will vary as $E \propto \rho_B(v\delta_B)^2 V \propto v^2\delta_B^2 \propto \delta_B^2 a^{-1} \propto \delta_B^2 t^{-2/3}$. In the case of $k v \gg H a$ there will be several acoustic oscillations within one expansion time scale, and hence, the expansion may be treated as (adiabatically) slow. In such a case, the quantity E/ν, which is an adiabatic invariant (see Vol. I, Chap. 2, Section 2.6), will remain constant. Because $\nu \propto t^{-1}$ and $E \propto t^{-2/3}\delta_B^2$, it follows that $t^{1/3}\delta_B^2 = $ constant or $\delta_B \propto t^{-1/6}$.

The preceding analysis shows that baryonic perturbations at moderate scales do not grow when baryons and photons are coupled. However, dark-matter perturbations, which are uncoupled to photons, have been growing since $a = a_{\rm eq}$. During the period $a_{\rm eq} < a < a_{\rm dec}$, dark-matter perturbations would have grown by a factor of $(a_{\rm dec}/a_{\rm eq}) \simeq 20\,\Omega h^2$. (The actual value is somewhat higher for two reasons: There is a small logarithmic growth in the amplitude of the dark matter for $a < a_{\rm eq}$ and some decay in the amplitude of the baryonic perturbations.) For $a > a_{\rm dec}$, baryonic perturbations can grow. During this epoch, baryons will feel the potential wells of the dark-matter perturbations, which would have already grown by a larger factor. It follows that baryonic perturbations at $a > a_{\rm dec}$ will be essentially driven by the dark-matter perturbations. We can then ignore the $\rho_B\delta_B$ term compared with the $\rho_{\rm DM}\delta_{\rm DM}$ term. In other words, we shall study the evolution of baryonic perturbations at $a > a_{\rm dec}$ and approximate Eq. (7.4) as follows: (1) The gravitational force is dominated by the dark matter, and we shall ignore $4\pi G\bar{\rho}_B\delta_B$ in comparison with $4\pi G\bar{\rho}_{\rm DM}\delta_{\rm DM}$. (2) We shall take $c_s^2 = (k_B T/m_p)$ with $T \propto a^{-1}$. In other words, the pressure support is provided by the gas, the temperature of which is maintained as equal to the radiation temperature. (3) The dark-matter density contrast grows as $\delta_{\rm DM} = Bt^{2/3} \propto a$.

Under these assumptions, Eq. (7.4) can be written as

$$\ddot{\delta}_B + \frac{4}{3t}\dot{\delta}_B + \left(\frac{k_B T}{m_p}\right)\frac{k^2}{a^3}\delta_B = \frac{2}{3}\frac{\delta_{\rm DM}}{t^2}, \qquad (7.14)$$

where δ_B stands for the baryonic density contrast corresponding to wave vector k; the subscript k is omitted to simplify the notation. Because the third term on the left-hand side varies as $a^{-3} = (t_0/t)^2$, the left-hand side is homogeneous in t and is driven by δ_{DM}. In that case we can find the solution by substituting the ansatz $\delta = Ct^{2/3}$. This leads to the relation

$$C\left[1 + \frac{3}{2}\left(\frac{k_B T_0}{m_p}\right)t_0^2\right] = B. \tag{7.15}$$

Therefore we get

$$\delta_{(B)}(t) = \frac{\delta_{DM}(t)}{1 + Ak^2}, \tag{7.16}$$

where

$$A = \frac{3}{2}\left(\frac{k_B T_0}{m_p}\right)\left(\frac{t^2}{a^3}\right) = 1.5 \times 10^{-6} h^{-2} \text{ Mpc}^2. \tag{7.17}$$

It is clear that at large scales (i.e., for small Ak^2) the dark-matter perturbations induce corresponding perturbations in the baryons. For $Ak^2 \gtrsim 1$, the baryonic perturbations are suppressed in amplitude with respect to the dark-matter perturbations because of the pressure support.

In the study of dark matter, we find that small-scale perturbations are wiped out because of the free streaming of the dark-matter particles. A somewhat similar phenomenon occurs for baryons as well. Photons can also undergo free streaming only if they are not tightly coupled to baryons. In the limit of infinitely tight coupling, no free streaming can occur. However, in reality, photons *can* diffuse from high-density to low-density regions, dragging baryons with them. This process can wipe out small-scale fluctuations in baryons. At $t \ll t_{\text{dec}}$, photons and baryons are very tightly coupled because of Thomson scattering. The proper length corresponding to the photon mean free path at some time t is

$$l(t) = \frac{1}{x_e n_e \sigma} \simeq 1.3 \times 10^{29} \text{ cm } x_e^{-1}(1 + z)^{-3}(\Omega_B h^2)^{-1}, \tag{7.18}$$

where x_e is the electron ionisation fraction. For wavelengths $\lambda \lesssim l$, the photon streaming will clearly damp any perturbation almost instantaneously. However, the damping effect is actually felt at even larger scales. Photons can slowly diffuse out of overdense to underdense regions, dragging the tightly coupled charged particles. Although not much decay occurs during one oscillation period of the perturbation, a significant loss can take place within an expansion time scale. The characteristic length scale for this process (called *Silk damping*) was obtained in Chap. 6 (see Eq. (6.18) as well as Exercise 6.7).

If we assume that baryons are tightly coupled to photons before t_{dec}, it follows that baryons will be dragged along with the photons. Then all perturbations at wavelengths $\lambda < l_{\text{diff}}$ [see Eq. (6.18)] will be wiped out. This length l_S

corresponds to the mass (called the *Silk mass*):

$$M_S \cong 6.2 \times 10^{12} \, M_\odot \left(\frac{\Omega}{\Omega_B}\right)^{3/2} (\Omega h^2)^{-5/4}. \tag{7.19}$$

Using the expression in Eq. (6.17) for an arbitrary instant of time t, we can easily work out the scaling of various quantities with a. We find that $l_S \propto a^{9/4}$ and $M_S \propto a^{15/4}$ during the period $a_{\rm eq} < a < a_{\rm dec}$. For $a > a_{\rm dec}$ this process, of course, ceases to exist. Thus the Silk mass rises very steeply to its final value at $a = a_{\rm dec}$.

In summary, the overall evolution of baryonic perturbations in the linear theory is as follows: During very early epochs, when the mode is outside the Hubble radius, pressure gradients have negligible influence on its dynamics. In that case, all components evolve in the same manner and baryonic perturbations also grow as a^2 (see Chap. 5, Section 5.3). At some time $t = t_{\rm enter}$ in the radiation-dominated epoch, a perturbation enters the Hubble radius. At this time, we may assume that baryons and dark-matter perturbations have equal amplitudes. Once inside the Hubble radius, processes like pressure support become operational and prevent the growth of baryonic perturbations. Hence δ_B oscillates as an acoustic wave with a decaying amplitude during the entire period $a_{\rm enter} < a < a_{\rm eq}$. (Note that dark-matter perturbations, unaffected by pressure gradients, grow logarithmically during $a_{\rm enter} < a < a_{\rm eq}$ and grow linearly during $a_{\rm eq} < a < a_{\rm dec}$.) Modes for which $M > M_J \simeq 10^{16} \, M_\odot$ could have grown after they entered the Hubble radius but these modes are not of much cosmological interest. For $a > a_{\rm dec}$, the baryonic perturbations can grow. During this epoch, its growth is driven by the already existing perturbations in the dark-matter component and $\delta_B \to \delta_{\rm DM}$ for large a. We have also seen that small-scale power in the baryonic component will be wiped out during $a < a_{\rm dec}$ because of Silk damping. However, this power can be regenerated for $a > a_{\rm dec}$ because of the driving by the dark matter. Thus, for $a \gtrsim a_{\rm dec}$, baryonic perturbations will closely follow that of dark matter.

Exercise 7.1
More on the Jeans mass: A more exact set of equations for the growth of dark matter and baryonic fluctuations, in the linear approximation, is given by

$$\ddot{\delta}_{\rm DM} + 2H\dot{\delta}_{\rm DM} = \frac{3}{2}H^2 \left(\Omega_B \delta_B + \Omega_{\rm DM}\delta_{\rm DM}\right), \tag{7.20}$$

$$\ddot{\delta}_B + 2H\dot{\delta}_B = \frac{3}{2}H^2 \left(\Omega_B \delta_B + \Omega_{\rm DM}\delta_{\rm DM}\right)$$
$$- \frac{k_B T_i}{\mu m_p}\left(\frac{k}{a}\right)^2 \left(\frac{a_i}{a}\right)^{(1+\beta)} \left(\delta_B + \frac{2}{3}\beta[\delta_B - \delta_{B,i}]\right), \tag{7.21}$$

where $\mu = 1.22$ is the mean molecular weight and the parameter β distinguishes between two limits of evolution for the gas temperature. $\beta = 1$ is the adiabatic limit, and $\beta = 0$ corresponds to a uniform baryonic temperature that is locked to the radiation temperature. (a) Convince yourself of the correctness of different terms in this equation. (b) The residual ionisation fraction can keep the gas temperature locked to the CMB temperature down to the redshift of z_t, where $(1 + z_t) = 137(\Omega_B h^2/0.02)^{2/5}$. Show that the Jeans mass during the epoch $z_t < z < z_{\rm rec}$ is given by

$$M_J = 1.35 \times 10^5 \left(\frac{\Omega_{\rm DM} h^2}{0.15}\right)^{-1/2} M_\odot, \tag{7.22}$$

whereas for $z < z_t$ it is given by

$$M_J = 5.73 \times 10^3 \left(\frac{\Omega_{\rm DM} h^2}{0.15}\right)^{-1/2} \left(\frac{\Omega_B h^2}{0.022}\right)^{-3/5} \left(\frac{1+z}{10}\right)^{3/2} M_\odot. \tag{7.23}$$

7.3 Nonlinear Collapse of Baryons

For the rest of this chapter we shall be concerned with the nonlinear collapse of baryons and the properties of the resulting structures. This is *the* key problem in the study of structure formation because all structures that are observed directly are baryonic and most of them are in their nonlinear phase. Unfortunately, this is also the area in which progress has been slow, mainly because of mathematical complexity. The brute-force method of simulating baryonic physics in computers should eventually yield dividends; but at present this approach is limited both in terms of input physics and in the resolution that is achievable. We shall therefore rely heavily on semianalytic approaches to understanding this phase of the universe.

Because there is at least ten times more dark matter than baryons in the universe, the former defines the pattern of gravitational potential wells in which structure formation essentially occurs. The study of dark matter, fortunately, is much simpler than the study of baryons. When the dark-matter fluctuation at a given mass scale M goes nonlinear, collapses, and virialises, it leads to a formation of a halo of mass M. The dark-matter particles, which would have undergone violent relaxation, can be described reasonably accurately as an isothermal sphere with a given velocity dispersion σ_v^2 and an asymptotic density distribution $\rho(r)$ given by

$$\rho(r) = \frac{\sigma_v^2}{2\pi G r^2}. \tag{7.24}$$

[To avoid possible misunderstanding we mention that different velocities are used in different contexts to describe the dark-matter halos. The first one is the mean-square velocity σ_v^2 of the particles, which is defined in such a way that the total kinetic energy of the system is $K = (1/2)M\sigma_v^2$, where M is the total mass of

Table 7.1. *Properties of collapsing structures*

	z = 5			z = 10		
	1σ	2σ	3σ	1σ	2σ	3σ
$M(M_\odot)$	1.8×10^7	3×10^{10}	7.0×10^{11}	1.3×10^3	5.7×10^7	4.8×10^9
T_{virial}(K)	2.0×10^3	2.8×10^5	2.3×10^6	6.2	7.9×10^3	1.5×10^5
v_c (km s^{-1})	7.5	88	250	0.41	15	65

the virialised halo. From virial theorem it follows that $2K + U = 0$, where U is the potential energy of the system that, in turn, will depend on the density profile. If the density profile is taken to fall as $1/r^2$, then the total mass M, cutoff radius R, and mean-square velocity σ_v^2 are related by $\sigma_v^2 = (GM/2R)$ and the density is related to r by Eq. (7.24). The second velocity often used is the *circular velocity* v defined through the centrifugal-force balance equation $v^2/r = GM(r)/r^2$. For the density profile in Eq. (7.24) we get $v^2 = 2\sigma_v^2$. Finally, we also occasionally use the line-of-sight velocity dispersion $\sigma_{\text{los}}^2 = \sigma_v^2/3$. Depending on the definition of the velocity used to characterise the halo, there could be a numerical factor of $\sim$2–3 in the density profile.] Table 7.1 summarises the properties of dark-matter halos arising out of the collapse of 1σ, 2σ, and 3σ fluctuations at $z = 5$ and 10.

Exercise 7.2

Timetable for galaxy formation: Let $\bar{\rho}$ be the density averaged over a radius r within a protogalaxy forming at a redshift z_f. Assume that such a structure can form provided that $\bar{\rho} \gtrsim f_c \rho_{\text{bg}}(z_f)$, where $f_c \approx 200$ and ρ_{bg} is the mean density of the background universe. Determine the redshift at which the mass inside the radius of $r = (10, 30, 100)h^{-1}$ kpc could have been assembled. Assume that the circular velocity of the protogalaxy is $v_c \approx 200$ km s^{-1}. [Answer: The given condition translates to

$$\frac{\bar{\rho}}{\rho_{bg}} = \frac{3}{4\pi r^3} \frac{v_c^2 r}{G} \frac{8\pi G}{3H_0^2 \Omega(1 + z_f)^3} = \frac{2v_c^2}{\Omega(H_0 r)^2(1 + z_f)^3} \gtrsim f_c \qquad (7.25)$$

Substituting the numbers, we find that $(1 + z_f) = (7.6, 3.0, 1.5)\Omega^{-1/3}$ for $r = (10, 30, 100)h^{-1}$ kpc.]

The central question, of course, is what happens to the baryons in this halo. The linear perturbation theory, described in the previous section, breaks down when $\delta_B \simeq 1$. The study of the nonlinear regime for the baryons is far more complicated than that of dark matter because of the need to take into account pressure gradients and radiative processes. Numerical simulations are also more complex for baryons as we need to use hydrodynamic codes. In this section, we shall examine some simple models that allow the estimate of nonlinear baryonic structures.

Let us begin by studying the evolution of the baryonic component located within a spherically symmetric dark-matter potential well by using the spherical top-hat model discussed in Section 5.10. When we ignore the effects of cooling, the dynamical evolution of baryons is similar to that of dark matter. The baryonic component expands more slowly compared with the background universe, turns around, and collapses. During the collapse, a gaseous mixture of H and He develops shocks and gets reheated to a temperature at which the pressure balance can prevent further collapse. At this stage the thermal energy will be comparable with the gravitational potential energy. The temperature of the gas, T_{vir}, is related to the velocity dispersion v^2 by $3\rho_{gas}T_{vir}/2\mu = \rho_{gas}v^2/2$, where ρ_{gas} is the gas density and μ is its mean molecular weight. This gives $T_{vir} = \mu v^2/3$. If the He fraction is Y by weight and the gas is fully ionised, then

$$\mu = \frac{(m_H n_H + m_{He} n_{He})}{(2n_H + 3n_{He})} = \frac{m_H}{2}\left(\frac{1+Y}{1+0.375\,Y}\right) \cong 0.57 m_H, \quad (7.26)$$

for $Y = 0.25$. Apart from the cosmological parameters, two other parameters need to be specified to determine the evolution. These may be chosen to be the mass M of the overdense region and the redshift of formation z_{coll}. Using the results of Section 5.10, we can express r_{vir}, v, and T_{vir} in terms of z_{coll} [or $\delta_0 = 1.68(1+z_{coll})$] and the mass M of the structure:

$$r_{vir} = 258(1+z_{coll})^{-1}\left(\frac{M}{10^{12}\,M_\odot}\right)^{1/3}h_{0.5}^{-2/3}\,\text{kpc} = 434\delta_0^{-1}h_{0.5}^{-2/3}M_{12}^{1/3}\,\text{kpc},$$

$$v = 100(1+z_{coll})^{1/2}\left(\frac{M}{10^{12}\,M_\odot}\right)^{1/3}h_{0.5}^{1/3}\,\text{km s}^{-1} = 77\delta_0^{1/2}M_{12}^{1/3}h_{0.5}^{1/3}\,\text{km s}^{-1},$$

$$T_{vir} = 2.32\times10^5(1+z_{coll})\left(\frac{M}{10^{12}\,M_\odot}\right)^{2/3}h_{0.5}^{2/3}\text{K} = 1.36\times10^5\delta_0 M_{12}^{2/3}h_{0.5}^{2/3}\text{K}.$$

$$(7.27)$$

We have used

$$x = 0.92(\Omega h^2)^{-1/3}(M/10^{12}\,M_\odot)^{1/3}\,\text{Mpc} \qquad (7.28)$$

to relate M and the comoving scale and have set $\Omega = 1$, $h = 0.5$. Also note that

$$t_{coll} = t_0(1+z_{coll})^{-3/2}, \quad (1+z_m) = 1.59(1+z_{coll}). \qquad (7.29)$$

The preceding results can be used to estimate the typical parameters of collapsed objects once we are given M and the collapse redshift. For example, if objects with $M = 10^{12}\,M_\odot$ (which is typical of galaxies) collapse at a redshift of, say, 2, then we get $r_{vir} \approx 86$ kpc, $t_{coll} \approx 1.2\times10^9$ yr, $v \approx 173$ km s^{-1}, and $T_{vir} \approx 7\times10^5$ K. The density contrast of the galaxy at present will be $(\rho_{coll}/\rho_0) \approx 170(1+z_{coll})^3 \approx 4.6\times10^3$.

The baryonic component will be heated to a rather high temperature if the preceding analysis is correct. At these temperatures, it is important to take into account the cooling processes operational in the gas. If the gas can radiate energy and cool effectively, it will be able to collapse further and form a more tightly bound object. The evolution will depend critically on the relative values of the dynamical time scale,

$$t_{\text{dyn}} \approx \frac{\pi}{2} \left[\frac{2GM}{R^3} \right]^{-1/2} = 5 \times 10^7 \, \text{yr} \left(\frac{n}{1 \, \text{cm}^3} \right)^{-1/2}, \tag{7.30}$$

and the cooling time scale,

$$t_{\text{cool}} = \frac{E}{\dot{E}} \approx \frac{3\rho kT}{2\mu \Lambda(T)}, \tag{7.31}$$

where $\Lambda(T)$ is the cooling rate. Several different processes can contribute to cooling, depending on the temperature range. For the sake of simplicity, we shall assume that the cooling is dominated by plasma bremsstrahlung and recombination. Let us recall that the energy loss that is due to bremsstrahlung is given by

$$\epsilon_B = 1.4 \times 10^{-27} \, T^{1/2} n_e^2 x_e^2 \tag{7.32}$$

in cgs units, where x_e is the ionisation fraction. Similarly, the atomic cooling that is due to collisions is given by

$$\epsilon_c = 7.5 \times 10^{-19} n_e^2 [x_e(1 - x_e)] \exp(-E_0/k_B T) \tag{7.33}$$

in cgs units, where $E_0 = 13.6$ eV. At low temperatures, the ionisation fraction starts dropping, and the cooling by either of the processes is suppressed for $T \lesssim 10^4$ K. In particular the collisional atomic cooling is inhibited exponentially by the $\exp(-E_0/k_B T)$ factor.

To obtain the actual temperature dependence of the cooling rate, we need to solve for the ionisation equilibrium and find $x_e(T)$, which can be done in an approximate manner as follows: The time scale for collisional ionisation is of the order of

$$t_i = (n_{\text{HI}} \sigma_{\text{coll}} v)^{-1} \approx \frac{1}{7.2 n_{\text{HI}} a_0^2} \left(\frac{m_e}{k_B T} \right)^{1/2} \exp\left(\frac{E_0}{k_B T} \right)$$

$$\approx \frac{2.2 \times 10^{14} \, \text{s}}{(1 - x_e)} T_5^{-1/2} e^{E_0/k_B T}, \tag{7.34}$$

and the time scale for recombination is

$$t_r = (n_e \sigma_{\text{PI}} v)^{-1} \approx \frac{5.8 \times 10^{19} \, \text{s}}{x_e} T_5^{2/3}. \tag{7.35}$$

Equating the two gives the ionisation fraction as

$$x_e = \left[1 + aT_5^{-7/6}\exp\left(\frac{E_0}{k_BT}\right)\right]^{-1}, \qquad a = 2.8 \times 10^{-6}. \qquad (7.36)$$

Using this in our expressions for ϵ_B and ϵ_c, we find that the total cooling rate can be expressed in the form $\epsilon_B + \epsilon_c = n_e^2\Lambda(T)$ with

$$\Lambda(T) = \frac{10^{-24}}{\left[1 + aT_5^{-7/6}\exp(E_0/k_BT)\right]^2}\left[2.1\,T_5^{-7/6} + 0.44\,T_5^{1/2}\right] \qquad (7.37)$$

in cgs units. This expression captures the behaviour of cooling rates at high temperatures ($T > 10^6$ K) and at low temperatures ($T < 10^4$ K), but is not very accurate in the intermediate regime. However, it clearly shows the origin of different patterns of behaviour of $\Lambda(T)$ in different temperature ranges. In the range $T \gtrsim 10^4$ K, another simpler fitting formula for the cooling rate is

$$\Lambda(T) = \left(A_BT^{1/2} + A_RT^{-1/2}\right)\rho^2, \qquad (7.38)$$

where the $A_B \propto (e^6/m_e^{3/2})$ term represents the cooling that is due to bremsstrahlung and the $A_R \simeq e^4mA_B$ term arises from the cooling that is due to recombination. This expression is valid for temperatures above 10^4 K; for lower temperatures, the cooling rate drops drastically because H can no longer be significantly ionised by collisions and Eq. (7.37) provides a better description. We shall use Eq. (7.38) in what follows; Eq. (7.37) is discussed later.

Introducing the numerical values appropriate for a H–He plasma (with a He abundance $Y = 0.25$ and some admixture of metals) the expression for $t_{\rm cool}$ becomes

$$t_{\rm cool} = \frac{E}{\dot{E}} \approx \frac{3\rho kT}{2\mu\Lambda(T)}$$

$$= 8 \times 10^6\,{\rm yr}\left(\frac{n}{1\,{\rm cm}^{-3}}\right)^{-1}\left[\left(\frac{T}{10^6\,{\rm K}}\right)^{-\frac{1}{2}} + 1.5\left(\frac{T}{10^6\,{\rm K}}\right)^{-\frac{3}{2}}\right]^{-1}. \qquad (7.39)$$

Here n is the number density of particles corresponding to a critical density; the number density of baryons will be smaller by a factor of Ω_B. We can see from Eq. (7.39) that there is a transition temperature $T^* \approx 10^6$ K. For $T > T^*$, bremsstrahlung dominates, whereas for $T < T^*$, line cooling dominates. We now determine the characteristic mass and length scales that emerge from these considerations.

The evolution of a gas cloud of mass M and radius R will depend crucially on the relative values of the cooling time scale, $t_{\rm cool}$, and the dynamical time scale

$$t_{\rm dyn} \approx \frac{\pi}{2}\left[\frac{2\,GM}{R^3}\right]^{-1/2} = 5 \times 10^7\,{\rm yr}\left(\frac{n}{1\,{\rm cm}^{-3}}\right)^{-1/2}. \qquad (7.40)$$

Note that we have taken t_{dyn} to be the free-fall time of a uniform-density sphere of radius R.

There are three possibilities that should be distinguished as regards the evolution of such a cloud. First, if t_{cool} is greater than the Hubble time, H_0^{-1}, then the cloud could not have evolved much since its formation. On the other hand, if $H_0^{-1} > t_{cool} > t_{dyn}$, the gas can cool; but as it cools the cloud can retain the pressure support by adjusting its pressure distribution. In this case the collapse of the cloud will be quasi static on a time scale of the order of t_{cool}. Finally there is the possibility that $t_{cool} < t_{dyn}$. In this case the cloud will cool rapidly (compared with its dynamical time scale) to a minimum temperature. This will lead to the loss of pressure support and the gas will undergo an almost free-fall collapse. Fragmentation into smaller units can now occur because, as the collapse proceeds, smaller and smaller mass scales will become gravitationally unstable.

The criterion $t_{cool} < t_{dyn}$ can determine the masses of galaxies. Only when this condition is satisfied can a gravitating gas cloud collapse appreciably and fragment into stars. Further, in any hierarchical theory of galaxy formation, unless a gas cloud cools within a dynamical time scale and becomes appreciably bound, collapse on a larger scale will disrupt it. In these theories, galaxies are the first structures that have resisted such disruption by being able to satisfy the preceding criterion.

Let us first consider the evolution of baryons, ignoring the dark-matter component. We are interested in the ratio $\mathcal{R} = (t_{cool}/t_{dyn})$ with the condition $\mathcal{R} = 1$ defining a curve on the ρ–T space that marks out the region of parameter space in which cooling occurs rapidly within a dynamical time from the region of weak cooling. (See Fig. 7.1; the thick solid curve is for $\mathcal{R} = 1$ in the region $T > 10^4$ K, which corresponds to the present discussion. The sharp rise of this line for $T < 10^4$ K, which arises from the use of Eq. (7.37), will be discussed below.)

For $T < T^*$, when line cooling is dominant, we have $t_{cool} \propto (T^{3/2}/\rho)$ and $t_{dyn} \propto \rho^{-1/2}$, giving $\mathcal{R} \propto (T^{3/2}/\rho^{1/2}) \propto M$; hence the $\mathcal{R} = 1$ curve will be parallel to the lines of constant mass in the ρ–T plane. In the figure, five lines of constant mass have been plotted, corresponding to $M = (10^{10}, 2 \times 10^{11}, 2 \times 10^{12}, 6 \times 10^{13}, 6 \times 10^{14}) M_\odot$. The thick curve is close to the $M = 2 \times 10^{12} M_\odot$ curve in the range 2×10^4 K $\lesssim T \lesssim 5 \times 10^5$ K. Substituting the numbers and using the expression for the cooling time from Eq. (7.39), we find that $\mathcal{R} = 1$ implies

$$f \left(\frac{T}{10^6 \text{ K}} \right)^{3/2} \left(\frac{n}{\text{cm}^{-3}} \right)^{-1/2} = 4.28. \tag{7.41}$$

Expressing the mass of the cloud as

$$M = \frac{5 \, RT}{G\mu} = 2.1 \times 10^{11} \, M_\odot \left(\frac{T}{10^6 \text{ K}} \right)^{3/2} \left(\frac{n}{\text{cm}^{-3}} \right)^{-1/2}, \tag{7.42}$$

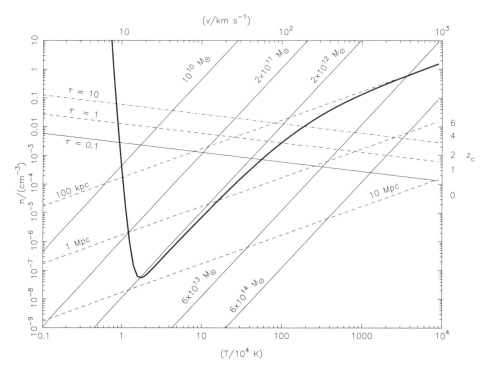

Fig. 7.1. Cooling diagram for collapsing gas. See text for discussion.

we can write

$$\mathcal{R} = \frac{t_{\text{cool}}}{t_{\text{dyn}}} \approx \frac{M}{9 \times 10^{11} \, M_\odot} \tag{7.43}$$

if $\mu = 0.57$. Thus the criterion for efficient cooling can be satisfied for masses below a critical mass of $\sim 10^{12} \, M_\odot$, provided that $T < 10^6$ K.

On the other hand, for $T > T^*$, when bremsstrahlung dominates the cooling process, $t_{\text{cool}} \propto (T^{1/2}/\rho)$ and $t_{\text{dyn}} \propto \rho^{-1/2}$. Therefore $\mathcal{R} \propto (T^{1/2}/\rho^{1/2}) \propto R$, and the curve $\mathcal{R} = 1$ will be parallel to the lines of constant radius in the ρ–T space. The curves corresponding to $R = 100$ kpc, 1 Mpc, and 10 Mpc are indicated by the dashed line. We now find that $\mathcal{R} = 1$ implies that

$$\left(\frac{T}{10^6 \, \text{K}} \right)^{1/2} \left(\frac{n}{\text{cm}^{-3}} \right)^{-1/2} = 6.43. \tag{7.44}$$

Expressing the radius of the cloud as

$$R = \frac{GM\mu}{5T} = 13 \, \text{kpc} \left(\frac{T}{10^6 \, \text{K}} \right)^{1/2} \left(\frac{n}{\text{cm}^{-3}} \right)^{-1/2}, \tag{7.45}$$

we get

$$\mathcal{R} = \frac{t_{\text{cool}}}{t_{\text{dyn}}} \cong \frac{R}{80 \, \text{kpc}}. \tag{7.46}$$

Therefore clouds with high temperature ($T > T^*$) have to shrink below a critical radius of $\sim 10^2$ kpc before being able to cool efficiently to form galaxies.

A gas cloud with constant mass evolves roughly along lines of constant M_J, with $T \propto \rho^{1/3}$, if it is pressure supported. Those outside the region enclosed by the thick curve, cool slowly (and undergo quasi-static collapse, with the pressure balancing gravity at each instant) until they enter the region where $\mathcal{R} < 1$. Gas clouds in the region above the thick curve can cool efficiently to form galaxies because they have masses below $10^{12} \, M_{\odot}$ or radii below 10^2 kpc. These masses and radii compare well with the scales characteristic of galaxies. On the right-hand side (vertical axis) we have indicated the collapse redshift for a structure if it should be frozen with the number density n. We compute this by assuming that, when a structure is formed, its density is a factor $f_{\text{coll}} \approx 178$ larger than the background density at the time of formation. This assumption leads to the relation $(1 + z_c)^3 = (n/n_0 f_{\text{coll}})$ that is used to arrive at the numbers indicated on the vertical axis on the right-hand side. Also note that the $1 - \sigma$ fluctuations containing the masses $(10^{10}, 2 \times 10^{11}, 2 \times 10^{12})M_{\odot}$ collapse at redshifts of $z = (2.1, 1.1, 0.47)$ whereas the $3 - \sigma$ fluctuations with the same mass scales collapse at $z = (8.4, 5.3, 3.4)$ (see Chap. 1, Section 1.4). The nearly horizontal lines marked by constant τ will be described in Section 7.4.

Let us now consider the effects of including the dark-matter component. The dynamical time scale is now determined by the total density of dark matter and baryons, whereas the cooling time still depends on only the density of the baryonic gas. In this case, the gas will not be at the virial temperature initially. It is only during collapse that the gas gets heated up by shocks produced when different bits of gas run into each other. If the cooling time scale of the shocked gas is larger than the dynamical time scale in which the cloud settles down to an equilibrium, then the gas will eventually get heated up to the virial temperature. On the other hand, if the cooling time is shorter, the gas may *never* reach such a pressure-supported equilibrium. Efficient cooling will result in the gas's sinking to the centre of the dark-matter potential well that is being formed until halted by rotation or fragmentation into stars.

Clearly it is again the ratio of the cooling time to the dynamical time of the object that governs the evolution. Further, note that smaller mass clumps are disrupted as larger masses turn around and collapse. However, if the gas component can cool efficiently enough, it may shrink sufficiently close to the centre of the dark-matter potential and thus resist further disruption. This process will break the hierarchy. Galaxies could be again thought of as the first structures that have survived the disruption that is due to hierarchical clustering.

The spherical model can be used to estimate the relevant dynamical time scale. We assume $t_{\rm dyn}$ to be comparable with $t_{\rm coll}/2$, the time it takes for a spherical top-hat fluctuation to collapse after turning around. This expression is the same as $t_{\rm dyn}$ given in Eq. (7.40), provided we identify R in Eq. (7.40) with the radius of turnaround r_m. Then

$$t_{\rm dyn} \approx \frac{t_{\rm coll}}{2} \approx 1.5 \times 10^9 \left(\frac{M}{10^{12}\, M_\odot} \right)^{-1/2} \left(\frac{r_m}{200\, {\rm kpc}} \right)^{3/2} \ {\rm yr.} \qquad (7.47)$$

For estimating the cooling time scale, we use Eq. (7.39) and assume that the gas makes up a fraction F of the total mass and is uniformly distributed within a radius $r_m/2$. The gas temperature is taken to be of the order of the virial temperature obtained in the spherical model; that is, $T_{\rm vir} \simeq (\mu v^2 / 3)$, where $v^2 \simeq (6\, G M / 5 r_m)$. This corresponds to the temperature achieved by heating by shocks, which have a velocity of the order of the virial velocity. In that case,

$$t_{\rm cool} \approx 2.4 \times 10^9 \left(\frac{F}{0.1} \right)^{-1} \left(\frac{M}{10^{12}\, M_\odot} \right)^{1/2} \left(\frac{r_m}{200\, {\rm kpc}} \right)^{3/2} \ {\rm yr.} \qquad (7.48)$$

We assumed that the line cooling dominates at the temperature $T = T_{\rm vir}$ relevant to the galaxies and adopted a typical value of $F \simeq (\Omega_B / \Omega) \simeq 0.1$. Note that the collapse, in general, is likely to be highly inhomogeneous and the preceding estimates are supposed to give only a rough idea of the numbers involved. From Eqs. (7.47) and (7.48) we get

$$\mathcal{R} = \left(\frac{t_{\rm cool}}{t_{\rm dyn}} \right) \approx 1.6 \left(\frac{F}{0.1} \right)^{-1} \left(\frac{M}{10^{12}\, M_\odot} \right) \qquad (7.49)$$

so that efficient cooling (with $\mathcal{R} < 1$) requires

$$M < M_{\rm crit} \approx 6.4 \times 10^{11}\, M_\odot \left(\frac{F}{0.1} \right). \qquad (7.50)$$

It is clear that masses of the order of galactic masses are again picked out preferentially, even when the dark matter is included.

The thick curve in Fig. 7.1 was obtained with expression (7.37) for low temperatures and an assumption of zero metallicity. It should also be stressed that cooling is fairly sensitive to the metallicity, and the forms of the curves change quite a bit even when small amounts of metals are added. We have not taken into account any photoionisation in this calculation; the role of photoionisation is discussed later on.

The procedure previously outlined can be used to analyse any particular theory of structure formation involving hierarchical clustering. The starting point will be the cooling diagram, in which the $\mathcal{R} = 1$ curve is plotted. Given the power spectrum of density fluctuations, we can work out density contrast that is due to ν–σ fluctuations at various mass scales $\delta_0 = \nu \sigma (M)$. Then the various properties, like ρ and T of the collapsed objects that are formed, can be estimated with the spherical model. We saw that these properties depend only on one parameter, M, once the density contrast $\delta_0(M)$ is fixed in terms of M. Thus, for each value of ν, we get a curve on the ρ–T plane, giving the properties of collapsed objects. These curves assume that the protocondensations have virialised but that the gas has not cooled and condensed. Cooling moves points on these curves to higher densities. In the same diagram we can also plot, for comparison, the observed positions of galaxies, groups, and clusters of galaxies.

The cooling arguments described above could be of particular relevance to clusters of galaxies. It is clear from Fig. 7.1 that the gas in the clusters has not yet cooled and is possibly in hydrostatic equilibrium. However, because the density of gas is higher near the centre of the halo, the cooling time will be lower in the central region. This could lead to a *cooling flow* in which the gas undergoes a slow subsonic inflow towards the centre because of the reduction in the pressure in the central rings. The flow rate should be such that the energy radiated by bremsstrahlung from the central region should be matched by the rate of flow of enthalpy towards the centre. Estimates indicate that the rate of deposit of mass inferred from such an argument can be fairly high: 50–100 $M_\odot$ per year. The observational evidence for such a flow is still somewhat controversial, although it seems very likely that it should exist.

Exercise 7.3

Cooling flows: (a) Assume that clusters of galaxies can be modelled as isothermal spheres with a velocity dispersion σ and mass M. Estimate the radius r_c at which the bremsstrahlung cooling time of the plasma is of the order of Hubble time. (b) Show that the mass flow rate $\dot{M}$ is given by $\dot{M}(r) = (2\mu m_H/5k_B T)L(< r)$, where μ is the mean molecular weight of the particles in the plasma. Estimate the mass flow rates for a typical cluster.

The baryons can also – in principle – cool through inverse Compton scattering with MBR photons. The cooling rate of a gas with electron density n_e and temperature T embedded in a blackbody radiation field of density ρ_R and temperature T_R is given by

$$\Lambda_{\text{Comp}} = \frac{4\sigma_T n_e \rho_R (T - T_R)}{m_e} \tag{7.51}$$

(see Vol. I, Chap. 6, Section 6.6). The cooling time for matter, because of inverse Compton scattering off the cosmic background photons, therefore will be

$$t_{\text{Comp}} = \frac{3m_p m_e (1 + z)^{-4}}{8\mu\sigma_T \Omega_R \rho_c} \approx 2.1 \times 10^{12} (1 + z)^{-4} \text{ yr.} \tag{7.52}$$

Here we have assumed that $T \gg T_R$ and have used $\rho_R(z) = \Omega_R \rho_c (1 + z)^4$ to take into account the expansion of the universe. Comparing t_{Comp} with the dynamical time in approximation (7.47) we get

$$\frac{t_{\text{Comp}}}{t_{\text{dyn}}} \approx 2 \times 10^2 (1 + z_{\text{coll}})^{-5/2}. \tag{7.53}$$

This ratio is less than unity for $z_{\text{coll}} > 7$, independent of the mass of the collapsing object. Therefore Compton cooling can efficiently cool an object only if it collapses at a redshift higher than $z \simeq 10$, whatever its mass is.

In an individual halo, baryons will originally follow the same broad distribution as dark matter. However, because they can radiate and cool, their energy will be dissipated and they will collapse to form smaller structures. Assuming that the angular momentum J of dark-matter particles remains an adiabatic invariant as the baryons accumulate in the centre and using $J = mvr$, we find $v^2 = GM/r$ so that $J \propto \sqrt{MR}$. We conclude that the quantity $r M(< r)$ is a constant. This gives

$$r[M_B(< r) + M_{\text{DM}}(< r)] = r_i M_i(< r_i). \tag{7.54}$$

If the final baryonic distribution is known from standard optical observations, etc., and the dark-matter distribution is known from rotation curves, this equation actually allows the determination of the initial density profile. Alternatively, we could determine the final distribution if the initial density profile is known. Suppose that initially $M(< r_i) = A r_i$ (corresponding to an isothermal sphere) and that a fraction α of the mass is baryonic. Equation (7.54) now gives

$$r[M_B(< r) + M_{\text{DM}}(< r)] = A r_i^2 \tag{7.55}$$

whereas the conservation of dark matter implies

$$M_{\text{DM}}(< r) = (1 - \alpha) M_i(< r_i) = (1 - \alpha) A r_i. \tag{7.56}$$

Eliminating r_i between these two equations and solving for M_{DM} gives

$$M_{\text{DM}}(< r) = A r \frac{(1 - \alpha)^2}{2} \left\{ 1 + \left[1 + \frac{4 M_B(< r)}{A r (1 - \alpha)^2} \right]^{1/2} \right\}. \tag{7.57}$$

This provides a simple but useful relation between the observable baryonic mass profile and the dark-matter profile.

Exercise 7.4

Halos of massive ellipticals: Consider a massive elliptical having the de Vaucouleur profile and assume that it has a constant M/L ratio. Determine the distribution of dark matter by using the result obtained above.

Exercise 7.5

Outer reaches of clusters: Assume that the Coma cluster is well described by an isothermal sphere containing a mass of $6 \times 10^{14} h^{-1} M_\odot$ within a radius of $1 h^{-1}$ Mpc. Show that the predicted turnaround radius for the Coma cluster in an $\Omega = 1$ universe is $r_{\text{ta}} \approx 9.8 h^{-1}$ Mpc. As a cluster turns around and collapses, the coherent infall in the outer parts will change to virialised random orbits near the centres. Can this effect be used to determine Ω?

7.4 Mass Functions and Abundances

The description developed so far can also be used to address an important question: What fraction of the matter in the universe has formed bound structures at any given epoch and what is the distribution in mass of these bound structures? We shall now describe a simple approach that answers these questions.

Gravitationally bound objects in the universe, like galaxies, span a large dynamic range in mass. Let $f(M)dM$ be the number density of bound objects in the mass range $(M, M + dM)$ (usually called the mass function) and let $F(M)$ be the number density of objects with masses *greater* than M. Because the formation of gravitationally bound objects is an inherently nonlinear process, it might seem that the linear theory cannot be used to determine $F(M)$. This, however, is not entirely true. In any one realisation of the linear density field $\delta_R(\mathbf{x})$, filtered by a window function of scale R, there will be regions with high density (i.e., regions with $\delta_R > \delta_c$, where δ_c is some critical value slightly greater than unity, say). It seems reasonable to assume that such regions will eventually condense out as bound objects. Although the dynamics of that region will be nonlinear, the process of condensation is unlikely to change the *mass* contained in that region significantly. Therefore, if we can estimate the mean number of regions with $\delta_R > \delta_c$ in a Gaussian random field, we will be able to determine $F(M)$.

One way of achieving this is as follows: Let us consider a density field $\delta_R(\mathbf{x})$ smoothed by a window function W_R of scale radius R. As a first approximation, we may assume that the region with $\delta(R, t) > \delta_c$ (when smoothed on the scale R at time t) will form a gravitationally bound object with mass $M \propto \bar{\rho} R^3$ by the time t. The precise form of the M–R relation depends on the window function used; for a step function, $M = (4\pi/3)\bar{\rho}R^3$, whereas for a Gaussian $M = (2\pi)^{3/2}\bar{\rho}R^3$. Here δ_c is a critical value for the density contrast that has to be supplied by theory. For example, $\delta_c \simeq 1.68$ in the spherical-collapse model. Because $\delta \propto t^{2/3}$ for an $\Omega = 1$ universe, the probability for the region to form a bound structure at t is the same as the probability $\delta > \delta_c(t_i/t)^{2/3}$ at some early epoch t_i. This probability can be easily estimated *because, at sufficiently early t_i*, the system is described by a Gaussian random field. Hence the fraction of bound objects with mass greater than M will be

$$F(M) = \int_{\delta_c(t,t_i)}^{\infty} P(\delta, R, t_i)\, d\delta = \frac{1}{\sqrt{2\pi}} \frac{1}{\sigma(R, t_i)} \int_{\delta_c}^{\infty} \exp\left(-\frac{\delta^2}{2\sigma^2(R, t_i)}\right) d\delta$$

$$= \frac{1}{2}\mathrm{erfc}\left[\frac{\delta_c(t, t_i)}{\sqrt{2}\sigma(R, t_i)}\right], \tag{7.58}$$

where $\mathrm{erfc}(x)$ is the complementary error function. The mass function $f(M)$ is just $\partial F/\partial M$; we can find the (comoving) number density $N(M, t)$ by dividing this expression by $(M/\bar{\rho})$. Carrying out these operations we get

$$N(M, t)\, dM = -\left(\frac{\bar{\rho}}{M}\right)\left(\frac{1}{2\pi}\right)^{1/2}\left(\frac{\delta_c}{\sigma}\right)\left(\frac{1}{\sigma}\frac{d\sigma}{dM}\right)\exp\left(-\frac{\delta_c^2}{2\sigma^2}\right) dM. \tag{7.59}$$

Given the power spectrum $|\delta_k|^2$ and a window function W_R we can explicitly compute the right-hand side of this expression.

There is, however, one fundamental difficulty with Eq. (7.58). The integral of $f(M)$ over all M should give unity; but it is easy to see that, for expression (7.58),

$$\int_0^\infty f(M)\,dM = \int_0^\infty dF = \frac{1}{2}. \tag{7.60}$$

This arises because we have not taken into account the underdense regions correctly. To see the origin of this difficulty more clearly, consider the interpretation of Eq. (7.58). If a point in space has $\delta > \delta_c$ when filtered at scale R, then that point should correspond to a system with mass greater than $M(R)$; this is taken care of correctly by Eq. (7.58). However, consider those points that have $\delta < \delta_c$ under this filtering. There is a *nonzero* probability that such a point will have $\delta > \delta_c$ when the density field is filtered with a radius $R_1 > R$. Therefore, to be consistent with the interpretation in Eq. (7.58), such points should *also* correspond to a region with mass greater than M. However, Eq. (7.58) ignores these points completely and thus *underestimates* $F(M)$ [by a factor of $(1/2)$]. To correct this, we "renormalise" the result by multiplying it by a factor 2. Then

$$dF(M) = \sqrt{\frac{2}{\pi}\,\frac{\delta_c}{\sigma_2}}\left(-\frac{\partial\sigma}{\partial M}\right)\exp\left(-\frac{\delta_c^2}{2\sigma^2}\right)dM \tag{7.61}$$

or

$$N(M)\,dM = -\frac{\bar{\rho}}{M}\left(\frac{2}{\pi}\right)^{1/2}\frac{\delta_c}{\sigma^2}\left(\frac{\partial\sigma}{\partial M}\right)\exp\left(-\frac{\delta_c^2}{2\sigma^2}\right)dM. \tag{7.62}$$

The quantity σ here refers to the linearly extrapolated density σ_L; the subscript L is omitted to simplify notation. The corresponding result for $F(M)$ is also larger by a factor of 2:

$$F(M, z) = \mathrm{erfc}\left[\frac{\delta_c}{\sqrt{2}\sigma_L(M, z)}\right] = \mathrm{erfc}\left[\frac{\delta_c(1+z)}{\sqrt{2}\sigma_0(M)}\right], \tag{7.63}$$

where $\sigma_0(M)$ is the linearly extrapolated density contrast today and we have used the fact that $\sigma_L(M, z) \propto (1+z)^{-1}$. In the case of models with $\Omega_{\mathrm{NR}} + \Omega_V = 1$, the growth law for the perturbations will not be $(1+z)^{-1}$ and we need to use the correct scaling. A reasonable fit for this purpose is given by

$$D(z) = \left[\frac{\Omega_{\mathrm{NR}} + 0.4545\,\Omega_V}{\Omega_{\mathrm{NR}}(1+z)^3 + 0.4545\,\Omega_V}\right]^{1/3}, \tag{7.64}$$

which should replace the $(1+z)^{-1}$ factor. It is convenient to quote the final results in the form

$$\Omega(M) = \frac{M^2 f(M)}{\rho_0} = \frac{dF}{d \ln M} = \left| \frac{d \ln \sigma}{d \ln M} \right| \sqrt{\frac{2}{\pi}} \, \nu \, \exp\left(-\frac{\nu^2}{2}\right), \quad (7.65)$$

where $\nu = [\delta_c / \sigma_0(M) D(z)]$. Physically, $\Omega(M)$ represents the fraction of critical density contributed by bound structures having mass M. We can obtain the baryonic density in the halos by multiplying this result by the factor $\Omega_B / \Omega_{\mathrm{NR}}$.

The simplest application of this formula will be for power-law spectra with $P(k) \propto k^n$. In this case, $\nu = (1 + z)(M/M_c)^{(n+3)/6}$, where M_c is the mass scale at which $\sigma_0(M_c) = \delta_c$; that is, M_c is the mass scale that is collapsing today. The evolution is now scale invariant and the abundance Ω depends on z only through a mass scale that varies as a power of $1 + z$. In models like CDM, the scale invariance will be broken because $\sigma(M)$ has an intrinsic scale. The results for standard CDM with $\Omega = 1$, $\sigma_8 \approx 0.52$, $h = 0.5$, and $\Omega_B h^2 = 0.02$ is shown in Fig. 7.2. (The broken curves arise when cooling of gas is taken into account and will be described below.) We see that, at any given mass scale M, the number

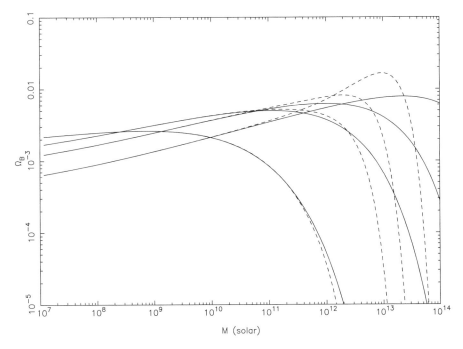

Fig. 7.2.　The density contributed by *baryons* in collapsed halos, having a specified total mass, is shown. The solid curves are for pure Press–Schecter theory and the dashed curves are for a model that incorporates the cooling of gas. The different curves are for $z = 0, 1, 2, 5$ with the peaks shifting to a larger mass at lower redshifts. The results are for $\Omega = 1$ CDM with $\sigma_8 \approx 0.52$, $h = 0.5$, and $\Omega_B h^2 = 0.02$.

density of collapsed objects first increases, reaches a maximum at a critical redshift, and then decreases, which could be interpreted as the formation of bound structures of a certain mass that later on merge to form still larger structures.

A simple argument shows that the merging of structures must scale rapidly with increasing redshift. The number of low-mass objects in any given epoch will scale as $n \propto [1/\sigma(M)]$; the merger rate will therefore scale as $\dot{n}$. Taking $\sigma \propto (1 + z)^{-1}$ and $t \propto (1 + z)^{-2/3}$, we find that $\dot{n} \propto (1 + z)^{5/2}$, which is a rapidly increasing function of redshift. This analysis, of course, is applicable only to the dark-matter halo. For the baryonic structures, merging is possible only if the typical velocities of the components,

$$\sigma_v \simeq \frac{GM}{r} \propto M^{2/3}\rho^{1/3} \propto M^{2/3}(1 + z), \tag{7.66}$$

are smaller than the relative escape velocity of ~ 200 km s^{-1}. Because the characteristic mass scale M_c varies rapidly with $(1 + z)$, σ_v is usually smaller in the past, encouraging the merger.

It is possible to obtain the formation and destruction rates of halos with a given mass from the Press–Schecter formula by fairly simple reasoning.[2] We first note that if $\dot{N}_{\rm for}$, $\dot{N}_{\rm dest}$, and $\dot{N}_{\rm PS}$ denote the rate of formation of halos, the rate of destruction of halos, and the rate of change of Press–Schecter abundance, respectively, then the conservation of halos gives the relation $\dot{N}_{\rm for} = \dot{N}_{\rm PS} + \dot{N}_{\rm dest}$. However, $\dot{N}_{\rm dest} = N_{\rm PS}(\dot{D}/D)$ as the destruction rate should be proportional to the number of halos present and can vary only because of cosmological expansion. Direct differentiation of $N_{\rm PS}$ shows that $\dot{N}_{\rm PS} = N_{\rm PS}(\dot{D}/D)(\nu^2 - 1)$. Using this in the conservation law, we immediately find that

$$\dot{N}_{\rm for} = \dot{N}_{\rm PS} + \dot{N}_{\rm dest} = N_{\rm PS}\left(\frac{\dot{D}}{D}\right)(\nu^2 - 1) + N_{\rm PS}\left(\frac{\dot{D}}{D}\right) = \nu^2 N_{\rm PS}\left(\frac{\dot{D}}{D}\right).$$
$$\tag{7.67}$$

Because Eq. (7.63) depends only on the linearly extrapolated $\sigma_0(M)$, it is possible to use this expression to obtain constraints on $\sigma_0(M)$ directly. Figure 7.3 gives $\Omega(M)$ as a function of $\sigma_0(M)$ for different z. The six different curves from top to bottom are for $z = 0, 1, 2, 3, 4$, and 5. The solid lines are for CDM with $\sigma_8 = 0.52$ and the dashed curves are for a model with $\Omega_{\rm NR} = 0.35$ and $\Omega_V = 0.65$. Both models have $h = 0.5$ and $\Omega_B h^2 = 0.02$.

As an application of this result, let us consider the abundance of Abell clusters. Let the mass of Abell clusters be $M = 5 \times 10^{14}\,\alpha M_{\odot}$, where α quantifies our uncertainty in the observation. Similarly, we take the abundance to be $\mathcal{A} = 4 \times 10^{-6}\beta h^3$ Mpc^{-3}, with β quantifying the corresponding uncertainty. The contribution of the Abell clusters to the density of the universe is

$$F = \Omega_{\rm clus} = \frac{M\mathcal{A}}{\rho_c} \approx 8\alpha\beta \times 10^{-3}. \tag{7.68}$$

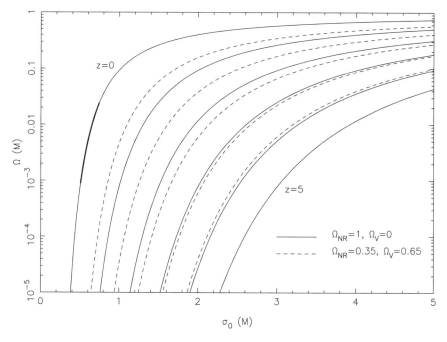

Fig. 7.3. The Ω contributed by collapsed halos containing mass greater than M is plotted against the linearly extrapolated density contrast $\sigma_0(M)$ at different values of z. The curves are for $z = 0, 1, 2, 3, 4,$ and 5, from top to bottom. The solid curves are for CDM with $\sigma_8 = 0.52$, and the dashed curves are for a model with $\Omega_{\mathrm{NR}} = 0.35$ and $\Omega_V = 0.65$. Both models have $h = 0.5$. The constraint arising from cluster abundance at $z = 0$ is marked.

Assuming that $\alpha\beta$ varies between 0.1 and 3, say, we get

$$\Omega_{\mathrm{clus}} \approx (8 \times 10^{-4} - 2.4 \times 10^{-2}). \tag{7.69}$$

This range is marked by thicker shading in Fig. 7.3 on the top curve for $z = 0$. The fractional abundance given in Eq. (7.69) at $z = 0$ requires $\sigma \approx (0.5$–$0.78)$ at the cluster scales. All we need to determine now is whether a particular model has this range of σ for cluster scales. Because this mass corresponds to a scale of $\sim 8h^{-1}$ Mpc, we conclude that the linearly extrapolated density contrast must be in the range $\sigma_L = (0.5$–$0.8)$ at $R = 8h^{-1}$ Mpc.

Note that observations of CMBR fixes the value of density contrast at $R \approx 3000h^{-1}$ Mpc whereas the abundance of clusters fixes its value at $R \approx 8h^{-1}$ Mpc. These observations taken together strongly constrain the shape of the power spectrum. For example, the $\Omega = 1$ CDM model, when normalised by COBE results at $R = 3000h^{-1}$ Mpc, leads to a value $\sigma_0(R = 8h^{-1}$ Mpc$) \equiv \sigma_8 \simeq 1.1$ and thus is ruled out by observations. Because of its convenience, we often use the value of σ_8 for normalising the power spectrum even though it is not as sound a procedure as using CMBR observations. Nevertheless, considering its widespread use, we quote here[3] a simple fitting function that relates σ_8 to Ω_{NR} in open models as well as in models with $\Omega_{NR} + \Omega_V = 1$:

$$\sigma_8 = \begin{cases} (0.52 \pm 0.04)\Omega_{NR}^{-0.46+0.10\Omega_{NR}} & (\text{if } \Omega_\Lambda = 0) \\ (0.52 \pm 0.04)\Omega_{NR}^{-0.52+0.13\Omega_{NR}} & (\text{if } \Omega_\Lambda = 1 - \Omega_{NR}) \end{cases}. \quad (7.70)$$

It is possible to obtain a more stringent but model-dependent constraint on σ_8 if we could use observationally determined mass function for clusters. There is some evidence that the cluster mass function is given by

$$n_{cl}(> M) = 4 \times 10^{-5} h^3 \left(\frac{M}{M^*}\right)^{-1} \exp\left(-\frac{M}{M^*}\right) \text{Mpc}^{-3}, \quad (7.71)$$

where $M^* = (1.8 \pm 0.3) \times 10^{14} h^{-1} M_\odot$ and M is the total gravitational mass within a sphere of radius $R = 1.5 h^{-1}$ Mpc. The integral of this mass function for objects with $M > 10^{14} h M_\odot$ gives the mean mass density in clusters as $\rho_{cl} \approx (7.7 \pm 2.5) \times 10^9 h^2 M_\odot \text{Mpc}^{-3}$ and the corresponding density parameter as $\Omega_{cl} \approx (0.028 \pm 0.009)$. This will lead to more stringent limits on σ_8 but is dependent on the mass function for clusters and the various cutoffs that are used.

The Press–Schecter formalism previously described does not take into account the fact that baryons can cool whereas dark matter cannot. If we are interested in the mass function of actual galaxies, characterised by certain M/L ratios, then we need the mass function for structures that can efficiently cool in a given halo. In the context of Press–Schecter theory, this is fairly easy to do and can be achieved as follows.

We recall from Eq. (7.39) that for a gas with hydrogen and helium with a helium fraction of $Y \approx 0.25$, the cooling time can be written as

$$\frac{t_{cool}}{\text{yr}} = 1.8 \times 10^{24} \left(\frac{\rho_B}{M_\odot \text{Mpc}^{-3}}\right)^{-1} \left(T_8^{-1/2} + 0.5 f_m T_8^{-3/2}\right)^{-1}, \quad (7.72)$$

where we have added a parameter f_m to take care of the metallicity; $f_m = 1$ for solar abundance and $f_m = 0.3$ if there are no metals. For the purpose of this analysis, the baryonic density ρ_B is treated as a fixed fraction Ω_B/Ω_{NR} of the total matter density. The matter density of the halo, in turn, can be expressed in the form $\rho_c = f_c \rho_0 (1 + z_c)^3$, with $f_c \simeq 178$ for an $\Omega_{NR} = 1$ model; a good fit for f_c for models with $\Omega_{NR} + \Omega_V = 1$ is given by $f_c \approx (178/\Omega_{NR}^{0.7})$. We determine the virial temperature T of the system by equating the gravitational potential energy $(3/5)(GM^2/r)$ to $(3M/\mu m_p)k_B T$. Using $M = (4\pi/3)\rho_c r^3$, $\mu = 0.59$, and $\rho_0 = 2.78 \times 10^{11} \Omega_{NR} h^2 M_\odot \text{Mpc}^{-3}$, we find the virial temperature as

$$T_{virial} = 1.26 \times 10^5 \text{ K} \left(\frac{M}{10^{12} M_\odot}\right)^{2/3} (f_c \Omega_{NR} h^2)^{1/3} (1 + z_c). \quad (7.73)$$

Given the temperature, we can determine t_{cool}; we obtain an upper bound on the structures that could have cooled by today by equating t_{cool} to $t_{\text{Hubble}} = (2/3)H_0^{-1}[1-(1+z_c)^{-3/2}]$, which is a slightly milder condition compared with equating $t_{\text{cool}} = t_{\text{coll}}$. If atomic cooling is the dominant process, then the equation $t_{\text{cool}} = t_{\text{Hubble}}$ gives

$$1+z_c = \left(1 + \frac{M}{M_{\text{cool}}}\right)^{2/3}, \qquad \frac{M_{\text{cool}}}{M_\odot} = 1.26 \times 10^{13} f_m f_c^{1/2} \Omega_B \Omega_{\text{NR}}^{-1/2}. \quad (7.74)$$

Because, in the original Press–Schecter formalism, the threshold for formation of bound structures is taken to be $\sigma(M) \geq \delta_c$ with the formation redshift determined by $(1 + z_{\text{for}}) \propto \delta_c^{-1}$, the only modification that is required for incorporating cooling is the replacement of the constant δ_c by a mass-dependent term

$$\delta_c \rightarrow \delta_c(M) = \delta_c[1 + z_{\text{cool}}(M)] = \delta_c[1 + (M/M_{\text{cool}})]^{2/3}. \quad (7.75)$$

The immediate effect of this replacement is to introduce two mass scales M_c and M_{cool} onto the Press–Schecter function (where M_c is the mass scale that is collapsing at a given epoch) so that the resulting form of the function depends on the relative values of these two. In general, the peaks become sharper, thereby reducing the *relative* number of low-mass and high-mass objects compared with the original distribution.

This could be important for the following reason: Consider a simple case of a power-law spectrum with $\sigma_0(M) \propto M^{-(n+3)/6}$. Then the low-mass end of the spectrum of the original Press–Schecter result (which should correspond to the low-luminosity sources in the universe) has the dependence $f(M) \propto (\nu/M^2) \propto M^{(n-9)/6}$. To relate the *mass* function (given by Press–Schecter) to the *luminosity* function of galaxies (which is directly observed), we need to make some ansatz regarding the M/L ratio of the galaxies. We saw, however, in Chap. 1, Section 1.7 that the normalisation of the Tully–Fisher relation requires very low dispersion on the M/L ratio of early-type spirals; we roughly require a constant $M/L \approx 5h$. Then, to match the faint end of the galaxy luminosity function (which varies as $\phi \propto L^{-1}$), we will require $n = 3$. Most of the realistic spectra, on the other hand, have $n \lesssim -2$, which would give a scaling of $M^{-11/6}$, suggesting the existence of far more numerous small-scale objects than seen in the universe. Cooling could change results in the correct direction as far as the *ratio* of high-mass objects to low-mass objects are concerned. Although this effect is in the right direction, it is not clear whether the problem can be completely solved along these lines. It is possible that the baryonic fraction that is retained in a halo is selectively suppressed at the faint end, which could arise, for example, because of mass loss from dwarf galaxies. It is not very clear whether these effects are adequate to provide the correct luminosity function for galaxies at all scales.

Equating the cooling time to the Hubble time is a valid procedure only if the system was allowed to cool undisturbed over the age of the universe. We obtain a better mass scale, valid for more realistic systems, by equating the cooling time and the free-fall collapse time, as was done in Section 7.3. This leads to the mass scale $M_{\text{coll}} = 3.16 \times 10^{13} \, M_\odot \, f_m \, \Omega_B \Omega_{\text{NR}}^{-1}$, which is of the same order of magnitude as M_{cool}; the modification of the Press–Schecter argument can be carried through with this criterion along the same lines.

In discussing the cooling of primordial gas, we have not taken into account the effect of molecular hydrogen H_2. It is possible that hydrogen molecular cooling is a significant effect in many scenarios of structure formation. H_2, however, is fragile and can be easily photodissociated by photons with the energy 11.3–13.6 eV to which the IGM is transparent even before it is ionised. The UV flux capable of dissociating H_2 in the collapsed environment is, in fact, lower by more than 2 orders of magnitude compared with the minimum flux necessary to ionise the universe. Hence, soon after trace amounts of stars form, the formation of additional stars that is due to H_2 cooling is suppressed. Further fragmentation is possible only through the atomic line cooling, which we have discussed.

There is, however, one further complication that needs to be taken into account in all these attempts. The cooling function used in Eq. (7.37) assumes that the gas is ionised collisionally and it is almost completely neutral for $T \lesssim 10^4$ K. If the gas is exposed to ionising radiation (which dominates over collisional ionisation) then the low-temperature peak in the cooling curve will be completely removed. To study this situation, we first need to determine the ionisation fraction x of the gas when it is immersed in a flux of radiation J_ν. For a hydrogen gas, we can determine x by equating the recombination rate per unit volume to the photoionisation rate:

$$x^2 n_B^2 \beta = (1 - x) n_B \int_{\nu_L}^\infty \frac{4\pi J_\nu}{h\nu} \sigma_{\text{PI}}(\nu) \, d\nu. \qquad (7.76)$$

The left-hand side is the recombination rate that is proportional to, $n_e n_p \propto x^2 n_B^2 = \beta x^2 n_B^2$, where n_B is the total number density of baryons (ions plus neutral atoms), and β is the recombination coefficient (see Vol. I, Chap. 6, Section 6.12). The right-hand side is proportional to the product of the ionising flux of photons and the number density of neutral hydrogen atoms [given by $(1 - x) n_B$]. In the integral, $(4\pi J_\nu / h\nu)$ is the flux of photons and $\sigma_{\text{PI}}(\nu) = \sigma_H (\nu/\nu_L)^{-3}$ is the photoionisation cross section for hydrogen with $\sigma_H = 8 \times 10^{-18}$ cm^2. The integral is over the range of frequencies at which ionisation is possible with $h\nu_L = 13.6$ eV. The recombination coefficient β has a weak dependence on temperature, which we will ignore for simplicity. In that case, this equation simplifies to

$$\frac{x^2}{1 - x} = \left(\frac{c\sigma_H}{\beta} \right) \left[\frac{n_\gamma^{\text{eff}}}{n_B} \right] \equiv 10^{5.2} \left(\frac{n_\gamma^{\text{eff}}}{n_B} \right) \equiv K \left(\frac{n_\gamma^{\text{eff}}}{n_B} \right),$$

$$n_\gamma^{\text{eff}} \equiv \frac{4\pi}{c} \int_{\nu_L}^\infty \frac{J_\nu}{h c} \left(\frac{\nu}{\nu_L} \right)^{-3} d\nu. \qquad (7.77)$$

The n_γ^{eff} can be thought of as an effective number of photons that can ionise the hydrogen atom and the ratio $(n_\gamma^{\text{eff}}/n_B)$ is usually called the *ionisation parameter*. If we take the ionising flux to be

$$J_\nu = J_{-21} \left(\frac{\nu}{\nu_L}\right)^{-\alpha} \times 10^{-21} \text{ ergs s}^{-1} \text{ Hz}^{-1} \text{ cm}^{-2} \text{ sr}^{-1}, \qquad (7.78)$$

where ν_L is the frequency of the Lyman line, then Eq. (7.77) gives the ionisation fraction x as

$$\frac{1-x}{x^2} \cong 10^{-7} (3+\alpha) (J_{-21})^{-1} \Omega_B h^2 (1+z)^3. \qquad (7.79)$$

We will see in Chap. 9 that an ionising flux of the form in Eq. (7.78) with $J_{-21} \approx 1$ does exist at high redshifts; in that case, the gas will be highly ionised, making it quite difficult for the galaxies to be formed at all. One possible way out of this dilemma is to assume that the protogalaxies are self-shielding and the outer layers of the material protect the central region from ionising radiation, thereby allowing it to cool. To determine the condition for this to happen, we need to determine the optical depth for the ionising photons, which is given by

$$\tau = N_{\text{HI}} \frac{\int (\sigma J_\nu / h\nu) \, d\nu}{\int (J_\nu / h\nu) \, d\nu} = N_{\text{HI}} \sigma_{\text{H}} \left(\frac{\alpha}{3+\alpha}\right), \qquad (7.80)$$

where $\sigma = \sigma_{\text{H}}(\nu/\nu_L)^{-3}$ is the photoionisation cross section and N_{H} is the column density of neutral hydrogen. Assuming that the critical size of the system is $2r$ and the density of neutral hydrogen is n_{H}, the column density of neutral hydrogen becomes

$$N_{\text{HI}} = 2r n_{\text{H}} = 2r n_B (1-x). \qquad (7.81)$$

The total background hydrogen density at redshift z is $n_{\text{H}}^{\text{tot}} = 8.4 \times 10^{-6} \Omega_B h^2 \times (1+z)^3$ cm^{-3}, whereas a collapsed object will have a density f_c times this value. Eliminating r in terms of density and $v^2 = GM/r$, we can easily estimate the optical depth to be

$$\tau = 6.31 \times 10^{-5} \alpha (J_{-21})^{-1} \left(\frac{v}{100 \text{ km s}^{-1}}\right) [f_c \Omega_B h^2 (1+z)^3]^{3/2}. \qquad (7.82)$$

For $f_c = 200$, $\Omega_B h^2 = 0.02$, the optical depth will be unity at $z = 3$ if $v \approx 3.8 \times 10^2$ km s^{-1}. Figure 7.1 has three diagonal lines corresponding to a constant optical depth of $\tau = 0.1$, 1, and 10 (from bottom to top). From the location of the intersection of the $M = 2 \times 10^{12} M_\odot$ line and the $\tau = 1$ line, we see that $M \gtrsim 10^{12} M_\odot$, $\tau > 1$ requires $v \gtrsim 10^2$ km s^{-1}. Thus galaxies can form only in halos with rotational velocities of $v \gtrsim 10^2$ km s^{-1}, and the formation of small galaxies is suppressed. Without this effect, it is possible that the first-generation low-mass structures that form could lock up most of the mass in the universe, which is not consistent with the observations.

Finally, we also mention another effect that works against the cooling of gas at high redshift, although this is not of equal significance. We saw in Chap. 4 that ionised gas embedded in the CMB feels a drag force with a characteristic time scale of $t_{\mathrm{drag}} \approx (3/4)(m_e c/\sigma_T)(U_{\mathrm{rad}} x)^{-1}$. Comparing this with the Hubble time $t_{\mathrm{Hubble}} = 6.7(\Omega_{\mathrm{NR}} h^2)^{-1/2}(1+z)^{-3/2}$ Gyr, we find that the radiation drag can be a significant effect on baryons until $(1+z) \lesssim 10(\Omega_{\mathrm{NR}} h^2)^{1/5} x^{-2.5}$. For $\Omega_B h^2 \approx 10^{-2}$, the residual ionisation fraction is $x \approx 10^{-3}$ [see relation (4.193)]. This suggests that the matter temperature is held equal to the radiation temperature all the way down to $z \approx 10^2$; it would be impossible for any cooling to take place until the radiation drag ceases to be effective. This effect, however, is not very important because other considerations (like the amplitude of the power spectrum) prevent structure formation at such high redshifts in any case.

7.5 Angular Momentum of Galaxies

One important feature of galactic systems, especially in the disklike systems, is their angular momentum. The angular momentum of a galaxy can be conveniently expressed in terms of the dimensionless parameter

$$\lambda \equiv \frac{L E^{1/2}}{G M^{5/2}}. \tag{7.83}$$

This parameter is the ratio between the *actual* angular velocity ω of the system and the *hypothetical* angular velocity ω_0 that is needed to support the system purely by rotation. It follows that a self-gravitating system with appreciable rotational support has a λ comparable with unity. Observations suggest that disk galaxies have $\lambda \simeq 0.4$ to 0.5.

One possible way of explaining the angular momentum of galaxies is as follows: During the initial collapse of the baryonic structures, tidal forces will be exerted on each protogalaxy by its neighbours. The tidal torque can spin up the protogalaxies, thereby providing them with some angular momentum. During the collapse of the gas that is due to cooling, the binding energy increases while mass and angular momentum remain the same. This will allow λ to increase as $\lambda \propto |E|^{1/2}$ and (possibly) reach observed values.

In this case, protogalaxies can acquire only initial angular momentum because of tidal torquing that is due to other collapsing material around it. If the typical mass, comoving radius, and density contrast are M, R, and δ, then the tidal acceleration of any one of a pair of blobs is $(\delta G M/a^3 R^3)(a R) = (\delta G M/a^2 R^2)$, where we have taken the separation between the blobs to be of the same order as the size of the blobs; if the masses involved are taken to be $M/2$ and the lever arm for the force is $aR/2$, then the net tidal torque is of the order of

$$T \simeq \frac{\delta G M}{a^2 R^2} \times \frac{M}{2} \times \frac{aR}{2} = \frac{\delta G M^2}{4Ra}. \tag{7.84}$$

In an $\Omega = 1$ universe, $\delta \propto a$, implying that the tidal torque is a constant and the angular momentum grows linearly with time: $J = Tt$. The total angular momentum acquired until the time of turnaround t_{ta} can now be estimated for a spherical top-hat model. Using the relation between energy and virial radius and the details of spherical top-hat model, we can easily show that the angular momentum parameter is

$$\lambda \simeq 0.025 \, \nu^{-1} \tag{7.85}$$

if the object has an overdensity that is ν times fractional density contrast. Once again simple arguments lead to meaningful final results, although several factors of the order of unity have been ignored in the analysis.

Further dissipational collapse of the baryons will increase the value of λ, especially in the case of disks embedded in a spherical halo. We shall, however, first examine this idea in the absence of dark-matter halos and assuming that a protogalaxy was just a self-gravitating cloud of baryonic gas. In this case, it turns out that the idea is not feasible. The binding energy of the protogalaxy will be $|E| \simeq GM^2/R$, where R is its characteristic radius. Because M is constant during collapse, $|E| \propto R^{-1}$ and so $\lambda \propto R^{-1/2}$. The gas cloud has to collapse by a factor of $(\lambda_d/\lambda_i)^2 \approx (0.5/0.025)^2 \simeq 400$ before it can spin up sufficiently to form a rotationally supported system, where λ_i is the initial value of λ produced by tidal torques. Therefore, to form a rotationally supported galactic disk of mass $10^{11} \, M_\odot$ and radius 10 kpc, matter needs to collapse from an initial radius of 4 Mpc. This process would take an inordinately long time of approximately $t_{\text{coll}} = (\pi/2)(R^3/2GM)^{1/2} \simeq 5.3 \times 10^{10}$ yr, much longer than the age of the universe. Note that even the material in the core, with a scale length of $r_c \simeq$ 3 kpc, would have to collapse from a distance of 1 Mpc and would take $\sim 10^{10}$ yr.

This difficulty is easily avoided if a massive halo exists. In the presence of a massive dark halo, the initial spin parameter of the system, before collapse of the gas, can be written as $\lambda_i = (L|E|^{1/2}/GM^{5/2})$ where the various quantities $L, E,$ and M refer to the *combined* dark-matter–gas system, although the contribution from the gas is negligible compared with that of the dark matter. After the collapse, the gas becomes self-gravitating and the spin parameter of the resulting disk galaxy is $\lambda_d = (L_d|E_d|^{1/2}/GM_d^{5/2})$, where the parameters now refer to the disk. Therefore we find that

$$\frac{\lambda_d}{\lambda_i} = \left(\frac{L_d}{L}\right) \left(\frac{|E_d|}{|E|}\right)^{1/2} \left(\frac{M_d}{M}\right)^{-5/2}. \tag{7.86}$$

The energy of the virialised dark-matter – gas system, assuming that the gas has not yet collapsed, can be written as $|E| = k_1(GM^2/R_c)$ and that of the disk is given by $|E_d| = k_2(GM_d^2/r_c)$. Here R_c and r_c are the characteristic radii associated with the combined system and the disk, respectively, and k_1 and k_2 are constants of the order of unity that depend on the precise density profile and geometry of the two systems. The ratio of the binding energy of the collapsed disk to that of the combined system is then

$$\frac{|E_d|}{|E|} = \frac{k_2}{k_1} \left(\frac{M_d}{M}\right)^2 \left(\frac{r_c}{R_c}\right)^{-1}. \tag{7.87}$$

Further, the total angular momentum (per unit mass) acquired by the gas, destined to form the disk, should be the same as that of the dark matter. This is because all the material in the system experiences the same external torques before the gas separates out because of cooling. Assuming that the gas conserves its angular momentum during the collapse, we have $(L_d/M_d) = (L/M)$. Hence,

$$\begin{aligned}
\frac{\lambda_d}{\lambda_i} &= \left(\frac{M_d}{M}\right)\left(\frac{k_2}{k_1}\right)^{1/2}\left(\frac{M_d}{M}\right)\left(\frac{R_c}{r_c}\right)^{1/2}\left(\frac{M}{M_d}\right)^{5/2} \\
&= \left(\frac{k_2}{k_1}\right)^{1/2}\left(\frac{R_c}{r_c}\right)^{1/2}\left(\frac{M}{M_d}\right)^{1/2},
\end{aligned} \tag{7.88}$$

where we have used Eq. (7.87) to simplify Eq. (7.86). The gas originally occupied the same region as the halo before collapsing and so had a precollapse radius of R_c. Hence the collapse factor for the gas is

$$\frac{R_c}{r_c} = \left(\frac{k_1}{k_2}\right)\left(\frac{M_d}{M}\right)\left(\frac{\lambda_d}{\lambda_i}\right)^2. \tag{7.89}$$

We see that the required collapse factor for the gas to attain rotational support has been reduced by a factor (M_d/M) from what was required in the absence of a dominant dark halo. For a typical galaxy with a halo that is 10 times as massive as the disk, we need a collapse by a factor of only ~ 40 or so before the gas can spin up sufficiently to attain rotational support.

It is also possible to extend these arguments and estimate the characteristic length scale R_d of an exponential disk profile $\exp(-R/R_d)$ if we assume that there is no exchange of angular momentum between the disk and the halo as the disk forms. For an exponential disk, $J/M = 2R_d v_g$, where v_g is the constant rotational velocity. Combining this with the virial theorem $|E| = K = (3/2)M\sigma_h^2$ (where σ_h^2 is the velocity dispersion of the halo), we can eliminate J and $|E|$ from λ in terms of M. Finally, expressing M in terms of the halo rotation velocity by $GM/R = v_h^2$ and assuming that the density was a factor of f_c larger than the background density $\rho_B(z) = 3\,\Omega_{NR}(z)H^2(z)/8\pi G$ at the time of the collapse redshift z, we get

$$R_d = \frac{\lambda}{2v_g}\left(\frac{3\sigma_h^2}{2}\right)^{-1/2} v_h^3 \left[\frac{f_c}{2}\Omega_{NR}(z)H^2(z)\right]^{-1/2}. \tag{7.90}$$

If we now use $f_c = 178/\Omega_{NR}^{0.7}$, $\lambda = 0.05$, $\sigma_h = v_h/\sqrt{2}$, and assume that, when the disk settles down, $v_g \simeq v_h$, we get

$$R_d \simeq 3h^{-1} \text{ kpc} \left(\frac{v_c}{100 \text{ km s}^{-1}} \right) \frac{H_0}{H(z_f)}. \tag{7.91}$$

This shows that, for reasonable scale lengths, the disk could not have formed at redshifts significantly higher than unity. The overall picture that arises from these considerations – that the spheroids and bulges form early on whereas the disks form at a somewhat later epoch – is consistent with the observations we discussed in Chap. 1, Section 1.7. It is, for example, known that metal-poor stars in a solar neighbourhood have highly eccentric orbits. This will arise naturally if such a spheroidal system of stars has formed early in a time that is short compared with the collapse time and their orbits ended up mixing through collisionless relaxation.

7.6 Galaxy Formation and Evolution

As has been emphasised several times, theoretical investigations of galaxy formation are severely hampered by the lack of a sufficiently large and accurate sample of galaxies – especially at high redshifts. A star-forming galaxy dominated by young stars and bursting at the rate of $2 \, M_\odot$ per year, will produce $2 \times 10^9 \, M_\odot$ of stars in a gigayear and will typically have a luminosity of $2 \times 10^9 \, L_\odot \simeq 7.6 \times 10^{35}$ W. [This corresponds to an efficiency $\epsilon \approx 1.5 \times 10^{-4}$, which is a factor ~ 100 lower than the efficiency of a p–p reaction. This arises because massive stars leave the main sequence after burning only a small fraction of their fuel (see Vol. II, Chap. 3, Section 3.4)]. We can estimate the *minimum* luminosity of such a galaxy by assuming that the star formation is spread over the time scale

$$t = \frac{2}{3 \, H(z)} = \frac{2}{3 \, H_0(1 + z)\sqrt{1 + \Omega z}}, \tag{7.92}$$

corresponding to the redshift z. (We assume that $\Omega_V = 0$ and $\Omega_{NR} = \Omega$.) For a $10^{11} \, M_\odot$ galaxy at redshift $z = 3$, this will lead to a minimum luminosity of approximately $L = 11h L_*$ (where $L_* \simeq 10^{10} h^{-2} \, L_\odot$.) There have been several attempts to detect such primeval galaxies (PGs) directly. Observational searches for PGs will need to cover sufficient volume of the universe and also have the capability of recognizing a PG amidst the background of older and normal galaxies. We will now discuss some broad characteristics that could be expected for a PG.

Let us start with an estimate of the expected abundance of PGs in the sky. If we assume that a significant fraction of the galaxies seen locally, with a density of $n_G \simeq 0.015h^3 \text{ Mpc}^{-3}$, have formed at a high redshift by gravitationally collapsing over a period of ~ 1 Gyr, then it is straightforward to determine the surface density of PGs in the sky. If the comoving density of galaxies n_G is conserved during $0 < z < z_f$, then the surface density

$$\frac{dN}{d\Omega} = n_G \int_0^{z_f} dz \frac{dV_{\text{comov}}}{d\Omega\, dz} \tag{7.93}$$

is essentially $n_G (c H_0^{-1})^3$ times a factor that depends on the cosmology and z_f. If the formation redshift is $z \simeq 3$, we obtain a surface density between 3×10^4 and $3 \times 10^5 h^{-1}$ deg^{-2}, depending on the cosmological model. The surface density increases with increasing redshift of formation.

The next important characteristic of a PG will be its angular size which, in turn, depends on when the brightest phase in the galaxy formation occurred relative to the gravitational collapse. If the star formation occurs soon after the onset of collapse, then the sizes will be large (with angular diameters of approximately 10–30 arcsec), leading to objects with low surface brightness. At $z_f > 5$, these will be virtually overlapping sources and will be difficult to detect as discrete entities. The best search strategy in this case will be to investigate the integrated background light arising from a population of such sources. At the other extreme, we may conjecture that the SFR is closely coupled to the gas density and will be highest in the central regions of the collapsing PGs. Then the angular diameters will be very small (0.1–0.5 arcsec) and detection from ground-based telescopes will be quite difficult.

A PG of mass $10^{11}\, M_\odot$ with a SFR of 100 $M_\odot$ yr^{-1} will have a net bolometric luminosity of approximately 6×10^{37} W. At a redshift of $z = 5$, this will lead to a flux of 10^{-18} W m^{-2} at the optical wavelength of $\lambda = 4500$ Å if the bolometric luminosity is interpreted as λI_λ. This is at least 100 times fainter than the night-sky background at these wavelengths and the situation is worse in the Near-IR (NIR) band, where the atmospheric background is nearly 1000 times brighter. However, because the luminosity is actually spread over a wide range of frequencies, the spectral-energy distribution needs to be computed from a detailed population synthesis code. Figure 7.4 shows the result of one such study. The data in Fig. 1.7 of Chap. 1, Section 1.6, are used to illustrate qualitatively the spectrum of a galaxy having a continuous SFR of 1 $M_\odot$ yr^{-1} Mpc^{-3}. The three thick solid curves in Fig. 7.4 show the spectra of a galaxy at $z = 1.45$ and $z = 1.85$ assuming that it forms at $z = 2$. The dotted curves give the spectra of a galaxy at $z = 4.24$ and $z = 2.76$ assuming that it formed at $z = 5$. Because observations suggest that the SFR in the universe is constant around this value from $z \approx 5$ (see Section 1.6) the appearance of two protogalaxies with the zero age set at $z = 5$ and $z \approx 2$ brackets a fair region of galaxy formation. One key feature that emerges from these studies is that young galaxies radiate almost constant energy per unit frequency in the wavelength range 0.1–2 μ. For a SFR of $\sim 1\, M_\odot$ yr^{-1}, this emission is $\sim 10^{20.6}$ W Hz^{-1} (see Chap. 1, Section 1.6). The relative amount of emission at larger wavelengths increases as the stellar

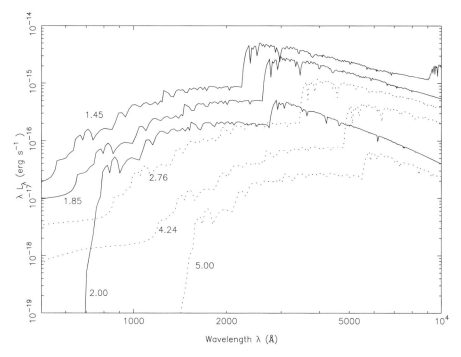

Fig. 7.4.   The spectra of two evolving galaxies, each with a continuous SFR of 1 $M_\odot$ yr^{-1} Mpc^{-3}. The solid curves give the spectrum at $z = 2$, 1.85, and 1.45 for a galaxy that has formed at $z = 2$. The dotted curves give the spectrum at $z = 5$, 4.24, and 2.76 for a galaxy that has formed at $z = 5$.

population ages through the death of massive hot stars that evolve into cooler red giants. In addition to such a continuum radiation, young star-forming regions will also emit intense line radiation especially Lyman-α and H-α. A relatively flat optical and NIR spectrum along with intense line emission could be one possible way of diagonalising such objects.

The search strategy for the Lyman-α (1215-Å) line from the PGs is based on the fact that, in unobscured young stellar population with normal IMF, a significant percentage of the total bolometric luminosity is contained in the Lyman-α line. Roughly speaking, approximately two Lyman-α photons are produced for three ionising ($\lambda < 912$ Å) photons. More precisely, a SFR of 1 $M_\odot$ yr^{-1} will lead to a Lyman-α line luminosity of $\sim 10^{41}$ ergs s^{-1} for typical IMFs.

The searches for objects with a single strong emission line within narrow passbands have not led to any detection of PGs. This, of course, could be due to several reasons. The measured flux F_λ^{line} around any strong line, like the Lyman-α line (1216 Å), will decrease as $(1 + z)^{-5}$ per square arcsecond of the sky. [The quantity λF_λ will decrease as $(1 + z)^{-4}$ and $\lambda_{\text{obs}} = \lambda_{\text{em}}(1 + z)$.] Further, to measure the Lyman-α line from a PG at a redshift of z we need observation at a redshift of $1216(1+z)$Å. The sky brightness (flux) per square arcsecond increases roughly as $F_\lambda^{\mathrm{bg}} \propto \lambda^{2.5}$ at optical and NIR wavelengths. It follows that $(F_\lambda^{\mathrm{line}}/F_\lambda^{\mathrm{bg}}) \propto (1+z)^{-7.5}$; it will be difficult to see Lyman-α emission unless it is very strong. Further, if the star formation takes place in smaller clumps that undergo repeated mergers, then the above calculation is not directly applicable. The second and possibly more likely situation is that the initial burst of star formation leads to radiation that is rapidly reprocessed by dust, thereby moving the emission into the IR or submillimeter band. If the dust grains in a star-burst galaxy radiate as a blackbody, then their luminosity will be Planckian: $L_\nu \propto \nu n_\nu$, where n_ν is the number of photons in the frequency interval $(\nu, \nu + \Delta\nu)$ in the Planck spectrum. However, because the dust grains are tiny, they do not radiate efficiently at long wavelengths (see Vol. I, Chap. 6, Section 6.4) and their luminosity varies as $L_\nu \propto \nu^\alpha n_\nu$ with $\alpha \approx 2$. When observations are made at, say, 100 GHz, the power $F_\nu \Delta\nu$ that is received in the interval $(\nu, \nu + \Delta\nu)$ would have been emitted in the interval $[\nu(1+z), (\nu + \Delta\nu)(1+z)]$. Because the energy received per square arcsecond decreases by a factor of $(1+z)^4$, the flux F_ν from a star-burst galaxy per square arcsecond is almost constant around this frequency for redshifts, $1 \lesssim z \lesssim 20$. With future detectors with sufficient sensitivity, we should be able to see extremely distant star-forming galaxies in this band.

Incidentally, the preceding discussion also brings to the forefront the strong selection effect that arises in the investigation of high-redshift phenomena that are due to different sensitivities in different wave bands. At present, we can reach $B \approx (29\text{--}30)$ in photometry and $B = 25$ in spectroscopy in the visible/UV bands. A star-burst galaxy with a SFR of 50 $M_\odot$ yr^{-1} should be detectable up to $z \lesssim 5$ under optimal circumstances. In the NIR, the corresponding limits are $K = 25$ (photometry) and $K = 21$ (spectroscopy) in VLT, for example; this will probably allow detection up to $z \approx 2.5$. In the mid-IR (ISOCAM; limit of 100 μJy at 15 μm) and in the FIR (ISOPHOT; 100 mJy at 175 μm) we could hardly reach $z \approx 1$. The situation is similar in radio bands (VLA; 10 μJy at 6 cm) but is quite hopeful in the submillimeter band, where SCUBA, for example, has a sensitivity of 3 mJy. The reprocessed IR emission in a broad redshift range of $z \approx (1\text{--}10)$ should be eventually detectable here.

Recent observational advances, that have helped us to probe more directly the nature of high-redshift galaxies, came from two fronts. The first is a technique called the Lyman break detection, and the second advance is from deep pencil-beam surveys from the HST called Hubble deep fields (HDFs).

We begin with a discussion of the Lyman break technique that is based on the fact that the broadband colours of galaxies could be used to select galaxies at high redshifts. Any star-forming galaxy will have a spectrum that is reasonably flat up to to the Lyman limit ($\lambda_c = 912$ Å), beyond which the luminosity will be cut off because of strong absorption. This drop in the luminosity at $\lambda < \lambda_c$ is usually called the *Lyman break*. For a galaxy at redshift z, the Lyman break will move to the wavelength $912(1 + z)$Å that will occur in the middle of the U band for $z = 3$. In a U–V diagram, this will lead to galaxies that are significantly red and thus could be picked up. This Lyman break technique turns out to be quite powerful. For example, in the redshift range $(2 < z < 3.5)$, the galaxies are identified with nearly 100% efficiency whereas in the range $(3.5 < z < 4.5)$, the detection rate is nearly 50%.

There is, however, one difficulty in carrying out this program. When $z \gtrsim 3.4$, the redshifted Lyman break can overlap with the 4000-Å break at low redshift, creating confusion in the identification (see Fig. 7.4). Hence we can pick out Lyman limit galaxies only in a narrow redshift window by identifying sources that are fairly red in U–V but have a relatively flat spectrum at longer wavelengths. At present we have observations that have detected galaxies in the redshift ranges of $(0 \lesssim z \lesssim 1)$ and $(2.5 \lesssim z \lesssim 5)$ with a gap in the middle-redshift range. One broad result that arises from these observations is that the SFR is not *very* high at $z \approx (3\text{–}3.5)$ and is typically $\sim 10\ M_\odot\ \mathrm{yr}^{-1}$. In other words, these observations have not revealed very luminous star-burst galaxies at high redshifts.

Even though the number of galaxies found in the redshift band $2.5 \lesssim z \lesssim 3.5$ is very small (compared with entire sample of galaxies that is currently available), the number is large enough to warrant some preliminary investigation of the clustering process. There is some tentative evidence that the distribution of high-redshift galaxies is fairly inhomogeneous and corresponds to a power spectrum with an apparent value of $\sigma_8 \approx 1$. This is to be compared with the present value of $\sigma_8 \approx 0.6$, which should have evolved from a value of $\sigma_8(1 + z)^{-1} \approx 0.13$ at a redshift of $z = 3.5$ in an $\Omega = 1$ universe. This suggests an apparent bias factor of $b \approx (1/0.13) \approx 7.7$ that can be related to the underlying density field along the following lines. We know that the number density of halos depend on the parameter $\nu = \delta_c/\sigma(M)$ as

$$\frac{dn}{d \ln M} \propto \nu \exp\left(-\frac{\nu^2}{2}\right). \tag{7.94}$$

This relation was derived for the population on the whole but we can try to use it for a more detailed modelling as follows. Consider a region that has a long-wavelength perturbation of magnitude ϵ. Any shorter-wavelength perturbation riding on top of it will need to reach a threshold of only $\delta_c - \epsilon$ for collapsing. The number density of halos will therefore be modulated to the form

$$n \to n + \frac{dn}{d\nu}\frac{d\nu}{d\epsilon}\epsilon = n\left[1 + \epsilon\frac{(\nu^2 - 1)}{\sigma\nu}\right], \tag{7.95}$$

where we have used $\nu = (\delta_c - \epsilon)/\sigma$. This obviously leads to a bias in the number density of halos in the form $\delta n/n = \bar{b}\epsilon$, where $\bar{b} = (\nu^2 - 1)/\nu\sigma$. The net density contrast (for $\epsilon \ll 1$) will be $\epsilon + \bar{b}\epsilon$ in the Eulerian space so that the effective bias will be

$$b = (1 + \bar{b}) = 1 + \frac{\nu^2 - 1}{\nu \sigma} = 1 + \frac{\nu^2 - 1}{\delta_c}. \qquad (7.96)$$

For $b = 7.7$, and $\delta_c = 1.68$, we get $\nu = 3.5$. If these objects correspond to $R \approx 1h^{-1}$ Mpc, the relation $\nu \sigma_0 [1h^{-1} \text{ Mpc}](1 + z)^{-1} = \delta_c$ shows that $\sigma_0 \approx 1.5$ on the scale of these objects. Assuming that $\sigma_0(R) \propto R^{-(n+3)/2}$ in the interval $1h^{-1}$ Mpc $\lesssim R \lesssim 8h^{-1}$ Mpc and using $\sigma_0[8h^{-1} \text{ Mpc}] \approx 0.6$, $\sigma_0[1h^{-1} \text{ Mpc}] \approx 1.5$ we can estimate n to be $n \approx -2$. This is broadly consistent with what we would have expected for galaxy scale fluctuations, suggesting that the high-redshift galaxies that are seen are indeed the first collapsed objects arising from the high peaks of the Gaussian random field.

The technique of using the Lyman break for detecting high-redshift galaxies can be complemented by deep pencil-beam surveys in single directions. The deepest available survey of this kind is the one based on HDFs in the north and the south. These surveys have led to ∼3000 galaxies at U_{300}, B_{450}, V_{606}, and I_{814} and 1700 galaxies at J_{110} and H_{160}. At still longer wavelengths, there are ∼300 galaxies in the K band, 50 at 6.7–15 μm and 9 at 3.2 μm. Nearly 200 of them now have redshifts, with the median redshift being approximately $z = 1$ at $R = 24$. The area of the sky covered by these observations is ∼5.3 arcmin^2 for HDF-N (which corresponds to 4.6 Mpc at $z = 3$ for a universe with $\Omega_{NR} = 0.3$ and $\Omega_V = 0.7$) and 0.7 arcmin^2 in the case of HDF-S.

We can convert the comoving volume $V(< z)$ covered by HDF-N to a characteristic number of L_* galaxies expected within the volume by multiplying it by $\phi_* = 0.017h^3$ Mpc^{-3}. At a redshift of $z = 1$, we find that $V(< z)\phi_* \simeq (10–30)$ depending on the cosmological model; the lower value is for $\Omega_{NR} = 1$, $\Omega_V = 0$ and the higher one is for $\Omega_{NR} = 0.3$, $\Omega_V = 0.7$. This increases to $V(< z)\phi_* \simeq (70–200)$ at $z = 3$ and $V(< z)\phi_* \simeq (100–300)$ at $z = 6$. A comparison of the actual number of galaxies detected with the expected number of L_* galaxies provides the measure of the distribution of galaxies with different luminosities. The actual galaxy counts in the K, I, B, and U bands are shown in Fig. 1.24 of Chap. 1. The smooth curves are based on a no-evolution model. It is obvious that although the K-band counts are reasonably consistent with a no-evolution model (in which the current galaxy population is moved to the high redshift without any change), there is a significant excess of faint galaxies in the B and U bands. This broadly indicates the existence of star-burst galaxies that have young stars. However, to maintain the overall number density, it is necessary to probably invoke a population of dwarf galaxies undergoing star bursts. Alternatively, a large amount of galaxy–galaxy merging could account for the decrease in the number of individual galaxies as we move from high to low redshifts. There have also been attempts to explain the counts by pure luminosity evolution as well as by invoking a new population of galaxies. Almost all these models run into difficulties either with the count of low-luminosity blue galaxies or with high-redshift

Lyman break galaxies. In short, these counts suggest that the galaxy evolution is a fairly complex process, and its details cannot be captured with simple models that have two parameters.

It is, of course, possible to obtain some important results directly from the HDF counts. If $n(F_\nu) \equiv (dN/d\Omega\, dF_\nu)$ denotes the number of sources with the flux densities between F_ν and $F_\nu + dF_\nu$ per solid angle in the sky, the total intensity of radiation that is due to all the sources is

$$I_\nu = \int_0^\infty F_\nu\, n(F_\nu)\, dF_\nu. \tag{7.97}$$

If apparent magnitudes are used instead of fluxes, they can be related by $F_\nu \propto 10^{-0.4m_\nu}$ with the constant of proportionality, depending on the passband in which observations are carried out. In this case, the integrand in Eq. (7.97) is proportional to $10^{-0.4m}(dN/d\Omega\, dm)$. The bottom panel of Fig. 1.24 of Chap. 1 plots these counts in a more physically relevant manner by multiplying by the factor $10^{-0.4m}$. The area under the curve in this figure indicates the total amount of intensity contributed by these galaxies. The figure suggests that when galaxies up to the limiting magnitudes are taken into consideration, the total integrated contribution of light has effectively converged.[4] Actual integration shows that νI_ν varies in the range 2.9–9.0 nW m^{-2} sr^{-1} in the wavelength band 3600–16,000 Å.

Exercise 7.6

Galaxy number counts: At the B magnitude of $m_B = 29$, the HDF results give $dN/dm = 3.47 \times 10^5$ galaxies per magnitude per square degree.

(a) Determine $dN/d\ln S$, where S is the blue bandpass flux of the HST, assuming that the universe is isotropic. Next obtain the total number of galaxies in the universe $N(S)$ visible to us with a limiting magnitude of $m_B = 29$ if the source count can be approximated to have the form $N \propto S^{-\beta}$.

(b) Suppose that the blue-band luminosity function for galaxies is a Schecter luminosity function with $\alpha = -1$, $\phi_* = 0.015h^3$ Mpc^{-3} and $M_* = -19.5 + 5\log h$. At $m_B = 29$, estimate the luminosity distance d_{29}^* (in h^{-1} Mpc) for a galaxy of luminosity L_*.

(c) Show that the differential count at $m_B = 29$ in a flat universe with $\alpha = -1$ is given by

$$-\frac{dN}{d\ln S} = 4\pi \left(\frac{c}{H_0}\right)^3 \int_0^{H_0\tau_0} \exp[-(d_L/d_{29}^*)^2]x^2\, dx, \tag{7.98}$$

where $x = H_0\chi/c$ and χ is the dimensionless radial coordinate. The luminosity distance d_L is treated as a function of χ in this expression. For the flat matter-dominated universe, $H_0 d_L/c = 4x/(2-x)^2$. Use this and integrate Eq. (7.98) numerically to determine the predicted source count at $m_B = 29$. How does it compare with the HDF measurement?

The UV emission from the galaxies is directly connected to the formation of metals as the same massive stars that produced the bulk of the UV photons also make the metals and disperse them through supernova. In contrast, the relation between UV emission and total SFR is lot more problematic as the low-mass stars of any standard IMF contains most of the mass whereas it is the high-mass stars that produce most of the UV emission and the metals. Nevertheless, there have been several attempts to connect the luminosity density seen in a HDF with the results of low-redshift surveys and obtain an overall picture of the SFR as a function of redshift. These approaches estimate the amount of stars produced in the universe at any given epoch from the abundance of UV luminosity of the star-forming galaxies. (The counts of Lyman break galaxies can be used to estimate the UV luminosity density that, in turn, can be converted to star-formation density.) One simple phenomenological relationship between the SFR and far-UV and H-α luminosities is given by

$$SFR(M_\odot \text{ yr}^{-1}) = 1.4 \times 10^{-28} L_{\text{FUV}}(\text{ergs s}^{-1} \text{ Hz}^{-1}), \qquad (7.99)$$

$$SFR(M_\odot \text{ yr}^{-1}) = 7.9 \times 10^{-42} L_{\text{H-}\alpha}(\text{ergs s}^{-1}). \qquad (7.100)$$

Both these relations apply to continuous star formation with a Salpeter IMF between 0.1 and 100 $M_\odot$. The key difficulty in using these relations arises from the unknown amount of dust obscuration. If the ISM of PGs contains a large amount of dust that absorbs UV and optical radiation and converts it into IR radiation, then the estimated SFR will be lower than the actual rate. In fact, the presence of dust in galaxies can lead to anything between 50% and 80% of UV light being absorbed. There is also some uncertainty arising from the fact that the stellar IMF is poorly known at the low-mass end where most of the mass is contained.

If we assume that the properties of the dust and intrinsic spectra of galaxies at high redshifts are similar to those of star-burst galaxies in the local universe, then the correction is approximately a factor of 5 at $z = 3$. Different models and assumptions could vary this correction factor anywhere from 2 to 15. This, in turn, suggests that the SFR observed in the universe is consistent with the assumption of near constancy at $\sim 10^{-0.8}$ $M_\odot$ yr^{-1} Mpc^{-3} in a redshift range of $(1 \lesssim z \lesssim 4.5)$. One convenient way of parameterising the SFR as a function of redshift is through a fit

$$\dot\rho_{\text{sfr}} = \frac{Ae^{az}}{e^{bz} + B} \qquad (7.101)$$

with some constants A, B, a, and b. Direct translation of UV luminosities suggests the parameters $a = 3.4, b = 3.8, B = 44.7$, and $A = 0.11$ $M_\odot$ Mpc^{-3} yr^{-1}. Correcting for the dust could change the fit parameters to $a = 2.2, b = 2.2, B = 6$, and $A = 0.13$ $M_\odot$ Mpc^{-3}. In the first case, the star formation decreases exponentially at high redshifts, whereas in the second case it stays nearly constant with redshift. (Of course, these fits are valid in only the range $z \lesssim 5$.)

Dividing $\dot{\rho}_{sfr}$ by $\rho_c = 2.8 \times 10^{11} h^2 \, M_\odot \, \text{Mpc}^{-3}$, we get the rate of change of Ω_{sfr} with respect to time. If we multiply this quantity by $H^{-1}(z)$, we will get a measure of the amount of matter that is being turned into stars over a Hubble time at any given redshift. (This illustrates the physical process more than the integral over the SFR that is constrained to be monotonically increasing.) These quantities were plotted in Fig. 1.11 for the two sets of fitting functions, covering an extreme range of assumptions regarding the dust.

These curves illustrate several interesting points. First, note that most of the star-formation activity is in the recent past; the peak activity over a Hubble time is around $z \approx 1.2$ (which corresponds to a cosmic time of $\sim 3 h_{50}^{-1}$ Gyr). This feature is more emphatically brought out by the time scales shown in the lower frames. The universe spends less time at high redshifts and is treated as a function of *time* rather than of redshift, so much of the star formation has occurred recently. Second, we can estimate the total star formation up to any given time by integrating the SFR. The results could vary by a factor of a few between the two curves but the typical value at $z = 0$ is $\Omega_* \approx 5 \times 10^{-3}$, which is adequate to account for the entire stellar content in disks and spheroids of the present-day universe. Half of this value is reached at a redshift of $z \approx 2$ in these models.

The stars will also produce the metals in the universe and we could ask for the amount of metal enrichment at any redshift z that is due to a given SFR. From standard stellar-evolution theory, we could make a rough estimate that the rate of production of metals that are ejected to enrich the surrounding medium as $\dot{\rho}_{metals} \approx 2.4 \times 10^{-2} \dot{\rho}_*$. For the SFR given above, the amount of metals produced by $z = 2.5$ is $\sim 0.04 \, Z_\odot$, which is in rough agreement with the measurement of metallicity in quasar absorption systems at high redshifts. We shall discuss the connection among SFR, metallicity, and the EBL in more detail in Chap. 9.

In fact, this case could also account for, quite naturally, the existence of the cosmic infrared background radiation (CIBR). Observations suggest that this background has an intensity of $\nu I_\nu \approx (24 \pm 5) \, \text{nW} \, \text{m}^{-2} \, \text{sr}^{-1}$ at $\lambda = 140 \, \mu\text{m}$ and $\nu I_\nu \approx (15 \pm 5) \, \text{nW} \, \text{m}^{-2} \, \text{sr}^{-1}$ at $\lambda = 240 \, \mu\text{m}$. The intensity is probably $\nu I_\nu \approx (40 \pm 5) \, \text{nW} \, \text{m}^{-2} \, \text{sr}^{-1}$ in the wavelength band $\lambda = (100\text{--}1000) \mu\text{m}$ (see Chap. 1, Section 1.14). This flux is larger than the "optical" background of $\nu I_\nu \approx 17 \, \text{nW} \, \text{m}^{-2} \, \text{sr}^{-1}$ that we obtain by counting all the galaxies detected by the HST in the band between 0.3 and 3 μm.

The general procedure for relating background radiation to any energy-generation process is fairly straightforward: The differential bolometric flux dF received from a comoving volume element dV located at a redshift z and containing a comoving luminosity density $\epsilon(z)$ is given by

$$dF = \frac{\epsilon(z) \, dV_c(z)}{4\pi \, d_L^2(z)}, \qquad (7.102)$$

where $d_L(z)$ is the luminosity distance to the volume element and $dV_c = a^{-1}c\,dtr_{\rm em}^2\delta\Omega$ is the comoving volume element along a solid angle $\delta\Omega$. [The *proper* volume element is given by Eq. (3.67); dividing by a^3 gives dV_c.] Using $d_L = a_0 r_{\rm em}(1+z)$, we get

$$\frac{dV_c}{4\pi\,d_L^2} = \left(\frac{\delta\Omega}{4\pi}\right)\left|\frac{c\,dt}{dz}\right|\frac{dz}{1+z}. \qquad (7.103)$$

The total intensity (integrated over all frequencies) received from the sources lying within a solid angle $\delta\Omega$ is given by the integral

$$I = \left(\frac{c}{4\pi}\right)\int_0^{z_*}\epsilon(z)\left|\frac{dt}{dz}\right|\frac{dz}{1+z}, \qquad (7.104)$$

where z_* is the redshift at which the process was turned on and

$$\left|\frac{dt}{dz}\right|^{-1} = H_0(1+z)[(1+z)^2(1+\Omega_{\rm NR}z) - z(2+z)\Omega_V]^{1/2}$$

$$\equiv H_0\mathcal{G}(\Omega_{\rm NR},\Omega_V,z). \qquad (7.105)$$

The corresponding energy density is given by $4\pi I/c$ and the contribution to the density parameter will be

$$\Omega_R = \left(\frac{4\pi}{c}\right)\frac{I}{\rho_c c^2} = 1.0\times 10^{-7}I(\mathrm{nW\ m^{-2}\ sr^{-2}})h_{50}^{-2}. \qquad (7.106)$$

Note that I has the units $\mathrm{nW\ m^{-2}\ sr^{-1}}$ with the conversion $1\ \mathrm{nW\ m^{-2}\ sr^{-1}} = 10^{-6}\ \mathrm{ergs\ cm^{-2}\ s^{-1}\ sr^{-1}}$; the frequency-integrated intensity I has the same dimensions as νI_ν and can be used interchangeably provided that the spectrum is sharply peaked or when the cumulative intensity is needed only approximately.

To see what these numbers imply, let us make some simple estimates based on fiducial star-formation cases. If a fraction f_* of the mass density in baryons,[5]

$$\rho_B = \frac{3H_0^2(1+z)^3}{8\pi G}\Omega_B \simeq 7\times 10^{10}(1+z)^3\,h_{50}^2\,\Omega_B\,M_\odot\,\mathrm{Mpc^{-3}}, \qquad (7.107)$$

undergoes processing, either through star formation or by gravitational accretion, with a radiative efficiency ϵ, then the locally observed energy density of photons will be

$$\rho_\gamma \stackrel{\simeq}{=} \rho_B\frac{c^2\epsilon f_*}{(1+z)^4}$$

$$\simeq 5\times 10^{-30}\left(\frac{\Omega_B h_{50}^2}{0.05}\right)\left(\frac{f_*}{0.1}\right)\left[\frac{2.5}{(1+z_*)}\right]^4\left(\frac{\epsilon}{0.001}\right)\mathrm{gm\ cm^{-3}}. \qquad (7.108)$$

The intensity corresponding to this photon density will be $I = (c/4\pi)(\rho_\gamma c^2)$;

approximating this total energy to νI_ν, we get

$$\nu\, I(\nu) \simeq 20 \left(\frac{\Omega_B h_{50}^2}{0.05}\right) \left(\frac{f_*}{0.1}\right) \left[\frac{2.5}{(1+z_*)}\right] \left(\frac{\epsilon}{0.001}\right) nW\ m^{-2}\ sr^{-1}, \quad (7.109)$$

which may be compared with the optical–UV background with a bolometric intensity of

$$\nu I_\nu|_{opt} \approx (17 \pm 3)\ nW\ m^{-2}\ sr^{-1}, \quad (7.110)$$

in the range $\lambda = 0.1$–$7\ \mu m$, which is due to distant galaxies. For a Salpeter IMF with a lower cutoff of $0.1\ M_\odot$, the efficiency is $\epsilon \approx 10^{-3}$.

In this picture, the optical emission arises from quiescent star formation in disk galaxies by intermediate- and low-mass stars. This is consistent with the assumption that 10% of all baryons are processed in the low-mass stars at $z_* \simeq 1.5$. The local density in low-mass stars will then be

$$\rho_B(stars) \simeq 7 \times 10^{10} f_* \Omega_B \simeq 3.4 \times 10^8\ M_\odot\ Mpc^{-3}, \quad (7.111)$$

which is consistent with observations based on photometric surveys and standard M/L ratios. Further, assuming solar metallicity, we can estimate the local density in metals to be

$$\rho_Z(stars) \simeq 1.6 \times 10^9 f_* \frac{Z}{Z_\odot} \Omega_B\ M_\odot\ Mpc^{-3} \simeq 7.7 \times 10^6\ M_\odot\ Mpc^{-3}. \quad (7.112)$$

However, if we further assume that the same process of star formation at $z_* \simeq 1.5$ has to produce CIBR, through reprocessing by dust, then we need a factor of ~ 2 increase in the efficiency. This is because the FIR background between 7 and $1000\ \mu m$ is $\nu I_\nu \approx 40\ nW\ m^{-2}\ sr^{-1}$, which is roughly twice as large as the optical background. (It is likely that most AGN have dusty gas that absorbs the power from the central source and emits in the FIR. However, this contributes only $\sim 10\%$–20% at most of the diffuse FIR background.) One possible way is to raise the lower cutoff of the Salpeter IMF to $2\ M_\odot$, thereby getting $\epsilon \simeq 2 \times 10^{-3}$ and

$$\nu\, I(\nu)|_{FIR} \simeq 40 \left(\frac{\Omega_B h_{50}^2}{0.05}\right) \left(\frac{f_*}{0.1}\right) \left[\frac{2.5}{(1+z_*)}\right] \left(\frac{\epsilon}{0.002}\right) nW\ m^{-2}\ sr^{-1}. \quad (7.113)$$

In other words, the star-bursting phase should have higher efficiency compared with the quiescent phase and the former should account for most of the CIBR.

A class of objects called *ultraluminous IR galaxies* are worth mentioning in this context. These systems are powered largely by star formation (rather than by embedded AGN) and have 100–$1000\ M_\odot\ yr^{-1}$ of star formation. Mapping in the NIR shows the de Vaucouleurs profile, and mapping with CO suggests a cold disk of molecular gas with $M \approx 10^{10}\ M_\odot$ within a few hundred parsecs. This star-burst population could account for a fair amount of CIBR. In such a case, we would expect a contribution of ~ 1 nW m^{-2} sr^{-1} to EBL at $\lambda \approx 1$ mm. The CMB flux at this wavelength is ~ 2000 nW m^{-2} sr^{-1} but because we can measure fluctuations at the level of 10^{-6} in the CMB, it may be possible to measure a population of ultraluminous FIR sources at high z.

The preceding discussion uses somewhat extreme values but it illustrates the close interplay among star-formation efficiency (which in turn is determined by the poorly known high-z IMF), the intensity of extragalactic background radiation, and the density of low-mass stars in the local neighbourhood. Any scenario should be able to account for all these observations. We will discuss these issues in greater detail in Chap. 9.

Finally, we mention a completely different approach to the detection of high-redshift galaxies. In the nearby star-burst galaxies, low-level rotational transitions of CO give luminosities of the order of $10^9 \, L_\odot$. This opens up the possibilities of detecting similar star-burst galaxies at high redshifts ($z = 10$–20) through the rotational transitions of CO in which a transition from level $(J + 1)$ to level J will lead to radiation at a frequency of $\omega \simeq 115(J + 1)$ GHz. If such star-burst galaxies, with a comoving density ρ similar to that of massive galaxies today, have existed at $z = 10$–20, we can easily compute their characteristic surface density in the sky by using Eq. (3.152) of Chap. 3:

$$\frac{dN}{d\Omega \, dz} = \frac{n(z) a_0^2 r_{\rm em}^2(z) \, d_H(z)}{(1 + z)^3}. \tag{7.114}$$

If these sources have a luminosity L, then the bolometric flux received from them will be

$$\frac{dS}{dz} = \frac{L}{d_L^2} \frac{dN}{d\Omega \, dz} = \frac{Ln(z)}{(1 + z)^3} \left(\frac{a_0 r_{\rm em}}{d_L} \right)^2 d_H(z) = Ln(z) \frac{d_H(z)}{(1 + z)^5}. \tag{7.115}$$

Specialising to an $\Omega = 1$ universe with a conserved population of sources having $n(z) = n_0(1 + z)^3$ and converting dz to the bandwidth of observation $d\nu$, we can easily show that the flux is given by

$$\Delta S_{\rm bol} = n_0 \, L H_0^{-1} (1 + z)^{-2.5} \frac{\Delta \nu}{\nu}. \tag{7.116}$$

If $n_0 = 0.01 h^3$ Mpc^{-3} and the bandwidth is 3×10^{-3}, there is approximately one object per 30 arcmin^2 of the sky. If the velocity width of the objects is σ, leading to a fractional linewidth ϵ, then we have the relation $S_\nu \simeq S_{\rm bol}/(\sqrt{2\pi}\nu\epsilon)$. Collecting together all these results and substituting in numbers, we find that

$$S_\nu \approx 500 \text{ mJy } (1 + z)^{-2.5} L_9 \, \sigma_{100}^{-1} \nu_{\rm GHz}^{-1}. \tag{7.117}$$

Although this requires submillijansky sensitivity for sources at $z \approx 10$, it is a potentially feasible observation. The difficulty, however, arises from the fact that the equivalent temperatures of the energy differences between the $J = 1, 2, \ldots,$ and $J = 0$ levels are approximately 5.5 K, 16.5 K, $\ldots$, etc. At low redshifts, the

CMB temperature is small enough to allow collisional effects to dominate the level populations. However, at $z = 10$, the CMB temperature ~ 30 K, thereby completely depopulating the lower levels and leaving only the weaker, higher-order transitions. This is, in fact, a general difficulty in observing many molecular transitions from high redshifts.

7.7 Galaxy Distributions in Projection

The collapse of overdense regions, the formation of structures, etc., are most conveniently described in terms of what happens in three-dimensional space at a given time t or equivalently at a given redshift z. Observationally, however, we probe structures along the backward light cone with objects that are farther away, being seen earlier on in the evolution. What is more, these observations usually use the redshift as a distance indicator, which is correct only if the peculiar velocities are ignored. In the study of inhomogeneities, we are interested in the density perturbations at least to linear order, and because peculiar velocities are of the same order as that of the density perturbations, we cannot ignore their effect in observations. We shall now see how this can be taken into account.

Let us consider the observations in a narrow cone of solid angle around some direction that we shall take to be the x axis. If the displacement field of particles is $\mathbf{x}$ and the velocity field is $\mathbf{u}$, then, as we have seen in Chap. 5, Section 5.6, in linear theory $\mathbf{u} = Hf(\Omega)\mathbf{x}$, where $f(\Omega) = \Omega^{0.6}$. Because the velocity v changes the redshift by replacing $(1 + z)$ with $(1 + z)(1 + v/c)$, the actual position vector of a particle $\mathbf{r}$ and its apparent position $\mathbf{r}_{\text{app}}$ (calculated from the observed redshift by the Hubble law), will be related by

$$\mathbf{r}_{\text{apparent}} = \mathbf{r} + \left(\frac{\hat{\mathbf{r}} \cdot \mathbf{u}}{H}\right) \hat{\mathbf{r}} = \mathbf{r} + \left(\frac{\mu u}{H}\right) \hat{\mathbf{r}}, \tag{7.118}$$

where $\cos^{-1} \mu$ is the angle between the velocity vector and the line of sight. If a plane-wave disturbance propagating at some angle to the line of sight produces a comoving displacement field $\mathbf{x}$ parallel to $\mathbf{k}$, then the apparent displacement produced will be $[\mathbf{x} + f(\Omega)\mu \, x\hat{\mathbf{r}}]$ and the component of this displacement along the wave vector will be $x[1 + f(\Omega)\mu^2]$. In the Zeldovich approximation we are using, the apparent density perturbation δ_{red} that is due to this mode is directly proportional to the amplitude of the apparent displacement. It follows that

$$\delta_{\text{red}}(k) = \delta_{\text{true}}(k)[1 + f(\Omega)\mu^2]. \tag{7.119}$$

The two power spectra differ by the factor $(1 + f(\Omega)\mu^2)^2$; the effect is to produce an overall enhancement and add a quadrupole term.

The preceding formula, of course, cannot be used at small scales where non-linearities are significant. The main effect at those scales is a modification of the density contrast because of the pairwise velocity dispersion along the line of sight. The actual form of the modifications will then depend on the distribution of relative radial pair velocities and will be a convolution in the Fourier space. If we take the distribution to be a Gaussian with some width σ, then we only need to make the replacement:

$$\delta_k \rightarrow \delta_k \, \exp\left(-\frac{k^2 \mu^2 \sigma^2}{2}\right). \tag{7.120}$$

An alternative expression often uses the distribution of relative pair velocities to be a pairwise exponential in real space, leading to the replacement

$$\delta_k \rightarrow \delta_k \left(1 + \frac{k^2 \mu^2 \sigma^2}{2}\right)^{-1/2}. \tag{7.121}$$

When the effects at both large and small scales are taken into account, the overall modification to the power spectrum is given by

$$\frac{P_s}{P_r} = \frac{[1 + f(\Omega)\mu^2]^2}{(1 + k^2 \mu^2 \sigma^2/2)}. \tag{7.122}$$

The power spectrum described here should correspond to the underlying mass distribution as the dynamical input used in the analysis, $u = Hf(\Omega)x$, assumes that all the gravitating matter is taken into account. The galaxy surveys, however, measure the visible matter, which is related in a fairly complicated manner to the distribution of underlying dark matter. The relationship between the two is totally uncertain, although it is often assumed in the literature that we can relate them linearly by taking $\delta^{\text{light}} = b\delta^{\text{mass}}$, where b is called the *bias parameter*. In that case, the expression in the linear limit will be modified, with $f(\Omega)$ replaced with $f(\Omega)/b$.

It is also possible to consider distribution functions of galaxies in the projected sky plane by integration along the redshift axis. For example, if the transverse and the radial separations are denoted by $r_\perp$ and $r_\parallel$, respectively, then the projected transverse correlation function $w_p(r_\perp)$ can be expressed in terms of the true correlation function $\xi(\mathbf{r})$ by

$$w_p(r_\perp) \equiv \int_{-\infty}^{\infty} \xi(r_\perp, r_\parallel) \, dr_\parallel = 2 \int_{r_\perp}^{\infty} \xi(r) \frac{r \, dr}{\left(r^2 - r_\perp^2\right)^{1/2}}. \tag{7.123}$$

This equation can be inverted to obtain

$$\xi(r) = -\frac{1}{\pi} \int_r^{\infty} w_p'(y) \frac{dy}{(y^2 - r^2)^{1/2}}. \tag{7.124}$$

Thus, in principle, the correlation function in the sky plane does contain complete information and is independent of redshift space distortions. Unfortunately, the integral inversion to obtain $\xi(r)$ is highly unstable and is difficult to use.

More generally, we may lack any distance information regarding the galaxy distribution and might know the distribution of galaxies in the celestial sphere only as a function of the angular coordinates θ and ϕ. This will allow us to compute the angular correlation function in the sky that for isotropic distribution, will be a function of only a single angular variable θ. To extract the actual power spectrum from the angular correlation function $w(\theta)$, we may proceed as follows: If the actual density contrast on the celestial sphere is $\delta(\hat{\mathbf{q}})$, where $\hat{\mathbf{q}}$ is a unit vector towards a particular direction in the sky corresponding to angles θ and ϕ, then we can expand it in spherical harmonics as

$$\delta(\hat{\mathbf{q}}) = \sum a_{lm}\, Y_{lm}(\hat{\mathbf{q}}) = \sum a_{lm}\, Y_{lm}(\theta, \phi). \qquad (7.125)$$

The inverse relation will be

$$a_{lm} = \int \delta(\hat{\mathbf{q}})\, Y_{lm}^*\, d^2q. \qquad (7.126)$$

We can now compute the correlation function $\langle \delta(\hat{\mathbf{q}})\delta(\hat{\mathbf{q}}')\rangle$ by integrating over the two-sphere with $\hat{\mathbf{q}} \cdot \hat{\mathbf{q}}' = \cos\theta$ held constant. This calculation is similar to the one carried out in Chap. 6. The analogue of the standard result between the correlation function and the power spectrum will now be

$$w(\theta) = \frac{1}{4\pi} \sum_l \sum_{m=-l}^{m=+l} |a_{lm}|^2 P_l(\cos\theta), \quad |a_{lm}|^2 = 2\pi \int_{-1}^{1} w(\theta) P_l(\cos\theta)\, d\cos\theta. \qquad (7.127)$$

This expression simplifies considerably when we look at a small patch in the sky and θ is small. Most of the contribution now comes from large l, and we can use the asymptotic expansions

$$P_l(\cos\theta) \simeq \sqrt{\frac{2}{\pi l \sin\theta}} \cos\left[\left(l + \frac{1}{2}\right)\theta - \frac{1}{4}\pi\right], \quad J_0(z) \simeq \sqrt{\frac{2}{\pi z}} \cos\left[z - \frac{1}{4}\pi\right]. \qquad (7.128)$$

The result will then be

$$w(\theta) = \int_0^\infty \Delta_\theta^2(K) J_0(K\theta) \frac{dK}{K}, \quad \Delta_\theta^2\left(K = l + \frac{1}{2}\right) = \frac{2l+1}{8\pi} \sum_m |a_{lm}|^2. \qquad (7.129)$$

What is of considerable interest is the relationship between the angular correlation function $w(\theta)$ and the spatial density perturbation $\delta(\mathbf{y})$ when the objects are selected according to a particular weightage given along the radial direction. (This could arise, for example, because of the sensitivity of the galaxy survey.)

Denoting the weightage function by $\phi(y)$, we define the angular density contrast through the relation

$$\delta(\hat{\mathbf{q}}) = \int_0^\infty \delta(\mathbf{y}) y^2 \phi(y)\,dy, \qquad \int_0^\infty y^2 \phi(y)\,dy = 1. \qquad (7.130)$$

Physically, the function ϕ represents the comoving density of objects in the survey we obtain by integrating the luminosity function all the way to the flux limit. Expanding $\delta(\mathbf{y})$ in Fourier series and expressing the plane waves in terms of spherical Bessel functions by

$$e^{ikr\cos\theta} = \sum_0^\infty (2l+1)\,i^l\,P_l(\cos\theta)\,j_l(kr), \qquad (7.131)$$

we can express the angular correlation function $w(\theta) = \langle\delta(\hat{\mathbf{q}})\delta(\hat{\mathbf{q}}')\rangle$ in terms of a_{lm}s. The use of the identity

$$P_l(\cos\theta) = \frac{4\pi}{2l+1} \sum_{m=-l}^{m=+l} Y_{lm}^*(\hat{\mathbf{q}})Y_{lm}(\hat{\mathbf{q}}') \qquad (7.132)$$

now leads to the result

$$\langle |a_{lm}|^2\rangle = 4\pi \int \Delta^2(k)\frac{dk}{k}\left[\int y^2\phi(y)\,j_l(ky)\,dy\right]^2. \qquad (7.133)$$

This result is exact and relates the three-dimensional power spectrum to the average of angular coefficients observed in the sky. At small angles, however, it is possible to obtain a much simpler expression along the following lines. From the definition, it follows that

$$\langle\delta(\hat{\mathbf{q}}_1)\,\delta(\hat{\mathbf{q}}_2)\rangle = \iint \langle\delta(\mathbf{y}_1)\,\delta(\mathbf{y}_2)\rangle\,y_1^2\,y_2^2\,\phi(y_1)\,\phi(y_2)\,dy_1\,dy_2. \qquad (7.134)$$

Changing the variables to $y = (y_1 + y_2)/2$ and $x = (y_1 - y_2)$, assuming that significant contributions arise only when $y_1 \cong y_2 \cong y$ (which is valid if the depth of the survey is larger than any correlation length), and extending the integration over x to infinite range (valid if ϕ is a slowly varying function so that the thickness of the shell being observed is also of the order of the depth), we get

$$w(\theta) = \int_0^\infty y^4\phi^2\,dy \int_{-\infty}^\infty \xi(\sqrt{x^2 + y^2\theta^2})\,dx. \qquad (7.135)$$

The corresponding result in terms of the power spectrum is

$$w(\theta) = \int_0^\infty y^4\phi^2\,dy \int_0^\infty \pi\,\Delta^2(k)J_0(ky\theta)\frac{dk}{k^2}. \qquad (7.136)$$

As an example, consider a survey with the selection function $\phi \propto y^{-1/2} \times \exp[-(y/y^*)^2]$, which is a reasonable approximation to a distribution of sources following the Schecter function. Evaluating the integrals, we get

$$w(\theta) = \frac{\pi}{2\Gamma^2(5/4)} \int_0^\infty \Delta^2(k) \exp\left[-\left(\frac{(k\theta y^*)^2}{8}\right)\right]\left[1 - \frac{1}{8}(k\theta y^*)^2\right]\frac{dk}{k^2 y^*},$$

$$(7.137)$$

which, when compared with Eq. (7.129), immediately gives the relationship between spatial and angular power spectra:

$$\Delta_\theta^2 = \frac{\pi}{K}\int \Delta^2\left(\frac{K}{y}\right) y^5\phi^2(y)\,dy.$$

$$(7.138)$$

Exercise 7.7

Effect of curvature: Generalise the results in the text for a universe with nonzero curvature. Show that if the selection function is normalised by

$$\int_0^\infty y^2\phi(y)C(y)\,dy = 1, \qquad C(y) = \frac{1}{\sqrt{1 - ky^2}},$$

$$(7.139)$$

then the relevant equations are

$$w(\theta) = \int_0^\infty [C(y)]^2\, y^4\phi^2\,dy \int_{-\infty}^\infty \xi(\sqrt{x^2 + y^2\theta^2})\frac{dx}{C(y)},$$

$$\Delta_\theta^2 = \frac{\pi}{K}\int \Delta^2\left(\frac{K}{y}\right) C(y)y^5\phi^2(y)\,dy.$$

$$(7.140)$$

Exercise 7.8

Effect of scaling: What happens to the angular power spectrum if the depth of the survey is increased by a factor D but the form of $\phi(y)$ is kept the same? Can you think of a use for this formula?

Exercise 7.9

Power laws: Show that, if the correlation function in three dimensions is given by $\xi(r) = Br^{-\gamma}$, then the angular correlation function is given by $w(\theta) = A\theta^{1-\gamma}$, where

$$\frac{A}{B} = \frac{\Gamma(1/2)\Gamma[(\gamma - 1)/2]}{\Gamma(\gamma/2)}\frac{\int_0^\infty x^{5-\gamma}\psi^2(x)\,dx}{\left[\int_0^\infty x^2\psi(x)\,dx\right]^2} D^{*-\gamma}$$

$$(7.141)$$

and D^* is the characteristic depth of the survey with the selection function $\psi(x)$.

7.8 Magnetic Fields in the Universe

Virtually all the baryonic structures seen in the universe possess some amount of energy density in the form of magnetic fields. Unfortunately, the origin of these magnetic fields is not clearly known at any scale, with the level of uncertainty increasing at galactic and extragalactic domains. We shall briefly comment on some of these issues in this section.

Accurate measurement of magnetic fields is not an easy task, and most of the observations require some extra input for the estimate. For example, synchrotron emission and rotation measures give a handle on the magnetic field only if an independent estimate of the electron density is available. If n_e is unknown, then synchrotron emission provides an estimate of $n_e B^2$ and we require some further assumption, such as equipartition, to make an estimate of B. These measurements lead to the following results regarding the magnetic-field strength at different scales:

(1) The interstellar magnetic field in the Milky Way is approximately 3–4 μG and corresponds to an energy equipartition among the magnetic field, the cosmic rays contained in the galaxy, and small-scale turbulent motion in the ISM. The magnetic energy density is also comparable with the energy density in CMBR, although we have no reliable explanation for any of these equalities. As discussed in Vol. II, Chap. 9, Section 9.7, the field maintains its direction over a few kiloparsecs and two reversals have been observed between the galactic arms. Magnetic fields are observed in other spiral galaxies as well, although their relative strength compared with the equipartition value is somewhat uncertain.

(2) The magnetic-field strength in the intracluster medium has been measured in several Abell clusters (because of rotation measure of background sources) and varies in the range of 1–10 μG. The observations seem to fit a phenomenological formula,

$$B_{\rm ICM} \simeq 2\,\mu G \left(\frac{L}{10\ {\rm kpc}} \right)^{-1/2} (h_{50})^{-1}, \qquad (7.142)$$

where $L \approx 10$–50 kpc is the typical scale over which the field maintains directionality. There is some evidence for the strong magnetic fields at the centres with a typical strength of 10–30 μG and peak values of $\sim$70 μG. At these field strengths, the magnetic pressure can be higher than the gas pressure derived from x-ray data, suggesting the dynamical importance of the field.

(3) High-resolution rotation measures of quasars show that they host a field with a strength of approximately 0.4–4 μG over a scale of $\sim$15 kpc. The best observations are from the radio emission of quasar 3C 191 at $z = 1.945$.[6] There is also a determination of the magnetic field in a high-redshift, young, spiral galaxy at $z = 0.395$ in the range of 1–4 μG.

(4) The radio emission from distant quasars can be used to constrain the magnetic fields in the IGM, provided that some assumptions are made regarding the structure of the IGM, its ionisation fraction, and the coherence length for the magnetic field. For example, if the field is assumed to be coherent over cosmological scales, then the rotation measure of distant quasars puts a bound $B_{\rm IGM} \lesssim 10^{-11}$ G for the IGM magnetic field. A more moderate assumption, that the field is coherent over scales of $\sim$1 Mpc, suggests a less stringent limit, $B_{\rm IGM} \lesssim 10^{-9}$ G.

Models for the generation of large-scale magnetic fields attempt to start with a small-seed magnetic field and amplify it by some well-defined mechanism.

One possible choice for an amplification mechanism is the dynamo described in Vol. I, Chap. 9, Section 9.6. Quite generally, the dynamo mechanism for amplification ceases to be effective when equipartition is reached between the kinetic energy density of the small-scale turbulence and the energy density of magnetic field. This equipartition field strength is approximately 2–8 μG and could be reached starting from a seed field of, say, 10^{-20} G in a time scale of approximately 10^9–10^{10} yr. The dynamo models, however, ignore the strong amplification of small-scale magnetic fields that can reach equipartition and stop the process before a coherent field can develop over galactic length scales. It is unlikely that dynamos can explain both the field strength and the length scales of the magnetic fields observed in the universe.

An alternative view is to assume that the galactic magnetic field, say, arises directly from a primordial field when the protogalactic cloud collapses. From flux conservation, it follows that

$$B_{\text{prim},0} = B_{\text{gal}} \left(\frac{\rho_{\text{bg}}}{\rho_{\text{gal}}} \right)^{2/3}. \tag{7.143}$$

Taking $\rho_{\text{bg}}/\rho_{\text{gal}} \approx 10^{-6}$ and $B_{\text{gal}} \approx 10^{-6}$ G, we need a primordial field strength of $B_0 \approx 10^{-10}$ G in the IGM. The spatial structure has to arise from differential rotation of the galaxy, and it is difficult to explain axisymmetric fields.

We could also estimate the effect of such a primordial magnetic field of strength B_0 at some length scale l_0 on the baryons during the evolution of perturbations. The magnetic force per unit area will be comparable with the magnetic energy density, which will scale as $(1 + z)^4$. Therefore we can take $F/A \approx B_0^2(1 + z)^4/4\pi$; this force acts over an area $A \approx l_0^2(1 + z)^{-2}$. The acceleration a induced on the mass $\rho_0 l_0^3$ will be $a = F/(\rho_0 l_0^3)$. This acceleration will produce a displacement of $\delta x \simeq at^2$, where $t = (2/3)(H_0^{-1}/\Omega^{1/2})(1 + z)^{-3/2}$. The ratio of this displacement to the characteristic length scale $l_0(1 + z)^{-1}$ will provide the density contrast induced by the magnetic force. Because the force acts on only the baryons, the final density contrast has to be reduced by a factor of Ω_B/Ω. Combining all these together, we get

$$\delta_i \simeq \frac{B_0^2(1 + z)^4}{4\pi} \frac{l_0^2}{(1 + z)^2} \frac{1}{\rho_0 l_0^3} \frac{4}{9} \frac{1}{\Omega H_0^2 (1 + z)^3} \frac{1 + z}{l_0} \frac{\Omega_B}{\Omega}. \tag{7.144}$$

Assuming that the density contrast grows as $\delta \propto (1 + z)^{-1}$, we can estimate the critical magnetic-field strength at, say, $z = z_{\text{dec}}$, which is capable of producing a density contrast δ_0 at a scale l_0 today. This is given by

$$B_0 \simeq \frac{l_0 H_0^2 \delta_0^{1/2}}{G^{1/2}(1 + z_{\text{dec}})^{1/2}} \left(\frac{\Omega}{\Omega_B^{1/2}} \right) \simeq 10^{-9} \delta_0^{1/2} l_{\text{Mpc}} (\Omega h^2) \Omega_B^{-1/2} \text{ G}, \tag{7.145}$$

which is quite comparable with the primordial field estimated with Eq. (7.143).

Thus if galactic fields are amplified by adiabatic compression of a primordial field in a collapsing protogalaxy, then such a field could be dynamically significant.

Much less is known regarding the origin of a primordial magnetic field in the universe. The ideas for its creation vary from a simple-minded battery mechanism based on conventional plasma physics to high-energy particle-physics models involving phase transitions in the early universe. None of these models are completely satisfactory in explaining the field in relation (7.145), although many mechanisms can produce the seed field, and at present this issue must be left as a totally open question.[7]

8

Active Galactic Nuclei

8.1 Introduction

A very special class of cosmic sources that has attracted a considerable amount of attention is the *active galactic nuclei* (AGN), which were described briefly in Chap. 1. In this chapter, we describe different physical phenomena associated with AGN in greater detail.[1] It depends on Chaps. 3 and 7 and, to a somewhat lesser extent, on Chaps. 2 and 6.

8.2 The Black Hole Paradigm

We begin with a simple calculation of the total amount of energy density that is due to AGN by using the observed number of AGN per steradian that have a flux in the range $(F_\nu, F_\nu + dF_\nu)$. Because the energy in the photons emerging from distant cosmic sources is rarely converted into a nonelectromagnetic form, the photons we receive set a (useful) lower bound to the total amount of electromagnetic energy that was radiated in our direction. If the universe is truly homogenous and isotropic, then the number of photons here, today, is a good indicator of their mean density in the universe. It follows that the local energy density of AGN photons at frequency ν is just

$$U_\nu = \frac{4\pi}{c} \int dF_\nu \left(\frac{dN}{dF_\nu} \right) F_\nu. \tag{8.1}$$

This quantity can be calculated in any band for which survey results are available. For example, we saw in Chap. 1 that the number $N(< B)$ of quasars in the B band can be expressed in the form $\log N = \alpha m_B + \beta$ [see Eq. (1.126)]. Computing dN/dm_B, converting F_ν to m_B, and evaluating the integral in Eq. (8.1) in the range, say, $m_B = (12, 24)$, we get the total B-band flux (over 4π sr) to be approximately $I_\nu = 40$ Jy. Multiplying by the B-band frequency and dividing by c gives the energy density per logarithmic band $\nu U_\nu \approx 0.8 \times 10^{-17}$ ergs cm^{-3}. To find the energy density in all the bands, we need to repeat the exercise

in each band. However, in the absence of good-quality data in other bands
we can attempt to integrate this expression over all frequencies by applying
a bolometric correction from, say, the B-band luminosity to the bolometric
luminosity. Because of redshift, the frequency ν that is observed will correspond to the frequency $\nu(1 + z)$ at the time of emission, and the bolometric
correction will therefore depend on the redshift. Taking these into account
by a factor of ~ 30, we find that the local energy density that is due to AGN
radiation is

$$\epsilon \simeq 2 \times 10^{-16} \left(\frac{\mathcal{B}}{30} \right) \text{ergs cm}^{-3}, \tag{8.2}$$

where $\mathcal{B}$ is the bolometric and redshift correction. For comparison, note that the
energy density of starlight is $\sim 10^4$ times larger, CMBR energy density is $\sim 10^3$
times larger, and the integrated light from normal galaxies (at $M_B = 20$) is ~ 10
times larger.

If the AGN radiation was emitted through a process of accretion with efficiency
η, then the cosmological mass density accumulated by all the quasars per unit
volume is given by

$$\rho_{\text{acc}} = \frac{E_{\text{rad}}}{\eta c^2} = 2 \times 10^{13} \eta^{-1} \ M_\odot \ \text{Gpc}^{-3}. \tag{8.3}$$

This is the mass density that, when converted through a process that has efficiency
η, will lead to the observed radiation from the AGN. If the process goes on in
a wide class of field galaxies with a luminosity $L \simeq L^*$, then we can obtain
the amount of mass that each of the galaxies should provide for this process
by dividing this expression by the space density of galaxies with luminosities
within, say, one magnitude of L^*. The latter is given by $2 \times 10^6 h_{0.75}^3$ Gpc^{-3} so
that the mass contributed per galaxy is

$$M_{\text{active}} \simeq 1.6 \times 10^7 \ M_\odot \left(\frac{\mathcal{B}}{30} \right) \left(\frac{\eta}{0.1} \right)^{-1} h_{0.75}^{-3}. \tag{8.4}$$

The scaling of this expression with respect to η is noteworthy. The efficiency of
chemical reactions corresponds to ~ 1 eV per atom so that the energy efficiency
per unit rest mass is $\eta_{\text{chem}} \simeq 10^{-10}$ if the mass of the atom is $\sim 10 m_p$. In stellar
nuclear reactions, the energy released in hydrogen burning is ~ 8 MeV$/m_p$, corresponding to $\eta_{\text{nucl}} \simeq 8 \times 10^{-3}$. If such processes are to be used for powering
AGN radiation, then the active mass that is involved will be 10^{16} $M_\odot$ in chemical
reactions and $\sim 10^9$ $M_\odot$ for nuclear reactions. In contrast, accretion onto a gravitational potential well from the lowest stable orbit can lead to efficiencies of the
order of $\eta_{\text{accr}} \approx 0.1$, which is the scaling used in the expression (8.4). In other
words, among all processes that assemble a certain amount of fuel for produc-
ing radiation from a compact region, accretion will require the least amount of
fuel for a given amount of radiation. Because assembling the fuel in a compact

region and converting it into radiation is a difficult process, the problems will be aggravated in any model that uses a process with a lower value of η.

Hence we shall follow the standard paradigm of AGN being powered by an accretion disk around a massive black hole. Some of the simple estimates that can be made in this case have already been discussed in Chap. 1, Section 1.10. The more detailed description of the AGN phenomena that is attempted in this chapter is necessarily more complex. One key reason is that the phenomena associated with AGN span a wide range of length scales – nearly 10 orders of magnitude from the compact dimension of the central black hole (10^{14} cm) to the length scales of radio jets originating from the central source and flowing through the IGM (10^{24} cm). Some of the length scales and related phenomena are shown in Fig. 8.1.

Some characteristic numbers can be immediately obtained based on this paradigm. The first one is the Eddington luminosity given by

$$L_E = \frac{4\pi G M m_p c}{\sigma_T} \simeq 1.3 \times 10^{46} \left(\frac{M}{10^8 \, M_\odot} \right) \text{ergs s}^{-1} \tag{8.5}$$

from which we can obtain the time scale for radiating an amount equivalent to

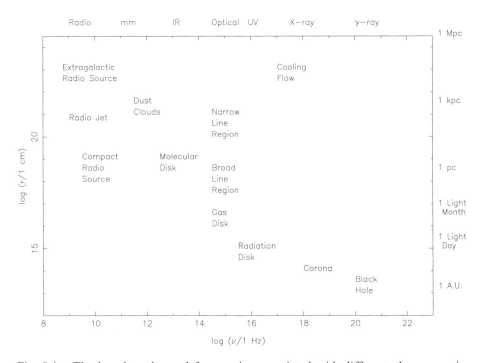

Fig. 8.1. The length scales and frequencies associated with different phenomena in AGN. (Adapted from a creation that is due to R.D. Blandford.)

the rest-mass energy M:

$$t_E = \frac{\sigma_T c}{4\pi G m_p} \simeq 4 \times 10^8 \text{ yr.}$$ (8.6)

The characteristic temperature, assuming thermal emission with the preceding luminosity from a region of radius $r_g = (GM/c^2)$, is

$$T_E = 5 \times 10^5 \left(\frac{M}{10^8 \ M_\odot}\right)^{1/2} \text{K.}$$ (8.7)

Further, if the radiation is in equipartition with a magnetic-field strength B_E, then we obtain a characteristic field of

$$B_E \simeq 4 \times 10^4 \left(\frac{M}{10^8 \ M_\odot}\right)^{-1/2} \text{G.}$$ (8.8)

The lifetime of electrons emitting synchrotron radiation in such a magnetic field will be

$$t_E^{\text{cool}} \simeq \left(\frac{m_e}{m_p}\right) \gamma^{-1} \left(\frac{r_g}{c}\right) \simeq 0.3 \gamma^{-1} \left(\frac{M}{10^8 \ M_\odot}\right) \text{s.}$$ (8.9)

Finally, we can also compute the probability of very high-energy processes like e^+–e^- pair creation taking place in such a system. The photon density in a volume scaled by r is $(L/r^2c)/\langle h\nu \rangle$. If a fraction f of the photons have energies $h\nu \approx m_e c^2$, then they can interact with a typical cross section σ_T to produce e^+–e^- pairs, provided that $f > (m_e/m_p)(L/L_E)^{-1}(r/r_g)$.

   These numbers imply that (1) thermal radiation from optically thick material at $r \approx r_g$ will be in the optical–UV region; (2) we expect high radio luminosity that is due to synchrotron emission; in fact, if most of the luminosity L is due to synchrotron radiation, then the characteristic frequency will be $\nu \simeq 2 \times 10^{14} (M/10^8 \ M_\odot)^{-5/14}$ Hz; (3) the synchrotron or Compton lifetimes of the electrons are fairly short and reacceleration mechanisms will be required; (4) if there is fair amount of γ-ray luminosity, then production of an e^+–e^- pair is a distinct possibility; and (5) The smallest time scale in the problem is the light travel time across r_g. If we assume that any variability in the radiation has to occur at a time scale larger than this, then it follows that

$$\Delta t_{\text{min}} \gtrsim \frac{2GM}{c^3} = 10^4 \left(\frac{M}{10^8 \ M_\odot}\right) \text{s} \approx 7.8 \left(\frac{L}{10^{44} \text{ergs s}^{-1}}\right) \text{s.}$$ (8.10)

We now describe the physical processes in the AGN in greater detail, starting from the optical–UV band.

8.3 Optical and UV Continua from AGN

The physics of an accretion disk around a compact object was discussed in Vol. II, Chap. 7, in the context of binary stars. The ideas can be directly generalised for a disk around a massive black hole, which is a popular model for AGN. We now review the emission of radiation from an accretion disk in the context of AGN and elaborate on some of the physical processes.

One of the key questions in modelling the disk is related to its thickness, which is measured by the dimensionless ratio (h/r), where h is the vertical thickness of the disk at the radial distance r. It was shown in Vol. II, Chap. 7, Section 7.5, that

$$h \simeq r \, \frac{c_s}{v_{\text{orb}}}, \tag{8.11}$$

where c_s is the sound speed and v_{orb} is the orbital speed of material at a radial distance r. The condition $(h/r) \ll 1$ can be expressed in the form

$$\left(\frac{k_B T}{\mu c^2}\right)\left(\frac{r}{r_g}\right) \ll 1, \tag{8.12}$$

where μ is the mean mass per particle and $r_g = (GM/c^2)$ is the Schwarzschild radius of the compact object. This condition is somewhat easy to meet in many contexts, and the simplest models for AGN are the ones in which the thin-disk approximation is valid.

If we further assume that the disk radiates locally as a blackbody with most of the radiation emerging from a region $r \simeq r_g$ then we must have $L \simeq \pi r_g^2 \sigma T^4$. This gives

$$T \simeq 10^6 \, L_{46}^{-1/4} \left(\frac{L}{L_E}\right)^{1/2} \text{K}, \tag{8.13}$$

where $L_E = 1.5 \times 10^{38}(M_{\text{BH}}/M_\odot)\,\text{ergs s}^{-1}$ is the Eddington luminosity. This is a rather moderate temperature even when $L \lesssim L_E$. Hence the blackbody emission from the disk will peak in the UV and could possibly account for the local maxima of νF_ν seen in the AGN around $h\nu \simeq 10$ eV (see Fig. 1.20). Also note that, for a fixed L/L_E, the characteristic temperature decreases with increasing luminosity – which is a feature we have already noted in Chap. 1.

In such a thin-disk approximation, it is possible to work out a model for accretion provided that some viscous dissipation is introduced. We saw in Vol. II, Chap. 7, Section 7.5 that microscopic models for disk viscosity are still very uncertain; nevertheless, it is possible to arrive at final expressions for the dissipation of energy and angular momentum, which are independent of the viscosity parameter. For example, the local rate of dissipation of energy into heat is given by

$$Q = \frac{3\,GM\dot{M}}{4\pi r^3}\,R(r), \qquad R(r) = 1 - \left(\frac{r_m}{r}\right)^{1/2}, \tag{8.14}$$

where r_m is a minimum radius at which we take the viscous stress to be zero. Equation (8.14) is valid for a Keplerian disk and was discussed in detail in Vol. II, Chap. 7, Section 7.5. The corresponding result for angular momentum flow is given by

$$\int dz\, T_{r\phi} = -\frac{\dot{M}\Omega(r)}{2\pi}\,R(r). \tag{8.15}$$

There is, however, one complication that needs to be mentioned before we proceed further. The results given above and derived in Vol. II, Chap. 7, Section 7.5, are based on a Newtonian analysis. In the proximity of a massive black hole, it is necessary to incorporate the effects of general relativity, which can be done through a fairly straightforward, but complicated, analysis.[2] In the case of a disk that is in steady state and has a rotation axis parallel to the rotation axis of the black hole, we can incorporate all the relativistic effects by modifying the function $R(r)$ in Eqs. (8.14) and (8.15) to $R_R(x)$ and $R_T(x)$, respectively, where $x = r/r_g$; the functions $R_R(x)$ and $R_T(x)$ are plotted in Fig. 8.2. The analytic

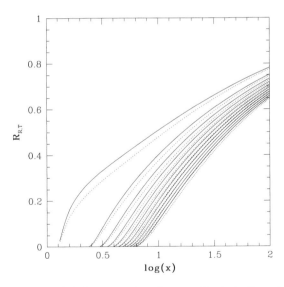

Fig. 8.2.   The general relativistic correction factors: R_R is indicated by solid curves and R_T is indicated by dotted curves. The top curve is for a maximally rotating curve black hole and the bottommost curve is for the Schwarzchild black hole. The curves in between correspond to a change of parameter values by $\Delta a = 0.1$. (Figure courtesy of Tapas Kumar Das.)

forms for these functions are

$$R_R(x) = \frac{C(x)}{B(x)}, \qquad R_T(x) = \frac{C(x)}{A(x)}, \tag{8.16}$$

with

$$A(x) = 1 - \frac{2}{x} + \frac{a_*^2}{x^2}, \qquad B(x) = 1 - \frac{3}{x} + \frac{2a_*^2}{x^{3/2}}, \tag{8.17}$$

$$C(x) = 1 - \frac{y_{ms}}{y} - \frac{3a_*}{2y} \ln\left(\frac{y}{y_{ms}}\right) - \frac{3(y_1 - a_*)^2}{yy_1(y_1 - y_2)(y_1 - y_3)} \ln\left(\frac{y - y_1}{y_{ms} - y_1}\right)$$

$$- \frac{3(y_2 - a_*)^2}{yy_2(y_2 - y_1)(y_2 - y_3)} \ln\left(\frac{y - y_2}{y_{ms} - y_2}\right) \tag{8.18}$$

$$- \frac{3(y_3 - a_*)^2}{yy_3(y_3 - y_1)(y_3 - y_2)} \ln\left(\frac{y - y_3}{y_{ms} - y_3}\right).$$

Here $y = x^{1/2}$ and a_* is the angular momentum of the black hole in normalised units; y_{ms} is the value of y at the marginally stable orbit and $y_{1,2,3}$ are the three roots of the equation $y^3 - 3y + 2a_* = 0$. The figure shows that the general relativistic correction could be a factor of few at all relevant ranges of radii. For this reason, the radius x_{max}, where the maximum dissipation per unit area occurs, is somewhat outside the marginally stable orbit. Taking the relativistic corrections into account, we can write the temperature of the disk in terms of the accretion rate as

$$T_s = \left(\frac{Q}{2\sigma}\right)^{1/4} = \left[\frac{3\,GM\dot{M}}{8\pi\sigma r^3} R_R(x)\right]^{1/4}$$

$$= 6.8 \times 10^5 \eta^{-1/4} \left(\frac{L}{L_E}\right)^{1/2} L_{46}^{-1/4} R_R^{1/4}(x) x^{-3/4} \text{ K}, \tag{8.19}$$

where $x = r/r_g$ and η is the efficiency ($\sim 10\%$) of converting the accreting rest-mass energy into radiation. The value of this temperature could differ by a factor of a few from the corresponding Newtonian value because of relativistic correction.

In the limit of a thin disk in local thermodynamic equilibrium, the corresponding total luminosity can be expressed as

$$L_\epsilon = 4\pi r_g^2 \int_{x_{max}} dx\, x \frac{2\epsilon^3/(h^3 c^2)}{\exp[\epsilon/k_B T_s(x)] - 1}$$

$$= \frac{32\pi}{3} \frac{r_g^2}{c^2 h^3} (k_B T_0)^{8/3} \epsilon^{1/3} \int_{u_{max}} du \frac{u^{5/3}}{e^u - 1}, \tag{8.20}$$

where $\epsilon = h\nu$ is the photon energy, $T_s(x) = T_0 x^{-3/4}$ [see Eq. (8.19)], and $u_{max} =$

$(\epsilon / k_B T_0) x_{\max}^{3/4}$. When $u_{\max} \ll 1$, the integral is dominated by contributions around $u \approx 1$ and can be extended from 0 to ∞; when $u_{\max} \gg 1$, the integrand is dominated by the lower bound. Taking both these limits into account, we can approximate the luminosity as

$$L_\epsilon \propto \epsilon^{1/3} \exp\left[-\frac{\epsilon}{k_B T_s(x_{\max})}\right]. \tag{8.21}$$

Thus we find that thermal radiation from a thin accretion disk should vary as $L \propto \epsilon^{1/3}$ at low frequencies.

The preceding analysis attributes a local temperature to each r, and hence specific (peak) photon frequencies translate into specific radii in the disk. If we define a radius r_ϵ such that $k_B T_s(r_\epsilon) = \epsilon$ then very little radiation at energy ϵ is emitted from larger radius $r \gtrsim r_\epsilon$. On the other hand, at smaller radii, the black body spectrum can be approximated by the Rayleigh–Jeans expression $B_\epsilon = (2k_B T_s \epsilon^2 / h^3 c^2)$, allowing us to write the total radiated energy at ϵ as

$$L_\epsilon \simeq \int_{r_{\max}}^{r_\epsilon} dr \, 2\pi r \, \frac{2\epsilon^2}{h^3 c^2} T_s(x_{\max}) \left(\frac{r}{r_{\max}}\right)^{-3/4}. \tag{8.22}$$

The integral is clearly dominated by the upper limit r_ϵ, showing that most of the radiation at energy ϵ is indeed emitted near the radius r_ϵ. If we consider, as an example, a $10^8 \, M_\odot$ black hole accreting at the Eddington rate, then the peak contribution to continuum emission at $\lambda = 1500 \, \text{Å}$ arises from $r = 50 r_g$ (which is $\sim$0.6 light days from the black hole), whereas the peak contribution to $\lambda = 5000 \, \text{Å}$ arises from $r = 240 r_g$ (which is $\sim$2.8 light days from the black hole).

Observationally, however, it is not clear whether the UV–optical continuum of AGN is consistent with the $\nu^{1/3}$ prediction. After emission lines and contamination of the continuum by starlight from the host galaxy are allowed for, power-law fits typically yield $F_\nu \propto \nu^{-0.3}$ that *decreases* with frequency. It is, however, possible to have acceptable fits to the spectrum including $\nu^{1/3}$ if it is assumed that there are other contributions to the UV continuum. For example, a power-law background continuum along with a thermal contribution to the blue bump can produce acceptable fits; similarly, a combination of a cool thermal spectrum in the IR plus an accretion disk spectrum in the UV–optical can successfully model the AGN continuum over a large spectral range.[3]

In the preceding analysis we assumed local thermodynamic equilibrium and took the radiation as emerging through the two surfaces of the thin accretion disk. At the next level, we need to understand (1) processes that could redistribute heat in the disk as well as (2) the force balance in the vertical direction. The redistribution of heat can arise from simple photon diffusion and more complicated thermal conduction and convection. The latter two processes depend crucially on the viscosity mechanism, which is quite uncertain. The exact modelling of photon diffusion is also complicated if the equations of radiative transfer are to be solved rigorously. We shall follow a simpler approach and describe the radiative diffusion when photon scattering and photon production are isotropic by (see Vol. 1, Chap. 6, Section 6.8)

$$\cos\theta \frac{\partial I}{\partial \tau} = J - I + \frac{\partial \mathcal{F}_{\text{rad}}}{\partial \tau}, \qquad (8.23)$$

where θ is the angle between the direction of photon and the normal to the disk, I is the intensity, $J = \langle I \rangle$ is the mean intensity, $\mathcal{F}_{\text{rad}}$ is the radiative flux, and τ is the optical depth that is taken to run from 0 at midplane to τ_{tot} at the outer edge of the disk. Multiplying this equation by $\cos\theta$ and integrating over angles give

$$\frac{\partial K}{\partial \tau} = -\mathcal{F}_{\text{rad}}(\tau) = -f(\tau)\mathcal{F}_{\text{rad}}(\tau_{\text{tot}}), \qquad (8.24)$$

where K is the second moment of I so that K/c is the radiation pressure. In arriving at the second equality, we have introduced a function $f(\tau)$ such that $f(0) = 0$ and $f(\tau_{\text{tot}}) = 1$. This is convenient as we know from Eq. (8.14) that the total flux can be expressed as

$$\mathcal{F}_{\text{rad}}(\tau_{\text{tot}}) = \frac{3}{8\pi} \frac{GM\dot{M}}{r^3} R_R(x). \qquad (8.25)$$

Actual determination of $f(\tau)$ will require solving the full radiative transfer problem, but even a simple extrapolation $f(\tau) = \tau/\tau_{\text{tot}}$ – which implies that dissipation per unit optical depth is a constant – will give us some insight into the problem. In this case, the solution to Eq. (8.24) is given by

$$K(\tau) = \frac{1}{2} \tau_{\text{tot}} \mathcal{F}_{\text{rad}} \left[1 - \left(\frac{\tau}{\tau_{\text{tot}}} \right)^2 \right]. \qquad (8.26)$$

That is, the radiation pressure at the disk centre is larger than the value near the outer edge by a factor τ_{tot}.

Let us next consider the vertical force balance against gravity that could be provided either by gas pressure or by radiation. In the case of gas-pressure-dominated vertical support, the equation for hydrostatic equilibrium in the vertical direction is given by

$$\frac{\partial p}{\partial z} = \rho g_z = -\frac{GM\rho z}{r^3} R_z(x). \qquad (8.27)$$

Such an equation with $R_z = 1$ was used in Vol. 2, Chap. 7, Section 7.5; the factor R_z is again a relativistic correction to the Newtonian gravitational force and is given by

$$R_z(x) = x^{-1}[L^2 - a_*^2(E_\infty - 1)], \qquad (8.28)$$

with

$$L(x) = x^{1/2}\frac{1 - 2a_*x^{-3/2} + a_*^2x^{-2}}{B^{1/2}(x)}, \qquad E_\infty(x) = \frac{1 - 2/x + a_*x^{-3/2}}{B^{1/2}(x)},$$

$$(8.29)$$

where the various terms are as defined in Eqs. (8.16). Unlike $R_{R,T}$, the function R_z is essentially unity for $x \gtrsim 10$. If the temperature is constant with height, then we can write $p = \rho k_B T/\mu$ and integrate Eq. (8.27) to get

$$\rho(z) = \rho(0)\exp\left(-\frac{GM\mu R_z}{2r^3k_B T}z^2\right). \qquad (8.30)$$

In this limit, the vertical scale height is given by

$$h = \left(\frac{2k_B Tr^3}{GMR_z\mu}\right)^{1/2} \simeq \frac{c_s}{\Omega} = 1.7 \times 10^9\, T_5^{1/2}\left(\frac{L}{L_E}\right)^{-1}L_{46}x^{3/2}R_z^{-1/2}\ \text{cm}.$$

$$(8.31)$$

The appropriate temperature to be used in this expression is, of course, $T_c \approx \tau_{\text{tot}}^{1/4}T_s$ rather than T_s.

It is possible to make a more direct estimate of the scale height if we assume that the viscosity is provided by a process parameterised by a variable α with $\nu = \alpha c_s h$ (see Vol. II, Chap. 7, Subsection 7.5.2). In this case, the vertically integrated pressure is constrained by

$$\int dz\, p(z) = \frac{\dot{M}\Omega}{2\pi\alpha}R_T(x) = \frac{2}{\eta\alpha}\frac{L}{L_E}\frac{c^2}{\kappa_T}x^{-3/2}R_T(x). \qquad (8.32)$$

Approximating the integrated pressure by $\Sigma(k_B T_c/\mu)$, where Σ is the surface density, and taking $\tau_{\text{tot}} = \kappa\Sigma/2$, we can easily show that

$$\Sigma = \kappa^{-1/5}\left[\frac{\mu\dot{M}\Omega R_T(x)}{2^{3/4}\pi\alpha k_B T_s}\right]^{4/5}. \qquad (8.33)$$

The opacity κ in the disk can be due to different physical processes in different regions. When Thomson opacity dominates over others, Eq. (8.33) becomes

$$\Sigma = 2.1 \times 10^6\alpha^{-4/5}\eta^{-3/5}\left(\frac{L}{L_E}\right)^{2/5}L_{46}^{1/5}R_T^{4/5}(x)R_R^{-1/5}(x)x^{-3/5}\ \text{gm cm}^{-2}.$$

$$(8.34)$$

Because $\kappa_T \simeq 0.33\ \text{cm}^2\ \text{gm}^{-1}$, gas-pressure-dominated disks are indeed optically thick for reasonable values of L/L_E and x. Taking $T_c \simeq \tau_{\text{tot}}^{1/4}T_s$ and using

Eq. (8.31), we get a scale height of

$$h = 2.2 \times 10^{10} \alpha^{-1/10} \eta^{-1/5} \left(\frac{L}{L_E} \right)^{-7/10}$$

$$\times L_{46}^{9/10} [R_R(x) R_T(x)]^{1/10} R_z^{-1/2}(x) x^{21/20} \text{ cm.} \qquad (8.35)$$

Given the surface density and scale height, we can also estimate the central density as

$$\rho_c = 5.6 \times 10^{-5} \alpha^{-7/10} \eta^{-2/5} \left(\frac{L}{L_E} \right)^{11/10} L_{46}^{-7/10}$$

$$\times R_T^{7/10}(x) R_R^{-3/10}(x) R_z^{1/2}(x) x^{-33/20} \text{ gm cm}^{-3}. \qquad (8.36)$$

The preceding analysis can also be used to estimate the gas pressure and the radiation pressure, and we will find that although the gas pressure increases inwards as $x^{-51/20}$, the radiation pressure increases at a faster rate of $x^{-18/5}$. (The procedure is identical to the one adopted in Vol. II, Chap. 7, Section 7.5, as regards accretion disks around stars.) Quite clearly, radiation pressure will dominate over the gas pressure near the central region. The equation for vertical pressure support in this region becomes

$$\frac{\partial p_g}{\partial z} - \frac{\kappa \rho}{c} \mathcal{F}_{\text{rad}} = -\frac{G M \rho z}{r^3} R_z(x). \qquad (8.37)$$

If we take $\mathcal{F}_{\text{rad}} = f \mathcal{F}_{\text{rad}}(\tau_{\text{tot}})$ as before and assume that radiation pressure dominates over the gas pressure, this equation reduces to

$$\frac{3}{8\pi} \frac{\kappa \dot{M}}{c} R_R(x) f(z) = z R_z(x). \qquad (8.38)$$

Note that, in this case, ρ disappears from the discussion because both radiation force and gravity are proportional to the density. This equation has the solution $f(z) = z/h$, with the vertical scale height determined as

$$h = \frac{3}{8\pi} \frac{\kappa \dot{M}}{c} \frac{R_R(x)}{R_z(x)} = 1.5 \times 10^{13} \frac{L_{46}}{\eta} \frac{\kappa}{\kappa_T} \frac{R_R(x)}{R_z(x)} \text{ cm.} \qquad (8.39)$$

In the final expression, we have scaled the opacity to κ_T, although, in general, other mechanisms could contribute. Note that this scale height depends only weakly on r through the relativistic correction; it is also independent of the uncertain viscous parameter α. The flaring angle of the radiation-pressure-dominated region is fixed by the ratio

$$\frac{h}{r} = \frac{3}{2} \frac{\kappa}{\kappa_T} \frac{L}{L_E} \frac{R_R(x)}{\eta x R_z(x)}, \qquad (8.40)$$

which can be $\mathcal{O}(1)$ if $L \approx L_E$.

We can also use this estimate to decide on the transition point between the radiation-dominated and the gas-pressure-dominated regions. Obviously, whichever process that produces larger value of h will be the dominant process at a given radius. Comparing the scale heights in the two cases, we find that the transition occurs around

$$x_{\mathrm{rg}} \simeq 1000 \, \alpha^{2/21} L_{46}^{2/21} \left(\frac{L}{L_E} \right)^{2/3} \left(\frac{\kappa}{\kappa_T} \right)^{20/21}. \tag{8.41}$$

This location is only weakly dependent on the total luminosity or α but moves outwards with increasing L/L_E.

The preceding estimates can also be used to determine the role of free–free and bound–free opacity mechanisms on the accretion disks. The free–free opacity is given by (see Vol. 1, Chap. 6, Subsection 6.9.4)

$$\kappa_{\mathrm{ff}} = \frac{256\pi^{5/2}}{3^{3/2}} Z^2 \alpha_{\mathrm{fs}} a_0^5 \left(\frac{I_H}{k_B T} \right)^{7/2} \frac{\rho g(\epsilon, T)}{\mu_e \mu_z} \epsilon^{-3} (1 - e^{-\epsilon})$$

$$= 4.90 \times 10^7 g(\epsilon, T) \rho T_5^{-7/2} \epsilon^{-3} (1 - e^{-\epsilon}) \mathrm{cm}^2 \, \mathrm{gm}^{-1}, \tag{8.42}$$

where $\epsilon = \hbar\omega/k_B T$, $\alpha_{\mathrm{fs}} = (e^2/\hbar c)$ is the fine-structure constant, a_0 is the Bohr radius, I_H is the ionisation potential for Hydrogen, μ_z is the mass per ion of atomic number Z, and g is the Gaunt factor that is generally of the order of unity in the relevant range. The numerical estimate is for a plasma with normal cosmic abundance in which both H and He are fully ionised. The optical depth evaluated with this opacity, for a radiation-dominated disk, is

$$\tau_{\mathrm{ff}} \simeq 1.1 \times 10^{-4} \frac{1 - \exp(-\epsilon)}{\epsilon^3} \frac{g(\epsilon, T)}{T_5^{7/2}} \frac{\eta^3}{\alpha^2} \left(\frac{L}{L_E} \right)^{-2} L_{46}^{-1} x^3 \frac{R_z^3 R_T^2}{T_R^5}. \tag{8.43}$$

The numerical value clearly shows that free–free opacity is subdominant to the electron-scattering opacity. In other words, in the inner portions of the accretion disks, photons with $\epsilon \approx 1$ in any local region will not be stopped by free–free absorption.

Let us next consider the bound–free opacity that is due to hydrogenlike atoms that have only a single electron. If the ionisation balance is in thermodynamic equilibrium, the cross section for ionisation from a state with quantum number n is (see Vol. II, Chap. 6, Section 6.12)

$$\sigma_{\mathrm{bf}} = \frac{64\pi}{3\sqrt{3}} g \frac{\alpha_{\mathrm{fs}}}{Z^2} n a_0^2 \left(\frac{E}{E_n} \right)^{-3}, \qquad E_n = \left(\frac{1}{2} \right) \left(\frac{m_e c^2}{n^2} \right) (Z\alpha_{\mathrm{fs}})^2. \tag{8.44}$$

The population of atoms in the energy level n is given by the Saha equation:

$$\frac{N}{N_Z} = n_e n^2 \frac{(2\pi)^{3/2} \hbar^3}{(m_e k_B T)^{3/2}} \exp\left(\frac{E_n}{k_B T} \right), \tag{8.45}$$

where N is the density of hydrogenlike ions (or neutral H if $Z = 1$) and N_Z is the density of bare nuclei with Z protons. We can now easily evaluate the total opacity as

$$\kappa_{bf} = 1.1 \times 10^7 \, g \frac{N_Z n_e}{N_H^2} Z^{-5} n^6 \left(\frac{E_n}{k_B T} \right)^{3/2}$$

$$\times \exp\left(\frac{E_n}{k_B T} \right) \rho \left(\frac{E}{E_n} \right)^{-3} (1 - e^{-\epsilon}) \text{cm}^2 \text{ gm}^{-1}, \qquad (8.46)$$

which has a dependence similar to κ_{ff}. In fact, the ratio between bound–free opacity for hydrogenlike ions with Z protons and free–free opacity including all ions whose total density is N_i will be given by

$$\frac{\kappa_{bf}}{\kappa_{ff}} = 1.7 \frac{g_{bf}}{g_{ff}} \left(\frac{N_Z}{N_i} \right) Z^4 n^{-3} T_5^{-1} \exp\left(\frac{1.58 Z^2}{T_5 n^2} \right). \qquad (8.47)$$

This shows that, in the inner regions of AGN, the two opacities are comparable whereas, in the outer region (having a lower temperature), bound–free opacity becomes progressively more important.

The scattering processes also have the effect of enhancing the chances of absorption of photons by increasing the effective path length. If τ_s and τ_{abs} are the scattering and the absorption optical depths, respectively, then the total path length of a photon to escape is the product of number of photon scatterings $(\tau_s + \tau_{abs})^2$ and the path length between each scattering, $h/(\tau_s + \tau_{abs})$; thus the effective absorptive opacity is increased by the factor $[1 + (\tau_s/\tau_{abs})]^{1/2}$. Hence the photosphere from which the radiation emerges with a Planckian spectrum should actually be determined by the condition $\tau_{eff} \simeq 1$ rather than by $\tau_{abs} = 1$. The column density Σ_{ph} above the photosphere is then given by

$$\Sigma_{ph} \simeq [\kappa_{abs}(\kappa_{abs} + \kappa_s)]^{-1/2}. \qquad (8.48)$$

The corresponding density at the photosphere is $\rho_{ph} \simeq \Sigma_{ph}/h$.

As a specific example, let us consider the case when the dominant absorption is due to a free–free process and scattering is due to a Thomson process. Then $\kappa_{scat} \gg \kappa_{abs}$ and

$$\rho_{ph} \simeq \frac{1}{h^{2/3} [(\kappa_{abs}/\rho) \kappa_T]^{1/3}}. \qquad (8.49)$$

Numerically,

$$\rho_{ph} \simeq 9 \times 10^{-12} h_{13}^{-2/3} T_5^{7/6} \frac{\epsilon}{g^{1/3} (1 - e^{-\epsilon})^{1/3}} \text{ gm cm}^{-3}. \qquad (8.50)$$

Because the scattering opacity is quite large in the radiation-pressure-supported part of the disk, the effective optical depth in this region will be significantly enhanced over the earlier estimate:

$$\tau_{\rm eff} \simeq 0.016[g\epsilon^{-3}(1 - e^{-\epsilon})]^{1/2}\frac{\eta^2}{\alpha^{3/2}}T_5^{-7/4}\left(\frac{L}{L_E}\right)^{-3/2}L_{46}^{-1/2}x^{9/4}\frac{R_z^2 R_T^{3/2}}{R_R^{7/2}}.$$

$$(8.51)$$

This implies that local thermalisation is a good assumption over the central regions of the accretion disks.

Although the numerical values for different expressions such as optical depth, opacity, etc., have been scaled to reasonable values of AGN parameters, we should also note that there is a residual frequency dependence (on ϵ) in these expressions. This leads to a frequency dependence in the conclusions we arrive at. For example, the location of the photosphere will be different at different frequencies. To understand these effects, it is useful to define certain characteristic frequencies, which can be done if the dominant absorption process is identified. We present the results assuming free–free opacity, although it is easy to generalise the results for any other case.

In this case, we can divide the spectrum into four different regions separated by three characteristic frequencies. The lowest of the three frequencies is ϵ_0, defined as the frequency below which free–free opacity is larger than the electron-scattering opacity. From $\kappa_{\rm ff} > \kappa_T$ we obtain

$$\epsilon_0 \simeq 5.2 \times 10^{-3}\eta\alpha^{-1/2}T_5^{-7/4}\left(\frac{L}{L_E}\right)^{-1/2}L_{46}^{-1/2}x^{3/4}\frac{R_z R_T^{1/2}}{R_R^{3/2}}$$

$$(8.52)$$

if we ignore the Gaunt factors. This suggests that, at sufficiently low frequencies, free–free opacity will become dominant and the emitted spectrum will become Planckian. It is also clear that near the peak of the thermal spectrum, free–free absorption is subdominant by a fair margin.

The second characteristic frequency is ϵ_t above which even the effect of electron scattering is not sufficient to increase the path length adequately for the disk to be optically thick to absorption. This is given by

$$\epsilon_t \simeq 1.6 \times 10^{-2}\frac{\eta^2}{\alpha^{3/2}}T_5^{-7/4}\left(\frac{L}{L_E}\right)^{-3/2}L_{46}^{-1/2}x^{9/4}\frac{R_z^2 R_T^{3/2}}{R_R^{7/2}}.$$

$$(8.53)$$

This ϵ_t is only a few times bigger than ϵ_0 for the relevant range of parameters, and hence there is only a limited range of frequencies over which the scattering enhances the path length in any interesting manner. However, because the ratio $\epsilon_t/\epsilon_0 \propto (L/L_E)^{-1}x^{3/2}$, this range increases as L decreases or x increases. In this range, the spectrum will be Planckian only if we go sufficiently deep inside the disk. If not, the spectrum will be distorted from the Planckian and the detailed form needs to be determined by solving equations of radiative transfer.

The third critical frequency, ϵ_{ic}, is the one beyond which inverse Compton scattering becomes important. We can define this through the relation

$$\kappa_{ff}(\epsilon_{ic}) = \left(\frac{4k_BT}{m_ec^2}\right)\kappa_T \qquad (8.54)$$

so that ϵ_{ic} is the frequency at which the small amount of energy lost by the free–free absorption is offset by the energy gained because of inverse Compton scattering. In the range $\epsilon_{ic} < \epsilon < 4$, photons gain energy by inverse Compton scattering, but when $\epsilon > 4$ they lose energy to electrons by Compton recoil. The emitted spectrum above ϵ_{ic} is essentially optically thin thermal bremsstrahlung.

Exercise 8.1

Scaling from stars to quasars: The parameters for a thin accretion disk based on the α prescription for the viscosity was derived in Chap. 7, Subsection 7.5.2 of Vol. II. Rescale those equations for parameters appropriate for the case of a quasar and compare the results with those obtained in this section. In particular, show that the surface density Σ, pressure scale height h, density ρ, and central temperature T_c are given by

$$\Sigma = 5.2 \times 10^6 \alpha^{-4/5} \dot{M}_{26}^{7/10} M_8^{1/4} R_{14}^{-3/4} f^{14/5} \text{ gm cm}^{-2},$$

$$h = 1.7 \times 10^{11} \alpha^{-1/10} \dot{M}_{26}^{3/10} M_8^{-3/8} R_{14}^{9/8} f^{7/5} \text{ cm},$$

$$\rho = 3.1 \times 10^{-5} \alpha^{-7/10} \dot{M}_{26}^{11/20} M_8^{5/8} R_{14}^{-15/8} f^{11/5} \text{ gm cm}^{-3}, \qquad (8.55)$$

$$T_c = 1.4 \times 10^6 \alpha^{-1/5} \dot{M}_{26}^{3/10} M_8^{1/4} R_{14}^{-3/4} f^{6/5} \text{ K},$$

where $f = 1 - (6\,GM/r_c^2)^{1/2}$. These formulas were obtained assuming that Kramers opacity dominates over electron-scattering opacity. Show that the condition for this is given by

$$r \geq 5.4 \times 10^{17} \dot{M}_{26}^{2/3} M_8^{1/3} f^{8/3} \text{ cm}. \qquad (8.56)$$

Finally, show that gas pressure will dominate over radiation pressure at radii with

$$r \gtrsim 5.2 \times 10^{14} \alpha^{8/30} \dot{M}_{26}^{14/15} M_8^{1/3} f^{56/15} \text{ cm}. \qquad (8.57)$$

Hence efficient subcritical accretion will make the disk largely gas pressure dominated.

Exercise 8.2

Dust in the AGN: There is some indirect evidence that at least some of the light that is scattered from AGN is by dust rather than by free electrons. (a) How can we observationally distinguish scattering by dust and scattering by free electrons? (b) Radiation pressure acting on a dust grain can expel it from a region of high intensity; estimate the size around the central source within which you expect this effect to be important for typical parameters of AGN.

Exercise 8.3

Seeing double: A disk emits isotropically flux of radiation $l(r)$ in an emission line with rest-frame frequency ν_0. The emitting material moves with a circular velocity of $v(r)$ in the plane of the disk. The normal to the disk is inclined at angle i to the line of sight. Show that the observed emission line will have a profile given by

$$L_v \, dv = \int \frac{l(r) \, r \, dr}{[(v^2/c^2) \sin^2 i - (v - v_0)^2]^{1/2}} \, dv, \qquad (8.58)$$

where the integral is over a region such that $(v/c) \sin i \geq |v - v_0|$. Explain why this profile can exhibit a double peak for a suitable class of functions $l(r)$. (Such double peak profiles of H-α and other lines have been seen in a few cases.)

8.4 High-Energy Spectra: X Rays and Gamma Rays

We now consider higher-energy radiation from the accretion disk, which is far more complex because several processes can contribute at these energies. The most important ones are relativistic and nonrelativistic bremsstrahlung, inverse Compton scattering, and electron–positron pair processes. We begin with a description of the radiation expected in the case of inverse Compton scattering.

8.4.1 Comptonisation

The basic physics of inverse Compton scattering has been discussed in Vol. I, Chap. 6, Section 6.6. Consider the scattering between an extremely relativistic electron and a photon of initial energy k_i. In the rest frame of the electron, the energy of the photon will be $\sim \gamma k_i$. (We will use units with $c = 1$.) The exact expression for the differential (Klein–Nishina) scattering cross section between electrons and photons (see Vol. I, Chap. 6, Section 6.4) is

$$\frac{\partial \sigma_{KN}}{\partial k_f} = \frac{3}{8} \sigma_T \frac{m_e}{k_i^2} \left[\frac{k_i}{k_f} + \frac{k_f}{k_i} + \left(\frac{m_e}{k_f} - \frac{m_e}{k_i} \right)^2 - 2m_e \left(\frac{1}{k_f} - \frac{1}{k_i} \right) \right], \quad (8.59)$$

which can be approximated by the Thomson scattering cross section when the photon energy (in the electron rest frame) γk_i is small compared with the rest energy of the electron, m_e. In this frame, when $\gamma k_i \ll m_e$, the scattering does not change the photon energy, and the final energy is also γk_i. Transforming back to the lab frame will introduce another factor γ, making $k_f \approx \gamma^2 k_i$. A more precise analysis, taking into account the angular dependence of scattering and averaging over the solid angle, gives $\langle k_f / k_i \rangle = (4/3)\gamma^2$, where k_f and k_i are the final and the initial energies, respectively, of the photons (see Vol. I, Chap. 6, Section 6.6).

When $\gamma k_i / m_e$ is of the order of unity or larger, it is important to take into account the corrections to the Thomson scattering cross section and use the full Klein–Nishina formula. In this limit, the scattering cross section is $\sigma_{KN} \propto (m_e / k_i)$, so that the final energy of the scattered photon is $\sim m_e$; transforming back to lab frame, we get $\langle k_f \rangle \simeq \gamma m_e$, which shows that almost all the energy of the electron goes into the photon at a single scattering.

If the electrons have a distribution of energy, then we can average over the velocity distribution in order to find the mean increase in the energy of the photon per scattering. This is straightforward (but tedious), and the final answer, in the case of nonrelativistic electrons, is given by

$$A \equiv \left\langle \frac{k_f}{k_i} \right\rangle = 1 + 4\Theta \frac{K_3(1/\Theta)}{K_2(1/\Theta)} - \left[3\Theta + \frac{K_1(1/\Theta)}{K_2(1/\Theta)} \right] \frac{k_i}{m_e}, \qquad (8.60)$$

where K_n is the modified Bessel function of the order of n and $\Theta = k_B T/m_e c^2$. When $\Theta \ll 1$, this reduces to the form

$$A = 1 + 4\Theta - \frac{k_i}{m_e}. \qquad (8.61)$$

This result has been obtained earlier in a different form in Vol. I, Chap. 6, Eq. (6.100). Note that $A = 1$ when $\langle k_i \rangle = 4\Theta m_e$; this relation defines the Compton temperature T_c. When $T \gg T_c$, the photon energy increases by the factor of $(1 + 4\Theta)$, on the average, per scattering. This allows the photon spectrum to retain the spectral shape, although it is shifted to higher energies by a constant factor.

In the relativistic case, the average photon energy is increased by the amount $(4/3)\gamma^2$ per collision as long as the scattering cross section can be approximated as $\sigma \approx \sigma_T$. If the input photon flux is J_k, then the output can be approximated as

$$j_\epsilon = \sigma_T \int d\gamma \int dk \frac{dn_e}{d\gamma} \frac{J_k}{k} 4\pi \delta_D(4\gamma^2 k/3 - \epsilon)\epsilon, \qquad (8.62)$$

where $dn_e/d\gamma$ is the spectrum of relativistic electrons. In this case, the output spectrum depends clearly on both the seed photon spectrum and the electron distribution. If we take $(dn_e/d\gamma) \propto \gamma^{-\beta}$ between γ_{min} and γ_{max} and the seed photon energy density per unit energy range to be $U_0 k^{-\alpha}$ in the range $k_{min} < k < k_{max}$, then Eq. (8.62) gives

$$j_\epsilon \propto \epsilon^{(1-\beta)/2} \int_{k_1}^{k_2} dk \, k^{-(2+\alpha)}, \qquad (8.63)$$

where

$$k_1 = \max\left[k_{min}, \frac{3\epsilon}{4\gamma_{max}^2} \right], \qquad k_2 = \min\left[k_{max}, \frac{3\epsilon}{4\gamma_{min}^2} \right]. \qquad (8.64)$$

If the initial photon spectrum is very narrow compared with the electron distribution, the integral does not contribute any ϵ-dependent terms, and the resulting power spectrum is a power law with index $(\beta - 1)/2$ over a dynamic range $(\gamma_{max}/\gamma_{min})^2$. On the other hand, if the initial range of electron energy is narrow, then we get essentially the input power spectrum scaled up in energy by a factor of γ^2.

When we proceed from single scattering to multiple scattering of photons, the results depend on whether the medium is optically thin or thick. A range of energies can exist just above $\epsilon \approx k_B T$ (when the electron spectrum is thermal) or $\epsilon \approx \gamma_{max} m_e$ (when the electron spectrum is a power law) in which the optical depth may be small but multiply scattered photons dominate the spectrum. From the previous analysis we know that a photon injected with energy k will reach an energy ϵ after m scatterings, where $A^m \simeq (\epsilon / k)$. When the Thomson optical depth τ_T is small compared with unity, the probability for making m scattering before escaping the region is $\sim \tau_T^m$ so that the emergent flux has the shape

$$\frac{dF}{d\epsilon} \propto \int dk \, J_k \frac{\epsilon}{k} \frac{dm}{d\epsilon} \tau_T^m \propto \int dk \, \frac{J_k}{k \ln A} \tau_T^{\ln(\epsilon/k)/\ln A} \propto \int dk \, \frac{J_k}{k \ln A} \left(\frac{\epsilon}{k}\right)^{\ln \tau_T / \ln A}.$$

$$(8.65)$$

Thus a power-law output spectrum arises, which is harder for higher temperatures and larger optical depths (as long as $\tau_T \ll 1$). The natural cutoff for this power law is at $\epsilon \simeq k_B T$ (for thermal electrons) or $\epsilon \simeq \gamma_{max} m_e$ (for power-law relativistic electrons), as no photons can be scattered to energies significantly greater than those of the electrons.

The case of multiple scattering by nonrelativistic thermal electrons in the optically thick limit is described by the Kompaneets equation (see Vol. I, Chap. 6, Section 6.7):

$$\frac{1}{n_e \sigma_T c} \frac{\partial \mathcal{N}}{\partial t} = \Theta \frac{1}{\epsilon^2} \frac{\partial}{\partial \epsilon} \left[\epsilon^4 \left(\frac{\partial \mathcal{N}}{\partial \epsilon} + \mathcal{N} + \mathcal{N}^2 \right) \right] + q(\epsilon), \qquad (8.66)$$

where $\Theta = k_B T / m_e c^2$, $q(\epsilon)$ is a source term in units of change in occupation number per unit interval in $(n_e \sigma_T ct)$ and ϵ is measured in units of $k_B T$ for convenience. As discussed in Vol. 1, Chap. 6, Section 6.6, the relevant measure of Comptonisation is the Compton y parameter defined as

$$y = 4\Theta \max\left(\tau_T, \tau_T^2\right). \qquad (8.67)$$

In the case of an accretion disk in which the absorption is dominated by a free–free process, this can be expressed as

$$y \simeq 0.05 h_{13}^{2/3} T_5^{10/3} \frac{\epsilon^2}{(1 - e^{-\epsilon})^{2/3}} = 0.08 \, L_{46}^{2/3} \left(\frac{R_R}{\eta R_z}\right)^{2/3} T_5^{10/3} \frac{\epsilon^2}{(1 - e^{-\epsilon})^{2/3}},$$

$$(8.68)$$

provided the scale height at the photosphere is taken to be the disk thickness.

The above version of Kompaneets equation (8.66) ignores spatial diffusion which, however, can be incorporated by the addition of a loss rate proportional to $\mathcal{N}/\tau_T^2$. Given this modification, it is easy to find stationary state solutions to the Kompaneets equation, as discussed in Vol. I, Chap. 6, Section 6.7. For energies much higher than the initial energy at which soft photons are injected, $\mathcal{N}(\epsilon)$ varies as $\exp(-\epsilon)$ for $\epsilon \gg 1$ and as ϵ^w, for $\epsilon \ll 1$. The index w is determined from the quadratic equation

$$w(w + 3) = \frac{4}{y},\qquad(8.69)$$

which has the solutions

$$w = \frac{3}{2}[-1 \pm \sqrt{1 + (16/9y)}].\qquad(8.70)$$

The solutions separate naturally into two classes. When $y \gg 1$, we have $0 \leq w \lesssim 0.5$ so that $\epsilon n_{ph}(\epsilon) \propto \epsilon^{\chi} \exp(-\epsilon/\Theta)$ with $3 \leq \chi \lesssim 3.5$. (Note that n_{ph} is the spatial density of photons and $\mathcal{N}$ is the phase-space density; hence $\chi = w + 3$.) In this limit, the spectrum approaches the Bose–Einstein distribution as the number of photons is conserved; the mean energy of the photons will be comparable with the mean thermal energy of the electrons. If ϵ_0 is the mean energy of the injected photons, then the Wien spectrum is reached for $y \gtrsim \ln(\Theta/\epsilon_0)$.

For $y \ll 1$, we need to take solution (8.70) of the quadratic equation with the negative sign so that $w \simeq -3/2 - (2/\sqrt{y})$ and $\chi \simeq 3/2 - (2/\sqrt{y})$. In this case, the photon spectrum is a power law all the way from the mean energy of the initial soft photons that were injected up to the electron temperature beyond which it is exponentially cut off. When y is not too small, the power-law index can be hard enough for most of the luminosity to reside in the highest-energy photons. It should be stressed that even though an average photon leaves the plasma with its energy increased by a factor of e^y, the output luminosity is *not* e^y times the input luminosity of the injected photons; this is because a minority of photons whose energy is enhanced by a large factor dominate the output even though the energy of most of the photons is enhanced only slightly.

For intermediate values of y, we need to solve the Kompaneets equation explicitly. However, the solution can be approximated by a sum of power law and Wien contributions with suitable relative proportionality. A good fit is given by

$$n_{ph} \propto \left[\left(\frac{\epsilon}{\Theta}\right)^{\chi} + Q\left(\frac{\epsilon}{\Theta}\right)^3\right]\frac{e^{-\epsilon/\Theta}}{\epsilon},\qquad(8.71)$$

where

$$Q \simeq \frac{\Gamma(-\chi)}{\Gamma(3 - 2\chi)}P(\tau_T), \qquad P(\tau_T) = 1 - \frac{3}{8\tau_T^3}\left[2\tau_T^2 - 1 + e^{-2\tau_T}(1 + 2\tau_T)\right].$$
$$(8.72)$$

Here Γ is the gamma function and χ is calculated from the negative square root of the quadratic solution. The form of the spectrum is plotted in Fig. 8.3.

When additional photons are injected into the hot plasma, the cooling rate of the electrons will increase monotonically with the injection rate of the photons. This suggests that, in the case of thermal electrons, an equilibrium will be reached at a temperature that will monotonically decrease with the injection rate of the

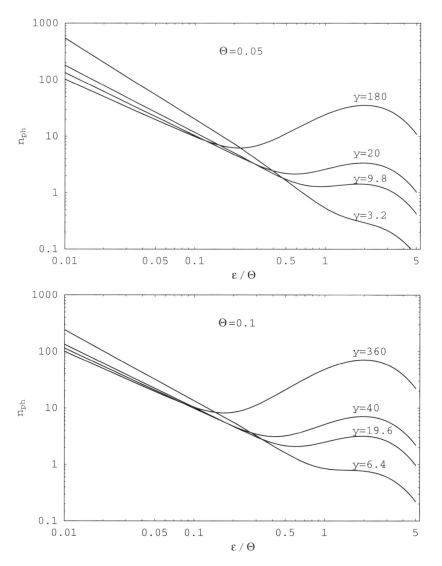

Fig. 8.3.   Output photon spectrum in the case of an optically thick Comptonised medium containing nonrelativistic thermal electrons. It is assumed that soft photons are injected at very low energies. The optical depth is 30, 10, 7 and 4 for the four curves and the *y*-axis is in arbitrary units.

photons. More formally, we can write a heat-balance equation in the form

$$F_h = \tau_T [A(\Theta) - 1] \, F_s f(\Theta, \tau_T), \qquad (8.73)$$

where F_h is the heating rate per unit area of a slab of an accretion disk, F_s is the seed photon flux, and $F_s f(\Theta, \tau_T)$ is the equilibrium photon flux density that takes into account the Compton scattering. The form assumes that, in the

Thomson limit of inverse Compton scattering, the photon energy is multiplied by a factor that depends on only electron distribution, not on the energy of the photon. It is clear from this equation that the equilibrium temperature Θ depends on only the ratio (F_h/F_s) and τ_T. Very often the disk geometry determines the ratio of heating rate to soft photon injection rate and hence we can think of (F_h/F_s) as more or less constant compared with either one of them. In such a case, the temperature depends on only τ_T and is independent of the actual heating rate. It is possible to solve for Θ with Eq. (8.60); it turns out that the result is very well approximated by the simple scaling law

$$\Theta \, \tau_T \simeq a\left(\frac{F_h}{F_s}\right)^{1/4} \tag{8.74}$$

over a wide range of parameter values provided that $\tau_T \lesssim 5$. The coefficient a depends on the geometry and is ~ 0.1 for spherical accretion and 0.06 for slabs. In the same limit, the spectral index can be approximated by a simple law, $\chi \approx -1.6(F_h/F_s)^{-1/4}$.

The situation is somewhat different when the electrons are not in thermal equilibrium but are described by a power-law spectrum. Inverse Compton scattering makes the electrons lose energy and thus changes the spectrum of the electrons when they are nonthermal. The continuity equation for electron density in the energy space has the form

$$\frac{\partial n_e(\gamma)}{\partial t} + \frac{\partial(\dot{\gamma} n_e)}{\partial \gamma} = Q(\gamma), \tag{8.75}$$

where $Q(\gamma)$ is the rate at which electrons are injected from external mechanisms at energy γm_e. From the basic theory of Compton scattering we also know that

$$\dot{\gamma} = \left(\frac{4}{3}\right)\left(\frac{\sigma_T U_{ph}}{m_e c}\right)\gamma^2, \tag{8.76}$$

which allows us to define a characteristic Compton cooling time for electrons $t_{cool} \equiv \gamma/\dot{\gamma} \propto \gamma^{-1}$. If the time scale that is due to external processes (like injection of electrons) is t_{evol} then, Eq. (8.75) can be easily solved in two limits of $t_{evol} \ll t_{cool}$ and $t_{evol} \gg t_{cool}$. In the first case, we can ignore the second term on the left-hand side of Eq. (8.75) and the solution is just

$$n_e \approx \int dt \, Q(\gamma). \tag{8.77}$$

This is appropriate at low energies for which the total number of electrons is essentially determined by the rate at which they are being injected from the external source. The more interesting situation is when $t_{cool} \ll t_{evol}$, which will arise in the high-energy limit in which we can ignore the first term on the left-hand side of Eq. (8.75). Then we can obtain a steady-state distribution given by

$$n_e(\gamma) = \frac{1}{\dot{\gamma}} \int_\gamma^\infty d\gamma' Q(\gamma') = \frac{3cm_e}{4\gamma^2 \sigma_T U_{\rm ph}} \int_\gamma^\infty d\gamma' Q(\gamma'). \qquad (8.78)$$

This shows that the number of electrons at γ is the ratio between the integrated production rate of all electrons at higher γ divided by the rate at which the electrons leave the energy corresponding to γ. Such a distribution implies that if there has been enough time for electrons to cool from, say, 2γ to γ, then there has been enough time for electrons to cool from any high energy to γ. In particular, if the electrons are injected with a power law $Q_0 \gamma^{-\psi}$, then

$$n_e(\gamma) = \frac{3cm_e}{4\sigma_T U_{\rm ph}} \frac{Q_0}{\psi - 1} \gamma^{-(\psi+1)}. \qquad (8.79)$$

Thus the cooling has steepened the power spectrum by one unit. We have seen earlier that a power-law distribution of electrons with the index β, interacting with a set of seed photons in a narrow window of energies, will lead to a resulting power-law spectrum of photons with spectrum $F_\epsilon \propto \epsilon^{(1-\beta)/2}$. Combining this with the above result, we conclude that when electrons are injected with an index β and remain in the medium long enough to cool, the resulting photon spectrum that is due to inverse Compton scattering is a power law with $F_\epsilon \propto \epsilon^{-\beta/2}$.

8.4.2 Pair Production

We have seen in Vol. I, Chap. 6, Section 6.4 that the quantum-electrodynamical interaction between two photons with energies ϵ_1 and ϵ_2 can lead to the production of $e^+ e^-$ pairs, provided that $\epsilon_1 \epsilon_2 \gtrsim (m_e c^2)^2$. In particular, this allows for pair production when high-energy gamma rays interact with, say, low-energy x-ray photons as long as this condition is satisfied. For a gamma ray of given energy, the cross section for pair production, $\sigma_{\gamma\gamma}$, is given by

$$\sigma_{\gamma\gamma}(s) = \frac{3}{8}\frac{\sigma_T}{s} \left[\left(2 + \frac{2}{s} - \frac{1}{s^2}\right) \cosh^{-1} s^{1/2} - \left(1 + \frac{1}{s}\right)\left(1 - \frac{1}{s}\right)^{1/2} \right], \qquad (8.80)$$

where $\sqrt{s}(m_e c^2)$ is the centre of mass energy for the two photons that are interacting. The limiting forms of this expression are

$$\sigma_{\gamma\gamma}(s) \to \frac{3}{8}\sigma_T \begin{cases} \sqrt{s-1} & (\text{for } (s-1) \ll 1) \\ [\ln(4s) - 1]/s & (\text{for } s \gg 1) \end{cases}, \qquad (8.81)$$

which suggests that we can approximate the cross section as a constant with $\sigma \simeq 0.2\sigma_T$ around a region centred at $\epsilon = \epsilon_{\rm th} = 2m_e^2/\epsilon_\gamma$ and having a width of $\sim 2\epsilon_{\rm th}$. Hence the optical depth for pair production for a given gamma ray of energy ϵ_γ located in a homogeneous region of size R containing softer photons of density n_ϵ is given by

$$\tau(\epsilon_\gamma) \simeq R \int_{\epsilon_{th}}^{\infty} n_\epsilon \sigma_{\gamma\gamma} \, d\epsilon \simeq 0.2\sigma_T n\epsilon_{th} R. \qquad (8.82)$$

Note that $n(\epsilon)\epsilon$ has the dimensions of inverse volume. Taking $n(\epsilon_{th})\epsilon_{th} = L(\epsilon_{th})/(4\pi R^2 c\epsilon_{th})$, where L is the luminosity of the central source, we can write the optical depth in the form

$$\tau \simeq \frac{0.2\sigma_T L(\epsilon_{th})}{4\pi \epsilon_{th} Rc}. \qquad (8.83)$$

Because the characteristic minimum threshold energy is $\epsilon_{th} \approx m_e c^2$, this suggests defining a dimensionless variable (called the compactness parameter) by

$$l \equiv \frac{L}{R} \frac{\sigma_T}{m_e c^3}. \qquad (8.84)$$

In general this definition can be used with the luminosity evaluated at any relevant energy of the hard photons involved in the pair creation. In terms of the compactness parameter, we have $\tau \simeq (l/60)$, showing that the optical depth for pair production exceeds unity when $l \gtrsim 60$. Expressing the definition for l in terms of the Eddington luminosity, we find

$$l = \frac{2\pi}{3} \frac{m_p}{m_e} \left(\frac{L}{L_E} \right) \frac{3r_g}{R}. \qquad (8.85)$$

For reasonable values of parameters, $l \gtrsim 60$, leading to copious pair production for $L \gtrsim 1.5 \times 10^{-2} L_E$.

We have seen earlier that the thermal properties of a hot Comptonised plasma depends on only τ_T and F_s/F_h. When both the soft photons as well as the heat flux are produced from the same source, we can express the flux ratio in terms of the corresponding compactness parameters by $(F_s/F_h) = l_s/l_h$. When $\Theta \approx 1$, there will be significant amount of pair production as well and the optical depth that is due to Thomson scattering can increase. The equilibrium density of pairs will now be determined by the balance between the pair production rate and the pair annihilation rate when the most dominant channels are chosen for each. If the pair production is dominated by photon–photon interaction, then the pair production rate is given by

$$\mathcal{R}_{\text{prod}} = \int d\epsilon \int_{m_e/\epsilon} d\epsilon' c\sigma_{\gamma\gamma}(\epsilon\epsilon') n_\gamma(\epsilon) n_\gamma(\epsilon'). \qquad (8.86)$$

The pair annihilation cross section can be obtained from that for pair production by use of the principle of detailed balance. In the centre-of-mass frame of the pair, the electron and the positron will have energies $\gamma_{cm} m_e$ and speed β_{cm}. The resulting photons (a pair of them with oppositely directed spins) will have energy

$\gamma_{cm} m_e$. The principle of detailed balance then gives

$$\sigma_{ann}(\gamma_{cm}) = \frac{1}{2\beta_{cm}^2}\sigma_{\gamma\gamma}\left(s = \gamma_{cm}^2\right), \tag{8.87}$$

which has the limiting forms

$$\sigma_{ann} = \frac{3}{16}\sigma_T \begin{cases} (1 + \beta_{cm}^2)/\beta_{cm} & (\alpha_{fs} \ll \beta_{cm} \ll 1) \\ [2\ln(2\gamma_{cm}) - 1]/\gamma_{cm}^2 & (\gamma_{cm} \gg 1) \end{cases}. \tag{8.88}$$

As noted in Vol. I, Chap. 6, Section 6.4, the fact that $\sigma_{ann} \propto \beta_{cm}^{-1}$ in the non-relativistic limit implies that the annihilation rate coefficient $\alpha_{ann} = \langle v\sigma_{ann} \rangle$ is a constant in the nonrelativistic limit: $\alpha_{ann} \simeq (3/8)\sigma_T c$. On the other hand, at high temperatures, we have the rate coefficient that is approximated by

$$\alpha_{ann} \simeq \frac{3}{16}\sigma_T c\, \Theta^{-2} \ln(2\eta_E \Theta), \tag{8.89}$$

with $\eta_E \simeq 0.56$. The pair annihilation rate can therefore be expressed in the form

$$\mathcal{R}_{ann} = n_+(n_+ + n_p)\alpha_{ann}(\Theta), \tag{8.90}$$

where n_+ is the number density of positrons (which is equal to the number density of pairs produced) and n_p is the number density of protons (which is equal to the original number density of electrons in the plasma). Equating $\mathcal{R}_{prod}$ with $\mathcal{R}_{ann}$ and assuming that the photon spectrum peaks at energies ϵ_s and ϵ_h and expressing everything in dimensionless form, we get the relation

$$z(1 + z)\Phi_{ann}(\Theta) = l_h l_s f(\tau_T)\frac{m_e^2}{\epsilon_s \epsilon_h}\left[\frac{\sigma_{\gamma\gamma}(\epsilon_h \epsilon_s)}{\sigma_T}\right], \tag{8.91}$$

where $z \equiv n_+/n_p$ is the ratio of pairs to net electrons $\Phi_{ann} \equiv (\alpha_{ann}/\sigma_T c)$ and $f(\tau_T)$ is a factor that accounts for the modification of the input photon spectrum by Comptonisation. It is clear that greater optical depth and seed photon density lead to smaller Θ at fixed l_h and thus postpone the onset of significant pair production – sometimes requiring $l_h \gg 1$.

The following point should be noted regarding the effect of pair production on the equilibrium temperature of the Comptonised plasma. When $l_s/l_h \ll 1$ and $l_h \lesssim 1$, there is paucity of photons available for inverse Compton scattering to higher energies. Hence, to maintain a given luminosity, most of the photons needs to be scattered up to Compton saturation, thereby forming a Wien spectrum. In such a case, the cooling rate is proportional to temperature and an increase in the heating rate (corresponding to an increase in l_h) leads to an increase in the temperature. However, when l_h is large (say, of the order of unity or higher), a significant number of pairs are created and the additional electrons can cool through thermal bremsstrahlung. Thus the cooling rate increases without an increase in temperature and, in fact, when the cooling rate is high enough, the equilibrium temperature actually decreases with increasing l_h. It follows that, for any particular rate of external soft photon injection and a given τ_T, there is a maximum equilibrium temperature. For example, the maximum temperature is ~ 1.5 MeV if $\tau_T = 1$ and $F_s \to 0$ so that the inverse Compton scattering acts only on locally created bremsstrahlung photons.

The possibility that even a small number of high-energy photons (or, for that matter, any other high-energy particle) can lead to pair production through interaction with softer photons gives rise to several interesting phenomena. In particular, it is possible for the pairs to be involved in further scattering of soft photons, which, in turn, could lead to a cascading production of pairs. We have seen earlier that if electrons are injected into a region with a power-law distribution of $\gamma^{-\psi}$, then the steady-state distribution arising through Compton scattering is γ^{-p}, where $p = \psi + 1$ and the resulting photon spectrum is $\epsilon^{-\psi/2}$. These photons can now lead to pair production, with the energy ϵ being shared by the e^+e^- pairs. The resulting pairs will have $\gamma \approx \epsilon/2m_e$ and hence the new generation of pairs will have a power spectrum $\gamma^{-\psi/2}$. These pairs will now get thermalised and the process can continue, leading to a limiting value of $\psi = 2$ and $p = 3$. The equilibrium density of pairs in the presence of hard photons of luminosity L_h and energy ϵ_h can be estimated by equating the pair production rate to the pair annihilation rate. This equation has the approximate form

$$(1 - e^{-\tau_{\gamma\gamma}})\frac{L_h}{\epsilon_h} = \frac{4\pi}{3}r^3 n_+ (n_+ + n_p)\,\Phi(\Theta)\sigma_T c. \tag{8.92}$$

The left-hand side is proportional to the probability for pair creation, and the right-hand side is the probability for pair annihilation; the factor $\Phi(\Theta) \approx (3/8)$ over a broad range of energies. In this limit, the Compton optical depth that is due to pairs is determined by

$$\tau_T(\tau_T - \tau_p) \simeq \frac{8}{3}(1 - e^{-\tau_{\gamma\gamma}})\frac{l_h m_e}{\epsilon_h}, \tag{8.93}$$

which reduces to

$$\tau_T \simeq \left(\frac{8l_h m_e}{3\epsilon_h}\right)^{1/2} \tag{8.94}$$

when $\tau_T \gtrsim 1$. In other words, a high value of compactness parameter automatically leads to a significant Compton depth even though originally it did not exist. This is an effect of pair cascading.

The high compactness parameter also makes it difficult for the pairs to cross the entire size of the region r before significant Compton cooling occurs. To see this, note that

$$\frac{t_{\rm cool}}{t_{\rm dyn}} \simeq \frac{m_e c^2 4\pi c r^2}{\gamma \sigma_T c L}\frac{c}{r} \simeq \frac{4\pi m_e c^3 r}{\gamma L \sigma_T} = \frac{4\pi}{\gamma l}, \tag{8.95}$$

implying a very short cooling time [compared with free-flight time (r/c)] for electrons with $\gamma \gg 1$ when $l \gtrsim 1$. Hence Comptonisation of low-energy photons are important when the compactness is large. Further, the downscattering of photons with energies of ~ 100 keV or more will also be important in this regime.

It is possible to derive a remarkably tight bound on the x-ray spectra produced by nonthermal pair plasmas by using these considerations. Over a wide range of conditions, the observed x-ray spectra has the form $F(\epsilon) \propto \epsilon^{-\alpha_x}$ with $0.5 \leq \alpha_x \leq 1$. To understand the lower limit, let us consider the hardest possible electron injection spectrum in the presence of pair creation that is monoenergetic with $Q(\gamma) = Q_0 \delta_D[\gamma - \epsilon_h/2m_e]$. Then the electron distribution function in steady state will be

$$n_e(\gamma) = \frac{3cm_e}{4\gamma^2 \sigma_T U_{\text{ph}}} \int_\gamma d\gamma' \, Q(\gamma') = \frac{3cm_e}{4\gamma^2 \sigma_T U_{\text{ph}}} Q_0 \qquad (8.96)$$

for $\gamma m_e < \epsilon_h/2$. We already know that if the electron spectrum is $\gamma^{-\psi}$ the resulting photon spectrum is $\epsilon^{-\alpha_x}$ with $\alpha_x = (\psi - 1)/2$; for $\psi = 2$ we get $\alpha_x = 0.5$, which is the lower bound, arising from the hardest possible input spectrum of electrons.

The upper bound to the index corresponds to the saturated limit, in which there is copious production of electrons and positrons at all energies below the energy of the first generation. In this case, we can take the total energy of the electrons to be conserved, although their number is not. The corresponding conservation law now becomes

$$\frac{\partial \dot{\gamma} n_e}{\partial \gamma} = \gamma \, Q(\gamma), \qquad (8.97)$$

which has the solution

$$n_e \propto \gamma^{-3} \int_\gamma d\gamma' \gamma' Q(\gamma'). \qquad (8.98)$$

At energies well below those of first-generation pairs, $n_e \propto \gamma^{-3}$ and the corresponding Compton scattered photon spectrum is proportional to ϵ^{-1}.

The whole preceding discussion pertains to the production of high-energy photons *at the source*. As was discussed in Vol. I, Chap. 7, Subsection 7.4.4, x rays have a fairly large opacity for absorption in cosmic gas. As a result the *observed* properties of x rays from a source will become modified because of propagation effects. The soft-x-ray opacity of our galaxy causes absorption of photons from any AGN whose line of sight is in the galactic plane. Because the column densities along a line of sight with galactic latitude b are approximately $(10^{20}/\sin b)$ cm^{-2}, it is often difficult to see intrinsic spectrum at energies much below 500 eV even for AGN well above the galactic plane. In fact, the local opacity makes the spectral range from 13.6 to 100 eV almost unobservable along any line of sight. If there is additional opacity that is due to the ISM of the host galaxy in the AGN, observations become still more difficult.

Observationally, radio-quiet AGN have very similar high-energy spectra described by a power-law cutoff of the form $f(\epsilon) \propto \epsilon^{-\alpha_x} \exp(-\epsilon/\epsilon_c)$. The index $\alpha_x \approx 0.9$ is reasonably well determined but ϵ_c is not; it is estimated to be ~ 100 keV or more. The thermal Comptonisation previously described can easily reproduce such a continuum shape of a power law with cutoff. The model, of course, need not be unique; for example, $l_s/l_h \simeq 0.5$ and $\tau_T \simeq 0.1$ is a feasible model and so is the one with $l_s/l_h = 0.1$ with any reasonable τ_T in the range of, say, 0.3–3. The first model will give $\Theta m_e c^2 \approx 400$ keV for the cutoff energy. Because $\Theta \propto \tau_T^{-1}$, smaller cutoff energies would require higher τ_T and will favour $l_s/l_h \simeq 0.1$. Because the cutoff energy is somewhat uncertain, it is difficult to determine the models uniquely.

8.4.3 Line Emission from Iron

The high-Z elements capable of producing line emission in the x-ray band are comparatively rare, leading to small equivalent widths. One fortunate exception is the Fe K-α line, which can be quite strong because of the relatively high abundance of Fe. The general features of x-ray line emission from Fe has been discussed in Vol. I, Chap. 7, Section 7.4. When Fe is ionised to a state with at least four remaining electrons, K-shell photoionisation will be followed by either Auger ionisation or fluorescence. The details of this process are complicated by uncertain atomic physics, especially for highly ionised states; but the final numbers do not change significantly. The line energy is, for example, 6.7 keV in heliumlike Fe and rises only to 6.9 keV for hydrogenlike Fe.

Some of the Fe fluorescence arises when x-ray photons are reflected off a cooler slab of material. To understand the issues involved, consider a plane-parallel semi-infinite slab exposed to x rays on the top surface. If the slab is completely ionised, then all the photons would be reflected but the high-energy photons would undergo a fair amount of energy loss. To begin with, photons more energetic than 100 keV can lose a significant part of their energy in a single scattering event; they cannot therefore be studied by normal tools of continuum evolution based on differential equations and need to be examined by Monte Carlo simulations. Such studies indicate that even a modest optical depth can wipe out the spectrum at such a high-energy range. At somewhat lower energies, we can try to estimate the energy loss by a simple approximation to Comptonisation described by

$$\frac{d\epsilon}{dt} = -n_e \sigma_T \frac{\epsilon^2}{m_e c^2}. \tag{8.99}$$

This has the solution

$$\epsilon(t) = \frac{\epsilon_0}{1 + n_e \sigma_T t \epsilon_0/(m_e c^2)} = \frac{\epsilon_0}{1 + (\epsilon_0/m_e c^2)N_{\text{scat}}}, \tag{8.100}$$

where the second expression uses the number of scatterings $N_{scat} = n_e \sigma_T ct$. This result shows that once the photon energy has been reduced by a factor of the order of unity, subsequent scatterings decrease the energy as N_{scat}^{-1}. Equivalently, when photons are random walking through a medium with $\tau_T > 1$, very few photons will emerge with energy greater than $m_e c^2 / \tau_T^2$. This feature affects the scattering of photons back from a fully ionised slab and, in practice, we find that incident photons of less than 30 keV are reflected with the unit reflection coefficient. Thus the reflected spectrum will be similar to the incident spectrum for $\epsilon < 30$ keV but will be depleted for $\epsilon \gtrsim 30$ keV with very few photons at $\epsilon \gtrsim 100$ keV.

The situation, however, is quite different if the slab is completely neutral. In this case, low-energy photons will suffer severe attenuation and, for any reasonable column density, the absorption depth is greater than unity for all photons with $\epsilon \lesssim 10$ keV. Even in the range of 20–30 keV, the reflection coefficient is only ~50%. The combination of photoelectric opacity at the low-energy end and Compton recoil loss at the high end eliminates both extremes of the spectrum, leaving a bump between 20 and 50 keV. Of these, the low-energy absorption in 7–20 keV is substantially contributed by the Fe K-edge opacity. Therefore a substantial amount of Fe K-α fluorescence will accompany the x-ray reflection. This model does reproduce some of the observed features, although the reflecting slab should have substantial covering fraction in the sky as seen from the x-ray source in order to produce the observed equivalent width of the K-α line.

Finally, we discuss an interesting application of the observations of the 6.5-keV Fe K line from some of the AGN. When the line emission occurs from a region close to the black hole, it is necessary to take into account the gravitational redshift in the strong field regime. For a general rotating Kerr black hole, this redshift is given by

$$1 + z = \sqrt{\frac{1 + a^2/r^2 + 2Ma^2/r^3}{1 + a^2/r^2 - 2M/r}} \tag{8.101}$$

(see Vol. II, Chap. 5, Section 5.9). The smallest stable circular orbit is at $r = 6\,M$ for the Schwarzchild black hole, with $a = 0$ and the corresponding redshift $(1 + z) = \sqrt{3/2}$ – which is not too large. In contrast, the redshift diverges for the extreme Kerr black hole with $a = M$ when emission occurs from $r = M$. Remarkably enough, we can see the signature of black hole rotation by comparing the profile of the 6.5-keV Fe line. Such a study in the Seyfert galaxies MCG-6-30-15 shows that the results match quite well with a nearly extreme Kerr black hole but not with a Schwarzchild black hole. In the future, such observations could provide a reliable strong field test of general relativity.[4]

8.5 Radio Emission from Quasars

In contrast to optical, UV, and x-ray spectra, the radio emission from AGN is better understood, at least as regards the basic mechanism. Strong linear polarisation as well as the nonthermal nature of radio emission suggest synchrotron emission as the possible mechanism and most models of AGN incorporate this. We first summarise some of the key features of synchrotron radiation that were discussed in detail in Vol. I, Chap. 6, Section 6.11.

The spectrum of radiation emitted by a single electron moving in a magnetic field can be expressed in the form

$$P_\nu = \frac{4}{3}\gamma^2 \sin\theta \left(\frac{\sigma_T c}{\nu_c}\right)\left(\frac{B^2}{8\pi}\right) F_s[\nu/\nu_c], \qquad (8.102)$$

where

$$\nu_c = \frac{3}{2}\gamma^2 \sin\theta \left(\frac{eB}{2\pi m_e c}\right), \qquad (8.103)$$

and the function F_s is given by

$$F_s[\nu/\nu_c] = \frac{3^{5/2}}{8\pi}\frac{\nu}{\nu_c}\int_{\nu/\nu_c}^{\infty} dy\, K_{5/3}(y)$$

$$\simeq \begin{cases} \dfrac{9}{2\Gamma(1/3)}\left(\dfrac{\nu}{2\nu_c}\right)^{1/3} & \text{(for } \nu \ll \nu_c) \\[2ex] \sqrt{\dfrac{243}{128\pi}}\left(\dfrac{\nu}{\nu_c}\right)^{1/2}\exp(-\nu/\nu_c) & \text{(for } \nu \gg \nu_c) \end{cases} \qquad (8.104)$$

The radiation is also highly polarised, and the magnitude of the polarisation for a bunch of electrons with different pitch angles (but same energy) is given by

$$P(\nu/\nu_c) = \frac{G_s(\nu/\nu_c)}{F_s(\nu/\nu_c)}, \qquad (8.105)$$

with

$$G_s[\nu/\nu_c] = \frac{3^{5/2}}{8\pi}\frac{\nu}{\nu_c} K_{2/3}[\nu/\nu_c]$$

$$\simeq \begin{cases} \dfrac{9}{4\Gamma(1/3)}\left(\dfrac{\nu}{\nu_c}\right)^{1/3} & \text{(for } \nu \ll \nu_c) \\[2ex] \sqrt{\dfrac{243}{128\pi}}\left(\dfrac{\nu}{\nu_c}\right)^{1/2}\exp(-\nu/\nu_c) & \text{(for } \nu \gg \nu_c) \end{cases} \qquad (8.106)$$

When $\nu \ll \nu_c$, $P \simeq 0.5$, whereas in the limit of $\nu \gg \nu_c$, $P \simeq 1$. We can compute the emissivity from a power-law spectrum of electrons with $dn/d\gamma = K\gamma^{-\xi}$ for

$\gamma_{\min} < \gamma < \gamma_{\max}$ in the optically thin limit by integrating the spectrum of a single electron over the distribution function of electrons. This gives the emissivity

$$j_\nu = E(\xi)\frac{Ke^3B}{m_ec^2}\left(\frac{2\pi m_ec\nu}{eB}\right)^{(1-\xi)/2}, \tag{8.107}$$

where

$$E(\xi) = \frac{3^{\xi/2}\pi^{1/2}}{2(\xi+1)}\frac{\Gamma\left(\frac{3\xi+19)}{12}\right)\Gamma\left(\frac{3\xi-1}{12}\right)\Gamma\left(\frac{\xi+5}{4}\right)}{\Gamma\left(\frac{\xi+7}{4}\right)}. \tag{8.108}$$

Usually $E(\xi)$ is of the order of unity and can be ignored in order-of-magnitude estimates.

It has also been pointed out in Vol. I, Chap. 6, Section 6.11, that there are striking similarities between the energy loss of electrons through synchrotron and the energy loss of electrons through inverse Compton radiation. Both processes lead to a rate of loss proportional to γ^2, and hence much of the discussion in Section 8.4 can be adopted for studying the effect of synchrotron energy loss on the electron spectrum. The cooling time varies as γ^{-1} and the distribution function at high energies (where cooling is significant) will be one unit steeper than at low energies [see Eq. (8.79)]. Another similarity between the two processes is that, for each output photon energy $h\nu$, there exists a narrow range of electron energies that is primarily responsible for emission of the radiation at that frequency. These two are related by

$$\gamma_s(\nu) \simeq \left(\frac{2\pi m_ec\nu}{eB}\right)^{1/2}. \tag{8.109}$$

The first feature in the model that can be compared with observations is the slope of the spectrum. The observed spectra from the extended radio-emission regions of AGN have spectral slopes of $\alpha \approx 0.5$–1 that match with the synchrotron spectra if $\xi \simeq 2$–3. Next, to compare the observed emissivity with the model prediction, we need to know the strength of the magnetic field present in these sources. Direct measurement of the magnetic field is extremely difficult, and we often resort to the minimum-energy argument (discussed in Vol. II, Chap. 4, Section 4.9). For a system from which the synchrotron power-law spectrum is observed in the range of frequencies $\nu_l < \nu < \nu_u$, the lowest- and the highest-energy electrons that contribute to emissivity have the γ factors given by $\gamma_s(\nu_l)$ and $\gamma_s(\nu_u)$, respectively, where the function $\gamma_s(\nu)$ is defined in relation (8.109). In the expression for the total energy density of the system,

$$\mathcal{E} = \frac{B^2}{8\pi} + \int d\gamma\,\frac{dn_e}{d\gamma}\gamma m_ec^2, \tag{8.110}$$

the second term is dominated by the contribution near $\nu = \nu_l$ for $2 < \xi < 3$. Normalising the electron power-law spectrum in terms of the observed emissivity at the lower cutoff, $j_\nu(\nu_l)$, we can easily determine the second term as

$$U_e = \frac{6 j_\nu(\nu_l)}{\sigma_T} \left(\frac{em_e c \nu_l}{2\pi} \right)^{1/2} B^{-3/2}. \tag{8.111}$$

Minimising the total energy with respect to B now gives the minimum magnetic strength as

$$B_m = \left[\frac{36\pi j_\nu(\nu_l)}{\sigma_T} \left(\frac{em_e c \nu_l}{2\pi} \right)^{1/2} \right]^{2/7}. \tag{8.112}$$

The corresponding minimum energy density is

$$\mathcal{E}_m = \left[1 + \left(\frac{9}{16} \right)^{4/7} \right] \frac{1}{\pi^{4/7}} \left[\frac{j_\nu(\nu_l)}{\sigma_T} \left(\frac{em_e c \nu_l}{2\pi} \right)^{1/2} \right]^{4/7}. \tag{8.113}$$

Numerically, the minimum magnetic field and the total energy stored in the radio-emission region are given by

$$B_m \simeq 2.4 \times 10^{-6} \left(\frac{L_{43}}{r_{100}^3} \right)^{2/7} \nu_{l9}^{(1-2\alpha)/7} \ \mathrm{G}, \quad E_m \simeq 2 \times 10^{57} L_{43}^{4/7} r_{100}^{9/7} \nu_{l9}^{(2-4\alpha)/7} \ \mathrm{ergs},$$
$$\tag{8.114}$$

where ν_{l9} is ν_l measured in gigahertz, r_{100} is the length scale of the source in units of 100 kpc, and L_{43} is the monochromatic luminosity at 1 GHz in units of 10^{43} ergs s^{-1}. The energy content is equal to the stellar luminosity of a typical galaxy over a period of a few million years. It should be stressed that this is merely a minimum energy and the actual field need not always reach the equipartition value. Further, the lowest frequency that is observed need not be the lowest frequency at which synchrotron radiation occurs. Because the total energy scales as $[j_\nu(\nu_l)\nu_l^{1/2}]^{4/7} \propto \nu_l^{(4/7)(0.5-\alpha)}$ with $\alpha > 0.5$, the equipartition energy observed from the computed flux is generally less than the possible true value.

It is easy to show from the preceding analysis that the minimum value B_{min} for the magnetic field is related to the equipartition value B_{eq} by $B_{min} = (3/4)^{2/7} B_{eq} \simeq 0.92 B_{eq}$. For numerical computations involving high-redshift sources, the equilibrium value can be expressed in the form

$$B_{eq} = 9.3 \times 10^{-6} \left[\frac{\nu_u^{(1/2+\alpha)} - \nu_l^{(1/2+\alpha)}}{(2\alpha + 1)f(\alpha)} \right]^{2/7}$$
$$\times \nu^{-2\alpha/7} S^{2/7} \theta^{-6/7} [1 - (1+z)]^{-2/7} (1+z)^{2(4-\alpha)/7} h_{50}^{2/7} \tag{8.115}$$

where all the frequencies are in megahertz, S is in jansky units, B is in gauss units, θ is the angular diameter of the emission region in arcseconds, and

$$f(\alpha) = \frac{3^{(1/2-\alpha)}}{16\sqrt{\pi}} \left(\frac{5-3\alpha}{1-\alpha}\right) \Gamma\left(\frac{1-3\alpha}{6}\right) \Gamma\left(\frac{5-3\alpha}{6}\right) \frac{\Gamma\left(\frac{3-\alpha}{2}\right)}{\Gamma\left(\frac{4-\alpha}{2}\right)}. \quad (8.116)$$

As a rule of thumb, this translates to the statement that an object of flux density S Jy at frequency ν MHz subtends an angle $\theta \simeq 2(S^{1/2}/\nu)''$.

If we take the field strength to be a few microgauss, then $\nu = 1$ GHz photons are emitted by highly relativistic electrons with $\gamma \approx 10^4$. The cooling time for these electrons will be

$$t_{\text{cool}} \simeq 2.4 \times 10^9 \gamma_4^{-1} B_{\mu G}^{-2} \text{ yr.} \quad (8.117)$$

There is some evidence that the ages of these systems are less than 10^8 yr, which shows that electrons producing radiation below 1 GHz would have lost very little of their original energy. Also, the mean total power injected into the source in the form of relativistic electrons is at least an order of magnitude larger than the observed radio luminosity. At somewhat higher frequencies, there must be reinjection or reacceleration of electrons in order to maintain the emission of radiation.

The cooling of the relativistic electrons will cause the spectrum to steepen by half a unit above the frequency at which the age of the source equals the cooling time for the electrons. Observations of the lobes of FRII radio galaxies seem to indicate such a steepening: The spectrum is flattest in the hot spots and steepens as we go away from them, which is what we would have expected if the electrons were accelerated in the hot spots and then streamed away.

Exercise 8.4
Direct measurements of the magnetic field: The electrons responsible for radio emission at $\nu = 1$ GHz have $\gamma \simeq 10^4$. Microwave photons can scatter off these electrons and acquire energies around 100 keV. Show that the ratio between x-ray luminosity produced by inverse Compton scattering and the radio luminosity produced by synchrotron radiation is

$$\frac{L_{\epsilon_x}}{L_{\epsilon_r}} \simeq \left(\frac{aT_{\text{CMB}}^4}{B^2}\right) \left(\frac{\hbar eB/m_ec}{k_BT_{\text{CMB}}}\right)^{(3-\xi)/2} \left(\frac{\epsilon_x}{\epsilon_r}\right)^{-(\xi-1)/2}, \quad (8.118)$$

where ϵ_x and ϵ_r are the typical energies at which x rays and radio fluxes are measured. Under what conditions can this relationship be used to measure the magnetic field of the source?

In the discussion so far we have assumed that synchrotron radiation originates from an optically thin region. As discussed in Vol. I, Chap. 6, Subsection 6.11.4, this is not correct at low frequencies at which we need to take into account the synchrotron self-absorption. The self-absorption opacity can be expressed in the form

$$\rho \kappa_\nu = \frac{c^2}{8\pi h\nu^3} \frac{1}{m_e c^2} \int d\gamma \, P_\nu(\gamma) \gamma^2 h\nu \frac{d}{d\gamma}\left(\gamma^{-2}\frac{dn_e}{d\gamma}\right) \quad (8.119)$$

[see Eq. (6.290) of Vol. I, Chap. 6, Section 6.11]. If $dn_e/d\gamma = K\gamma^{-\xi}$, this integral can be easily computed for an isotropic pitch-angle distribution, giving

$$\rho \kappa_\nu = A(\xi) K \frac{e}{B}\left(\frac{2\pi\nu}{eB/m_e c}\right)^{-(\xi+4)/2}, \quad (8.120)$$

with

$$A(\xi) = \frac{\pi^{3/2}}{12} 3^{(\xi+3)/2} \frac{\Gamma\left(\frac{3\xi+2)}{12}\right)\Gamma\left(\frac{3\xi+22)}{12}\right)\Gamma\left(\frac{\xi+6)}{4}\right)}{\Gamma\left(\frac{\xi+8)}{4}\right)}. \quad (8.121)$$

[The coefficient $A(\xi)$ is typically ~ 10 and hence cannot be ignored; for example, $A(2.5) = 10.41$.] Given this expression, we can define a synchrotron self-absorption cross section per electron by

$$\sigma_e = \frac{(d/d\gamma)(\rho\kappa_\nu)}{dn_e/d\gamma} \approx \frac{4\pi^2}{9}(\xi+2)\frac{e}{B}\gamma^{-5}\left(\frac{\nu}{\nu_c}\right)^{-2}F\left(\frac{\nu}{\nu_c}\right) \propto \gamma^{-6}r_L r_e, \quad (8.122)$$

where r_L is the Larmor radius and r_e is the classical electron radius. The result depends on the shape of the electron distribution function because the process depends on the stimulated emission that, in turn, depends on the electron distribution function. For typical FRII radio galaxies, $\sigma_e \approx 10^{-24}$ cm^2, which is comparable with σ_T. The optical depth, τ_{eq} across a region of size r can be estimated if we assume that the magnetic field is given by the equipartition value:

$$\tau_{eq} \simeq \frac{\pi^2}{36}(\xi+2)\frac{r}{c}\left(\frac{eB_m}{2\pi m_e c}\right)^4 v_l^{-3}\left(\frac{\nu}{\nu_l}\right)^{(4+\xi)/2}$$

$$\simeq 5\times 10^{-11}L_{43}^{8/7}r_{100}^{-17/7}v_{l9}^{-25/7}\left(\frac{\nu}{\nu_l}\right)^{-3.25} \quad (8.123)$$

for $\xi = 2.5$. This shows that extended lobes of giant radio galaxies are optically thin. However, the steep dependence of the preceding result in r makes the situation different for compact cores. These regions have luminosities comparable with extended sources, but their sizes are smaller by a factor 10^4. This increases the equipartition field strength by $\sim 10^{3.4}$ and the energy of the relevant electrons by $10^{1.7}$; $\tau_{eq}(\nu_l)$ now rises to ~ 0.3 with the same scaling as before on L, r, and ν_{l9}. If the assumption of equipartition is valid, then synchrotron self-absorption should play a dominant role in these compact sources. Assuming that the source is optically thick, we can estimate the intensity deep inside the source as

$$I_\nu = \frac{j_\nu}{4\pi\rho\kappa_\nu} = \frac{1}{4\pi}\frac{E(\xi)}{A(\xi)}\frac{e^2 B^2}{m_e c^2}\left(\frac{2\pi m_e c\nu}{eB}\right)^{5/2} = \pi\frac{E(\xi)}{A(\xi)}\frac{\gamma_s(\nu)m_e c^2}{\lambda^2} \quad (8.124)$$

(see the discussion in Vol. I, Chap. 6, Subsection 6.11.4). The same result applies to the surface brightness of the synchrotron source, which we expect to scale as $B^{-1/2}\nu^{5/2}$. The last equality shows that the brightness temperature at $\lambda = (c/\nu)$ is of the order of the energy of the electrons responsible for radiating at frequency ν.

Because the spectrum rises as $\nu^{5/2}$ at low frequencies and falls as $\nu^{-\alpha}$ at high frequencies, it should exhibit a distinct peak near a turnaround frequency ν_T, at which the optical depth is unity. This frequency is given by

$$\nu_T = \frac{eB}{2\pi m_e c}\left[\frac{(\xi-1)A(\xi)e\tau_T'}{B\sigma_T}\right]^{2/(\xi+4)} = 2.4\times 10^{11}\,B^{0.692}\tau_T'^{0.308}\ \text{Hz} \quad (8.125)$$

if $\xi = 2.5$. (This value will be used in all numerical estimates.) The quantity τ_T' is the electron-scattering optical depth across the radius of the source for the electron distribution function continued all the way to $\gamma = 1$. Equating the surface brightness to the source function, we can estimate the brightness temperature at ν_T to be approximately

$$T_{bT} = \frac{\pi}{2}\frac{(\xi)}{A(\xi)}\frac{m_e c^2}{k_B}\left[\frac{(\xi-1)A(\xi)e\tau_T'}{B\sigma_T}\right]^{1/(\xi+4)} = 3.8\times 10^{11}\left(\frac{\tau_T'}{B}\right)^{0.154}\ \text{K}, \quad (8.126)$$

which has only a weak dependence on the parameters. This suggests that most compact sources should have similar peak brightness temperatures.

The observed spectrum of the compact radio core is indeed flatter than those of extended sources but they do not exhibit a spectrum rising as sharply as $\nu^{5/2}$. We usually tackle this discrepancy by assuming that the observed spectrum comes from the superposition of different components, each of which peaks at different values of ν_T. It is fairly easy to produce spectra that are flat over a decade of frequency by this procedure.

One possible way of cross checking this assumption is as follows. Equations (8.125) and (8.126) can be solved for magnetic field and optical depth in terms of ν_T and the brightness temperature T_{bT} at the turnaround frequency:

$$B = \frac{2\pi m_e c\nu_T}{e}\left[\frac{\pi m_e c^2}{2k_B T_{bT}}\frac{E(\xi)}{A(\xi)}\right]^2 = 5.8\times 10^{-4}\nu_{T\text{GHz}}T_{bT12}^{-2}\ \text{G}, \quad (8.127)$$

$$\tau_T' = \frac{16\pi^2}{3(\xi-1)A(\xi)}\frac{e^2\nu_T}{m_e c^3}\left[\frac{2k_B T_{bT}}{\pi m_e c^2}\frac{A(\xi)}{E(\xi)}\right]^{\xi+2} = 0.34\,T_{bT12}^{4.5}\nu_{t\text{GHz}}. \quad (8.128)$$

Most core-dominated sources have a relatively flat spectrum in the range 1–100

GHz, so that each of the components that contributes to the flat region of the spectrum should have v_T in this range. Because $T_b \propto F_v/v^2$ the largest value for the brightness temperature in a flat spectrum arises at the lowest frequencies, and observations suggest that the distribution of the observed peak brightness temperature lies between 10^{11} and 3×10^{12} K. This agrees fairly well with theory and leads to $\tau_T' \approx 0.01$–10 and $B \approx 10^{-4}$–10^{-3} G.

Exercise 8.5

Energetics of self-absorbed synchrotron sources: Show that, if the electron distribution function has the same slope down to a lower energy cutoff γ_{min}, then the total energy density of the electrons is

$$U_e = 0.40 \frac{\xi - 1}{\xi - 2} \tau_T' r_{pc}^{-1} \gamma_{min}^{2-\xi} \text{ ergs cm}^{-3},\tag{8.129}$$

and the ratio between magnetic energy density and electron energy density is given by

$$\frac{U_m}{U_e} = 3.3 \times 10^{-8} T_{bT12}^{-8.5} v_{tGHz} r_{pc} \gamma_{min}^{0.5}.\tag{8.130}$$

By what factor should T_{bT} decrease in order for equipartition to be achieved? Also show that the ratio between photon energy density and magnetic energy density is given by

$$\frac{U_{ph}}{U_m} = \frac{256}{\pi^4} \frac{e^2 v_T}{m_e c^3} \left[\frac{A(\xi)}{E(\xi)}\right]^4 \left(\frac{k_B T_{bT}}{m_e c^2}\right)^5 \left\{\frac{2}{7} + \frac{2}{3 - \xi}\left[\left(\frac{v_{max}}{v_T}\right)^{(3-\xi)/2} - 1\right]\right\},\tag{8.131}$$

where v_{max} is the high-frequency cutoff for the optically thin synchrotron spectrum. Hence show that, for reasonable values of parameters, inverse Compton luminosity will dominate the synchrotron luminosity when $T_{bT} \gtrsim 10^{12}$ K.

8.6 Radio Jets

The most impressive feature about the radio galaxies, which has attracted significant amount of attention, is the well-collimated jets that these objects exhibit. In spite of extensive research, we still do not understand clearly the mechanism for the production and collimation of these jets. The energetics of the jets demand continuous feeding of energy and relativistic particles, but even the physical content of the jet material is not well known. (For example, it is not clear whether it is a normal plasma or an e^+e^- plasma.) The most popular schemes involve magnetohydrodynamic (MHD) winds driven by rapidly rotating black holes, but the details of the model are fairly speculative. We briefly describe some of the issues involved, although conclusive answers are not yet available.

Let us begin by considering the actual content of the jet. For a given (observed) radio emission in the jet the least energy is required if the jet is made of only electrons and positrons. To be precise, we need only a particular kind of charged particle orbiting in a magnetic field to produce the observed radiation. If we take these to be electrons, then neutralising the total electric charge of the jets by adding protons will increase the total energy by a factor $\mathcal{O}(m_p/m_e)$ compared with having a jet made of electrons and positrons. However, if we freeze the estimated total momentum and energy fluxes of the jets – rather than the observed radio emission – then the total energy is independent of the positron-to-proton ratio. So, at a fundamental level, it is not clear whether the jet is made of a pair plasma or normal plasma. (A pair plasma will emit e^+e^- annihilation radiation but the predicted fluxes are well below the current upper bounds.) Further, we also do not have a complete picture of how such high-energy particles are created and collimated.

Similar uncertainties exist as regards the magnetic field, which is another ingredient in explaining the synchrotron radiation. Even if the magnetic field in the inner part of the jet is somewhat tangled, the magnetic flux conservation in a jet with opening angle θ will imply that the component of the magnetic field parallel and transverse to the jet will scale as

$$B_{\parallel} \propto (r\theta)^{-2}, \qquad B_{\perp} \propto r^{-1}, \qquad (8.132)$$

where r is the distance along the jet. Hence, at large distances, we would expect the field to be predominantly perpendicular to the jet axis, which is consistent with observations in radio jets of low luminosity. However, in the case of jets with higher luminosities, the field seems to be significantly longitudinal far away from the source, which requires more complicated modelling for its explanation. The Faraday rotation (see Vol. I, Chap. 9, Section 9.5) of the plane of polarisation of the synchrotron radiation could provide another independent and direct diagnostic of the magnetic field strength. Unfortunately, this will give only a lower limit to the field strength (even assuming that the electron density is known independently) as the field can double back and forth along the line of sight. In a pair plasma, positrons and electrons will lead to a near cancellation of the Faraday rotation, so that we need some independent estimate of the positron content before any definite conclusion can be reached.

The second set of issues connected with the jets is related to their acceleration and collimation, as any suggested mechanism should be able to accelerate the material to Lorentz factors of ~ 10 and collimate the flow to ~ 0.1 rad. One general feature about jet propagation is that it will develop shock fronts as it pushes its way out through the host galaxy and into the IGM. A primary shock front will develop ahead of the jet and will move through the external medium. Material at the tip of the jet is continously losing momentum because of direct contact with the external medium, and hence the tip cannot move outward as fast as the jet material itself. This will lead to the formation of a second shock (called *reverse shock*) that will be moving backwards into the jet, slowing it down. In general, a relativistic jet can be characterised by its mass, momentum, and energy outputs:

$$\dot{M} = A\gamma \rho_r v,$$

$$T = A\gamma^2(\rho + p/c^2)v^2 + p, \tag{8.133}$$

$$K = A\gamma^2(\rho + p/c^2)c^2 v - \dot{M}c^2,$$

where A is the cross-sectional area of the jet, ρ_r is the rest-mass energy density, ρ is the total rest-frame energy density, including the internal energy density $u = (\rho - \rho_r)c^2$ of the fluid, and p is the pressure. Because a jet is likely to expand sideways with an opening angle $\theta \approx \mathcal{M}^{-1}$, where $\mathcal{M}$ is the Mach number, it is likely that a cold supersonic jet will be required at least far away from the source to maintain collimation. In this case, $T \approx A\rho v^2$ and $K = (1/2)\dot{M}v^2$. Equating the internal pressure of the radiating plasma in the frame of the hot spot to the ram pressure of the incoming IGM, we get

$$2\left(\frac{B^2}{8\pi}\right) = \rho_i v_h^2 = \rho_h(v_j - v_h)^2, \tag{8.134}$$

where ρ_i and ρ_j are the densities of IGM and the jet and v_j and v_h are the velocities of the jet and the hot spot. Writing this equation in the form

$$\frac{v_j}{v_h} = 1 + \left(\frac{\rho_i}{\rho_j}\right)^{1/2}, \tag{8.135}$$

we see that the jet velocity will be comparable with the hot-spot velocity unless the jet is much lighter than the IGM. In the case of Cygnus-A, for example, we require $v_j \approx c$ and hence $\rho_j < 10^{-3}\rho_i$. The equipartition field in the hot spot is $B \simeq 5 \times 10^{-4}h^{2/7}$G, implying an external proton density of $n \simeq 10^{-2}h^{12/7}$ cm^{-3}. This is comparable, within an order of magnitude, to the density prevalent in the IGM of a cluster.

We can also attempt to estimate the jet density and velocity by equating the total kinetic energy delivered within the characteristic lifetime,

$$\tau = \frac{E}{|\dot{E}|} = 2.45 \times 10^{11}\gamma^{-1}\left(\frac{B}{10^{-5}\,\text{G}}\right)^{-2} \text{yr}, \tag{8.136}$$

to the sum of total energy radiated, the energy resident in relativistic particles and fields, and the work done in pushing the IGM. Of these three, the last one is the dominant term and can be estimated along the following lines. If the source is taken to be a cylinder of length L and radius R (for example, $L \simeq 88h^{-1}$ kpc and $R \simeq 0.05\,L$ for Cygnus-A), with the cylinder expanding uniformly over the source lifetime, the integrated PdV work done against the ram pressure is

$$E \simeq \frac{\pi R^2 L\rho_i}{12}v_h^2. \tag{8.137}$$

Equating this to $K\tau$ and using the values appropriate for Cygnus-A, we get $K \simeq 3 \times 10^{38}h^{-2}$ W, $v_j/c \simeq 0.22h^{-4/7}$, and $\dot{M} \simeq 1.5h^{-6/7}M_\odot$ yr^{-1}. This is marginally relativistic and implies that the efficiency of radio AGN in converting directed kinetic energy into relativistic particles and fields is only ~ 0.01. Direct proper motion measurements of M87, for example, give a speed of $0.35h^{-1}c$ at larger distances from the source.

Exercise 8.6

Jet power: Show that the flux Q of energy of the jet (usually called the power of the jet) can be expressed in the form $Q = \Phi v P$, where v is the speed of the jet, P is the momentum flux of the jet, and

$$
\Phi = \begin{cases} \left(\dfrac{5\mathcal{M}^2 + 15}{10\mathcal{M}^2 + 6}\right) & \text{(supersonic, nonrelativistic)} \\[4mm] \left(\dfrac{4}{1 + 3\beta^2}\right) & \text{(extreme relativistic)} \end{cases}, \tag{8.138}
$$

where $\mathcal{M}$ is the Mach number of the flow. Show that the maximum value for the product Φv is $(2c/\sqrt{3})$. Explain how this can be used to put an upper bound on Q in terms of the parameters of the hot spot.

Let us next consider the issue of collimating the jet. Simple ideas of collimation, based on gas-pressure gradients and supersonic flow through a de Laval nozzle, require the gas to have very high pressure (see Vol. 1, Chap. 8, Subsection 8.9.1) if the jets are collimated at VLBI scales. Such a gas will cool very rapidly, and reheating will lead to observable effects in the spectrum, which are not seen. Purely radiative forces are also inefficient unless the angular distribution of the radiation is such that the directed component is nearly 10^4 times larger than the random component. This is quite unlikely in the AGN. One surviving possibility is the collimation through MHD wind. This is a fairly generic feature, although it requires a certain ad hoc assumption regarding the field geometry. If the footprints of the magnetic-field lines are anchored in the accretion disk, then the field lines will rotate with the orbital frequency of the footprints, creating a toroidal field. Once the field is toroidal, with, say, $\mathbf{B} = B(r, z)\mathbf{e}_\phi$ in the cylindrical (r, ϕ, z) coordinate system, the MHD force equation becomes

$$
\rho\left[\frac{\partial \mathbf{v}}{\partial t} + \mathbf{v}\cdot\nabla\mathbf{v}\right] = -\frac{B^2}{4\pi r}\left[\left(\frac{1}{2} + \frac{\partial \ln B}{\partial \ln r}\right)\mathbf{e}_r + \frac{r}{z}\frac{\partial \ln B}{\partial \ln z}\mathbf{e}_z\right] - \nabla p + \mathbf{F}_{\text{ext}}. \tag{8.139}
$$

It is clear that even in the absence of field gradients there is a component of the field exerting a force towards the axis that could lead to collimation. Such a toroidal field, however, requires the flow of a net current $I(r) = (cr/2)B_\phi(r)$ within a cylindrical radius r. Hence there must exist a return current somewhere, and a complete model should correctly account for the flow pattern, which is a difficult task. Further, such a return current flowing back with the central engine via different field lines (compared with the outgoing current) must cross the field lines, thereby allowing dissipation and breakdown of flux-freezing conditions. A similar dissipative region will exist near the base of the flow, which possibly could be associated with the event horizon of the black hole.

We can develop a scheme involving a rotating black hole that – under certain circumstances – can work as a battery and produce jets. This is one possible case, although it must be stressed that it uses features outside the context of pure MHD. We shall now briefly describe this process (known as the *Blandford–Znajek process*). The key idea behind this process is based on the fact that there are some remarkable analogies between a rotating black hole and an electrically charged conductor with some resistance. This analogy is best seen by consideration of a series of thought experiments that are now described.

To begin with, we note that the solution of Maxwell's equations in a Schwarzchild metric shows that the electric-field lines cross the event horizon normally suggesting that the event horizon behaves as an equipotential surface of a conductor (see Exercise 8.7). To determine the resistivity of this conductor, we consider a thought experiment in which a magnetised cloud of plasma is engulfed by the black hole. After the matter has crossed the horizon, the electromagnetic fields will decay in a time scale of the order of $t_{\text{decay}} \simeq GM/c^3$. On the other hand, Maxwell's equations suggest that

$$\frac{\partial \mathbf{B}}{\partial t} = -c\nabla \times \mathbf{E} \simeq -\frac{c^3}{G}\frac{E}{M} \simeq -\frac{c^3 J R_H}{GM} \simeq -\frac{c^3 B R_H}{4\pi GM}. \qquad (8.140)$$

Equating this to $(-B/t_{\text{decay}})$, we find that the effective resistance of the black hole is given by $R_H = (4\pi/c) = 377\Omega$. More rigorous analysis shows that this result is exact.

This finite resistance will allow us to spin up the black hole like an electric motor by the following process. Suppose we embed a nonrotating black hole in a uniform magnetic field B and arrange a battery of voltage V to be connected between the pole and the equator, thereby generating a current. Clearly, $I \approx V/R_H$ is the current that crosses the magnetic-field lines. (In a physical context, the current could be generated by, say, electrons falling onto a region around the pole and protons onto a region around the equator.) Such a current in a magnetic field will exert a torque $\tau \simeq IB$ on the black hole, thereby spinning it up. The reverse of this process will involve spinning a black hole at a an angular velocity Ω_H in a constant magnetic field that – treating the event horizon as a conducting surface – leads to the idea that a potential difference will arise because of unipolar induction. The magnitude of this potential difference between the pole and the equator will be $V \simeq \Omega_H M^2 B \simeq \Omega_H \Phi$, where Φ is the magnetic flux threading the hole.

To make an order-of-magnitude estimate of this phenomenon, let us consider a black hole accreting mass at a rate $\dot{M}_E = L_E/c^2$, where $L_E \simeq 1.3 \times 10^{46} M_8$

ergs s^{-1} is the Eddington luminosity. If the matter is falling with speed $v \approx c$ near the horizon, then the number density is $n_E \simeq (c^2/\sigma_T GM) \simeq 6 \times 10^{10} M_8^{-1}$ cm^{-3}. The characteristic magnetic-field strength is given by the equipartition value $B^2/8\pi = n_E m_p c^2$, giving

$$B_E = \left[\frac{8\pi c^2 m_p}{\sigma_T GM}\right]^{1/2} = 4 \times 10^4 \, M_8^{-1/2} \, \text{G}. \qquad (8.141)$$

In a maximally rotating black hole embedded in such a magnetic field with the electromagnetic power extracted, the gyro frequency is

$$\Omega_g = \frac{e B_e}{m_e c} = 3 \times 10^{11} M_8^{-1/2} \, \text{rad s}^{-1}. \qquad (8.142)$$

This will lead to the maximum electromotive force of

$$\frac{\Omega_g r_g}{c} \frac{J}{J_{\max}} m_e c^2 \simeq 10^{21} M_8 \left(\frac{J}{J_{\max}}\right) \text{V}, \qquad (8.143)$$

where $J_{\max}$ is the maximum possible angular momentum for the black hole. The corresponding current will be $\sim 10^{18}$ A, which is high enough to induce electron–positron pair creation on the black hole's magnetosphere and supply energy to accelerate a jet.

Exercise 8.7

Field of a point charge in the Schwarzchild metric: Consider a charge at rest at the coordinates $r = r', \theta = 0$ in a Schwarzschild space–time with the metric

$$ds^2 = \left(1 - \frac{2M}{r}\right) dt^2 - \frac{dr^2}{(1 - 2M/r)} - r^2 (d\theta^2 + \sin^2\theta \, d\phi^2). \qquad (8.144)$$

(The charge is located along the z axis at a distance r' from the origin.) (a) Show that the differential equation satisfied by the scalar potential A_0 is

$$\frac{1}{r^2}\partial_r(r^2 \partial_r A_0) + \frac{1}{(1 - 2M/r)} \frac{1}{r^2 \sin\theta} \partial_\theta(\sin\theta \, \partial_\theta A_0) = -4\pi e \delta_D(r - r')\delta_D(\cos\theta - 1).$$

$$(8.145)$$

(b) Solve this equation and show that the solution with proper boundary conditions is given by

$$A_0 = \frac{e[(r - M)(r' - M) - M^2 \cos\theta]}{rr'\sqrt{(r - M)^2 + (r' - M)^2 - 2(r - M)(r' - M)\cos\theta - M^2 \sin^2\theta}}. \qquad (8.146)$$

(c) Sketch the electric-field lines for such a charged particle as r' is varied and interpret the results.

Let us now turn to some of the consequences of a jet model based on prespecified content and field geometry. To parameterise the different variables, let us assume that the electron density varies as $n_e \propto r^{-a}$ and $B \propto r^{-b}$. (If the number of electrons is conserved and flux freezing occurs, then $a = 2$ and $b = 1$.) In such a nonrelativistic jet, the synchrotron luminosity at the optically thin region with $(\nu > \nu_T)$ will be given by

$$L_\nu \propto \int_{r_{\min}} dr \; \Delta\Omega(r)\, n_e(r)\, B^{(1+\xi)/2}(r)\, \nu^{(1-\xi)/2}, \qquad (8.147)$$

where $r_{\min}$ is defined by the relation $\nu = \nu_T(r_{\min})$ and $\Delta\Omega(r)$ is the solid angle of the jet as a function of radial distance. This integral is essentially a superposition of different components of synchrotron radiation described earlier. Because

$$\nu_T(r) \propto B^{(\xi+2)/(\xi+4)}(n_e r)^{2/(\xi+4)}, \qquad (8.148)$$

we find that

$$r_{\min} \propto \nu^{1/[2(b-a+1)/(\xi+4)-b]}. \qquad (8.149)$$

For the range of values $\xi \simeq (2\text{–}3)$, $B \simeq 1$, $r_{\min}$ generally decreases as ν increases. For a conical jet, with $\Delta\Omega$ independent of r, we can easily work out the integral and obtain the spectral index as

$$\frac{d \ln L_\nu}{d \ln \nu} = s = \frac{1-\xi}{2} - \frac{3 - a - b(1+\xi)/2}{b - 2(b-a+1)/(\xi+4)}. \qquad (8.150)$$

This has the interesting feature that, for a conical jet with $a = 2$ and $b = 1$, we get $s = 0$ – implying a flat spectrum – independently of ξ. Such a jet has several other special features. For example, the minimum radius at which a given frequency is radiated varies as ν^{-1}, which scales with the frequency exactly as the resolution of an interferometric image. If we take the smallest dimension of the jet segment at a distance r as r_T, then the peak brightness temperature $T_{bT} \propto (n_e r_T / B)^{1/(\xi+4)}$ is independent of radius because $r_T \propto r$ in a conical jet. These results are consistent with observations, although the model is by no means unique.

Exercise 8.8

Radiative acceleration of jets: (a) Consider a beam of radiation with intensity I_ν that is hitting an electron and making it accelerate. Show that, in the Thomson scattering limit, the energy of the electron changes according to

$$m_e c^2 \frac{d\gamma}{dt} = \int d\nu \int d\Omega \, I_\nu(\nu, \hat{\mathbf{n}}_i) \sigma_T \, v \, \gamma^2 (1 - \mathbf{v} \cdot \hat{\mathbf{n}}_i)(\hat{\mathbf{n}}_i \cdot \mathbf{v} - v). \qquad (8.151)$$

Argue from this equation that, unless the photons are directed closely along the direction of the velocity of the electron, they decelerate the electrons rather than accelerate. Give a physical reason for this effect. (b) Specialise to a radiation field containing a well-collimated component along with a more isotropic component in the form

$$I(\hat{\mathbf{n}}_i) = \frac{F}{2\pi} \delta_D(\hat{\mathbf{n}}_i - \mathbf{z}) + J_*. \tag{8.152}$$

Work out the angular integrations in Eq. (8.151) and take the limit $v \to 1$ to obtain

$$m_e c^2 \frac{d\gamma}{dt} = \sigma_T \left(\frac{1}{4\gamma^2} F - \frac{16\pi}{3} \gamma^2 J_* \right). \tag{8.153}$$

Hence show that the equilibrium value of γ is $\gamma = (3F/64\pi J_*)^{1/4}$. Explain why this creates difficulties for radiative acceleration as an efficient mechanism.

Exercise 8.9
Compton scattering of synchrotron photons: Show that, in a nonrelativistic jet, the photon energy density at any given location is given by

$$U_v(r) \propto \int_{r_{min}(v)}^{r_{max}(v)} dr' \min[1, (r'/r)^2] n_e(r') \gamma_{min}^{\xi-1} B^{(1+\xi)/2}(r') v^{(1-\xi)/2}, \tag{8.154}$$

where $r_{max}(v)$ is defined through the relation $v = 4\gamma_{max}^2(eB(r)/2\pi m_e c)$. When $a + b(1 + \xi)/2 > 2$, the energy density is dominated by photons produced near r_{min} and not by photons produced locally. In this limit, the emissivity for an inverse Compton photon is

$$J_\epsilon(r) \propto \sigma_T \int d\gamma n_e(r) \gamma_{min}^{\xi-1} \gamma^{-\xi} \int dv \frac{U_v(r)}{hv} \epsilon \delta_D \left(\frac{4}{3} \gamma^2 hv - \epsilon \right). \tag{8.155}$$

Assume further that $U_v(r) = [U_0(r)/v_0](v/v_0)^{-l}$. Evaluating the integral, show that

$$J_\epsilon(r) = (\xi - 1)\gamma_{min}^{\xi-1} n_e \sigma_T c \frac{U_0(r)}{hv_0} \left(\frac{\epsilon}{hv_0} \right)^{-s} \int_{\gamma_{min}}^{\gamma_{max}} d\gamma \gamma^{2s-\xi} \tag{8.156}$$

in the range

$$\gamma_{min}^2 hv_{min} \leq \epsilon \leq \min[\gamma_{max} m_e c^2, \gamma_{max}^4 \hbar e B_{max}/(m_e c)]. \tag{8.157}$$

Under what conditions will the upscattered spectrum have the same shape as that of the original synchrotron spectrum? What will be the consequence of this?

Let us next consider the energetics of the electrons emitting the radiation. A stream of electrons injected into the jet and emitting synchrotron radiation will continuously lose energy, thereby changing the spectral shape. If $N(t, E)$ is the number density of electrons with energy E at time t, we can write the conservation law for electrons in the E–t space as

$$\frac{\partial N(E)}{\partial t} = -\frac{\partial}{\partial E}\left[\dot{E} N(E)\right] + Q, \tag{8.158}$$

where $\dot{E} = -bE^2$ represents the synchrotron energy loss and Q is the rate of injection of electrons with energy E. When $Q \propto E^{-x}$, the steady-state solution is $N \propto E^{-(x+1)}$, showing that the ageing will steepen the spectrum [see Eq. (8.79)].

To obtain a more general solution, we need to first determine the Green's function for this equation. Any given electron loses energy at the rate $\dot{E} = -bE^2$ so that

$$E(t) = \frac{E_0}{1 + E_0 bt}. \tag{8.159}$$

If the initial *cumulative* energy distribution is $\mathcal{N}(> E_0) \propto E_0^{-(x-1)}$, then the cumulative distribution at any other later time will be the same, with E_0 expressed as a function of E through Eq. (8.159). Hence

$$\mathcal{N}(> E, t) \propto \left(\frac{E}{1 - Ebt} \right)^{-(x-1)}. \tag{8.160}$$

Differentiating this with respect to E will give the Green's function for the problem. Consider now a source that had the form $Q \propto E^{-x}$ during $0 < t < T$. The energy distribution at any time t is given by

$$N(E, t) = \int E^{-x} [1 - Eb(t - t_0)]^{x-2} \, dt_0. \tag{8.161}$$

The source is alive in the interval $0 < t < T$, but the solution is applicable only for $t_0 > t - (Eb)^{-1}$. Hence there are two separate solutions in two separate regimes corresponding to $Ebt > 1$ and $Ebt < 1$. In the simplest case of a continuously active source, we get

$$N = \begin{cases} \dfrac{E^{-x}}{(x-1)Eb}[1 - (1 - Ebt)^{x-1}] & \text{(for } E < 1/bt) \\[4mm] \dfrac{E^{-x}}{(x-1)Eb} & \text{(for } E > 1/bt) \end{cases}. \tag{8.162}$$

This gives a continuously changing slope, making the gradual transition between x and $x + 1$. Such a signature can, in principle, be used to estimate the age of the electrons emitting the radiation.

The relativistic electrons in the jet also lose energy through inverse Compton scattering of CMB photons. Let us assume that the electrons are injected at a constant rate into the hot spot of a double radio source with a power-law spectrum $\dot{N}(\gamma)d\gamma \propto \gamma^{-p} \, d\gamma$, reside there for a fixed time t_r before being advected, and are taken away by the backflow. In this case, it is fairly easy to work out the effects of inverse scattering with the microwave background photons of energy density U_{CMB}. The key effect is to change the steady-state spectrum of the relativistic electrons from γ^{-p} to γ^{-p+1} at the break value

$$\gamma \simeq \gamma_b \simeq \frac{m_e c^2}{\sigma_T c t_r U_{\text{CMB}}}. \tag{8.163}$$

The corresponding radio spectrum of synchrotron radiation steepens from $I_\nu \propto \nu^{-(p-1)/2}$ to $I_\nu \propto \nu^{-p/2}$ at a corresponding break frequency ν_b.

Exercise 8.10

Filling in the blanks: Prove relation (8.163) and determine the redshift dependence of ν_b. Can this phenomenon explain why radio sources with steep spectra turn out to be at high redshifts whereas similar ones with flat spectra turn out to have low redshifts?

8.7 Effects of Bulk Relativistic Motion

For luminous jets moving with relativistic speeds, we need to consider several effects of special relativity to relate the observed quantities to the intrinsic properties of the source. Although all these effects arise from a fairly straightforward application of special relativity, some of them can be counterintuitive and deserve a discussion.

The first nontrivial effect that is due to relativistic motion is that the apparent proper motion of a jet in the sky could be at a speed greater than that of light. This arises as follows: Consider a blob of plasma that moves with speed v at an angle ψ to the line of sight. Two signals emitted by the blob in a time interval dt_e in the rest frame of the source will be detected during a time interval dt_0 in the rest frame of the observer. There are two effects that make $dt_e \neq dt_0$. The first is the cosmological redshift that scales the time intervals by the factor $1 + z$; it turns out, however, that the really interesting features do not arise from this factor, and hence we shall temporarily ignore it. The second effect is due to the fact that the blob moves a distance $v\, dt_e \cos \psi$ in the direction of the observer between the emission of two signals so that the second signal has to travel less distance compared with that of the first. This reduction in the distance implies that the signals will arrive in a time interval (in units of $c = 1$)

$$dt_0 = (1 - v \cos \psi)\, dt_e. \tag{8.164}$$

The distance travelled by the blob in the plane of the sky, transverse to the line of sight, is $dl_\perp = v\, dt_e \sin \psi$, so that the apparent transverse velocity measured by an observer is

$$v_a = \frac{dl_\perp}{dt_0} = (v \sin \psi)\frac{dt_e}{dt_0} = \frac{v \sin \psi}{1 - v \cos \psi}. \tag{8.165}$$

Figure 8.4 plots v_a as a function of ψ for different values of v. It is obvious that v_a can be greater than unity even for large inclination angles, provided that γ is high enough. The following results can be obtained directly from Eq. (8.165). (1) For a given value of v, the apparent velocity v_a is a maximum when $\psi = \cos^{-1} v$, with the maximum value being γv. (2) In the limit of $v \to 1$, the maximum occurs for $\psi \simeq (1/\gamma)$. (3) It also follows from this analysis that

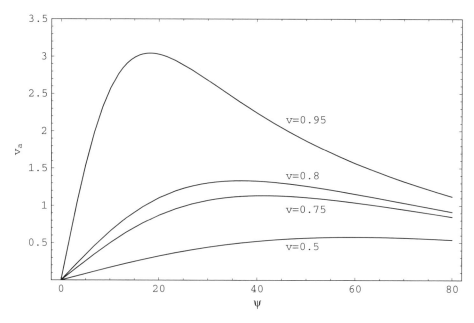

Fig. 8.4. Apparent transverse speed as a function of the inclination angle of the jet with respect to the line of sight.

superluminal motion occurs at some angle, provided that $v_a^{\max} > 1$, corresponding to $v > (1/\sqrt{2})$. (4) The same analysis can be repeated in terms of a fixed observed value for v_a. The minimum speed v required for producing a given value of v_a occurs at $\psi = \cot^{-1}(v_a)$, with $v_{\min} = v/\sqrt{1 + v^2}$. (5) When the cosmological redshift is taken into account, Eq. (8.164) needs to be modified by a factor of $1 + z$ on the right-hand side, which does not change the condition for observing the superluminal motion.

Exercise 8.11

Getting the numbers: (a) Consider a source, with $\mu = d\theta/dt_0$ being the angular proper motion in units of milliarcseconds per year, in a matter-dominated universe. Show that

$$\frac{v_a}{c} = 94.4h^{-1}\left(\frac{\mu}{1 \text{ milliarcsec yr}^{-1}}\right)\left[\frac{\Omega z + (\Omega - 2)(\sqrt{1 + \Omega z} - 1)}{\Omega^2(1 + z)}\right]. \quad (8.166)$$

(b) Assume that a source emits blobs of matter with speed v in a random direction. Show that the probability that the apparent transverse speed for the source will exceed a value v_a is given by

$$p(> v_a) = \frac{1}{1 + v_a^2}\sqrt{1 - \frac{v_a^2}{\gamma^2 - 1}}. \quad (8.167)$$

The relativistic motion also affects the flux of the source. If cosmological factors are ignored, then the constancy of I_ν/ν^3 implies that the intensity (or surface brightness) of the source will scale as

$$I_\nu(\mathcal{D}\nu_0) = \mathcal{D}^3 I_\nu(\nu_0),\qquad(8.168)$$

where

$$\mathcal{D} = \frac{1}{\gamma(1 - v\cos\psi)}.\qquad(8.169)$$

For an unresolved optically thin source or an optically thick spherically symmetric source, the flux $F(\nu)$ transforms in the same way as the intensity. In particular, for a source with power-law spectrum $F(\nu) \propto \nu^{-\alpha}$, we have the result

$$F(\nu) = \mathcal{D}^{3+\alpha} F'(\nu),\qquad(8.170)$$

where the prime quantities refer to the rest frame. It is clear that there is a boosting of the observed flux over the rest-frame flux. When the cosmological redshift cannot be ignored, the observed frequency ν and the rest frame ν' are related by

$$\nu = \frac{\mathcal{D}\nu'}{1+z},\qquad(8.171)$$

so that the flux transforms to

$$F(\nu) = \left(\frac{\mathcal{D}}{1+z}\right)^{3+\alpha} F'(\nu).\qquad(8.172)$$

In particular, consider a source that emits two blobs of material moving in the opposite directions ψ and $\pi + \psi$ with the same speed v. If F_{in} and F_{out} are the fluxes from the sources moving towards the observer and away from the observer, respectively, then we have

$$\frac{F_{\mathrm{in}}}{F_{\mathrm{out}}} = \left(\frac{1 + v\cos\psi}{1 - v\cos\psi}\right)^{3+\alpha},\qquad(8.173)$$

which shows that there is a preferential enhancement of fluxes in the component that is moving towards the observer.

The results described above are valid for a single blob of material moving with a given velocity. In the case of a continuous jet formed out of a series of blobs, the number of blobs observed at any given instant scales as $\mathcal{D}^{-1}$, and hence the indices in the formulas for the flux change from $(3 + \alpha)$ to $(2 + \alpha)$. So the ratio of the flux arising from advancing jet to the identical receding jet will be

$$\frac{F_{\mathrm{in}}}{F_{\mathrm{out}}} = \left(\frac{1 + v\cos\psi}{1 - v\cos\psi}\right)^{2+\alpha}.\qquad(8.174)$$

Such a relativistic beaming effect is often invoked to explain the fact that several observed radio sources have only a one-sided jet. If the ratio of the fluxes is

$R = F_{in}/F_{out}$, we can obtain an upper limit on the angle of inclination as

$$\psi \leq \frac{R^{1/(2+\alpha)} - 1}{R^{1/(2+\alpha)} + 1}. \tag{8.175}$$

The relativistic effects also play a role in determining the time scale of variability of the sources and hence in putting bounds on the size of the source. If the observed variability is in a time scale Δt_{var}, then the intrinsic time scale of variability in the rest frame of the source is $\Delta t'_{var} = \mathcal{D}\Delta t_{var}$. The minimum source size corresponding to this variability time scale is $R \simeq c\Delta t'_{var} \simeq c\mathcal{D}\Delta t_{var}$. The solid angle subtended at the source is $\Omega \simeq (c\mathcal{D}\Delta t_{var}/D)^2$, where D is the distance to the source. The observed flux is then given by

$$F(\nu) = I(\nu)\left(\frac{c\mathcal{D}\Delta t_{var}}{D}\right)^2, \tag{8.176}$$

where $I(\nu)$ is the intensity measured in the observer's frame. We can now easily relate the brightness temperature T' corresponding to a given intensity in the rest frame and the brightness temperature as determined in the observer's frame T_{var} corresponding to the angle θ subtended by the source where $\Omega \approx \theta^2$. Simple algebra now gives

$$T_{var} = \mathcal{D}^3 T'. \tag{8.177}$$

The physical origin of the $\mathcal{D}^3$ factor can be understood as follows: The brightness temperature is related to the intensity by $T \propto (I_\nu/\nu^2)\theta^2$. If θ is treated as a fixed quantity, then T will scale as $(I_\nu/\nu^2) = (I_\nu/\nu^3)\nu$; because (I_ν/ν^3) is invariant, this will bring in one factor of $\mathcal{D}$. If we further estimate θ^2 by the time variability (rather than treat it as fixed), then we will get an additional $\mathcal{D}^2$ factor from this, thereby producing the $\mathcal{D}^3$ factor. This shows that there is an enhancement of the brightness temperature that is due to relativistic beaming.

The relativistic beaming, because it amplifies the observed luminosity of the source, will also change the luminosity function, if we have a class of sources all having the same luminosity l but in relativistic motion over randomly distributed directions. The observed luminosity L is then given by

$$L = \mathcal{D}^m l, \quad \mathcal{D} = \frac{1}{\gamma(1 - \beta \cos \psi)}, \tag{8.178}$$

where $m = 2 + \alpha$ for a continuous jet and $m = 3 + \alpha$ for a blob. As the angle ψ varies from 0 to $\pi/2$, the observed luminosity varies in the range $[(2\gamma)^{-m}l, (2\gamma)^m l]$. For a collection of sources with the same γ but randomly oriented beaming directions, the probability distribution for $\mathcal{D}$ is given by $P(\mathcal{D})d\mathcal{D} = \sin \psi \, d\psi$; from this it is easy to work out the conditional probability of observing a luminosity L for a source with luminosity l:

$$P(L|l)\, dl = \left(\frac{1}{\beta\gamma m}\right) l^{1/m} L^{-(m+1)/m}. \tag{8.179}$$

If we further assume that the intrinsic luminosity is distributed according to some luminosity function $\Phi_{\text{int}}(l)$, then the observed luminosity function is given by

$$\Phi_{\text{obs}}(L) = \int dl\, P(L|l)\, \Phi_{\text{int}}(l). \tag{8.180}$$

The limits of integration will be fixed by the cutoffs on the intrinsic luminosity function as well as the range of L and γ. If $\Phi_{\text{int}}(l) = K l^{-q}$ with $q > 1, l > l_{\text{min}}$, then the observed luminosity function is given by

$$\Phi_{\text{obs}}(L) = \left(\frac{K}{\beta\gamma m}\right) L^{-(m+1)/m} \int_{l_1}^{(2\gamma)^m L} dl\, l^{-q+1/p},$$

$$l_1 = \max[l_{\text{min}}, L(2\gamma)^{-m}], \tag{8.181}$$

that is,

$$\Phi_{\text{obs}}(L) \propto \begin{cases} L^{-q} & (\text{for } L > (2\gamma)^m l_{\text{min}}) \\ L^{-(m+1)/m} & (\text{for } L < (2\gamma)^m l_{\text{min}}) \end{cases}. \tag{8.182}$$

The physical reason for this scaling should be obvious: At high luminosities a whole range of l contributes to Φ_{obs}, thereby maintaining the original shape, whereas for low l, the contribution comes mainly from the sources near l_{min}. The effect of beaming is thus to produce a power-law luminosity function with a break that is flattened (compared with the intrinsic luminosity function) at low luminosities.

Exercise 8.12
Useful relations: Prove that γ, $\mathcal{D}$, ψ, and v_a are related by

$$\gamma = \frac{v_a^2 + \mathcal{D}^2 + 1}{2\mathcal{D}}, \qquad \tan\psi = \frac{2v_a}{v_a^2 + \mathcal{D}^2 - 1}, \qquad \frac{1 + v\cos\psi}{1 - v\cos\psi} = v_a^2 + \mathcal{D}^2. \tag{8.183}$$

Exercise 8.13
Numerical estimates in variability: The brightness temperature T_b, for a source with brightness S at wavelength λ and subtending an angle θ, is given by $T_b \simeq \lambda^2 S/(2k_B\theta^2)$. Suppose the flux changes by an amount ΔS (in jansky units) in a time scale τ (in years), show that

$$T_b \geq 2 \times 10^{12} \, K\lambda_{\text{cm}}^2 \left(\frac{\Delta S}{\tau^2}\right) \tag{8.184}$$

if we ignore relativistic effects. Estimate T_b for the two sources, 3C 454.3 ($\Delta S \simeq 5\,\text{Jy}, \tau \simeq$ 1 yr, $\lambda \simeq 74$ cm) and AO 0235 ($\Delta S \simeq 1$ Jy, $\tau \simeq 30$ days, $\lambda \simeq 11$ cm). What is the range of the Lorentz factor γ and the direction of motion θ (with respect to the line of sight) for which relativistic effects can bring the actual brightness temperature down to 10^{12} K?

Exercise 8.14

Jet scaling: Consider a jet with constant opening angle and assume that the surface brightness (luminosity per unit projected surface area) scales as $r^{-2.5}$, where r is the distance along the jet.

(a) Determine the scalings for (i) the equipartition pressure in the jet, (ii) the equipartition magnetic-field strength, and (iii) lifetime that is due to synchrotron cooling.

(b) Suppose the jet has a constant bulk Lorentz factor and that the synchrotron radiation is due to relativistic electrons with energy distribution $N(\gamma) \propto \gamma^{-2}$. Show that if the electron gas expands rapidly and the jet is dominated by a transverse magnetic field, then the surface brightness will scale as $r^{-19/6}$.

[Answers: (a) Constant opening angle $\theta = R/r$ implies that $r \propto R$. The volume emissivity is then $\epsilon_\nu \propto I_\nu/R \propto r^{-3.5}$. The volume emissivity in synchrotron radiation scales as $\epsilon_\nu \propto n_e(\gamma^2 B^2/\nu) \simeq p_e(B^2\gamma/\nu)$, where $p_e \simeq n_e\gamma mc^2$ is the pressure of the electron gas. Assuming that most of the radiation is at $\nu \simeq \nu_c \propto \gamma^2 B$, we get $\epsilon_\nu \propto p_e B^{3/2}\nu_c^{-1/2} \propto B_{eq}^{7/2}\nu_c^{-1/2}$ if $p_e \propto B_{eq}^2$ under the assumption of equipartition. Comparing $\epsilon_\nu \propto r^{-3.5}$ with $\nu_c^{1/2}\epsilon_\nu \propto B_{eq}^{3.5}$, we find that the equipartition field $B_{eq} \propto r^{-1}$ and the equipartition pressure scales as r^{-2}. The synchrotron cooling time varies as $t_{cool} = E/\dot{E} \propto \gamma/\gamma^2 B^2 \propto \nu^{-1/2}B^{-3/2} \propto r^{3/2}$. (b) For a relativistic gas expanding adiabatically, $p \propto \rho^{4/3} \propto R^{-8/3}$. Equating this to $p_e \propto n_e\gamma \propto \gamma^2(dn/d\gamma)$, we get $\gamma^2 dn/d\gamma \propto R^{-8/3}$. The emissivity varies as $\epsilon_\nu \propto (\gamma^2 dn/d\gamma)B^{3/2}\nu^{-1/2} \propto B^{3/2}\nu^{-1/2}R^{-8/3}$. Using the result $B \propto R^{-1}$, we get $\epsilon_\nu \propto \nu^{-1/2}R^{-25/6}$. The corresponding surface brightness is $I_\nu \propto \epsilon_\nu R \propto \nu^{-1/2}R^{-19/6} \propto r^{-3.1}$. (Because the observed distribution scales as $I_\nu \propto r^{-2.5}$, there must be a process of continuous acceleration to restore the energy lost because of adiabatic expansion.)]

8.8 The Broad-Line and Narrow-Line Regions

Quasars display a wide variety of emission lines among which there exist two specific categories, usually called *broad lines* and *narrow lines*. There is fair amount of evidence to suggest that these lines originate from physically different regions, called the *broad-line region* (BLR) and the *narrow-line region* (NLR). The BLR has a typical size of 10–100 light days in Seyfert I galaxies and could be up to a few light years in bright quasars. The linewidths for lines originating from a BLR correspond to approximately $(3\text{–}10) \times 10^3$ km s^{-1}. In contrast, the NLR emits lines with widths of $(3\text{–}10) \times 10^2$ km s^{-1}. We now discuss some aspects of the modelling of these regions.

8.8.1 Broad-Line Regions

To obtain a linewidth of $\sim$5000 km s^{-1} from purely thermal motion would require temperatures in excess of 10^9 K, which are clearly ruled out in BLRs. It is generally assumed that the linewidth arises because of the bulk motion of the cloud rather than from thermal motion. The actual physical parameters of the BLR can be estimated by the study of different lines that are detected, and we shall begin with some simple examples of such a diagnosis.

We recall from Vol. II, Chap. 9, that the population equilibrium between two ionic levels 1 and 2 is represented by an equation of the form

$$N_1 N_e \Gamma_{12} = N_2 A_{21} + N_2 N_e \Gamma_{21}, \tag{8.185}$$

where $\Gamma = \langle v\sigma(v) \rangle$ is the collisional excitation rate, A_{21} is the spontaneous emission coefficient, and N_i are the number densities at the two levels. Solving this for N_1/N_2 and noting that the line flux j_ν is proportional to $N_2 A_{21} h\nu$, we get

$$j_\nu \propto \frac{N_e}{1 + A_{21}/(N_e\Gamma_{21})} \propto \frac{N_e}{1 + (N_c/N_e)}, \qquad N_c \equiv \left(\frac{A_{21}}{\Gamma_{21}}\right), \tag{8.186}$$

where N_c is the critical density for the particular line transition. The critical density for the [OIII]λ4363 line, for example, is $\sim 10^8$ cm^{-3} and the fact that this line is generally absent (or very weak) in the BLR gives a lower bound to the electron density in these regions; $N_e \gtrsim N_c \approx 10^8$ cm^{-3}. Similarly, we can use the presence of semiforbidden line CIII]λ1909 (with a critical density of $\sim 10^{10}$ cm^{-3}) to set an upper limit to the electron density, provided the emission is from the same region. (We shall see later, however, that there is some evidence from a procedure called reverberation mapping to suggest that the CIII] emission arises from a different region and the electron densities are probably closer to 10^{11} cm^{-3}.)

Given the temperature and electron density, it is easy to estimate the mass of the BLRs using the intensity of any convenient line. For example, the emissivity (ergs s^{-1}cm^{-3} sr^{-1}) of collisionally excited CIV λ1549 is given by

$$j(\text{CIV}) = n_e n_{C^{3+}} q(2s\,^2S, 2p\,^2P)\frac{h\nu}{4\pi}, \tag{8.187}$$

where $q \simeq 2.6 \times 10^9$ cm^3 s^{-1} (at $T \simeq 20{,}000$ K) is the collisional excitation rate. If the radius of the emitting region is r and the filling factor – which gives the fraction of the emitting volume that actually contains line-emitting material – is ϵ, then the total luminosity is

$$L(\text{CIV}) \simeq 4.6 \times 10^{-23}\epsilon r^3 n_e^2 \tag{8.188}$$

in cgs units. We can estimate the size r of the BLR by noting that the region responds very quickly (in ~ 8 days) to the variations in the intensity of the continuum radiation. This suggests that $r \lesssim 8$ light days $\cong 2 \times 10^{16}$ cm. Using this and $n_e \simeq 10^{11}$ cm^{-3}, we get the filling factor of the BLR as

$$\epsilon = 2.7 \times 10^{-7} L_{42}^{-1/2}(\text{CIV}). \tag{8.189}$$

This small value shows that the BLR structure is very clumpy and the emission linewidth could indeed arise from the bulk motion of individual clouds. If there are N_c clouds, each of radius l, then $\epsilon = N_c(l/r)^3$ and the mass of the BLR is

$$M_{\text{BLR}} = \frac{4\pi}{3}\ell^3 N_c n_e m_p = \frac{4\pi}{3}\epsilon r^3 n_e m_p \approx \frac{1.5 \times 10^{41}}{n_e} L_{42}(\text{CIV}) \text{ gm}$$
$$\approx 10^{-3} M_\odot L_{42}(\text{CIV}), \tag{8.190}$$

where we have used relation (8.188) in arriving at the third equality and used $n_e = 10^{11} \text{ cm}^{-3}$ to arrive at the last step. It is clear that the total mass in the BLR is fairly small.

The filamentary structure of the BLR also suggests that it will absorb only a fraction of the continuum ionising photons emitted by the central source. If the solid angle subtended by the BLR at the origin is Ω, then the covering factor f is defined by $f = (\Omega/4\pi)$. One way of estimating this covering factor is to begin by noting that in optically thick hydrogen nebula every photoionisation ultimately results in one Lyman-α photon. We take the ionising flux from the central source as

$$F(\nu) = F(\nu_0)\left(\frac{\nu}{\nu_0}\right)^{-\alpha} \tag{8.191}$$

(where we take ν_0 to be the frequency corresponding to $\lambda_0 = 1216$ Å for convenient normalisation). The total number of ionising photons is given by

$$\Phi = \int_{\nu_1}^\infty \frac{F(\nu)\,d\nu}{h\nu} = \frac{F(\nu_0)}{h\alpha}\left(\frac{\nu_1}{\nu_0}\right)^{-\alpha}, \tag{8.192}$$

where ν_1 is the frequency corresponding to the Lyman edge at 912 Å. If the covering factor is f, then a fraction f of these photons will be absorbed by the BLR. Because the number of ionising photons absorbed is equal to the number of Lyman-α photons eventually emitted, the flux in the Lyman-α emission line is

$$F(\text{Ly }\alpha) = f\Phi h\nu_0 = f\frac{F(\nu_0)}{\alpha}\left(\frac{\nu_1}{\nu_0}\right)^{-\alpha}\left(\frac{c}{\lambda_0}\right). \tag{8.193}$$

By definition, the equivalent width of the Lyman-α emission line is

$$W(\text{Ly }\alpha) = \left[\frac{F(\text{Ly }\alpha)}{F_\lambda(\lambda_0)}\right] = \left[\frac{F(\text{Ly }\alpha)}{F(\nu_0)}\right]\left(\frac{\lambda_0^2}{c}\right) = f\frac{\lambda_0}{\alpha}\left(\frac{\nu_1}{\nu_0}\right)^{-\alpha}. \tag{8.194}$$

For $\alpha \approx 1.4$, the equivalent width is $\sim 580f$ Å whereas the observed equivalent width is ~ 70 Å, suggesting that $f \approx 0.1$; thus the covering factor of the BLR is only $\sim 10\%$.

The energy source behind the BLR emission lines is essentially the central source that photoionises the cloud. The photoionisation equilibrium of such a system depends on several factors, but the most important parameter that governs the equilibrium is the ratio between the number density of ionising photons and the number density of H atoms. Taking the total number of photons emitted by the central source per second (which is capable of ionising H) as

$$Q(\text{H}) = \int_{\nu_1}^{\infty} \frac{L_\nu}{h\nu}\, d\nu, \tag{8.195}$$

we can define the *ionisation parameter U* as

$$U = \frac{Q(\text{H})}{4\pi r^2\, c\, n_{\text{H}}}. \tag{8.196}$$

The behaviour of the system in equilibrium can then be understood along the lines similar to those used in the study of the ISM in Vol. II, Chap. 9, Section 9.6. In particular, photoionisation calculations show that the temperature and the ionisation state of a photoionised gas depends primarily on the ionisation parameter and only weakly on the shape of the ionising spectrum or the pressure. If the spectrum has no sharp cutoffs between 10 and 10^4 eV, then the ionisation states of different elements are also essentially determined by the ionisation parameter. The variation of temperature as a function of the ionisation parameter is shown in Fig. 8.5. The freezing of the temperature at $T \simeq 10^4$ K is an effect that we had already seen in the case of the ISM (see Vol. II, Chap. 9, Section 9.5). The strongest atomic and ionic lines have energies in the range of a few

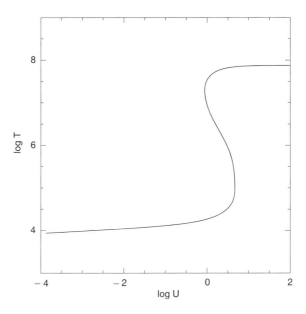

Fig. 8.5. A schematic diagram showing the typical variation of the temperature with the ionisation parameter.

electron volts that act as a thermostat to maintain the temperature around 10^4 K. A dramatic change occurs when the ionisation parameter approaches ~ 10. Around this stage, even at relatively low temperatures, the ionisation balance of C, N, and O becomes dominated by H- and He-like species. Because the energies involved go up, the cooling rate is sharply suppressed until the temperature rises close to 10^6 K. Depending on the shape of the spectrum, we may get a multivalued dependence of T on the ionisation parameter, as seen in Fig. 8.5. Of the three equilibrium temperatures, only the lowest and the highest are stable, for the reasons discussed in Vol. II, Chap. 9, Section 9.6. When the ionisation parameter is much higher than 10, the only equilibrium possible is Compton equilibrium, in which the cooling is dominated by Compton processes.

Given these ideas, it is possible to obtain a broad physical picture of a cloud that is radiated on one side by a power-law spectrum of ionising photons.[5] For the sake of definiteness, let us consider a plane-parallel geometry with an ionisation parameter $U = 1$, $p \simeq 0.01$ dyn cm^{-2}, a column density of $N_H = 3 \times 10^{22}$ cm^{-2}, and a solar abundance of metals. At the illuminated edge of the slab, the temperature will reach approximately 2.5×10^4 K because of the effects that we had already discussed. H will be almost completely ionised in this region, and most of the heating will be actually due to photoionisation of HeII. The cooling will be due to the $O_{VI}1034$ line, with some amount being contributed by $N_V 1240$. C is also highly ionised, with C_V being the most abundant ionisation state. In fact, the relative population of an element in different ionisation states can be a useful probe of the physical conditions in the cloud (see Vol. II, Chap. 9, Section 9.7). The ionisation balance between two ionisation states I and $I+1$ of an element (say, for example, C) is determined by the relation

$$n_{I+1} n_e \alpha_R = n_I \int_{\nu_I}^{\infty} \frac{F_\nu}{h\nu} \sigma_{PI}(\nu) \, d\nu, \qquad F_H = \int_{\nu_I}^{\infty} \frac{F_\nu}{h\nu} \, d\nu, \qquad (8.197)$$

where α_R is the relevant recombination coefficient. If $\sigma_{PI} \propto \nu^{-3}$ and $F_\nu \propto \nu^{-1}$, then we get

$$\frac{n_{I+1}}{n_I} \simeq \frac{\sigma_{PI}(\nu_I)}{4\alpha_R} \left(\frac{F_H}{n_e} \right). \qquad (8.198)$$

Using the numbers relevant for CII and CIV, we find that

$$\frac{n(\text{CII})}{n(\text{CIV})} \simeq 2.08 \times 10^{13} \left(\frac{F_H}{n_e} \right)^{-2}. \qquad (8.199)$$

If the ionisation parameter $F_H/(n_e c)$ varies from 1.3×10^{-2} to 1.3×10^{-4}, this ratio varies from 1.3×10^{-4} to 1.3, showing that the relative abundances in the two states can cross over depending on the ionisation parameter. Because the number of ionising photons decreases as we go into the slab, the relative ratios of atoms in different states of ionisation will also show a corresponding spatial variation.

As we move into the cloud, the HeII edge becomes optically thick and other species (with ionisation potentials greater than 55 eV, like O_{VI} and N_V) disappears. The cooling will now be mainly due to $C_{IV}1549$. The zone, a little deeper, presents the transition between ionised and neutral H. The sharpness of this zone can be estimated as follows: If we take the photoionisation cross section to vary as $\sigma \propto \nu^{-3}$, then an approximate radiative transfer equation for the ionising continuum will be

$$\frac{\partial I_\nu}{\partial \tau_L} = -\left(\frac{\nu_H}{\nu}\right)^3 I_\nu, \tag{8.200}$$

where $h\nu_H$ is the ionisation potential for Hydrogen and τ_L is the optical depth at the Lyman edge. This integrates to give

$$I_\nu(\tau_L) = I_\nu(0) \exp\left[-\tau_L\left(\frac{\nu}{\nu_H}\right)^{-3}\right]. \tag{8.201}$$

The variation of the optical depth with distance can be expressed by

$$\frac{d\tau_L}{dz} = n\,\sigma_{ph}(\nu_H) = \frac{n_e^2 \alpha_{rec}}{\int_{\nu_H} d\nu\,\sigma_{ph}(\nu)I_\nu/(h\nu)}, \tag{8.202}$$

provided the neutral fraction is reasonably small. For an incident power-law spectrum $I_\nu \propto \nu^{-\alpha}$, the photoionisation rate will be given by

$$\int_{\nu_H} d\nu \left(\frac{\nu}{\nu_H}\right)^{-3} \frac{I_\nu}{h\nu} = h^{-1} \int_1 dx\,x^{-4-\alpha} I_\nu(\nu_H)\exp\left(-\frac{\tau_L}{x^3}\right)$$

$$= \frac{I_\nu(\nu_H)}{h(1+\alpha/3)}\tau_L^{-1-\alpha/3}\gamma(1+\alpha/3,\tau_L), \tag{8.203}$$

where $\gamma(a,t)$ is the incomplete gamma function that varies slowly for $a > 0$ and $t > 1$. Ignoring the slow variation of this function, we find that the photoionisation rate is dominated by the power-law dependence on τ_L. This is in contrast with the situation in an ISM ionised by massive OB stars in which the spectral dependence is exponential for the incoming photons. In such a case, in which the gas is ionised by a thermal spectrum, the photoionisation rate drops exponentially with τ_L, whereas in the present case it drops only as a power law in τ_L. The actual variation of optical depth with z is now governed by the equation

$$\frac{d\tau_L}{dz} \propto \tau_L^{1+\alpha/3}, \tag{8.204}$$

which integrates to give

$$\tau_L \propto (z_0 - z)^{-3/\alpha}. \tag{8.205}$$

Thus the distance between the point at which the Lyman edge becomes optically thick to the point z_0 at which the optical depth is formally infinite, τ_L rises fairly rapidly. In other words, the ionisation transition is still very sharp in terms of physical length scales.

This argument can be presented in a somewhat different manner that would allow us to relate directly the column density of ionised electrons to the ionisation parameter. The ionisation transition occurs at a depth z_0 at which the integrated recombination rate is equal to the rate at which ionising photons impinge on the slab. Hence,

$$\mathcal{N}_{\text{ion}} = \int_0^{z_0} dz \, n_e^2 \, \alpha_{\text{rec}}, \tag{8.206}$$

where $\mathcal{N}_{\text{ion}}$ is the number flux of incoming photons. Treating the variables as approximately constant we can relate the column density of ionised electrons in this region $N_e = n_e z_0$ to the ionisation parameter $\Gamma \equiv (\mathcal{N}_{\text{ion}}/n_e c)$. We get

$$N_e \simeq \frac{\mathcal{N}_{\text{ion}}}{n_e \, \alpha_{\text{rec}}} = \frac{c\Gamma}{\alpha_{\text{rec}}} \simeq 1.5 \times 10^{23} \, U \, \text{cm}^{-2}. \tag{8.207}$$

Hence the column density that can be kept ionised is proportional to the incident ionisation parameter.

The fact that ionising spectra of photons from quasars is a power law, whereas the ionising flux from O stars involved in the case of Stromgen spheres in the ISM is exponential, has another effect. We saw in Vol. II, Chap. 9, Section 9.3, that the Stromgen radius $r_s(\text{He})$ for helium is related to the Stromgen radius for hydrogen r_s (H) by

$$\left[\frac{r_s(\text{He})}{r_s(\text{H})} \right]^3 = 2.5 \, \frac{n_\gamma^{\text{ion}}\left(E_0^{\text{He}}\right)}{n_\gamma^{\text{ion}}\left(E_0^{\text{H}}\right)}. \tag{8.208}$$

In the case of stars with steep spectra, this number on the right-hand side is small and hence the length scales are fairly different. Interpreting the radii of Stromgen spheres as the typical length scales over which ionisation takes place even in the context of planar geometry, we find that the ratio on the right-hand side may not be small for a power-law ionising spectrum. In other words, $r_s(\text{He}) \ll r_s(\text{H})$ in the stellar context, whereas $r_s(\text{He}) \simeq r_s(\text{H})$ in the context of quasar radiation ionising a slab of gas. The different ionisation regions will now be very narrowly crammed in the gas cloud.

Even though the transition region from ionised to neutral H (and He) occurs in a short length scale, this transition zone is of considerable significance. This is the location where most of the continuum energy of photons with 13.6 eV $\lesssim h\nu \lesssim$ 55 eV is absorbed and hence – because of local energy conservation – this is also the location where a large part of line emission occurs. Most of the Lyman$-\alpha$, Balmer, CIII] 1909 and MgII 2800 lines arise from this zone. Once this transition zone is passed, the material becomes optically thick to photons from 13.6 eV to several hundred electron volts. At these optical depths, the Lyman$-\alpha$ is very nearly thermalised so that the population of H atoms in an $n = 2$ excited state becomes comparable with the abundance of heavy elements:

$$\frac{n(n = 2)}{n(n = 1)} = 4\exp\left(-\frac{11.84}{T_{\mathrm{ex},4}}\right) \approx 2.9 \times 10^{-5}, \qquad (8.209)$$

where the second equality is valid for $T_{\mathrm{ex},4} = 1$, with $T_{\mathrm{ex},4}$ denoting the excitation temperature for Lyman-α. It follows that the collisional excitation of Balmer and other excited H lines can be quite important.

The preceding discussion shows that the line emission from the BLR is a complex but somewhat localised process. Given the broad features of the cloud, the properties of ionising radiation, and other physical parameters, we can work out the photoionisation equilibrium structure for the cloud. Because of the varying physical conditions, different lines are produced at different locations and with different intensities. Figure 8.6 shows the relative fluxes from a BLR cloud with the ionisation parameter of unity, $p = 0.01$ dyn cm^{-2} and $N_{\mathrm{H}} = 3 \times 10^{22}$ cm^{-2}.

From such a simplified model in which the majority of the lines are created within a narrow range of physical conditions (usually called *single-zone models*), we can relate the line ratios to the physical parameters and obtain an estimate

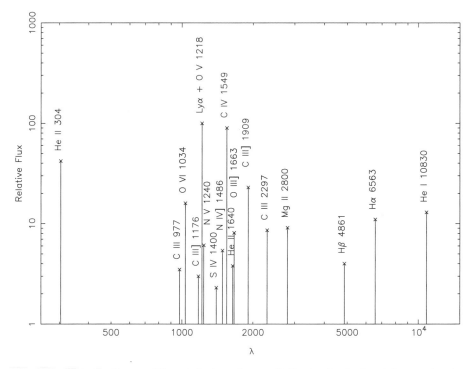

Fig. 8.6.   Relative fluxes of line emissions from a BLR cloud calculated from a simple photoionisation model.

of the latter. The following constraints are fairly easy to obtain if a power-law spectrum and a plane-parallel slab geometry are assumed: (1) To begin with, the existence of high ionisation states like C^{+3} and N^{+4} shows that the ionisation parameter cannot be too small. When combined with the line ratios of N_V 1240 and O_{VI} 1034 with respect to Lyman$-\alpha$, we can estimate the ionisation parameter to be in the range 0.1–1. (2) The density can be constrained at the lower end from the [OIII] 5007 line, which becomes thermalised at densities above $N_c \simeq 3 \times 10^6$ cm^{-3}. The upper bound on the density comes from the collisional deexcitation of CIII] 1909 with a critical density of 6×10^9 cm^{-3}. The ratio of the He lines, He I 5976 and He I 10830, is also a very sensitive indicator of density, and the observed ratio favours a value of $\sim 10^9$ cm^{-3}. (3) These pieces of information can be combined to estimate the distance of BLR clouds from the central source, and we get

$$ r = \left(\frac{L_{\mathrm{ion}}}{4\pi c n_H k_B T U} \right)^{1/2} = 0.067 \, L_{\mathrm{ion},45}^{1/2} \, U^{-1/2} n_{\mathrm{H},10}^{-1/2} \left(\frac{T_4}{2} \right)^{-1/2} \quad \mathrm{pc.} \quad (8.210) $$

This region is quite far away from the relativistic accretion flow zone, and we expect $r \propto L^{1/2}$. (4) Finally, the mean column density can be determined from either the strengths of low ionisation lines like MgII 2800 or the relative strengths of the Balmer and Lyman-α lines. Because the former has an ionisation potential of only ~ 15 eV, it cannot maintain substantial abundance unless the Lyman continuum photons have been depleted significantly. This requires a substantial optical depth at the Lyman edge. If we take the ionisation parameter to be ~ 0.3, the minimum column density that is required is $\sim 10^{22}$ cm^{-2}.

The diagnostic procedures described are all indirect in the sense that we need to model the photoionisation equilibrium of the BLR and then compare the model predictions with observations. There is an alternative method, known as *reverberation mapping*, that potentially allows a more direct determination of the conditions in the BLR. The key idea behind this technique is as follows. From the estimate in Eq. (8.210), it is clear that photons take $\sim 80 \, L_{45}^{1/2}$ days to travel from the central source to the line-emitting gas. The time scale for readjustment of the gas condition, in response to the continuum radiation, is very short and could be thought of as almost instantaneous. Because the line radiation emitted by the cloud depends on the continuum radiation emitted by the central source monotonically, we will expect the line radiation received by us to reflect the variations of the continuum *with a time delay*. By monitoring both the line radiation and the continuum radiation over a stretch of time, we should be able to probe the geometry of the line-emitting region.

Let a line-emitting cloud be located at (r, θ) with respect to the central point source located at the origin. For a distant observer along the positive z axis, the difference in the time of arrival in the photons from the cloud and the source will be

$$\tau = \frac{r}{c}(1 - \cos\theta). \tag{8.211}$$

The relation between the emissivity j_l in any given line from the cloud and the continuum flux F_c is fairly complicated. However, if the change in the continuum is small, then we can relate the changes in line and continuum radiation by a Taylor series expansion of the form

$$j_l(F_c) = j_l(\langle F_c \rangle) + \frac{\partial j_l}{\partial F_c}(F_c - \langle F_c \rangle) + \cdots. \tag{8.212}$$

The linear relationship between δj_l and δF_c immediately leads to a linear relationship between the fluctuating line luminosity and the continuum luminosity, with a time delay. Hence we can write, quite generally,

$$\delta L_l(t) = \int_0^\infty d\tau \, \Psi(\tau)[L_c(t - \tau) - \langle L_c \rangle], \tag{8.213}$$

where $\Psi(\tau)$ is the response function of the system. We can express this quantity as an integral over a surface of constant τ, with the integrand depending on the local volume-filling factor $f(\mathbf{r})$:

$$\Psi(\tau) = c \int_S dA \, \frac{f(\mathbf{r})}{4\pi r^2} \frac{\partial j_l}{\partial F_c}. \tag{8.214}$$

Equation (8.213) is a convolution in time coordinates and can be solved trivially in Fourier space. Inverting back, we get

$$\Psi(\tau) = \int \frac{d\omega}{2\pi} e^{-i\omega\tau} \frac{\delta L_l(\omega)}{L_c(\omega)}. \tag{8.215}$$

To get accurate results with such a Fourier inversion technique, we need a sampling interval that is smaller than $\tau/2$ and the duration of the sampling should be significantly larger than τ. Further, in such a technique, the signal-to-noise ratio will *not* be the ratio of mean flux to rms error but the ratio of the rms fluctuations on the time scales of interest to the rms error. This is because the AGN flux varies significantly on the relevant time scales during the monitoring. Hence we require a considerable amount of effort in practical situations to extract high-quality information from this study.

This procedure can, in fact, be used for several emission lines, thereby allowing us to simultaneously obtain a mapping of the marginal emissivity $(\partial j_l/\partial F_c)$ for a number of lines; photoionisation models can now be used to predict the ratios of marginal emissivities, thereby allowing us to get a more detailed picture of the physical conditions on each isodelay surface. In fact, one of the most important results of reverberation mapping is the demonstration that the response functions for different lines have different time lags. Lines that are prominent in highly ionised gases respond to a continuum faster than lines that are prominent at lower ionisation levels, suggesting that the BLR is radially stratified.

We can also generalise this procedure by replacing the total line flux with the specific flux in a number of frequency bins and converting the frequencies to corresponding line-of-sight velocities. This will provide, in principle, a function $\Psi(\tau, v)$, depending on the velocity structure of the region.

One disadvantage of reverberation mapping has to do with the fact that it probes $(\partial j_l/\partial F_c)$ rather than j_l. There are lines that are insensitive to continuum fluctuations because an increase in the ionising flux causes an almost exactly compensating decrease in the neutral fraction and hence in the optical depth. Any line (for example, CIV 1549) that is energised primarily by the Lyman continuum will not vary, even for large continuum changes. This also shows that the lines that are selected in this process may not be the ones that may be of interest in other contexts.

Exercise 8.15

Escape from optically thick lines: The cross section for scattering in a line is given by $\sigma(v) = (h v_0/4\pi)B\phi(v)$, where B is the Einstein coefficient and $\phi(v)$ is the line profile for Doppler broadening, given by

$$\phi(v) = \frac{1}{\Delta v_D \sqrt{\pi}} \exp\left[-\frac{(v - v_0)^2}{(\Delta v_D)^2}\right]. \tag{8.216}$$

Assume that as the photons are scattered across the line there is no correlation between the incoming and the scattered photons and that they are independently distributed over the line profile. Consider a bunch of photons trapped in a system of scatterers with high optical depth at the centre of the line; that is, $\tau_0 \equiv \tau(v_0) \gg 1$. (a) Prove that the mean number of scatterings before the photon escapes is $N \simeq \tau_0(\ln \tau_0)^{1/2}$. How does this compare with the escape of photons from a region in which electron scattering (rather than line scattering) is the dominant source of opacity? (b) Estimate the energy radiated from the surface of a uniform spherical cloud of radius r within which the line scattering is important. Assume that the excitation temperature of the line is T_e and take $(\ln \tau_0)^{1/2}$ to be of the order of unity. [Hint: Compute the net rate at which the photons (i) are created in spontaneous decay and (ii) escape from the cloud.]

8.8.2 Narrow-Line Regions

The NLRs are the largest spatial scales wherein the ionising radiation from the central source dominates. It is also the region that can be spatially resolved in the optical. Just like the BLR, the NLR is also clumpy with a filling factor ϵ. However, in contrast to the BLRs, the NLRs have a low number density of electrons ($\sim 10^4 \, \text{cm}^{-3}$), which is comparable with the HII regions of our galaxy. The low number density of electrons allows us to obtain fairly detailed information about the density and the temperature of a NLR by studying suitably chosen line intensities. We now illustrate some of the diagnostic methods used in the study of a NLR.

Let us consider, for example, the Hβ luminosity from a NLR, which is given by

$$j_{H\beta} = n_e n_p \alpha_{H\beta}^{\text{eff}} \frac{h\nu_{H\beta}}{4\pi} = n_e^2 \alpha_{H\beta}^{\text{eff}} \frac{h\nu_{H\beta}}{4\pi} = 1.24 \times 10^{-25} \frac{n_e^2}{4\pi} \text{ ergs s}^{-1} \text{ cm}^{-3} \text{ sr}^{-1},$$

(8.217)

so that a volume of radius r emits an amount

$$L(H\beta) = \int \int j_{H\beta} \, d\Omega \, dV = 1.24 \times 10^{-25} \frac{4\pi \epsilon n_e^2}{3} r^3 \text{ ergs s}^{-1}. \quad (8.218)$$

This allows us to relate the size of the emitting region to the luminosity by

$$r \approx 19 \left[\frac{L_{41}(H\beta)}{\epsilon n_3^2} \right]^{1/3} \text{ pc}, \quad (8.219)$$

where $n_3 \equiv (n_e/10^3 \text{ cm}^{-3})$. In nearby AGN, the NLR is often (at least partially) resolved, implying that $r \gtrsim 100 \text{ pc}$ so that $\epsilon \lesssim 10^{-2}$. Estimating the combination $(4\pi \epsilon r^3/3)$ from Eq. (8.218) and (8.219), we also get

$$M_{\text{NLR}} = 8.1 \times 10^{59} \frac{L_{41}(H\beta)}{n_3^2} n_e m_p = 1.4 \times 10^{39} \frac{L_{41}(H\beta)}{n_3} \text{ gm}$$

$$= 7 \times 10^5 \frac{L_{41}(H\beta)}{n_3} M_\odot. \quad (8.220)$$

It is obvious that a NLR is several orders of magnitude more massive than a BLR, even though the amounts of line emission from both are comparable. This is because the emissivity of recombination lines is proportional to n_e^2 and the denser BLR acts as a more efficient emitter.

The electron densities in the NLR are low enough that many forbidden transitions are not collisionally suppressed, allowing them to be used as a probe. Let us consider two simple examples of this technique. We saw in Vol. II, Chap. 9, Section 9.8, that the behaviour of line intensities produced by transitions among three separate energy levels can depend sensitively on temperature and density. Consider, for example, a system with three relevant energy levels, E_1, E_2, and E_3. We will assume that only collisional excitation, deexcitation, and spontaneous decay occur between the ground and either excited state. Further, we will neglect direct transitions between levels 2 and 3. Then, it was shown in Vol. II, Chap. 9, Section 9.8, that the line intensities are determined by the expression

$$\frac{j(3 \to 1)}{j(2 \to 1)} = \frac{n_3 h\nu_{31} A_{31}}{n_2 h\nu_{21} A_{21}} = \frac{g_3 A_{31} \nu_{31}}{g_2 A_{21} \nu_{21}} \frac{[1 + (A_{21}/\mathcal{R}_{21})]}{[1 + (A_{31}/\mathcal{R}_{31})]} e^{-(E_{13} - E_{12})/k_B T}.$$

(8.221)

There are two limiting forms of this expression that are of interest. If the element is such that E_{12} and E_{13} are quite different, then the argument of the exponent,

$(E_{13} - E_{12})/k_B T = E_{23}/k_B T$, will vary rapidly with the temperature T. Hence the main dependence of the line strength on the temperature will arise from the exponential factor, and the ratio of the line strengths can be used as a probe of the temperature. On the other hand, if $E_{13} \approx E_{12}$ so that $E_{23} \ll k_B T$, the exponential factor is irrelevant and the temperature dependence of the line ratios essentially arises from $\mathcal{R}_{ij} \propto n T^{-1/2}$. This corresponds to a situation in which the upper energy level is split by a small amount, having $E_3 \approx E_2$, with the excited-state energies satisfying $(E_3, E_2) \gg E_1$.

An example of the first situation is provided by the transitions in OIII that are due to $^1S_0 \to {}^1D_2(\lambda = 4363 \,\text{Å})$, $^1D_2 \to {}^3P_2(\lambda = 5007 \,\text{Å})$, and $^1D_2 \to {}^3P_1(\lambda = 4959 \,\text{Å})$. Then the line ratios turn out to be (see Eq. (9.134) of Vol. II, Chap. 9, Section 9.8)

$$\frac{j(4363)}{j(5007) + j(4959)} = 0.12 \, \exp\left(-\frac{3.29 \times 10^4 \, \text{K}}{T}\right). \tag{8.222}$$

The assumption that collisional deexcitation is subdominant is valid for $n_e \lesssim 10^5 \, \text{cm}^{-3}$. Because of the exponential factor, the ratio varies by $\sim$2 orders of magnitude when the temperature varies from 6000 to 20,000 K and thus can be a sensitive probe of the temperature.

As an example of the second case, let us consider the transitions in OII that are due to $^2D_{3/2} \to {}^4S_{3/2}(\lambda = 3726 \,\text{Å})$ and $^2D_{5/2} \to {}^4S_{5/2}(\lambda = 3729 \,\text{Å})$. In this case, $E_{23} \ll k_B T$ and the line intensity ratio is given by

$$\frac{j(3729)}{j(3726)} = \frac{g_3 A_{31}}{g_2 A_{21}} \left[\frac{1 + (A_{21}/\mathcal{R}_{21})}{1 + (A_{31}/\mathcal{R}_{31})}\right]. \tag{8.223}$$

We saw in Vol. II, Chap. 9, Section 9.8, that, in the dilute regime (as $n_e \to 0$), the line intensity takes the limiting value

$$\frac{j(3729)}{j(3726)} \approx \frac{g_3 \mathcal{R}_{31}}{g_2 \mathcal{R}_{21}} = \frac{\Omega_{31}}{\Omega_{21}} \approx 1.5. \tag{8.224}$$

In the high-density limit (as $n_e \to \infty$), the line ratio drops to

$$\frac{j(3729)}{j(3726)} \approx \frac{g_3 A_{31}}{g_2 A_{21}} \approx 0.35. \tag{8.225}$$

The transition occurs over a band of 2 orders of magnitude when $n_e T^{-1/2}$ varies from 1 to 10^2 in cgs units. In this range the line ratio can be a sensitive probe of the parameter $n_e T^{-1/2}$. Similar results can be obtained with [SII] $\lambda6716$, $\lambda6731$ lines.

We must, however, keep in mind that the NLR is characterised by a reasonably broad range of densities, and hence a measurement, say, of T from the [OIII] line is not necessarily representative of the temperature of, say, [SII] emitting clouds. Because of the stratification of the region, only spatially resolved spectrophotometry can give more detailed information.

The ionisation structure of NLRs can be understood by photoionisation models in a manner similar to the study of BLRs. Such an analysis usually gives an ionisation parameter of $\sim 10^{-2}$ and a Stromgen depth of $z_0 \simeq 10^{18}(n/10^3 \text{ cm}^{-3})^{-1}$. Taking this to be a lower limit on the size of the clouds, we can estimate the number of narrow-line clouds to be $N_c \lesssim 10^5 n_3^2 L_{41}(H\beta)$.

8.9 Intrinsic Absorbers in AGN

Any gaseous material along the line of sight to AGN will absorb light from the source and could be potentially detected by its absorption signature in the AGN spectrum. A significant component of such absorption arises from neutral H in the IGM along the line of sight to the quasar; we will discuss this in Chap. 9. There are also signatures in the spectrum related to the absorbing gas that is intrinsic to the AGN in the sense that the absorber resides broadly within the radius of AGN's host galaxy. A prime example is the broad-absorption-line (BAL) systems that are possibly due to energetic outflows of gas from the central engines. These lines are blue shifted relative to AGN emission lines, implying velocities that vary anywhere from a few kilometres per second to 6×10^4 km s^{-1}. A representative linewidth will be $\sim 10^4$ km s^{-1}, although there is considerable variation in this value. In contrast, narrow absorption lines (NALs) have widths less than few hundred kilometres per second. The evidence for the intrinsic origin of specific NALs come from different spectroscopic indicators such as time-variable line strengths, well-resolved line profiles that are smooth and broad compared with the thermal velocity, multiplet ratios in the lines that imply partial coverage of the background source, and high space densities. All these features are most easily understood in terms of the dense environments near the AGN.

The theoretical picture is based on a geometry in which the absorbing gas resides near the equatorial plane of the accretion disk of the AGN. The variation in the percentage of polarisation seen in BALs can be then understood in terms of light scattering in the disk geometry. Quasars viewed nearly edge-on will exhibit a BAL; they will also be more polarised because the direct unpolarised light through the BAL region is somewhat attenuated. Several other statistical correlations (like the observation that quasars with reddened spectra are more likely to have BALs) can be naturally explained in this model.

In such a geometry, the two most important derived parameters will be the covering fraction and the column density of the absorbers. The partial coverage leads to the filling of the bottoms of measured absorption troughs so that the observed line intensities are given by

$$I_v = (1 - C)I_0 + C I_0 \, e^{-\tau_v}, \tag{8.226}$$

where C is the line-of-sight coverage fraction, I_v and I_0 are the observed and the unabsorbed intensities, respectively and τ_v is the optical depth. (It is conventional to use the velocity v as the independent variable rather than the corresponding frequency.) In the case of a large optical depth, we get $C = 1 - (I_v/I_0)$. If this is not the case, we can still determine C by using lines like CIV $\lambda\lambda 1548$–1550 for which the ratios of the true optical depths are known. In the case of CIV, the optical depth ratio is ~ 2, and if the observed line intensities are I_1 and I_2, normalised by I_0 at the same velocity in the weaker and stronger line troughs, we can obtain from Eq. (8.226) the result

$$C = \frac{I_1^2 - 2I_1 + 1}{I_2 - 2I_1 + 1}. \tag{8.227}$$

Such an analysis, applied to well-resolved multiplets in NALs, leads to covering fractions ranging in few tens of percentage. The corresponding column densities in many of the ions are usually in the range 10^{13}–10^{16} cm^{-2}.

These systems are usually highly ionised, and the lowest levels of ionisation are characterised by CIII, NIII, SiIV, CIV, etc. The typical value for the ionisation parameter is between 0.01 and 1. The density of material can again be obtained from the absence of broad or blue-shifted forbidden emission lines and from the fact that a certain minimum density is required for the gas to adjust its ionisation structure within the time scale of variability. These suggest densities in a wide range of 7–10^6 cm^{-3}. When combined with the estimates for the ionisation parameter, this gives a distance scale varying between 10 pc to 300 kpc in different sources.

The actual mechanisms for the outflow of gas are quite uncertain. However, to illustrate the possibilities, consider a simple case in which the radial acceleration of the wind is produced by the radiation pressure from the central source with luminosity L and mass M. In that case,

$$\frac{dv}{dt} = v \frac{dv}{dR} = f_L \left(\frac{L}{4\pi R^2} \right) \frac{1}{c m_H N_H} - \frac{GM}{R^2}, \tag{8.228}$$

where f_L is the fraction of the luminosity contributing to the momentum transfer. Integrating this equation from R to infinity gives the terminal velocity as

$$v_\infty \approx 10^4 R_{0.1}^{-1/2} \left(\frac{f_{0.1} L_{46}}{N_{22}} - 0.8 M_9 \right)^{1/2} \text{ km s}^{-1}, \tag{8.229}$$

where $R_{0.1}$ is the radial distance in units of 0.1 pc and other quantities have the usual meanings. Because $f_{0.1}$ could be a factor of a few and N_{22} can vary between 0.01 and unity, radiative acceleration can produce the observed velocities if $R_{0.1}$ is not too large. The same radial scale also determines the total mass in the wind at any instant as $M_w \approx 1.1\ Q_{0.1} N_{22} R_{0.1}^2\ M_\odot$, where $Q_{0.1}$ is the global covering factor relative to the fiducial 10%. Such a flow can carry $\sim 0.1\ M_\odot$ yr^{-1} with a kinetic-energy flow that is equivalent to a luminosity of approximately 4×10^{42} ergs s^{-1}. During a quasar's lifetime of 10^8 yr, these outflows will eject a total mass of $10^7 M_\odot$ with a kinetic energy of 10^{58} ergs. It is not very clear whether the required energetics can be easily achieved in many sources. A serious challenge to many of these models is that none of the intrinsic absorbers have changed their velocity (that is, exhibited accelerated motion) in the observations that now span 10–20 yr. In one particular case, there is an upper bound on the acceleration of ~0.02 cm s^{-2}. The outflow speed of 10^4 km s^{-1} is therefore stable to less than 0.2% over this time scale. This is particularly surprising as the crossing time for a flow with 2×10^4 km s^{-1} in a region of 0.1 pc is only ~5 yr.

8.10 Quasar Luminosity Function

The simplest statistical description of the abundance of quasars – or for that matter any other source of a particular kind – is based on the distribution of source counts that measure the number of sources as a function of their brightness. To see the importance of a source count, let us consider a survey in which the apparent magnitude of all the sources brighter than a limiting magnitude $m_{\max}$ are measured in a given patch of the sky, subtending a solid angle $\Delta \Omega$. Translating the magnitudes into fluxes by using the relation $F = F_0 10^{0.4(m_0 - m)}$, we can determine the number of sources $N(> F)$ brighter than a given flux that exists in the survey. If the sources are distributed in a spherically symmetric manner with a luminosity function dn/dL (which is independent of distance), then it is easy to find an expression for $N(> F)$ in terms of (dn/dL). We note that for each luminosity L there exists a limiting luminosity distance given by $d_L(L) = (L/4\pi F)^{1/2}$. Hence the number of sources that we can see with flux greater than F will be

$$N(> F) = \int dL \, V[d_L(L, F)] \frac{dn}{dL}, \tag{8.230}$$

where $V[d_L(L, F)]$ is the volume within which a source of luminosity L could be seen with a minimum flux of L. In a Euclidean universe, $V = (\Delta\Omega/3)d_L^3$; this gives

$$N_{\text{Eucl}}(> F) = \frac{\Delta\Omega}{3(4\pi)^{3/2}} F^{-3/2} \int dL \, L^{3/2} \frac{dn}{dL}. \tag{8.231}$$

Thus, for a spherically symmetric homogeneous population in a Euclidean universe, $N(> F) \propto F^{-3/2}$ *independently* of the shape of the luminosity function. An important conclusion is that the source counts of such a population in the Euclidean universe cannot be used to obtain any information about the shape of the luminosity function.

The situation, of course, is different if the geometry of the universe is not Euclidean so that V does not vary as d_L^3. In this case, the correct expression for

$N(>F)$ will be

$$N(>F) = \int dL \int_0^{r_1(d_L)} dr \, \Delta\Omega \frac{r^2}{(1-kr^2)^{1/2}} \frac{dn}{dL}(r), \qquad (8.232)$$

which takes into account the geometry of the universe. This result also implies that we cannot really distinguish the effects of genuine evolution of sources from the effects arising from the geometry of the universe as both factors are convolved into the expression for $N(>F)$. Some amount of effort was spent in the past under the assumption that $N(>F)$ could allow us to determine the geometry of the universe. This is clearly an impossibility, given the fact that any realistic source population will evolve over cosmic time and the final results will depend on the assumptions made for this evolution. Although the source counts can act as a consistency check, it is very unlikely that they will ever determine the geometry of the universe reliably. Fortunately, there are better and more reliable techniques available to determine the geometrical parameters of the universe (like the CMBR anisotropies described in Chap. 6). Once such methods are used to determine the geometry of the universe, the study of source counts will tell us about the physical evolution of the population and will allow us to constrain different models for the sources.

Exercise 8.16
K correction: While observing sources like quasars, with redshifts comparable with unity, it is necessary to take into account two simple kinematic effects: (i) The frequency at which the radiation is detected is not the same as the frequency at which it is emitted and (ii) the bandwidth over which the observations are carried out will not be the same as the bandwidth over which the radiation is emitted because of redshift. It is conventional to write the apparent magnitude of the object in the form $m_{\text{intrinsic}} = m_{\text{observed}} - K(z)$, where $K(z)$ is called the K correction. (a) Show that $K(z) = K_1 + K_2$, where $K_1 = 2.5 \log(1 + z)$,

$$K_2 = 2.5 \log \frac{\int_0^\infty F(\lambda) S(\lambda) \, d\lambda}{\int_0^\infty F(\lambda/[1+z]) S(\lambda) \, d\lambda}, \qquad (8.233)$$

$F(\lambda)$ is the spectral-energy distribution of the source, and $S(\lambda)$ is the filter function determining the wave band over which observations are carried out. (b) Consider a power-law source with $S_\nu = C\nu^{-\alpha}$ and show that, in this case, the total K correction is given by

$$K(z) = 2.5(\alpha - 1) \log(1 + z). \qquad (8.234)$$

Exercise 8.17
Practice with luminosities: Consider a class of quasars with intensity $F_\nu \propto \nu^{-0.5}$ in a matter-dominated universe with $h = 0.5$, $\Omega = 1$.

(a) What is the absolute B magnitude of a quasar that is observed to have an apparent B magnitude 19.5 and redshift $z = 2$?

(b) There are two quasars per square degree of the sky with redshifts between 1.75 and 2.25 having B magnitudes brighter than 19.5. Estimate the comoving number density of quasars around the redshift $z \simeq 2$.

[Answers: (a) The luminosity distance for the relevant universe is given by $d_L = 12(1 + z) - \sqrt{1 + z}]$ Gpc $\simeq 15.2$ Gpc at $z = 2$. Using the relation $M = m - 5\log(d_L/10\text{pc})$, we find that $M = -26.4$ if $M = 19.5$. However, M is not the absolute B magnitude, as the cosmological redshift implies that the light we see was emitted at $4400(1 + z)^{-1}$ Å $\simeq 1460$ Å. To find the correct magnitude, we can use the fact that $M = \text{constant} - 2.5\log(F_{\text{band}}/F_\nu)$, where $F_{\text{band}} = \Delta\nu F_\nu = (\Delta\nu/\nu)\nu F_\nu$ is the flux within a given band. Because $\Delta\nu/\nu$ is independent of the redshift for a given band, we can write

$$M_B' - M_B = -2.5\log\left(\frac{\nu' F_{\nu'}}{\nu F_\nu}\right). \tag{8.235}$$

Using $F_\nu \propto \nu^{-0.5}$ and $\nu'/\nu = (1 + z)^{-1}$, we find that $M_B' = M_B + 1.25\log(1 + z) = -25.8$. (b) For the given cosmology, the comoving volume between redshifts z_1 and z_2 is

$$\Delta V = \frac{8}{3}\Omega\left(\frac{c}{H_0}\right)^3\left[\left(1 - \frac{1}{\sqrt{1 + z_2}}\right)^3 - \left(1 - \frac{1}{\sqrt{1 + z_1}}\right)^3\right], \tag{8.236}$$

where the solid angle Ω is taken to be 1 square degree $= 3 \times 10^{-4}$ sr. Given $z_1 = 1.75$, $z_2 = 2.25$, we get $\Delta V = 4.5 \times 10^6$ Mpc^3. Because there are two quasars in this volume, the space density is $(2/\Delta V) \simeq 4.4 \times 10^{-7}$ Mpc^{-3}.]

Because we expect quasars also to evolve over cosmic time scales, the luminosity function of quasars will depend on the redshift as well. Careful observations suggest that this luminosity function can be expressed in the form (see Chap. 1, Section 1.12)

$$\phi(L, z)\,dL = \phi^*\left\{\left[\frac{L}{L^*(z)}\right]^{-\alpha} + \left[\frac{L}{L^*(z)}\right]^{-\beta}\right\}^{-1}\frac{dL}{L^*(z)}. \tag{8.237}$$

Using the relationships

$$M_B = M_B^*(z) - 2.5\log\left(\frac{L}{L^*(z)}\right), \qquad \phi(M_B, z)\,dM_B = \phi(L, z)\left|\frac{dL}{dM_B}\right|dM_B, \tag{8.238}$$

we can express this in a more convenient form in terms of the blue magnitudes as follows:

$$\phi(M_B, z)\,dM_B = \frac{\phi^*\,dM_B}{10^{0.4[M_B - M_B^*(z)](\alpha+1)} + 10^{0.4[M_B - M_B^*(z)](\beta+1)}}, \tag{8.239}$$

where

$$M_B^*(z) = \begin{cases} A - 2.5k \, \log(1+z) & (\text{for } z < z_{\max}) \\ M_B^*(z_{\max}) & (\text{for } z > z_{\max}) \end{cases}, \qquad (8.240)$$

with

$$\alpha = -3.9, \quad \beta = -1.5, \quad A = -20.9 + 5 \log h, \quad k = 3.45,$$
$$z_{\max} = 1.9, \quad \phi^* = 5.2 \times 10^3 h^3 \, \text{Gpc}^{-3} \text{mag}^{-1}. \qquad (8.241)$$

Equation (8.239) gives the number density of quasars with magnitudes between M_B and $M_B + dM_B$ at a redshift z (the inverse relation is based on $L = L_\odot \, 10^{0.4[M_\odot - M]}$ with $M_\odot = 5.48$ and $L_\odot = 2.36 \times 10^{33}$ ergs s^{-1}) and is plotted in Fig. 1.27 of Chap. 1 for different redshifts. There has been a significant amount of effort to understand the luminosity function of quasars at $z > 3$; however, the situation is still very inconclusive.

One obvious feature of Fig. 1.27 is that the luminosity function of quasars evolves in time although the actual direction of evolution is not very easy to determine. If the shape of the intrinsic luminosity function of quasars did not change with time but the comoving density of AGN varied with z, then the luminosity function will move vertically in Fig. 1.27 (this is usually called *pure density evolution*). The other extreme situation corresponds to the assumption that the space density of AGN is constant in time but the luminosity of the population changes with redshift. (This is usually called *pure luminosity evolution*.) In both the cases, we can explain the absence of very high-luminosity quasars in the local region: In the case of density evolution, this is because they are spatially very rare; in the case of luminosity evolution, this occurs because the sources are much fainter today than they were, say, at $z = 2$.

There is no physical basis to assume that either extreme assumption should hold, and we must treat them as purely convenient parameterisations for more complicated situations. Nevertheless, it is possible to work out some physical consequences of these assumptions that are illuminating. Let us begin with pure density evolution in which the probability for AGN to turn off is independent of luminosity. Because there are fewer AGN now than in the past in this case, some of the AGN in the past must have turned off and the dead remnants of quasars must reside at the centres of galaxies, which appear to be normal at present. Taking the luminosity function for normal galaxies as

$$\phi_G(L) \, dL = \phi_G^* \left(\frac{L}{L^*}\right)^{-1.1} \exp\left(\frac{-L}{L^*}\right) d\left(\frac{L}{L^*}\right), \qquad (8.242)$$

with L^* corresponding to $M_B^* = -19.7 + 5 \log h$ and $\phi_G^* = 1.6 \times 10^7 h_0^3 \, \text{Gpc}^{-3}$, and defining the bright galaxies to be the ones with, say, $L \geq 0.1 \, L^*$ (which is equivalent to $M_B \leq -17.5 + 5 \log h$) we get the space density of bright galaxies as

$$n_G(L \geq 0.1\,L^*) = \phi_G^* \int_{0.1}^{\infty} \left(\frac{L}{L^*}\right)^{-1.1} \exp\left(\frac{-L}{L^*}\right) d\left(\frac{L}{L^*}\right)$$
$$\approx 5.7 \times 10^7 h_0^3 \; \text{Gpc}^{-3}. \tag{8.243}$$

We can obtain the accreted mass per dead AGN by dividing the accumulated mass density of Eq. (8.3) by the space density of host galaxies:

$$M_{\text{AGN}} = \frac{2 \times 10^{13} \eta^{-1} M_\odot \; \text{Gpc}^{-3}}{5.7 \times 10^7 h^3 \; \text{Gpc}^{-3}} \approx 3.5 \times 10^5 h^{-3} \eta^{-1} \; M_\odot. \tag{8.244}$$

Although this is not very high, it is at a ballpark value that should be detectable in nearby galaxies.

The space density of AGN today can be determined from the luminosity function and is approximately $1.1 \times 10^6 h^3 \; \text{Gpc}^{-3}$. A comparison with Eq. (8.243) shows that the fraction of bright galaxies containing active AGN today is $f \simeq 0.019$. If all the bright galaxies contained AGN then the density of quasars would have been higher by a factor of 50 or so. If pure density evolution is valid, then a vertical downward translation by factor of 50 of the $z = 2$ curve in Fig. 1.27 should make it coincide with the $z = 0$ curve. Such a translation, however, seems, to overpredict the number of high-luminosity quasars at low redshifts. For example, calculating the number density of quasars brighter than $M_B = -23 + 5 \log h$ at $z = 2$ and rescaling by a factor of 50 will lead to a local space density of $\sim 440 h^3 \; \text{Gpc}^{-3}$. In a sky survey, extending upto $z = 2$ and covering $\sim 25\%$ of the entire sky, we would expect to find ~ 100 such objects whereas only ~ 8 were found. It is obvious that pure density evolution will not work. This is also, of course, clear from the fact that the break point in the luminosity function does evolve with z at least for $z < 1.9$.

It is also possible to see that pure luminosity evolution leads to somewhat unacceptable values for the parameters. In this case, we can take the current space density of all Seyfert galaxies ($1.1 \times 10^6 h^3 \; \text{Gpc}^{-3}$) to be representative of all the AGN that ever existed. This immediately suggests that the accreted mass per active galactic nuclears is higher compared with our previous estimate and is approximately

$$M_{\text{AGN}} = \frac{2 \times 10^{13} \eta^{-1} M_\odot \; \text{Gpc}^{-3}}{1.1 \times 10^6 h^3 \; \text{Gpc}^{-3}} \approx 1.8 \times 10^7 h^{-3} \eta^{-1} \; M_\odot. \tag{8.245}$$

This will make the detection of dead AGN remnants somewhat easier.

A more quantitative estimate of accretion phenomena over the lifetime of the universe can be obtained as follows: Consider a quasar at a redshift of $z = 2$ with luminosity $L = 10^{48} \; \text{ergs s}^{-1}$. The parameterisation of the luminosity function in Eq. (8.237) shows that the characteristic luminosity $L^*(z)$ varies as $(1 + z)^k$ with $k \simeq 3.45$. Taking $L^*(z) = L^*(0)(1 + z)^{3.45}$, we can estimate the total amount of energy radiated by this quasar during $0 < z < 2$. We get

$$E_{\text{rad}} = \int_{t(z=2)}^{t_0} L(t)\,dt = \int_{z=0}^{z=2} \frac{L(z)\,dz}{H_0(1+z)^2(1+\Omega z)^{1/2}}$$

$$= \int_{z=0}^{z=2} \frac{L(z=0)(1+z)^{3.45}\,dz}{H_0(1+z)^{5/2}} = 2.5 \times 10^{64} h_0^{-1} \text{ ergs.}$$

(8.246)

The total mass accreted during this process is given by

$$M_{\text{acc}} = \int \dot{M}\,dt = \frac{1}{\eta c^2} \int_{t(z=2)}^{t_0} L(t)\,dt = \frac{E_{\text{rad}}}{\eta c^2} = 1.4 \times 10^{10} \eta^{-1} h_0^{-1} \, M_\odot,$$

(8.247)

which is comparable with that of a galaxy. Such objects probably should have been detected in the local neighbourhood. The problem cannot be avoided by introducing a finite lifetime for the quasar phase; this will decrease the mass per quasar but will increase the number of galaxies involved.

Exercise 8.18
Density of dead black holes: Calculate the density of relic black holes in dead AGN in terms of the luminosity function. [Answer: The comoving density of mass in black holes is given by

$$\rho_{\text{BH}} c^2 = \int n(L, z)(BL/\epsilon)\,dt\,d\ln L,$$

(8.248)

where B is the bolometric correction, ϵ is the efficiency, and $n(dL/L)$ is the comoving number density of sources. With $dt = a_0\,dr/(1+z)$, $L = S(a_0\chi)^2(1+z)^{1+\alpha}$, and $n(S, z)\,d\ln S\,dz = n(L, z)\,d\ln L\,dV$, it follows that

$$\rho_{\text{BH}} c^2 = \frac{B}{\epsilon} \int n(S, z)S(1+z)^\alpha\,d\ln S\,dz.$$

(8.249)

This shows that the result can be expressed entirely in terms of the observable quantity $n(S, z)$.]

One important application of the abundance of quasars is to check whether it is consistent with standard structure-formation models. Although the results are not as discriminatory as those that are based on abundance of clusters, it is still interesting from the point of view that it probes much higher redshifts. To obtain such a constraint, we begin by noting that the luminosity of a system accreting at a maximum possible value can be related to the mass of the central object by

$$L_E = \frac{4\pi G m_p c\, M_{\text{BH}}}{\sigma_T} \approx 1.3 \times 10^{47} \left(\frac{M_{\text{BH}}}{10^9\,M_\odot}\right) \text{ergs s}^{-1}.$$

(8.250)

We can infer a characteristic black hole mass from Eq. (8.250) by using the observed luminosity of the quasar. The quasars with $z > 4$ have typical luminosities of 10^{47}ergs s^{-1} in a universe with $\Omega = 1$ and $h = 0.5$. This requires black holes of mass $M_{BH} \simeq 10^9\, M_\odot$.

From the luminosity we can also estimate the amount of fuel that must be present to power the quasar for a lifetime t_Q. If ϵ is the efficiency with which the rest-mass energy of the fuel is converted into radiation, then the fuel mass is

$$M_f = \frac{Lt_Q}{\epsilon} \approx 2 \times 10^9\, M_\odot \left(\frac{L}{10^{47}\text{ergs s}^{-1}}\right)\left(\frac{t_Q}{10^8\text{ yr}}\right)\left(\frac{\epsilon}{0.1}\right)^{-1}. \quad (8.251)$$

So if the lifetime is $\sim 10^8$ yr and the efficiency is $\sim 10\%$, the required fuel mass is comparable with the mass of the central black hole.

The mass previously estimated corresponds to that involved in the central engine of the quasar. This mass will, in general, be a small fraction F of the mass of the host galaxy. We can write F as a product of three factors: a fraction f_b of matter in the universe that is baryonic; some fraction f_{ret} of the baryons originally associated with the galaxy that was retained when the galaxy was formed (the remaining mass could be expelled by a supernova-driven wind); and a fraction f_{hole} of the baryons retained that participates in the collapse to form the compact central object. In standard CDM models, $f_b \simeq 0.1$ for an $\Omega = 1$ universe whereas f_{ret} will depend crucially on the depth of the potential well of the galaxy or equivalently on the circular velocity v_c; $f_{ret} \simeq 0.1$ for $v_c \lesssim 100\,\text{km s}^{-1}$. The quantity f_{hole} depends on the way the central mass accumulates and how efficiently the system can lose angular momentum and sink to the centre, and it is difficult to estimate reliably. A very optimistic estimate for $F = f_b f_{ret} f_{hole}$ will be $F \simeq 0.01$; $F \simeq 10^{-3}$ is more likely. We then get, for the mass of the host galaxy,

$$M_G = 2 \times 10^{11}\, M_\odot \left(\frac{L}{10^{47}\text{ergs s}^{-1}}\right)\left(\frac{t_Q}{10^8\text{ yr}}\right)\left(\frac{\epsilon}{0.1}\right)^{-1}\left(\frac{F}{0.01}\right)^{-1}. \quad (8.252)$$

For $(F/0.01) \simeq (0.1 - 1)$, with the earlier value being more likely, Eq. (8.252) implies a mass for the host galaxy in the range of 10^{11}–$10^{12}\, M_\odot$ if other dimensionless parameters in Eq. (8.251) are of the order of unity. Therefore the existence of quasars at $z > 4$ suggests that a reasonable number of objects with galactic masses should have formed before this redshift.

An estimate of the fraction of the mass density, $f_{coll}(> M_G, z)$, in the universe that should have collapsed into objects with mass greater than M_G at redshift z can be made with the luminosity function of quasars. We have

$$\rho(z) f_{coll}(> M, z) = \int_{-\infty}^{M_B(M)} \tau M_{host}(M_B)\phi(M_B, z)\, dM_B. \quad (8.253)$$

On the right-hand side, $M_B(M)$ is the bolometric magnitude of a quasar situated in a host galaxy with a total mass M; $\phi(M_B, z)\, dM_B$ gives the number density of quasars in the magnitude interval dM_B. Therefore the product $M_{host}(M_B)\phi\, dM_B$

gives the collapsed mass density around quasars with magnitude in the range $(M_B, M_B + dM_B)$. We integrate this expression up to some limiting magnitude M_B corresponding to a host mass M. Because magnitude decreases with increasing luminosity, this picks out objects with luminosity higher than some particular luminosity or – equivalently – a mass higher than M. Finally, the factor,

$$\tau = \max\left[\frac{t(z)}{t_Q}, 1\right] \tag{8.254}$$

takes into account the possibility that the typical lifetime of a quasar t_Q may be shorter than the age of the universe at the redshift z. In that case, typically $[t(z)/t_Q]$ generations of quasars can exist at the particular redshift, and we need to enhance the abundance by this factor. Given the luminosity function of the quasars and the relations previously derived, we can calculate the fraction by doing the integral numerically. We find that

$$f_{\text{coll}}(> 10^{12}\, M_\odot, z \gtrsim 2) \simeq 1.4 \times 10^{-3}. \tag{8.255}$$

From the results of Chap. 7, Section 7.4, we can write the collapse fraction as

$$f_{\text{coll}}(> M, z) = \text{erfc}\left[\frac{\delta_c(1 + z)}{\sqrt{2}\sigma(M)}\right], \tag{8.256}$$

where erfc(z) is the complementary error function and $\sigma(M)$ is the linearly extrapolated density contrast today at the mass scale M. Figure 7.3 of Chap. 7 plots the fraction f as a function of σ for various redshifts. This figure shows that at the scales corresponding to the host mass of quasars (which could be above $10^{12}\, M_\odot$), we must have $\sigma \simeq 1.4$–2.6. Any theoretical model for structure formation should produce at least these values of σ to be feasible. It turns out that most models that satisfy the abundance criteria arising from clusters or damped Lyman-α systems also satisfy this criterion; hence it is not very stringent.

9

Intergalactic Medium and Absorption Systems

9.1 Introduction

In this chapter we deal with the physical processes that take place in the diffuse IGM and their signatures in the quasar spectra in the form of hydrogen absorption lines.[1] It uses material developed in several previous chapters, especially Chap. 7.

9.2 Gunn–Peterson Effect

We have seen in Chap. 4 that the formation of neutral matter and the decoupling of radiation occurred at $z \simeq 10^3$. Because it is unlikely that the formation of structures at lower redshifts could have been 100% efficient, we would expect at least a fraction of the neutral material, especially hydrogen, to remain in the IGM with nearly uniform density. This neutral hydrogen could, in principle, be detected by an examination of the spectrum of a distant source, like a quasar. Neutral hydrogen absorbs Lyman-α photons, which are photons of wavelength 1216 Å that corresponds to the energy difference between the $n = 1$ and the $n = 2$ states of the H atom. Because of the cosmological redshift, the photons that are absorbed will have a shorter wavelength at the source and the signature of the absorption will be seen at longer wavelengths at the observer. We expect the spectrum of the quasar to show a dip (known as the *Gunn–Peterson effect*) at wavelengths on the blue side (shortwards) of the Lyman-α emission line if neutral hydrogen is present between the source and the observer. The magnitude of this dip depends on the neutral-hydrogen density and can be calculated with the optical depth for such absorption.

The procedure for calculating the optical depth τ has been described in Chap. 3, Section 3.5, where it was shown that

$$\tau[\nu_0; z] = \int_0^z \sigma(\nu) n(z) \left| \frac{c\,dt}{dz} \right| dz, \qquad \nu = \nu_0(1 + z). \tag{9.1}$$

518

In the case of a Lyman-α photon, the absorption cross section is

$$\sigma(\nu) = \frac{\pi e^2}{m_e c} f \delta_D(\nu - \nu_\alpha), \qquad (9.2)$$

where $f = 0.416$ is the oscillator strength and ν_α is the frequency corresponding to the Lyman-α photon. The Dirac delta function in $\sigma(\nu)$ allows us to do the integration easily, and we get

$$\tau_{GP}(z) = \left(\frac{\pi e^2 f}{m_e c \nu_\alpha}\right) \frac{n_{HI}(z)}{(1+z)} d_H(z), \qquad (9.3)$$

where

$$d_H(z) = c H_0^{-1} [\Omega_{NR}(1+z)^3 + \Omega_K(1+z)^2 + \Omega_V]^{-1/2} \qquad (9.4)$$

in an arbitrary cosmology with $\Omega_K = 1 - \Omega_{NR} - \Omega_V$. Here z is related to the observing frequency ν_o by $(1+z) = (\nu_\alpha/\nu_o) = (\lambda/\lambda_\alpha)$. Clearly all $\nu > \nu_\alpha$ will contribute to the absorption and will lead to an absorption trough in the continuum blueward of the Lyman-α emission line from the source.

To estimate the numbers, let us consider $\Omega_{NR} = 1$, $\Omega_V = 0$ universe. In this case, we get

$$\tau_{GP}(z) = \frac{\pi e^2 f}{m_e H_0 \nu_\alpha} \frac{n_{HI}(z)}{(1+z)^{3/2}} = 6.6 \times 10^3 h^{-1} \left(\frac{\Omega_B h^2}{0.019}\right) \frac{n_{HI}}{\bar{n}_H} (1+z)^{3/2} \quad (9.5)$$

where we have expressed the answer in terms of the mean density of H nuclei at a redshift z, given by

$$\bar{n}_H = \frac{\rho_c}{m_H}(1 - Y)\Omega_B(1+z)^3 = (1.6 \times 10^{-7} \text{ cm}^{-3}) \left(\frac{\Omega_B h^2}{0.019}\right)(1+z)^3. \quad (9.6)$$

Equation (9.5) shows that even if neutral H is only a small fraction of the total H density, the optical depth should be fairly high.

No such absorption troughs are seen in the continuum of the quasar. The current upper bound at $z \simeq 5$ is approximately $\tau_{GP} < 0.1$, implying that $(n_{HI}/\bar{n}_H) < 10^{-6} h$. Even if 99% of all the baryons are confined to structures and only 1% remained as a smoothly distributed component, this result requires the diffuse IGM to be ionised to better than 1 part in 10^4. Assuming that the bound on the optical depth is typically $\tau \lesssim 0.1$ in the redshift range $0 < z < 5$, we may write the above bound as

$$n_{HI} \lesssim 2 \times 10^{-12} \Omega_{NR}^{1/2} h(1+z)^{3/2} \text{ cm}^{-3}, \qquad (9.7)$$

or equivalently as

$$\Omega_{NR}(n_{HI}) \lesssim 2 \times 10^{-7} \Omega_{NR}^{1/2} h^{-1}(1+z)^{-3/2}. \qquad (9.8)$$

In arriving at the last two expressions, we have assumed that $\Omega_V = 0$ and used the approximation

$$(1 + z)(1 + \Omega_{NR}z)^{1/2} \approx \Omega_{NR}^{1/2}(1 + z)^{3/2}. \tag{9.9}$$

This is, of course, exact if $\Omega_{NR} = 1$ but remains a reasonable approximation for $z \gtrsim 1$ if $\Omega_{NR} \gtrsim 0.3$.

Because the high-redshift universe contains ~25% He, we could also perform a similar analysis with regard to the absorption of radiation by neutral He. The sources that reionise H are likely to have caused the single reionisation of He from HeI to HeII. The ionisation threshold of neutral He is 24.6 eV, and its recombination rate is comparable with that of H. On the other hand, the threshold for HeII ionisation is 54.4 eV and fully ionised He recombines at a rate nearly five times faster than that of H. This suggests that the reionisation of HeII should occur later than the reionisation of H, even though the number of He atoms is smaller by approximately an order of magnitude. Indeed, the Lyman-α absorption by intergalactic HeII (at $\lambda = 304$ Å) has been observed in a few quasars in the redshift range $2.4 < z < 3.2$. The HeII optical depth is probably $\tau \approx 4$ and is indicative of a HeII Gunn–Peterson absorption that is due to a diffuse IGM. This also suggests that the background photon spectrum should have a spectral shape that ionises H much more thoroughly than it does HeII.[2]

Note that expression (9.5) for the optical depth is also valid for absorption arising from the distribution of structures (like discrete clouds of neutral H) as long as we interpret n_{HI} as the average neutral density of individual clouds multiplied by the volume-filling factor.

In the standard evolutionary history of the universe, we expect the electrons and the protons to have recombined around $z = 1000$, forming an essentially neutral medium except for a small fraction $(< 10^{-4})$ of residual ionisation, whereas the observations at $z < 5$ discussed above show that the IGM is ionised to a high degree. This raises two central questions: (1) How has the IGM been ionised so efficiently? and (2) When did it happen?

Let us assume that the reionisation occurred at $z \simeq z_{ion}$ because of the energy input from the first structures. Because the evolutionary model for the universe suggests that most of the structures formed around $z \lesssim 20$, it is interesting to ask whether we can directly probe the neutral phase of the universe by, say, the 21-cm emission. If the universe remained neutral for $z > z_{ion}$ and was rapidly ionised at $z = z_{ion}$ then, in principle, we will see redshifted 21-cm radiation at all frequencies $\nu < \nu_{21}(1 + z_{ion})^{-1}$. The intensity can again be calculated with the formulas in Chap. 3. The observed flux in a matter-dominated universe will be

$$F(\nu, z = 0) = \frac{1}{4\pi} \int_{z_1}^{z_2} \frac{J(\nu_{em}; z)}{(1 + z)^3} \frac{dl}{dz} dz, \qquad \nu_{em} = \nu(1 + z) \tag{9.10}$$

[see Eq. (3.155) of Chap. 3]. For 21-cm emission, $J(\nu) = (3/4)Ah\nu n_{HI}\delta_D(\nu - \nu_{21})$.

Using the Dirac delta function, we can easily evaluate the integral; for a matter-dominated universe we get

$$F(\nu, z = 0) = \frac{cH_0^{-1}P_0}{4\pi} \int n_{HI}(z) \frac{\delta([1+z]\nu - \nu_{21})}{(1+z)^5 \sqrt{1+\Omega_{NR}z}} \frac{d([1+z]\nu)}{\nu}$$

$$= \frac{cH_0^{-1}P_0 n_{HI}(z)}{4\pi \nu(1+z)^5 \sqrt{1+\Omega_{NR}z}}, \qquad (9.11)$$

where $P_0 = (3/4)Ah\nu \simeq 2 \times 10^{-32}$ ergs s^{-1} is the average power emitted by a single neutral-H atom at 21 cm. For a universe with $\Omega_{NR} = 1$, this gives

$$F(\nu) \approx \left(\frac{cH_0^{-1}P_0}{4\pi}\right)\left(\frac{\rho_c\Omega_B}{m_p}\right)\left(\frac{1}{\nu_{21}}\right)\left(\frac{\nu}{\nu_{21}}\right)^{3/2} \qquad [\text{for } \nu < \nu_{21}(1+z_{ion})^{-1}].$$

$$(9.12)$$

Clearly, F_ν increases with ν up to the maximum limit and then drops to zero. For $z_{ion} \simeq 9$, $\Omega_B h^2 \approx 0.02$, the maximum flux is ~ 0.01 mJy arcmin^{-2}. The corresponding brightness temperature is $\sim 0.2\Omega_{HI}$ K. Unfortunately, this is too small a signal to detect in the presence of other backgrounds, even though the spectral signature of this signal will be quite different from everything else.

Having decided that the emission from the neutral phase is not directly detectable, we next turn our attention to emission from the ionised phase. The most obvious candidate here is the thermal bremsstrahlung of a hot plasma. We have seen in Vol. I, Chap. 6, Section 6.9, that the emissivity for a pure H plasma can be expressed as

$$\epsilon_\nu = 6.8 \times 10^{-31} T^{-1/2} n_e^2 e^{-h\nu/k_B T} G(\nu, T) \text{ ergs cm}^{-3} \text{ s}^{-1} \text{ Hz}^{-1}, \qquad (9.13)$$

where the Gaunt factor G can be approximated as $G \simeq 1 + \log(k_B T/h\nu)$ in the relevant range. This should be compared with the diffuse x-ray background that is actually observed. Current observations suggest that the diffuse x-ray background (obtained after known point sources are subtracted) is given approximately by

$$\nu I_\nu = 1.1 \times 10^{-8} \left(\frac{E}{3 \text{ keV}}\right)^{1-\alpha} \exp\left(-\frac{E}{k_B T}\right) \text{ ergs cm}^{-2} \text{ s}^{-1} \text{ sr}^{-1}, \qquad (9.14)$$

with $0 < \alpha < 0.2$. This background can indeed be produced by a plasma[3] with temperature varying as $T = T_0(1+z)^2$, with $T_0 \simeq 36$ keV and $\Omega_B h^2 \simeq 0.24$ if the reionisation occurred at $z > 6$. However, it must be noted that an ionised IGM will strongly couple to CMBR through inverse Compton scattering and will distort the CMBR spectrum, which can be described by the standard Compton y parameter:

$$y \equiv \int \sigma_T n_e \frac{k_B T}{m_e c^2} dl. \qquad (9.15)$$

For a fully ionised plasma with 25% He and 75% H, the electron density is

$$n_e = 0.88 \left(\frac{\rho_{bg}}{m_H} \right) = 9.84 \times 10^{-6} \Omega_B h^2 (1+z)^3 \text{ cm}^{-3}. \tag{9.16}$$

Taking the temperature to vary as $T = T_0(1+z)^2$, we find that the Compton y parameter now becomes

$$y = 1.2 \times 10^{-4} \Omega_B h^2 \left(\frac{T_0}{\text{keV}} \right) \int_0^{z_{max}} \frac{(1+z)^3}{\sqrt{1+\Omega_{NR}z}} \, dz. \tag{9.17}$$

Observations of CMBR spectral distortions[4] give $y < 1.5 \times 10^{-5}$ (see Chap. 6), requiring $z_{max} < 0.1$. This suggests that the ionised IGM could not contribute significantly to the diffuse x-ray background.

The above result was based on the assumption that the IGM is smooth. The x-ray emission goes up by a factor of $\langle n_e^2 \rangle / \langle n_e \rangle^2$ if the IGM is clumped while the Compton y parameter is independent of the clumping factor. Even with clumping, we require a clumping factor greater than $\sim 10^4$ if the x-ray background has to receive a significant contribution from the diffuse IGM. Recent observations from the x-ray satellite Chandra seem to confirm the view that most of the x-ray background is indeed due to individual point sources in the sky when observed with high resolution.

Finally, it must be noted that direct distortion of the temperature anisotropies of CMBR is also not very significant if the reionisation occurred at $z_{ion} \lesssim 20$. We have seen in Chap. 6 that the angular scale at which reionisation will make its presence felt is

$$\theta_{max} \simeq 3° \Omega_{NR}^{1/2} \left(\frac{z_{ion}}{100} \right)^{-1/2}, \tag{9.18}$$

provided the optical depth τ_T for Thompson scattering is at least unity at $z = z_{ion}$. The τ_T is given by

$$\tau \simeq \int \sigma_T n_e c \, dt \simeq \begin{cases} 0.026(\Omega_B z) & (\Omega_{NR}z \ll 1) \\ 0.017\left(\frac{\Omega_B}{\Omega_{NR}^{1/2}} \right) z^{3/2} & (\Omega_{NR}z \gg 1) \end{cases}. \tag{9.19}$$

It is easy to see from these expressions that, for realistic values of the parameters, there is no strong effect on $\Delta T / T$ if $z_{ion} \lesssim 16$.

Thus it is difficult to detect directly either the ionised IGM at $z < z_{ion}$ or the neutral phase at $z > z_{ion}$ for reasonable values of z_{ion}. The ionisation history and the evolution of the IGM need to be studied by indirect physical considerations. We now turn our attention to this task.

9.3 Ionisation of the IGM

The Gunn–Peterson effect described in the last section shows that the IGM has to be predominantly ionised at redshift $z \lesssim 5$. We can think of collisional ionisation and photoionisation as two possible means of achieving the ionisation.

In the case of photoionisation, the source of photons could be quasars or stars with $M > 10\ M_\odot$. We will see in the next section that a high degree of ionisation requires an energy input of $\sim 13.6(1 + t/t_{\text{rec}})$eV per H atom, where t_{rec} is the volume-averaged H recombination time scale and t is the cosmic time. The factor t/t_{rec} is large compared with unity, even at $z \simeq 10$, and thus photoionisation requires significantly higher energy than 13.6 eV per H atom.

The collisional ionisation also requires an IGM temperature of 10^5–10^6 K or equivalently ~ 25 eV per atom. The evolution of stellar population, leading to supernova explosions (which reheat the ISM and contaminate it by metals), injects both mechanical energy as well as UV photons into the surroundings. However, during the evolution of a typical stellar population, more energy is injected in the form of UV photons than in mechanical form. This is because the energy radiated per baryon (in nuclear burning) that eventually leads to solar metallicity, $Z_\odot \simeq 0.02$, is $\sim 0.007\ Z_\odot\ m_H c^2$ – of which approximately one-third is radiated as UV light. The massive stars, when they eventually explode as supernova, inject most of the metals as well as $\sim 10^{59}$ ergs of energy per explosion into the surroundings. For a Salpeter IMF, there is approximately one supernova per $150\ M_\odot$ of baryons in the form of stars. The mass fraction in mechanical energy works out to be approximately 4×10^{-6}, which is 10 times lower than the energy released in ionising photons. This suggests that photoionisation could be a major effect in the physics of the IGM, which we shall now discuss.

9.3.1 Photoionisation Equilibrium of the IGM

If the mean intensity of ionising radiation (in units of energy per unit area, time, solid angle, and frequency interval) is J_ν, then the photoionisation rate per unit volume is given by

$$\mathcal{R} = n_{\text{HI}} \int_{\nu_L}^{\infty} \frac{4\pi J_\nu \sigma_H(\nu)}{h\nu}\, d\nu, \tag{9.20}$$

where the photoionisation cross section for H is given by (see Vol. I, Chap. 6, Section 6.12)

$$\sigma_I = \sigma_0 \left(\frac{\omega_0}{\omega}\right)^{7/2}, \tag{9.21}$$

with $\sigma_0 \simeq 8 \times 10^{-18}$ cm^2 and $\omega_0 \simeq 2 \times 10^{16}$ Hz. This ionising flux has to come from a variety of sources at high redshift; and we first provide the general prescription for calculating the same.

Let us consider a class of sources with specific luminosity $L(\nu)$ (in units of ergs s^{-1} Hz^{-1}) and the luminosity function $\phi(L, z)$. Then $L(\nu)\phi(L, z)$ (in units of ergs s^{-1} Hz^{-1} cm^{-3}) gives energy radiated at a frequency ν per unit comoving volume by all the sources. The *proper* volume emissivity will then be

$$\epsilon(\nu, z) = \int_0^\infty (1 + z)^3 L_\nu(\nu)\phi(L, z)\, dL. \tag{9.22}$$

We next consider the specific intensity of the radiation field $J(\nu_0, z_0)$ (in units of ergs s^{-1} Hz^{-1} cm^{-2} sr^{-1}) at some frequency ν_0 and redshift z_0 arising from all the sources at $z > z_0$. The frequency of the emitted radiation will undergo a redshift by a factor of $(1 + z)/(1 + z_0)$ on propagation from z to z_0. Further, the light arriving from a source at a higher redshift z would also have been attenuated by some amount τ_{eff} because of IGM absorption, which will depend on the specific model of the IGM. Taking both these effects into account, we can now express the mean specific intensity in the form

$$
\begin{aligned}
J_\nu(\nu_0, z) &= \int_{z_0}^\infty \left(\frac{1 + z_0}{1 + z}\right)^3 \epsilon[\nu(z), z]\exp\{-\tau_{\text{eff}}[\nu(z), z_0, z]\}\frac{dl}{dz}\, dz \\
&= \frac{c}{4\pi H_0} \int_{z_0}^\infty \left(\frac{1 + z_0}{1 + z}\right)^3 \epsilon[\nu(z), z]\frac{\exp[-\tau_{\text{eff}}]\, dz}{(1 + z)^2(1 + \Omega_{\text{NR}}z)^{1/2}}.
\end{aligned}
\tag{9.23}
$$

The first factor is due to expansion, the exponential is due to absorption, and the rest of the factors have the usual geometric significance. The second equality is valid in a matter-dominated universe and can be easily generalised for any other model if dl/dz is replaced with the appropriate expression.

The actual form of τ_{eff} will depend on the structure of the IGM. For example, if the IGM contains clumps of H clouds, with each cloud having an opacity $\tau_{\text{cloud}}(N_{\text{H}}, \nu_0, z')$ that depends on the column density of hydrogen N_{H} and on the frequency and redshift, then the mean opacity that is due to all the clouds averaged over all directions will be

$$\tau_{\text{eff}} = \int_{z_0}^z \int_0^\infty \frac{\partial^2 N}{\partial N_{\text{H}}\partial z'}\{1 - \exp[-\tau_{\text{cloud}}(N_{\text{H}}, \nu_0, z')]\}\, dN_{\text{H}}\, dz'. \tag{9.24}$$

The first factor gives the number of clouds with H column densities in the range $(N_{\text{H}}, N_{\text{H}} + dN_{\text{H}})$ in the redshift interval dz'; the second factor takes into account the attenuation in a single cloud. Our discussion of different absorption systems later in this chapter will show that even if bulk of the baryons in the universe reside in the well-ionised diffuse IGM at $z \lesssim 5$, the residual neutral H that is present in higher-density clouds can provide a significant amount of optical depth to the continuum radiation. We now provide a derivation of Eq. (9.24).

The effective optical depth τ_{eff} is most conveniently defined by the average

$$\langle e^{-\tau}\rangle \equiv e^{-\tau_{\text{eff}}}. \tag{9.25}$$

To derive an explicit formula for τ_{eff}, we first consider a situation in which all absorbers have the same optical depth τ_0. Let the mean number of systems along the path be ΔN, which in turn can be obtained by integration of (dN/dz) over z. Then the Poisson probability of encountering an optical depth $k\tau_0$ along the line of sight will be

$$P(k\tau_0) = e^{-\Delta N} \left[\frac{(\Delta N)^k}{\tau_0 k!} \right]. \tag{9.26}$$

In this case, the mean transmitted flux will be

$$\langle e^{-\tau} \rangle = \sum_{k=0}^{\infty} e^{-k\tau_0} P(k\tau_0) = \exp[-\Delta N(1 - e^{-\tau_0})]. \tag{9.27}$$

This gives

$$\tau_{\mathrm{eff}} = \Delta N(1 - e^{-\tau_0}) = \int dz \frac{dN}{dz}(1 - e^{-\tau_0}). \tag{9.28}$$

When different column densities and redshifts are taken into account, this equation generalises to a τ_{eff} at frequency ν_0 at redshift z_0:

$$\tau_{\mathrm{eff}}(\nu, z_0, z) = \int_{z_0}^{z} dz' \int_0^{\infty} dN_{\mathrm{HI}} f(N_{\mathrm{HI}}, z)(1 - e^{-\tau}), \tag{9.29}$$

where $\tau = N_{\mathrm{H}} \sigma_{\mathrm{I}}(\nu)$ is the H continuum optical depth of an individual cloud at frequency $\nu = \nu_0(1 + z)/(1 + z_0)$. We shall see later that $f(N_{\mathrm{HI}}, z)$ is well approximated by

$$f(N_{\mathrm{HI}}, z) = A N_{\mathrm{HI}}^{-1.5}(1 + z)^{\gamma}, \qquad \gamma \approx 1.5. \tag{9.30}$$

In this case, we can explicitly compute τ_{eff} in closed form and obtain

$$\tau_{\mathrm{eff}}(\nu_0, z_0, z) = \frac{4}{3} \sqrt{\pi \sigma_0} A \left(\frac{\nu_0}{\nu_L} \right)^{-1.5} (1 + z_0)^{1.5}[(1 + z)^{1.5} - (1 + z_0)^{1.5}]. \tag{9.31}$$

This shows that, because of a rapid increase in the number of absorbers with look-back time, the mean free path of photons with $\nu > \nu_L$ becomes so small beyond the redshift of $z \approx 2$ that the radiation field is largely local. Expanding τ_{eff} in a Taylor series around some redshift z, we get $\tau_{\mathrm{eff}}(\nu_L) \simeq 0.36 \times (1 + z)^2 \Delta z$; that is, photons around $z = 3$ giving $\tau_{\mathrm{eff}}(\nu_L) = 1$ should have originated within a narrow local redshift window of $\Delta z = 0.18$.

Thus the ionising flux of radiation J can be explicitly computed for any specific class of sources (described by L and ϕ) and in a given model for the IGM (described by τ_{eff}) through Eq. (9.23). Because $\epsilon \propto L\phi$ with $L \propto h^{-2}$ and $\phi \propto h^3$ usually, the intensity $J \sim (\epsilon/h)$ is independent of h. The specific models can also give information regarding the variation of J with z, which is important for understanding the evolution of the IGM.

Later we shall see examples of different ionising sources and their spectrum. Right now, for the sake of illustration, we consider a spectrum of the form

$$J_\nu = J_0 \frac{\nu_L}{\nu} \times 10^{-21} \text{ ergs cm}^{-2} \text{ s}^{-1} \text{ Hz}^{-1} \text{ sr}^{-1}, \tag{9.32}$$

where $\nu_L = 3 \times 10^{15}$ Hz is the ionisation threshold corresponding to $\lambda \simeq 912$ Å. It is assumed that J_0 is of the order of unity. The magnitude of the intensity – with $J_0 \approx 1$ – is typically what arises in most models involving star-forming galaxies or quasars as the source. The spectral dependence is more uncertain and is essentially modelled by the continuum spectrum of quasars. The rate of ionisation per H atom that is due to the flux J_ν will be

$$\mathcal{R} = \int_{\nu_L}^\infty \frac{4\pi J_\nu}{h\nu} \sigma_I \, d\nu \simeq 3 \times 10^{-12} J_0 \text{ s}^{-1}. \tag{9.33}$$

In this case, the equilibrium between photoionisation and recombination will lead to the condition

$$\mathcal{R} n_H = \alpha n_e n_p = \alpha n_p^2 \tag{9.34}$$

if $n_e = n_p$. In reality, we also have to take into account the fact that the plasma may not fill space uniformly but could be clumped. If we assume that a fraction $1/C$ of space is uniformly filled by plasma, with the rest of the region empty, we need to multiply the right-hand side of Eq. (9.34) by a factor C. (Note that, in this case, the mean-square density will be $\langle n^2 \rangle = (n^2/C) = \langle n \rangle^2 C$.) Then we obtain

$$\frac{n_H}{n_p} = \frac{\alpha C n_p}{\mathcal{R}} \simeq 10^{-6} \frac{C \Omega_{IGM} h^2 (1+z)^3}{J_0} \left(\frac{T}{10^4 \text{ K}} \right)^{-1/2}. \tag{9.35}$$

Using the Gunn–Peterson bound on n_p/n_H, we can translate this equation to

$$\Omega_{IGM} \lesssim 0.4 \frac{\Omega_{NR}^{1/4} J_0^{1/2}}{h^{3/2} C^{1/2}} \left(\frac{T}{10^4 \text{ K}} \right)^{1/4} (1+z)^{-9/4}. \tag{9.36}$$

Although the preceding bound by itself is not very restrictive, the detailed mechanism for photoionisation as well as the issue of generating the necessary background intensity turns out to be quite nontrivial. The number density of photons per logarithmic interval in frequency can be related to the flux J_ν as

$$n_\gamma \equiv \frac{dN}{dV d(\ln \nu)} = \nu \frac{dN}{dA(c\,dt)d\nu} = \frac{1}{hc} \frac{dE}{dAdt\,d\nu} = \frac{4\pi}{hc} \left(\frac{dE}{dA\,dt\,d\nu\,d\Omega} \right) = \frac{4\pi}{hc} J_\nu. \tag{9.37}$$

Numerically, for J_ν in Eq. (9.32),

$$n_\gamma = 6 \times 10^{-5} J_0 \left(\frac{\nu_L}{\nu}\right) \text{cm}^{-3}. \tag{9.38}$$

Comparing Eq. (9.38) with Eq. (9.6), we find that $(n_\gamma/n_B) \simeq 0.1 J_0 (\Omega_B h^2)^{-1}$ at $z = 3$, $\nu = \nu_L$. For $\Omega_B h^2 \simeq 1$, $J_0 = 1$, this is hardly sufficient to ionise the IGM. For lower values of $\Omega_B h^2$, this may be marginally sufficient, although more detailed modelling suggests that it may not be quite adequate. (Also note that the ionisation of the IGM has to be reasonably uniform over sufficiently large scales.) We now discuss some of the details of the process of photoionisation arising from the emission of UV photons by discrete sources that could have formed at high redshifts.

Exercise 9.1
Photoionisation in a ΛCDM model: When H atoms are photoionised, their equilibrium can be described by the relation $n_{\text{HI}} \beta = n_e n_p \alpha(T)$ or, equivalently,

$$\frac{1 - x_e}{x_e^2} = \frac{n_H \alpha(T)}{\beta}, \tag{9.39}$$

where x_e is the ionisation fraction, $n_H = (1 - Y)\rho_B/m_H$, and

$$\beta = \int_{\nu_L}^{\infty} 4\pi J_\nu \sigma_I(\nu) \frac{d\nu}{h\nu} \approx 3 \times 10^{-12} J_{21} \text{ s}^{-1}, \tag{9.40}$$

$$\alpha = 2.06 \times 10^{-11} T^{-1/2} \phi_2(T) \text{ cm}^3 \text{ s}^{-1}, \qquad \phi_2(T) = 0.448 \ln\left(1 + \frac{h\nu_L}{k_B T}\right). \tag{9.41}$$

Here $h\nu_L = 13.6$ eV is the ionisation energy of H, σ_I is the photoionisation cross section, Y is the He fraction, and T is the temperature in degrees kelvin. Assuming $T \simeq 10^4$ K, determine the value of J_{21} needed to account for the Gunn–Peterson effect at $z = 3$ in a ΛCDM cosmology. Assume that $\Omega_B h^2 = 0.019$, $h = 0.65$, $Y = 0.24$, $\Omega_V = 0.65$, and $\Omega_{NR} = 0.35$.

9.3.2 Photoionisation of the IGM by Discrete Sources

The ionisation of the IGM by discrete sources like quasars – which could have formed at high redshift – leads to a situation that is similar to the ionisation of the ISM by OB stars. As an illustrative example, consider a point in an intergalactic HII region at, say, $z \simeq 6$, around a quasar with luminosity $L_\nu \simeq 10^{30}(\nu_L/\nu)$ ergs s^{-1} Hz^{-1}. The mean density of neutral H at this redshift will be $\bar{n}_H \simeq 1.6 \times 10^{-7}(1 + z)^3$ cm$^{-3} \simeq 5.5 \times 10^{-5}$ cm^{-3}. The intensity of radiation J_ν from the quasar can be approximated by $4\pi J_\nu = L_\nu/(4\pi r^2)$, where we will take $r \simeq 3$ Mpc, say. Given these expressions, we can calculate the photoionisation time scale

$$t_{ion} = \left[\int_{\nu_L}^{\infty} \frac{4\pi J_\nu \sigma_H(\nu)}{h\nu} d\nu \right]^{-1} = 5 \times 10^{12} \text{ s} \qquad (9.42)$$

and the recombination time scale

$$\bar{t}_{rec} = \frac{1}{n_e \alpha} = 5 \times 10^{16} \text{ s} \frac{\bar{n}_H}{n_e}, \qquad (9.43)$$

where α is the recombination coefficient. In photoionisation equilibrium, we would expect $(n_{HI}/t_{ion}) = (n_p/\bar{t}_{rec})$ so that we get $(n_{HI}/n_p) \simeq 10^{-4}$, implying nearly complete ionisation.

It is, of course, impossible for any given source to ionise an infinite region of space, and we will expect the formation of structures similar to Stromgen spheres of the ISM, which need to expand and overlap completely. The mean free path for an ionising photon is $(\sigma_0 n_H)^{-1} \approx 1$ kpc (at the threshold), which is much smaller than the radius of the ionised region. If the source spectrum is steep enough so that very little energy is carried out by higher-energy soft-x-ray photons, then we will have inhomogeneous structures in the IGM similar to the Stromgen sphere of the ISM. In an example for photoionisation, triggered by UV photons from an early generation of stars and quasars that were formed in the collapsed galaxy halos, we expect an expanding HII region to originate around individual sources of photons, inside of which H is fully ionised and He is either singly or doubly ionised. The reionisation will be complete when these HII regions overlap and every point in the IGM is illuminated by a nearly uniform Lyman continuum background. The equation governing the propagation of the ionisation front is similar to the one we used in the study of Stromgen spheres in the ISM in Vol. II, Chap. 9, Section 9.3. In terms of the proper volume V_I of the ionised zone, this equation is

$$\frac{dV_I}{dt} - 3HV_I = \frac{\dot{N}_{ion}}{\bar{n}_H} - \frac{V_I}{\bar{t}_{rec}}, \qquad (9.44)$$

where $\dot{N}_{ion}$ is the number of ionising photons emitted by the central source per second and $\bar{n}_H(0) = 1.7 \times 10^{-7}(\Omega_B h^2/0.02) \text{ cm}^{-3}$ is the mean H density today. When $\bar{t}_{rec} \ll t$, we can ignore the cosmological expansion and the solution is essentially given by

$$V_I = \frac{\dot{N}_{ion}\bar{t}_{rec}}{\bar{n}_H}\left(1 - e^{-t/\bar{t}_{rec}}\right). \qquad (9.45)$$

Note that, although the volume of the ionised region depends on the luminosity of the central source, the time taken to produce effective ionisation is decided by t_{rec} alone.

Although Eq. (9.45) describes the behaviour of a Stromgen sphere around a given source, what we are really interested in is a statistical description of the ionisation state of the IGM. This can be done in terms of a volume-filling factor Q for the ionised regions. (To be precise, we needs different Q to describe HII, HeII, and HeIII regions but we will concentrate on just HII.) In the early epochs, Q will be far less than unity and the radiation sources will be randomly distributed. The Stromgen spheres will be isolated from each other and every UV photon will be absorbed somewhere in the IGM with a highly inhomogeneous UV radiation field. As time goes on, Q increases and the percolation of ionised regions occurs around $Q = 1$. Because the mean free path of Lyman continuum radiation is much smaller than the Hubble radius, the filling factor Q can be easily estimated by the following argument. The fact that every UV photon is either absorbed by a newly ionised H atom or by a recombining one implies that $Q(t)$ must be equal to the total number of ionising photons emitted per H atom until that moment t, minus the total number of radiative recombinations per atom. That is, we must have

$$Q(t) = \int_0^t dt' \frac{\dot{n}_{\text{ion}}(t')}{n_{\text{H}}} - \int_0^t dt' \frac{Q}{\bar{t}_{\text{rec}}}. \tag{9.46}$$

Differentiating, we get

$$\frac{dQ}{dt} = \frac{\dot{n}_{\text{ion}}}{\bar{n}_{\text{H}}} - \frac{Q}{\bar{t}_{\text{rec}}}, \tag{9.47}$$

which describes statistically the transition of the universe from neutral to ionised form. Given a model for $\dot{n}_{\text{ion}}$, this equation can be integrated to describe the filling factor. Initially, when $Q \ll 1$, recombination can be neglected and the ionised volume increases at a rate $\dot{n}_{\text{ion}}/n_{\text{H}}$. At late times, in the limit of a fast recombining IGM, the asymptotic value of Q will be

$$Q \lesssim \frac{\dot{n}_{\text{ion}}}{\bar{n}_H} \bar{t}_{\text{rec}}. \tag{9.48}$$

This shows that the volume-filling factor is less than the Lyman continuum photons emitted per H atom within one recombination time. In other words, of all the photons with $\nu > \nu_I$, only a fraction $(\bar{t}_{\text{rec}}/t) \ll 1$ is actually used to ionise the new IGM material. The condition for complete reionisation of the IGM will then be

$$\dot{n}_{\text{ion}} \bar{t}_{\text{rec}} \gtrsim \bar{n}_{\text{H}}, \tag{9.49}$$

which says that the rate of emission of UV photons must exceed the rate of recombination.

The result also shows that clumpiness in the universe can have a significant effect on the conclusions. If the ionised gas with density n_p filled a fraction $1/C$ of the available volume uniformly, then we will have $\langle n_p^2 \rangle = (n_p^2/C) = \langle n_p \rangle^2 C$. More generally, if f_m is the fraction of baryonic mass in photoionised gas with density contrast δ and the underdense region is distributed uniformly, the fractional volume occupied by the denser component will be $f_v = f_m/\delta$

whereas the density of the diffuse component will be $\bar{n}_p[(1 - f_m)/(1 - f_v)]$. In this case, the recombination rate will be larger than that of a uniform universe by a factor of

$$C = f_m \delta + \frac{(1 - f_m)^2}{1 - f_v}.$$

(9.50)

Numerically, we have

$$\bar{t}_{\text{rec}} = [(1 + 2Y)\bar{n}_p \alpha_B C]^{-1} = 0.06 \text{ Gyr} \left(\frac{\Omega_B h^2}{0.02}\right)^{-1} \left(\frac{1+z}{10}\right)^{-3} \frac{\bar{n}_H}{\bar{n}_p} C_{10}^{-1}$$

(9.51)

where Y is the He–H abundance ratio and $C_{10} = (C/10)$. Unfortunately, it is difficult to estimate C theoretically, although it is likely to be fairly large compared with unity. For example, virialised halos with $T = 10^4$ K and a mass fraction of 0.04 will have a large comoving space density of $(dn/d \ln M) \simeq 10^3 h^3$ Mpc^{-3} at $z = 9$, corresponding to a mean proper distance of only $6h^{-1}$ kpc. These will contribute a factor $\sim$7 to clumping. It should also be stressed that the volume-averaged clumping factor is not a good statistic to use in the early epochs of ionisation. The halos previously described have a mean separation of $d = 6h^{-1}$ kpc but their virial radius is only $r_v \simeq 0.4h^{-1}$ kpc. It is only at scales larger than $(d^3/r_v^2) \simeq 2h^{-1}$ Mpc that we can use an average clumping factor as a useful statistic. Detailed modelling of these processes, therefore, has to depend on numerical simulations.

Assuming that the average clumping factor is a good descriptor and that $\bar{t}_{\text{rec}} \ll t$, we can use relation (9.49) to arrive at a critical rate of emission for ionising photons, $\dot{N}_c$ per unit comoving volume, independently of the previous history. This is given by

$$\dot{N}_c(z) = \frac{\bar{n}_H(0)}{\bar{t}_{\text{rec}}(z)} = (2.51 \times 10^{51} \text{ s}^{-1} \text{ Mpc}^{-3})C_{10} \left(\frac{1+z}{10}\right)^3 \left(\frac{\Omega_B h^2}{0.02}\right)^2.$$

(9.52)

To see the implications of this number, it is useful to determine an effective SFR that will provide this rate of photon emission. With the usual IMF, for each 1 $M_\odot$ star, a fraction, 0.08, goes into massive stars with $M > 20 M_\odot$. In these massive stars, at the end of C burning, nearly half the initial mass is converted into He and C, with a nuclear-energy-generation efficiency of 0.007. Further, $\sim$25% of the luminosity is in the form of Lyman continuum photons with $E > 20$ eV. Thus, for every 1 $M_\odot$ star that is formed per year, we expect an emission rate of

$$N_{\text{photons}} \approx 0.08 \times 0.5 \times 0.007 \times 0.25 \times \left(\frac{M_\odot c^2}{20 \text{ eV yr}}\right) \approx 10^{53} \text{ photons s}^{-1}.$$

(9.53)

Using this, we can convert the critical UV photon flux to an equivalent minimum SFR per unit comoving volume, $\dot{\rho}_* = (\dot{N}_c/N_{\text{photons}} f_{\text{esc}})$, where f_{esc} is a factor that determines what fraction of the photons escapes out of the star-forming region. Numerically, for $\Omega_B h^2 = 0.02$, we have

$$\dot{\rho}_* \approx (0.12 \, M_\odot \, \text{yr}^{-1} \, \text{Mpc}^{-3}) \left(\frac{0.5}{f_{\text{esc}}} \right) C_{10} \left(\frac{1+z}{10} \right)^3. \qquad (9.54)$$

This is a fairly high rate and, as we shall see, the energy injected into the IGM is comparable with (or even larger than) the energy radiated by quasars at the peak of their activity – which raises the question as to whether it is indeed possible to provide such a high rate of ionising photons in the universe. It should be stressed that this estimate depends sensitively on the value assumed for f_{esc} that denotes the fraction of high-energy ionising photons that can escape the star-forming regions and reach the IGM. Direct observations of nearby galaxies at the present epoch suggest that this fraction can be very low – possibly only a few (3–14) percent. If this is indeed true for the high-redshift protogalaxies as well, then we would require a significantly higher SFR in order to make this model work.

A more formal way of doing this calculation is as follows: The galaxy survey data (from ground-based telescopes as well the HDF) for star-forming galaxies at $2 \lesssim z \lesssim 4$ provide direct information about the UV flux of these galaxies, as the rest-frame UV continuum at 1500 Å will be redshifted into the visible band for a source at $z = 3$. There have been several detailed studies of UV luminosity functions of Lyman break galaxies spanning a factor of ~40 in luminosity. Integrating the luminosity functions over the range $L > 0.1 \, L_*$ and using the fact that, for a Salpeter mass function, the luminosity at 1500 Å is ~6 times the luminosity at 912 Å, we can determine the comoving emissivity at $\nu = \nu_I$ for these galaxies.[5] The emissivity is approximately

$$I = \begin{cases} (9 \pm 2) \times 10^{25} h \text{ ergs s}^{-1} \text{ Hz}^{-1} \text{ Mpc}^{-3} & \text{(for } z \approx 3) \\ (7 \pm 2) \times 10^{25} h \text{ ergs s}^{-1} \text{ Hz}^{-1} \text{ Mpc}^{-3} & \text{(for } z \approx 4) \end{cases} . \qquad (9.55)$$

This is roughly four times the quasar contribution at $z = 3$. There is, however, still the uncertainty as to what fraction of the photons actually escapes the galaxy, which needs to be addressed. If we assume that ~50% of the photons escape (which is far higher than the fraction that escapes our galaxy today), then we get approximately 1.58×10^{51} photons s^{-1} Mpc^{-3} at $z = 4$, which is high enough to provide the necessary ionisation. The situation at higher redshifts, of course, is quite uncertain because of the lack of direct observational evidence.

The second obvious source of photons is the quasars. The optical surveys of quasars seem to suggest that there exists clear evidence for a turnover in the QSO counts with the space density of quasars having a maximum somewhere around $z \approx 2.5$. Given the luminosity function for the quasars, it is possible to obtain the emission rate of photons per unit comoving volume by convolving with a fiducial quasar spectrum. Although this is straightforward in principle, there are several uncertainties in actual implementation, the most important of which is the incompleteness of the sample at high z. If we assume that the luminosity function varies as $L^{-\beta}$, with $\beta = 1.64$ at the faint end, then the total contribution from faint high-redshift quasars converges[6] only like $L^{0.36}$ as $L \to 0$. In such a case, it is possible to obtain a peak emission rate of approximately 1.26×10^{51} photons s^{-1} Mpc^{-3} at $z = 2$. However, at higher redshifts, the luminosity function suggests that $\dot{N}_{photons}$ will fall fairly rapidly as

$$\dot{N}_{photons} \simeq 3.66 \times 10^{51} \exp[-0.368(1 + z)] \qquad \text{(for } z > 2.2\text{).} \quad (9.56)$$

This falls below the critical rate needed to ionise the IGM at a redshift somewhere between $z = 3$ and $z = 4$ for clumping factors in the range of $C = 20$–40; if the ionisation of the IGM has to be complete before a redshift of 5, then quasars may not be the best candidates.

It is possible to circumvent this difficulty by boosting the luminosity function. If we take $\beta = 2$ and a comoving space density that remains constant above $z = 2.5$, we can increase the rate of UV photon emission. It must, however, be stressed that down to a 50% completeness limit, no quasar candidates at $z > 4$ have been found in the unresolved faint objects of the HDF, whereas the boosted luminosity function predicts roughly 10 objects.[7]

9.3.3 Collisional Ionisation

Even though it has been previously argued that photoionisation could be a major source of energy injection into the IGM, it must be remembered that the relative efficacy of the two processes – photoionisation and mechanical injection of energy – will depend on how easily radiation and mechanical energy penetrate through the IGM. This can be illustrated by the following simple argument. Consider an early generation of baryonic objects that form, through spherical top-hat collapse, a virialised structure with the parameters

$$v_c \approx 50\,\text{km s}^{-1}, \quad T = \frac{\mu m_p v_c^2}{2k_B} \simeq 2 \times 10^5\,\text{K}, \quad M \simeq 10^9 h^{-1} \left[\frac{(1+z)}{10}\right]^{-3/2} M_\odot.$$
$$(9.57)$$

Cooling by atomic H will allow rapid star formation in these structures within a dynamic time scale, converting a significant fraction $f\Omega_B$ into stars. If $f \approx 0.05$, $\Omega_B h^2 \simeq 0.02$, and $h = 0.5$, then the explosive output of 5×10^4 supernova will inject an energy $E_0 \simeq 5 \times 10^{55}$ ergs into the medium. The hot gas will escape the host and will appear as a cosmological blast wave, shocking the IGM, and it can be modelled by a Sedov solution. Assuming that the explosion occurs at $t = 4 \times 10^8$ yr (corresponding to $z = 9$), the radius of the shock wave after a time $\Delta t \approx 0.2t$ will be

$$R_s \approx \left(\frac{12\pi G E_0}{\Omega_B} \right)^{1/5} t^{2/5} \Delta t^{2/5} \approx 23 \text{ kpc}. \tag{9.58}$$

The corresponding shock velocity will be $v_s \approx (2R_s/5\Delta t) \approx 110 \text{ km s}^{-1}$, which is significantly higher than the escape velocity from the halo. The gas temperature behind the shock, $T = (3\mu m_p v_s^2/16 k_B) \simeq 4 \times 10^5$ K, is sufficiently high to ionise the incoming H. The Press–Schecter formalism shows that (see Chap. 7, Section 7.4) the comoving abundance of collapsed dark halos is approximately $(dn/d \ln M) = 5h^3 \text{ Mpc}^{-3}$ for $M = 10^9 h^{-1} \, M_\odot$ at $z = 9$. The corresponding mean proper distance between the neighbouring halos is $\sim 40 h^{-1}$ kpc, which should be compared with R_s obtained above. Even though only $\sim 1\%$ of the stars seen today might have formed at these early epochs, the blast waves from such a population of pregalactic objects could overlap and heat the IGM to a higher temperature, $T \gtrsim 10^5$ K, and pollute the IGM with metals. Although this case cannot be ruled out, it depends crucially on several input assumptions regarding star formation at high redshifts.

If the IGM could indeed be heated to 10^5 K, then we need to take the effect of collisional ionisation seriously. The cross section for collisional ionisation is maximum when the electron energy is ~ 100 eV, that is, when the temperature is $\sim 10^6$ K. In the case of collisional ionisation, we need to maintain equilibrium between the reaction $H + e \rightarrow p + e + e$ and $p + e \rightarrow H + \gamma$. This gives

$$\langle \sigma_{\text{coll}} v \rangle n_e n_H = \alpha n_p n_e, \tag{9.59}$$

where n_e, n_p, and n_H are the number densities of free electrons, protons, and H atoms, respectively, $\sigma_{\text{coll}} \simeq 2a_0^2$ is the collisional cross section (where a_0 is the Bohr radius), v is the typical velocity of the electron at $T \simeq 10^6$ K, and α is the recombination coefficient. We have seen in Vol. I, Chap. 6, Section 6.12, that

$$\alpha \simeq 2 \times 10^{-13} \left(\frac{T}{10^4 \text{ K}} \right)^{-1/2} \text{ cm}^3 \text{ s}^{-1}. \tag{9.60}$$

For $T \approx 10^6$ K, we have $\sigma_{\text{coll}} v \simeq 3 \times 10^{-8} \text{ cm}^3 \text{ s}^{-1}$, and we get

$$\frac{n_H}{n_p} \simeq 5 \times 10^{-7} \left(\frac{T}{10^6 \text{ K}} \right)^{-1/2}. \tag{9.61}$$

When the Gunn–Peterson bound in relation (9.8) is used, this becomes

$$\Omega_{\text{IGM}} = \frac{n_p}{n_{\text{crit}}} \lesssim 0.4(1 + z)^{-3/2} \Omega_{\text{NR}}^{1/2} h^{-1} \left(\frac{T}{10^6 \text{ K}} \right)^{1/2}. \tag{9.62}$$

The preceding result shows that unless the IGM could be heated to $\sim 10^6$ K, the collisional processes are not very efficient. On the other hand, photoionisation could be significant even at lower temperatures of $\sim 10^4$ K.

9.4 Background Radiation from High-Redshift Sources

The different sources that could possibly contribute to the ionisation of the IGM will also contribute to the EBL at different wavelengths, especially in the IR–UV range. Because we have direct observation as well as limits on the latter, this could provide some amount of consistency check on the different models.

The calculation of total background intensity from a class of sources is fairly straightforward. If the spectrum of emitted radiation from a class of objects is given in the form $\nu B(\nu)$ then the luminosity in a small-redshift cell Δz will be given by

$$\Delta L(\nu_{\mathrm{em}}, z) = \nu_{\mathrm{em}} B(\nu_{\mathrm{em}}) G(z) \frac{\Delta V}{\Delta z} \Delta z \qquad (9.63)$$

where the various factors are as follows: $(\Delta V / \Delta z)$ is the volume per one square arcsecond, corresponding to the redshift interval Δz given by

$$\Delta V = \frac{4\pi r_{\mathrm{em}}^2 c \Delta z}{\omega H_0 (1 + z)(1 + \Omega_0 z)^{1/2}},$$

$$r_{\mathrm{em}} = \left(\frac{2c}{H_0} \right) \frac{\Omega_0 z + (\Omega_0 - 2)[(1 + \Omega_0 z)^{1/2} - 1]}{\Omega_0^2 (1 + z)} \qquad (9.64)$$

where $\omega = 5.3464 \times 10^{11}$ is the number of square arcseconds on the sphere and $G(z)$ in Eq. (9.63) represents the time history of the sources. If the source is present only in a redshift interval $(z_c, z_c + \Delta z)$, we could, for example, take $G(z)$ to be a Gaussian peaked at z_c with a width Δz. The flux corresponding to this luminosity in Eq. (9.63) is

$$\Delta I_\nu(z) = \frac{L(\nu_{\mathrm{em}}, z)}{4\pi(1 + z)r_{\mathrm{em}}^2}, \qquad (9.65)$$

so that the total observed flux at a given frequency is

$$I_\nu = \int_{z_{\mathrm{min}}}^{z_{\mathrm{max}}} \frac{\nu_{\mathrm{em}} B(\nu_{\mathrm{em}}) G(z)}{4\pi(1 + z)r_{\mathrm{em}}^2} \frac{dV}{dz} \, dz. \qquad (9.66)$$

If the physical processes describing the evolution of the source are given in terms of time t rather than redshift z, we need to convert to redshift space by the general result

$$t(z) = H^{-1} \left[\frac{\sqrt{\Omega_0 z + 1}}{(1 - \Omega_0)(1 + z)} + \frac{\Omega_0}{2(1 - \Omega_0)^{-3/2}} \ln \left(\frac{\sqrt{\Omega_0 z + 1} - \sqrt{1 - \Omega_0}}{\sqrt{\Omega_0 z + 1} + \sqrt{1 - \Omega_0}} \right) \right]. \qquad (9.67)$$

A simple example of this calculation for a class of sources with spectra of the form $B(\nu) \propto \nu^{-\alpha}$ has already been given in Chap. 3, Section 3.5. There we found that if the sources are distributed uniformly then the intensity is given by

$$I(\nu_0) = \frac{c}{H_0} \frac{L(\nu_0)N_0}{4\pi} \int_0^\infty \frac{dz}{(\Omega_0 z + 1)^{1/2}(1+z)^{2+\alpha}}. \tag{9.68}$$

Assuming that $\alpha > -1.5$, which should be true for any realistic spectrum, we find that the total intensity is given by

$$I(\nu_0) = \frac{c}{(n+\alpha)H_0} \frac{N_0 L(\nu_0)}{4\pi}, \tag{9.69}$$

where $n = 1$ for an $\Omega_0 \to 0$ universe and $n = 1.5$ for an $\Omega_0 = 1$ universe. This expression shows that the emission is essentially due to all the sources in a volume $(cH_0^{-1})^3$ falling off with an inverse square factor $(cH_0^{-1})^2$. Also note that for an $\Omega_0 = 1$, $\alpha = 1$ model, the total background intensity from all sources up to a redshift z is given by

$$I(\nu_0) = \frac{2c}{5H_0} L(\nu_0)N_0 \left[1 - (1+z)^{-5/2}\right]. \tag{9.70}$$

The z dependence shows that half of the background intensity actually originates at $z < 0.31$; for an $\Omega_0 \to 0$ model, half the intensity comes from redshifts $z < 0.42$. Although high-redshift sources do contribute to background radiation unlinking the effects of more nearby sources is not going to be easy.

There are direct observations of background light intensity at 3.5 μm, 3000 Å, 5500 Å, and 8000 Å, all of which seem to lie a factor of a few above the contribution from the observed galaxy counts. In the FIR band there are measurements that yield $I_{FIR} \approx 14$ nW m^{-2} sr^{-1} in the 125- to 2000-μm range. Taking all these together, we can estimate the total background light as $I_{EBL} \simeq (55 \pm 20)$ nW m^{-2} sr^{-1}. We shall use a fiducial figure of 50 I_{50} nW m^{-2} sr^{-1}, with I_{50} parameterising the uncertainty.

Among the various sources that contribute to background light, galaxies have attracted considerable attention because of the availability of the HST and ground-based data. Given the differential number counts in various wavelength bands, we can directly integrate the counts multiplied by the emitted flux all the way down to the detection threshold to obtain the total background radiation. In the 0.2 to 2.2-μm range, the available data indicate an intensity of $I_{opt} \simeq 15$ nW m^{-2} sr^{-1}. This, of course, should be treated as a lower bound as much of the contribution in the past could not be observed directly and had to be inferred by theoretical analysis.

The direct estimate of background light from quasars is uncertain because of unknown bolometric correction, the behaviour of luminosity function at the faint end, and the space density of objects at high redshifts. The possible existence of a population of dusty AGN with strong intrinsic absorption will also add to the uncertainty. One possible way of bypassing some of these difficulties is to consider the mass density of black hole remnants of erstwhile AGN. We saw in

Chap. 1 that there is a direct correlation between the mass M_{sph} of the spheroidal bulge and the mass of a remnant black hole M_{BH}, given by $M_{BH} \simeq 0.006\, M_{sph}$. The mass density of a spheroidal population today[8] is approximately $\Omega_{sph}h = 0.0018^{+0.0012}_{-0.001}$. Combining these, we get the mean mass density of quasar remnants today as

$$\rho_{BH} = (3 \pm 2) \times 10^6 h\, M_\odot\, \text{Mpc}^{-3}. \tag{9.71}$$

Because the observed comoving energy density of radiation emitted by all quasars will be the total emitted energy divided by the mean redshift, we can estimate the contribution to background radiation as

$$I_{BH} = \frac{c}{4\pi} \frac{\eta \rho_{BH} c^2}{\langle 1+z \rangle} \simeq (4 \pm 2.5) \left(\frac{\langle 1+z \rangle}{2.5} \right)^{-1} \left(\frac{\eta}{0.05} \right) \text{nW m}^{-2}\, \text{sr}^{-1} \tag{9.72}$$

for $h = 0.5$, with η denoting the efficiency of the accretion process. This is not more than $\sim 20\%$ of the observed background radiation.

Let us next consider the contribution to background light from star-forming galaxies. The bolometric emissivity at time t that is due to stars with a SFR per comoving cosmological volume $\dot\rho_s$, bolometric luminosity L, and formation epoch t_F is given by the convolution integral

$$\rho_{bol} = \int_0^t L(\tau)\dot\rho_s(t-\tau)\,d\tau. \tag{9.73}$$

The total background light at $t = t_H \propto H_0^{-1}$ is given by

$$I_{EBL} = \frac{c}{4\pi} \int_{t_F}^{t_H} \frac{\rho_{bol}(t)}{1+z}\,dt, \tag{9.74}$$

where the $1 + z$ factor takes into account the redshift that is due to cosmological expansion. To compute I_{EBL} we need to know the SFR, the IMF, and the luminosity evolution. A simple model for estimating I_{EBL} can be constructed along the following lines.

Stellar-evolution theory suggests that the bolometric luminosity of a single generation of homogeneous stars having total mass M and solar metallicity can be expressed as a function of stellar age by a simple fitting function for $L(\tau)$ (valid for $\tau \gtrsim 3$ Myr):

$$L(\tau) = 1.3\, L_\odot \left(\frac{M}{M_\odot} \right) \left(\frac{\tau}{T} \right)^{-n}, \qquad (n \simeq 0.8, \quad T \simeq 1\ \text{Gyr}). \tag{9.75}$$

(This fitting function assumes that the luminosity scales in proportion to the mass of the stars produced, which is an approximation; for a more precise calculation, we need to integrate over an IMF.) For the SFR, we consider two extreme limits: (1) In the first case, we assume that the SFR is a constant, $\dot\rho_*(t) = \dot\rho_*$. (2) The other extreme corresponds to the formation of all the stars in a single burst at some time t_F so that $\dot{\rho}_*(t) = C\delta(t - t_F)$. In the first case, the total amount of mass per unit volume converted into stars within the age of the universe is $\dot{\rho}_* t_0$, which can be equated to $\Omega_* \rho_c$, where Ω_* is the fraction of the critical density contributed by the stars. Hence we can parameterise the SFR by $\dot{\rho}_* = \Omega_* \rho_c t_0^{-1}$. Similarly, in the second case, the mass density of the stars is $C = \Omega_* \rho_c$. In terms of this parameterisation, I_{EBL} for the first case can be easily estimated as

$$I_{\text{EBL}}^{(1)} = \frac{3.9\Omega_*}{(1-n)(8-3n)} \frac{c}{4\pi} \left(\frac{t_0}{T}\right)^{1-n} \left(\frac{\rho_c L_\odot T}{M_\odot}\right)$$

$$\approx 1.58 \times 10^4 (\Omega_* h^2) \text{ nW m}^{-2} \text{ sr}^{-1}. \tag{9.76}$$

This is equivalent to the relation (with $\Omega = 1$, $h = 0.5$),

$$I_{\text{EBL}} = 1460 \text{ nW m}^{-2} \text{ sr}^{-1} \left(\frac{\dot{\rho}_s}{M_\odot \text{ yr}^{-1} \text{ Mpc}^{-3}}\right). \tag{9.77}$$

The observed background radiation therefore requires a mean SFR of $\dot{\rho}_s = 0.034 \, I_{50} \, M_\odot \text{ yr}^{-1} \text{ Mpc}^{-3}$. If we ignore the recycling of gas returned to the ISM in the new stars, then the visible mass density at the present epoch is

$$\rho_{g+s} = \int_0^{t_H} \dot{\rho}_s(t) \, dt = 4.4 \times 10^8 \, I_{50} \, M_\odot \text{ Mpc}^{-3}, \tag{9.78}$$

corresponding to $\Omega_{\text{gs}} h^2 = 1.6 \times 10^{-3} \, I_{50}$.

We compute the second case most easily by estimating the ratio $I_{\text{EBL}}^{(2)}/I_{\text{EBL}}^{(1)}$, which is given by

$$\frac{I_{\text{EBL}}^{(2)}}{I_{\text{EBL}}^{(1)}} = \frac{(1-n)(8-3n)}{3} \int_{t_F/t_0}^1 dx \, x^{2/3} \left(x - \frac{t_F}{t_0}\right)^{-n}$$

$$= 0.37 \int_{t_F/t_0}^1 dx \, x^{2/3} \left(x - \frac{t_F}{t_0}\right)^{-0.8}. \tag{9.79}$$

The $I_{\text{EBL}}^{(2)}$ is plotted in the top frame of Fig. 9.1 as a function of redshift. The three curves are for $\Omega_* h^2 = 0.0018, 0.0013$, and 0.0008, corresponding to $0.09, 0.07$, and $0.04 \, \Omega_B h^2$. The figure shows that the time evolution of the luminosity makes the dependence on z_F much shallower than the $(1 + z_F)^{-1}$ factor (which is plotted as a dashed curve) because the energy emission is spread over the lifetime of the star. The bottom frame of Fig. 9.1 plots the ratio $I_{\text{EBL}}^{(2)}/I_{\text{EBL}}^{(1)}$ as a function of z_F for the same amount of total star formation. Clearly the star formation has to be of recent origin if the EBL has to be greater than what is produced by a constant rate of star formation. Further, to generate background light of $\sim 50 \, I_{50}$ nW m^{-2} sr^{-1}, we require $\Omega_{\text{gs}} h^2 > 1.3 \times 10^{-3} \, I_{50}$ for an $\Omega = 1$, $h = 0.5$ universe. This will correspond to a mean mass-to-(blue)light ratio of $(M/L_B)_{\text{gs}} > 3.5 \, I_{50}$ if we take the present-day luminosity density to be

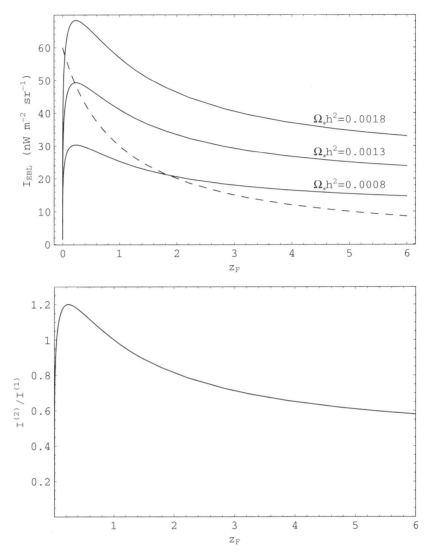

Fig. 9.1.   Top frame: The intensity of EBL produced at a star burst occurring at a redshift z_F. The three curves are for $\Omega_* h^2 = 8 \times 10^{-4}$, 1.3×10^{-3}, and 1.8×10^{-3}. The dashed curve is $(1 + z)^{-1}$, which shows that the dependence of EBL intensity on z_F is shallower than the $(1 + z_F)$ dependence. Bottom frame: The ratio of the EBL intensity produced when a certain amount of star formation takes place in a burst at z_F compared with the same amount of star formation occurring at a constant rate throughout the age of the universe. Unless the star formation has occurred recently, the latter process produces higher EBL.

$\mathcal{L}_B = 2 \times 10^8 h \, L_\odot \, \mathrm{Mpc}^{-3}$. Although such a case can explain the sky brightness, it requires most of the stars to form at a fairly low redshift (at $z_F \lesssim 0.5$). It is possible that the observed evolution of the UV flux rules out this possibility, although the situation is not yet completely clear.

A more precise calculation would require considering a realistic IMF, luminosity evolution, and SFR; we shall comment briefly on each of these ingredients. The IMF has a fair amount of uncertainty at the low-mass end. Recent HST observations suggest that it is possible to fit the IMF of the galactic disk in the range $0.1 < m < 1.6$ (where m is the mass in solar units) by[9]

$$\log \phi(m) = \text{constant} - 2.33 \log m - 1.82 (\log m)^2, \qquad (9.80)$$

which agrees fairly well with the Salpeter mass function for $m > 1$. It has been suggested that a combination of the Salpeter mass function for $m \geq 1$ and the preceding form for $m < 1$ matching to the Salpeter slope will provide an adequate description of the IMF. A somewhat better fit for the bolometric luminosity [than the one in Eq. (9.75)] of a population of stars is given by the fitting function

$$L(\tau) = \begin{cases} 1200 \, L_\odot \dfrac{M}{M_\odot} & (\tau \leq 2.6 \text{ Myr}) \\[2ex] 0.7 \, L_\odot \dfrac{M}{M_\odot} \left(\dfrac{\tau}{1 \text{ Gyr}} \right)^{-1.25} & (2.6 \leq \tau \leq 100 \text{ Myr}) \\[2ex] 2.0 \, L_\odot \dfrac{M}{M_\odot} \left(\dfrac{\tau}{1 \text{ Gyr}} \right)^{-0.8} & (\tau \geq 100 \text{ Myr}) \end{cases} \qquad (9.81)$$

The results of evaluating Eq. (9.74) for a model in which all the stars were formed in a single burst at $z = z_F$ are very similar to the ones shown in Fig. 9.1.

We saw in Chap. 1, Section 1.6, that measurements of the UV continuum suggest a star-formation density that evolves with redshift as[10]

$$\dot{\rho}_s(z) = \frac{0.11 e^{3.4z}}{e^{3.8z} + 44.7} M_\odot \text{ yr}^{-1} \text{ Mpc}^{-3}. \qquad (9.82)$$

Integrating the Eq. (9.74) with this fit gives approximately the correct amount of background radiation if $I_{50} \simeq 1$. The resulting visible mass density is $\Omega_{\text{gs}} h^2 = 3 \times 10^{-3} I_{50}$, corresponding to $(M/L_B) = 8.5 I_{50}$, which is almost twice as large in the constant SFR approximation.

Thus, depending on the star-formation history and the IMF, we require between 0.07 and 0.16 of the baryon density to be in the form of stars, processed gas, and their remnants if the observed background light arises from stars. Observations suggest that the current mass density in stars and their remnants is $\Omega_s h = (2.45 \pm 1) \times 10^{-3}$, corresponding to $(M/L)_* \simeq 3.4$ for $h = 0.5$. This is a factor of 2.5 smaller than that calculated from Eq. (9.82). Efficient recycling of ejected material could reduce the apparent discrepancy somewhat, but it is not clear whether the overall picture is completely consistent.

It was mentioned earlier (see Chap. 1) that there is an interesting connection between the amount of background radiation contributed by the stars and the mean metallicity in the universe, as the same process of nucleosynthesis in the stellar interior generates both the light and the metals. This connection can be made precise along the following lines. Let $\rho(L, z)\, d \ln L$ denote the comoving density of galaxies per logarithmic interval of luminosity. The total metal density is then given by

$$\rho Z = \frac{1}{\epsilon c^2} \int\!\!\int \rho(L, t) L \frac{dL}{L}\, dt, \tag{9.83}$$

where $\epsilon = 0.007$ is the efficiency of nuclear-energy conversion. If the bolometric luminosity is S and we assume, for simplicity, that the galaxies have flat spectra (that is, F_ν is independent of frequency in the relevant range), then we also have the relation $L = (4\pi \mathcal{B} S D^2)(1 + z)$, where $\mathcal{B}$ is the bolometric correction, D is the comoving distance, and the factor 4π takes into account the full solid angle. The observed distribution of galaxies in the redshift–flux space $N(S, z)$ is related to ρ by

$$\rho(L, z) \frac{dL}{L} D^2\, dr = N(S, z) \frac{dS}{S}\, dz. \tag{9.84}$$

Using $c\, dt = dr/(1 + z)$, we can write $cL\, dt = (4\pi S D^2)\, dr$ and convert the integrals in Eq. (9.83) to the form

$$\rho Z = \frac{4\pi \mathcal{B}}{\epsilon c^3} \int\!\!\int N(S, z) S \frac{dS}{S}\, dz = \frac{4\pi \mathcal{B}}{\epsilon c^3} I_\nu. \tag{9.85}$$

In arriving at the last equality, we have used the fact that we obtain the observed surface brightness of the population by integrating over all fluxes and redshifts. This result shows that the total metallicity is directly related to I_ν of the background light. Numerically,

$$I_\nu = 6.5 \times 10^{-22} \left(\frac{\rho Z}{10^{-34}\ \text{gm cm}^{-3}} \right) \text{ergs s}^{-1}\ \text{Hz}^{-1}\ \text{cm}^{-2}\ \text{sr}^{-1}. \tag{9.86}$$

A similar analysis can also be performed to relate the intensity of light that is expected in IR–submillimeter bands after being reprocessed by dust from primeval galaxies. If we assume that all the energy is produced at a mean redshift z, then the energy density of radiation is $U = \epsilon \rho Z c^2 (1 + z)^3$, leading to a specific intensity of radiation $I_{\text{tot}} = (c/4\pi) U$. Evolving this intensity to the current redshift will dilute it by a factor of $(1 + z)^{-4}$ so that the present background will be

$$I_{\text{tot}} = \frac{\epsilon \rho Z c^3}{4\pi(1 + z)} = \frac{2.1 \times 10^{-6}}{1 + z} \left(\frac{\rho Z}{10^{-34}\ \text{gm cm}^{-3}} \right) \text{ergs s}^{-1}\ \text{cm}^{-2}\ \text{sr}^{-1}. \tag{9.87}$$

9.5 Lyman-α Absorption by a Diffuse IGM

Even though quasar spectra do not show any absorption that is due to smoothly distributed neutral H in the IGM (see Section 9.2), it does exhibit several other absorption features that are due to H. We shall first describe some of the broad features seen in a typical quasar spectrum and will then discuss each category of the absorption systems in more detail.

9.5.1 Classification of Lyman-α Absorption Lines

A typical quasar spectrum is shown in Fig. 9.2. The absorption lines at wavelengths shortward of the Lyman-α emission line are thought to be Lyman-α absorption lines arising from clumps of gas along the line of sight, with each line being contributed by a particular region of neutral H in the IGM. The key feature that allows us to distinguish different Lyman-α absorption systems is the column density of neutral H present in them, which in turn we can determine by fitting the absorption line to a theoretically known profile and determining the equivalent width. We shall briefly recall the procedure involved in this analysis that was described in detail in Vol. I, Chap. 7, Section 7.3.

The equivalent width of an absorption line is defined by the integral across the line

$$w_{\mathrm{obs}} = \int \frac{I_c - I}{I_c}\, d\lambda = \int \left(1 - e^{-\tau(\lambda)}\right) d\lambda, \qquad (9.88)$$

where I is the observed intensity, I_c is the estimate of the absorption-free continuum across the line, and $\tau(\lambda)$ is the optical depth. The definition makes it clear that the equivalent width w_{obs} does not depend on the spectral resolution, which is a useful property; also note that, for a redshifted absorption line,

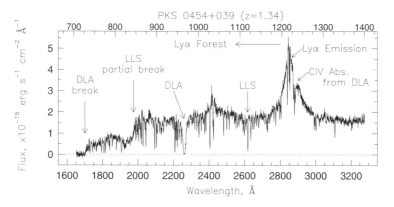

Fig. 9.2. The spectrum of a quasar at $z = 1.34$ showing the different kinds of absorption systems discussed in the text. The rest-frame wavelength is shown on the top and the observed wavelength is shown at the bottom. (Figure courtesy of C.W. Churchill.)

$w_{obs} = w_{rest}(1 + z_{abs})$. We have seen in Vol. I, Chap. 7, Section 7.3, that the optical depth can be expressed in terms of the Voigt profile, which is a convolution of natural linewidth and thermal linewidth, through the formula

$$\tau(v) = N\frac{1}{\sqrt{\pi}b}\int_{-\infty}^{\infty}\sigma(v')e^{-\frac{(v-v_0)^2}{b^2}}\,dv, \tag{9.89}$$

where $v' = v(1 - v/c)^{-1}$ and the parameter $b = \sqrt{2}v_{rms} = (FWHM)/2\sqrt{\ln 2}$ can be related to the temperature by

$$b = \sqrt{\frac{2kT}{m}} = 12.8 \text{ (km s}^{-1})\sqrt{\frac{T_4}{\mu}}. \tag{9.90}$$

Here μ is the molecular weight of the absorbing gas and T_4 is the temperature in units of 10^4 K. The cross section $\sigma(v)$, given by

$$\sigma = \left(\frac{\pi e^2 f}{m_e c}\right)\left[\frac{1}{\pi}\frac{(\gamma/4\pi)}{(v - v_0)^2 + (\gamma/4\pi)^2}\right], \tag{9.91}$$

incorporates the natural linewidth γ and the oscillator strength f. Equation (9.89) can now be rewritten in the form

$$\tau(\lambda) = (1.498 \times 10^{-2})\left(\frac{Nf\lambda}{b}\right)H(a, u), \tag{9.92}$$

where

$$H(a, u) = \frac{a}{\pi}\int_{-\infty}^{\infty}\frac{e^{-y^2}}{(u - y)^2 + a^2}\,dy, \quad a = \frac{\lambda\gamma}{4\pi b}, \quad u = -\frac{c}{b}\left[\left(1 + \frac{v}{c}\right) - \frac{\lambda}{\lambda_0}\right]. \tag{9.93}$$

The function $H(a, u)$ describes the Voigt profile. As we discussed in Vol. I, Chap. 7, Section 7.3, the core of the line is dominated by the Gaussian profile arising from the thermal width whereas the wings are dominated by the Lorentzian profile arising from the natural width. The optical depth at the centre of the line is approximately

$$\tau_0 = 1.497 \times 10^{-15}\frac{N(\text{cm}^{-2})f\lambda_0(\text{Å})}{b(\text{km s}^{-1})} \tag{9.94}$$

and essentially determines the nature of the absorption line profile that is observed. This is usually represented in the form of a curve of growth that is a relation between the column density N and the equivalent width w. There are three distinct regions in the curve of growth (see Vol. I, Chap. 7, Section 7.3). First, when the column density is small ($\tau_0 < 0.1$), the absorption line is optically thin and w is independent of b. In this linear part of curve of growth, we have

$$N(\text{cm}^{-2}) = 1.13 \times 10^{20} \frac{w_r(\text{Å})}{\lambda^2(\text{Å}) f}. \tag{9.95}$$

Given the equivalent width of a given transition, it is straightforward to determine the column density in this regime. Second, when the column density has an intermediate value, N depends strongly on b for a given w and the relationship is approximately logarithmic:

$$\frac{w}{\lambda_0} = 2\frac{b}{c}\sqrt{\ln(\tau_0)}. \tag{9.96}$$

The determination of column density is now more difficult because of the weak dependence of w on N. Finally, when τ_0 is large, the line gets saturated and the wings of the line are determined by the natural width. The equivalent width is again independent of b, and the relations become

$$\frac{w}{\lambda_0} = 2.64\frac{b\sqrt{a}}{c} \times \sqrt{\tau_0}, \qquad N(\text{cm}^{-2}) = 1.88 \times 10^{18} w_{\text{rest}}^2(\text{Å}), \tag{9.97}$$

where the second relation is valid for the HI λ 1215 line with $\gamma = 6.27 \times 10^8 \text{ s}^{-1}$. For realistic values of b encountered in the IGM, the HI λ 1215 line is in the saturated regime for approximately $N > 10^{19} \text{ cm}^{-2}$. This suggests that a natural limiting column density could be around $N \simeq 10^{20} \text{ cm}^{-2}$. Systems with higher column density are usually called *damped Lyman-α absorbers*. Because the photoionisation cross section for H is $\sigma_{\text{I}} \approx 10^{-17} \text{ cm}^2$, a column density of $N \simeq \sigma_{\text{I}}^{-1}$ provides a natural limit for photoionisation shielding. Lines with column densities between $N \simeq 10^{17}$ and $N \simeq 10^{20} \text{ cm}^{-2}$ are usually called *Lyman limit* systems. Systems with low column density – usually below $N \simeq 10^{17} \text{ cm}^{-2}$ – are called *Lyman-α forests*. The term "forest" is used to denote a large number of such lines that are typically seen in such absorption systems. It must be stressed that these limits merely reflect historical nomenclature and the division is not rigid.

The number of absorbing systems with different column densities of H seem to follow a fairly simple power law over a wide range, as shown in Fig. 9.3. This result can be expressed in a more useful form in terms of the probability dP that a random line of sight intercepts a cloud with a column density in the range $(\Sigma, \Sigma + d\Sigma)$ at the redshift interval $(z, z + dz)$. This is given by

$$dP = g(\Sigma, z)d\Sigma \, dz \simeq 91.2 \left(\frac{\Sigma}{\Sigma_0}\right)^{-\beta} \frac{d\Sigma}{\Sigma_0} dz, \tag{9.98}$$

with $\Sigma_0 \simeq 10^{14} \text{ cm}^{-2}$ and $\beta \cong 1.46$. This formula is valid around $z = 3$ for $10^{13} \text{ cm}^{-2} \lesssim \Sigma \lesssim 10^{22} \text{ cm}^{-2}$. We now consider these absorption systems in detail.

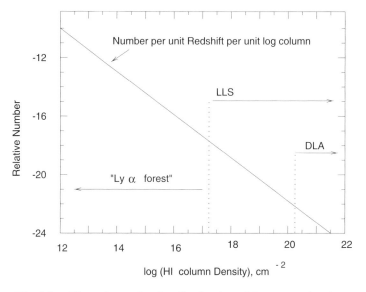

Fig. 9.3.   The column density distribution of Lyman-α clouds.

9.5.2 Lyman-α Forest and a Diffuse IGM

Let us begin by studying the Lyman-α forest that has the lowest column densities. These systems could naturally arise because of the fluctuations in the neutral-H density in the IGM in the following manner. Consider a region of diffuse low-density IGM in which the matter is in ionisation equilibrium under the action of an ionising flux J_ν and recombination. Because the density contrast in the IGM will be fluctuating and evolving, we will obtain transient absorption systems because of all the low-column-density neutral-H regions along the line of sight to a quasar. The neutral-H fraction in these systems can be estimated exactly as in Section 9.2 and we will get

$$\frac{n_H}{n_p} = \frac{\alpha f n_p}{\mathcal{R}} \simeq 10^{-6}\frac{\Omega_{IGM}h^2(1+z)^3}{J_0}\left(\frac{T}{10^4\,\mathrm{K}}\right)^{-0.76}[1+\delta(z)]^2. \qquad (9.99)$$

This is the same as Eq. (9.35) with two modifications. We have introduced a factor $[1+\delta(z)]^2$ to take into account the density variations – where δ is the density contrast compared with the background density – in the IGM; we have changed the exponent of temperature dependence from $-(1/2)$ to a more accurate value of -0.76. The neutral-H density in Eq. (9.99) will correspond to an optical depth of

$$\tau(z) = 6.5 \times 10^{-4}\left(\frac{\Omega_B h^2}{0.019}\right)^2\left(\frac{h}{0.65}\right)^{-1}\frac{(1+z)^6}{H_0}d_H(z)\frac{T_4^{-0.76}}{(\mathcal{R}/10^{-12}\,\mathrm{s}^{-1})}[1+\delta(z)]^2,$$
$$(9.100)$$

where $\mathcal{R}$ is the photoionisation rate defined in Eq. (9.33). This equation gives the typical opacity that is due to a Lyman-α forest in a diffuse IGM.

Observations suggest that the number of lines per unit redshift range is given by

$$\frac{dN}{dz} = B(1+z)^\gamma, \tag{9.101}$$

where $\gamma \simeq 2.78 \pm 0.71$ and B is in the range of 3–4 in the redshift range $2 < z < 3.5$; the evolution seems to be stronger at higher redshifts whereas HST observations[11] indicate a much weaker evolution with $\gamma = 0.48 \pm 0.62$ for $z < 1$. Another way of presenting this result is in terms of the mean proper distance between the Lyman forest clouds along the line of sight. This is given by

$$L = \frac{c\,dt}{dN} = \frac{c\,dt}{dz}\frac{dz}{dN} \simeq \frac{c}{H_0\,\Omega_{NR}^{1/2}B(1+z)^{\gamma+5/2}}$$

$$= 10^3(\Omega_{NR}h^2)^{-1/2}(1+z)^{-5.25}\,\mathrm{Mpc}$$

$$\simeq 0.6\,(\Omega_{NR}h^2)^{-1/2}\,\mathrm{Mpc} \qquad (\text{at } z = 3). \tag{9.102}$$

[We have used approximation (9.9) to get the Ω_{NR} dependence and set $B = 3.5$; approximation (9.9) will be used in the rest of the section whenever convenient.]

The amount of mass density contributed by these absorption systems is of considerable interest and – fortunately – can be estimated without knowing the size of the neutral-H patch involved in the absorption. Consider a region of IGM from which Lyman-α forest absorption takes place. Let us assume that the typical overdense fluctuation has a size of l, neutral-H mass M_H, column density in the range $(\Sigma, \Sigma + d\Sigma)$, and number density of overdense regions $n(z)$. Then we have

$$\frac{dN}{d\Sigma\,dz}d\Sigma \equiv g(\Sigma, z)\,d\Sigma \simeq \pi l^2 n(z)\frac{c\,dt}{dz}, \tag{9.103}$$

$$M_H \simeq m_p \pi l^2 \Sigma. \tag{9.104}$$

The total density contributed by neutral H in these clouds is

$$\rho_H = n(z)M_H = \left[g\,d\Sigma\frac{1}{\pi l^2}\frac{dz}{c\,dt} \right][m_p\pi l^2\Sigma] = m_p\Sigma g(\Sigma, z)\left(\frac{dz}{c\,dt} \right)d\Sigma. \tag{9.105}$$

Note that this result is independent of the size l of the clouds. We can obtain the total ρ_H contributed by clouds with column densities in the range $(\Sigma_{min}, \Sigma_{max})$ by integrating this expression between the two limits. Dividing by $\rho_c(1+z)^3$, we can find the Ω_H contributed by neutral H in these clouds:

$$\Omega_H(z) = \frac{m_p}{\rho_c(1+z)^3}\left(\frac{dz}{c\,dt}\right)\int_{\Sigma_{min}}^{\Sigma_{max}} \Sigma g(\Sigma, z)\,d\Sigma$$

$$= \left(\frac{m_p}{\rho_c}\right)\left(\frac{H_0\Omega}{c}\right)^{1/2}\frac{1}{(1+z)^{1/2}}\int_{\Sigma_{min}}^{\Sigma_{max}} \Sigma g(\Sigma, z)\,d\Sigma. \quad (9.106)$$

Taking $g = 91.2(\Sigma/\Sigma_0)^{-\beta}\Sigma_0^{-1}$ with $\Sigma_0 = 10^{14}$ cm^{-2}, $\beta = 1.46$, $\Sigma_{max} \approx 10^{22}$ cm^{-2}, and $\Sigma_{min} \approx 10^{14}$ cm^{-2}, we get, at $z = 3$,

$$\Omega_H \simeq 2\times 10^{-3}\Omega_{NR}^{1/2}h^{-1}\left(\frac{\Sigma_{max}}{10^{22}\text{ cm}^{-2}}\right)^{2-\beta}. \quad (9.107)$$

For $\Sigma_{max} \approx 10^{15}$ cm^{-2}, $\Sigma_{min} \approx 10^{13}$ cm^{-2}, which is appropriate for Lyman forest clouds, the corresponding result is

$$\Omega_H(\text{Lyman forest}) \simeq 3\times 10^{-7}\Omega_{NR}^{1/2}h^{-1} \quad (9.108)$$

at $z = 3$.

Let us next consider the Ω_{NR} contributed by the ionised component. In a cloud with column density Σ, the plasma contributes a mass that is F^{-1} larger than that of neutral H. This fraction F can be determined from Eq. (9.34) rewritten in the form

$$n_p = \left(\frac{Rn_H}{\alpha}\right)^{1/2} = \left(\frac{R\Sigma}{\alpha l}\right)^{1/2}, \quad (9.109)$$

where l is the characteristic scale of the system. This gives

$$F = \frac{n_H}{n_p} = \frac{\alpha n_p}{R} = \left(\frac{\alpha\Sigma}{Rl}\right)^{1/2}. \quad (9.110)$$

So the $d\Omega_{Bary}$ that is due to clouds with column density in the range $(\Sigma, \Sigma + d\Sigma)$ will be

$$d\Omega_{Bary} = \frac{d\Omega_H}{F} = \left(\frac{Rl}{\alpha\Sigma}\right)^{1/2}\left(\frac{d\Omega_H}{d\Sigma}\right)d\Sigma. \quad (9.111)$$

The result in Eq. (9.106) is now modified to

$$\Omega_{Bary} = \left(\frac{Rl}{\alpha}\right)\left(\frac{m_p}{\rho_c}\right)\left(\frac{H_0\Omega_{NR}^{1/2}}{c}\right)\frac{1}{(1+z)^{1/2}}\int_{\Sigma_{min}}^{\Sigma_{max}}\Sigma^{1/2}g(\Sigma, z)\,d\Sigma.$$

$$(9.112)$$

Because the exponent β of $g \propto \Sigma^{-\beta}$ is close to 1.5, the integral is close to a logarithm. Taking $\beta = 1.5$ for simplicity, we get, at $z = 3$,

$$\Omega_{Bary} \approx 6.3 \times 10^{-3} \left(\frac{J_0 l_0 \Omega}{h^3}\right)^{1/2} \left(\frac{T}{10^4 \, K}\right)^{1/4} \ln\left(\frac{\Sigma_{max}}{\Sigma_{min}}\right). \qquad (9.113)$$

If we take $\Sigma_{max} = 10^{15} \, cm^{-2}$ and $\Sigma_{min} = 10^{13} \, cm^{-2}$, the logarithm is ~4.6, and we find that

$$\Omega_{Bary} \approx 3 \times 10^{-2} \left(\frac{J_0 l_0 \Omega}{h^3}\right)^{1/2} \left(\frac{T}{10^4 \, K}\right)^{1/4}. \qquad (9.114)$$

So far we have been treating the Lyman-α forest as being made of primordial gas. Observations, however, reveal that ~50% of the Lyman-α lines with column densities higher than $3 \times 10^{14} \, cm^{-2}$ and nearly all lines with column densities higher than $10^{15} \, cm^{-2}$ originate from regions that also show CIV $\lambda\lambda$ 1548, 1550 absorption. Although this shows that stellar nucleosynthesis has already distributed the metals in the IGM at $z \approx 3$, the actual abundance of metals is difficult to determine. Estimates center around $Z \approx 10^{-2.5} \, Z_{\odot}$ with a large (approximately an order of magnitude) scatter around this value.

9.6 Damped Lyman-α Clouds

Historically, damped Lyman-α clouds (DLA hereafter) were defined as systems with column densities higher than $2 \times 10^{20} \, cm^{-2}$. This was essentially because of observational convenience when surveys were performed at low resolution and because of a theoretical prejudice that DLAs were progenitors of galactic disks. Because current observations clearly show damped wings, even at $10^{19} \, cm^{-2}$, it may be worthwhile to have a somewhat more general definition of what a DLA is.

The high column densities make the optical depth at the Lyman limit sufficiently large for the H to be mostly neutral. The gas has a temperature of $T \lesssim 3 \times 10^3 \, K$ and could even contain molecules. The distribution of column densities follows a slope of -1.5 even at these high column densities (see Fig. 9.3), but there seems to be virtually no evolution in these systems. The number density of these systems per unit redshift interval does not show any significant z dependence and is given by $(dN/dz) \simeq 0.3$ around $z \simeq 2.5$. The cosmic density of neutral H in a DLA at $z \approx 3$ is comparable with that of stars today. All these suggests that DLAs can be thought of as individual clouds, unlike the low-column-density Lyman-α forest, which is most likely to have originated from a fluctuating IGM.

Scaling the size of the clouds by $l = 10 \, l_0 h^{-1}$ kpc, we can estimate different parameters corresponding to these clouds. To begin with, $(dN/dz) = B \approx 0.3$ implies that the mean proper distance between the clouds along the line of sight is aprroximately

$$L = \frac{c\,dt}{dz}\frac{dz}{dN} = \frac{c}{H_0\Omega_{NR}^{1/2}B(1+z)^{5/2}} = 10^4(\Omega_{NR}h^2)^{-1/2}(1+z)^{-5/2}$$

$$= 300(\Omega_{NR}h^2)^{-1/2}\,\text{Mpc} \qquad (\text{at } z = 3).$$

(9.115)

This is considerably larger than the mean separation between Lyman-α forest systems, as to be expected. The density of H atoms and the mass of neutral H in these systems are given by

$$n_H \simeq \frac{\Sigma}{l} = 0.3hl_0^{-1}\,\Sigma_{22}\,\text{cm}^{-3},$$

$$M_H \approx m_p\Sigma l^2 \approx 10^{10}l_0^2\,\Sigma_{22}h^{-2}\,M_\odot,$$

(9.116)

where Σ is the column density. The mass in neutral hydrogen is fairly high allowing these systems to be interpreted as progenitors of the disks.

The fraction f of the space filled by these clouds is also of interest and can be estimated as follows. If $n(z)$ is the number density of clouds, then

$$\frac{dN}{dz} = (\pi l^2)n(z)\frac{c\,dt}{dz}.$$

(9.117)

The fraction of space filled by clouds in some large volume V is

$$f = \frac{(4\pi/3)l^3n(z)V}{V} = \frac{4\pi}{3}n(z)l^3.$$

(9.118)

Substituting for $n(z)$ from Eq. (9.117) we get

$$f = \frac{4\pi}{3}l^3\left(\frac{c\,dt}{dz}\right)^{-1}\left(\frac{dN}{dz}\right)\left(\frac{1}{\pi l^2}\right) \approx \left(\frac{dN}{dz}\right)\left(\frac{dz}{c\,dt}\right)l = \frac{l}{L},$$

(9.119)

where L is the mean distance between the clouds along the line of sight. The fraction of space filled by DLAs is

$$f = \frac{l}{L} = 3.3 \times 10^{-5}l_0\,\Omega_{NR}^{1/2} \qquad (\text{at } z = 3).$$

(9.120)

The number density of the clouds will be

$$n_{\text{cloud}} = \frac{f}{l^3} = 33h^3\,\Omega_{NR}^{1/2}l_0^{-2}\,\text{Mpc}^{-3},$$

(9.121)

with a mean intercloud separation of $n_{\text{cloud}}^{-1/3} \approx 0.3h^{-1}l_0^{2/3}$ Mpc. This is nearly 30 times larger than the size of damped Lyman-α systems (DLASs) if $l_0 \approx 1$.

The high column density of these clouds implies that they will be mostly neutral. The metagalactic flux of photons will be able to ionise only a thin layer of matter on the surface. Near the surface, equilibrium between ionisation and recombination,

$$\frac{n_{\mathrm{H}}}{n_p} = \frac{\alpha n_p}{\mathcal{R}}, \tag{9.122}$$

will lead to a proton density of

$$n_p = \left(\frac{\mathcal{R}}{\alpha} n_{\mathrm{H}}\right)^{1/2} = \left(\frac{\mathcal{R}}{\alpha} \frac{\Sigma}{l}\right)^{1/2}$$

$$\approx 2 \left(\frac{T}{10^4 \,\mathrm{K}}\right)^{1/4} \left(\frac{\Sigma}{10^{22} \,\mathrm{cm}^{-2}}\right)^{1/2} \left(\frac{J_0 h}{l_0}\right)^{1/2} \mathrm{cm}^{-3}, \tag{9.123}$$

which is approximately six times the neutral-H density of n_{H} of relation (9.116). Further, the penetration length λ of photons at such high column densities is low; we have

$$\lambda \approx \frac{1}{n_{\mathrm{H}} \sigma_{\mathrm{I}}} \approx 0.14 \, l_0 h^{-1} \,\mathrm{pc} \left(\frac{\Sigma}{10^{22} \,\mathrm{cm}^{-2}}\right)^{-1}. \tag{9.124}$$

This length is much smaller than the size of the system. Note that $(\lambda/l) \approx (\Sigma \sigma_{\mathrm{I}})^{-1}$, so $\lambda \approx l$ at column densities of $\Sigma \approx 10^{17} \,\mathrm{cm}^{-2}$. For $\Sigma \gg 10^{17} \,\mathrm{cm}^{-2}$, $\lambda \ll l$, and the system is mostly neutral. Thus most of the mass in DLASs is contributed by the neutral H:

$$M_{\mathrm{Baryons}} \approx M_{\mathrm{H}} \approx 10^{10} l_0^2 h^{-2} \, M_\odot. \tag{9.125}$$

The Ω_{NR} contributed by DLASs is quite large and comparable with that of luminous matter in galaxies. We have, from relation (9.107),

$$\Omega_{\mathrm{H}} \approx 2 \times 10^{-3} \Omega_{\mathrm{NR}}^{1/2} h^{-1} \quad (\text{for } \Sigma \approx 10^{22} \,\mathrm{cm}^{-2}; z \approx 3). \tag{9.126}$$

The dark matter associated with such a cloud will be higher by a factor of Ω_B^{-1}. If $\Omega_B \simeq 0.06$, this factor will be ~ 18. Thus the total Ω_{NR} associated with DLASs could be approximately 0.05–0.1 in the redshift range of $z \approx 2$–3.

These observations show that in the redshift range $1 \lesssim z \lesssim 4$, the universe contained both star-forming galaxies as well as DLASs. Because both populations are made of baryons that have settled down in dark-matter potential wells, we can attempt to relate their abundances along the following lines. Consider a dark-matter halo parameterised by a total mass M or circular velocity v_c. The abundance of these halos, as well as the density of baryons Ω_B in these halos, at any given redshift can be determined from the Press–Schecter theory developed in Chap. 7, Section 7.4. Some of these halos will host baryons that evolve into early star-forming galaxies. A remaining fraction will appear as protogalactic gaseous structures of which DLAS is a prime example. If the baryonic abundance of star-forming galaxies is subtracted from the total baryonic abundance in collapsed halos, we should be able to obtain the baryonic abundance contributed by DLASs. Because the SFR can be independently determined from the UV

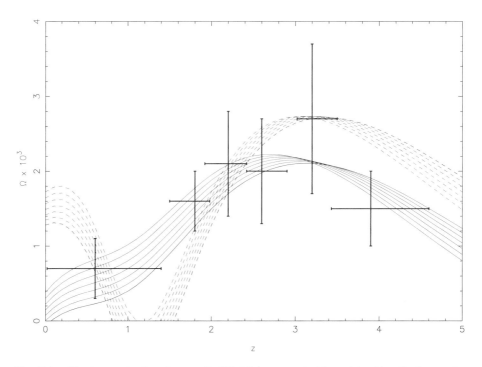

Fig. 9.4.   The baryonic abundance of a DLAS is computed by subtracting the baryonic abundance contributed by star formation from the baryonic abundance in collapsed halos at different redshifts. The baryonic fraction contributed by star formation is computed as in Chap. 1, Section 1.6, with the fitting functions in Eq. (1.59), and the baryonic abundance in collapsed halos is computed with the Press–Schecter theory. The two extreme forms of the fitting function for the SFR lead to the two sets of solid and dashed curves. The solid curves are for the SFR with dust absorption corrected, leading to nearly constant SFR in the redshift range $1 \lesssim z \lesssim 4$, and the dashed curves are for the SFR without dust absorption corrected. Each of the seven curves can be parameterised by the total dark-matter mass or the velocity dispersion of the halo. The masses used are 7.4×10^9, 8.7×10^9, 1.0×10^{10}, 1.2×10^{10}, 1.4×10^{10}, 1.7×10^{10}, and 2×10^{10} $M_\odot$; the corresponding velocity dispersions vary between 28 and 36 km s^{-1}.

luminosities, we could directly compute the baryonic abundance of a DLAS as a function of redshift and compare it with observations. The free parameters in the calculation are the mass of the collapsed halo and the fitting function used to describe the SFR. Because both of these are fairly well constrained by other observations as well, this approach serves as a good cross check on the scenario.

Figure 9.4 shows the results of such an analysis. The data points[12] give the Ω_B contributed by the DLAS with column densities above 3×10^{20} cm^{-2}. The solid curves are the theoretical expectation for the baryonic abundance in a DLAS obtained when the baryonic abundance contributed by star formation is substracted from the total baryonic abundance in collapsed halos. The baryonic fraction contributed by star formation is computed precisely as in Chap. 1, Section 1.6, with the fitting functions in Eq. (1.59), and the baryonic abundance in collapsed halos is computed with the Press–Schecter theory. The two extreme forms of the fitting function for the SFR lead to the two sets of solid and dashed curves. The solid curves are for the SFR with dust absorption corrected, leading to nearly constant SFR in the redshift range $1 \lesssim z \lesssim 4$, and the dashed curves are for the SFR without dust absorption corrected. Each of the seven curves can be parameterised by the total dark-matter mass or the velocity dispersion of the halo. In the figure, the masses used are 7.4×10^9, 8.7×10^9, 1.0×10^{10}, 1.2×10^{10}, 1.4×10^{10}, 1.7×10^{10}, and 2×10^{10} $M_{\odot}$; the corresponding velocity dispersions vary between 28 and 36 km s^{-1}. The figure shows that, within the somewhat large error bars, the theoretical predictions agree quite well with the observations. Alternatively, we could say that the sum of the baryonic abundance in DLASs and star-forming regions matches reasonably well with the total amount of baryons in collapsed halos.

This result gives credence to the standard assumption that DLASs are the progenitors of present-day disk galaxies observed at a time when most of their mass is still in gaseous form. If we take $\Omega_{\mathrm{DLAS}} \approx 2.5 \times 10^{-3}$, around $z = 3$, then we have the approximation

$$\Omega_{\mathrm{DLAS}}(z = 3) \approx \Omega_{\mathrm{diskstars}}(z = 0) \approx 5\Omega_{\mathrm{HI}}(z = 0). \qquad (9.127)$$

Although this conclusion is interesting, it must be stressed that the numerical values are highly uncertain and that the disk stars contribute less than 5% of the total baryons today. In fact, it is quite possible to argue that the known DLASs trace a population of which the galaxies that form the bulk of the stars comprise only a minor component at high redshifts. Given a range of surface densities for disks, the most compact galaxies will be the sites of rapid star formation whereas the more diffuse and slowly evolving systems will dominate absorption of quasar light by providing a larger cross section.

Understanding DLASs is further complicated by the fact that they contain a fair amount of metals compared with Lyman-α forest systems. For example, the abundance of Zn in ~40 DLASs in the redshift range 0.5–3.5 is approximately $Z_{\odot}/13$. (Once again the scatter on the abundance is fairly large and could vary by ~2 orders of magnitude at worst.) The metals in DLASs contribute $\Omega_Z \approx 2 \times 10^{-3}$ $\Omega_B Z_{\odot}$ whereas the Lyman-α forest contributes $\Omega_Z \approx (2-3) \times 10^{-3}$ $\Omega_B Z_{\odot}$. In units of solar metallicity, $Z_{\mathrm{DLAS}} \approx 0.07$ $Z_{\odot}$ whereas $Z_{\mathrm{Lyforest}} \approx 3 \times 10^{-3}$ $Z_{\odot}$. It is not very clear whether there is a problem with missing metals in this accounting. Although the Lyman-α forest may account for a large fraction of the baryons, its metal content is 1 order of magnitude too low compared with that of a DLAS. On the other hand, DLASs have the right amount of metallicity but they contain a smaller fraction of the total baryons. It is possible that when all the metals that have been measured are added up they account for no more than 10% of what we expect to have been produced and released by $z = 2.5$.

10

Cosmological Observations

10.1 Introduction

The purpose of this chapter is to give an overview and summarise several observations that are of cosmological relevance. Many of the results (like the existence of dark matter) described here were taken for granted in the earlier chapters, and we shall make an attempt to provide the description of the evidence in support of these results. This chapter also provides a summary of the current knowledge of different parameters of cosmological significance.[1]

10.2 Cosmic Distance Scale

The measurement of distances to different celestial bodies is of primary importance not only in understanding their properties but also – for objects at cosmological distances – in determining the geometry of the universe and the cosmological parameters. Obviously the technique used to measure the distance will depend on the properties of the object that is being studied; we shall concentrate on the measurement of extragalactic distances.

The procedures used for distance measurements can be divided into two natural classes. The first one uses what could be called the *absolute distance estimator*. These estimators are certain properties of (or features in) an object that can be used to directly measure the distance to the object. The second approach uses a *relative distance estimator* that allows us to determine the ratio between the distances to two different objects. By choosing a wide class of overlapping relative distance estimators, we can build what is known as a *cosmic distance ladder*. If an absolute distance estimator is used to fix the distance corresponding to the lowest rung of the ladder, then all other distances can be fixed. We now discuss several examples of these estimators.

10.2.1 Examples of Direct Distance Estimates

As the first example of an absolute distance estimator, we consider the Sunyaev–Zeldovich (SZ) effect discussed in Chap. 6, Section 6.7. Observation of the spectral distortion of CMBR in the vicinity of a cluster allows us to estimate the (inverse) Compton scattering optical depth in a cluster, given by

$$\tau_{SZ} = \sigma_T \int ds\, n_e(s), \qquad (10.1)$$

where n_e is the electron density and the integral is along the line of sight. This quantity is obviously independent of the distance to the cluster. If we can compare a distance-dependent feature of the same cluster with this optical depth, we will be able to estimate the distance to the cluster. Such a feature is provided by the x-ray bremsstrahlung emission from the cluster that has the luminosity

$$\epsilon(\nu) = A n_e^2 T^{-1/2} e^{-h\nu/k_B T}, \qquad (10.2)$$

where A is a known constant. By measuring the x-ray spectrum and fitting it to Eq. (10.2) we can determine the temperature T. The observed flux then depends on only the electron density n_e and the distance D to the cluster. Once n_e is determined from the SZ effect, we can determine D.

To illustrate the procedure, we consider a simple model for the cluster in which the gas is distributed with uniform density inside a sphere of radius R. The optical depth through the middle of the cluster is $\tau_{SZ} = 2\sigma_T R n_e$, and the x-ray flux is $f_X = (4/3)\pi R^3 [\epsilon(\nu)/4\pi d_L^2]$, where d_L is the luminosity distance. The angular extent of the x-ray-emitting region, which can also be directly measured, is $\theta = 2(R/d_A)$, where d_A is the angular diameter distance. Combining these relations and using the relation $d_A = (1+z)^{-2} d_L$, we get

$$D = \frac{A}{24\sigma_T^2} T^{-1/2} e^{-h\nu/k_B T} \, \tau_{SZ}^2 \frac{\theta}{f_X(\nu)} (1+z)^{-2}. \qquad (10.3)$$

All the quantities on the right-hand side can be directly measured, and hence D can be determined. The model can be easily generalised to include radial variations of density and temperature, provided they are independently measured.

The key weakness of this approach lies in the assumption of spherical symmetry made in the preceding analysis, which is important to relate the optical depth (measured along the line of sight) to the angular scale (measured across the sky). Because images of x-ray clusters show highly distorted, nonspherical shapes with an ellipticity ratio of $(a/b) \simeq 1.5$, this assumption is clearly questionable. In the case of an elliptical cluster, the results will change by a factor of $\mathcal{O}(a/b)$, depending on the orientation. We cannot easily circumvent this problem by observing a large sample of clusters and averaging, as we need to pay attention to the following selection effect: In a random sample, x-ray emission will be most detectable from those clusters that lie with the longest axis along the line of sight (as gas will appear to be concentrated in the smallest region in such a case); they will also have the largest SZ effect, which is due to a longer path length through the systems. Unless this selection bias is corrected, we cannot use the averaging procedure for a more accurate determination.

Because extragalactic distances scale with the Hubble constant as h^{-1}, determination of the former leads to an estimate of h that is of cosmological significance. The first result for h from the SZ effect in cluster A 665 ($z = 0.182$) gave $h = (0.4 - 0.5) \pm 0.12$. Combining it with the data on A 2218 ($z = 0.171$) raised it to $h = (0.55 \pm 0.17)$. Since then there have been several other measurements of clusters at different redshifts, allowing us to make a tentative plot of the angular diameter distance versus the distance. A fit to these data suggests that $h = 0.63 \pm 0.03$ in a universe with $\Omega_m = 0.3$ and $\Omega_V = 0.7$. The error bar is a formal statistical uncertainty that is not of much significance as the systematics contributes at least at the 30% level. It is therefore fair to say that currently available data[2] indicate $h = 0.6$ with a scatter of ~ 0.2.

The SZ effect can also be used to estimate the gas fraction in the clusters, although the result will depend on the model assumed for the intracluster medium. Several such studies suggest that the gas mass fraction is approximately $f_g = (0.081 \pm 0.01)h^{-1}$ in a universe with $\Omega_m = 0.3$ and $\Omega_V = 0.7$.

As the second example, let us consider the time-delay measurements of gravitational lens systems. We have seen in Chap. 3, Section 3.6, that, in the case of a multiply imaged quasar, light takes different amounts of time to go from the source to the observer along different paths. This time delay is due to a combination of differences in the geometrical path length and the effect of the gravitational potential. If the image of a background source at redshift z_s appears shifted by an angle α because of a lens at a redshift z_L, then the extra time delay is

$$t_L = (1 + z_L) \left[\frac{1}{2c} \frac{D_{\text{OL}} D_{\text{OS}}}{D_{\text{LS}}} \alpha^2 - \frac{2}{c^3} \int ds \, \Phi(s) \right], \qquad (10.4)$$

where the quantity on the right-hand side has to be evaluated separately for each of the images [see Chap. 3, Eq. (3.175)]. Given a detailed model for the lens, we can obtain the deflection α and the integral over the potential for each of the images. The time delay between, say, two images can be directly measured with the variation in the intrinsic brightness of the quasar as seen in the two images. Hence the only unknown quantities are the distances to the source and the lens. If the redshifts to the source and the lens are known, then these distances depend on only the cosmological parameters such as h, Ω, etc., and could be used to determine the latter.

The difficulties in this method are the following: The primary measurement of the time delay, although straightforward in principle, can be uncertain in practice, especially because all the features of one image are not often detected in the other. This leads to a fundamental doubt regarding the accuracy of the model and also produces disparate time delays, depending on how the data are analysed. For example, in the case of QSO $0957 + 561$, which was monitored by two groups for over 10 years, the time-delay estimates by the two groups have varied between 410 ± 20 days and 540 ± 10 days.[3]

A more recent analysis seems to support the shorter time scale, and, combining all available data, the two different groups concluded that $h = (0.64 \pm 0.13)$ and $h = (0.62 \pm 0.07)$. A further complication arises from the fact that the positions and relative brightnesses of the images do not often uniquely specify the lens model and thus can lead to very different estimates for the distance estimates and the Hubble constant. Attempts have also been made to use this technique with quadrapole image lenses, like PG $1115 + 080$ and B $1606 + 656$. Unfortunately, these results are plagued by large systematic effects that are due to uncertainties in the modelling of the lens.

10.2.2 Development of a Cosmic Distance Ladder

Let us now consider the second approach, which involves building up a cosmic distance ladder. The most direct measurements of distances are possible to the nearby stars in our galaxy. As the Earth goes around the Sun, the angular position of nearby stars will shift compared with those of more distant stars in the sky. This parallax, combined with the knowledge of the semimajor axis of Earth's orbit, allows us to measure distances to nearby stars. (In fact, the unit parsec is so defined that a star that has a parallax of 1 s will be at a distance of 1 pc.) This method works for distances up to about $\sim$30 pc or so.

The procedure adopted for objects in the next layer of distances uses a technique based on stellar evolution. To do this successfully, we need some objects for which both the methods can be applied. The nearest star cluster (called *Hyades*) can be used for this purpose. The distance to the Hyades cluster can be determined by geometrical methods to be approximately 46 ± 2 pc. [To be precise, this star cluster is at a distance that makes the parallax measurements unreliable. The distance is actually determined by a variant known as statistical parallax. This cluster is moving away from us in such a way that there is a measurable decrease $(\Delta\theta)$ in the angular size (θ) with time (t). It is easy to show that the distance is $r = (vt)(\theta/\Delta\theta)$, where the velocity v could be obtained from the Doppler shift.] The known distance to Hyades can then be used to calibrate the zero-age main-sequence stars. From the theory of stellar evolution, we can obtain a relation between the luminosity and the surface temperature for stars that have just begun the conversion of H into He. This relation is parameterised by the stellar mass and also depends on the abundance of elements heavier than He. We can measure these abundances from the study of the spectra of the stars without knowing the distance to the stars; the observed flux has an inverse square dependence on the distance. Therefore comparing the zero-age main-sequence stars in the Hyades with those in a more distant cluster will give the ratio of the distances to the two clusters. The measured distance to Hyades can be used to determine the distance to the second cluster. This method is used to cover the stars at distances in the range of 50 pc to 10 kpc.

To measure still larger distances, we can make use of a class of variable stars like RR Lyrae or Cepheids. Cepheid variables, whose outer atmospheres pulsate with periods in the range of 2–100 days (see Vol. II, Chap. 3, Section 3.7), are bright young stars that exist in abundant quantities in nearby spiral and irregular galaxies and have been studied extensively. Cepheids have absolute magnitudes $M_V \approx -3$ so that they can be studied up to a distance modulus of $m - M \simeq 25$ from ground-based telescopes and up to $m - M \simeq 28$ with the HST. RR Lyraes are much fainter (with $M_V \simeq 0.6$) and hence can be studied to only comparatively shorter distances. In absolute terms, it is impossible to find RR Lyrae outside the Local Group, whereas Cepheids have been seen even at Virgo cluster.

We have seen in Vol. II, Chap. 3, Section 3.7, that there exists a very tight relation between their periods of variation and the intrinsic luminosities (calculated either at maximum luminosity or based on an average over one period), with brighter Cepheids having longer periods. The correlation between period and luminosity of a Cepheid variable in the I band (~ 8000 Å) is accurate to $\sim 20\%$ in the luminosity. When the luminosity is used through an inverse square law to obtain the distance, this will lead to a 10% uncertainty in the distance to a single Cepheid; if a sample of $N \simeq 25$ Cepheids is available in a galaxy, we can reach a statistical accuracy of $\sim 10/\sqrt{N} = 2\%$ in the distance measurement. Thus, to use this method successfully, we require many Cepheids located at a common distance. The reach of the Cepheid variables is ~ 20 Mpc or so, beyond which we would require brighter objects. Further, we need to know the absolute distances to a few Cepheids to calibrate the relation. The LMCs and SMCs, two dwarf satellite galaxies of the Milky Way, are usually used to provide such a sample of Cepheids.

This is the best procedure currently available, but it suffers from several intrinsic errors. The distance to the Andromeda galaxy, ~ 740 kpc, is determined in this manner; the uncertainty in this value turns out to be $\sim 20\%$. Although the statistical uncertainty is fairly low ($\sim 5\%$), there are many systematic efforts that need to be corrected for. Dust grains along the line of sight will make the objects redder and fainter and hence – no correction is made for dust intervention – they will appear farther away than they actually are. The second difficulty in using any stellar standard candle has to do with the range of metallicities that stars have. Metals in the stellar atmosphere act as a source of opacity, and the actual calibration will depend on the value of metallicity, which is difficult to determine accurately.

To gauge larger distances, it is possible to adopt a method based on the luminosity function of a certain well-defined class of objects. If it is known that the luminosity function of any particular class of object does not vary from system to system, then comparing the characteristic apparent luminosities of the distribution of objects in two different systems will provide the ratio of the distances to these systems. An advantage of this method is that even if the absolute magnitudes of the objects have a spread σ, the average magnitude of the sample with n such objects will have a smaller scatter of $\sigma/\sqrt{n}$. One such example is provided by globular clusters (GCs) in which the number of stars per unit apparent magnitude is well described by

$$\phi_{GC}(m) \propto \exp\left[-\frac{(m - \bar{m}_{GC})^2}{2\sigma_{GC}^2}\right]. \tag{10.5}$$

Fitting this form to observed luminosity distribution of globulars in external galaxies, we can estimate $\bar{m}_{GC}$ and thus compare the distances to different galaxies. For accurate use, we need to determine the distribution of magnitudes to a level much fainter than $\bar{m}$ so that the peak of the Gaussian is well determined. In the Milky Way, the peak is at the absolute V-band magnitude of $\bar{M} = -7.4 \pm 0.1$ and it is possible to use the HST to observe the GC population down to approximately $m_V = 27.6$. A comparison shows that we should be able to detect the Gaussian peak all the way out to distances of $\sim 10^2$ Mpc.

Another example of the same technique uses the luminosity function for planetary nebula from which line emission can be detected, even against a bright stellar continuum of a galaxy, by use of narrow-band pass filters. They also occur with sufficient abundance in all types of galaxies and hence are not limited in their use. Observations centred at the 5007-Å O line from the planetary nebula in nearby galaxies suggest that their luminosity function can be fitted by

$$\phi_{PN}(m) \propto e^{0.307m}\left[1 - e^{3(m_c - m)}\right]. \tag{10.6}$$

The sharp cutoff in this luminosity function allows us to measure the cutoff magnitude m_c accurately, even with limited data. Further, because the cutoff occurs at the bright end, we need to measure only that part accurately, provided, of course, that the luminosity function is universal. The calibration is usually done by using the cutoff magnitude in the bulge of M31, which occurs at $m = 19.77 \pm 0.04$. Taking the distance to M31 as 740 ± 40 kpc, the absolute magnitude of the cutoff is $M = -4.6 \pm 0.1$, which can now be used to obtain distances to other systems.

Another class of standard candles based on stellar physics is provided by novae and supernovae. The idea in these methods is to use some robust feature of the light curve of these sources and relate it to their absolute magnitude. Comparison with apparent magnitude could then provide an estimate of the distance. In the case of novae, unfortunately, the absolute magnitudes vary quite a bit from $M_V = -4.8$ to $M_V = -8.9$ – a result known from observations of galactic novae at measured distances. Novae, however, seem to show a correlation between the rate of decay of the light curve and the maximum brightness, which can be quantified as

$$M_V(\text{max}) = -10.7 + 2.3 \log(t_2/\text{day}), \tag{10.7}$$

where t_2 is the time taken by the novae to decline by 2 magnitudes from the maximum. This fit has an error of $\sim$0.5 magnitude so that an accurate measurement of t_2 will allow us to determine the distance to an accuracy of 25%.

Type IA supernovae seem to fare better as standard candles and exhibit remarkably similar light curves. Recent HST observations detected Cepheid variables in the galaxies that also host type Ia supernovae, allowing us to calibrate the absolute magnitudes of the latter. It was found that $M_V = -19.52 \pm 0.07$ and $M_B = -19.48 \pm 0.07$. Theoretical models of supernovae also predict an absolute magnitude around $M_B = -19.4 \pm 0.3$. If we take these systems as standard candles, they provide another useful distance indicator.

The procedures outlined above were all based on properties of stellar systems. Much larger distances can be probed if galaxies themselves are used as secondary indicators. We have seen in Chap. 2 that the disk components of spiral galaxies are supported against gravity by centrifugal force. From radio and optical measurements, we can determine the quantity $v \sin i$, where v is the rotational speed of the stars and i is the angle between the axis of rotation of the galaxy and the line of sight. The speed v is related to the luminosity L by Eq. (1.67) of Chap. 1, Section 1.7, viz., $L \propto v^4$ (called the Tully–Fisher relation). This relation allows us to determine L with $\sim$40% uncertainty – the main source of scatter being the determination of $\sin i$, which has to be inferred from the shape of the galaxy. Given L, the distance can be determined with the observed luminosity. An important feature of this method is that the measured distances have a constant *fractional* error; therefore the absolute error increases linearly with distance. To decrease the fractional error and make sensible predictions at large distances, it would be convenient if we have a large number N of galaxies in some region of space close to each other, as in a cluster. In such a case, the statistical error can be reduced by the usual $\sqrt{N}$ factor.

A corresponding method exists for elliptical galaxies. Ellipticals are characterised by a velocity dispersion σ that seems to be well correlated with the characteristic size D by the relation

$$D \propto \sigma^{1.2}. \tag{10.8}$$

Once the intrinsic diameter D is found, the observed diameter can be used to estimate the distance. In practice, the scatter in the data leads to $\sim$20% error. The last two methods are potentially very powerful and can be used to very large distances (say, 10^2 Mpc). At present, however, they suffer from statistical bias and lack of good calibrators.

Figure 10.1 summarises the key results in the measurement of extragalactic distances obtained by use of several of the methods previously outlined. (The entry "surface-brightness fluctuations" is described in Exercise 10.1.) We shall

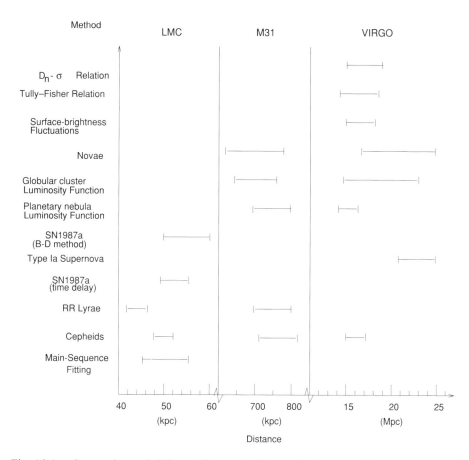

Fig. 10.1. Comparison of different distance indicators in the case of measurements of distance to LMC, M31, and Virgo cluster.

now briefly summarize some of the results. The distances to the LMC is comparatively well determined with an error of ~5%–7%. At the next stage, the distance to M31 has an error budget that is essentially dominated by the uncertainty in the zero-point calibration. Comparison of the Cepheids in the LMC and M31 show that the ratio of distances is 15.3 ± 0.8, which is accurate to ~5%. A mean distance of 50 ± 2 kpc for the LMC will place M31 at 760 ± 50 kpc. The uncertainties and scatter among the results from different methods increase when we move to Virgo cluster. Cepheids are supposed to give the most reliable estimate of distance to Virgo quoted in Fig. 10.1; but there is clearly a fair amount of variation between different measurements.

Given the distance to an extragalactic object that is far away and its line-of-sight velocity (which can be obtained from the redshift), we can attempt to determine the Hubble constant $H \simeq (v/d)$. The accurate determination of the Hubble constant from these distance measurements, however, is still problematic.

Virgo is far too close to participate in the Hubble flow, and the next obvious candidate will be Coma cluster. When different analyses are combined, the best estimate for the ratio of distances between Coma and Virgo turns out to be approximately 5.5 ± 2. Using the observed Hubble flow velocity of Coma allows us to express the Hubble constant as

$$H_0 = (63 \pm 2) \left(\frac{D_{\text{virgo}}}{20 \text{ Mpc}} \right)^{-1} \text{ km s}^{-1} \text{ Mpc}^{-1}. \qquad (10.9)$$

If we take the Virgo cluster distance to be 16 ± 1 Mpc, we get $H_0 = 79 \pm 6$ km s^{-1} Mpc^{-1}, whereas if we take the Virgo cluster distance to be 23 ± 2 Mpc, the Hubble constant becomes $H_0 = 55 \pm 5$ km s^{-1} Mpc^{-1}.

One of the main aims of the HST H_0 key project is to use the high resolving power of the HST to establish a primary calibration of Cepheid variables (and an accurate local extragalactic distance scale) and use this calibration in conjunction with several secondary distance indicators in order to determine H_0. As a part of this project, Cepheid distances were obtained for 17 galaxies located between 3 and 25 Mpc. Along with this, significant progress has been achieved for measuring relative distances to the galaxies by use of secondary techniques by calibration of supernova type IA light curves, rotational linewidths, I-band magnitudes for Tully–Fisher relations, and fundamental plane characteristics for elliptical galaxies. Table 10.1 summarises the results obtained for H_0 through several of these methods.

The largest sources of uncertainty in any individual determination of H_0 arise from the number of Cepheid calibrators per method, the effects of metallicity, and the velocity fields at large scales. However, there are two major sources of systematic uncertainty in all these measurements. The first one has to do with the adopted distance to the LMC that provides the fiducial Cepheid period luminosity relationship for the distance scale; the uncertainty to the LMC distance is $\sim 7\%$. The second source of systematic uncertainty is the photometric calibration of

Table 10.1. Summary of key project results on H_0

Method	H_0
Local Cepheid galaxies	$73 \pm 7 \pm 9$
Surface-brightness fluctuations	$69 \pm 4 \pm 6$
Tully–Fisher clusters	$71 \pm 4 \pm 7$
FP/$D_N - \sigma$ clusters	$78 \pm 7 \pm 8$
Type Ia supernovae	$68 \pm 2 \pm 5$
SNII	$73 \pm 7 \pm 7$
Combined	$71 \pm 3 \pm 7$

Note: The systematic errors are ± 5 (LMC), ± 3 ([Fe/H]), ± 3 (global), ± 4 (photometry).

HST magnitudes, which is approximately ± 0.09 mag.[4] The HST key project adopts a value of 18.5 mag for the distance modulus to the LMC. The distance to the LMC has been studied by several authors who obtained values in the range of 18.1–18.7 mag with a median of 18.45 and a 1σ error of ± 0.13 mag. The situation is still unsatisfactory in the sense that LMC distance leads to the largest uncertainty in determination of the Hubble constant. It has been emphasised in the literature that the zero points of the Cepheid and RR Lyrae calibrations differ at the level 0.15–0.30 mag, which could lead to 8%–15% uncertainty in distance.

Finally, we mention a method, based on the statistical fluctuations of the surface brightness of a galaxy, that could act as a distance indicator. To illustrate the idea, let us assume that all the stars in a distant galaxy are of the same intrinsic luminosity. If the galaxy is far away, then there will be $N(\gg 1)$ stars within one resolution element of the instrument. Hence there will be fluctuations in the brightness between the resolution elements, which will vary as $N^{-1/2}$. Measuring these fluctuations allows us to count the stars and thus determine the luminosity; knowing the total apparent brightness of the galaxy, we can determine its distance (see Exercise 10.1).

The development of the cosmic distance ladder based on different standard candles is probably one of the best available procedures if we attempt to measure distances directly. However, because each step of the ladder introduces certain errors, it is very unlikely that such a procedure will help in determining the Hubble constant accurately. Consider, for example, a set of four standard candles used to reach distances at which the Hubble flow is important. If each of the standard candle carries an uncertainty of 0.2 magnitudes, then recalibration at each step within the allowed range will make the final value for the Hubble constant uncertain by a factor in the range 0.7–1.4.

Exercise 10.1

Surface-brightness fluctuations: It is possible to develop a distance indicator based on the fluctuations of surface brightness in the galaxy. Consider a surface of a galaxy intercepted by a given solid angle. Let this surface be divided into a large number (M) of pixels. Let the total flux density from this area be I and the expectation number of stars in the region be m. Show that

$$\frac{\sigma_I}{\langle I \rangle} = \frac{1}{\sqrt{m}} \left(1 + \frac{\sigma_S^2}{\langle S \rangle^2} \right)^{1/2}, \tag{10.10}$$

where $\sigma_I^2 = \langle I^2 \rangle - \langle I \rangle^2$; $\langle S \rangle$ is the mean stellar flux density averaged over the stellar luminosity function and σ_S^2 is the variance in the flux. Show that the unknown number of stars m can be eliminated by the total flux so as to obtain

$$\left(\frac{\sigma_I}{\langle I \rangle} \right)^2 \langle I \rangle = \frac{\langle S^2 \rangle}{\langle S \rangle} = D^{-1} \frac{\langle L^2 \rangle}{\langle L \rangle}, \tag{10.11}$$

where D is the distance to the galaxy.

10.3 Age of the Universe

In the standard cosmological model, the current age of the universe $t_0(h, \Omega_{DM},$ $\Omega_V)$ is a function of the Hubble constant and the density parameter of matter and vacuum energy present in the universe. This age is obviously constrained to be larger than the age of the oldest structures seen in the universe. Given the latter, we get a constraint on the important cosmological parameters h, Ω_{DM}, and Ω_V. Further, if we have direct observations of the Hubble constant, it will allow us to combine these constraints to demark allowed regions in the Ω_{DM}–Ω_V plane. We shall now discuss some of the relevant results regarding this analysis.

In the standard example for the origin of the Milky Way, it is generally assumed that the halo of the GCs formed first, followed by the disk. There have been several attempts to determine the ages of objects in the disk as well as in the halo. Much of the work regarding disk age determination concentrated on the stellar ages of the open clusters (OCs) in the disk or the white dwarf luminosity function (WDLF). The results of several such recent observations are given in Table 10.2. The average value (excluding the age of the Sun) is $t_{disk} = 8.7 \pm 0.4$ Gyr, although there is significant variation around this mean. It should also be stressed that most of the error estimates are conservative and the possibility of unknown systematics cannot be completely ruled out.[5]

The corresponding results for the halo of the galaxy, based mostly on the stellar ages of halo GCs, is given in Table 10.3. The average in this case is $t_{halo} = 12.2 \pm 0.5$ Gyr with comparatively lower scatter.

The age determinations in both the tables, of course, are quite independent of any cosmological model. The age of the universe t_0 must be offset from the age t_{halo} by the relation $t_0 = t_{halo} + \Delta t$, where Δt is the time difference between the formation of the galaxy and the formation of the universe. An independent estimate of Δt will allow us to tighten the constraints on cosmological models.

Table 10.2. *Age of the Milky Way disk*

Object	Technique	Age (Gyr)
Sun	Isotopes[a]	4.53 ± 0.04
disk OC	Stellar ages[b]	8.0 ± 0.5
disk WD	WDLF[c]	8.0 ± 1.5
disk OC	Stellar ages[d]	9.0 ± 1
disk WD	WDLF[e]	$9.7^{+0.9}_{-0.8}$
disk OC	Stellar ages[f]	$12.0^{+1.0}_{-2.0}$
t_{disk} (ave)		8.7 ± 0.4

Note: The superscript letters in this table and Table 10.3 refer to individual references given in Ref. 5 for Chap. 10.

Table 10.3. *Age of the Milky Way halo*

Object	Technique	Age(Gyr)
Halo GC	Stellar ages[g]	11.5 ± 1.3
Halo GC	Stellar ages[h]	$11.8^{+1.1}_{-1.3}$
Halo GC	Stellar ages[i]	12 ± 1
Halo GC	Stellar ages[j]	12 ± 1
Halo GC	Stellar ages[k]	12.5 ± 1.5
Halo GC	Stellar ages[l]	13.5 ± 2
Halo GC	Stellar ages[m]	$14.0^{+2.3}_{-1.6}$
t_{Gal} (ave)		12.2 ± 0.5

Unfortunately, this result is highly model dependent and estimates vary in the range 0.1–5 Gyr.

It must be stressed that some of the techniques for age determination (like those based on main-sequence turn-off for GCs) depends crucially on the absolute magnitude calibration of the main sequence. If this changes by ϵ magnitudes, then the age estimate for globular cluster will change by $\Delta t_{gc} \approx 0.85 t_{gc} \epsilon$ (see Vol. II, Chap. 10, Section 10.2). If the main-sequence fitting is used in one of the steps of the distance ladder, the error ϵ will change the age estimate for the universe by approximately $\Delta t_{univ} = -0.46 t_{univ} \epsilon$. This shows that a difference of, say, 3 Gyr between t_{gc} and t_{univ} can occur if the main-sequence calibration is wrong by $\epsilon = -0.2$ magnitude.

Using the light curves of a supernova as a standard candle, we can make an independent estimate of the Hubble constant (given a cosmological model) or obtain best-fit results for all three parameters (h, Ω_{DM}, Ω_V). It is, for example, possible to use the supernova measurements to constrain Ω_{DM} and Ω_V and obtain a value for $H_0 t_0$. The same data will provide a best-fit value for H, using which we can determine t_0.[5] One such study has a best-fit value of $h = 0.63$ and $t_0 = 14.5 \pm 1.0(0.63/h)$ Gyr. Another attempt based on supernova data gives $h = 0.65 \pm 0.02$, and a best-fit analysis marginalising over all the parameters gives $t_0 = 14.2 \pm 1.7$ Gyr. These groups have also obtained the best fits, imposing the constraints that $\Omega_{DM} + \Omega_V = 1$. This leads to the values $t_0 = 14.9^{+1.4}_{-1.1}(0.63/h)$ Gyr and $t_0 = 15.2 \pm 1.7$ Gyr.

Although these observations have enjoyed significant popularity, certain key points that underlie these analyses need to be stressed. (1) The basic approach uses the supernova type I light curve as a standard candle. Although this is generally accepted, it must be remembered that we do not have a sound theoretical understanding of the underlying emission process. (2) The supernova data and fits are dominated by the region in the parameter space around (Ω_{DM}, Ω_V) $\approx$ (0.8, 1.5), which is strongly disfavoured by several other observations. If this disparity is due to some other unknown systematic effect, then this will have an effect on the estimate previously given. (3) The statistical issues involved in the analysis of these data to obtain best-fit parameters are nontrivial and certain subjectivity in the choice of method for the analysis cannot be completely ruled out.

The situation regarding the Hubble constant and the age of the universe can be summarised as follows. Observationally, it seems reasonable to assume that $H_0 t_0$ is constrained by the bounds $0.37 < H_0 t_0 < 1.47$. Using a reasonably good fit for $H_0 t_0$ [see Chap. 3, Section 3.5, relation (3.128)], we can obtain the bound

$$0.37 < \frac{2}{3}(0.7\Omega_m - 0.3\Omega_V + 0.3)^{-0.3} < 1.47. \qquad (10.12)$$

From this formula, we can immediately draw several conclusions: (1) If $\Omega_m > 0.1$, then Ω_V is nonzero if $H_0 t_0 > 0.9$ and – in fact – $\Omega_m + \Omega_V > 1$ if $H_0 t_0 > 1.33$. A more reasonable assumption of $\Omega_m > 0.3$ will require vacuum energy density if $H_0 t_0 > 0.82$. (2) If we take $\Omega_m = 1$ and $\Omega_V = 0$, then $t_0 > 12$ Gyr (which is a conservative lower bound from stellar ages) will require $h < 0.54$. Thus a purely matter-dominated $\Omega = 1$ universe would require a low Hubble constant, which is contradicted by most of the observations. (3) An open model with $\Omega_m \approx 0.2$ and $\Omega_V = 0$ will require $H_0 t_0 \approx 0.85$. This still requires ages on the lower side but values like $h \approx 0.6$ and $t_0 \approx 13.5$ Gyr are acceptable within error bars. (4) A straightforward interpretation of observations suggests a maximum likelihood for $h \approx 0.65$ and $t_0 \approx 14$ Gyr, giving $H_0 t_0 = 0.93$. This can be consistent with a $\Omega = 1$ model only if $\Omega_m \approx 0.3$ and $\Omega_V \approx 0.7$.

Given the best-fit values for the parameters h, Ω_{DM}, and Ω_V, as well as for t_0, t_{disk}, and t_{halo}, we can convert the time scales to redshifts at which the disk and the halo of the galaxy form. We typically get $z_{disk} \simeq 0.8$–2.8 and $z_{halo} \simeq 6 \pm 4.3$. The errors are fairly large at present, but it does indicate the existence of a phase in the universe dominated by diskless galaxies.

10.4 Observational Evidence for Dark Matter

Any form of energy density that makes its presence felt only by its gravitational effects may be called dark matter. For the purpose of this section, we shall not make any further distinction about the nature of the dark matter. (For example, planets like Jupiter that may exist around other stars will fall under the category of dark matter as far as the discussion in this chapter is concerned. This is in contrast to the discussion in earlier chapters where the term dark matter was used, essentially, to denote the *nonbaryonic* dark matter.) Observations suggest that there exists a significant amount of dark matter in our universe.

The quantity of dark matter in a system is usually determined in the following way: The gravitational field of the dark matter affects the motion of other objects (stars, clouds of gas, etc.) that are visible. If v is the typical velocity of test particles in a system with *total* gravitating mass $M(R)$ and size R, then

$$v^2 = k \frac{GM(R)}{R},$$
(10.13)

where k is a numerical constant that depends on the geometry of the system and the density distribution. The velocities of test particles can be inferred from the measurement of the redshifts. The size R is determined by different techniques for different objects; for the extragalactic objects, distances are usually determined from the redshift by Hubble's law. Once v and R are known from the observations and k from a theoretical model, M can be determined as $M = (v^2 R/kG)$.

It is conventional to quote the results of such an analysis in terms of a quantity Q, called the *mass-to-light* ratio: $Q = (M/L)$, where L is the luminosity of the system. This quantity can be expressed in units of the mass-to-light ratio for the sun $Q_\odot = (M_\odot/L_\odot) \simeq 0.5$ gm ergs^{-1} s. The existence of dark matter is signalled if the observed value of Q, based on luminous matter, is smaller than the value of Q inferred from the motion of test particles. Quite often it is convenient to determine the mass-to-light ratio of the system from the ratio of mass *density* ρ to the luminosity *density* J.

The following scaling may be noted as regards the mass-to-light ratio of the extragalactic objects. If R is measured with the mean redshift and the angular size, then it will scale with the Hubble constant, i.e., $R \propto h^{-1}$, making $M \propto R \propto h^{-1}$. Similarly, the absolute luminosity of the system L will be proportional to the square of the distance to the object and hence will scale as h^{-2}. Therefore $Q = (M/L) \propto (h^{-1}/h^{-2}) \propto h$. The inferred density ρ of the system M/R^3 will scale as $(h^{-1}/h^{-3}) = h^2$; because the critical density ρ_c of the universe also varies as h^2, the estimated value of the density parameter $\Omega = (\rho/\rho_c)$ will be *independent* of h.

We shall briefly summarise the observed results before describing them in detail. The observed distribution of galaxy luminosities $\phi(L)$, when integrated over the whole luminosity range, leads to an approximate total B-band luminosity density of $L \simeq 2 \times 10^8 h\, M_\odot$ Mpc^{-3}. This corresponds to a mass-to-light ratio of $Q = 1350\, \Omega_m h$ in solar units. On the scale of 0.1–20 kpc, the total gravitating mass can be measured from the rotation curve of spiral galaxies. These observations suggest that M/L increases up to the scales of binary pairs of galaxies ($r \simeq 100 h^{-1}$ kpc) and even to the scales of groups of galaxies ($r \simeq 1 h^{-1}$ Mpc). At still larger scales, the situation is more uncertain, although it is best probed by rich clusters of galaxies. The observed velocity dispersion of galaxies in the clusters range from 500 to 1500 km s^{-1}. The simple application of the virial theorem with a characteristic size of $R \simeq 1 h^{-1}$ Mpc leads to a B-band M/L ratio in the range 200–400h, corresponding to $\Omega_m \simeq 0.15$–0.30. We shall now discuss in detail a series of methods that are used in different contexts, from the smallest scales to the largest.

10.4.1 Solar Neighbourhood

The mass density of visible stars near the Sun in our galaxy is $\sim 0.044\ \dot{M}_\odot\ \mathrm{pc}^{-3}$. Almost an equal amount of density, $0.042\ M_\odot\ \mathrm{pc}^{-3}$, is contributed by gas. Further, the standard theory of stellar evolution suggests that the mass density of stellar remnants with negligible luminosity should be $\sim 0.028\ M_\odot\ \mathrm{pc}^{-3}$. Thus the total mass density near the Sun is $\sim 0.114\ M_\odot\ \mathrm{pc}^{-3}$. The luminosity is contributed mainly by the stars and amounts to $\sim 0.067\ L_\odot\ \mathrm{pc}^{-3}$. Dividing the mass density by the luminosity density, we find that the mass-to-light ratio near the Sun is

$$Q_{\mathrm{solar}}(\mathrm{observed}) \cong 1.7\ Q_\odot. \tag{10.14}$$

This value is based on the *volume* density of mass and luminosity. Because different stellar populations are distributed differently in the direction perpendicular to the galactic plane, it is probably more meaningful to consider the projected *surface* density of mass and luminosity. The surface mass density near the Sun (obtained by integration of the mass density within a size of 700 pc perpendicular to the galactic plane) is $\sim 50\ M_\odot\ \mathrm{pc}^{-2}$ and the surface brightness is $15\ L_\odot\ \mathrm{pc}^{-2}$. This leads to a slightly larger value for Q:

$$Q_{\mathrm{solar}}(|z| < 700\ \mathrm{pc},\ \mathrm{observed}) \cong 3.3\ Q_\odot. \tag{10.15}$$

To estimate the mass-to-light ratio that is due to the total gravitating mass in the solar neighbourhood, we first have to model the galaxy and obtain a relation between the mass density and the stellar velocities. The basic principle behind this estimate can be understood as follows: Consider a class of stars (say, the K giants) near the Sun. Let ρ be their density in the plane of the disk ($z \simeq 0$) and v_z be the mean random velocity in the z direction. If H is the mean scale height of the stars and g_z is the z component of gravitational force, then "vertical equilibrium" requires $\rho g_z H = \rho v_z^2$ or $g_z \simeq (v_z^2/H)$. For a plane slab of gravitating matter with surface mass density μ, Gauss theorem implies that $\mu = (g_z/2\pi G)$. Hence $\mu = (v_z^2/2\pi G H)$. Measuring v_z and H, we can determine the density of the gravitating mass.

To do the calculation more precisely, we use the model for our galaxy based on the CBE (see Chap. 2). The methods used to determine the amount of matter near the solar neighbourhood use the relation between the gravitational force **F** and the mass density ρ arising from the Poisson equation: $\nabla \cdot \mathbf{F} = -4\pi G\rho$. Assuming that the galaxy is axisymmetric and using a cylindrical coordinate system (R, ϕ, z), we can write this equation in the form

$$\rho = -\frac{1}{4\pi G}\left(\frac{\partial F_z}{\partial z} - \frac{1}{R}\frac{\partial v_c^2}{\partial R}\right), \tag{10.16}$$

where we have used the fact that $F_R = -(v_c^2/R)$. Observationally, v_c is approximately independent of R and the second term within the parentheses in any case does not vary much with z for $z \ll R$. Hence we can integrate Eq. (10.16) along the z axis to obtain the surface mass density:

$$\Sigma(R, z) \equiv 2 \int_0^z dz' \rho(R, z') \simeq -\frac{1}{2\pi G} \left[F_z(R, z) - \frac{z}{R} \frac{\partial v_c^2}{\partial R} \right]. \quad (10.17)$$

We have assumed that the galaxy is symmetric about the $z = 0$ plane so that $F_z(R, 0) = 0$. It follows from this equation that an accurate determination of Σ requires an accurate determination of F_z. To do this, we usually select a population of stars and determine their random velocities σ_z along the z direction and their number density ν along the z axis. We can then determine F_z from the Jeans equation,

$$\nu F_z = \frac{\partial \nu \sigma_z^2}{\partial z} + \frac{1}{R} \frac{\partial}{\partial R} (R \nu \sigma_{Rz}^2), \qquad \sigma_{ij}^2 \equiv \langle v_i v_j \rangle, \quad (10.18)$$

provided the second term can be ignored.

If we can measure ν and $\langle v_z^2 \rangle$ as functions of z, then this equation can be used in Eq. (10.16) to estimate the mass density ρ. Unfortunately, such a procedure leads to large observational errors because it involves two differentiations. (In reality, we need three differentiations because the quantity that is usually observed is not ν but the cumulative stellar density). To minimise the uncertainties, we can try to obtain independent estimates of ν for different stellar populations like K giants, F stars, etc., and average the results. Such an analysis, by use of Eq. (10.18), leads to the estimate $\rho \cong 0.15 \, M_\odot \, \mathrm{pc}^{-3}$ for the volume density. The integrated (surface) mass density Σ can be determined somewhat more accurately as it involves only two differentiations. A calculation similar to the preceding one gives $\Sigma(z \leq 700 \, \mathrm{pc}) \simeq 90 \, M_\odot \, \mathrm{pc}^{-2}$.

It is possible to increase the reliability of these measurements by more elaborate statistical techniques. Such a study leads to the estimates $\rho \simeq 0.18 \pm 0.03 \, M_\odot \, \mathrm{pc}^{-3}$ and $\Sigma(700 \, \mathrm{pc}) \simeq 75 \, M_\odot \, \mathrm{pc}^{-2}$. Because the luminosity density (near the Sun) is $\sim 0.067 \, L_\odot \, \mathrm{pc}^{-3}$, we get the mass-to-light ratio as

$$Q = 2.7 \, Q_\odot. \quad (10.19)$$

Similarly, using the fact that the surface brightness is $\sim 15 \, L_\odot \, \mathrm{pc}^{-2}$, we find the integrated mass-to-light ratio to be $\sim 5 \, Q_\odot$. Both these values exceed the corresponding mass-to-light ratios, based on luminous matter, by $\sim 50\%$. Thus nearly one-third of the material in the solar neighbourhood must be considered dark matter.

A more accurate estimate would require handling the quantity $\sigma_{Rz}^2 = \langle v_R v_z \rangle$ in Eq. (10.18) by some modelling. One idea is to assume that the gravitational potential is separable in the form $\Phi(R, z) = \Phi_R(R) + \Phi_z(z)$. In that case, we can separate the part of the distribution function $f(z, v_z)$ that depends on the z coordinate from the rest. In steady state, this will be a function $f(E_z)$ of the variable:

$$E_z \equiv \frac{1}{2}v_z^2 + \Phi_z(z). \tag{10.20}$$

The stellar number density is

$$\nu(z) = \int dv_z f(z, v_z) = 2 \int_{\Phi_z(z)}^{\infty} dE_z \frac{f_z(E_z)}{\sqrt{2[E_z - \Phi_z(z)]}}, \tag{10.21}$$

which, being an Abel integral equation, can be inverted to give

$$f_z(E_z) = -\frac{1}{\pi} \int_{E_z}^{\infty} d\Phi_z \frac{(d\nu/d\Phi_z)}{\sqrt{2[\Phi_z - E_z]}}. \tag{10.22}$$

If we can determine the form of $\Phi_z(z)$, then we can eliminate E_z between Eqs. (10.20) and (10.22) and predict the distribution of stars in the z–v_z plane. However, because $\Phi_z(z)$ is not directly observed, it is necessary to proceed in a statistical manner. If we can choose a suitable form for $\Phi_z(z)$, containing some undetermined constants, then we can compare the theoretically predicted form of distribution of stars in the z–v_z plane with the observed distribution and calculate the likelihood for any given set of parameters that appears in $\Phi_z(z)$. The set of parameters for which the likelihood is maximum is chosen as the correct one, thereby fixing $\Phi_z(z)$. Differentiation now provides F_z, from which ρ can be computed. One particularly simple form for $\Phi_z(z)$ and the corresponding force F_z are given by

$$\frac{\Phi_z(z)}{2\pi G} = (\sqrt{z^2 + D^2} - D)K + Fz^2, \qquad \frac{|F_z|}{2\pi G} = \frac{Kz}{\sqrt{z^2 + D^2}} + 2Fz. \tag{10.23}$$

The first term in the force varies as $K(z/D)$ for $z \ll D$ and is a constant for $z \gg D$. This is the form that we expect for a disk with a scale height of the order of D. The second term is the contribution from the spheroidal components of the disk. We can now use this form of Φ_z – parameterised by the constants D, F, and k – and the observed density $\nu(z)$ to determine $f(z, v_z)$. Comparing this distribution with the observed one, we can determine the optimum values for the parameters D, F, and k.

Exercise 10.2
Being upfront: Consider a disk galaxy in which the motion of stars perpendicular to the disk is described by a distribution function $f(z, v_z)$ that has decoupled from the rest of the directions. If the disk is seen face on, we can spectroscopically determine the quantity

$$F(v_z) = \int_{-\infty}^{\infty} f(z, v_z) \, dz. \tag{10.24}$$

What we are usually interested in is the density:

$$v(z) = \int_{-\infty}^{\infty} f(z, v_z) \, dv_z. \tag{10.25}$$

Show that, if the gravitational potential is of the form $\phi(z) = (1/2)Cz^2$, then $v(z) = \sqrt{C}F(v_z = \sqrt{C}z)$. For a more challenging task, relate $v(z)$ and $F(v_z)$ for a completely general self-consistent potential.

An analysis along these lines has yielded a value for the surface density of material within 1.1 kpc of the plane as

$$\Sigma_{1.1}(R_0) = 71 \pm 6 \, M_{\odot} \, \mathrm{pc}^{-2}. \tag{10.26}$$

It should, however, be stressed that although the data constrains the force F_z around $z \approx 1.1$ kpc, they permit variations in the parameters D, F, and k that are consistent with this value. Because uncertainty in k directly leads to the uncertainty in the surface density of the disk, the result is subject to uncertainties of modelling. It is possible to circumvent this difficulty by more sophisticated modelling of the galaxy that will provide further constraints on the parameter, but even these approaches are still plagued by serious uncertainties. One such modelling gives the total surface density of local disk as[6]

$$\Sigma_d(R_0) = 48 \pm 9 \, M_{\odot} \, \mathrm{pc}^{-2}. \tag{10.27}$$

The corresponding mass-to-light ratio is $Q \simeq 2.3 \, Q_{\odot}$. These results are fairly sensitive to the statistical analysis and modelling that have been used. There have also been reports in the literature claiming that the amount of dark matter in the solar neighbourhood is much lower than the figure previously quoted. Hence the numbers previously quoted should still be considered tentative.

We can also estimate the surface mass density of all gravitating matter in our galaxy from the rotational speed of stars in the disk. Let us assume, for simplicity, that the stars are moving in circular orbits with velocity $v(r)$. If the surface mass density of the disk is $\Sigma(r)$, then the balance between centrifugal force and gravitational force gives the relation

$$\frac{v^2(r)}{r} = G \int_0^{\infty} dx \, x\Sigma(rx) \int_0^{2\pi} d\phi \frac{(1 - x \cos \phi)}{(1 + x^2 - 2x \cos \phi)^{3/2}}. \tag{10.28}$$

Observations of $v^2(r)$ in our galaxy suggest that it is constant for a wide range of values of r. A constant rotational velocity v can be obtained from the surface density $\Sigma(r) \propto r^{-1}$. The integrals can be easily evaluated for this case, giving $v^2 = 2\pi G\Sigma(r)r$. Because $v \simeq 220$ km s^{-1} near the location of the Sun, $r = R_0 = 8.5$ kpc, we get

$$\Sigma = \frac{v^2}{2\pi G R_0} \simeq 210 \, M_{\odot} \, \mathrm{pc}^{-2}. \tag{10.29}$$

Note that this value is nearly three times larger than the value $\Sigma(700 \text{ pc})$ determined from the analysis of the z component of velocities of the stars. Clearly a significant fraction of the dark matter must be distributed with a scale height much greater than 700 pc. The same result can be stated in a different manner. If all the matter were confined to the disk and had a surface density of only $75 \, M_\odot \, \text{pc}^{-2}$ (as determined before) then the rotational velocity of stars would be lower than 220 km s^{-1}. Thus all the dark matter cannot be confined to the disk.

The existence of dark matter near the Sun raises the following question: How far does the dark-matter halo extend? It is difficult to answer this question precisely because we do not have reliable observations at distances far greater than the distance of the Sun from the galactic center, R_0. However, a rough estimate can be made along the following lines: Let us assume that the rotational velocity of our galaxy has a constant value V_0 up to a radius $r = L$ and falls as $r^{-1/2}$ beyond L. This is equivalent to assuming that the mass distribution of the halo is given by

$$M(r) = \begin{cases} (V_0^2 r/G) & (\text{for } r < L) \\ M_0 & (\text{for } r > L) \end{cases}. \tag{10.30}$$

Corresponding to this mass distribution, we get the gravitational potential ϕ:

$$\phi = \begin{cases} V_0^2 \left[\ln(r/L) - 1\right] & (\text{for } r < L) \\ -\left(V_0^2 L/r\right) & (\text{for } r > L) \end{cases}. \tag{10.31}$$

All the stars in the solar neighbourhood with speeds significantly higher than the escape velocity in this potential would have escaped by now. So the maximum stellar velocity v_{max} we expect in solar neighbourhood will be given by the condition

$$\frac{1}{2} v_{\text{max}}^2 + \phi(R_0) < 0. \tag{10.32}$$

Observations in the solar neighbourhood suggest that the velocity distribution of stars shows a sharp cutoff around $v_{\text{max}} = 500 \text{ km s}^{-1}$. Substituting $V_0 = 220 \text{ km s}^{-1}$ and $v_{\text{max}} = 500 \text{ km s}^{-1}$ into Eq. (10.31), we find that $L \gtrsim 4.9 \, R_0$. Or, because $R_0 \simeq 8.5 \text{ kpc}$, $L \gtrsim 41 \text{ kpc}$, corresponding to a total mass of $M_0 \gtrsim 4.6 \times 10^{11} \, M_\odot$. Because the total luminosity of our galaxy is $L_{\text{total}} \simeq 1.4 \times 10^{10} \, L_\odot$, the mass-to-light ratio for our galaxy is bounded by the inequality $Q \gtrsim 33 \, Q_\odot$. This value is at least six times larger than the value in the solar neighbourhood, suggesting that Q increases with the increasing scale.

An independent estimate of the extent of our galactic halo can be made with the Magellanic clouds and the dynamics of the satellite galaxies that are gravitationally bound to the Milky Way. These procedures lead to somewhat larger values of L, in the range of 50–80 kpc.

10.4.2 Rotation Curves of Other Disk Galaxies

One of the most striking and reliable pieces of evidence for the presence of dark matter in the galactic systems comes from the study of rotation curves of disk galaxies. A rotation curve is a plot of the rotational velocity $v(r)$ of some suitable test particle at a distance r from the centre of the spiral galaxy. The velocities are measured optically from the emission lines in the HII regions or from the radio measurements of the 21-cm line of neutral H. Because the neutral-H clouds exist even at a large radius from the centre of the galaxy, they serve as good tracers of the mass distribution for distances much larger than the visible extent of the galaxy.

For a spiral galaxy with finite mass, we would expect the rotation curve to fall off at sufficiently large distances. Such a behaviour is not seen. We now have measurements for over 70 spiral galaxies, and in almost all of them the rotation curve is either flat or slowly rising. The simplest interpretation of this result is that the spiral galaxies, like ours, contain massive spherical halos, with the halo mass increasing linearly with radius.

If both the rotation curve and accurate photometry of the galaxy are available, then we can attempt to model the distribution of the dark halo. This has been done, for example, for the Sc galaxy NGC 3198. The luminosity profile of this galaxy can be fitted by an exponential disk of scale height $2h^{-1}$ kpc and total luminosity $4 \times 10^9 h^{-2} \ L_\odot$. The dark-matter halo is well described by a density profile of the form $\rho(r) = \rho_0[1 + (r/a)^n]^{-1}$. If we assume a constant mass-to-light ratio throughout the galaxy, then the parameters that will provide the least mass-to-light ratio turn out to be $\rho_0 = 0.013h^2 \ M_\odot \ \mathrm{pc}^{-3}$, $a = 6.4h^{-1}$ kpc, and $n = 2.1$. For this set of values, $Q = 5.8h \ Q_\odot$. The total mass inside a radius of $22h^{-1}$ kpc (which is the distance up to which rotation curve has been determined) is approximately $1.1 \times 10^{11} h^{-1} \ M_\odot$, corresponding to a total mass-to-light ratio of $28h \ Q_\odot$. Although these results seem reasonable, they are – unfortunately – not unique. It is possible to model this system with the mass distributed differently between the disk and the halo, leading to a fair variation in these parameters, as discussed in Chap. 2, Section 2.3.

10.4.3 Cores of Spiral Galaxies and Dwarf Spheroidals

There are two systems that can be modelled theoretically by an isothermal sphere with some level of accuracy. These are the dwarf galaxies and the core regions of the galaxies. Using such a theoretical model, we can estimate the amount of dark matter in these systems.

The isothermal sphere is parameterised by two variables: the velocity dispersion, σ^2 and the central density ρ_0. Using these two variables, we can define a core radius $r_0 = (9\sigma^2/4\pi G\rho_0)^{1/2}$. The central *surface* density Σ_0 of the isothermal sphere is given by $\Sigma_0 \simeq 2.018\rho_0 r_0 \simeq 2\rho_0 r_0$. To determine the mass-to-light ratio

for such a system – say, the core of an elliptical galaxy – we proceed as follows:
The observed luminosity profile of the elliptical core is fitted to an isothermal
sphere, allowing us to determine the best-fit values for r_0 and central brightness I_0. The central emissivity is therefore $j_0 \cong (I_0/2r_0)$. On the other hand,
the central density is $\rho_0 = (9\sigma^2/4\pi Gr_0^2)$. Thus the central mass-to-light ratio
is

$$Q = \frac{M}{L} = \frac{\rho_0}{j_0} = \frac{9\sigma^2}{2\pi G I_0 r_0}. \tag{10.33}$$

Because all the quantities on the right-hand side are known from observation, Q
can be found.

   This procedure has been applied to cores of elliptical galaxies, bulges of spiral
galaxies, and dwarf spheroidal galaxies. The elliptical cores and spiral bulges
give the mass-to-light ratio of $\sim 12h\, Q_\odot$. This is consistent with the mass-to-
light ratio we would obtain in solar neighbourhood if the young stars and the gas
are excluded.

   The results of the analysis for dwarf spheroidals have been somewhat controversial. The main uncertainty is in the determination of the velocity dispersion
of the stars. The analysis gives $Q/Q_\odot$ as 1.7, 6.8, 12, 71, and 220 for the satellite galaxies Fornax, Sculptor, Carina, Draco, and Ursa Minor, respectively. The
wide variation in the Q values, as well as the large value for Ursa Minor, makes
these observations somewhat suspect. An anomalous value for σ^2 will be obtained if the stars for which the observations are carried out are members of a
binary system; in that case, part of the velocity that is due to orbital motion will
be erroneously attributed to σ^2. A detailed statistical study, however, suggests
that this effect is small and could possibly change the results at most by 20%.
Another possibility (which is difficult to test) is that Draco and Ursa Minor are
not gravitationally bound systems. The situation regarding the dwarf galaxies is
still somewhat unclear.

   There exists one dwarf irregular galaxy (called DDO 154) that is very rich in
gas. Gaseous H has been detected in this galaxy up to a distance of 8 kpc and the
visible extent is ~ 2 kpc. Up to this point, the rotation curve is flat, implying that
90% of the matter is dark. The estimated mass-to-light ratio is greater than ~ 75.
One of the simplest theoretical models gives $M_{vis} \simeq 5 \times 10^7\, M_\odot$ and $M_{dark} \simeq 4 \times 10^9\, M_\odot$.

   Except in very few cases – such as NGC 1052, NGC 4278, and NGC 5128,
which have rings of neutral H that allow the determination of the mass just as
in the case of spiral galaxies – the elliptical galaxies do not have systematic
rotation curves and their dark-matter content needs to be ascertained by other
methods like the ones previously described. Another possibility is to use the
velocity dispersion of the stars in the ellipticals and determine the total mass
by using virial theorem. This procedure, however, has the difficulty indicated in

Chap. 2, Section 2.4 – namely, we need a handle on the velocity anisotropy in order to obtain a reliable estimate of the mass.

Some ellipticals have a hot-x-ray-emitting halo of gas extending up to $\sim$50 kpc in a few cases. If hydrostatic equilibrium can be assumed for the gas and the radial variations of density and temperature can be obtained from observations, then the mass can be estimated, for a spherical halo, from the formula

$$M(< R) = \frac{k_B T R}{G m_p} \left[\frac{d \ln \rho}{d \ln R} + \frac{d \ln T}{d \ln R} \right], \tag{10.34}$$

where ρ is the density of the gas. The difficulties with this method are twofold. First, the assumption of hydrostatic equilibrium may not be valid if the central regions cool faster than the outer regions, initiating a flow of gas towards the center. Second, the $T(R)$ profile is not very accurately known in many cases.

Exercise 10.3
Filling in the gaps: Derive Eq. (10.34) for the mass of the x-ray-emitting gas, assuming hydrostatic equilibrium and spherical symmetry.

Finally, it is worth mentioning that star clusters, both open and globular, do not contain any amount of dark matter. This is of some importance because both GCs and dwarf spheroidals have comparable sizes and masses.

10.4.4 Dark-Matter Estimates from the Dynamics of the Local Group

The Local Group consists of two dominant spiral galaxies, the Milky Way and Andromeda, each surrounded by a number of smaller galaxies. The next nearest groups of large galaxies are Sculptor and M81, which are $\sim$3 Mpc away. Therefore, to first approximation, we may consider the Local Group to be a gravitationally isolated system. By studying the dynamics of the Local Group, we can estimate the total gravitational mass in it and thus its mass-to-light ratio. The basic idea is extremely simple and can be understood as follows: The relative velocity v between Andromeda and the Milky Way, the most dominant members of the Local Group, is $\sim 10^2$ km s^{-1} and their relative separation r is $\sim 10^3$ kpc. Taking the age of the universe t_0 to be 1–2 $\times 10^{10}$ yr, we can form the dimensionless combination $(v t_0 / r) \simeq (1\text{–}2)$. The fact that this ratio (which could have had any value a priori) is close to unity suggests that the Andromeda–Milky Way system is gravitationally bound. In that case, the mass may be estimated to be $M \simeq (v^2 r / G) \simeq 10^{12}$ $M_\odot$.

To do a more precise job, we proceed as follows: The line-of-sight velocity that will be measured by a hypothetical observer located on the Sun (called the heliocentric velocity) is known for all the members of the Local Group. Because we know the motion of the Sun, we can compute the line-of-sight velocity with respect to the galactic centre. If an external galaxy is at the galactic angular coordinates (l, b), then the line-of-sight velocity in a frame at rest with respect to the galaxy centre is given by

$$v_G = v_{he} + v_{LSR} + v_\odot, \tag{10.35}$$

where v_{he} is the heliocentric velocity that is observed, v_{LSR} is the velocity of a hypothetical star in a circular orbit at the location of the Sun, and $v_\odot$ is the deviation of the velocity of the Sun from v_{LSR}. The latter two quantities, when projected along the radial direction, have the values

$$v_{LSR} = (220 \, \text{km s}^{-1}) \sin l \cos b, \tag{10.36}$$

$$v_\odot = (16.5 \, \text{km s}^{-1})[\cos b \cos 25^o \cos(l - 53^o) + \sin b \sin 25^o]. \tag{10.37}$$

Because all the members of the Local Group are at distances much greater than R_0, the line-of-sight velocity is approximately radial from the centre of the galaxy. In particular, the line-of-sight velocity of Andromeda is $-297 \, \text{km s}^{-1}$. Because Andromeda is located at $l = 121.2$ and $b = -21.6$, Eq. (10.35) gives $v_G = -119 \, \text{km s}^{-1}$. The most natural explanation for this motion of the Milky Way and Andromeda towards each other (as indicated by the negative sign) is that the relative Hubble expansion between the two galaxies has been halted by their mutual gravitational attraction. In that case, we may treat Andromeda and the Milky Way as a gravitationally bound system, with each galaxy moving in a high-eccentricity Keplerian orbit. For such an orbit, the separation of the galaxies r as a function of the cosmic time t is given by the Kepler equation:

$$r = \alpha(1 - e \cos \eta), \qquad t = \left(\frac{\alpha^3}{GM}\right)^{1/2}(\eta - e \sin \eta) + T, \tag{10.38}$$

where α is the semimajor axis, e is the eccentricity, and M is the total mass of the Local Group. Using the conditions that, at $t = 0$, $r = 0$, we may set $e = 1$ and $T = 0$, thereby obtaining a radial orbit. (The orbit may not be precisely radial because the tidal torques between the galaxies can lead to the exchange of orbital angular momentum and spin. However, such a transfer cannot produce *large* deviations from radial velocity.) From these equations we find the radial velocity to be

$$\frac{dr}{dt} = \frac{dr/d\theta}{dt/d\theta} = \left(\frac{GM}{\alpha}\right)^{1/2}\left(\frac{\sin \eta}{1 - \cos \eta}\right) = \frac{r}{t}\frac{\sin \eta(\eta - \sin \eta)}{(1 - \cos \eta)^2}. \tag{10.39}$$

At the present epoch, $t_0 = 1-2 \times 10^{10}$ yr, we have $(dr/dt) = -119 \, \text{km s}^{-1}$ and $r = 730$ kpc. Substituting these values and solving the resulting equation, we find that η must be between 4.11 and 4.46 rad. Substituting this value for $r(\eta)$ into Eq. (10.38) we find that α is between 0.47 and 0.58 Mpc. Finally, using this information on the expression for $t(\eta)$, we get the mass of the Local Group to be between 3.2×10^{12} $M_\odot$ and 5.5×10^{12} $M_\odot$, with the larger age leading to smaller mass. The luminosity of the Milky Way is 1.4×10^{10} $L_\odot$ whereas that of Andromeda is approximately 2.8×10^{10} $L_\odot$. Thus we get the mass-to-light ratio of the Local Group to be between 76 $Q_\odot$ and 130 $Q_\odot$. This value is consistent with our earlier observation that the Milky Way halo must be extending beyond 50 kpc.

Exercise 10.4

Magellanic clouds and dark matter: (a) The LMC ($d_{\text{LMC}-\text{MW}} \simeq 52$ kpc, $M_{\text{LMC}} \simeq 2 \times 10^{10}$ $M_\odot$) and the SMC ($d_{\text{SMC}-\text{MW}} \simeq 63$ kpc; $M_{\text{SMC}} \simeq 2 \times 10^{9}$ $M_\odot$) are the nearest satellite galaxies to us. There exists a long, circular trail of neutral H (the "Magellanic stream") connecting the clouds and our galaxy. The material at the tip of the stream is falling on the Milky Way with a speed of ~ 220 km s^{-1}. The stream may be interpreted as being due to the material torn out of the clouds in the past encounter with our galaxy. Make a reasonable, physical model for the observations and argue that they suggest an extended halo for the Milky Way.

(b) Our galaxy is surrounded by several satellite galaxies and GCs. Assume that they are outside our halo and feel the potential $\phi = -GM/r$. Let the radial velocities and galactocentric distances of these objects be v_i and r_i respectively; $i = 1, 2, \ldots, N$. Show that the time average of $v^2 r$ for any one satellite in Kepler orbit is

$$\langle v^2 r \rangle = \frac{GM}{4}.$$

(Assuming that the time average is the same as the "ensemble average," we can estimate the mass of our galaxy to be

$$M = \frac{4}{GN} \sum_{i=1}^{N} v_i^2 r_i.$$

This gives a value of $M \simeq 3.8 \times 10^{11}$ $M_\odot$.)

10.4.5 Groups of Galaxies

Groups of galaxies contain typically 10–100 galaxies with a mean separation that is much smaller than the typical intergalactic separation in the universe. The masses of these groups can be estimated if we assume that they are gravitationally bound systems in steady state. For such systems, virial theorem gives the relation $2T + U = 0$, where T is the kinetic energy and U is the potential energy of the system. For a group of N galaxies, treated as point masses with masses m_i, positions $\mathbf{r}_i$, and velocities $\mathbf{v}_i$, the virial theorem can be written as

$$\sum_{i=1}^{N} m_i v_i^2 = \sum_{i \neq j}^{N} \frac{G m_i m_j}{|\mathbf{r}_i - \mathbf{r}_j|}. \tag{10.40}$$

Rigorously speaking, we should treat the quantities appearing in Eq. (10.40) as time averages; we shall assume that the instantaneous values will be close to the time-averaged values for systems with a sufficiently large number of galaxies. Observationally, we can determine only the line-of-sight velocity dispersion σ^2 and the projection $\mathbf{R}_i$ of $\mathbf{r}_i$ onto the plane of the sky. On the average, we expect $v_i^2 = 3\sigma^2$; we can relate the projected inverse separation to the true inverse separation by averaging over all possible orientations of the group. It is easy to show that (see Exercise 10.5):

$$\langle(|\mathbf{R}_i - \mathbf{R}_j|)^{-1}\rangle = \frac{\pi}{2}(|\mathbf{r}_i - \mathbf{r}_j|)^{-1}. \tag{10.41}$$

Finally, we also assume that the mass-to-light ratios Q of all the member galaxies are the same. Substituting $m_i = QL_i$ into the virial equation and using Eq. (10.41) we get

$$Q_{\text{group}} = \frac{3\pi}{2G} \frac{\sum_{i=1}^{N} L_i \sigma_i^2}{\sum_{i \neq j}^{N} L_i L_j(|\mathbf{R}_i - \mathbf{R}_j|^{-1})}. \tag{10.42}$$

Because all the quantities on the right-hand side can be determined from the observations, we can find Q for the group. This procedure will give reasonably reliable results for groups with sufficiently large number of galaxies.

Several groups have been studied by use of Eq. (10.42). The median value of Q is $\sim 260h\ Q_\odot$ but there is a wide spread in the values. It is believed that the spread is mostly statistical. In spite of the statistical uncertainty, it is obvious that the mass-to-light ratios in the groups are much larger than the values obtained in the luminous part of the galaxies. (The procedure needs to be modified if the groups are dominated by a single large galaxy or if the the dark-matter halo forms a uniform background around all the galaxies; these modifications are discussed in Exercise 10.6.)

Exercise 10.5
Averaging the radii: Let $\mathbf{r}_i$ be the position of a galaxy and let $\mathbf{R}_i$ be the projection of this vector on the sky. Show that

$$\langle|\mathbf{R}_i - \mathbf{R}_j|^{-1}\rangle = \frac{\pi}{2}|\mathbf{r}_i - \mathbf{r}_j|^{-1},$$

where the average is taken over all possible orientations of the members of a set.

Exercise 10.6
Weightage for galaxies in a group: In estimating Q for a group of galaxies in the text, we have treated all the members on the group on the same footing. The equation for Q needs to be modified when this is not the case. Suppose that the group is dominated by a massive galaxy of luminosity L. Let v_i and R_i be the line-of-sight velocity and projected

distance relative to the dominant galaxy, respectively. Show that

$$Q \simeq \frac{3\pi}{2GL} \left(\sum_{i=1}^{N} v_i^2 \right) \left(\sum_{i=1}^{N} R_i^{-1} \right)^{-1}.$$

Suppose the dark matter is distributed over the size of the group, with the galaxies embedded in it. Further, assume that the spatial distribution of galaxies is similar to that of dark matter. Show that, in this case,

$$Q \simeq \frac{3\pi N}{2G} \left(\sum_{i=1}^{N} v_i^2 \right) \left(\sum_{i=1}^{N} L_i \right)^{-1} \left(\sum_{i=1}^{N} \sum_{j<i} |\mathbf{R}_i - \mathbf{R}_j|^{-1} \right)^{-1}.$$

If the dark-matter distribution is more extended than that of galaxies, will Q be larger than Q_{true} or smaller?

10.4.6 Clusters of Galaxies

For a rich cluster containing several hundred galaxies, like the Coma cluster, we can use a detailed model of the cluster for determining the mass-to-light ratio. One such approach uses the moments of the Boltzmann equation in the spherical coordinates. For a spherically symmetric system, the velocity moment of the CBE becomes

$$\frac{d}{dr}\left(n\langle v_r^2 \rangle\right) + \frac{2n}{r}\left[\langle v_r^2 \rangle - \langle v_\phi^2 \rangle\right] = -n\frac{GM(r)}{r^2}. \qquad (10.43)$$

Introducing the anisotropy parameter $\beta(r) \equiv 1 - (\langle v_\phi^2 \rangle / \langle v_r^2 \rangle)$, we can rewrite Eq. (10.43) as

$$\frac{d}{dr}\left(n\langle v_r^2 \rangle\right) + \frac{2n}{r}\langle v_r^2 \rangle \beta(r) = -n\frac{GM(r)}{r^2}. \qquad (10.44)$$

Among the four quantities $\beta(r)$, $M(r)$, $\langle v_r^2 \rangle$, and $n(r)$, we can observationally determine $n(r)$ and the line-of-sight velocity dispersion. These two functions, along with the moment equation, still constitute an underdetermined set. To proceed further, we have to make some assumption about the structure of the cluster. The mass that is calculated will depend on the nature of the assumption we make.

As an example of the procedures that are adopted, let us consider the Coma cluster. Among the several assumptions made in the literature to determine the mass of the Coma cluster, the following three attempts seem to be reasonably reliable: (1) We may assume that n is proportional to ρ, i.e., the number density traces the actual mass density. This provides the extra relation needed to close the set of equations. The calculated mass for the Coma cluster turns out to be approximately $1.8 \times 10^{15} h^{-1}\ M_\odot$. (2) It is possible to use suitable forms of the density profile $\rho(r)$, containing some free parameters, and choose these parameters so as to obtain minimum mass for the cluster. This often leads to fairly implausible density configurations but does constrain the mass from below. Such attempts have yielded a minimum mass of approximately $0.7 \times 10^{15}h^{-1}\ M_\odot$. (3) It is also possible to estimate the mass by choosing suitable functional forms for the anisotropy $\beta(r)$. In particular, the form $\beta(r) = [r^2/(r^2 + a^2)]$, where a is a free parameter, has several attractive features. [We can show, for example, that this anisotropy parameter arises from a distribution function of the form $f = f(E + L^2/2a^2)$]. Such a procedure also leads to masses slightly greater than $2 \times 10^{15}h^{-1}\ M_\odot$.

The total luminosity of the Coma cluster is approximately $5 \times 10^{12}h^{-2}\ L_\odot$. Our estimates therefore suggest a mass-to-light ratio of ~$400h\ Q_\odot$ for the Coma cluster. A similar analysis of the Perseus cluster gives a value of $600h\ Q_\odot$. Quite clearly, these clusters contain a large amount of dark matter.

An indirect way of determining the amount of dark matter around the clusters is by using clusters as gravitational lenses for distant objects. In one such study, the lensing of faint blue galaxies (with redshifts between 0.8 and 3) by 23 foreground clusters (in the redshift range of 0.2–0.4) was investigated. The images of the faint galaxies were used to determine the mass distribution around the clusters. Such an analysis indicates that the dark-matter content of the clusters is 10–25 times larger than the visible mass. The central density of dark matter in these clusters is approximately $2 \times 10^5 \rho_c$ and the core radius of the dark-matter distribution is approximately 30–50 kpc. These observations also suggest that the dark-matter distribution in these clusters is extremely smooth to the level of approximately 10^{10}–$10^{11}\ M_\odot\ \mathrm{kpc}^{-1}$.

During the past decade, x-ray measurements of clusters have supplied a more accurate method for mapping the mass distribution. This approach assumes that the hot gas is in hydrostatic equilibrium in the cluster potential and derives the mass distribution from the mean temperature of the gas and radial gradients of temperature and density. Such an analysis of the Perseus cluster of galaxies and other clusters has led to the conclusion that the radial distributions of total mass, gas mass, and galaxy mass are similar. Intracluster hot gas contributes approximately $14\% \pm 2\%$ of the total mass whereas visible galaxies contribute only ~3% of the total mass. The resulting mass-to-luminosity ratio is approximately $M/L = (213 \pm 60)h\ M_\odot/L_\odot$.[7] The x-ray–emitting gas within $1.5h^{-1}$ Mpc contributes a fraction $0.06h^{-3/2}$ of the cluster virial mass and luminous stars contribute another fraction of 0.02. This gives a lower bound (as it is possible for some baryons as a dark component) to the baryon fraction as

$$\frac{\Omega_B}{\Omega_m} \geq 0.06h^{-3/2} + 0.02, \qquad (10.45)$$

which can be combined with the nucleosynthesis bound on $\Omega_B h^2$ to provide a bound on Ω_m if h is known. We find that $\Omega_m < 0.40$ if $h = 0.5$ whereas

$\Omega_m < 0.30$ if $h = 0.75$. A somewhat similar result is obtained from the study of the SZ effect in 18 different clusters, which leads to an estimate of $\Omega_m h = 0.22^{+0.05}_{-0.08}$; this gives $\Omega_m \simeq 0.31$ for $h = 0.7$.

In summary, the studies indicate that (1) the radial distributions of total mass, gas mass, and mass in galaxies are similar; (2) intracluster hot gas contains ~15% of the total mass of the cluster; and (3) the mass in the visible population of galaxies is only ~3% of the total mass of the cluster. The mass-to-luminosity ratio averaged over several clusters by x-ray data is $M/L_V \simeq 150h(M_\odot/V_\odot)$; for comparison, the data based on velocity dispersion give $M/L_V = (213 \pm 60)h(M_\odot/V_\odot)$.[8]

10.4.7 Virgo-Centric Flow and Velocity Fields

At least part of the velocity of the Local Group (discussed in Chap. 6, Section 6.6) must be due to the gravitational attraction of the Virgo cluster. This fact can be used to estimate the amount of mass contained in a sphere, centred at Virgo and having a radius equal to the distance between Virgo and the Local Group. Although this method has now become somewhat obsolete, we will discuss it because it illustrates an interesting method for determining the Ω of the universe.

The recession velocity between Virgo and the Local Group will be $H_0 r_{V_0}$ (where r_{V_0} is the distance to Virgo) if we ignore the gravitational influence of Virgo on the Local Group. To take into account the gravitational influence, we may proceed as follows: We assume that the mass concentration of the Virgo supercluster is spherically symmetric and that, at any time t, the total mass contained inside a sphere of radius $r_V(t)$ is M. We have seen in Chap. 5 that any spherical region in a Friedmann universe behaves as another isolated Friedmann universe with its own parameters. Thus a sphere, centred on Virgo with the Local Group at its periphery, can be treated as a Friedmann universe with the density $\rho_V(t) = [3M/4\pi r_V^3(t)]$ and the Hubble parameter $H_V(t) = (\dot{r}_V/r_V)$. This sphere can also be assigned a density parameter $\Omega_V = (8\pi G\rho_V/H_V^2)$. We shall assume that Ω_V is less than unity. Using the standard equations for the $k = -1$ Friedmann universe, we can show that

$$H_{V_0} t_0 = \frac{\sinh \eta_{V_0}(\sinh \eta_{V_0} - \eta_{V_0})}{(\cosh \eta_{V_0} - 1)^2} \equiv F_1(\eta_{V_0}), \tag{10.46}$$

$$\rho_{V_0} = \left(\frac{3}{4\pi G t_0^2}\right) \frac{(\sinh \eta_{V_0} - \eta_{V_0})^2}{(\cosh \eta_{V_0} - 1)^3} \equiv \frac{3}{4\pi G t_0^2} F_2(\eta_{V_0}), \tag{10.47}$$

$$\Omega_{V_0} = 2(1 + \cosh \eta_{V_0})^{-1}. \tag{10.48}$$

To proceed further, we compare these values with that of a background universe that, we assume, has $\Omega < 1$. Because the spherical region should match the background universe at the initial singularity, the constant t_0 can be eliminated by such a comparison. It follows that

$$\frac{H_{V_0}}{H_0} = \frac{F_1(\eta_{V_0})}{F_1(\eta_0)}, \qquad \frac{\rho_{V_0}}{\rho_0} = \frac{F_2(\eta_{V_0})}{F_2(\eta_0)}. \tag{10.49}$$

These equations give two relations among the four variables: η_{V_0}, η_0, and the ratios ρ_{V_0}/ρ_0 and H_{V_0}/H_0. If two of these quantities, viz., (ρ_{V_0}/ρ_0) and (H_{V_0}/H_0), can be determined observationally, then these equations can be used to fix η_0; once η_0 is known, Ω_{V_0} can be determined.

The quantities H_{V_0}/H_0 and ρ_{V_0}/ρ_0 can be measured in the following way. Suppose the mean velocity of the galaxies in the core of the Virgo cluster is u_1 and the mean velocity of the galaxies in some very distant cluster (located at distance l) is u_2. Let the observed brightness of the Virgo cluster and the distant cluster be L_1 and L_2, respectively. Because $u_1 = H_{V_0} r_{V_0}$, $u_2 = H_0 l$, and $L_1/L_2 = (l^2/r_{V_0}^2)$ it follows that

$$\frac{H_{V_0}}{H_0} = \frac{u_1}{u_2}\left(\frac{L_1}{L_2}\right)^{1/2}. \tag{10.50}$$

Because all quantities on the right-hand side are measurable, we can determine this ratio. Similarly, we can determine ρ_{V_0}/ρ_0 by estimating the ratio of luminosity distances, provided we assume that the average mass-to-light ratio on a sphere centred on Virgo is the same as that of the universe as a whole. Given these ratios and Eqs. (10.49), Ω can be determined.

Estimates of H_{V_0}/H_0 are in the range 0.7–0.8, with most of the scatter originating because of (1) uncertainties in the velocity of the Virgo cluster and (2) the differences in the definition used for the volume on which the sampling is made. The estimates for ρ_{V_0}/ρ_0 range between 3 and 4. Using these values, we obtain $\Omega = 0.25 \pm 0.1$. The mean value of Q is $\sim 400h\, Q_\odot$.

It must be noted that this procedure is plagued by several uncertainties. The procedure adopted for measuring ρ_{V_0}/ρ_0 requires knowledge of the absolute luminosities of nearby galaxies, which in turn requires knowledge of the distance to these galaxies. The distance indicators are calibrated against a set of galaxies whose distances are usually determined by Hubble's law. If the Virgo cluster affects these galaxies (as it will), this primary calibration needs to be corrected in an iterative manner. This can be done if more sophisticated models are used for the velocity flow in the local region. There have been several attempts to map the velocity field of the universe and compare it with the observed mass distribution. Each of these comparisons depends on the combination $(b^{-1}\Omega^{0.6})$, where b is the biasing factor, as a free parameter and can be used to determine the best-fit value for this parameter. The QDOT-IRAS galaxy survey, for example, suggests that $b^{-1}\Omega^{0.6} = (0.81 \pm 0.15)$. Assuming $b = 1$ gives Ω to be approximately

Table 10.4. Mass density in the universe

System	Method	Scale	$Q/Q_\odot$	Ω
Milky Way	Near sun	100 pc	5(?)	$0.003h^{-1}$(?)
	Escape velocity	20 kpc	30	$0.018h^{-1}$
	Satellites	100 kpc	30	$0.018h^{-1}$
	Magellanic stream	100 kpc	>80	$>0.05h^{-1}$
Ellipticals	Core fitting	2 kpc	$12h$	0.007
	X-ray halo	100 kpc	>750	$>0.46h^{-1}$
Spirals	Rotation curves	50 kpc	$>30h$	>0.018
Groups	Local group	800 kpc	100	$0.06h^{-1}$
	Other groups	1 Mpc	$260h$	0.16
Clusters	Coma	2 Mpc	$400h$	0.25

$0.7(\pm 0.3, -0.2)$. On the other hand, $\Omega = 1$ would correspond to a biasing factor of $b = 1.23 \pm 0.23$.

The results discussed so far are summarised in Table 10.4, in which we have given both the mass-to-light ratio and the corresponding value of Ω. We have also provided a rough "length scale" at which each of the observations is relevant.

For comparison, note that the density parameter that is due to luminous matter in the universe is $\Omega_{lum} \lesssim 0.01$, whereas that due to baryons is constrained by the primordial nucleosynthesis to the range $0.011 < \Omega_b < 0.12$.

10.5 Nature of Dark Matter

It is clear that the density parameter Ω is at least ~ 0.2 over scales at which reasonably accurate dynamical measurements are possible. There is also some evidence that suggests that Ω increases with the scale and is possibly as high as unity at the largest scales. The luminous matter contributes only approximately one hundredth of this value.

It is therefore important to ask what the dark matter is composed of. *There is no a priori reason for the dark matter in different objects to be made of the same constituent.* For example, our analysis of dark matter near the Sun suggests that roughly half the mass in the solar neighbourhood is not visible. This component of dark matter is *confined* to the galactic disk and hence must have undergone dissipation. It is therefore reasonable to conclude that this component must be baryonic. In fact, there *must* be baryonic dark matter even at a larger scale. This is because the dark matter in the solar neighbourhood contributes only $\Omega \simeq 0.003$, which is at least a factor of 4 less than the *lower* limit on the baryonic density arising from the nucleosynthesis bounds. This conclusion is strengthened by the observations of baryons in the Lyman$-\alpha$ forest systems, which show that the baryons in the diffuse IGM could contribute $\Omega_{gas} \approx 0.04$. This goes a long way in filling the baryonic budget from nucleosynthesis. However, at $z \approx 0$, the luminous baryon component amounts to only $\Omega_* \approx 0.003$ in stars. Rich clusters contain a significant fraction of baryons, but the clusters themselves account for only 5% of the stellar component of the universe. Conservatively, we would say that $\Omega_{gas} \approx 0.005$. Hence it is obvious that we are missing quite a fraction of baryons seen at $z \approx 3$ by the time we reach $z \approx 0$.

At the same time, it is unlikely that *all* the dark matter in the universe is contributed by baryons. There are two reasons for this conclusion: (1) We have seen in Chap. 6 that baryonic universes with $\Omega_B \simeq 0.2$ (and adiabatic initial fluctuations) violate the MBR anisotropy observations. (2) We saw in Chap. 4 that $\Omega = 1$ is the only stable value in the standard Friedmann cosmology. Universes with $\Omega \neq 1$ evolve to $\Omega = 0$ or $\Omega = \infty$ within a very short time, unless we resort to extreme fine-tuning. This argument favours the value $\Omega = 1$, which cannot be contributed entirely by baryons, because such a case would violate the upper bounds on Ω_B arising from the nucleosynthesis.

The preceding arguments, taken together, suggest that *both* baryonic and nonbaryonic dark matter exist in the universe, with nonbaryonic matter being dominant. We now discuss these components separately.

10.5.1 Baryonic Dark Matter

Baryonic dark matter can exist in several forms. Interstellar clouds with masses of less than $\sim 0.08\ M_\odot$ will not be able to reach the central temperatures that are needed for nuclear ignition. Such clouds can end up as "brown dwarfs" (see Vol. II, Chap. 3, Subsection 3.3.4). However, the standard theory of star formation predicts very few stars with masses of less than $0.08\ M_\odot$. Therefore, in order for this model to be feasible, the IMF governing the star formation must have a nonstandard shape. Because the process of star formation is not quite well understood, such a possibility cannot be ruled out at this stage.

The baryonic dark matter could also be in the form of white dwarfs, neutron stars, or black holes, all of which are remnants of stellar evolution. There is a constraint on the density that can be contributed by such remnants that arises from the fact that stellar evolution should not contribute too much to the background radiation. This constraint gives $\Omega \lesssim 0.03$. In fact, we require a contrived IMF between 2 and 8 $M_\odot$ to avoid excessive production of light or metals. The halo fraction of white dwarfs is also constrained to be less than 10% to avoid luminous precursors that will contradict upper limits from galaxy counts. Further, a large fraction of white dwarf precursors in binaries will also lead to an excessive number of type IA supernovae. More recent constraints come from CNO, He, and D production, as well as from limits on the EBL; these have strengthened the belief that white dwarfs are unlikely candidates.

Another possibility is for the baryons to exist in the form of primordial black holes. This can happen in scenarios in which there existed an epoch of star formation *before* the onset of galaxy formation. In such a case, the primordial stars, heavier than a few hundred solar masses, can collapse directly, forming a black hole. This is to be contrasted with the black holes produced in standard stellar evolution, in which much of the stellar mass is expelled during the black hole formation. The latter case is severely constrained by the fact that the concentration of heavy elements in the ISM should not be too large (the resulting bound is $\Omega < 10^{-4}$). The bounds are much less severe on the formation of black holes directly from supermassive stars with masses in the range $10^2 \, M_\odot < M < 10^6 \, M_\odot$. (The upper bound of $10^6 \, M_\odot$ comes from the fact that gravitational perturbation from the black holes should not heat up the disk stars too much.) There are, however, several other indirect constraints on the existence of such supermassive black holes.

It may be possible to detect some of the compact objects previously discussed (usually called *massive astrophysical compact halo objects*, or machos for short) by the gravitational microlensing effect of these objects on the stars in the LMC. This process operates as follows: When we look at the stars in the LMC through our galactic halo, every once in a while, a macho will cross the line of sight. This macho will act as a lens for the starlight and will enhance the intensity of the star by some amplification factor that can be calculated. We saw in Chap. 3 that in the case of lensing by a point source, the total amplification is given by

$$\mu = \frac{u^2 + 2}{u\sqrt{u^2 + 4}}, \tag{10.51}$$

where $u = r/r_E$ with the Einstein radius r_E being given by

$$r_E \approx \left(\frac{GMD}{c^2}\right)^{1/2} \approx 9 \, \text{AU} \left(\frac{M}{M_\odot}\right)^{1/2} \left(\frac{D}{10 \, \text{kpc}}\right)^{1/2}. \tag{10.52}$$

The characteristic time scale of the lensing event is that corresponding to the crossing time of r_E and is given by

$$t_E = 78 \, \text{days} \left(\frac{M}{M_\odot}\right)^{1/2} \left(\frac{D}{10 \, \text{kpc}}\right)^{1/2} \left(\frac{v}{200 \, \text{km s}^{-1}}\right)^{-1}. \tag{10.53}$$

Figure 3.8 of Chap. 3, Section 3.6, shows the expected light curve in this context. From the observed light curve, we can obtain primarily four parameters: (1) the baseline magnitude of the lensed object, (2) the peak magnification, (3) the peak time, and (4) the crossing time for the Einstein radius, t_E. Of these, only t_E contains information about the lens and only the combination MD/v^2 can be determined. Further, to proceed from an individual microlensing light curve to a sample population on the whole, we require a statistical description. This can be given in terms of the optical depth,

$$\tau = \int_0^{D_s} n\left(\pi r_E^2\right) dD_d, \tag{10.54}$$

which is the probability that there is a lens located inside the Einstein radius at any given distance along the line of sight. For a self-gravitating system with a constant rotational velocity, V, $\tau \propto (V^2/c^2)$ is approximately 5×10^{-7} for the Milky Way. Because τ depends on only the total mass of all lenses, it can be used (only) to infer the overall mass distribution of the lenses. We can do better with the event rate (ie., the number of lensing events per unit time), which is given by

$$\Gamma \equiv \frac{d N_{\text{event}}}{dt} = \frac{2}{\pi} N_* \tau \frac{1}{t_E} \epsilon\,(t_E), \tag{10.55}$$

where N_* is the number of stars monitored in the experiment and $\epsilon(t_E)$ is the efficiency of the experiment in detecting events with a characteristic time scale. (This factor is added because most experiments miss events varying at time scales less than a day or more than a few years.) The event rate depends not only on the total mass but also on the mass function through t_E. Thus the study of the event distribution function offers a possible way of determining the mass function.

By continuously monitoring $\sim 10^6$ stars in the LMC for a period of ~ 4 months (10^7s), we can either find or impose constraints on the machos in our galactic halo. For a macho with a mass of $10^{-2}\ M_\odot$, we expect approximately five events in such a search, each lasting ~ 9 days. For objects with a mass of $10^{-6}\ M_\odot$, there will be ~ 500 events, each lasting ~ 2 h (see Chap. 3, Section 3.6).

The experiments looking for intensity variations for stars in the Magellanic clouds and the galactic bulge have now been underway for a decade. This method is sensitive to lens masses in the range 10^{-7}–$10^2\ M_\odot$ but the optical depth for an individual star being lensed is only $\tau \simeq 10^{-6}$. The estimated duration and likely event rate are

$$P \simeq 0.2(M/M_\odot)^{1/2} \text{ yr}, \qquad \Gamma \simeq N\tau/P \simeq (M/M_\odot)^{-1/2} \text{ yr}^{-1}, \tag{10.56}$$

where $N \simeq 10^6$ is the number of stars. The macho group currently has 13–17 LMC events, with the duration spanning 34–230 days. For a standard halo model, the data suggest a lens mass in the range 0.15–0.9 $M_\odot$ with a halo fraction of 0.08–0.5 at 95% confidence level. Although the data do not conclusively favour any specific candidate, they have already excluded objects in the range of $(5 \times 10^{-7}$–$0.002)M_\odot$ from providing more than 0.2 of the halo density. Because the number of events scale as $M^{-1/2}$, we expect many short-duration events if the halo is dominated by low-mass objects. Because none were found, it was possible to rule out a halo dominated by objects in the mass range $\sim 10^{-7}$–10^{-2} $M_\odot$. In the worst case, objects in this mass range could contribute only less than $\sim 20\%$ of the mass to the halo.

The analysis of the data also gives an optical depth of $\tau \simeq 2.9^{+1.4}_{-0.9} \times 10^{-7}$. To understand this result, we could compare it with the optical depths contributed by other known populations. Galactic disk populations give $\tau \simeq 1.5 \times 10^{-8}$ whereas the self-lensing of LMC stars by themselves will be $\tau \approx (1\text{–}5) \times 10^{-8}$. Because the observed optical depth is larger, the most straightforward interpretation is that it is contributed by a new class of macho population. A Salpeter mass function with $(dN/dM) \propto M^{-2.35}$ fits the observed distribution of lensing duration reasonably well, whereas the one dominated by brown dwarfs, for example, predicts too many short events. On the other hand, the mass functions for the disk and the bulge determined from the HST seem to be too flat to explain the observed distribution. This suggests that the mass functions at different places in the galaxy (in particular the disk and the bulge) may be different.[9]

Finally, we shall summarise the constraints on many other forms of baryonic dark matter. For example, small objects like dust grains, asteroids, comets, etc. (which are dominated by molecular forces rather than by gravitational forces) cannot contribute much to dark matter. This is because such bodies will be composed of heavy elements like Si, C, etc., which are always much less abundant (lower by a factor of at least 100) than H. H in the form of solid snowballs is also not feasible because these would evaporate too fast.

Elliptical galaxies and clusters do contain a hot gas, made of baryons, which is detectable from x-ray emission. However, the mass in the gaseous component is not dynamically significant. For example, in M87, the total mass within 100 kpc is $\sim 10^{12}$ $M_\odot$ but the mass of gas is only approximately 3×10^{11} $M_\odot$. Similarly, observations from the 21-cm line of neutral-H clouds show that the neutral gas present around galaxies is also not dynamically significant.

Neutral H at extragalactic distances can be constrained from the observations of the quasar absorption spectra. These bounds were discussed in Chap. 9, where it was shown that the baryonic density, in either uniformly distributed neutral H or in compact Lyman-α clouds, should be much smaller than the critical density. Ionised matter in the IGM can be constrained by indirect methods similar to the ones discussed in Chap. 9. These constraints show that the IGM cannot have too much baryonic dark matter.

The big bang nucleosynthesis abundances – especially the D abundance – are strong constraints on the total amount of baryons that are present in the universe.[10] Although there have been some controversial measurements of anomalously high D abundance in intergalactic Lyman-α clouds, most of the recent studies suggest a D abundance of 3.3×10^{-5}, which would correspond to a constraint of $0.018 < \Omega_B h^2 < 0.020$. Three conclusions are apparent from this bound. First, the upper bound rules out the possibility of $\Omega_B = 1$, corresponding to a critical density universe entirely made of baryons. Second, the upper limit allowed for Ω_B is significantly higher than the density of visible baryons in the universe. The latter is not very well determined but is generally believed to be in the range $\Omega_{\text{visible}} \simeq (2.2 + 0.6h^{-1.5}) \times 10^{-3} \simeq 0.003$ for reasonable values of h. (This estimate is tentative because baryons could exist as warm intergalactic gas, cold gas in GCs, etc., which are difficult to observe and account for.) Third, it is possible for all the matter in the galactic disk to be baryonic. Once again, it is not very clear what fraction of matter in a galactic disk is dark; recent observations suggest that this could be below 10%.

A more interesting question would be whether the dark-matter halo of the Milky Way contributing to $\Omega_h \simeq 0.03h^{-1}(R_{\text{halo}}/100 \text{ kpc})$ could be baryonic. The nucleosynthesis upper bound shows that all dark-matter halos can be baryonic provided that $R_{\text{halo}} < 70h^{-1}$ kpc. Even if halos are larger than this, it could still be that a large fraction of the matter could be baryonic and invisible.

10.5.2 Nonbaryonic Dark Matter

Any massive neutral fermion with weak interactions provides an a priori candidate for dark matter and would have existed in large numbers in an early universe. As the universe cools, the number density of such particles will change, depending on the details of their interaction. Given a specific particle-physics model, we can compute, in a rather straightforward manner, the relic abundance of any such particle today. Knowing their mass, we can work out the density contributed by these particles.

Such particles are of primary interest only if they can contribute significantly to the closure density. We saw in Chap. 4 that, for fermions with weak interactions, three mass ranges are of interest. Particles that decouple when they are relativistic will contribute closure density if their total mass is $\sim$100 eV; more precisely, $\Omega h^2 \simeq (m/92 \text{ eV})$. Particles that are in the mass range 100 eV $\ll m < 100$ GeV and decouple after becoming nonrelativistic can contribute closure density if their masses are a few giga-electron-volts; $\Omega h^2 \simeq 3(m/1 \text{ GeV})^{-2}$. This result depends on the estimate of annihilation cross sections, which is valid provided that the mass of the particle is less than that of Z_0. For particles with m greater than 100 GeV the annihilation cross section decreases as m^{-2} and Ωh^2 can again increase to approximately unity for $m \simeq 1$ TeV. In this connection, it must be mentioned that the annihilation cross section for any such structureless fermionic particle is bounded from above by certain field-theoretical considerations. This bound implies that particles with masses higher than $\sim$340 TeV need not be considered at all. Thus the cosmologically interesting mass ranges are $m < 100$ eV, $m = (1\text{–}3)$GeV, and $m \simeq 1$ TeV.

The minimal particle-physics model, based on $SU(3)_C \otimes SU(2)_L \otimes U(1)_Y$, provides *no* such dark-matter candidate. The only neutral fermion in such a theory – the neutrino – is modelled to have zero mass. However, almost all extensions of the standard model provide several massive candidates. These involve generalisations of the standard model with massive neutrinos, grand unified theories, theories incorporating supersymmetry, etc. We now discuss some of these candidates and the constraints on their masses.

There exist several simple generalisations of the standard model in which the neutrino becomes massive with a low mass: $m_\nu < 100$ eV. All such models can provide a hot dark-matter candidate. Such a light neutrino is the most difficult candidate to constrain experimentally.

Although the minimal standard model of electroweak interactions has only zero-mass neutrinos, this result, in some sense, is put in by hand in the standard model. Because all known experimental results are consistent with a left-handed neutrino (ν_L) and a right-handed antineutrino ($\bar{\nu}_R$), the *minimal* model for electroweak interaction does not use ν_R and $\bar{\nu}_L$ and sets the neutrino masses to zero. In particular, note that the vanishing of the neutrino mass does *not* arise from any fundamental physical symmetry (unlike, for example, the gauge invariance that makes the photon massless).

There are several theoretical and observational motivations for generalising the electroweak model to include massive neutrinos. On the theoretical side, the absence of any guiding principle that makes neutrinos massless, coupled with the fact that the experimental bounds on the masses of the neutrinos are too weak, suggests that it is worth examining models with massive neutrinos. The direct and most reliable experimental bounds on the neutrino masses are – unfortunately – somewhat uninteresting: $m(\nu_e) < 15$ eV, $m(\nu_\mu) < 0.19$ MeV, and $m(\nu_\tau) < 18.2$ MeV. These bounds, however, suggest that the neutrino masses are much smaller than the corresponding charged lepton and quark masses. This theoretical puzzle – namely that leptons have widely different mass scales – is often addressed by invoking some kind of a mixing between the masses. It is then possible to introduce a very heavy-mass scale, $M \simeq 10^{15\pm2}$ GeV, into the theory and produce a light-mass m_L for a neutrino, say, through a relation of the form $m_L = m_H^2/M$, where m_H is the corresponding charged lepton mass. In such models, the neutrino masses will scale as the square of the corresponding charged lepton masses[11] [$m(\nu_e) \propto m_e^2$, etc.].

There are now three experiments that actually indicate a nonzero neutrino mass: (1) Studies of the ratio ν_μ and ν_e produced by the decay of pions in the atmosphere suggest that there are neutrino oscillations between ν_μ and ν_τ corresponding to a mass difference of $\Delta m_{\tau\mu}^2 = 2 \times 10^{-3}$ eV^2; (2) the deficit of electron neutrinos coming from the Sun suggests an oscillation between ν_e and either of the other two species (denoted by X), with a mass difference given by $\Delta m_{eX}^2 = 2 \times 10^{-5}$ eV^2; (3) there is a claim for laboratory detection of oscillations between v_μ and v_e in the neutrino beams coming from accelerators, corresponding to a mass difference of $0.2 \text{ eV}^2 < \Delta m_{\mu e}^2 < 10 \text{ eV}^2$. If the last result is confirmed, it requires $m(v_\mu) > 0.5$ eV, giving $\Omega_v > 0.005$, which is already comparable with the density of visible matter. Because (1) implies $\Delta m_{\tau e}^2 \gg 10^{-5} \text{ eV}^2$, it is likely that the X neutrino of (2) will most naturally be identified with v_μ. In that case, we would expect the masses to be

$$m(v_e) \ll 5 \times 10^{-3} \text{ eV}, \qquad m(v_\mu) \simeq \Delta m_{\mu e} \simeq 5 \times 10^{-3} \text{ eV},$$
$$m(v_\tau) \simeq \Delta m_{\tau \mu} \simeq 0.05 \text{ eV}. \tag{10.57}$$

Another possibility – which is also consistent with (3) – is $m(v_\tau) \simeq m(v_\mu) \gg \Delta m_{\tau \mu}$. This could lead to fairly large values of Ω_v but will fail to provide a natural solution to the solar neutrino problem.

Much better bounds are available on heavier wimps, including heavy neutrinos, from the decay width of the Z_0 boson at Stanford Linear Collider (SLC) and Large Electron-Positron (LEP) collider. The essential idea behind these experiments is as follows: The decay width Γ of the Z boson can be written in the form

$$\Gamma = \Gamma_{\text{hadron}} + 3\Gamma_{\text{lepton}} + N_v \Gamma_{v\bar{v}}, \tag{10.58}$$

where the first two terms represent the width that is due to the hadronic channel and due to $e\bar{e}$, $\mu\bar{\mu}$, and $\tau\bar{\tau}$ channels, and the last term incorporates contributions that are due to all "neutrinolike" particles that couple to Z. In the standard model, the first two terms can be computed quite reliably; we can also compute $\Gamma_{v\bar{v}}$ from first principles. Because all the quantities on the right-hand side except N_v are known, the experimentally observed value of Γ can be used to put bounds on N_v. (It turns out that a change of N_v by unity will change Γ by $\sim 13\%$; thus this is a fairly sensitive test.) The LEP measurements, for example, give $N_v = 2.95 \pm 0.11$; all other results are also consistent with the value $N_v = 3$. Because three neutrino species are known to exist, this does not allow for any other neutral wimp to which Z_0 can decay. Any other wimp, if it exists, must be very massive – so that Z_0 cannot decay by that channel; or they must be coupled to the Z_0 much more weakly compared with the "standard" neutrinos. For example, a heavy neutrino, if it exists, must have a mass greater than ~ 35 GeV.

Further constraints are available from semiconductor ionisation detectors designed to observe the wimps in our halo. As the Earth moves through the halo, the wimp strikes the nucleus of the detector, depositing a small recoil energy (~ 10 keV). A fraction of this energy 0.3–0.2 goes into ionisation, giving a signal that is typically 1 count $\text{keV}^{-1} \text{ kg}^{-1} \text{ day}^{-1}$. The best bounds currently available are from the University of California at Santa Barbara, the Lawrence Berkeley Laboratory, and the University of California at Berkely groups. Assuming that the halo density of wimps on the location of the Earth is ~ 0.3 GeV cm^{-3} (i.e. $\rho_{\text{DM}} \simeq 5 \times 10^{-25}$ gcm^{-3}) and that the rms velocity of halo particles is

~300 km s^{-1}, these groups find that Dirac fermions in the mass range 10 GeV to 3 TeV can be excluded. Because SLC-LEP rules out the 4- to 30-GeV range and the cosmological constraints eliminate the 30-eV to 4-GeV range, we have excluded Dirac fermions in the entire range of interest (30 eV–1 TeV). This bound rules out *all* dark-matter candidates that have coherent interaction with the nuclei and a coupling strength that is comparable with that of standard weak interaction.

These results also constrain several other particles. In the models incorporating supersymmetry, every particle will have a supersymmetric partner (called a *sparticle*). In most supersymmetric models there exists a quantum number called *R parity* that is +1 for ordinary particles and −1 for supersymmetric particles. The conservation of R parity makes the lightest supersymmetric particle (LSP) stable, thereby making it an ideal candidate for cold dark matter. If the stable LSP with a mass of 1–10^4 GeV has an electric charge or strong interaction, then it would be detectable as an anomalous heavy isotope in geological studies.[12] The observed bounds rule out this possibility, suggesting that the LSP must be a neutral, weakly interacting particle; and hence, LSP is a feasible candidate for dark matter. Some of these LSPs are constrained by the LEP-SLC result and the others are not. For example, if the LSP is the "sneutrino," the superpartner to a neutrino, then the decay width of Z_0 constrains its mass to be higher than ~35 GeV. [Note that a sneutrino is a boson.] Moreover, snuetrinos in the halo (with $m \gtrsim 3$ GeV) would have been captured by the Sun. The annihilation of solar snuetrinos will lead to high-energy neutrinos from the Sun that should have been noticed in the experiments designed to detect the proton decay.[13] The absence of such signals rules out the range $m > 3$ GeV. These two results together rule out the sneutrino as a LSP and as a dark-matter candidate.

One of the most popular candidates is the lightest *neutralino*, which is a mixture of the supersymmetric partners of the photons, the Z bosons, and two other Higgs bosons that are present in the minimal extension of the standard model. If the LSP is a neutralino – denoted by χ – an analysis similar to the one performed in Chap. 4 shows that $\Omega_\chi \simeq 1$ if 15 GeV $< m_\chi < 1$ TeV. The coupling of χ to Z_0 is complicated and somewhat model dependent. However, in most cases, much of the parameter space for $m_\chi < 30$ GeV is excluded by current observations. The range 30 GeV $< m_\chi < 1$ TeV is still open.[14] A wide variety of models lead to neutralino abundances in the present day universe which is cosmologically interesting. The difficulty with this candidate is that virtually nothing is known about the supersymmetry breaking so that any given supersymmetric model contains several (even ~100) undetermined parameters. It is impossible to scan the entire parameter space without making further simplifying assumptions; then the results, naturally, depend on the model that is chosen.

The search for neutralino at CERN, Fermilab, etc., constitutes one of the largest current experimental endeavours being undertaken for the verification of supersymmetry. In addition to this, there exists the possibility of detecting the neutralinos directly if they constitute the dark-matter particles in the galactic halo. One of the most promising techniques involves looking for recoil effects in the elastic scattering of neutralinos by normal nuclei in a detector. A particle with mass $m_X \simeq 100$ GeV and electroweak scale interaction will have a scattering cross section of $\sigma \simeq 10^{-38}$ cm^2 for interaction with a nucleus. If the galactic halo has a density of 0.4 GeV cm^{-3} and the particles are moving with velocities $v \simeq 300$ km s^{-1}, then the rate of elastic scattering of these particles by a Ge nucleus ($m_n \simeq 70$ GeV) will be

$$\mathcal{R} \frac{\rho_0 \sigma v}{m_X m_n} \simeq 1 \text{ event kg}^{-1} \text{ yr}^{-1}. \tag{10.59}$$

If a 100-GeV wimp moving with $v/c \simeq 10^{-3}$ scatters with a nucleus of similar mass, it will impart a recoil energy up to 100 keV to the nucleus. Hence a detector with 1 kg of Ge will see $\sim$1 nucleus per year, recoiling spontaneously with an energy of $\sim$100 keV.

Although this description provides the basic idea, there are several uncertainties that plague its actual implementation. The supersymmetric models have numerous parameters that make unique predictions very difficult. For example, the event rates could vary from 10^{-4} to 10 events per kilogram per day for the plausible region of supersymmetric parameter space. There are also several details that need to be attended to as regards the actual interpretation of the signal. The recoil of the nucleus could cause dislocation of the crystal structure, vibration of the crystal lattice, or ionisation. All these effects could arise from many other backgrounds that deposit a comparable amount of energy on the detector at a higher rate. Special precautions (underground operation, low-temperature cooling, radioactive shielding) are necessary to obtain a sensible signal-to-noise ratio. One method is to use the fact that the Earth's orbit is aligned with the motion of the Sun in the galaxy in June whereas it is antialigned in December, thereby leading to a 6-month periodicity in the signal. The current detectors have a threshold of $\sim$1 event per kilogram per day whereas the next-generation detectors could go down to 10^{-2} events per kilogram per day. As mentioned earlier, the current results rule out Dirac neutrinos and other simple variants related to Dirac neutrinos.

If a wimp, while passing through the Sun or the Earth, undergoes a scattering event such that its velocity is reduced to a value less than the escape velocity, it will get gravitationally bound to the Sun (or the Earth). This will lead to an enhancement of density of wimps at the centre of the Earth that will eventually annihilate through different channels. Among the decay products will be energetic muon neutrinos that can be detected by neutrino telescopes like IMB, Kamiokande, MACRO, AMANDA, or NESTOR. Because the energies of these neutrinos will be in tens or hundreds of giga-electron-volts, they can be easily distinguished from solar neutrinos. Theoretical predictions for the flux of muons fall in the range of 10^{-6}–1 events m^{-2} yr^{-1}. At present, several detectors have put constraints on the flux of such energetic neutrinos at less than ~0.02 m^{-2} yr^{-1}.

There is even a preliminary report of a tentative detection of a neutralino in the NaI experiment DAMA with a tentative mass estimate of $m_w = 59^{+17}_{-14}$ GeV.[15] There are, however, some experimental uncertainties in this result, and it may be premature to draw strong conclusions at this stage. Even if the detection is true, the predicted neutralino density seems to be somewhat lower than what is required for being identified as a cosmological dark-matter candidate. Figure 10.2 summarises some of the efforts for laboratory detection of a neutralino and the parameter space that is being explored.[16] The different curves described in the legend and the figure caption correspond to existing and predicted regions of parameter space that can be covered, and the crosshatched region denotes possible theoretical values. It is clear that although a detection will be a major breakthrough in physics, upper bounds will not conclusively rule out all the possible models.

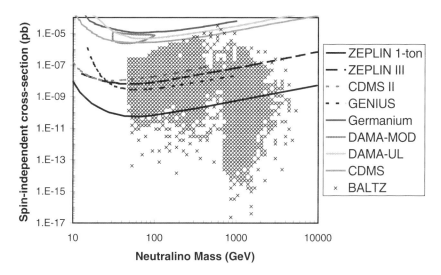

Fig. 10.2.   Current and predicted statuses of some of the experiments for direct detection of a wimp candidate. The region marked BALTZ indicates the allowed neutralino parameter space within MSSM allowing for existing cosmological constraints. The other curves are (1) CDMS, upper-limit result from the Berkeley/Stanford Cryogenic Dark Matter Search; (2) DAMA-UL, upper limit from pulse-shape analysis of the first year of DAMA NaI data; (3) DAMA-MOD, claimed annual modulation result (4 years) (enclosed area is allowed); (4) ZEPLIN 1-ton, the prediction of what could be done with a 1-ton two-phase Xe detector; (5) ZEPLIN III, the prediction of the performance of the 6-kg two-phase Xe detector being built in Great Britain; (6) CDMS II, prediction of the performance of the CDMS upgrade plus deeper site; (7) Germanium, combined best limits from all Ge ionisation detectors; (8) GENIUS, predicted sensitivity of dedicated 100-kg Ge detector system. (Figure courtesy of T. Sumner.[16])

Finally, we mention a general constraint that can be imposed on the mass of any collisionless wimp based on the scale at which it is clustering.[17] Consider a fermionic wimp that was a relic from the early universe. At some time in the past, when the universe was homogeneous, the wimps were described by the Fermi–Dirac distribution $f_{FD}(\mathbf{p})$. The maximum phase-space density (i.e., the maximum value for number $\mathcal{N}$ of neutrinos per unit phase-space volume) is $\mathcal{N}_{max} = f_{FD} (\mathbf{p} = 0) = 2g/(2\pi\hbar)^3$ for each neutrino species. The phase-space density, of course, decreases with larger $|\mathbf{p}|$. During the subsequent evolution, the distribution function evolved to some fairly complicated, space-dependent form $f(\mathbf{x}, \mathbf{p})$. Let the *average* phase-space density of wimps in some system (galaxy, cluster) today be $\mathcal{N}$. Because collisionless evolution *only* mixes up regions of high phase density with those of low phase density, the final, average phase density $\mathcal{N}$ cannot be larger than the initial maximum phase density $\mathcal{N}_{max}$. (This is a special case of a more general result; see Vol, I, Chap. 10, Section 10.6.) We can use this fact to obtain a lower bound on the mass m of the wimp. Suppose N wimps, each of mass m, have clustered in a scale R. The phase volume is approximately $\mathcal{V} = (4\pi/3)R^3 \cdot (4\pi/3)(mv_{max})^3$ where $v_{max}^2 = (2GM/R) \cong 2\sigma^2$; so $\mathcal{V} \propto R^3 m^3 \sigma^3$. The average phase-space density is $\mathcal{N} = (N/\mathcal{V}) \propto (M/m)\mathcal{V}^{-1} \propto (\sigma^2 R/m)\mathcal{V}^{-1} \propto \sigma^{-1}R^{-2}m^{-4}$. The condition $\mathcal{N} < \mathcal{N}_{max}$, on putting in the numbers, leads to the constraint

$$m > 30 \text{ eV} \left(\frac{\sigma}{220 \text{ km s}^{-1}}\right)^{-1/4} \left(\frac{R}{10 \text{ kpc}}\right)^{-1/2}. \qquad (10.60)$$

Clearly, the wimps should have a higher mass to cluster at smaller scales. If, for example, it is confirmed that the wimps cluster at the scale of dwarf galaxies that have masses of $M \simeq 10^6 M_\odot$ and radius $R \simeq 1$ kpc, then $m > 10^2$ eV, ruling out the possibility of hot dark matter in these systems.

10.6 Axions

Another possible candidate for the dark matter is a particle called an *axion*. Because this candidate is somewhat peculiar compared with the rest, we shall discuss it in some detail.

Axions arise in the theory of quantum chromodynamics, which is a very successful model for explaining the strong interactions. The basic constituents of this model are quarks (which make up the baryons) and gluons (which mediate the strong interactions). The gluons are described by a vector field $A_\mu^i(x)$, where μ denotes the usual space–time index and i denotes an "internal" symmetry space index. This field A_μ^i is analogous to the vector potential in electrodynamics. We can also construct a field tensor $F_{\mu\nu}^i$ from A_μ^i in a well-defined manner; $F_{\mu\nu}^i$ will denote the field strength of the gluon field and is analogous to the electromagnetic field.

In electromagnetism, the ground state of the theory is characterised by the vanishing of the electromagnetic field tensor. Any vector potential that leads to a vanishing electromagnetic field can itself be made to vanish by a suitable gauge transformation. It turns out that the corresponding transformation is *not* possible in quantum chromodynamics. There can exist different configurations $A^i_\mu(x)$, all of which lead to the vanishing of $F^i_{\mu\nu}(x)$ but are *not* connected by gauge transformation. Any one such configuration of the gauge field is a valid vacuum state in quantum chromodynamics. It can be shown that the inequivalent vacuum states can be classified by the set of integers.[18] The most general ground state can therefore be expressed as a superposition:

$$|\theta\rangle = \sum_{n=0}^{\infty} \exp(-in\theta)|n\rangle, \qquad (10.61)$$

where θ is a free parameter and $|n\rangle$ is a ground state labelled by the integer n. In other words, the nontrivial nature of the vacuum introduces an extra parameter θ into the theory, whose value needs to be determined by experiments.

The nonzero value of θ leads to an electric dipole moment[19] for the neutron of the order of $d_n \simeq 10^{-15}\theta$e cm; comparing this value with the experimental bound that $d_n < 10^{-25}$ e cm, we see that $\theta < 10^{-10}$. (To be precise, the parameter θ that is constrained in the above manner is related to the θ in relation (10.61) by the addition of a quark mass determinant. We have ignored this complication for simplicity.) We are therefore led to the existence of a dimensionless parameter that is extremely small; the question arises as to why θ is so small.

The simplest solution could be to ignore the nontrivial structure of the vacuum state. Unfortunately, this will lead to contradictions with the observed mass spectrum of mesons.[20] It is therefore necessary to explain the smallness of θ by some other means. We can do this by introducing a new dynamical field (called the axion field) and arranging the potential for the axion field in such a way that the effective value of θ is zero at the minimum of the potential. In such a model, the axion field will acquire a mass that depends on the energy scale f at which the symmetry of the new potential is broken. If f is taken to be anywhere between 100 and 10^{19} GeV, the corresponding axion mass m_a will range between 10^{-12} eV and 1 MeV, with the larger f leading to smaller m_a.

Given a specific model for the axion field, we can also work out the coupling of axions to other matter fields.[21] It turns out that, in general, the axion couples to the electrons, nucleons, and photons. The strengths of these couplings are somewhat model dependent; however, the smaller the axion mass, the weaker the coupling. In particular, the axion can decay into two photons, with the lifetime

$$\tau \cong 6 \times 10^{24} \text{ s } q^{-2} \left(\frac{m_a}{1 \text{ eV}}\right)^{-5}, \qquad (10.62)$$

where the value of q ranges between 0.05 and 0.7 and depends on the model.

The interaction between the axion and other matter fields can be used to put severe constraints on the possible range of m_a. To begin with, laboratory observations[22] of the decay of the K meson and quarkonium allow us to constrain the coupling constant by $f > 10^3$ GeV, giving $m_a \lesssim 6$ keV. The remaining window (10^{-12} eV to 6 keV) can be constrained from the study of stellar evolution.

The key idea behind such constraints can be understood as follows: Consider a star like the Sun. The nuclear energy produced in the solar core is transported to the surface by photon diffusion. Because photons are very strongly coupled to the solar material, the energy transport to the surface is a very slow process (i.e., "photon cooling" is quite inefficient); this is the key reason for the long lifetime of the main-sequence star. The situation changes considerably when the axion is introduced into the picture. Because axions couple very weakly to matter, they can transport energy rapidly from the core. Such an efficient cooling will accelerate several phases of the stellar evolution. From the observed properties of the stars, we can put bounds on the axion coupling and hence on the axion mass.

The constraints from the evolution of the main-sequence stars give a maximum bound for the axion mass in the range of 1–20 eV, depending on the model. A similar analysis of the evolution of a red giant phase gives a bound of 10^{-2}–2 eV. Thus the possible mass range for the axion is reduced[23] to the range of 10^{-12} to ~ 1 eV.

Further constraints on the mass of the axion can be imposed from cosmological considerations. At very early phases of the universe, axions will be in thermal equilibrium with other particles. Depending on the relative values of the interaction rate and expansion rate, the axions will decouple from the rest of the matter at some temperature. Axions with masses greater than 10^{-3} eV decouple while they are relativistic (just like neutrinos) and hence have a relic number density that is comparable with that of photons. From our earlier discussion, we know that, for such particles,

$$\Omega_a h^2 \simeq 10^{-2}(m_a/1 \text{ eV}). \tag{10.63}$$

Therefore these axions can close the universe only if their mass is $\sim 100 h^2$ eV. However, we see from relation (10.62) that such massive axions have a lifetime that is much shorter than the age of the universe. The lifetime can be larger for axions with masses in the range of a few electron volts. However, the decay of such axions in the halo of our galaxy will produce a detectable line signal of photons at a frequency that depends on the mass of the axion.[24] The absence of such signals excludes axions of mass greater than ~ 4 eV. Thus the relic axions, even if they exist, cannot contribute significantly to the mass density of the universe.

There is, however, another way by which axions can be produced. We saw that axions arise when a symmetry breaking drives the value of the parameter θ to zero. When this symmetry breaking occurs, θ does not instantaneously relax to zero. Instead, it oscillates around the value $\theta = 0$. These oscillations of a scalar field behave like a concentration of energy density with the equation of state $p = 0$. The *effective* mass of these condensates made of coherent oscillations of the scalar field will be quite different from the mass of the axion. For $m_a \simeq 10^{-5}$ eV, the effective mass can be as high as a few giga-electron-volts, making these axions an ideal CDM candidate. A careful analysis shows that this process contributes to the density parameter

$$\Omega_a h^2 = \mathcal{O}(1) (\Lambda/200 \text{ MeV})^{-0.7} (m_a/10^{-5} \text{ eV})^{-1.18}, \qquad (10.64)$$

where Λ is a parameter that arises in the quantum chromo dynamics (QCD) models. Note that Ω_a decreases with increasing mass of the axions. Because Ωh^2 must be less than approximately unity, it follows that $m_a \gtrsim 10^{-6}$ eV. Thus we are left with a small window in the range 10^{-6}–10^{-3} eV for the mass of the axion. These and similar observations have now restricted the range of the axion mass considerably; the only available window is around 10^{-5}–10^{-2} eV.[25] An axion produced at the symmetry breaking, having a mass of $\sim 10^{-5}$ eV, is thus a feasible candidate for CDM.

It is possible to detect axions in this mass range using their electromagnetic coupling.[26] For example, an axion can be converted into a microwave photon in the presence of a strong magnetic field. In a properly tuned microwave cavity, we may be able to detect these signals.

10.7 Cosmological Constant

One important feature that must be stressed in the determination of dark-matter densities is the following: Gravitational effects acting on a system of size l are usually insensitive to energy densities that are distributed smoothly over scales much larger than l. If the universe contains such a smooth component of energy density it may not be detectable by the conventional probing of dark matter through gravitational effects. Nevertheless, it will contribute to the expansion of the universe and, in particular, can close the universe. By introducing such a dark-matter candidate we can reconcile the observational result of $\Omega \simeq 0.3$ with the theoretical prejudice for $\Omega = 1$ in a natural fashion.

For this idea to be feasible we must prevent the smooth component from condensing on smaller scales. One way of doing this will be to use a smooth component made of particles that are relativistic at present. This idea, however, runs into several difficulties; for example, the age of a radiation-dominated universe is smaller than that of a matter-dominated universe and can lead to difficulties with the age of GCs.

Another alternative for a smooth energy density is a cosmological constant. The source term for Einstein's equations can be any conserved stress tensor. If this stress tensor has the form $T_k^i = \Lambda \delta_k^i$ with a constant Λ (corresponding to an "equation of state" $p = -\rho = -\Lambda$, implying that either pressure or density is negative), then the quantity Λ is called the *cosmological constant*. Such a term was originally postulated by Einstein and has an interesting history.[27] We shall briefly discuss some general features of the cosmological constant that were not adequately stressed earlier.

We should realise that the existence of a nonzero cosmological constant in Einstein's equations is a feature of deep and profound consequence. (Among other things, it implies that flat space–time is not a solution to Einstein's equation, even in the absence of any other matter.) This issue is of fundamental significance to *all of physics* and transcends the limited domain of cosmology. In fact, if cosmological observations lead to the detection of such a cosmological constant, then it would be a major contribution made by cosmology to fundamental physics. The reason for this significance arises from the fact that the value of such a constant term is severely constrained by cosmological observations. Because Λ contributes $\Omega_\Lambda = (8\pi G\Lambda/3\ H^2)$ to the critical density, we can safely conclude (in spite of any astronomical uncertainties) that

$$|\Lambda| \lesssim 10^{-29}\,\mathrm{g\,cm^{-3}} \approx 10^{-47}(\mathrm{GeV})^4. \tag{10.65}$$

The smallness of this value is a deep mystery because we do not know of any symmetry mechanism that requires it to be zero. In fact, we know of several independent, unrelated phenomena that contribute to Λ. To produce such a small Λ, these terms have to be fine-tuned to a bizarre accuracy.

To begin with, nothing prevents the existence of a Λ term (say, Λ_0) in the Einstein equations themselves. This will make the gravitational part of the Lagrangian dependent on two fundamental constants Λ_0 and G, *which differ widely in scale*; the dimensionless combination made of fundamental constants $(G\hbar/c^3)^2 \Lambda_0$ has a value of less than 10^{-124}, which appears to be very unaesthetic.

Quantum field theory provides a wide variety of contributions to Λ. For example, consider a scalar field with a potential $V(\phi)$. The particle-physics predictions do not change (except possibly in theories with exact supersymmetry, which is anyway broken at some energy scale) if we add a constant term V_0 to this potential. Potentials like $V_1(\phi) = (1/2)\mu^2\phi^2 + (\lambda/4)\phi^4$ and $V_2 = (\lambda/4)(\phi^2 - \mu^2/\lambda)^2$ will lead to the same particle physics even though they differ by the constant term $(\mu^4/4\lambda)$. However, such a shift in the energy density will contribute to Λ. According to currently accepted cases, the value of the constant term in $V(\phi)$ changes in every phase transition by the amount E^4, where E is the energy scale at which the phase transition occurs: At grand unified theory (GUT) transition this change is $10^{56}(\mathrm{GeV})^4$; at the Salam–Weinberg transition it changes by $10^{10}(\mathrm{GeV})^4$. These are enormous numbers compared with the present value of $10^{-47}(\mathrm{GeV})^4$. It is not clear how a physical quantity can change by such a large magnitude and finally adjust itself to be zero to such a fantastic accuracy.

Last, we should not forget that the "zero-point energy" of quantum fields will also contribute to gravity.[28] Each degree of freedom contributes an amount

$$\Lambda \cong \int_0^{k_{max}} \frac{4\pi k^2 \, dk}{(2\pi)^3} \sqrt{k^2 + m^2} \cong \frac{k_{max}^4}{8\pi^2}, \tag{10.66}$$

where k_{max} is an UV cutoff. If we take general relativity to be valid up to the Planck energy, then we may take $k_{max} \approx 10^{19}$ GeV and the contribution to Λ will be $10^{74}(\text{GeV})^4$.

If we assume that all the contributions are indeed there, then they have to be fine-tuned to cancel each other, for no good reason. Before the entry of GUTs into cosmology, we needed to worry only about the first and the last contributions, both of which could be tackled in an ad hoc manner. We arbitrarily set $\Lambda_0 = 0$ in the Lagrangian defining gravity and try to remove the zero-point contribution by complicated regularisation schemes. (Neither argument is completely satisfactory but appears plausible.) However, with the introduction of GUTs and inflationary scenarios, the cosmological constant became a dynamical entity and the situation has become more serious. Note that it is precisely the large change in the $V(\phi)$ that leads to a successful inflation; it has to be large to inflate the universe and change to a small value in the end for a graceful exit from the inflationary phase. Several mechanisms have been suggested in the literature to make the cosmological constant zero. Supersymmetry, complicated dynamical mechanisms, and probabilistic arguments from quantum gravity are only a few of them. None of these seems to provide an entirely satisfactory solution. If the smallness (or vanishing) of Λ is because of some general symmetry consideration, it may be necessary for Λ to be small (or vanish) identically at all epochs. This can, for example, wipe out the entire inflationary picture.

In summary, combining the considerations of quantum field theory and general relativity, we are led to *only* two natural values for Λ: (1) If there is an unknown symmetry principle operating in nature, then Λ should be strictly zero; and (2) if the vacuum fluctuations of energies up to Planck energy contribute to gravity, then Λ should be of the order of 10^{74} GeV^4, which is huge and unacceptable.

Until recently, most cosmologists preferred models with $\Lambda = 0$, hoping that – eventually – a quantum gravitational explanation would be found for the vanishing of this quantity. Several recent observations (discussed in various chapters of this book) seem to suggest that our universe actually has a small but finite value of Λ with $\Omega_V \approx 0.65$, with $\Omega_{tot} = \Omega_V + \Omega_{NR} \approx 1$. At present, the most promising techniques for measuring Ω_m (and Ω_V) are based on the CMBR temperature anisotropy and type IA supernovae. The location of the first acoustic peak at $l = 200$ in the MAXIMA and BOOMERANG data favours a $\Omega_m + \Omega_V = 1$ universe with $\Omega_m \simeq 0.3$. The analysis of the first seven high-redshift SN Ia gave a value for Ω_{NR} that is consistent with unity: $\Omega_{NR} = (0.94^{+0.34}_{-0.28})$. However, adding a single $z = 0.83$ supernova for which good HST data was available lowered the value to $\Omega_{NR} = (0.6 \pm 0.2)$. More recently, the analysis of a larger data set of 42 high-redshift SN Ia gives $\Omega_m = (0.28^{+0.09+0.05}_{-0.08-0.04})$, where the first errors are statistical and the second are due to identified systematic errors. These results assume that $\Omega_m + \Omega_V = 1$. Taken together, we are led to a model with $\Omega_m \approx (1/3)$ and $\Omega_V \approx (2/3)$.

Although the above results from CMBR and supernovae enjoy considerable popularity, it would be nice if these results are confirmed by other observations as well. Unfortunately, other direct determinations of Ω_m from different dynamical considerations are more inaccurate compared with the CMBR and supernovae measurements. Here is a rapid overview of some of these observations: (1) The analysis of bulk motion of galaxies gives a result of $0.3 \lesssim \Omega_m \lesssim 3$, having a range of factor 10. (2) The measurement of the cluster M/L ratio of 16 clusters at $z \lesssim 0.3$ gives $\Omega_m = (0.19 \pm 0.06)$. The study of x-ray gas in clusters suggests $\Omega_m = (0.3 \pm 0.1)$ if we take $h = 0.65 \pm 0.08$. A more direct determination of Ω_B from the SZ effect when combined with the x-ray observations of clusters suggests that $\Omega_m \simeq 0.25h^{-1} \simeq (0.38 \pm 0.1)$. (3) An independent constraint on the cosmological constant arises from the statistics of gravitational lensing events.[29] The original result showed that $\Omega_V < 0.70$ whereas more recent data suggest that $\Omega_V < 0.66$ at a 95% confidence interval if $\Omega_{NR} + \Omega_V = 1$.[30] These results could depend on the extinction by dust in the E/S0 galaxies responsible for lensing as well as the evolution of galaxies. A more stringent bound that is not affected by dust comes from radio observations and gives $\Omega_V < 0.73$ at the 2σ level.[31] (4) Finally, given the likelihood range of Ω_V and Ω_m, we can determine the range of $H_0 t_0$. The 95% confidence limit on this parameter turns out to be $0.85 \leq H_0 t_0 \leq 1.13$. If we take $h = 0.71 \pm 0.06$, this leads to, at the 95% confidence limit, $10.8 \leq t_0/(1 \text{ Gyr}) \leq 16.7$ on the age of the universe, which is consistent with other determinations of the age.

It is obvious that such a model with $\Omega_V \approx 0.7$ is aesthetically quite ugly. As mentioned before, the only two natural values for the vacuum energy density are 0 and $(10^{19} \text{ GeV})^4$. Any small, nonzero value for this parameter requires extreme fine-tuning of the model so that just around the present epoch vacuum energy begins to dominate over matter. In spite of this fact, such cosmological models are being investigated purely because they seem to agree with observations. If these observations stand the test of time, then the smallness of the cosmological constant will be *the* most important problem that needs to be settled in physics.

Notes and References

General References

The following references are relevant to several chapters in this volume:

a. B. W. Carroll and D. A. Ostlie, *Modern Astrophysics* (Addison-Wesley, New York, 1996).
b. R. L. Bowers and T. Deeming, *Astrophysics* (Jones and Bartlett, Boston, 1984), Vols. I and II.
c. P. A. Charles and F. D. Seward, *Exploring the X-Ray Universe* (Cambridge Univ. Press, New York, 1995).
d. *Encyclopedia of Astronomy and Astrophysics* (Institute of Physics Publishing and Nature Publishing Group, London, England), Vols. 1–4.
e. J. A. Peacock, *Cosmological Physics* (Cambridge Univ. Press, New York, 1999).
f. P. J. E. Peebles, *Principles of Physical Cosmology* (Princeton Univ. Press, Princeton, NJ, 1993).
g. T. Padmanabhan, *Structure Formation in the Universe* (Cambridge Univ. Press, New York, 1993).
h. T. Padmanabhan, *Cosmology and Astrophysics through Problems* (Cambridge Univ. Press, New York, 1996).
i. J. Binney and M. Merrifield, *Galactic Astronomy* (Princeton Univ. Press, Princeton, NJ, 1998).

Chapter 1

1. The following general references are relevant to this chapter: a, b, d, and i.
2. For a review, see R. C. Kennicutt, *Ann. Rev. Astron. Astrophys.* **36**, 189 (1998).
3. The data are provided by C. C. Steidel; see C. C. Steidel et al., *Astrophys. J.* **519**, 1 (1999).
4. J. Maggorian et al., *Astrophys. J.* **115**, 2285 (1998).
5. The data on galaxies, etc., given in this chapter, in Figures and Tables, are taken from different sources cited in the General References, especially from a, c, d, and i.

6. A. Udalski et al., *J. Acta. Astron.* **44**, 165 (1994); C. Alcock et al., *Astrophys. J.* **109**, 1653 (1995); B. Paczynski et al., *Astrophys. J.* **435**, L113 (1994).

7. Most of the data related to AGN can be found in Ref. 1 of Chap. 8.

8. L. Woltjer, in R. D. Blandford, H. Netzer, and L. Woltjer, eds., *Active Galactic Nuclei* (Springer-Verlag, Berlin, 1990), p. 1.

9. For more details, see the following review article: H. C. Ferguson et al., astro-ph/0004319, *The Hubble deep fields*.

10. P. Madau and L. Pozzetti, *Mon. Not. R. Astron. Soc.* **312**, L9–15 (2000).

11. See Ref. 3 of this chapter.

12. Some of the quasar surveys are described in the following articles: M. Schmidt and R. F. Green, *Astrophys. J.* **269**, 352 (1983); B. J. Boyle, T. Shanks, and B. A. Peterson, *Mon. Not. R. Astron. Soc.* **235**, 935 (1988); F. D. A. Hartwick and D. Schade, *Ann. Rev. Astron. Astrophys.* **28**, 437 (1990).

13. For a visual display of galaxy distributions in the nearby region, see R. B. Tully and J. R. Fisher, *Nearby Galaxies Atlas* (Cambridge Univ. Press, New York, 1987). Detailed descriptions of various structures can also be found in J. Audouze and G. Israel, eds., *The Cambridge Atlas of Astronomy*, 2nd ed. (Cambridge Univ. Press, New York, 1988).

14. A nontechnical description of the Local Supercluster can be found in R. B. Tully, *Sky Telesc.* **63**, 550 (1982).

15. Some of the older galaxy surveys are described in the following papers: V. de Lapparent, M. J. Geller, and J. Huchra, *Astrophys. J.* **302**, L1 (1986); V. de Lapparent, M. J. Geller, and J. Huchra, *Astrophys. J.* **332**, 44 (1988); M. J. Geller, J. Huchra, and V. de Lapparent, in A. Hewitt, G. Burbidge, and L.-Z. Fang, eds., *Observational Cosmology*, IAU Symposium Series No. 124 (Reidel, Dordrecht, The Netherlands), p. 301; M. Seldner et al., *Astron. J.* **82**, 249 (1977); R. P. Kirshner et al., *Astrophys. J.* **248**, L57 (1981); M. Rowan-Robinson et al., *Mon. Not. R. Astron. Soc.* **247**, 1 (1990); W. Saunders et al., *Nature (London)* **349**, 32 (1991); S. J. Maddox et al., *Mon. Not. R. Astron. Soc.* **242**, 43 (1990). All the recent galaxy surveys have well-maintained web pages; some of the relevant ones are http://www.roe.ac.uk/japwww/2df; http://www.manaslu.astro.utoronto.ca/~lin/lcrs.html; http://www.sdss.org; http://www.astro.utoronto.ca/~lilly/CFRS.

16. M. Davis and P. J. E. Peebles, *Astrophys. J.* **267**, 437 (1983); M. Davis and P. J. E. Peebles, *Astrophys. J.* **267**, 465 (1983); P. J. E. Peebles, *Large Scale Structure of the Universe* (Princeton Univ. Press, Princeton, NJ, 1980).

17. N. A. Bahcall, *Ann. Rev. Astron. Astrophys.* **26**, 631 (1988); R. C. Nichol, astro-ph/0110231; L. Guzzo, astro-ph/9911115.

18. V. de Lapparent, M. J. Geller, and J. Huchra, *Astrophys. J.* **302**, L1 (1986). For a good, popular account, see M. J. Geller and J. P. Huchra, *Science* **246**, 897 (1989).

19. This figure is compiled and adapted from different sources: G. F. Smoot, astro-ph/9705101, G. Lagache et al. (1999), astro-ph/9901059, 9910255; M. G. Hauser et al., *Astrophys. J.* **508**, 25 (1998); C. H. Leinert et al., *Astron. Astrophys. Suppl.* **127**, 1 (1998); E. Dwek et al., *Astrophys. J.* **508**, L9 (1998); L. Pozzetti et al., *Mon. Not. R. Astron. Soc.* **298**, 1133 (1998); T. Miyaji et al., *Astron. Astrophys.* **334**, L13 (1998); P. Sreekumar et al., *Astrophys. J.* **494**, 523 (1998); D. Scott, astro-ph/9912038.

Chapter 2

1. The following general references are relevant to this chapter:

   i. J. Binney and S. Tremaine, *Galactic Dynamics* (Princeton Univ. Press, Princeton, NJ, 1987).
   ii. R. C. Kennicutt et al., eds., *Galaxies: Interactions and Induced Star Formation*, Vol. 26 of the Proceedings of the Saas Fee Advanced Course (Springer-Verlag, New York, 1998).
   iii. J. Binney et al., eds., *Morphology and Dynamics of Galaxies*, Vol. 12 of the Proceedings of the Saas Fee Advanced Course held at Geneva Observatory (Switzerland, 1982).
   iv. There are several web-based courses dealing with galactic dynamics. See, for example, the following web sites: http://www.ifa.hawaii.edu/~barnes; http://www.astrua.edu; http://www.usm.uni-muenchen.de/people/bender.
   v. D. M. Elmegreen, *Galaxies and Galactic Structures* (Prentice-Hall, Saddle River, NJ, 1998).

2. See, e.g., Eq. (10.36) of i in the General References.
3. See Ref. 1(i) of this chapter, p. 69.
4. See, e.g., F. H. Shu, *The Physics of Astrophysics*-Volume 2 (University Science, Mill Valley, CA, 1992).
5. The chemical evolution of galaxies is covered in the following books: D. Arnett, *Supernova and Nucleosynthesis* (Princeton Univ. Press, Princeton, NJ, 1996); B. E. J. Pagel, *Nucleosynthesis and Chemical Evolution of Galaxies* (Cambridge Univ. Press, New York, 1997); J. E. Beckman and T. J. Mahoney, eds., *The Evolution of Galaxies on Cosmological Timescales*, Vol. 187 of ASP Conference Series (Astronomical Society of the Pacific, San Francisco, CA, 1999).
6. See, e.g., F. Combes et al., eds., *Dynamics of Galaxies: Proceedings of the Fifteenth IAP Meeting*, Vol. 197 of the ASP Conference Series (Astronomical Society of the Pacific, San Francisco, CA, 2000).
7. Several such codes are available on the Web; also see (a) of General References and Ref. 1(i). The results given here are based on a code written by the author, adapting these resources.

Chapter 3

1. The following general references are relevant to this chapter: e, f, and g.
2. One example of this result is given in T. Padmanabhan, *Gen. Relativ. Gravit.* **19**, 927 (1987).
3. P. Schneider, J. Ehlers, and E. Falco, *Gravitational Lenses* (Springer, New York, 1992).

Chapter 4

1. The following general references are relevant to this chapter:

   i. P. Coles and F. Lucchin, *Cosmology: The Formation and Evolution of Cosmic Structure* (Wiley, New York, 1995).

   ii. P. J. E. Peebles, *Physical Cosmology* (Princeton Univ. Press, Princeton, NJ, 1971).

   iii. References e, f, and g of the General References.

2. See, e.g., K. A. Olive et al., *Phys. Lett. B* **236**, 454 (1990).

3. See, e.g., the books by Pagel and Arnett cited in Ref. 5 of Chap. 2.

4. S. Burles and D. Tytler, *Astrophys. J.* **499**, 699 (1998); S. Burles and D. Tytler, *Astrophys. J.* **507**, 732 (1998); K. A. Olive et al., *Phys. Rep.* **333**, 389 (2000).

5. V. V. Smith et al., *Astrophys. J.* **408**, 262 (1992); L. Hobbs and J. Thorburn, *Astrophys. J.* **428**, L25 (1994).

6. This analysis is based on the following paper: R. Esmailzadel et al., *Astrophys. J.* **378**, 504 (1991).

7. See, e.g., A. R. Liddle and D. H. Lyth, *Phys. Rep.* **231**, 1 (1993); A. R. Liddle and D. H. Lyth, *Cosmological Inflation and Large-Scale Structure* (Cambridge Univ. Press, New York, 2000).

Chapter 5

1. The following general references are relevant to this chapter:

   i. References e, f, g, and h of the General References and Ref. 7 of Chap. 4.

   ii. R. H. Brandenberger, *Rev. Mod. Phys.* **57**, 1 (1985); V. F. Mukhanov, H. A. Feldman, and R. H. Brandenberger, *Phys. Rep.* **215**, 203 (1992).

   iii. T. Padmanabhan, in R. Mansouri and R. H. Brandenberger, eds., *Large Scale Structure Formation*, Vol. 247 of Astrophysics and Space Science Library Series (Kluwer, Dordrecht, The Netherlands, 2000).

   iv. P. J. E. Peebles, *Large Scale Structure of the Universe* (Princeton Univ. Press, Princeton, NJ, 1980).

2. The discussion is based on E. Bertschinger, in R. Schaffer et al., eds., *Cosmology and Large Scale Structure, Les Houches LX* (Elsevier Science, Amsterdam, 1996), pp. 273–347, as well as on some unpublished work by the author.

3. These fitting functions are described in the book by Liddle and Lyth, cited in Ref. 7 of Chap. 4.

4. See, e.g., A. N. Taylor and A. J. S. Hamilton, *Mon. Not. R. Astron. Soc.* **282**, 767 (1996).

5. See, e.g., J. S. Hamilton, P. Kumar, L. E., and A. Mathews, *Astrophys. J.* **374**, L1 (1991); T. Padmanabhan, *Mon. Not. R. Astron. Soc.* **278**, L29 (1996); T. Padmanabhan, in S. V. Dhurandhar and T. Padmanabhan, eds., *Gravitation and Cosmology*, Proceedings of the ICGC '95 Conference (Kluwer, Dordrecht, The Netherlands), p. 37; J. S. Bagla and T. Padmanabhan, *Mon. Not. R. Astron. Soc.* **286**, 1023 (1997).

6. See J. S. Bagla, S. Engineer, and T. Padmanabhan, *Astrophys. J.* **495**, 25 (1998).

Chapter 6

1. The following general references are relevant to this chapter:

   i. References e, f, and g of the General References and Ref. 7 of Chap. 4.

   ii. M. Zaldarriaga, *Astrophys. Space Sci.* **247**, 213 (2000).

iii. P. Binetroy et al., eds., *Primordial Universe, Les Houches LXXL* (EDP Sciences, Berlin, 2000).

2. W. Hu and N. Sugiyama, *Astrophys. J.* **444**, 489 (1995); W. Hu and N. Sugiyama, *Phys. Rev. D* **51**, 2599 (1995).

3. The theoretical curves are computed with the code CMBFAST; see U. Seljak and M. Zaldarriaga, *Astrophys. J.* **469**, 437 (1996).

4. W. Hu and N. Sugiyama, *Phys. Rev. D* **51**, 2599 (1995).

5. W. Hu et al., *Astrophys. J.* **549**, 669 (2001).

6. B. E. Corey and D. T. Wilkinson, *Bull. Am. Astron. Soc.* **8**, 351 (1976); G. F. Smoot, M. V. Gerenstin, and R. A. Muller, *Phys. Rev. Lett.* **39**, 898 (1977); E. K. Conklin, *Nature (London)* **222**, 971 (1969); G. F. Smoot et al., *Astrophys. J.* **371**, L1 (1991).

7. For a review of cosmic velocity fields, see S. Courteau and A. Dekel, astro-ph/0105490.

8. For a discussion of the results from various "dipoles," see the following papers: O. Lahav, *Mon. Not. R. Astron. Soc.* **225**, 213 (1987); D. Lynden, O. Lahav, and D. Burstein, *Mon. Not. R. Astron. Soc.* **241**, 325 (1989); A. Yahil, D. Walter, and M. Rowan-Robinson, *Astrophys. J. Lett.* **301**, 1 (1986); O. Lahav, M. Rowan-Robinson, and D. Lynden-Bell, *Mon. Not. R. Astron. Soc.* **234**, 677 (1988).

9. R. B. Tully and J. R. Fisher, *Astron. Astrophys.* **54**, 661 (1977); A. Dressler et al., *Astrophys. J.* **313**, 42 (1987); S. Djorgovski and M. Davis, *Astrophys. J.* **313**, 59 (1987). The estimates for the scatter in the IRTF relation quoted in the literature vary quite a bit. For a discussion of this issue, see, e.g., D. Burstein and S. Raychaudhury, *Astrophys. J.* **343**, 18 (1989); M. Han and J. Mould, *Astrophys. J.* **360**, 448 (1990).

10. D. Burstein et al., in B. F. Madore and R. B. Tully, *Galaxy Distances and Deviations from Universal Expansion* (Reidel, Boston, 1986), p. 123; see also A. Dressler, *Astrophys. J.* **317**, 1 (1987); P. B. Lilje, A. Yahil, and B. J. T. Jones, *Astrophys. J.* **307**, 91 (1986); D. Lynden-Bell et al., *Astrophys. J.* **326**, 16 (1988).

11. This is discussed in detail in the review by D. Burstein, *Rep. Prog. Phys.* **53**, 421 (1990). See also the review in Ref. 7 of this chapter.

12. These are discussed, e.g., in the book by Liddle and Lyth cited in Ref. 7 of Chap. 4.

13. This is based on T. Padmanabhan and S. Sethi, *Astrophys. J.* **555**, 125 (2001).

14. See, e.g., E. T. Vishniac, *Astrophys. J.* **322**, 597 (1987).

15. Ya. B. Zeldovich and R. A. Sunyaev, *Astrophys. Space Sci.* **4**, 301 (1969); K. L. Chan and B. J. T. Jones, *Astrophys. J.* **195**, 1 (1975); J. G. Barlett and A. Stebbins, *Astrophys. J.* **371**, 8 (1991); also see Ref. 1(ii) of Chap. 4 as well as S. Sarkar in *Large Scale Structure Formation*, Vol. 247 of Astrophysics Space Science Library Series (Kluwer, Dordrecht, The Netherlands, 2000), p. 37.

16. For a review of this effect, see R. A. Sunyaev and Ya. B. Zeldovich, *Ann. Rev. Astron. Astrophys.* **18**, 537 (1980) and references cited therein; M. Birkinshaw, *Phys. Rep.* **310**, 97 (1999).

Chapter 7

1. The following general references are relevant to this chapter:

i. A. J. Bunker and W. J. M. Van Breugel, *The Hy-Redshift Universe: Galaxy*

Formation and Evolution at High Redshift, Vol. 193 of ASP Conference Series (Astronomical Society of the Pacific, San Francisco, CA, 1999).

ii. See Ref. 1(i) of Chap. 4 and Refs. e, f, g, and h of the General References.

2. See, e.g., S. Sasaki, *Proc. Astron. Soc. Jpn.* **46**, 427 (1994).
3. V. R. Ike et al., *Mon. Not. R. Astron. Soc.* **282**, 263 (1996).
4. P. Madau and L. Pozzetti, *Mon. Not. R. Astron. Soc.* **312**, L9–15 (2000).
5. A. Franceschini (2000), *High-Redshift Galaxies: The Far-Infrared and Sub-Millimeter View*, astro-ph/0009121.
6. P. P. Kronberg, *Rep. Prog. Phys.* **57**, 325 (1994) and references therein.
7. G. Grasso and H. R. Rubinstein (2000), *Magnetic fields in the early universe*, astro-ph/0009061.

Chapter 8

1. The following general references are relevant to this chapter:

    i. G. V. Bicknell et al., *The First Stromlo Symposium: The Physics of Active Galaxies*, Vol. 54 of ASP Conference Series (Astronomical Society of the Pacific, San Francisco, CA, 1994).

    ii. G. Ferland and J. Baldwin, *Quasars and Cosmology*, Vol. 162 of ASP Conference Series (Astronomical Society of the pacific, San Francisco, CA, 1999).

    iii. J. H. Krolik, *Active Galactic Nuclei* (Princeton Univ. Press, Princeton, NJ, 1999).

    iv. B. M. Peterson, *An Introduction to Active Galactic Nuclei* (Cambridge Univ. Press, New York, 1997).

    v. R. D. Blandford, H. Netzer, and L. Woltjer, in T. J. L. Courvoisier and M. Mayor, eds., *Active Galactic Nuclei*, Vol. 20 of Proceedings of Saas Fee Advanced Course (Springer-Verlag, New York, 1990).

    vi. J. Frank, A. King, and D. Raine, *Accretion Power in Astrophysics* (Cambridge Univ. Press, New York, 1992).

    vii. D. Emerson, *Interpreting Astronomical Spectra* (Wiley, New York, 1996).

    viii. D. E. Osterbrock, *Astrophysics of Gaseous Nebulae and Active Galactic Nuclei* (University Science, Mill Valley, CA, 1989).

2. I. D. Novikov and K. S. Thorne, in C. DeWitt and B. DeWitt, eds., *Blackholes* (Gordon & Breach, New York, 1973), p. 343; D. N. Page and K. S. Thorne, *Astrophys. J.* **191**, 499 (1974).
3. See, e.g., M. A. Malcolm, in F. Mayer et al., eds., *Theory of Accretion Disks* (Kluwer, Dordrecht, The Netherlands, 1989).
4. K. Iwasawa et al., *Mon. Not. R. Astron. Soc.* **282**, 1038 (1996); Y. Dabrowski et al., *Mon. Not. R. Astron. Soc.* **288**, L11 (1997).
5. These processes are described in several textbooks; see, in particular, the book by Krolik cited in Ref. 1(iii) of the chapter.

Chapter 9

1. The following general references are relevant to this chapter:

   i. See Refs. f and h of the General References.

   ii. S. M. Viegas et al., eds., *Young Galaxies and QSO Absorption Line Systems*, Vol. 114 of ASP Conference Series (Astronomical Society of the Pacific, San Francisco, CA, 1997).

2. P. Jacobson, *Nature (London)* **370**, 35 (1994); A. F. Davidson et al., *Nature (London)* **380**, 47 (1996); C. J. Hogan et al., *Astron. J.* **113**, 1495 (1997); D. Reimers et al., *Astron. Astrophys.* **327**, 890 (1997); S. F. Anderson et al., *Astron. J.* **117**, 56 (1999); S. R. Heap et al., *Astrophysics J.* **534**, 69 (2000).

3. P. W. Guilbert and A. C. Fabian, *Mon. Not. R. Astron. Soc.* **220**, 439 (1986).

4. D. J. Fixsen et al., *Astrophys. J.* **473**, 576 (1996).

5. C. C. Steidel et al., *Astrophys. J.* **519**, 1 (1999).

6. Y. C. Pei, *Astrophys. J.* **438**, 623 (1995).

7. A. Conti et al., *Astron. J.* **117**, 645 (1999).

8. G. Magorrian et al., *Astron. J.* **115**, 2285 (1998); M. Fukugita et al., *Astrophys. J.* **503**, 518 (1998).

9. A. Gould et al., *Astrophys. J.* **503**, 798 (1998).

10. This form of the fit was used, e.g., in the following paper: P. Madau, in M. Colless, ed., *Cosmology and the Brightness of the Night Sky*, Proceedings of the Second Coral Sea Cosmology Conference Dunk Island, 24–28 Aug, 1999; http://mso.anu.edu.au/Dunk Island/proceedings. The numerical values are chosen here to fit the current data.

11. T. Kim, *Astron. J.* **114**, 1 (1997); C. D. Impey, *Astrophys. J.* **463**, 473 (1996); J. N. Bahcall et al., *Astrophys. J. Suppl.* **87**, 1 (1993).

12. The data are from L. Storrie-Lombardi, in Ref. 1(ii) of Chap. 8.

Chapter 10

1. The following general references are relevant to this chapter:

   i. See Ref. i in the General References.

   ii. S. Webb, *Measuring the Universe: The Cosmological Distance Ladder* (Springer, Chichester, 1999).

2. More details can be found in the following papers: M. Birkinshaw et al., *Astrophys. J.* **420**, 33 (1994); M. Birkinshaw et al., *Astrophys. J.* **379**, 466 (1991); M. Birkinshaw et al., *Phys. Rep.* **310**, 97 (1999); J. E. Carlstrom et al., astro-ph/0103480.

3. See, e.g., p. 409 of Ref. i in the General References; T. Kundic et al., *Astrophys. J.* **482**, 75 (1997); M. Serra-Ricart et al., *Astrophys. J.* **526**, 40 (1999); E. E. Falco et al., *Astrophys. J.* **484**, 70 (1997).

4. For a summary, see W. L. Freedman, *Phys. Rep.* **333**, 13 (2000).

5. References for the two tables (the letters correspond to those given as italic superscripts in column 2 of the tables) are as follows:

  a. D. B. Guenther and P. Demarque, *Astrophys. J.* **484**, 937, (1997).

  b. B. Chaboyer, E. M. Green, and J. Liebert, astro-ph/9812097.

  c. S. K. Leggett, M. T. Ruiz, and P. Bergeron, *Astrophys. J.* **497**, 294 (1998).

  d. G. Carraro et al., *Astron. Astrophys.* **343**, 825 (1999).

  e. T. D. Oswalt et al., *Nature (London)* **382**, 692 (1996).

  f. R. L. Phelps, *Astrophys. J.* **483**, 826 (1997).

  g. B. Chaboyer et al., *Astrophys. J.* **494**, 96 (1998).

  h. R. G. Gratton et al., *Astrophys. J.* **491**, 749 (1997).

  i. I. N. Reid, *Astron. J.* **114**, 161 (1997).

  j. M. Salaris and A. Weiss, *Astron. Astrophys.* **327**, 107 (1997).

  k. F. Grundahl, D. A. VandenBerg, and M. T. Andersen, *Astrophys. J.* **500**, L179 (1998).

  l. R. Jimenez, astro-ph/9810311.

  m. F. Pont et al., *Astron. Astrophys.* **329**, 87 (1998). For a summary, see C. Lineweaver, astro-ph/9911493.

6. K. Kuijken and G. Gilmore, *Astrophys. J.* **367**, L9 (1991).

7. See e.g., J. Einasto et al. (1999), astro-ph/9909437.

8. L. Grego et al. (2000), astro-ph/0003085; L. P. David et al., *Astrophys. J.* **473**, 692 (1996); R. G. Callberg et al., *Astrophys. J.* **478**, p. 462 (1997).

9. See e.g., S. Mao (1999), astro-ph/9909302.

10. For a review of dark-matter candidates, especially the baryonic dark matter, see B. Carr (2000), gr-qc/0008005.

11. B. Zeitnitz et al., *Prog. Part. Nucl. Phys.* **40**, 69 (1998); K. S. Hirata et al., *Phys. Lett. B* **205**, 416 (1988); D. Casper et al., *Phys. Rev. Lett.* **66**, 2561 (1991); M. Ambrosio et al., *Phys. Lett. B* **434**, 451 (1998); Y. Fukuda et al., *Phys. Rev. Lett.* **82**, 2644 (1998).

12. S. Wolfram, *Phys. Lett. B* **82**, 65 (1979); C. B. Dover, T. K. Gaisser, and G. Steigman, *Phys. Rev. Lett.* **42**, 1117 (1979).

13. J. Silk, K. A. Olive, and M. Srednicki, *Nucl. Phys. B* **279**, 804 (1987).

14. J. Ellis, J. S. Hagelin, and D. V. Nanopoulous, *Phys. Lett. B* **159**, 26, (1985); M. Oreglia, in *Proceedings of the Fourteenth International Conference on Neutrino Physics and Astrophysics* (CERN, Geneva, 1990); J. Silk and M. Srednicki, *Phys. Rev. Lett.* **53**, 624 (1984); K. Griest, *Phys. Rev. Lett.* **61**, 666 (1988); K. Griest, *Phys. Rev. D* **38**, 2357 (1988); J. Ellis et al., *Nucl. Phys. B* **238**, 453 (1984); K. Griest, M. Kamiokowski, and M. S. Turner, *Phys. Rev. D* **41**, 3565 (1990); K. Olive and M. Srednicki, *Phys. Lett. B* **230**, 78 (1989).

15. R. Bernabei et al., *Phys. Lett. B* **450**, 448 (1999); A. Bottino et al., *Phys. Lett.* **423**, 109 (1998); R. Arnowitt and P. Nath, *Phys. Rev. D* **60**, 044002 (1999).

16. T. Sumner, in *Living Reviews in Relativity* (Potsdam, in press); T. Sumner, paper presented at Rencontres de Blois–2001, 17–23 June, 2001, Blois, France.

17. S. Tremaine and J. E. Gunn, *Phys. Rev. Lett.* **42**, 407 (1979); S. Tremaine, M. Henon, and D. Lynden-Bell, *Mon. Not. R. Astron. Soc.* **219**, 285 (1986).

18. See, e.g., P. Ramond, *Field Theory: A Modern Primer* (Benjamin-Cummings, Reading, MA, 1981).

19. V. Baluni, *Phys. Rev. D* **19**, 2227, (1979); R. Crewther et al., *Phys. Lett. B* **88**, 123

(1979); N. Ramsey, *Phys. Rep.* **43**, 409 (1977); I. S. Alterev, *Phys. Lett. B* **136**, 327 (1984).

20. S. Weinberg, *Phys. Rev. D* **11**, 3583 and references cited therein (1975).

21. R. D. Peccei and H. R. Quinn, *Phys. Rev. Lett.* **38**, 1440 (1977); R. D. Peccei, and H. R. Quinn, *Phys. Rev. D* **16**, 1791 (1977); S. Weinberg, *Phys. Rev. Lett.* **40**, 223 (1978); F. Wilezek, *Phys. Rev. Lett.* **40**, 279 (1978); K. Choi, K. Kangand, and J. E. Kim, *Phys. Rev. Lett.* **62**, 849 (1989); D. Kaplan, *Nucl. Phys. B* **260**, 215 (1985); R. Mayle et al., *Phys. Lett. B* **203**, 188 (1988); G. G. Raffelt and D. Seckel, *Phys. Rev. Lett.* **60**, 1793 (1988).

22. For a more complete review of laboratory searches for axions, see, e.g., J. E. Kim, *Phys. Rep.* **150**, 1 (1987).

23. S. Dimopoulos et al., *Phys. Lett. B* **176**, 223 (1986); J. Frieman, S. Dimopoulos, and M. S. Turner, *Phys. Rev. D* **36**, 2201, (1987); G. G. Raffelt and D. S. P. Dearborn, *Phys. Rev. D* **36**, 2211 (1987); D. S. P. Dearborn, D. N. Schramm, and G. Steigman, *Phys. Rev. Lett.* **56**, 26 (1986); N. Iwamoto, *Phys. Rev. Lett.* **53**, 1198 (1984); S. Truruta and K. Nomoto, in A. Hewitt, G. Burbridge, and L.-Z. Fang, *Observational Cosmology*, Proceedings of IAU Symposium Vol. 124 (Kluwer, Dordrecht, The Netherlands, 1987), p. 713; D. E. Morris, *Phys. Rev. D* **34**, 843 (1986); G. G. Raffelt and L. Stodolsky, *Phys. Rev. D* **37**, 1237 (1988); R. P. Brinkmann and M. S. Turner, *Phys. Rev. D* **38**, 2338 (1998).

24. T. Kephart and T. Weler, *Phys. Rev. Lett.* **58**, 171 (1987); M. S. Turner, *Phys. Rev. Lett.* **59**, 2489 (1987); A. L. Broadfoot and K. R. Kendall, *J. Geophys. Res. [Space Phys.]* **73**, 426 (1968).

25. G. Raffelt, *Nucl. Phys. Proc. Suppl.* **77**, 456 (1999).

26. See, e.g., S. De Panfills et al., *Phys. Rev. Lett.* **59**, 839 (1987).

27. See, e.g., S. Weinberg, *Rev. Mod. Phys.* **61**, 1 (1989).

28. Ya. B. Zeldovich, *JETP Lett.* **6**, 316 (1967); T. Padmanabhan, *Int. J. Mod. Phys. A* **4**, 4735 (1989).

29. L. Bergstrom (2000), hep-th/0002126.

30. C. S. Kochanek, *Astrophys. J.* **419**, 12 (1993); C. S. Kochanek, *Astrophys. J.* **466**, 638 (1996); D. Maoz and H. W. Rix, *Astrophys. J.* **416**, 425 (1993).

31. E. E. Falco et al., *Astrophys. J.* **484**, 70 (1997).

Ithi the gyanamakhyatham guhyathguhyatharam maya
Vimrushya asheshena yadhechasi thatha kuru.
(Thus has Knowledge more secret than all secrets been declared
to you by me. Ponder over it fully and then act as you like.)
– Bhagavat Gita, Chapter 18, Verse 63.

Index